Springer Protocols

Shirley Pease • Thomas L. Saunders
Editors

Advanced Protocols for Animal Transgenesis

An ISTT Manual

Springer

ISTT www.transtechsociety.org
International Society for Transgenic Technologies

Editors
Shirley Pease
California Institute of Technology
Div. Biology
Pasadena California
USA
ssp@its.caltech.edu

Ph.D. Thomas L. Saunders
University of Michigan
Transgenic Animal Model Core
Ann Arbor Michigan
USA
tsaunder@umich.edu

ISBN 978-3-642-20791-4 e-ISBN 978-3-642-20792-1
DOI 10.1007/978-3-642-20792-1
Springer Heidelberg Dordrecht London New York

Library of Congress Control Number: 2011936518

Cover illustration: Laser assisted injection with a beveled injection pipette; see Fig. 11b in Chap. 17 "Combining ES cells with Embryos", Elizabeth Williams, Wojtek Auerbach, Thomas M. DeChiara, and Marina Gertsenstein (Photo by Michael Brown)

Printed on acid-free paper

Springer is part of Springer Science+Business Media (www.springer.com)

Foreword

The International Society for Transgenic Technologies (ISTT) is proud to present this book entitled "Advanced Protocols for Animal Transgenesis", edited by Shirley Pease and Thom Saunders, whose commitment and perseverance have led this initiative successfully to term.

The ISTT, founded in 2006, established among its aims "*to foster and encourage knowledge generation, discussion, training and education, and the diffusion of the technologies and specific research used for the genetic modification of animals, in particular those aimed at generating and/or analysing transgenic and mutant animals as particularly useful experimental models in the biology, biomedicine and biotechnology disciplines*", as stated in the ISTT bylaws, Article 3a. According to this aim, a number of activities were identified to achieve the objectives, including: "*to publish information bulletins, abstracts, monographs, books, protocols and other specialised texts aimed at promoting knowledge generation, discussion, training and education, and the diffusion of the technologies and specific research associated with animal transgenesis*" (ISTT bylaws, Article 4c). Therefore, this book on transgenic technology represents one of the reasons why ISTT was founded, and we all hope it will become a useful reference manual for anyone interested in the generation and the analysis of transgenic animals.

The manual, "Advanced Protocols for Animal Transgenesis", includes chapters covering almost all current methods that can be applied to the generation and analysis of genetically modified animals. All chapters have been written and contributed by experts in their corresponding fields. Most of them are members of the ISTT. Topics range from standard and classical techniques, such as pronuclear microinjection of DNA, to more sophisticated and modern methods, such as the derivation and establishment of ES cell lines, with defined inhibitors in cell culture medium. In addition, topics that are not usually included in this type of book have been addressed in this manual, because of their relevance to the field. Such topics include global web-based resources, legal issues, colony management, shipment of mice and embryos, and the three Rs, refinement, reduction, and replacement. From the ISTT, we hope that this book will be useful in the daily work of any Transgenic Facility or laboratory, providing information and references that help to keep us up to date in the field.

Finally, from the ISTT, we would like to specifically thank Springer, for their sincere interest and most generous support for this book. Springer and ISTT have a close, and we hope, mutually rewarding relationship. ISTT is associated with the scientific journal "Transgenic Research", published by Springer. With this book, "Advanced Protocols for Animal Transgenesis", presented by ISTT and published by Springer, our relationship is strengthened and we anticipate that the whole field will benefit as a result.

The ISTT Council.[1]

[1]The ISTT Council is currently formed by: Lluis Montoliu (President), Thom Saunders (Vice-President), Carlisle Landel (Secretary), Manuel Sánchez-Martín (Treasurer), Shirley Pease (Officer), Jan Parker-Thornburg (Officer), Boris Jerchow (Officer), Wojtek Auerbach (Officer), Tom Fielder (Officer), Aimee Stablewski (Officer), and Elizabeth Williams (Officer).

Preface

Two years passed between conception and publication of this book. A lengthy time, perhaps, but came about because all those contributing to and working on the manual were fully occupied professionals Although this took more time than we would have liked, the strength of the final product lies in the fact that all those who contributed *are* leaders in their particular field of interest and are actively using the technology in their laboratories. The methodology has been written up by those with first-hand experience, and as a result, in most cases, we have been able to include a troubleshooting section at the end of each chapter, which will help those new to the technology to overcome the difficulties that sometime plague such detailed procedures.

We owe a tremendous vote of thanks to all our contributors. Thank you all so much for making time during your busy workweek to contribute your knowledge and experience to this project. Additional thanks must go to Belen Pintado, whose idea it was, in the first instance, that ISTT should embark upon such a project and to the ISTT Council of January of 2008, for the assembly of a putative list of contents for this manual. Thanks, too, to Kristina Nagy, who managed to write up her contribution while moving her family half way around the world for a sabbatical! Thanks to Jenny Nichols and Marina Gerstenstein, who jumped in with additional contributions at the last minute. Thanks to Lluis Montoliu, whose indefatigable energy levels keep us all on our toes, at ISTT Council! Thanks to Jan Parker Thornberg for imbuing her contribution with the passion she has for teaching and sharing of information. To Karen Brennan, whose wry sense of humor made it easy to thrash out the details of Colony Management. To Jorge Sztein and many others, who fielded our questions with great patience. To Elizabeth Williams, who managed to finish off her contribution while dealing with the effects of flooding in Brisbane. To Tom Fielder who was able to summarize the work so far on the international survey of pronuclear injection, even though he was still buried in data for the project. Thanks to those who have made multiple contributions, Anna and Wojtek Auerbach, Lluis Montoliu, Marina Gertstenstein, and Thom Saunders. And to all those not mentioned by name here, your contribution is equally important and we thank you sincerely.

Finally, thanks to my co-ed, Thom Saunders, a veritable rock in a stormy sea, always ready with information, insight, and references to back it all up!

Also to Jutta Lindenborn and Sabine Schwarz of Springer publishing, for guiding us through this process and to Springer Publishing Company, for providing us with the means to bring this project to fruition.

Pasadena, CA, USA Shirley Pease

Contents

Contributors

Ignacio Anegon Platform Rat Transgenesis IBiSA-CNRS, Nantes, France; INSERM, UMR643, Nantes 44093, France; CHU Nantes, Institut de Transplantation et de Recherche en Transplantation, ITERT, Nantes 44093, France; Faculté de Médecine, Université de Nantes, Nantes 44093, France

Anna B. Auerbach Mouse Genetics Core Facility, Sloan-Kettering Institute, Memorial Sloan-Kettering Cancer Center (MSKCC), 1275 York Avenue, New York, NY 10065, USA, auerbacb@mskcc.org

Wojtek Auerbach Velocigene, Regeneron Pharmaceutical Inc., 777 Old Saw Mill River Road, Terrytown, NY 10591, USA, wojtek.auerbach@regeneron.com

Darius Balciunas Department of Biology, Temple University, Philadelphia, PA, USA

Katja Becker Max Delbrueck Center for Molecular Medicine, Transgenic Core Facility, Berlin, Germany

María Blasco Telomeres and Telomerase Group, Molecular Oncology Program, Spanish National Cancer Research Centre (CNIO), C/Melchor Fernández Almagro 3, 28029 Madrid, Spain

Karen Brennan Victor Chang Cardiac Research Institute, Lowy Packer Building, 405 Liverpool Street, Darlinghurst, NSW 2010, Australia, k.brennan@victorchang.edu.au

Karen S. Canady canady + lortz LLP, 4201 Wilshire Blvd., Suite 622, Los Angeles, CA 90010, USA, karen@canadylortz.com

Karen M. Chapman The Department of Pharmacology and the Cecil H. & Ida Green Center for Reproductive Biology Sciences, University of Texas Southwestern Medical Center, 6001 Forest Park Dr., Dallas, TX 75390, USA

Thomas M. DeChiara Velocigene, Regeneron Pharmaceutical Inc., 777 Old Saw Mill River Road, Terrytown, NY 10591, USA, thomas.dechiara@regeneron.com

Verónica Domínguez Transgenic Mice Unit, Biotechnology Program, Spanish National Cancer Research Centre (CNIO), C/Melchor Fernández Almagro 3, 28029 Madrid, Spain

Almudena Fernández Department of Molecular and Cellular Biology, Centro Nacional de Biotecnología (CNB), Consejo Superior de Investigaciones Científicas (CSIC), Campus de Cantoblanco, Darwin 3, 28049 Madrid, Spain; Centro de Investigación Biomédica en Red de Enfermedades Raras (CIBERER) ISCIII, Madrid, Spain

Thomas J. Fielder University of California, Irvine, CA 92697–1310, USA; University Laboratory Animal Resources, Irvine, CA 92697–1310, USA, tjf@uci.edu

Marina Gertsenstein Toronto Centre for Phenogenomics and Samuel Lunenfeld Research Institute, Mount Sinai Hospital, Toronto, ON, Canada; Transgenic Core, Toronto Centre for Phenogenomics, 25 Orde Street, Toronto, ON, Canada M5T 3H7, marina.gertsenstein@phenogenomics.ca

Aron Geurts Department of Physiology, Human and Molecular Genetics Center, Cardiovascular Research Center, Medical College of Wisconsin, Milwaukee, WI, USA, ageurts@mcw.edu

F. Kent Hamra The Department of Pharmacology and the Cecil H. & Ida Green Center for Reproductive Biology Sciences, University of Texas Southwestern Medical Center, 6001 Forest Park Dr., Dallas, TX 75390, USA, Kent.Hamra@UTSouthwestern.edu

J. Hourcade Dpto Reproducción Animal, INIA Ctra de La Coruña Km- 5,9, 28040 Madrid, Spain, jhourcade@cib.csic.es

Elizabeth D. Hughes University of Michigan Transgenic Animal Model Core, Room 2570B, MSRB II, 1150 W. Med. Center Drive, Ann Arbor, MI 48109, USA

Anne-Laure Iscache Platform Rat Transgenesis IBiSA-CNRS, Nantes, France; INSERM, UMR643, Nantes 44093, France; CHU Nantes, Institut de Transplantation et de Recherche en Transplantation, ITERT, Nantes 44093, France; Faculté de Médecine, Université de Nantes, Nantes 44093, France

Boris Jerchow Transgenic Core Facility, Max Delbrueck Center for Molecular Medicine, Berlin, Germany, jerchow@mdc-berlin.de

R. J. Kastenmayer Comparative Medicine Branch, National Institute of Allergies and Infectious Diseases, National Institute of Health, ARTiC-CMB-NIAID-NIH, TwinBrook II, Room 201 D, 12441 Parklawn Drive, Rockville, MD 20852, USA

Han Li Tumor Suppression Group, Molecular Oncology Program, Spanish National Cancer Research Centre (CNIO), C/Melchor Fernández Almagro 3, 28029 Madrid, Spain

Ping Li Eli and Edythe Broad Center for Regenerative Medicine and Stem Cell Research at USC, Department of Cell and Neurobiology, Keck School of Medicine, University of Southern California, 1425 San Pablo Street, BCC 512, Los Angeles, CA 90033, USA

Carlos Lois Department of Neurobiology, University of Massachusetts Medical School, Aaron Lazare Medical Research Building, Worcester, MA 01605-2324, USA, carlos.lois@umassmed.edu

Willie H. Mark Mouse Genetics Core Facility, Sloan-Kettering Institute, Memorial Sloan-Kettering Cancer Center (MSKCC), 1275 York Avenue, New York, NY 10065, USA

Javier Martín Transgenic Mice Unit, Biotechnology Program, Spanish National Cancer Research Centre (CNIO), C/Melchor Fernández Almagro 3, 28029 Madrid, Spain

Lajos Mates Biological Research Centre, Hungarian Academy of Sciences, Szeged, Hungary

Séverine Ménoret Platform Rat Transgenesis IBiSA-CNRS, Nantes, France, Severine.Menoret@univ-nantes.fr

Eduardo Moltó Department of Molecular and Cellular Biology, Centro Nacional de Biotecnología (CNB), Consejo Superior de Investigaciones Científicas (CSIC), Campus de Cantoblanco, Darwin 3, 28049 Madrid, Spain; Centro de Investigación Biomédica en Red de Enfermedades Raras (CIBERER), ISCIII, Madrid, Spain

Lluis Montoliu Department of Molecular and Cellular Biology, Centro Nacional de Biotecnología (CNB), Consejo Superior de Investigaciones Científicas (CSIC), Campus de Cantoblanco, C/Darwin 3, 28049 Madrid, Spain; Centro de Investigación Biomédica en Red de Enfermedades Raras (CIBERER), ISCIII, Madrid, Spain, montoliu@cnb.csic.es

Diego Muñoz Department of Molecular and Cellular Biology, Centro Nacional de Biotecnología (CNB), Consejo Superior de Investigaciones Científicas (CSIC), Campus de Cantoblanco, Darwin 3, 28049 Madrid, Spain; Centro de Investigación Biomédica en Red de Enfermedades Raras (CIBERER) ISCIII, Madrid, Spain

Kristina Nagy Samuel Lunenfeld Research Institute, Mount Sinai Hospital, Toronto, ON, Canada, knagy@lunenfeld.ca

Jennifer Nichols Wellcome Trust Centre for Stem Cell Research, University of Cambridge, Tennis Court Road, Cambridge CB2 1QR, UK

Sagrario Ortega Transgenic Mice Unit, Biotechnology Program, Spanish National Cancer Research Centre (CNIO), C/Melchor Fernández Almagro 3, 28029 Madrid, Spain, sortega@cnio.es

Jan Parker-Thornburg University of Texas – M. D. Anderson Cancer Center, 6767 Bertner Avenue Unit 1000, Houston, TX 77030, USA, jpthorn@mdanderson.org

Shirley Pease Director Genetically Engineered Mouse Services, Division of Biology, California Institute of Technology, 156-29, Pasadena, CA 91125, USA, ssp@its.caltech.edu

K. A. Perdue Comparative Medicine Branch, National Institute of Allergies and Infectious Diseases, National Institute of Health, ARTiC-CMB-NIAID-NIH, TwinBrook II, Room 201 D, 12441 Parklawn Drive, Rockville, MD 20852, USA

B. Pintado Transgenic Service. Centro Nacional de Biotecnología, CSIC, C/Darwin 3, 28040 Madrid, Spain, bpintado@cnb.csic.es

Séverine Remy Platform Rat Transgenesis IBiSA-CNRS, Nantes, France; INSERM, UMR643, Nantes 44093, France; CHU Nantes, Institut de Transplantation et de Recherche en Transplantation, ITERT, Nantes 44093, France; Université de Nantes, Faculté de Médecine, Nantes 44093, France, Severine.Remy@univ-nantes.fr

Peter J. Romanienko Mouse Genetics Core Facility, Sloan-Kettering Institute, Memorial Sloan-Kettering Cancer Center (MSKCC), 1275 York Avenue, New York, NY 10065, USA

Dalia Saidley-Alsaadi The Department of Pharmacology and the Cecil H. & Ida Green Center for Reproductive Biology Sciences, University of Texas Southwestern Medical Center, 6001 Forest Park Dr., Dallas, TX 75390, USA

Thomas L. Saunders Transgenic Animal Model Core, University of Michigan Medical School, 2570 MSRB II SPC 5674, 1150 West Medical Center Drive, Ann Arbor, MI 48109, USA; Division of Molecular Medicine and Genetics, Department of Internal Medicine, University of Michigan Medical School, Ann Arbor, MI 48109, USA, tsaunder@umich.edu

Eric N. Schulze Eli and Edythe Broad Center for Regenerative Medicine and Stem Cell Research at USC, Department of Cell and Neurobiology, Keck School of Medicine, University of Southern California, 1425 San Pablo Street, BCC 512, Los Angeles, CA 90033, USA

Manuel Serrano Tumor Suppression Group, Molecular Oncology Program, Spanish National Cancer Research Centre (CNIO), C/Melchor Fernández Almagro 3, 28029 Madrid, Spain

James R. Shirley The Department of Pharmacology and the Cecil H. & Ida Green Center for Reproductive Biology Sciences, University of Texas Southwestern Medical Center, 6001 Forest Park Dr., Dallas, TX 75390, USA

Katerina Strati Telomeres and Telomerase Group, Molecular Oncology Program, Spanish National Cancer Research Centre (CNIO), C/Melchor Fernández Almagro 3, 28029 Madrid, Spain

Andrew E. Syvyk The Department of Pharmacology and the Cecil H. & Ida Green Center for Reproductive Biology Sciences, University of Texas Southwestern Medical Center, 6001 Forest Park Dr., Dallas, TX 75390, USA

J. M. Sztein Comparative Medicine Branch, National Institute of Allergies and Infectious Diseases, National Institute of Health, ARTiC-CMB-NIAID-NIH, TwinBrook II, Room 201 D, 12441 Parklawn Drive, Rockville, MD 20852, USA, szteinj@niaid.nih.gov

Laurent Tesson Platform Rat Transgenesis IBiSA-CNRS, Nantes, France; INSERM, UMR643, Nantes 44093, France; CHU Nantes, Institut de Transplantation et de Recherche en Transplantation, ITERT, Nantes 44093, France; Université de Nantes, Faculté de Médecine, Nantes 44093, France

Lindsay M. Thompson The Department of Pharmacology and the Cecil H. & Ida Green Center for Reproductive Biology Sciences, University of Texas Southwestern Medical Center, 6001 Forest Park Dr., Dallas, TX 75390, USA

Chang Tong Eli and Edythe Broad Center for Regenerative Medicine and Stem Cell Research at USC, Department of Cell and Neurobiology, Keck School of Medicine, University of Southern California, 1425 San Pablo Street, BCC 512, Los Angeles, CA 90033, USA

Claire Usal Platform Rat Transgenesis IBiSA-CNRS, Nantes, France; INSERM, UMR643, Nantes 44093, France; CHU Nantes, Institut de Transplantation et de Recherche en Transplantation, ITERT, Nantes 44093, France; Université de Nantes, Faculté de Médecine, Nantes 44093, France

Margaret L. Van Keuren Transgenic Animal Model Core, University of Michigan, Medical School, 2526 MSRBI, 1150 West Medical Center Drive, 48109 Ann Arbor, MI, USA, mvkeuren@umich.edu

Nguyen VanThuan Center for Developmental Biology RIKEN Kobe, 2-2-3 Minatojima-minamimachi, Kobe 650-0047, Japan; Department of Animal Biotechnology, Konkuk University, 1 Hwayang-dong, Gwangjin-gu, Seoul 143-701, Korea

Cristina Vicente-García Department of Molecular and Cellular Biology, Centro Nacional de Biotecnología (CNB), Consejo Superior de Investigaciones Científicas (CSIC), Campus de Cantoblanco, Darwin 3, 28049 Madrid, Spain; Centro de Investigación Biomédica en Red de Enfermedades Raras (CIBERER), ISCIII, Madrid, Spain

Sayaka Wakayama Center for Developmental Biology RIKEN Kobe, 2-2-3 Minatojima-minamimachi, Kobe 650-0047, Japan

Teruhiko Wakayama Center for Developmental Biology RIKEN Kobe, 2-2-3 Minatojima-minamimachi, Kobe 650-0047, Japan, teru@cdb.riken.go.jp

Elizabeth Williams Transgenic Animal Service of Queensland. University of Queensland Biological Resources, University of Queensland, Brisbane 4072, QLD, Australia, elizabeth.williams@uq.edu.au

Qi-Long Ying Eli and Edythe Broad Center for Regenerative Medicine and Stem Cell Research at USC, Department of Cell and Neurobiology, Keck School of Medicine, University of Southern California, 1425 San Pablo Street, BCC 512, Los Angeles, CA 90033, USA, qying@usc.edu

Michael G. Zeidler Transgenic Animal Model Core, University of Michigan, Medical School, 2574 MSRBII, 1150 West Medical Center Drive, 48109 Ann Arbor, MI, USA, zeidler@umich.edu

Chapter 1

Patent and Licensing Issues in Transgenic Technology

Karen S. Canady

Abstract

The use and study of transgenic organisms raises legal issues, particularly the potential for patent infringement if a necessary license is not obtained. Scientists, regardless of whether they practice in an academic or corporate setting, should be aware of patents relating to the research tools in use. An understanding of the intellectual property issues involved enables one to avoid unwanted legal obstacles. An overview of patent and licensing issues provides an introduction to why and how inventions are patented. A summary of some exemplary patents in the field of transgenic technology and their claims provides guidance in navigating the relevant patent landscape.

Abbreviations

BAC	Bacterial artificial chromosome
cDNA	Complementary DNA
DNA	Deoxyribonucleic acid
EST	Expressed sequence tag
IRES	Internal ribosome entry site
KCTT	Karolinska Center for Transgene Technologies
MTA	Materials transfer agreement
NAIL	Natural killer cell activation inducing ligand
NLS	Nuclear localization signal
PAIR	Patent application information retrieval
PTO	Patent and Trademark Office
U.S.	United States
USPTO	United States Patent and Trademark Office

S. Pease and T.L. Saunders (eds.), *Advanced Protocols for Animal Transgenesis*, Springer Protocols,
DOI 10.1007/978-3-642-20792-1_1, © Springer-Verlag Berlin Heidelberg 2011

1.1 Introduction

The use and study of transgenic organisms raises legal issues, particularly the potential for patent infringement. Scientists, regardless of whether they practice in an academic or corporate setting, should be aware of patents relating to the research tools in use. An understanding of the intellectual property issues involved enables one to avoid unwanted legal obstacles. This chapter provides an overview of patent and licensing issues, as well as some guidance on navigating the patent landscape and a review of some exemplary patents in the field of transgenic technology. The discussion is presented from the perspective of the U.S. patent system. While other jurisdictions have their own patent systems, many of the same concepts and principles apply to patents in other countries.

1.2 The Patent System

The power to grant patents was given to Congress by the framers of the U.S. Constitution as a means to promote economic development through progress in science and technology. A patent grants an inventor the right to exclude others from making or using the claimed invention for a limited period of time, in exchange for disclosing how to make and use the invention. Ensuring that inventors are able to reap the rewards of their efforts, rather than have others with more power or resources benefit from the inventor's discovery, encourages investment in research and development.

One common misconception about patents is the belief that a patent grants its owner the right to practice the claimed invention. A patent is not an affirmative right to the invention, but a right to exclude others. Some patents dominate others such that a broad patent covering a new technology might dominate a patent to an improvement on that core technology. For example, the first inventor to figure out how to get snakes to grow fur might obtain a patent covering all snakes genetically modified to grow fur. A second party might later improve on that technology to engineer a snake that grows angora. The second party would be able to patent the improvement, but could not practice that invention without a license from the owner of the earlier, dominating patent.

A second common misconception is that patents do not affect academic or nonprofit institutions. Although patent law allows for an experimental use exception, this exception has been interpreted quite narrowly. The courts have made it abundantly clear that academic research per se is not exempt from patent infringement. The work that begins in an academic setting is often later patented and licensed to outside entities. Even the enhancement of an academic institution's reputation in a field of study is considered furtherance of the commercial purpose of the institution.

The Bayh-Dole Act was passed in 1984 for the purpose of facilitating the patenting of inventions developed by research funded by the federal government. In part, this was a recognition that encouraging academic institutions to patent their inventions would benefit the objectives of federally funded research. By patenting and thereby acquiring the right to negotiate licenses on terms that favor bringing the nascent technology to market, universities could both further their mission to benefit the public with their developments and cultivate a revenue stream that could, in turn, be used to fund more research.

The challenge lies in understanding what activities are covered by a given patent. Understanding the scope encompassed by a patent's claims is not straightforward. First, the language of the claims themselves must be construed in light of both the language of the patent's text, known as the "specification," and the file history. The file history refers to the record of communications between the patentee and the Patent and Trademark Office (PTO) that occurred from the time the patent application was filed through the issuance of the patent. Second, the process of examining and issuing patents is imperfect. The mere existence of a patent does not necessarily mean that all of its claims are valid and therefore enforceable. It is therefore helpful to understand how to evaluate patentability and patent validity.

1.2.1. Patentability

To be patented, an invention must comprise patentable subject matter, must be new, useful and nonobvious, and must be supported by an adequate written description. Patentable subject matter is defined broadly in the U.S. Code as "any new and useful process, machine, manufacture, or composition of matter, or any new and useful improvement thereof" [1]. The U.S. Supreme Court has identified just a few categories of subject matter as outside this definition of patentable subject matter: laws of nature, physical phenomena, and abstract ideas [2]. For example, a new mineral discovered in the earth or a new plant found in the wild is not patentable subject matter [3]. A natural product is patentable, however, in its purified form (e.g., prostaglandins, adrenaline). Examples of patentable subject matter in the biotechnology field include: DNA that has been newly identified and claimed in purified or isolated form (rather than in its naturally occurring state), new organisms or parts of organisms, such as genetically modified seeds, cells, or animals, as well as vectors containing modified DNA.

The requirement for utility means the invention must be "useful," in keeping with the statutory language. This means a use for the claimed invention must be either apparent or explicitly stated in the patent application. The rare instances in which an invention is found lacking in utility typically involve subject matter whose use has not yet been identified. In the biotechnology field, this concern arose primarily in connection with attempts to patent expressed sequence tags (ESTs) without identifying what the ESTs would encode. An EST, although representing patentable subject matter, can only be patented if a credible and substantial utility specific to the claimed EST has been described.

The novelty criterion requires that the invention not be known prior to the filing of the patent application. The applicant cannot have made the invention known to the public through a published article or have sold the invention more than 1 year before the filing date. An invention cannot be patented if it was previously described in another's patent application, nor can the applicant patent the invention of another who has not suppressed, abandoned, or concealed the invention [4]. Many inventions developed in academia fail to meet the novelty requirement because of journal publications, abstracts presented at meetings, or funded grant applications that disclosed the invention more than 1 year before the U.S. patent application was filed. The date of disclosure for this analysis is the first date on which the item was available to the public, e.g., via online publication or via request from a public agency, regardless of the actual publication date that may appear on a corresponding printed publication of the same item.

The nonobviousness criterion means the invention must differ from what was previously known by more than just obvious modifications. An invention based on such a modification is obvious if "the differences between the subject matter sought to be patented and the prior art are such that the subject matter as a whole would have been obvious at the time the invention was made to a person having ordinary skill in the art to which the subject matter pertains" [5]. In biotechnology, the issue of obviousness has arisen in the context of claims to DNA encoding a known protein. When sequencing technology was nascent, claims to a cDNA encoding a known protein were considered nonobvious and therefore patentable because the mere availability of cloning and sequencing technology did not make apparent the specific nucleotide sequence claimed [6]. The same Court of Appeals for the Federal Circuit came to a different conclusion in an April 2009 decision that held claims to the cDNA encoding natural killer cell activation inducing ligand (NAIL) obvious and unpatentable because of a publication describing p38, a protein later found to be encoded by the NAIL cDNA [7]. The description of the encoded protein, together with the cloning manual published by Sambrook et al., was considered enough to show that it would have been obvious to isolate the claimed NAIL cDNA. It appears that claims to DNA sequences encoding a known protein cannot be patented if standard cloning techniques are used to identify the sequence. Such claims may be patentable, however, where the applicant can show that discovering the particular sequence required overcoming a technical challenge that made achieving that end result unpredictable.

Finally, the patent statute requires a written description of the invention that includes enough particularity that those persons with ordinary skill in the art can make and use the invention [8]. There are two key aspects to this disclosure requirement. The written description requirement is most concerned with ensuring that the application not claim subject matter beyond what was described at the time the application was filed. The other aspect is referred to as the enablement requirement, which requires that the description be sufficient to enable one skilled in the relevant art to be able to make and use the invention as claimed. In addition, the description must include the best mode contemplated by the inventor of carrying out the invention. This latter "best mode" requirement rarely raises a problem during the process of examination to obtain issuance of a patent, but has been used as a basis for seeking to invalidate an issued patent during litigation.

1.2.2. Patent Rights and Infringement

A patent does not confer an affirmative right to practice the claimed invention. Rather, the patent grants a right to exclude others from making or using the invention as claimed. This means that a party who practices the invention must have permission (referred to as a license) from the patent owner, or risk being sued for patent infringement. Where infringement is found, the remedy can be an injunction (preventing the activity) and/or monetary compensation based on a "reasonable royalty." If the infringement is found to be willful, the damage award can be tripled. The threat of treble damages serves to motivate parties to negotiate a license with the patent owner rather than await the outcome of litigation.

Academics should note that university research often fails to qualify for the experimental use defense to infringement. Acts "in furtherance of the alleged infringer's legitimate business" and "not solely for amusement, to satisfy idle curiosity, or for strictly philosophical inquiry" do not qualify for this exception [9]. The courts have considered nonprofit status to carry little weight in the analysis, nor is it determinative that the activity is not being pursued for commercial gain. In *Madey v. Duke University*, the Court of Appeals for the Federal Circuit noted that university research has been regarded as an activity in furtherance of the legitimate business of "educating and enlightening students and faculty" (*Id.*).

If a researcher is using patented materials or methods in the course of research, it is likely the owner of the patent rights expects to extract revenue from such use. This is especially true in the case of materials whose primary utility is in a research setting. In the case of products sold by a commercial vendor, the sale may well be authorized by the patent owner and notification of any restrictions on its use would accompany the product at the time of purchase. Where the research tool can be produced and/or practiced by the researcher without involvement of a commercial vendor, the researcher proceeds at his or her own peril.

A patent owner may become aware of the unauthorized practice of the claimed invention when the work is published or presented at a conference. Because damages for patent infringement can be trebled if it is shown to have been willful, the patent owner is likely to provide written notice of the suspected infringing activity accompanied by an offer of a license or a request to cease and desist the activity. At such point, the researcher should either seek professional advice as to the validity of the potential infringement or negotiate terms with the patent owner. Merely receiving a cease and desist letter, however, does not necessarily mean that infringement has occurred. It would be unwise to respond to such a letter without first obtaining advise of counsel, as the actions taken at this point can have significant consequences.

1.3 Licensing

A license can be granted in exchange for a fixed fee or for a royalty based on items sold, amount of use, or a portion of revenue. When considering whether to take a license or how to negotiate fair value, relevant factors include the amount of damages at stake and the likelihood of being sued successfully for infringement. The greater the potential for extracting commercial value from the activity, such as selling a product or service, the greater the risk of being sued for infringement and the higher the potential damage award. On the other hand, some purely academic pursuits would be unlikely to be the target of an infringement suit, even if it did not qualify for the narrowly construed "experimental use" exception to patent infringement. The key, however, is likely the extent to which a patent owner's commercial interests are threatened or compromised by the activity. If the contemplated activity could generate significant revenue, or otherwise threatens a source of revenue for the patent owner, then it may be worthwhile for the patent owner to seek a legal remedy.

One example of licensing in transgenic technology is the sublicense offered by the Biomolecular Research Facilities at the University of Missouri-Columbia [10]. Their web site explains that:

> Methods of pronuclear microinjection are patented under United State Patent No. 4,873,191, entitled "Genetic Transformation of Zygotes," assigned to Ohio University and licensed to Xenogen Biosciences. The University of Missouri currently holds a Sublicense Agreement which provides for the generation of transgenic animals for faculty at the University of Missouri, as well as for individuals at other organizations. For individuals external to the University of Missouri, a specific MTA form must be completed before service can be provided.

The "MTA" referenced in the above text is a materials transfer agreement. This is a contract often used to govern the terms under which one party will provide materials, such as biological material (antibodies, vectors, etc.) to another. MTAs are often used to clarify the permitted uses of the materials to be provided, as well as other intellectual property issues, such as who will own the rights to any new inventions developed by the party receiving the transferred material.

The referenced patent, U.S. Patent No. 4,873,191 (the "'191 patent"), was issued on October 10, 1989. Based on this issue date, the patent would have expired in October of 2006. A check of the status of this patent at http://www.uspto.gov shows that the required maintenance fees were paid, the twelfth year and final payment having been made on March 7, 2001. The continuity data for this patent show that a related patent, U.S. Patent No. 6,872,868 (the '868 patent) entitled "Transgenic Mammals," issued on March 29, 2005, was filed as a continuation of this same '191 patent. The '868 patent is based on an application filed in May 1995, and is therefore set to expire in March 2022, or 17 years from the issue date. No pending patent applications claiming benefit to either of these patents are identified in the patent records.

Thus, although one does not need a license to practice the invention claimed in the '191 patent mentioned in the sublicense offered by the University of Missouri-Columbia, the invention claimed in the '868 patent is currently covered by a patent. One interested in practicing this technology, therefore, would want to know what was claimed in each of these patents. The invention claimed in the expired patent is now in the public domain, while that covered by the claims of the '868 patent is not.

The '191 patent claimed a "method of obtaining a mammal characterized as having a plurality of cells containing exogenous genetic material, said material including at least one gene and a control sequence operably associated therewith, which, under predetermined conditions, express said gene under the control of said control sequence in a cell of said mammal, which comprises (a) introducing exogenous genetic material into a pronucleus of a mammalian zygote by microinjection" to obtain a genetically transformed zygote, (b) transplanting an embryo derived from such a zygote into a pseudopregnant female, and (c) allowing the embryo to develop to term.

Claim 1 of the '868 patent is to a "nonhuman transgenic mammal whose somatic and germ cells contain exogenous genetic material, wherein said material does not include any virus-specific DNA and includes at least one heterologous gene and a transcriptional control sequence operably associated therewith, wherein said gene is expressed at a detectable level in a plurality of said somatic cells or said germ cells, where said genetic material is selected so that the normal development of the embryo to term is not prevented by said material, where said mammal is selected from the group consisting of rodents, rabbits, goats, pigs, cattle, and sheep." Claim 6 of this patent is also directed to such a transgenic mammal, but explicitly recites that the mammal is produced by the steps of "introducing exogenous genetic material into a pronucleus of a mammalian zygote by microinjection" to obtain a genetically transformed zygote, transplanting an embryo derived from such a zygote into a pseudopregnant female, and allowing the embryo to develop to term. Claim 8 of this patent is directed to a method of producing a polypeptide or protein using such a transgenic mammal.

Although the sublicense terms set forth by the University of Missouri-Columbia reference the expired '191 patent, it may be difficult to proceed with the same technology without infringing the '868 patent, unless the exogenous genetic material in use includes virus-specific DNA. While the agreement should be updated to exclude reference to the expired patent, the University can continue to offer its services subject to its own terms of use.

The terms clarify that the sublicense granted thereby is limited to making transgenic animals for nonprofit organizations, provided that the transgenic animals and materials are used solely for internal noncommercial research purposes in the designated research field. The final term of the agreement explicitly states that no implied right or license is granted to the sublicensee to utilize the technology, animals, or materials in a manner not expressly included within the scope of the agreement. This document makes it clear that, even if you pay the University to prepare transgenic materials for you, you are not free to do as you wish with that material.

This example illustrates how licensing patented material differs from purchasing patented goods. The first sale doctrine, also known as the doctrine of patent exhaustion, applies to the latter. Under the doctrine, the first unrestricted sale of patented goods exhausts the patent owner's control over the item. Where the goods are provided under a license that restricts the terms of use of the patent goods, as is the case with the University of Missouri-Columbia's Pronuclear Microinjection Sublicense Terms, the patent owner retains control over the licensee's use of the goods.

Another example of an agreement governing pronuclear injection can be found in the Appendix to the pronuclear injection order form offered by the Karolinska Center for Transgene Technologies (KCTT, see http://www.kctt.ki.se). This appears to relate to pronuclear injection of DNA constructs into fertilized mouse eggs, an activity that, on its face, may run afoul of the above-mentioned '868 patent if mammals in accordance with the '868 patent were brought into the United States.

1.4 Freedom to Operate and Clearance Searching

While there is no duty to conduct a patent search, there are circumstances for which a search of patent databases may be advisable. This type of searching is referred to as a freedom to operate search or clearance searching. If one is about to invest significant resources in an endeavor for which a later judgment of patent infringement would be devastating, a clearance search can offer either peace of mind or an early warning. An example of such a circumstance would be starting up a new business venture that relies on exploitation of a core technology. If one or more blocking patents are turned up by the search, one can investigate options such as the availability and cost of a license or the feasibility of working around the patent.

Patent databases of most countries are searchable via the internet and accessible to the public. The United States Patent and Trademark Office (USPTO) publishes all issued patents as well as pending applications that were filed on or after November 1, 2000, and which have been pending for at least 18 months (or 18 months have elapsed since the earliest priority date). Searchable databases, one for issued patents and one for published pending applications, can be accessed at http://www.uspto.gov. Careful selection of a variety of search terms is advised, as one can easily miss relevant patent documents. The terminology used in connection with the area of technology may have changed since the technology was in its infancy or the drafter of the patent document used unconventional language in the description and/or claims. Professional searches can be ordered for those seeking a greater degree of confidence in a well-conducted and thorough search, typically costing in the range of several hundred dollars.

Reviewing and interpreting patent search results can be the greater challenge. Titles and abstracts can provide initial clues in identifying which patent documents deserve further review. When evaluating freedom to operate in a given technological area, the critical determinant is the language of the patent claims (or potential future claims, in the case of a pending patent application). It is not uncommon for a patent to describe the technology in terms much broader than the scope of its claims. It is the claims, however, that define the metes and bounds of what is covered by the patent. The patent specification, that is the description of the invention that precedes the claims, is used to inform the meaning of the terms used in the claims and is the primary source for resolving any ambiguities in claim language.

If a search turns up a patent whose claims encompass the contemplated activity, this does not necessarily mean that a problem with patent infringement exists. First, one must assess whether the patent is valid and enforceable. A patent may have expired or lapsed due to failure to pay required maintenance fees, or it may have issued with claims that would not be found valid if subject to a legal challenge. To determine whether a patent has expired, one must first look to the filing date and any claimed priority dates. U.S. patents expire 20 years from the earliest filing date, or 17 years from the date of issue, if the patent was filed prior to November 1, 2000. Even if the patent's term has not yet expired, the patent may have been abandoned for failure to pay maintenance fees required of all patents at designated intervals during the patent's term. Records indicating maintenance fee status for issued patents can be found at the USPTO web site, http://www.uspto.gov. Finally, one might be able to modify the technical approach employed to avoid infringement.

1.5 Patent Validity

Evaluating whether a patent's claims would hold up to a legal challenge is far less straightforward, and generally requires the expertise of a patent attorney. An issued patent is presumed valid, meaning that the burden of proving invalidity rests with the challenger. A patent can be invalidated if it is shown, with clear and convincing evidence, that the claims fail to meet any of the requirements for utility, novelty, nonobviousness, definiteness, written description, and enablement. Invalidating a patent for lack of utility is rare and unlikely to succeed. While there are many examples of patents that have been successfully challenged for failure to satisfy the requirements for definiteness, written description, or enablement, the subjective nature of these standards make the outcome of such challenges difficult to predict. These types of challenges succeed in the biotechnology area most often in cases in which the patent issued prior to developments in the case law that defined these standards more stringently. For example, claims that appear broad in scope relative to the embodiments described in detail in the patent specification are more likely to be invalidated for lack of adequate written description and/or sufficient enabling disclosure.

A stronger case for invalidating a patent typically requires identifying a piece of prior art that anticipates the patent claims. To mount an effective challenge, the prior art should be one that was not cited during examination of the patent in question. A list of prior art references considered during examination is listed on the front page of issued patents. Proving invalidity by anticipation means showing that a single document (journal article, publicly available grant application, published patent, or patent application) that predates the earliest priority date of the patent discloses each element of the claim. The teaching of each element must be present, either explicitly or inherently, in a single reference.

Where multiple prior art documents can be combined to show the invention, it can be argued that the claimed invention was obvious and therefore not patentable. To be successful challenging a patent for obviousness, however, one should have a strong argument that one skilled in the relevant art would have been motivated to combine and/or modify the elements required to arrive at the claimed invention with a reasonable expectation of success. It is much more risky to base a patent challenge on obviousness as compared to anticipation. Obviousness is more easily established where the invention involved a relatively predictable technology.

Advice of qualified patent counsel should be sought before proceeding to invest substantial sums in practicing in an area where there is risk of patent infringement. Preliminary searching and evaluation can be done prior to contacting an attorney, to clarify concerns and focus the task. A good attorney will guide the client in determining whether a formal opinion or further searching is warranted under the circumstances. The scope of such an evaluation can range from a few hours to a few months, with the corresponding range of expense, making it important to have a clear understanding of the client's needs and expectations at the outset.

1.6 Selected Patents in Transgenic Technology

Each endeavor involves the use of a unique combination of technological tools, making it impractical to provide here a thorough review of the patent landscape in transgenic technology. Some representative patents in the field are summarized below to provide a few examples of patents that have issued in this area and a sense for differences in patent claim scope. No representations are made regarding the validity of the patents discussed herein, nor regarding the applicability of their claims to a particular circumstance. Analyses of patent validity and infringement are not only specific to individual circumstances, but also they involve considerations too complex to elaborate in a review of this nature.

While the following provides analysis of a number of patents listed by first inventor name and patent number, these examples are used to exemplify how patent records can be reviewed and analyzed. The first patents listed relate to transgenic avian technology, but also provide an introduction to reviewing patent abstracts and claims. In some examples, the abstract is not a good indicator of claim scope and content. The fourth patent listed (Sect. 1.6.4) provides an example of continuation-in-part filings and the significance of applications having multiple filing and priority dates.

Section 1.6.5 deals with a patent relating to transgenic fish and also exemplifies relatively narrow patent claims. Section 1.6.6 presents a patent in the area of nuclear transfer and one which is part of an active family of related patents, all of which should be reviewed by one employing these techniques. A patent in the field of intracytoplasmic sperm injection is addressed in Sect. 1.6.7. This case illustrates broad claim coverage. Finally, in Sect. 1.6.8, a relatively mature patent relating to transgenic bovines is presented, as is the topic of divisional patent applications.

1.6.1. Patent No. 7,534,929: Avians Expressing Heterologous Protein

U.S. Patent No. 7,534,929, entitled "Avians Expressing Heterologous Protein," issued May 19, 2009, to Robert D. Ivarie et al. Based on the information printed on the front page of the patent, this technology is assigned to (owned by) Synageva BioPharma Corporation of Waltham, Massachusetts and University of Georgia Research Foundation, Inc. of Athens, Georgia.[1] The front of each patent also includes an abstract that summarizes the invention disclosed in the patent. The abstract of this patent reads:

> This invention provides vectors and methods for the stable introduction of exogenous nucleic acid sequences into the genome of avians in order to express the exogenous sequences to alter the phenotype of the avians or to produce desired proteins. In particular, transgenic avians are produced which express exogenous sequences in their oviducts and which deposit exogenous proteins into their eggs. Avian eggs that contain exogenous proteins are encompassed by this invention. The instant invention further provides novel forms of interferon and erythropoietin which are efficiently expressed in the oviduct of transgenic avians and deposited into avian eggs.

While the abstract provides a useful guide to the contents of a patent, it is the claims that define the scope of the intellectual property rights conferred by the patent. It is not uncommon for the abstract to describe the invention in very broad terms. The claims may well be limited to some narrower embodiments of the invention. Note also that the original patent application may have been restricted by the examiner, meaning the applicant was

[1]To verify the current status of a patent's ownership, one can search the records of the Assignment Branch of the USPTO on their web site (http://www.uspto.gov). The assignee named on the printed patent was the owner at the time the patent issued, but this could have changed subsequent to issuance.

required to pursue some of their claims in a divisional application, which would give rise to a separately issued patent. If the abstract describes technology for which you would like to know if a patent covers it, check for other patents issued to the same inventor, same assignee, or claiming priority to the same application.[2]

1.6.2. Patent No. 7,521,591: Transgenic Chickens

Just 1 month earlier, another patent issued to Ivarie, No. 7,521,591, entitled, "Transgenic Chickens That Lay Eggs Containing Exogenous Proteins" (the "'591 patent"). Although the abstract for this patent emphasizes the provision of eggs which contain exogenous proteins, the abstract also mentions vectors and methods for stable introduction of sequences into the genome of a bird.[3] Based on the abstract, the difference between these two patents is the further provision of "novel forms of interferon and erythropoietin which are efficiently expressed in the oviduct of transgenic avians and deposited into avian eggs." Looking at the claims, however, shows that the invention also encompasses a method of purifying such proteins as interferon and erythropoietin from a transgenic egg produced by a transgenic chicken of the invention.[4] While the claims of this patent do not cover the proteins interferon and erythropoietin per se, a party using the claimed method, or the claimed transgenic chickens, to

---

[2]In addition to noting related patents and applications listed on the front page of the patent, one can search the USPTO records for other patents and applications referencing the same application serial number to identify additional related patents. Using the "quick search" or "Boolean search" options, one can search by inventor name, assignee, serial number, title, as well as by keyword. In addition, related patents can be identified using the Patent Application Information and Retrieval (PAIR) available via http://www.uspto.gov (look for "Public PAIR"). In Public PAIR, one can retrieve the records of a particular patent (issued patent or published pending application). The tab "continuity data" provides links to related patents and pending applications.

[3]The abstract reads:

This invention provides eggs which contain exogenous proteins. The invention further provides transgenic chickens which express exogenous sequences in their oviducts, and vectors and methods for the stable introduction of exogenous nucleic acid sequences into the genome of a bird for expressing said exogenous sequences to alter the phenotype of the bird or to produce desired proteins.

[4]Selected claims (of the 43 total) read:

1. A germ-line transgenic chicken whose genome contains a transgene introduced by a replication-deficient retroviral vector which lays an egg containing an exogenous protein encoded by the transgene wherein the exogenous protein is produced in the transgene chicken oviduct at a detectable level.

6. A transgenic chicken whose genome contains a transgene introduced by a replication-deficient retroviral vector which lays an egg containing a cytokine encoded by the transgene wherein the cytokine is produced in the transgenic chicken oviduct at a detectable level and is exogenous to the chicken.

7. The transgenic chicken of claim 6 wherein the cytokine is interferon.

8. The transgenic chicken of claim 6 wherein the cytokine is GM-CSF.

9. The transgenic chicken of claim 6 wherein the cytokine is G-CSF.

10. The transgenic chicken of claim 6 wherein the cytokine is erythropoietin.

11. A method comprising purifying an exogenous protein from an egg of a transgenic chicken whose genome contains a transgene introduced by a replication-deficient retroviral vector which lays the egg containing the exogenous protein at a detectable level wherein the exogenous protein is encoded by the transgene.

12. The method of claim 11 wherein the exogenous protein is interferon.

13. The method of claim 11 wherein the exogenous protein is G-CSF.

14. The method of claim 11 wherein the exogenous protein is erythropoietin.

produce these proteins would infringe the '591 patent. This party would also infringe any patents covering these proteins and/or their use for particular methods, if applicable to their activities.

1.6.3. Patent No. 7,527,966: Gene Regulation in Transgenic Animals

Patent number 7,527,966 (the "'966" patent) issued to Cooper et al. in May 2009, and is entitled "Gene Regulation in Transgenic Animals Using a Transposon-Based Vector." The abstract describes administration of vectors to create transgenic animals, animals producing transgenic progeny, production of large quantities of molecules encoded by the transgene, as well as transgenic egg-laying animals that produce large quantities of these molecules and deposit them in the egg. This abstract implies an intent to patent transgenic animals, both broadly and more specifically, transgenic egg-laying animals, as well as vectors and methods of using such vectors to create the transgenic animals.

Most likely, the patent application was filed with claims directed to these various aspects of the invention. The patent as issued, however, contains only two claims, both directed to very specifically identified vectors.[5] This suggests that a look at the file history of this patent would show that the applicants had to elect vectors as one category of invention for prosecution in this case in response to a requirement for restriction. The extensive limitations recited in claim 1, further suggest the applicant may have been required to amend the claims to add these limitations in order to overcome rejections by the patent examiner.

---

[5]The abstract and claims:

Administration of modified transposon-based vectors has been used to achieve stable incorporation of exogenous genes into animals. These transgenic animals produce transgenic progeny. Further, these transgenic animals produce large quantities of desired molecules encoded by the transgene. Transgenic egg-laying animals produce large quantities of desired molecules encoded by the transgene and deposit these molecules in the egg.

1. A vector comprising a nucleic acid sequence as set forth in SEQ ID NO: 31, SEQ ID NO: 32, SEQ ID NO: 42, or SEQ ID NO: 43, wherein each sequence comprises: (a) a prokaryotic transposase gene operably linked to a first promoter, wherein the nucleic acid sequence 3′ to the first promoter comprises the Kozak sequence as set forth in SEQ ID NO: 13, the Kozak sequence being positioned so as to include at least the first codon of the transposase gene, wherein the transposase gene is modified such that a plurality of the codons of the transposase gene that encode for amino acids 2–10 of a transposase protein encoded by the transposase gene are individually modified from the wild-type sequence of cytosine or guanine at the third base position of the codon to an adenine or a thymine, such that the modification does not change the amino acid encoded by the modified codon, and wherein the first promoter is a viral or a eukaryotic promoter; (b) one or more genes of interest operably-linked to one or more additional promoters, wherein at least one of the genes of interest encodes for proinsulin or human growth hormone; and (c) insertion sequences recognized by a transposase encoded by the modified transposase gene, wherein the transposon insertion sequences are positioned to flank the one or more genes of interest and their operably-linked promoters.

2. A vector comprising the nucleic acid sequence as set forth in SEQ ID NO: 31, SEQ ID NO: 32, SEQ ID NO: 41, SEQ ID NO: 42, or SEQ ID NO: 43.

1.6.4. Patent No. 7,550,650: Production of a Transgenic Avian by Cytoplasmic Injection

Patent No. 7,550,650, entitled "Production of a Transgenic Avian By Cytoplasmic Injection," issued on June 23, 2009, to Rapp and, per the printed patent, is assigned to Synageva BioPharma Corp. of Waltham, Massachusetts. According to the abstract, this patent provides methods for the stable introduction of heterologous coding sequences into the genome of a bird and expressing the coding sequences to produce desired proteins or to alter the phenotype of the bird. It describes methods for introducing a transgene into the cytoplasm of avian embryonic cells by cytoplasmic microinjection. The embryo then develops into a transgenic adult capable of expressing a heterologous protein and/or capable of generating a line of transgenic birds through breeding. The abstract also mentions synthetic vectors and gene promoters useful in the methods, as well as transgenic birds that express heterologous protein and avian eggs containing heterologous protein.

The front page of the patent also lists related patent applications. Under the heading "Cross Reference To Related Applications," it states:

> The application is a continuation-in-part and claims the benefit of U.S. application Ser. No. 10/251,364, filed Sep. 18, 2002, now issued U.S. Pat. No. 7,312,374, issued Dec. 25, 2007, which claims the benefit of U.S. Provisional Application No. 60/322,969, filed Sep. 18, 2001, and U.S. Provisional Application No. 60/351,550, filed Jan. 25, 2002, all of which are incorporated by reference herein in their entireties.

1.6.4.1. Continuation-in-Part Patents

The indication that this is a "continuation-in-part application" means that this patent is based on the contents of both an earlier, "parent" filing submitted in September 2002, and a further filing that added more disclosure to the application that was submitted in October 2003. Both of these filings claim priority back to the two provisional application filing dates in September 2001 and January 2002. Each of these filing dates, and the contents of the applications at each of those points in time, carries significance for the validity (and thus enforceability) of the claims of the patent that issued on the basis of this continuation-in-part application.

Each claim of a patent is entitled to the priority date of the earliest filing in a series of related filings, such as the series described previously, as long as the contents of that filing provides an adequate written description and enabling disclosure to support that claim. The priority date to which a claim is entitled determines the relevant date used for determining whether the claimed invention is novel and nonobvious over the prior art.

For example, two claims of a patent, one claim to Protein X, and another claim to Protein X fused with Tag Q, might be entitled to two different priority dates. Suppose that a provisional application was filed on May 12, 2001, which described Protein X by molecular weight and other characteristics, but not amino acid sequence. On May 12, 2002, a regular, nonprovisional patent application was filed, claiming the benefit of the provisional application filing date and further disclosing the amino acid sequence of Protein X. In 2004, a continuation-in-part application was filed, claiming priority to both the provisional and the "parent" filing of May 12, 2002, and adding a description of Tag Q and its amino acid sequence.

In this example, the claim to Protein X would likely be entitled only to the May 12, 2002, filing date. In most circumstances, a protein is not considered to have an adequate written description without an amino acid sequence.[6] The claim to Protein X plus Tag Q, however, would only be entitled to the filing date of the continuation-in-part application in 2004. This means that the relevant state of the art for evaluating validity of the claim to Protein X is May 12, 2002. If another party filed a patent application or published a journal article in April of 2001 that disclosed Protein X, the claim to Protein X would not be valid, unless the Patentee could successfully argue that the identifying characteristics of Protein X that were disclosed in the provisional application were sufficient to meet the written description requirement.

We can further suppose for this example that Tag Q was first described by another scientist in a poster presentation at a meeting in 2003. Validity of the claim to Protein X fused to Tag Q would then depend on whether it would have been obvious to one skilled in the art of protein chemistry to combine the known Protein X with Tag Q to create this fusion protein. Even if that were the case, the Patentee might argue that the disclosure of Tag Q in the poster of 2003 was not sufficient to teach one skilled in the art how to make and use Tag Q, perhaps because the sequence was not described at that time. Alternatively, the Patentee might argue that the nature of the disclosure of Tag Q in the 2003 poster did not provide any suggestion or motivation to use it in conjunction with a protein like Protein X. Thus, each filing date in the series can be important to the patent analysis.

1.6.4.2. Broad Independent Claims

Returning to the Rapp patent, we can next look at the claims. Claim 1 is rather broadly directed to producing transgenic birds via microinjection of an embryonic cell:

> A method of producing a transgenic avian, the method comprising: microinjecting into a cell of an avian embryo a DNA molecule comprising a transgene containing a nucleotide sequence encoding a heterologous

[6]This would depend on the facts surrounding a particular protein and how specifically it could be identified by characteristics other than amino acid sequence. For example, antibodies are typically and adequately described by their ability to specifically bind a given antigen. In such an instance, however, the structure of the antigen must be known (e.g., by amino acid sequence). ***Noelle v. Lederman***, 355 F.3d 1343, 69 U.S.P.Q.2d 1508 (Fed. Cir. 2004). Regarding claims to a protein not supported by description of the amino acid sequence, see ***In re Wallach***, 378 F.3d 1330, 71 U.S.P.Q.2d 1939 (Fed. Cir. 2004).

polypeptide; introducing the microinjected avian embryo into an oviduct of a recipient hen, such that the recipient hen lays a shelled egg containing the microinjected avian embryo; and incubating the shelled egg containing the microinjected avian embryo until the shelled egg hatches; testing a hatched chick for the presence of the transgene; and developing a chick that tests positive for the transgene to sexual maturity, thereby producing a transgenic avian containing the transgene.

This claim is rather broad in that it applies to use of cellular microinjection with any avian species, and to using a transgene encoding any heterologous protein. A broad claim such as this can be advantageous for the patentee, as it would dominate the field of transgenic bird production via cytoplasmic injection. On the other hand, such broad claims can be vulnerable to a validity challenge based on either prior art that was not considered during examination of the patent application, or evidence that the application on which the patent is based did not enable the full breadth of the claim. For example, if one were to find examples of avian species that respond differently to cytoplasmic injection, suggesting the ability to adapt the technique exemplified in chickens to other species is impractical or unpredictable. Dependent claim 4 of the Rapp patent requires that the avian be a chicken. This dependent claim provides an important back-up position for the Patentee, as it would survive a challenge based on insufficient enablement for applying the invention to other species.

1.6.4.3. Dependent Claim Strategies

Similarly, claim 2 provides a back-up position should one successfully argue that claim 1 improperly encompasses microinjecting an avian embryo at an embryonic stage that would not support successful practice of the invention. Claim 2 requires that the avian embryo be a stage I embryo. Another purpose served by dependent claims that recite more specific features is that it provides a claim that may be more closely directed to a competitor's infringing activity. A competitor who is injecting DNA containing a transgene into chicken embryos at stage I is infringing all three of claims 1, 2, and 4 (and perhaps others).

The Rapp patent includes additional dependent claims directed to the incorporating into the transgene a transcriptional regulatory element that can direct gene expression in one or more cells of the transgenic avian, such as a promoter region of an avian gene which encodes ovalbumin, lysozyme, ovomucoid, ovomucin, conalbumin, or ovotransferrin. Other dependent claims relate to use of a nuclear localization signal (NLS) peptide, an internal ribosome entry site (IRES), or a bacterial artificial chromosome (BAC). Also claimed is use of at least two nucleotide sequences each encoding a heterologous polypeptide that are introduced into the avian embryo, such as heavy and light chains of an antibody.

One dependent claim specifies that the DNA molecule is not a eukaryotic viral vector. This may have been included to avoid potential validity issues relating to the use of such viral vectors. This would be an example of a limitation a patentee might prefer not to have to include in the broadest independent claim for fear of needlessly narrowing the claim scope. On the other hand, it may be recognized as an area that the patentee might be willing to concede if necessary to obtain or retain the patent.

The additional element added in dependent claim 14, further comprising isolating the heterologous peptide from the transgenic avian or an egg laid by the transgenic avian, contemplates an obvious step that would be employed in commercial applications of the invention. This adds an additional claim that would likely be infringed by an entity found to be infringing any of the other claims.

The Rapp patent contains an additional series of claims that include a slightly different independent claim and a parallel set of dependent claims. The different independent claim requires a nonavian protein be injected into the avian embryo, and includes the step of isolating the nonavian protein from an egg of a resultant chick. This reflects a common strategy of including a separate independent claim that more closely tracks a likely commercial application of the invention.

1.6.5. Patent No. 7,525,011: Transgenic Cancer Models in Fish

Patent No. 7,525,011 issued on April 28, 2009, to Look et al., entitled "Transgenic Cancer Models in Fish." The abstract describes an invention relating to transgenic fish whose genome contains an expressible oncogene as well as methods of using such transgenic fish in a variety of research-oriented methods (drug screening, identifying mutations or agents that modulate sensitivity to chemotherapy or radiation therapy).[7] A look at the claims shows that the patent is limited to a transgenic fish with the cMYC oncogene linked to a RAG2 promoter.[8] The broadest claim is directed to the transgenic fish, meaning that any production, use, or sale of a transgenic fish whose genome contains cMYC linked to RAG2, provided the oncogene is expressed in T-lymphocytes and induces T-cell acute lymphoblastic leukemia, would infringe the patent. The method claims that issued in this particular patent are

---

[7]The abstract:

The present invention provides transgenic fish whose genome has stably-integrated therein an oncogene operably linked to a promoter. Methods of making the transgenic fish and methods for their use are also provided. In one embodiment, the transgenic fish may advantageously be utilized in methods of screening for drugs or agents that modulate oncogene-mediated neoplastic or hyperplasic transformation, or that modulate sensitivity to chemotherapy or radiation therapy. In another embodiment, the transgenic fish may be used methods of identifying mutations that modulate oncogene-mediated neoplastic or hyperplastic transformation, or that modulate sensitivity to chemotherapy or radiation therapy.

[8]Claim 1:

1. A transgenic fish whose genome comprises a transgene encoding cMYC oncogene operably linked to a RAG2 promoter, wherein the oncogene is expressed in T-lymphocytes and induces T-cell acute lymphoblastic leukemia.

directed to screening drugs or agents that suppress cMYC-induced leukemia using this type of fish.[9]

The Look et al. patent provides an example of a patent that may seem broad based on an initial review of the abstract. A closer look at the claims shows a fairly specific invention. A check of the records for this patent on the public patent application information retrieval (PAIR) system at the USPTO's web site shows that no related applications were filed. One can also view this application's file history on PAIR and see that the original claims filed in September 2003 were much broader, seeking to cover transgenic fish having any oncogene stably integrated into the genome. The applicant was not able to obtain such broad patent coverage, at least not via this filing.

1.6.6. Patent No. 7,524,677: Mammalian Cultured Inner Cell Mass Cell Culture

"Mammalian Cultured Inner Cell Mass Cell Culture Using a G1 Cell As Nuclear Donor" is the title of Patent No. 7,524,677 to Campbell et al., dated April 28, 2009. The abstract indicates that the invention relates to a method of reconstituting an animal embryo. The method involves transferring a diploid nucleus into an oocyte which is arrested in the metaphase of the second meiotic division. The oocyte is not activated at the time of transfer, so that the donor nucleus remains exposed to the recipient cytoplasm for a period of time. The diploid nucleus can be donated by a cell in either the G0 or G1 phase of the cell cycle at the time of transfer. Subsequently, the reconstituted embryo is activated. Correct ploidy is maintained during activation, for example, by incubating the reconstituted embryo in the presence of a microtubule inhibitor such as nocodazole. The reconstituted embryo may then give rise to one or more live animal births. The invention is useful in the production of transgenic animals as well as nontransgenics of high genetic merit.

The broadest claim is a little narrower than this abstract. It reads:

> A method for producing a mammalian cultured inner cell mass cell by nuclear transfer comprising: (i) inserting a nucleus of a diploid mammalian differentiated cell in the G1 phase of the cell cycle into an unactivated, enucleated metaphase II-arrested mammalian oocyte of the same species to reconstruct an embryo; (ii) activating the resultant reconstructed embryo;

[9]The independent (broadest) method claim:

8. A method of screening test drugs or agents that suppress cMYC oncogene-induced leukemia, comprising: contacting or otherwise exposing a transgenic fish to a test drug or agent, wherein the transgenic fish has a genome that comprises a transgene encoding mouse cMYC oncogene operably linked to a RAG2 promoter, and wherein expression of the oncogene in T-lymphocytes induces T-cell acute lymphoblastic leukemia; comparing the leukemia in said transgenic fish after contact or exposure to said test drug or agent relative to the leukemia of said fish prior to contact or exposure with said test drug or agent; wherein suppression of the leukemia in said transgenic fish after contact or exposure to said test drug or agent relative to the leukemia of said fish prior to contact or exposure with said test drug or agent is indicative of a test drug or agent that suppresses cMYC-induced leukemia.

(iii) culturing said activated, reconstructed embryo; and (iv) isolating and culturing inner cell mass cells obtained from said cultured, activated, reconstructed embryo to obtain a cultured inner cell mass cell.

1.6.6.1. Patent Family Re Nuclear Transfer

This patent is an example of one which, when checking the USPTO web site for continuity data, one finds a related patent, number 6,252,133, that issued in 2001. This earlier patent is entitled, "Unactivated oocytes as cytoplast recipients of quiescent and nonquiescent cell nuclei, while maintaining correct ploidy." Its first claim is to:

A method of reconstructing an embryo of a nonhuman mammal, comprising:

(a) Transferring the nucleus of a diploid donor cell in the G0 phase of the cell cycle into an unactivated, enucleated metaphase II oocyte, without concomitantly activating the oocyte so as to form a reconstructed embryo, wherein the donor cell and the oocyte are from the same nonhuman mammalian species.

(b) Maintaining the reconstructed embryo without activation such that correct ploidy is maintained, wherein the reconstructed embryo subsequently can develop to term.

(c) Activating the reconstructed embryo under conditions that maintain correct ploidy.

In addition, a further related patent application is still pending, which means that additional subject matter that is described in the application but does not appear in the claims of either of these issued patents could be claimed in a later-issuing patent.

1.6.7. Patent No. 6,376,743: Transgenesis by Intracytoplasmic Sperm Injection

Patent No. 6,376,743, "Mammalian Transgenesis By Intracytoplasmic Sperm Injection," issued April 23, 2002, to Yanagimachi, and is assigned to the University of Hawaii. The abstract indicates that the invention relates to coinjection of unfertilized mouse oocytes with sperm heads and exogenous nucleic acid encoding a transgene. This results in transgene-expressing embryos, reflecting nucleic acid-sperm head association before coinjection. Nonselective transfer to surrogate mothers of embryos resulting from coinjection produced offspring expressing the integrated transgene.

Claim 1 is directed to a method for obtaining a transgenic embryo. The claimed method comprises the steps of (1) incubating a nucleic acid that is exogenous to the embryo with a membrane-disrupted sperm head or a demembranated sperm head for a period of time; (2) co-inserting the exogenous nucleic acid and sperm head into an unfertilized oocyte to form a transgenic fertilized oocyte; and (3) allowing the transgenic fertilized oocyte to develop into a transgenic embryo.

1.6.7.1. Broad Coverage of ICSI Technology

Note that this claim does not limit the class of organism from which the embryo or oocyte are derived. No limitation is placed on the nature or content of the nucleic acid other than being exogenous to the embryo. The specification of this patent indicates that "[m]embrane-disrupted sperm heads suitable for use in the invention can be obtained from frozen-thawed spermatozoa or rehydrated freeze-dried spermatozoa. A method for preserving spermatozoa by freeze-drying and using the resulting reconstituted freeze-dried spermatozoa to fertilize oocytes in vitro to produce embryos and live offspring is the subject of our copending U.S. patent application, Ser. No. 09/177,391, filed Oct. 23, 1998."

Searching the USPTO patent database with this serial number shows that Patent No. 6,641,526 issued from this related application on November 4, 2003. This patent claims a "membrane-damaged freeze-dried spermatozoon or freeze-dried spermatozoon head, having a moisture content of less than 1% and comprising a nucleus that has retained genetic integrity, wherein after rehydration and insertion of the nucleus into an isolated oocyte, the nucleus fertilizes the oocyte, and the retained genetic integrity of the nucleus is sufficient to result in the development of an embryo and the production of a live offspring." Thus, if one were to employ this type of spermatozoon material in the method of intracytoplasmic sperm injection, a license would be required under both patents.

1.6.8. Patent No. 5,741,957: Transgenic Bovine

Patent No. 5,741,957, entitled, "Transgenic Bovine," issued April 21, 1998, to Deboer et al. and is assigned to Pharming B.V. This patent is based on an application that was filed June 5, 1995. This filing date is significant because it was the last date on which a patent application could be filed under prior rules that provided for patents having a 17-year term calculated from the date of issuance. This patent, therefore, barring other restrictions or extensions affecting its term, would be in force until April 21, 2015.

This patent to Deboer et al. indicates also that it is based on a divisional application from "application Ser. No. 08/154,019 filed Nov. 16, 1993, now U.S. Pat. No. 5,633,076 issued May 27, 1997, which is a continuation-in-part of U.S. patent application Ser. No. 08/077,788, filed Jun. 15, 1993, now abandoned, which is a continuation-in-part of U.S. patent application Ser. No. 07/898,956, filed Jun. 15, 1992, now abandoned, which is a continuation-in-part of U.S. patent application Ser. No. 07/619,131 filed Nov. 27, 1990, now abandoned, which is a continuation-in-part of U.S. patent application Ser. No. 07/444,745 filed Dec. 1, 1989 now abandoned. Each of the above applications is incorporated by reference in its entirety for all purposes." This latter sentence is a safeguard against having failed to include material from any of the prior related applications. By stating their contents are incorporated by reference, the patentee has effectively made the contents of all applications in the series part of the present patent.

1.6.8.1. Divisional Patents

A divisional application means that the patentee pursued this application on the basis of material that was disclosed but not claimed in the parent filing. Divisional applications are typically filed in response to a requirement by the patent examiner to restrict the application to a single invention. Most patent applications in the biotechnology field (as well as some in other fields) are subjected to such a restriction requirement because they are regarded as containing claims to multiple inventions, each requiring a separate search and consideration of differing issues relating to the patentability analysis.

For example, claims to DNA, protein, antibodies, methods of using these materials, methods of treatment, and methods of diagnosis are often included in a single filing but each of these types of claims may be regarded by the USPTO as a separate invention. In response to a restriction requirement, the applicant must elect a single invention for examination. To pursue any of the other inventions, the applicant must file one or more divisional applications. Each divisional filing is entitled to the same filing date for priority purposes as the parent application, and must use the same specification and drawings.

1.6.8.2. Transgenic Bovine Claims

The abstract of the Deboer et al. patent states:

> A transgenic bovine is disclosed whose somatic and germ cells contain a transgene, wherein the transgene comprising [sic] a mammary gland specific promoter, a mammary gland specific enhancer, a DNA sequence encoding a signal sequence functional in bovine mammary gland secretory cells and a DNA sequence encoding a heterologous polypeptide of interest wherein the transgenic bovine expresses the transgene such that the polypeptide of interest is detectable in milk produced by the transgenic bovine.

Claim 1 of the Deboer et al. patent is directed to a transgenic bovine whose somatic and germ cells contain a transgene, wherein the transgene comprises in operable association: a mammary gland specific promoter; a mammary gland specific enhancer; a DNA sequence encoding a signal sequence functional in bovine mammary gland secretory cells; and a DNA sequence encoding a heterologous polypeptide of interest; wherein the transgenic bovine or a female descendant of the transgenic bovine expresses the transgene in mammary secretory cells such that the polypeptide of interest is detectable in milk produced by the transgenic bovine or a female descendant of the transgenic bovine.

This patent also claims a method of producing a polypeptide, comprising recovering milk from the transgenic bovine or female descendent of claim 1, wherein the milk contains the polypeptide.

Thus, both the transgenic bovine itself and the recovery of the transgene's product from its milk are covered by this patent. Further claims also relate to transplanting a preimplantation stage embryo comprised of the transgene into a recipient female bovine, wherein the female bovine gestates the embryo to give birth to the transgenic bovine. Additional claims are included that cover formulating the recovered polypeptide into a pharmaceutical, to a transgenic bovine that produces human lactoferrin as well as other specific embodiments of the invention.

1.7 Conclusion

Patents relating to the production and use of transgenic organisms can be found online via the U.S. Patent and Trademark Office's web site (http://www.uspto.gov). One can hire a professional search service to assist in identifying patents that may block a particular area of endeavor, or search the patent database oneself.

The identification of relevant patents is but a starting point in the analysis of freedom to operate. One must then review the claims and the related file history. Retaining the services of a qualified professional is strongly recommended prior to the investment of significant sums in practicing a technology. If the desired activity is blocked by one or more patent holders, one can explore the feasibility of working around the patent claims, seeking appropriate licenses, or entering into a joint development agreement.

References

1. §101 of Title 35 of the *United States Code* (35 U.S.C. §101)
2. *Diamond v. Chakrabarty*, 447 U.S. 303 (1980)
3. 35 U.S.C. §161, which provides for a special type of plant patent available to one who invents or discovers and asexually reproduces a distinct and new variety of plant
4. 35 U.S.C. §102
5. 35 U.S.C. §103
6. *In re Deuel*, 51 F.3d 1552 (Fed. Cir. 1995)
7. *In re Kubin*, 561 F.3d 1351 (Fed. Cir. 2009)
8. 35 U.S.C. §112, first paragraph
9. *Madey v. Duke Univ.* 307 F.3d 1351, 64 U.S. P.Q.2d 1737 (Fed. Cir. 2002) *cert. denied*, 123 S.Ct. 2639 (2003)
10. University of Missouri-Columbia, Pronuclear Microinjection Sublicense Terms (2009) http://biotech.rnet.missouri.edu/tac/sublicense-terms.html. Cited July 15, 2009

Chapter 2

Global Resources: Including Gene Trapped ES Cell Clones – Is Your Gene Already Knocked Out?

Lluis Montoliu

Abstract

The design of any new mouse genetic modification today should start with careful scrutiny of the resources that are already available, through the internet, for information relating to your gene of interest. International mouse consortia are constantly providing new genetically modified alleles of virtually any gene in the mouse genome. Therefore, unless a very specific knock-in allele is required, it is more than likely that the envisaged mutation has already been obtained somewhere and made available in the form of embryonic stem (ES) cell clones, live animals, or cryopreserved sperm or embryos. In this chapter, I will review the current (November 2010) global resources that are available through the internet, where the most updated information about any given mouse gene should be examined, before any new experiment is planned or conducted. The knowledge and adequate use of all these global resources should speed up the acquisition of knowledge in the fields of biology, biomedicine, and biotechnology, while avoiding the redundant use of animals for experimentation and optimizing the use of limited funding resources. In this chapter, I will try to respond to two basic questions: where is my mouse? and what is known about my gene?

Abbreviations

CMMR	Canadian mouse mutant repository
CREATE	Coordination of resources for conditional expression of mutated mouse alleles
EBI	European Bioinformatics Institute
EMAP	Edinburgh Mouse Atlas Project
EMBL	European Molecular Biology Laboratory
EMMA	European Mouse Mutant Archive
EMPReSS	European mouse phenotyping resource of standardized screens
ENSEMBL	A joint project between EMBL – EBI and the Wellcome Trust Sanger Institute to develop a software system which produces and maintains automatic annotation on selected eukaryotic genomes
ES	Embryonic stem

S. Pease and T.L. Saunders (eds.), *Advanced Protocols for Animal Transgenesis*, Springer Protocols,
DOI 10.1007/978-3-642-20792-1_2,

ESPCR	European Society of Pigment Cell Research
EUCOMM	European Conditional Mouse Mutagenesis
EuMMCR	European mouse mutant cell repository
EUMODIC	European Mouse Disease Clinic
EUMORPHIA	European Union Mouse Research for Public Health and Industrial Applications
EUROPHENOME	Open source project to develop a software system for capturing, storing, and analyzing raw phenotyping data from SOPs contained in EMPReSS
FP6	Framework Programme 6
ICS	Institut Clinique de la Souris
IGTC	International Gene Trap Consortium
IKMC	International KnockOut Mouse Consortium
IMSR	International Mouse Strain Resource
ISTT	International Society for Transgenic Technologies
JAX	The Jackson Laboratory
KOMP	Knock-Out Mouse Project
KORC	Knock-Out Rat Consortium
MGI	Mouse Genome Informatics
MMRRC	Mouse Mutant Regional Resource Centres
NBRP	National BioResource Project for the Rat
NCBI	National Center for Biotechnology Information
NIH	National Institutes of Health
NorCOMM	North-American Conditional Mouse Mutagenesis
OMIM	Online Mendelian Inheritance in Man
RGD	Rat genome database
RRRC	Rat Resource & Research Centre
SNP	Single nucleotide polymorphism
SOP	Standard operating procedures
TIGM	Texas A&M Institute for Genomic Medicine
UCSC	University of California, Santa Cruz
ZFIN	Zebrafish model organism database
ZGC	Zebrafish gene collection
ZIRC	Zebrafish International Resource Center

2.1 Has My Favorite Gene Already Been Knocked-Out? Where Should I Start?

After sequencing of the human [1] and mouse [2] genomes, strategies were needed to reveal gene function. Since human and mouse genes share 95% homology, it was established that mouse genes could serve as tools for understanding human gene function. This can be achieved due to the ease by which the mouse genome can be genetically manipulated with the available genetic toolbox, by knocking-out the corresponding murine homologous loci and interpreting the associated phenotypes generated. Globally, this process is known as mouse functional genomics.

Several approaches were initiated with intent to produce embryonic stem (ES) cell lines carrying gene mutations. At first, several gene trap consortia were arranged worldwide, with collaborative intent to saturate the mouse genome with gene trap vector insertions in mouse ES cells. This was based on the proposition that most genes could be mutated and the corresponding mouse mutants derived from these ES cell clones, carrying such random insertions. Eventually, all gene trap projects merged into the International Gene Trap Consortium (IGTC) [3] (Fig. 2.1).

Independently, three additional consortia were organized in Europe, USA, and Canada. In Europe, the European Conditional Mouse Mutagenesis (EUCOMM) project [4] was formed; in Canada the North-American Conditional Mouse Mutagenesis (NorCOMM) project came to be, and in the USA, the Knock-Out Mouse Project (KOMP) [5] was set up. Their orchestrated purpose was to systematically knockout all mouse genes using gene targeting approaches. These consortia used different approaches to vector design. Eventually, all three projects merged under the umbrella of the International KnockOut Mouse Consortium [6]. Later, the Texas A&M Institute for Genomic Medicine (TIGM) joined in as the fourth project of this type [7].

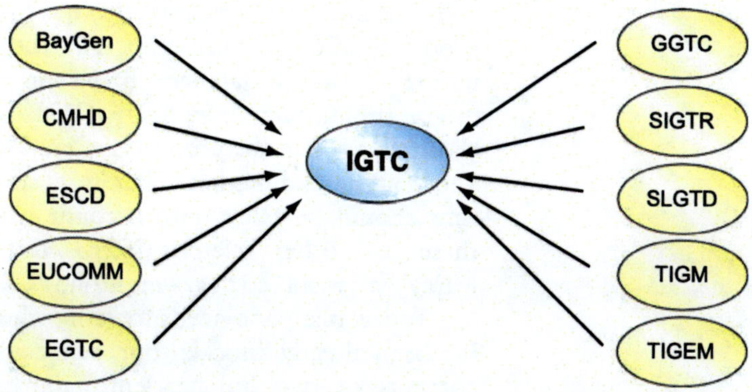

Fig. 2.1 The *International Gene-Trap Consortium* (IGTC) is constituted by the following ten members: [BayGen] BayGenomics (USA); [CMHD] Centre for Modeling Human Disease (Toronto, Canada); [ESCD] Embryonic Stem Cell Database (University of Manitoba, Canada); [EUCOMM] European Conditional Mouse Mutagenesis (European Union); [EGTC] Exchangeable Gene Trap Clones (Kumamoto University, Japan); [GGTC] German Gene Trap Consortium (Germany); [SIGTR] Sanger Institute Gene Trap Resource (Cambridge, UK); [SLGTD] Soriano Lab Gene Trap Database (Mount Sinai School of Medicine, New York, USA); [TIGM] Texas Institute for Genomic Medicine (USA); and [TIGEM] TIGEM-IRBM Gene Trap (Naples, Italy). The entire contents of the IGTC database can be browsed and searched via http://www.genetrap.org

2.1.1. Recommended Web Sites

All available ES cell clones from the various gene-trap consortia can be searched and browsed, at once, from IGTC at: http://www.genetrap.org/. Simply typing in the gene of interest will give an indication of whether there are any gene-trapped ES cell clones already generated for that gene and from where they can be obtained.

The EUCOMM project can be accessed at: http://www.eucomm.org and all the associated EUCOMM ES cell clones and vectors can be searched for and ordered from the European Mouse Mutant Cell Repository (EuMMCR) at: http://www.eummcr.org/. Live mice and cryopreserved embryos derived from EUCOMM ES cell lines can be searched and ordered through the European Mouse Mutant Archive (EMMA) [8] at: http://www.emmanet.org.

The NorCOMM Project is available at: http://www.norcomm.org/, the KOMP Project from: http://www.nih.gov/science/models/mouse/knockout/, and the TIGM Project from: http://www.tigm.org/. Global resources made available by the merging of EUCOMM, NorCOMM, and KOMP and the formation of IKMC are available from: http://www.knockoutmouse.org/. Biological material from KOMP (ES cell clones, live mice, and cryopreserved embryos) can be obtained through the Mouse Mutant Regional Resource Centres (MMRRC) http://www.mmrrc.org/. Similarly, biological material from NorCOMM is available through the Canadian Mouse Mutant Repository (CMMR) at: http://www.cmmr.ca/. The description of the ES cells used by IKMC has been reported [9] and details are available from: http://www.eummcr.org/products/wild_type_cells.php. All the international knockout mouse consortia data are based on the C57BL/6N inbred mouse strain, in contrast to the C57BL/6J inbred mouse strain, classically used in the previous generation of many transgenic and knockout animal models. Therefore, specific genetic polymorphisms should be taken into account where they differ between these and other related C57BL/6 mouse substrains ([10]; http://www.cnb.csic.es/~montoliu/C57/).

Today, if anyone needs to verify whether a given mouse gene has been already knocked out, one could start by searching the contents of two independent databases: the IGTC database (http://www.genetrap.org/) and the IKMC database (http://www.knockoutmouse.org/).

However, there may be other previously made mouse models or spontaneous mutants available relating to the gene of interest, not necessarily hit by the IGTC and/or not included by the IKMC. How could we look for them? The best global resource to find any mouse mutant strain, to browse whether a given mouse gene has been mutated or not, to eventually obtain biological material, in the form of ES cell clones, cryopreserved embryos, cryopreserved sperm, or live animals, is the International Mouse Strain Resource (IMSR), available from Mouse Genome Informatics (MGI), within The Jackson Laboratory (JAX) web site (http://www.jax.org), at: http://www.findmice.org/. Searching IMSR does, in one single step, a systematic search of most available databases, the contents of which have been merged. This includes IKMC, EMMA, MMRRC, CMMR, JAX, and all major mouse archives worldwide. The only exception would be the contents of the IGTC database (gene-traps), which is not entirely directly searchable through the IMSR database (Fig. 2.2). However, gene-trapped ES cell clones from some IGTC members are already included in the IMSR, such as those distributed by TIGM.

Therefore, submitting search requests through the IGTC (http://www.genetrap.org) and the IMSR databases (http://www.findmice.org) should be the first two steps in any experimental planning for a new mouse mutation, in order to explore whether mouse strains, ES cell clones (targeted or gene-trapped), or cryopreserved material already exist for the envisaged mutation in our favorite mouse gene.

Where a gene symbol has been used as search term, a typical IGTC search would bring up a list of ES cell lines where the

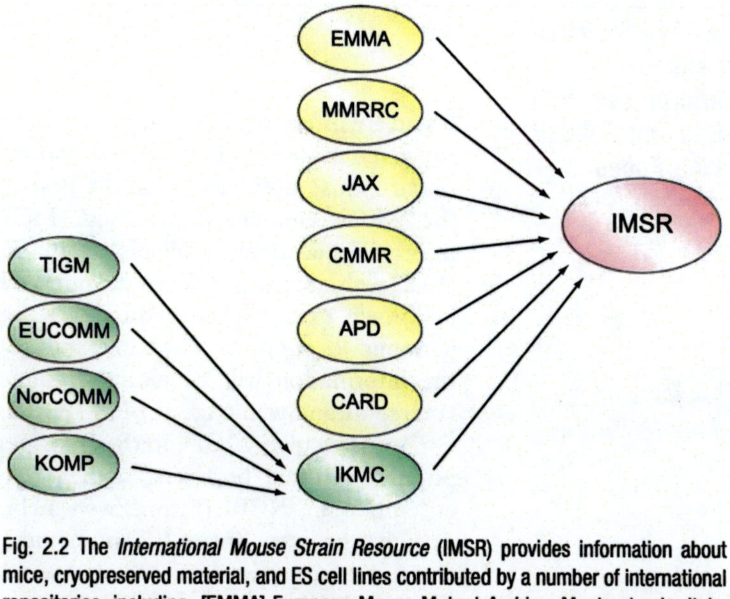

Fig. 2.2 The *International Mouse Strain Resource* (IMSR) provides information about mice, cryopreserved material, and ES cell lines contributed by a number of international repositories, including: [EMMA] European Mouse Mutant Archive, Monterotondo, Italy; [MMRRC] Mutant Mouse Regional Resource Centers, USA; [JAX] The Jackson Laboratory, Bar Harbor, Maine, USA; [CMMR] Canadian Mouse Mutant Repository, Toronto, Canada; [APD] Australian Phenome Bank, Acton, Australia; and [CARD] Centre for Animal Resources and Development, Kumamoto, Japan. In addition, the IMSR includes ES cell lines produced through the International KnockOut Mouse Consortium (IKMC). Additional repositories contributing to IMSR can be identified at the corresponding web site: http://www.findmice.org

gene-trapped locus is the gene of interest. The list indicates which gene-trap repository banks each ES cell line and from where the cells can be obtained. All potentially useful ES cell lines should be explored, and their gene-trap events understood in great detail, prior ordering any clone, for assessment of whether the insertion is likely to result in a knockout or knockdown effect. Each ES cell line is associated with plenty of genetic and mapping information that is absolutely required for analysis of the relevance of each gene-trap event. An example of IGTC output is shown in Fig. 2.3. A typical IMSR search would produce a list of mouse strains, in the form of (1) ES cell clones, indicating the particular project within the mouse consortium that has generated the biological resource; (2) live mice; or (3) cryopreserved embryos or sperm, linked to the repository where the mouse line is held. All suggested mouse mutant strains should be explored in detail. Usually, there will be several genetic backgrounds available, out of which the best suitable strain for our purposes should be ordered. In addition, not all mutant mouse strains will be available in the form of live mice. Most strains will be cryopreserved, as embryos or sperm, making them specially suited for shipping purposes. An example of IMSR output is shown in Fig. 2.4.

2.2 The Mouse Genome Informatics Web Site and Related Web Pages

If you are interested in exploring all that is currently known about any given mouse gene, its corresponding mutant alleles and associated mouse mutant strains, the best starting point is currently the "Mouse Genome Informatics" (MGI) web site (http://www.informatics.jax.org), available from The Jackson Laboratory (JAX) web site (http://www.jax.org). Whether you are interested in known gene alleles at this locus, gene expression patterns, genomic location, or associated mouse mutant strains, etc., all the information will be nicely arranged and organized on the corresponding web page at MGI (Fig. 2.5).

In particular, MGI interfaces its genomic information with popular genome browsers, such as ENSEMBL (http://www.ensembl.org), NCBI (http://www.ncbi.nlm.nih.org), or UCSC genome browser (http://genome.ucsc.edu), where a greater amount of genetic detail can be searched for, researched and downloaded.

There are many sections with information and useful links in every single gene card (Fig. 2.6). Information about the corresponding human disease associated with each gene is linked through the OMIM (Online Mendelian Inheritance in Man) database (http://www.ncbi.nlm.nih.gov/omim). Information about all known alleles and mouse strains available which carry

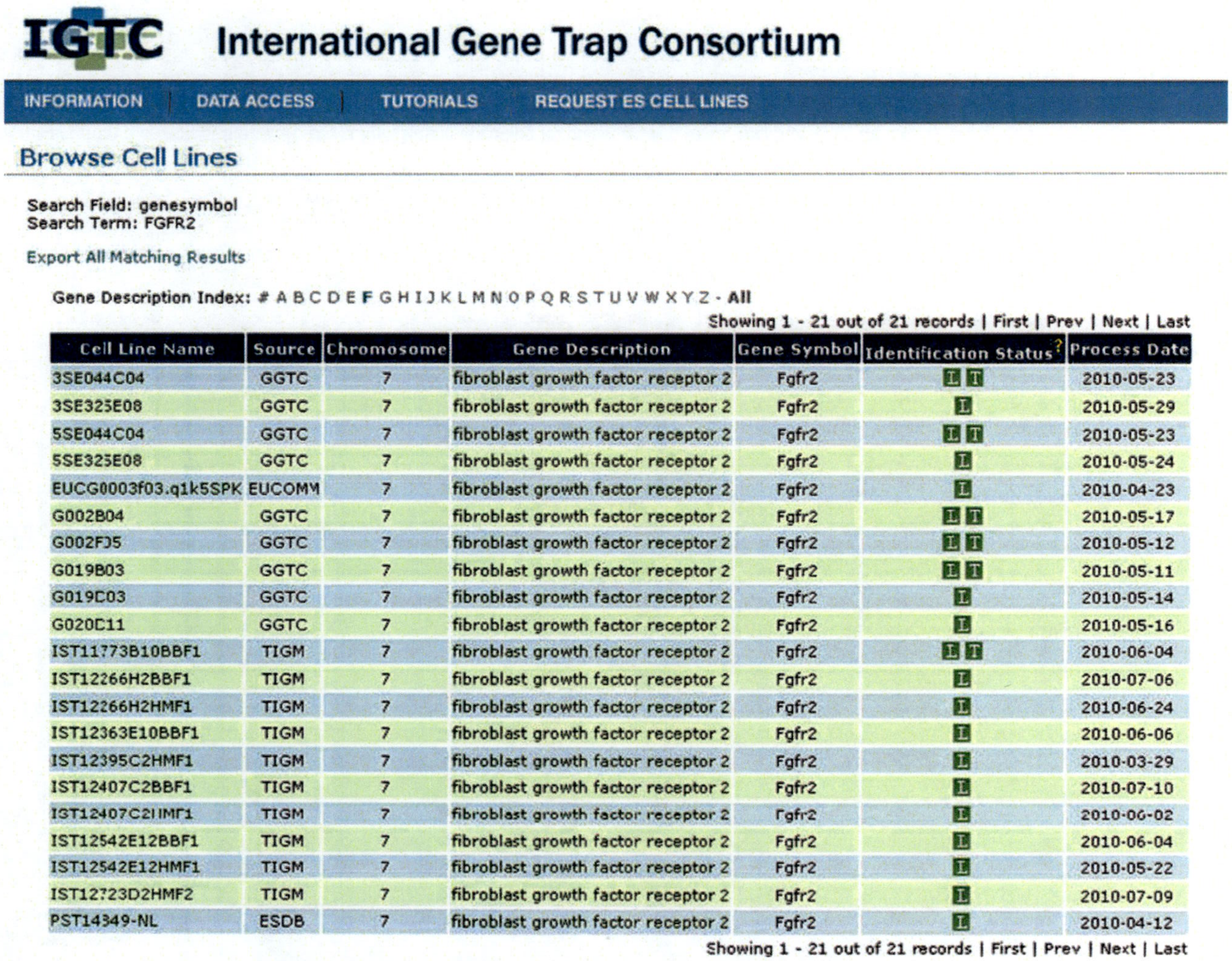

Cell Line Name	Source	Chromosome	Gene Description	Gene Symbol	Identification Status?	Process Date
3SE044C04	GGTC	7	fibroblast growth factor receptor 2	Fgfr2	L T	2010-05-23
3SE325E08	GGTC	7	fibroblast growth factor receptor 2	Fgfr2	L	2010-05-29
5SE044C04	GGTC	7	fibroblast growth factor receptor 2	Fgfr2	L T	2010-05-23
5SE325E08	GGTC	7	fibroblast growth factor receptor 2	Fgfr2	L	2010-05-24
EUCG0003f03.q1k5SPK	EUCOMM	7	fibroblast growth factor receptor 2	Fgfr2	L	2010-04-23
G002B04	GGTC	7	fibroblast growth factor receptor 2	Fgfr2	L T	2010-05-17
G002F05	GGTC	7	fibroblast growth factor receptor 2	Fgfr2	L T	2010-05-12
G019B03	GGTC	7	fibroblast growth factor receptor 2	Fgfr2	L T	2010-05-11
G019C03	GGTC	7	fibroblast growth factor receptor 2	Fgfr2	L	2010-05-14
G020C11	GGTC	7	fibroblast growth factor receptor 2	Fgfr2	L	2010-05-16
IST11773B10BBF1	TIGM	7	fibroblast growth factor receptor 2	Fgfr2	L T	2010-06-04
IST12266H2BBF1	TIGM	7	fibroblast growth factor receptor 2	Fgfr2	L	2010-07-06
IST12266H2HMF1	TIGM	7	fibroblast growth factor receptor 2	Fgfr2	L	2010-06-24
IST12363E10BBF1	TIGM	7	fibroblast growth factor receptor 2	Fgfr2	L	2010-06-06
IST12395C2HMF1	TIGM	7	fibroblast growth factor receptor 2	Fgfr2	L	2010-03-29
IST12407C2BBF1	TIGM	7	fibroblast growth factor receptor 2	Fgfr2	L	2010-07-10
IST12407C2HMF1	TIGM	7	fibroblast growth factor receptor 2	Fgfr2	L	2010-06-02
IST12542E12BBF1	TIGM	7	fibroblast growth factor receptor 2	Fgfr2	L	2010-06-04
IST12542E12HMF1	TIGM	7	fibroblast growth factor receptor 2	Fgfr2	L	2010-05-22
IST12723D2HMF2	TIGM	7	fibroblast growth factor receptor 2	Fgfr2	L	2010-07-09
PST14349-NL	ESDB	7	fibroblast growth factor receptor 2	Fgfr2	L	2010-04-12

Fig. 2.3 Typical results from a search at the IGTC. Using *Fgfr2* (gene encoding fibroblast growth factor receptor 2) as the search term, up to 21 different gene-trap ES cell lines are listed, from various programs and centers (GGTC, TIGM, ESDB, EUCOMM). Clicking on each of the ES cell line names will provide additional useful information of the gene-trap event.

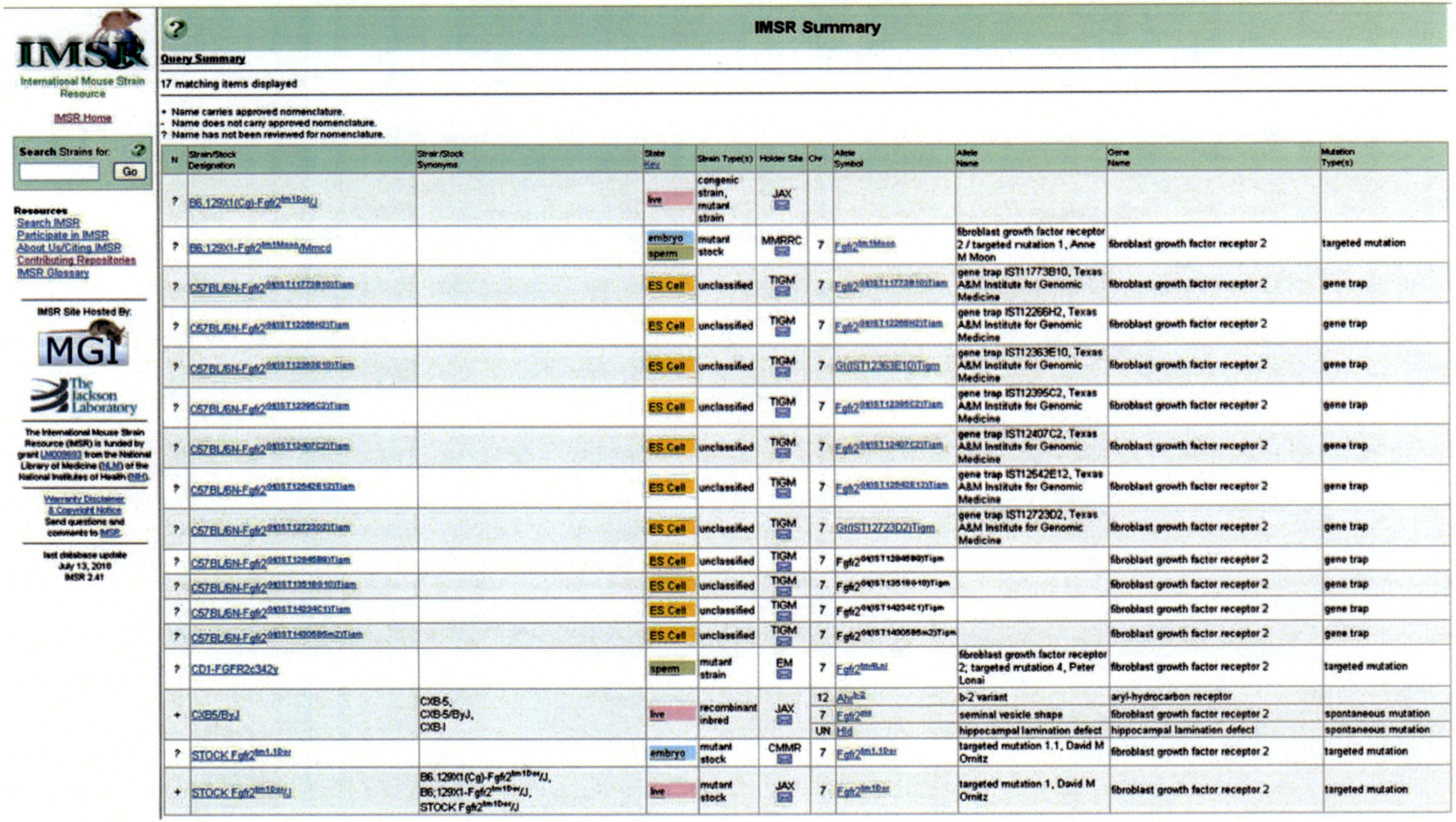

Fig. 2.4 Typical results from a search at the IMSR. Using *Fgfr2* (gene encoding fibroblast growth factor receptor 2) as the search term, up to 17 different mouse strains appear as available, in the form of live mice, frozen embryos, frozen sperm, or ES cell lines, from various repositories (JAX, MMRRC, TIGM, EM, CMMR) and on different genetic backgrounds. Clicking on each of the mouse strain names will provide additional useful information of the associated mutation. Please note that some (but not all) of the IGTC ES cell lines (i.e., from TIGM) are also included in the IMSR.

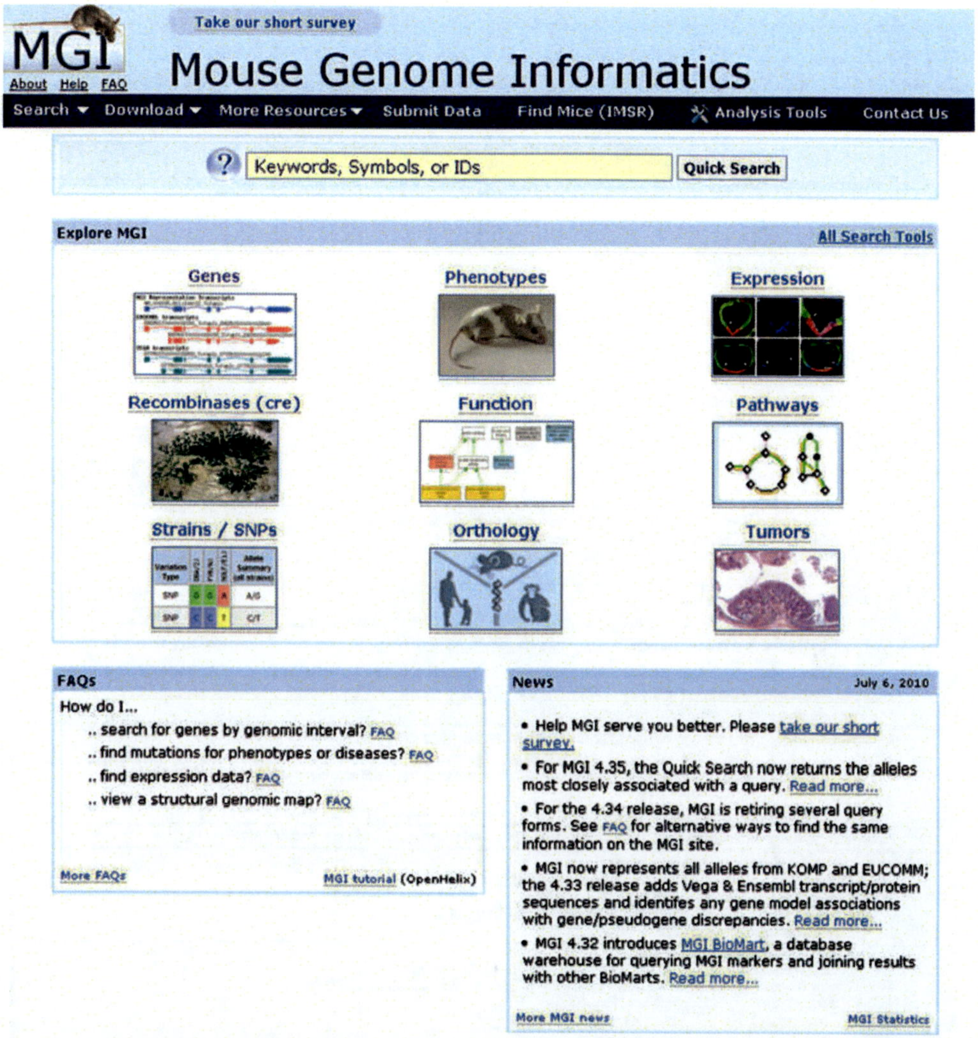

Fig. 2.5 The *Mouse Genome Informatics* (MGI) web site (http://www.informatics.jax.org) at The Jackson Laboratory. Main menu of the MGI web pages leading to various sections with different, but linked, types of information. If you are interested in statistics and would like to see the progress of mouse genome coverage in the form of gene targeting events, number of mouse models created, etc., simply click on "MGI statistics" (*bottom right corner* of this main menu page).

an allele at the locus of interest is also linked through the IMSR database, as described before, or through the Phenotypic Alleles summary. If you are interested in genetic polymorphisms (i.e., single nucleotide polymorphisms, SNPs) that could be used to differentiate the same gene in different mouse genetic backgrounds these are also indicated. The best collection of known mouse genome SNPs is found at the Mouse Phenome Database (http://phenome.jax.org) where a whole section is devoted to SNPs (http://phenome.jax.org/SNP). Regarding expression data there are various links to resources detailing where this gene is expressed. In this regard, complementary information can be

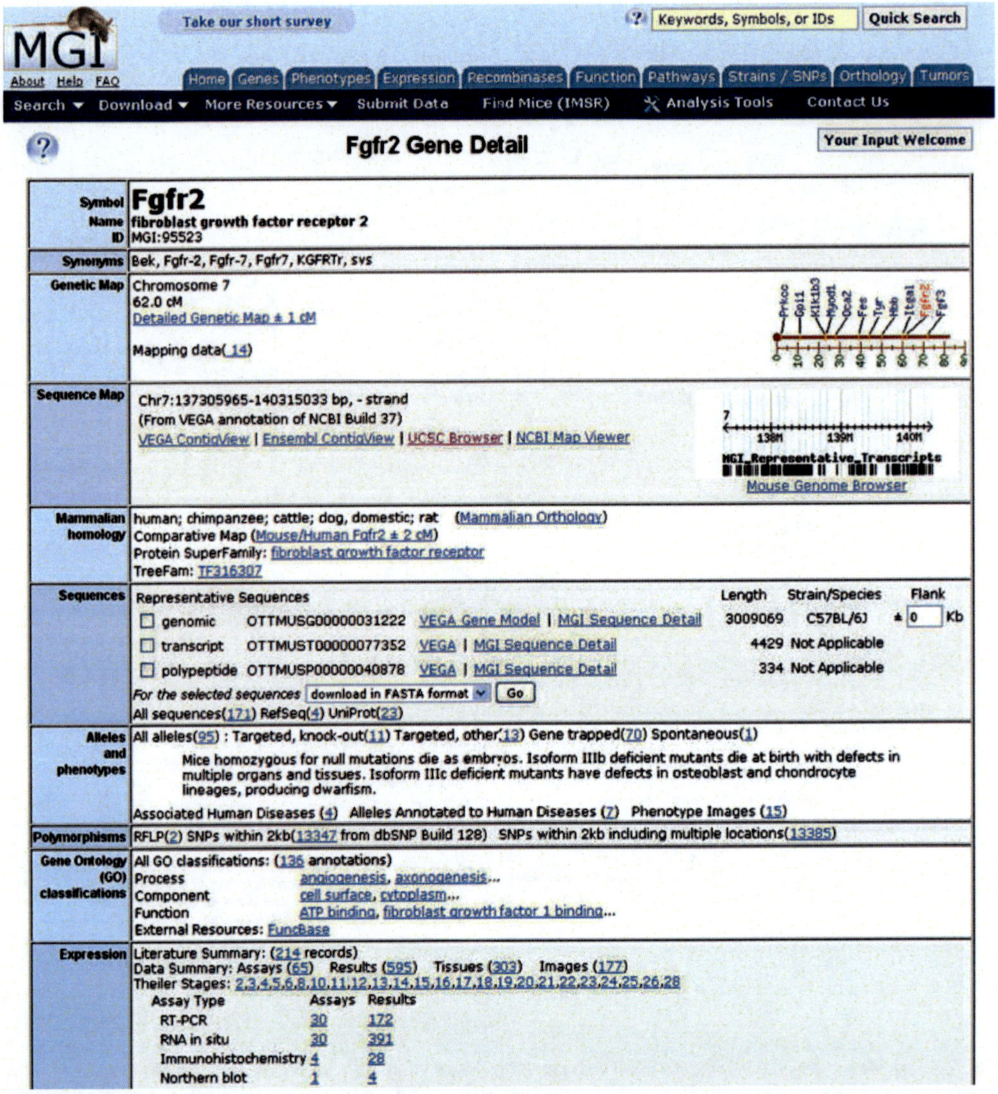

Fig. 2.6 Typical *example of MGI* search results regarding available information on the *Fgfr2* gene (list of topics covered is longer than shown, but it has been truncated for illustrative purposes). Each section is linked to additional sources of information.

obtained from Genepaint (http://www.genepaint.org), a digital atlas of gene expression patterns in mice, determined by nonradioactive in situ hybridization on serial tissue sections and associated with each gene, all available through their web site. Particularly interesting, for neuroscientists, are the new links to the Allen Brain Atlas (http://www.brain-map.org) for adult mouse brain and developing mouse brain sections, where the expression of every gene is annotated.

MGI collects and annotates expression and activity data for cre recombinase-containing transgenes and knock-in alleles. All these very useful cre-mouse lines can be browsed and searched through a specific site (http://www.creportal.org).

At MGI, they also provide links to complete reference books in the mouse field that are out of print. These valuable online books include: "The Biology of the Laboratory Mouse" Earl L. Green (ed.) (http://www.informatics.jax.org/greenbook); "Mouse Genetics" by Lee Silver (http://www.informatics.jax.org/silverbook); "The Anatomy of the Laboratory Mouse" by Margaret J. Cook (http://www.informatics.jax.org/cookbook); "The coat colors of mice" by Willys K. Silvers (http://www.informatics.jax.org/wksilvers); and the "Origins of Inbred Mice" Herbert C. Morse III (ed.) (http://www.informatics.jax.org/morsebook). Additional information on the genetics of pigmentation, or genes whose function affect coat color pigmentation, can be obtained from the "Color Genes" web site, at the European Society of Pigment Cell Research (ESPCR), at: http://www.espcr.org/micemut/.

One of the most useful sections within MGI is the "Mouse Nomenclature Home Page" (http://www.informatics.jax.org/mgihome/nomen), where the guidelines for nomenclature of genes, genetic markers, alleles, and mutations in the mouse and rat are found. The Mouse Genome Informatics (MGI) Database is the authoritative source of official names for mouse genes, alleles, and strains. Nomenclature follows the rules and guidelines established by the International Committee on Standardized Genetic Nomenclature for Mice. Recently, from the International Society for Transgenic Technologies (ISTT) (http://www.transtechsociety.org) and the scientific journal Transgenic Research (Springer) (http://www.springer.com/biomed/molecular/journal/11248), a combined position paper has been recently published, encouraging the use of standard nomenclature to adequately name transgenes, knockout gene alleles, and any mutation associated to a genetically modified mouse strain [11].

The MGI is fully interconnected with ENSEMBL and NCBI. At NCBI, one all-in-one bioinformatic resource can complement the information obtained from a given mouse gene. This is the "all databases" feature of NCBI (global query: http://www.ncbi.nlm.nih.gov/gquery/) that provides all the known information about a gene, interfacing with all NCBI databases, including published articles from PubMed.

2.3 Additional Databases for Mouse Transgenesis

Besides MGI, the reference for all mouse databases, there are additional bioinformatic resources available which are worth being aware of, since they also provide useful information.

Most of these additional databases are already compiled at the "General Links" page of the ISTT web site (http://www.transtechsociety.org/link.php).

Of outstanding interest are several independent databases that account for different Cre-transgenic mouse lines created for use in combination with mice carrying *floxed* (flanked-by-loxP-sites) alleles, for mouse conditional gene mutagenesis. Besides the creportal at MGI already mentioned, an additional database for cre-mouse lines is an initiative pioneered by Andras Nagy, the Cre-X-Mice database (http://nagy.mshri.on.ca/cre/). Other transgenic mouse cre lines can be obtained from the crezoo database (http://bioit.fleming.gr/crezoo/), originating at the Fleming Institute (Vari, Greece) and from the MouseCre database (http://www.ics-mci.fr/mousecre/), at the *Institut Clinique de la Souris* (ICS, Strasbourg-Illkirch, France). All worldwide databases collecting Cre transgenic mouse lines are coordinated through the CREATE consortium (http://creline.org/), a Cre recombinase portal organized by the European Bioinformatic Institute (EBI, Hinxton-Cambridge, UK).

Information on existing ES cell lines (name and mouse strain of origin) can be downloaded from MGI (ftp://ftp.informatics.jax.org/pub/reports/ES_CellLine.rpt). The diverse 129 mouse substrains follow revised nomenclature, indicated by Simpson et al. [12] and now are available through a useful web site at the MGI (http://www.informatics.jax.org/mgihome/nomen/strain_129.shtml).

Specific details on the use of the popular R1 mouse ES cell line [13] is available from a web site devoted to the topic (http://www.mshri.on.ca/nagy/r1.htm).

With regard to web sites oriented toward phenotyping of mice, those from the EUMORPHIA European Project (http://www.eumorphia.org/) should be mentioned, since that led to the EMPReSS initiative (http://empress.har.mrc.ac.uk/), a database of Standard Operating Procedures (SOPs) for procedures that can be used to characterize the phenotype of a mouse, and to EUROPHENOME (http://www.europhenome.org/), a database for collection of phenomic data obtained from the EMPReSS SOPs. The interaction of these phenotyping projects with the international knockout consortia can be followed with EUMODIC (http://www.eumodic.org/), a new project funded by the European Commission under Framework Program 6 (FP6) to generate phenome data on 650 mutant mice generated by EUCOMM, using the EMPReSS SOPs.

The Edinburgh Mouse Atlas Project (EMAP, http://genex.hgu.mrc.ac.uk) is another great resource for a 3D-mouse embryo anatomy atlas and its corresponding expression database. Again, for those focused on neuroscience, you will find The Mouse Brain Library (MBL, http://www.mbl.org) a very useful resource, consisting of high-resolution images and databases of brains from several inbred mouse strains.

On the subject of mouse welfare issues, several projects have been initiated associated with their corresponding web sites, including "Mouse Welfare Terms" (http://www.mousewelfareterms.org/), a site dedicated to standardizing the way different characteristics which may impact on the welfare of laboratory mice, are described. Also the COST B24 Action on "Laboratory Animal Science and Welfare" (http://www.cost.esf.org/domains_actions/bmbs/Actions/B24-Laboratory-Animal-Science-and-Welfare-End-date-April-2009) that recently published The COST Manual of Laboratory Animal Care and Use. Refinement, Reduction and Research.

Finally, from the ISTT web site, it is possible to reach many transgenic cores, facilities, and/or units producing genetically modified mice and rats in many countries all over the world (http://www.transtechsociety.org/linkstg.html).

2.4 Resources on Additional Animal Models

Mice are the most frequently used animal models in vertebrate functional genomics and for experiments involving mammalian genetic modification, but they are not the only species that might be used. Other species to consider as candidates for genetic modification are rats, zebrafish, flies, worms, etc., and, correspondingly, web sites listing such resources provide lots of interesting and useful information about these alternative and additional animal models. In this section, I will review some of these web sites, the most important for each species, where additional global resources can be readily explored and information obtained.

2.4.1. Rats

Some might still consider rats to be "bigger" mice, but this is not so. Rats are truly a different rodent species, with a specific reproductive system physiology that has precluded their routine use in most transgenic facilities for many years. Fortunately, several recent efforts and methods of investigation have resulted in the establishment of robust protocols that allow the generation of transgenic rats with efficiency comparable to that currently obtained in mice [14, 15]. Rats are the animal model of choice for most toxicological and pharmacological studies. For many years, the rat genome was not available to investigators for gene targeting, as is often used in mice. Despite the initial excitement generated with the cloning of rats [16], the nuclear-transfer technique has proven to be difficult to reproduce in this species [17]. Recently, true rat ES cells were obtained [18, 19] providing the tools for the generation of future knockout rats through standard gene targeting in ES cells. The first gene knockout engineered by homologous recombination in rat ES cells has been published [20]. However, a totally different method, using Zinc-finger nucleases, has been reported to produce the first gene-specific knockout rats [21, 22].

2.4.1.1. Recommended Web Sites

The reference entry point for almost anything related to rat genetic and genomic research is the Rat Genome Database (RGD), at: http://rgd.mcw.edu/. Complementary resources can be obtained from the NIH Rat Genomics and Genetics web site (http://www.nih.gov/science/models/rat/). In addition, rat genome information can also be obtained from the specific ENSEMBL (http://www.ensembl.org/Rattus_norvegicus/) and NCBI (http://www.ncbi.nlm.nih.gov/genome/guide/rat/) project web sites. Specific archives for obtaining rat strains are also available, such as The National BioResource Project for the Rat in Japan (NBRP: http://www.anim.med.kyoto-u.ac.jp/nbr/) [23], the Rat Resource & Research Centre (RRRC) at the University of Missouri (http://www.nrrrc.missouri.edu/), or the Michael Festing's collection of rat inbred strains (http://www.informatics.jax.org/external/festing/search_form.cgi). Finally, the standard nomenclature rules and guidelines to name genes, alleles, or strains are also available for rats at the Mouse Genome Informatics web site of The Jackson Laboratory (http://www.informatics.jax.org/mgihome/nomen/). A few transgenic core facilities are also producing transgenic rats by request, such as the University of Michigan Transgenic Core (http://www.med.umich.edu/tamc/rats.html) and the Transgenic Rats common facility of IFR26 and Biogenouest in Nantes, France (http://www.ifr26.nantes.inserm.fr/ITERT/transgenese-rat/).

Recently established, the Knock-Out Rat Consortium, (KORC;http://www.knockoutrat.org) is pledged to the creation of knockout mutations in rats by means of multiple technologies.

KORC is a consortium, with goals similar to that of KOMP. Additional rat global resources can be found linked to any of these web sites.

2.4.2. Other Mammals

Global information on genetic, genomic, and biological resources relating to various other mammalian species are available from NCBI. They include the following. For the pig, at (http://www.ncbi.nlm.nih.gov/projects/genome/guide/pig/), for sheep (http://www.ncbi.nlm.nih.gov/genome/guide/sheep/), for the cow (http://www.ncbi.nlm.nih.gov/projects/genome/guide/cow/), the rabbit (http://www.ncbi.nlm.nih.gov/projects/genome/guide/rabbit/), the goat (http://www.ncbi.nlm.nih.gov/projects/genome/guide/goat/), and the horse (http://www.ncbi.nlm.nih.gov/projects/genome/guide/horse/). These are a few among other mammals where genetic modification methods can be applied.

2.4.3. Zebrafish

Zebrafish have become a reference animal model for early vertebrate genomic research. The ease by which genetic modification can be accomplished and the visual transparency and short duration of zebrafish embryo development make them unique for many exploratory experiments or genetic screening. The genetic toolbox available for zebrafish includes standard transgenesis, through the use of *Tol2* transposon-mediated methods [24], gene targeting in zebrafish ES cells [25], site-specific recombination using the Cre/lox [26], or Flp/frt technologies [27], among other techniques.

Furthermore, most of the mammalian genes have their homologous counterpart in the zebrafish genome. The essential functions of most loci, especially if they are relevant during embryo development, are evolutionarily conserved, hence genetic studies in zebrafish are of value and provide a more efficient approach to understanding corresponding gene function in mammals [28].

2.4.3.1. Recommended Web Sites

The reference gate to access to all zebrafish biological and genetic resources is ZFIN, the Zebrafish Model Organism Database [29], available at: http://zfin.org. The ZFIN database is interconnected with many other useful resources for zebrafish, such as the specific web site for the Zebrafish genome project within ENSEMBL, at: http://www.ensembl.org/Danio_rerio/ or its equivalent web site at the NCBI server: http://www.ncbi.nlm.nih.gov/genome/guide/zebrafish/. Additional web sites with helpful information are the NIH Zebrafish Gene Collection (ZGC) database, at: http://zgc.nci.nih.gov/ and the Zebrafish International Resource Center (ZIRC), at: http://zebrafish.org/zirc. Additional zebrafish resources can be found linked to any of these web sites already mentioned.

2.4.4. Flies

The fruit fly, *Drosophila melanogaster*, has been a classical animal model for genetic studies for more than a century. Even though flies and mice are very distantly evolutionary related, many fundamental gene functions have proven to be surprisingly similar [30, 31], therefore genetic modification studies conducted in *Drosophila* have been, and will continue to be, instrumental for the understanding of mammalian genomes.

2.4.4.1. Recommended Web Sites

The essential reference entry point for all genetic, genomic, and biological information and resources currently available for *Drosophila* is the FlyBase (http://flybase.org/) [32]. This impressive resource offers links to almost everything in existence relating to *Drosophila* genetics. The corresponding *Drosophila* genome web sites in ENSEMBL (http://www.ensembl.org/Drosophila_melanogaster/) and NCBI (http://www.ncbi.nlm.nih.gov/projects/genome/guide/fly/) can also be used to access supplementary information.

2.4.5. Worms

The nematode *Caenorhabditis elegans* (*C. elegans*) was introduced by Sydney Brenner in 1974 as a new model organism for biology and genetic studies. Due to its apparent simplicity and rapid and transparent embryo development, the entire fate map for the approximately thousand cells that constitute an adult individual was known quite soon. The sequencing of this genome triggered many comparative studies of genomes and the use of worm models in the study of complex biological processes such as ageing [33].

2.4.5.1. Recommended Web Sites

Essential global resources for genetic, genomic, and biological information about *C. elegans* are WormBase (http://www.wormbase.org/) and WormBook (http://www.wormbook.org/). Additional helpful information on behavioral and structural anatomy can be obtained from the WormAtlas (http://www.wormatlas.org/).

References

1. Venter JC, Adams MD, Myers EW, Li PW, Mural RJ, Sutton GG et al (2001) The sequence of the human genome. Science 291:1304–1351
2. Mouse Genome Sequencing Consortium et al (2002) Initial sequencing and comparative analysis of the mouse genome. Nature 420:520–562
3. Skarnes WC, Von Melchner H, Wurst W, Hicks G, Nord AS, Cox T, Young SG, Ruiz P, Soriano P, Tessier-Lavigne M, Conklin BR, Stanford WL, Rossant J, International Gene Trap Consortium (2004) A public gene trap resource for mouse functional genomics. Nat Genet 36:543–544
4. Auwerx J, Avner P, Baldock R, Ballabio A, Balling R, Barbacid M, Berns A, Bradley A, Brown S, Carmeliet P, Chambon P, Cox R, Davidson D, Davies K, Duboule D, Forejt J, Granucci F, Hastie N, de Angelis MH, Jackson I, Kioussis D, Kollias G, Lathrop M, Lendahl U, Malumbres M, von Melchner H, Müller W, Partanen J, Ricciardi-Castagnoli P, Rigby P, Rosen B, Rosenthal N, Skarnes B, Stewart AF, Thornton J, Tocchini-Valentini G, Wagner E, Wahli W, Wurst W (2004) The European dimension for the mouse genome mutagenesis program. Nat Genet 36:925–927
5. Austin CP, Battey JF, Bradley A, Bucan M, Capecchi M, Collins FS, Dove WF, Duyk G, Dymecki S, Eppig JT, Grieder FB, Heintz N, Hicks G, Insel TR, Joyner A, Koller BH, Lloyd KC, Magnuson T, Moore MW, Nagy A, Pollock JD, Roses AD, Sands AT, Seed B, Skarnes WC, Snoddy J, Soriano P, Stewart DJ, Stewart F, Stillman B, Varmus H, Varticovski L, Verma IM, Vogt TF, von Melchner H, Witkowski J, Woychik RP, Wurst W, Yancopoulos GD, Young SG, Zambrowicz B (2004) The knockout mouse project. Nat Genet 36:921–924
6. International Mouse Knockout Consortium, Collins FS, Rossant J, Wurst W (2007) A mouse for all reasons. Cell 128:9–13
7. Collins FS, Finnell RH, Rossant J, Wurst W (2007) A new partner for the International Knockout Mouse Consortium. Cell 129:235
8. Wilkinson P, Sengerova J, Matteoni R, Chen CK, Soulat G, Ureta-Vidal A, Fessele S, Hagn M, Massimi M, Pickford K, Butler RH, Marschall S, Mallon AM, Pickard A, Raspa M, Scavizzi F, Fray M, Larrigaldie V, Leyritz J, Birney E, Tocchini-Valentini GP, Brown S, Herault Y, Montoliu L, de Angelis MH, Smedley D (2010) EMMA–mouse mutant resources for the international

scientific community. Nucleic Acids Res 38 (Database issue):D570–D576
9. Pettitt SJ, Liang Q, Rairdan XY, Moran JL, Prosser HM, Beier DR, Lloyd KC, Bradley A, Skarnes WC (2009) Agouti C57BL/6N embryonic stem cells for mouse genetic resources. Nat Methods 6:493–495
10. Zurita E, Chagoyen M, Cantero M, Alonso R, González-Neira A, López-Jiménez A, López-Moreno JA, Landel CP, Benítez J, Pazos F, Montoliu L (2010) Genetic polymorphisms among C57BL/6 mouse inbred strains. Transgenic Res 2011, 20:481–489
11. Montoliu L, Whitelaw CB (2011) Using standard nomenclature to adequately name transgenes, knockout gene alleles and any mutation associated to a genetically modified mouse strain. Transgenic Res 20(2):435–440
12. Simpson EM, Linder CC, Sargent EE, Davisson MT, Mobraaten LE, Sharp JJ (1997) Genetic variation among 129 substrains and its importance for targeted mutagenesis in mice. Nat Genet 16:19–27
13. Nagy A, Rossant J, Nagy R, Abramow-Newerly W, Roder JC (1993) Derivation of completely cell culture-derived mice from early-passage embryonic stem cells. Proc Natl Acad Sci USA 90:8424–8428
14. Filipiak WE, Saunders TL (2006) Advances in transgenic rat production. Transgenic Res 15:673–686
15. Charreau B, Tesson L, Soulillou JP, Pourcel C, Anegon I (1996) Transgenesis in rats: technical aspects and models. Transgenic Res 5:223–234
16. Zhou Q, Renard JP, Le Friec G, Brochard V, Beaujean N, Cherifi Y, Fraichard A, Cozzi J (2003) Generation of fertile cloned rats by regulating oocyte activation. Science 302:1179
17. Popova E, Bader M, Krivokharchenko A (2009) Efficient production of nuclear transferred rat embryos by modified methods of reconstruction. Mol Reprod Dev 76: 208–216
18. Buehr M, Meek S, Blair K, Yang J, Ure J, Silva J, McLay R, Hall J, Ying QL, Smith A (2008) Capture of authentic embryonic stem cells from rat blastocysts. Cell 135:1287–1298
19. Li P, Tong C, Mehrian-Shai R, Jia L, Wu N, Yan Y, Maxson RE, Schulze EN, Song H, Hsieh CL, Pera MF, Ying QL (2008) Germline competent embryonic stem cells derived from rat blastocysts. Cell 135:1299–1310
20. Tong C, Li P, Wu NL, Yan Y, Ying QL (2010) Production of p53 gene knockout rats by homologous recombination in embryonic stem cells. Nature 467:211–213
21. Geurts AM, Cost GJ, Freyvert Y, Zeitler B, Miller JC, Choi VM, Jenkins SS, Wood A, Cui X, Meng X, Vincent A, Lam S, Michalkiewicz M, Schilling R, Foeckler J, Kalloway S, Weiler H, Ménoret S, Anegon I, Davis GD, Zhang L, Rebar EJ, Gregory PD, Urnov FD, Jacob HJ, Buelow R (2009) Knockout rats via embryo microinjection of zinc-finger nucleases. Science 325:433
22. Rémy S, Tesson L, Ménoret S, Usal C, Scharenberg AM, Anegon I (2010) Zinc-finger nucleases: a powerful tool for genetic engineering of animals. Transgenic Res 19:363–371
23. Serikawa T, Mashimo T, Takizawa A, Okajima R, Maedomari N, Kumafuji K, Tagami F, Neoda Y, Otsuki M, Nakanishi S, Yamasaki K, Voigt B, Kuramoto T (2009) National BioResource Project-Rat and related activities. Exp Anim 58:333–341
24. Burket CT, Montgomery JE, Thummel R, Kassen SC, LaFave MC, Langenau DM, Zon LI, Hyde DR (2008) Generation and characterization of transgenic zebrafish lines using different ubiquitous promoters. Transgenic Res 17:265–279
25. Fan L, Moon J, Crodian J, Collodi P (2006) Homologous recombination in zebrafish ES cells. Transgenic Res 15:21–30
26. Pan X, Wan H, Chia W, Tong Y, Gong Z (2005) Demonstration of site-directed recombination in transgenic zebrafish using the Cre/loxP system. Transgenic Res 14: 217–223
27. Wong AC, Draper BW, Van Eenennaam AL (2011) FLPe functions in zebrafish embryos. Transgenic Res 20:409–415
28. Haga Y, Dominique VJ 3rd, Du SJ (2009) Analyzing notochord segmentation and intervertebral disc formation using the twhh:gfp transgenic zebrafish model. Transgenic Res 18:669–683
29. Sprague J, Bayraktaroglu L, Clements D, Conlin T, Fashena D, Frazer K, Haendel M, Howe D, Mani P, Ramachandran S, Schaper K, Segerdell E, Song P, Sprunger B, Taylor S, Van Slyke C, Westerfield M (2006) The Zebrafish Information Network: the zebrafish model organism database. Nucleic Acids Res 34:D581–D585
30. Mercader N, Leonardo E, Azpiazu N, Serrano A, Morata G, Martínez C, Torres M (1999) Conserved regulation of proximodistal limb axis development by Meis1/Hth. Nature 402:425–429
31. Giraldo P, Martínez A, Regales L, Lavado A, García-Díaz A, Alonso A, Busturia A, Montoliu L (2003) Functional dissection of the mouse tyrosinase locus control region

identifies a new putative boundary activity. Nucleic Acids Res 31:6290–6305
32. Tweedie S, Ashburner M, Falls K, Leyland P, McQuilton P, Marygold S, Millburn G, Osumi-Sutherland D, Schroeder A, Seal R, Zhang H, The FlyBase Consortium (2009) FlyBase: enhancing *Drosophila* Gene Ontology annotations. Nucleic Acids Res 37: D555–D559
33. Kenyon CJ (2010) The genetics of ageing. Nature 464:504–512

Chapter 3

Designing Transgenes for Optimal Expression

Eduardo Moltó, Cristina Vicente-García, and Lluis Montoliu

Abstract

In theory, designing a DNA construct to be used for transgene purposes, for standard pronuclear microinjection, would seem a rather easy task. The combination of a given promoter and some regulatory elements of choice, driving the expression of the construct to the desired tissue, with a suitable coding region of the gene of interest, and finishing the construct with an adequate transcription terminator would appear to be a straightforward process. However, chromosomal position effects, variegated expression, non-expressing transgenic mouse lines or those displaying ectopic and unexpected patterns of transgene expression are not uncommon. Therefore, great care should be invested in the design of the transgene, with optimal transgene expression the goal. With very few exceptions, there is no reliable catalogue of plasmid-based promoters that one could refer to when looking for robust tissue-specific transgene expression. Instead, BAC- and YAC-based transgenes have proven to produce optimal results, thus suggesting that genomic-type constructs may be more reliable as promoters than standard plasmid-type constructs. This and other observations will be discussed in this chapter. Three golden rules must be applied when designing a transgene (1) transgenes should not contain vector sequences; (2) transgenes should not contain DNA sequences derived from prokaryotic genomes; and, most importantly, (3) the more a transgene resembles the corresponding endogenous locus, the better it will behave in terms of expression levels and pattern. These very basic rules should be taken into account when preparing a DNA construct to be used as transgene, enabling easy removal of vector and prokaryotic sequences that are no longer required (and will normally have a detrimental effect upon transgene expression) and allowing the inclusion of genomic sequences that are fundamental for the faithful regulation of the locus.

3.1 Standard Approach: Adding Elements to a Basic DNA Construct

The simplest transgene should contain, in this order (1) a promoter, capable of driving the transcription of the DNA construct; (2) a given cDNA or coding sequence from the gene of interest; and (3) a suitable transcription termination signal. However, different functional elements (different promoter and terminator sequences) will need to be used depending whether RNA polymerase II (i.e. as in the case of the vast majority of DNA constructs used in transgenic mice) or RNA polymerase III (i.e. as in the case

S. Pease and T.L. Saunders (eds.), *Advanced Protocols for Animal Transgenesis*, Springer Protocols, DOI 10.1007/978-3-642-20792-1_3, © Springer-Verlag Berlin Heidelberg 2011

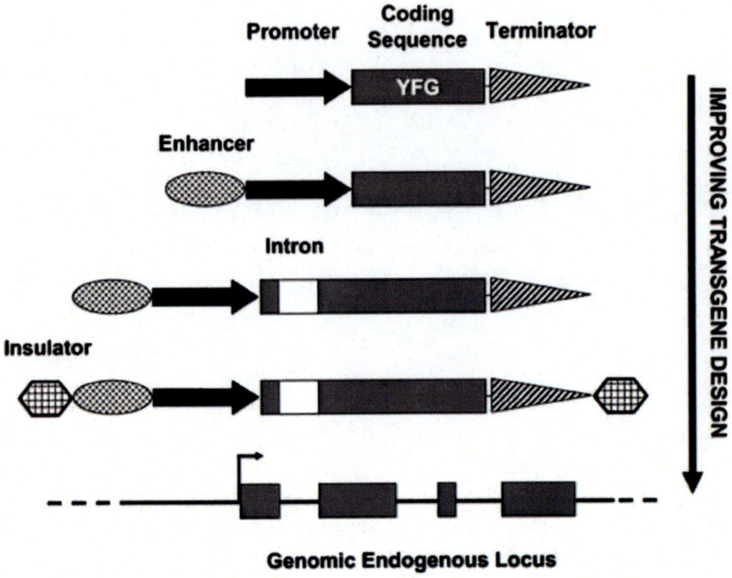

Fig. 3.1 *Improving transgene design.* From *top* to *bottom*, the drawing illustrates several fundamental steps towards improving transgene design and increasing transgene expression efficiency. The sequential addition of enhancers, intronic sequences and flanking boundaries or insulators have substantially improved transgene expression. However, optimal results would be always obtained using the corresponding endogenous genomic locus, where all natural regulatory elements that are required for the expression domain to function, are present. YFG stands for "your favourite gene".

of shRNA or RNAi-producing DNA constructs) transcription is envisaged (Fig. 3.1).

Transgenes lacking a promoter (promoter-less constructs) will only be expressed if integrated in the vicinity of an endogenous promoter, thus normally adopting the expression pattern of the interrupted locus. This strategy has been applied in the so-called gene-trap (or promoter-trap) DNA approaches, whose aim is to "trap" endogenous loci, randomly, whose expression pattern might be relevant or of interest [1–3]. In contrast, transgenes lacking proper transcription termination signals will usually generate longer RNA transcripts, because they will run along the genome until the next available termination signal is encountered. These unusually long RNA transcripts are likely to display altered expression patterns, associated with eventual gene expression regulatory elements trapped in these long 3′ transcribed and untranslated sequences. Standard and universal terminators can be used, harbouring functional polyadenylation signals. One of the most popular terminators, present in many plasmids made for eukaryotic gene expression [4], is derived from the simian virus 40 (SV40) genome and contains a small intron, which is useful for stabilisation and for monitoring transgene expression. This has been used, successfully, in many different transgenic constructs (i.e. [5–8]). However, whenever possible, endogenous 3′ untranslated regions (UTR) should be included, derived from the locus whose expression pattern is to be replicated [9]. If endogenous terminator elements are available they should be selected because they can efficiently contribute to the optimal expression of transgenes.

This basic and simple scheme has been used, usually with success, from the first transgenes that were created and reported [10] to date [11, 12]. However, it soon became obvious that promoter elements, normally short DNA sequences within 1–2 kb from the start of the transcription, were not sufficient to drive the expression of transgenes robustly to the expected tissues or cell-types of interest. The addition of "tissue-specific" regulatory elements (enhancers) was reported to greatly improve the tissue specificity of transgene expression in mice [13]. These enhancers were found both at the 5′ [13] and the 3′ [14] regions of the endogenous loci, and even within intronic sequences [15, 16].

The addition of intronic sequences to transgenes was soon reported to increase transgenic expression in mice [17]. This beneficial effect was observed both with homologous and heterologous introns [18–20], and specifically when using the first intron and when inserting it between the promoter and the first exon or the coding region of the gene to be expressed in transgenic mice [18]. These were the first observations suggesting that genomic-type constructs (i.e. sequences including endogenous DNA elements, like introns, enhancers, etc.) could be expressed more efficiently in transgenic animals. This led to the common use of the so-called "minigene" constructs, where genomic and cDNA-derived fragments of a locus are fused to include some intronic sequences, notably from the 5′ area, encompassing the first endogenous intron, while still maintaining the overall size of the construct within a manageable size (usually <20 kb, cloned in standard plasmids) [5, 8].

These groundbreaking observations were eventually confirmed a couple of years later, with the use of yeast-artificial chromosome (YAC) type constructs, where optimal transgene expression was first reported ([21]; reviewed in [11]). The use of genomic-type constructs will be described in more detail later in this chapter.

Further additional progress was made with several proposals aiming to improve the efficiency of transgene expression [22–24]. Among those, the transgene rescue strategy, where co-integration of a poorly expressed transgene with a full transcription unit carrying the genomic locus of interest resulted in increased transgene expression levels [25].

Transgenic constructs carrying enhancers and introns had to be assembled and cloned into plasmid-type vectors. However, it was also discovered that the presence of vector sequences negatively influenced transgene expression [26]. In general, all DNA sequences of prokaryotic origin (such as vector sequences but also some common reporter genes: *lacZ*, *cre*, *CAT*, *neo*) will fail to organise chromatin appropriately around them. In this case, they will trigger an altered transcription of the locus, normally observed as variegation, although unexpected ectopic expression domains can also be observed [27, 28]. Variegated expression can be interpreted, at the cellular level, as some (but not all) cells of the expected tissue eventually expressing the transgene, whereas the rest of the cells fail to express it, even though they also carry the transgene integrated in their genome. Variegated expression has been reported, for example, with *lacZ*-expressing transgenes [6, 29, 30] or associated with the lack of relevant regulatory elements [28, 31–33]. Variegation occurs in transgenic founder animals but also, characteristically, in transgenic individuals from subsequent generations. Variegation is not equivalent to mosaicism, though the terms "mosaic transgene expression" or "variegated transgene expression" have sometimes been used indistinctly. The term "mosaic" is best applied to transgenic founder animals that fail to transmit the transgene to their progeny according to Mendelian ratios, indicating that not all their cells carry the transgene (in particular, not all cells in their gonads will produce gametes carrying the transgene). Mosaic transgenic founder animals may be generated when the DNA construct does not integrate at the 1-cell embryo stage (which would ensure that all cells in the developing embryo would carry the transgene). Rather, the DNA integrates at the 2-, 4-, 8-cell stages or later and therefore not all cells within the developing embryo eventually carry the transgene [34]. Most transgenic founder animals have been reported to be mosaic [35].

Variegation is just one manifestation of many altered behaviours of transgenes that are dependent on the sequences included in the construct and, most importantly, on the chromosomal insertional location. The presence of neighbouring enhancers, silencers or the existence of heterochromatic regions at or around the transgene insertion site can impair expected transgene expression patterns, resulting in the generation of transgenic mouse lines that do not express the construct at all, or which have variegated expression patterns or ectopically express the transgenes in unexpected tissues [36, 37]. This undesired behaviour of transgenes, associated with their integration site within the host genome, is globally known as "chromosomal position effects" [38, 39].

A subset of regulatory elements of gene expression was found to effectively counteract chromosomal position effects, thereby facilitating position-independent and copy-number-dependent transgene expression. The first of these elements was described in the human beta-globin locus and was initially called a "dominant control region" (DCR) [40]. The inclusion of DCR in transgenic beta-globin constructs resulted in optimal expression patterns, irrespective of their chromosomal position. Eventually, analogous elements were described in other loci, such as the human CD2 gene [41] or the mouse tyrosinase gene [42, 43]. These elements are now best known as "locus control regions" (LCR). LCRs have been shown to contain powerful tissue-specific enhancers that can efficiently overcome heterochromatin-induced transgene inactivation in mammals [31] while having a specific developmental role in their corresponding genomic loci [32, 44].

LCRs are not the only class of regulatory element shown to correlate with position-independent and copy-number-dependent expression. Scaffold/matrix-attachment regions (S/MARs) are just another subset of DNA sequences that have been shown to overcome chromosomal position effects in transgenic animals [45–47] although in some cases not to a full extent or with a significant impact [48, 49]. S/MARs have also been reported to contribute to the adequate transcription of the endogenous loci they are associated with [50, 51].

Taking into account all these various elements described so far, all devised to improve transgene expression, we can illustrate the progress made in transgene design as depicted in Fig. 3.1. From the initial basic transgene construct, tissue-specific enhancers were added to increase expression in the cell-type of interest. Later, some of these enhancers were shown to display peculiar properties of the LCR, thus resulting in transgenes with optimal expression levels. Furthermore, the inclusion of intronic sequences in the 5′ region of the construct greatly improved its stability, and hence, the resulting expression level. Eventually, a new class of regulatory elements, called chromatin or genomic boundaries (also known as insulators), were described and reported to confer position-independent and copy-number-dependent transgene expression [7]. Insulators will be described in more detail in the next section of this chapter.

Currently, very sophisticated DNA constructs can also accommodate more than one transcriptional unit, either driven by the same promoter, aiming to be expressed in the same cellular types (i.e. [8]), or combining different promoters and transcriptional units, where the addition of insulators is specifically suited to shield each transcription unit and to avoid unwanted cross-interactions [52, 53].

Combining several encoding DNA fragments (cistrons) to operate under the same promoter and regulatory elements can be best achieved through the use of internal ribosomal entry sites (IRES) that allow polycistronic mRNAs to be efficiently translated [54]. These are short DNA elements, usually derived from viral genomes, that are capable of retaining the cellular translational machinery attached and operative, once a first protein has been made, therefore triggering the translation of a subsequent second protein encoded by a second linked transcriptional unit [1]. Lower or even non-existing expression of the second transcriptional unit has been reported with the use of IRES elements in transgenic animals [55, 56]. Therefore great attention must be given to determine which IRES element will be used [57] and to the correct spacing and length of the intercistronic sequences between consecutive transcriptional units [58], in order to achieve optimal results and best IRES performance in your transgenic construct [8, 59].

3.2 The Use of Boundary Elements in Transgene Design

Insulators were first described in the Drosophila genome [60] and later reported in vertebrates, the first being identified within the 5′ region of the chicken beta-globin locus [61]. Thereafter, various additional insulators have been reported in other vertebrate loci [62, 63], including the mouse tyrosinase gene [7], the mouse growth-hormone locus [64] and the chicken alpha-globin locus [65], as reviewed in Moltó et al. [53]. Their presumed endogenous role, insulating the protected expression domains from the undesired interaction of distal DNA sequences, was rapidly explored to test their potential benefit in protecting transgenes from chromosomal position effects [66, 67]. However, in spite of these initially promising results, several reports described the absence of the robust benefits expected for transgenes after the addition of insulator sequences (i.e. [68, 69]) strongly indicating the absence of a universal beneficial effect, and the need of novel insulator elements, that should be identified and carefully analysed before their function could be efficiently exploited in heterologous constructs [53, 70]. In this regard, new retrotransposon elements of the SINE B1 family have been recently reported to display the properties of insulators [71], as was described previously with SINE B2 repetitive elements [64] and with the analogous Alu repetitive elements from the human genome [72, 73]. All these short DNA elements, shown to behave as insulators in transgenic animals, can now be easily functionally validated in vivo, using a simple assay, which was developed with transgenic zebrafish [52], and is expected to become useful in heterologous constructs [53].

The results accumulated so far in transgenic animals seem to indicate that the addition of insulator sequences, although they may not always result in the expected benefits, they generally do no harm and hence, would not impair the otherwise characteristic expression pattern of the transgene. Therefore, the inclusion of genomic boundaries or insulators, to improve transgene expression, should be always carefully considered and regarded as potentially beneficial [53, 69, 74, 75] (Fig. 3.1). Ideally, your DNA construct should be protected from chromosomal position effects that potentially may arise at both sides of your transgene. However, due to frequent multicopy integration of most transgenes at one genomic locus, it would normally be sufficient to place one insulator inserted at one end of the DNA construct, to efficiently protect the transgene once inserted as a tandem array in the host genome [7, 65].

3.3 New Approaches: Transgenes Based on Genomic-Type DNA Constructs

In 1993, several groups, independently, demonstrated that large genomic DNA fragments, encompassing entire mouse and human loci cloned within YACs, correctly reproduced the expression pattern of the corresponding endogenous genes at their genomic locations [21, 76, 77]. These observed optimal performance of large genomic-type transgenes were rapidly replicated and extended to other mammalian loci, as the human gene encoding APP (>450 kb), whose enormous size had prevented its direct use in previous gene transfer attempts [78, 79]. The use of YACs and the large size of the heterologous genomic DNA inserts they could accommodate (>1 Mb) supported the inclusion of the entire set of regulatory elements that identify a given expression domain, thereby allowing faithful transgene expression in the ectopic host genome locations where these large transgenes would eventually integrate [11]. The benefits of YACs in transgenesis could be illustrated with the results obtained with transgenic mice generated with the mouse tyrosinase locus, whose mutations are associated with albinism [80]. Tyrosinase minigene-type DNA constructs had been used before with intermediate success, resulting in a variety of transgenic founder animals with variable, and sometimes variegated, expression [5, 81]. However, the use of a YAC harbouring 250 kb of mouse tyrosinase DNA sequences, resulted in transgenic mice that were undistinguishable from wild-type pigmented individuals [21], strongly supporting that the entire mouse tyrosinase expression domain was contained within those 250 kb. Subsequent experiments demonstrated that this genomic fragment included a LCR [43] and a boundary element [7] that were absent in the previous tyrosinase minigene constructs, thereby explaining the good performance observed with YAC tyrosinase transgenes [80]. Recently, updated methods have been described to generate transgenic mice with YACs (and BACs) using intracytoplasmic sperm injection (ICSI), resulting in an increased number of transgenic founder animals produced [82, 83].

A few years later, in 1997, the bacterial artificial chromosomes (BACs), capable of harbouring heterologous genomic DNA sequences up to 300 kb, were also shown to sustain optimal transgene expression [84], for analogous reasons as described in the case of YACs. The inclusion of large genomic regions, surrounding the locus of interest, guaranteed the correct expression of the gene according to the endogenous pattern [11]. Since then, due to the ease by which they can be grown and manipulated in the laboratory, BACs have become the most popular genomic-type transgene, normally reproducing the expected optimal performance when applied for the generation of expressing transgenic animals [85–87].

Both BACs and YACs can be easily manipulated through homologous recombination approaches, using a wide variety of useful toolboxes now available in bacteria and yeast cells [11, 43, 88–90]. Through specific homologous recombination events, the insertion of cDNAs, minigenes, polycistronic constructs or, simply, reporter genes, can be easily targeted to predetermined sequences within the BAC or the YAC, thereby allowing the inserted DNA construct to be expressed under optimal conditions [91–95].

For all the reasons expressed above, nowadays the first option when attempting to prepare a transgene, a DNA construct to be used for the production of transgenic mice, should always be a genomic-type transgene, a BAC or a YAC, depending on the size of the locus and on the availability of the BAC or YAC clones. Normally, BAC clones are easier to obtain than YACs from public repositories (i.e. BACPAC resources: http://bacpac.chori.org/) and, moreover, for each genomic region there is usually a large number of BAC clones encompassing the locus of interest, in an overlapping fashion. Which BAC or YAC clones should be then selected? Genomic comparative approaches should be best applied for selecting the optimal BAC clone. This will be the BAC clone that not only carries the known coding part of your favourite gene but also includes all those flanking sequences, proximal and distal, that might have been evolutionarily conserved, and therefore, might contain crucial regulatory elements for the faithful expression of the locus [96–98].

3.4 Recommendations

This is a list of simple recommendations to be taken into account when preparing a new DNA construct to be used for the generation of transgenic animals. Carefully adhering to this list of tips, as much as reasonably possible, will normally result in better and more robust transgene expression levels, according to the expected pattern.

1. First, and foremost, try to use the largest available genomic DNA fragment harbouring your locus of interest, the one you plan to use for driving the expression of your transgene. This will normally be a BAC or a YAC clone.
2. Aim to select a BAC or YAC clone that includes your locus of interest and most of the neighbouring regulatory elements that appear to be evolutionarily conserved and associated to this gene. The closer you will be to the correct endogenous expression domain, the better the performance of your derived transgene will be expected.
3. If a BAC or a YAC are not available or the use of these large genomic-type transgenes is not feasible in your project, try to include enough genomic regions, such as for minigene constructs, including the promoter and known tissue-specific enhancers of your locus of interest.
4. If a BAC or a YAC are not available or the use of these large genomic-type transgenes is not feasible in your project, try also to include boundary elements or insulators in, at least, one side of your DNA construct, to potentially overcome any chromosomal position effect associated with the site of integration of the transgene. Normally, only one insulator would be sufficient since, usually, plasmid-type transgenes integrate as multicopy tandem arrays, thereby resulting in the boundary element flanking either side of most of the integrated transgene units.
5. If you can obtain intronic sequences available from the gene to be expressed, try including some of these introns in the DNA construct, specially if the 5′ most distal intron is available. If endogenous introns are not available, try using a generic heterologous intron cloned in the equivalent 5′ UTR.
6. If you have the endogenous 3′ UTR, including terminators, available, try including the own polyadenylation signals in the DNA constructs. If the endogenous transcription termination signals are not available, try using a generic terminator proven to work in heterologous contexts.
7. If you need to co-express two (or more) transcriptional units under the control of the same promoter consider using IRES elements to link the consecutive encoding sequences. However, pay attention to the length of the inter-cistronic sequences and to the IRES element selected, that must have been proven to work in the expected cellular-type where you plan your transgene to be functional.
8. In general, if reasonably possible, try to avoid the inclusion of coding sequences derived from prokaryotic genomes in your DNA constructs. They are normally prone to impaired expression patterns and, hence, are usually associated to variegation,

unless these negative properties can be compensated with the use of large genomic-type transgenes (i.e. BACs or YACs), where the variegation and expression impairment will normally not be observed.

9. In general, try to avoid the inclusion of vector sequences, usually derived from prokaryotic genomes, in your DNA constructs to be used as transgenes. The presence of vector sequences is normally associated with impaired expression patterns. However, exceptionally, vector sequences from BACs or YACs do not appear to negatively influence the otherwise optimal expression patterns derived from the use of large genomic heterologous inserts.

Acknowledgements

This work was supported by grants to L.M. from the Spanish Ministry of Science and Innovation (MICINN; BIO2009-1297). E.M. is supported by CIBERER (ISCIII) and C.V-G. by the CSIC JAE Predoc program.

References

1. Chowdhury K, Bonaldo P, Torres M, Stoykova A, Gruss P (1997) Evidence for the stochastic integration of gene trap vectors into the mouse germline. Nucleic Acids Res 25:1531–1536
2. Kitajima K, Takeuchi T (1998) Mouse gene trap approach: identification of novel genes and characterization of their biological functions. Biochem Cell Biol 76:1029–1037
3. Stanford WL, Cohn JB, Cordes SP (2001) Gene-trap mutagenesis: past, present and beyond. Nat Rev Genet 2:756–768
4. Boshart M, Klüppel M, Schmidt A, Schütz G, Luckow B (1992) Reporter constructs with low background activity utilizing the cat gene. Gene 110:129–130
5. Beermann F, Ruppert S, Hummler E, Bosch FX, Müller G, Rüther U, Schütz G (1990) Rescue of the albino phenotype by introduction of a functional tyrosinase gene into mice. EMBO J 9:2819–2826
6. Montoliu L, Blendy JA, Cole TJ, Schütz G (1995) Analysis of perinatal gene expression: hormone response elements mediate activation of a lacZ reporter gene in liver of transgenic mice. Proc Natl Acad Sci USA 92:4244–4248
7. Giraldo P, Martínez A, Regales L, Lavado A, García-Díaz A, Alonso A, Busturia A, Montoliu L (2003) Functional dissection of the mouse tyrosinase locus control region identifies a new putative boundary activity. Nucleic Acids Res 31:6290–6305
8. Lavado A, Jeffery G, Tovar V, de la Villa P, Montoliu L (2006) Ectopic expression of tyrosine hydroxylase in the pigmented epithelium rescues the retinal abnormalities and visual function common in albinos in the absence of melanin. J Neurochem 96:1201–1211
9. Overbeek PA, Aguilar-Cordova E, Hanten G, Schaffner DL, Patel P, Lebovitz RM, Lieberman MW (1991) Coinjection strategy for visual identification of transgenic mice. Transgenic Res 1:31–37
10. Palmiter RD, Brinster RL, Hammer RE, Trumbauer ME, Rosenfeld MG, Birnberg NC, Evans RM (1982) Dramatic growth of mice that develop from eggs microinjected with metallothionein-growth hormone fusion genes. Nature 300:611–615
11. Giraldo P, Montoliu L (2001) Size matters: use of YACs. BACs and PACs in transgenic animals. Transgenic Res 10:83–103
12. Zhang Q, Triplett AA, Harms DW, Lin WC, Creamer BA, Rizzino A, Wagner KU (2010) Temporally and spatially controlled expression of transgenes in embryonic and adult tissues. Transgenic Res 19:499–509

13. Hammer RE, Swift GH, Ornitz DM, Quaife CJ, Palmiter RD, Brinster RL, MacDonald RJ (1987) The rat elastase I regulatory element is an enhancer that directs correct cell specificity and developmental onset of expression in transgenic mice. Mol Cell Biol 7:2956–2967
14. Kollias G, Hurst J, deBoer E, Grosveld F (1987) The human beta-globin gene contains a downstream developmental specific enhancer. Nucleic Acids Res 15:5739–5747
15. Aronow B, Lattier D, Silbiger R, Dusing M, Hutton J, Jones G, Stock J, McNeish J, Potter S, Witte D, Wiginton D (1989) Evidence for a complex regulatory array in the first intron of the human adenosine deaminase gene. Genes Dev 3:1384–1400
16. Brooks AR, Nagy BP, Taylor S, Simonet WS, Taylor JM, Levy-Wilson B (1994) Sequences containing the second-intron enhancer are essential for transcription of the human apolipoprotein B gene in the livers of transgenic mice. Mol Cell Biol 14:2243–2256
17. Brinster RL, Allen JM, Behringer RR, Gelinas RE, Palmiter RD (1988) Introns increase transcriptional efficiency in transgenic mice. Proc Natl Acad Sci USA 85:836–840
18. Palmiter RD, Sandgren EP, Avarbock MR, Allen DD, Brinster RL (1991) Heterologous introns can enhance expression of transgenes in mice. Proc Natl Acad Sci USA 88:478–482
19. Whitelaw CB, Archibald AL, Harris S, McClenaghan M, Simons JP, Clark AJ (1991) Targeting expression to the mammary gland: intronic sequences can enhance the efficiency of gene expression in transgenic mice. Transgenic Res 1:3–13
20. Petitclerc D, Attal J, Théron MC, Bearzotti M, Bolifraud P, Kann G, Stinnakre MG, Pointu H, Puissant C, Houdebine LM (1995) The effect of various introns and transcription terminators on the efficiency of expression vectors in various cultured cell lines and in the mammary gland of transgenic mice. J Biotechnol 40:169–178
21. Schedl A, Montoliu L, Kelsey G, Schütz G (1993) A yeast artificial chromosome covering the tyrosinase gene confers copy number-dependent expression in transgenic mice. Nature 362:258–261
22. Houdebine LM (2000) Transgenic animal bioreactors. Transgenic Res 9:305–320
23. Houdebine LM (2002) Animal transgenesis: recent data and perspectives. Biochimie 84:1137–1141
24. Houdebine LM (2007) Transgenic animal models in biomedical research. Methods Mol Biol 360:163–202
25. Yull F, Binas B, Harold G, Wallace R, Clark AJ (1997) Transgene rescue in the mammary gland is associated with transcription but does not require translation of BLG transgenes. Transgenic Res 6:11–17
26. Palmiter RD, Brinster RL (1986) Germ-line transformation of mice. Annu Rev Genet 20:465–499
27. Giménez E, Lavado A, Giraldo P, Montoliu L (2003) Tyrosinase gene expression is not detected in mouse brain outside the retinal pigment epithelium cells. Eur J Neurosci 18:2673–2676
28. Montoliu L, Larue L, Beermann F (2004) On the use of regulatory regions from pigmentary genes to drive the expression of transgenes in mice. Pigment Cell Res 17:188–190
29. Montoliu L, Chávez S, Vidal M (2000) Variegation associated with lacZ in transgenic animals: a warning note. Transgenic Res 9:237–239
30. Ramírez A, Milot E, Ponsa I, Marcos-Gutiérrez C, Page A, Santos M, Jorcano J, Vidal M (2001) Sequence and chromosomal context effects on variegated expression of keratin 5/lacZ constructs in stratified epithelia of transgenic mice. Genetics 158:341–350
31. Kioussis D, Festenstein R (1997) Locus control regions: overcoming heterochromatin-induced gene inactivation in mammals. Curr Opin Genet Dev 7:614–619
32. Giménez E, Giraldo P, Jeffery G, Montoliu L (2001) Variegated expression and delayed retinal pigmentation during development in transgenic mice with a deletion in the locus control region of the tyrosinase gene. Genesis 30:21–25
33. Giménez E, Lavado A, Jeffery G, Montoliu L (2005) Regional abnormalities in retinal development are associated with local ocular hypopigmentation. J Comp Neurol 485:338–347
34. Wilkie TM, Brinster RL, Palmiter RD (1986) Germline and somatic mosaicism in transgenic mice. Dev Biol 118:9–18
35. Whitelaw CB, Springbett AJ, Webster J, Clark J (1993) The majority of G0 transgenic mice are derived from mosaic embryos. Transgenic Res 2(1):29–32
36. Pravtcheva DD, Wise TL, Ensor NJ, Ruddle FH (1994) Mosaic expression of an Hprt transgene integrated in a region of Y heterochromatin. J Exp Zool 268:452–468
37. Koetsier PA, Mangel L, Schmitz B, Doerfler W (1996) Stability of transgene methylation patterns in mice: position effects, strain specificity and cellular mosaicism. Transgenic Res 5:235–244
38. Wilkins AS (1990) Position effects, methylation and inherited epigenetic states. Bioessays 12:385–386

39. Grosveld F, de Boer E, Dillon N, Gribnau J, McMorrow T, Milot E, Trimborn T, Wijgerde M, Fraser P (1998) The dynamics of globin gene expression and position effects. Novartis Found Symp 214:67–79, discussion 79–86
40. Grosveld F, van Assendelft GB, Greaves DR, Kollias G (1987) Position-independent, high-level expression of the human beta-globin gene in transgenic mice. Cell 51:975–985
41. Lang G, Mamalaki C, Greenberg D, Yannoutsos N, Kioussis D (1991) Deletion analysis of the human CD2 gene locus control region in transgenic mice. Nucleic Acids Res 19:5851–5856
42. Ganss R, Montoliu L, Monaghan AP, Schütz G (1994) A cell-specific enhancer far upstream of the mouse tyrosinase gene confers high level and copy number-related expression in transgenic mice. EMBO J 13:3083–3093
43. Montoliu L, Umland T, Schütz G (1996) A locus control region at −12 kb of the tyrosinase gene. EMBO J 15:6026–6034
44. Grosveld F, Antoniou M, Berry M, De Boer E, Dillon N, Ellis J, Fraser P, Hanscombe O, Hurst J, Imam A, Lindenbaum M, Philipsen S, Pruzina S, Strouboulis J, Raguz-Bolognesi S, Talbot D (1993) The regulation of human globin gene switching. Philos Trans R Soc Lond B Biol Sci 339:183–191
45. Bonifer C, Vidal M, Grosveld F, Sippel AE (1990) Tissue specific and position independent expression of the complete gene domain for chicken lysozyme in transgenic mice. EMBO J 9:2843–2848
46. McKnight RA, Shamay A, Sankaran L, Wall RJ, Hennighausen L (1992) Matrix-attachment regions can impart position-independent regulation of a tissue-specific gene in transgenic mice. Proc Natl Acad Sci USA 89:6943–6947
47. McKnight RA, Spencer M, Wall RJ, Hennighausen L (1996) Severe position effects imposed on a 1 kb mouse whey acidic protein gene promoter are overcome by heterologous matrix attachment regions. Mol Reprod Dev 44:179–184
48. Gutiérrez-Adán A, Pintado B (2000) Effect of flanking matrix attachment regions on the expression of microinjected transgenes during preimplantation development of mouse embryos. Transgenic Res 9:81–89
49. Moreira PN, Pérez-Crespo M, Ramírez MA, Pozueta J, Montoliu L, Gutiérrez-Adán A (2007) Effect of transgene concentration, flanking matrix attachment regions, and RecA-coating on the efficiency of mouse transgenesis mediated by intracytoplasmic sperm injection. Biol Reprod 76:336–343
50. Porter SD, Hu J, Gilks CB (1999) Distal upstream tyrosinase S/MAR-containing sequence has regulatory properties specific to subsets of melanocytes. Dev Genet 25:40–48
51. Whitelaw CB, Grolli S, Accornero P, Donofrio G, Farini E, Webster J (2000) Matrix attachment region regulates basal beta-lactoglobulin transgene expression. Gene 244:73–80
52. Bessa J, Tena JJ, de la Calle-Mustienes E, Fernández-Miñán A, Naranjo S, Fernández A, Montoliu L, Akalin A, Lenhard B, Casares F, Gómez-Skarmeta JL (2009) Zebrafish enhancer detection (ZED) vector: a new tool to facilitate transgenesis and the functional analysis of cis-regulatory regions in zebrafish. Dev Dyn 238:2409–2417
53. Moltó E, Fernández A, Montoliu L (2009) Boundaries in vertebrate genomes: different solutions to adequately insulate gene expression domains. Brief Funct Genomic Proteomic 8:283–296
54. Mountford P, Zevnik B, Düwel A, Nichols J, Li M, Dani C, Robertson M, Chambers I, Smith A (1994) Dicistronic targeting constructs: reporters and modifiers of mammalian gene expression. Proc Natl Acad Sci USA 91:4303–4307
55. Gorski JA, Jones KR (1999) Efficient bicistronic expression of cre in mammalian cells. Nucleic Acids Res 27:2059–2061
56. Shaw-Jackson C, Michiels T (1999) Absence of internal ribosome entry site-mediated tissue specificity in the translation of a bicistronic transgene. J Virol 73:2729–2738
57. Attal J, Théron MC, Rival S, Puissant C, Houdebine LM (2000) The efficiency of different IRESs (internal ribosomes entry site) in monocistronic mRNAS. Mol Biol Rep 27:21–26
58. Attal J, Theron MC, Puissant C, Houdebine LM (1999) Effect of intercistronic length on internal ribosome entry site (IRES) efficiency in bicistronic mRNA. Gene Expr 8:299–309
59. Attal J, Théron MC, Houdebine LM (1999) The optimal use of IRES (internal ribosome entry site) in expression vectors. Genet Anal 15:161–165
60. Udvardy A, Maine E, Schedl P (1985) The 87A7 chromomere. dentification of novel chromatin structures flanking the heat shock locus that may define the boundaries of higher order domains. J Mol Biol 185:341–358
61. Chung JH, Whiteley M, Felsenfeld G (1993) A 5′ element of the chicken beta-globin domain serves as an insulator in human

erythroid cells and protects against position effect in Drosophila. Cell 74:505–514
62. Bell AC, Felsenfeld G (1999) Stopped at the border: boundaries and insulators. Curr Opin Genet Dev 9:191–198
63. Bell AC, West AG, Felsenfeld G (2001) Insulators and boundaries: versatile regulatory elements in the eukaryotic genome. Science 291:447–450
64. Lunyak VV, Prefontaine GG, Núñez E, Cramer T, Ju BG, Ohgi KA, Hutt K, Roy R, García-Díaz A, Zhu X, Yung Y, Montoliu L, Glass CK, Rosenfeld MG (2007) Developmentally regulated activation of a SINE B2 repeat as a domain boundary in organogenesis. Science 317:248–251
65. Furlan-Magaril M, Rebollar E, Guerrero G, Fernández A, Moltó E, González-Buendía E, Cantero M, Montoliu L, Recillas-Targa F (2011) An insulator embedded in the chicken {alpha}-globin locus regulates chromatin domain configuration and differential gene expression. Nucleic Acids Res 39:89–103
66. Taboit-Dameron F, Malassagne B, Viglietta C, Puissant C, Leroux-Coyau M, Chéreau C, Attal J, Weill B, Houdebine LM (1999) Association of the 5′HS4 sequence of the chicken beta-globin locus control region with human EF1 alpha gene promoter induces ubiquitous and high expression of human CD55 and CD59 cDNAs in transgenic rabbits. Transgenic Res 8:223–235
67. Potts W, Tucker D, Wood H, Martin C (2000) Chicken beta-globin 5′HS4 insulators function to reduce variability in transgenic founder mice. Biochem Biophys Res Commun 273:1015–1018
68. Rival-Gervier S, Pantano T, Viglietta C, Maeder C, Prince S, Attal J, Jolivet G, Houdebine LM (2003) The insulator effect of the 5′HS4 region from the beta-globin chicken locus on the rabbit WAP gene promoter activity in transgenic mice. Transgenic Res 12:723–730
69. Giraldo P, Rival-Gervier S, Houdebine LM, Montoliu L (2003) The potential benefits of insulators on heterologous constructs in transgenic animals. Transgenic Res 12: 751–755
70. Recillas-Targa F, Valadez-Graham V, Farrell CM (2004) Prospects and implications of using chromatin insulators in gene therapy and transgenesis. Bioessays 26:796–807
71. Román AC, González-Rico FJ, Moltó E, Hernando H, Neto A, Vicente C, Ballestar E, Gómez-Skarmeta JL, Vavrova J, White RJ, Montoliu L, Fernández-Salguero PM (2011) Dioxin receptor and SLUG transcription factors regulate the insulator activity of B1 SINE retrotransposons via an RNA polymerase switch. Genome Res 21(3):422–432
72. Thorey IS, Ceceña G, Reynolds W, Oshima RG (1993) Alu sequence involvement in transcriptional insulation of the keratin 18 gene in transgenic mice. Mol Cell Biol 13:6742–6751
73. Willoughby DA, Vilalta A, Oshima RG (2000) An Alu element from the K18 gene confers position-independent expression in transgenic mice. J Biol Chem 275:759–768
74. Montoliu L (2002) Gene transfer strategies in animal transgenesis. Cloning Stem Cells 4:39–46
75. Recillas-Targa F (2006) Multiple strategies for gene transfer, expression, knockdown, and chromatin influence in mammalian cell lines and transgenic animals. Mol Biotechnol 34:337–354
76. Jakobovits A, Moore AL, Green LL, Vergara GJ, Maynard-Currie CE, Austin HA et al (1993) Germ-line transmission and expression of a human-derived yeast artificial chromosome. Nature 362:255–258
77. Strauss WM, Dausman J, Beard C, Johnson C, Lawrence JB, Jaenisch R (1993) Germ line transmission of a yeast artificial chromosome spanning the murine alpha 1(I) collagen locus. Science 259:1904–1907
78. Lamb BT, Sisodia SS, Lawler AM, Slunt HH, Kitt CA, Kearns WG, Pearson PL, Price DL, Gearhart JD (1993) Introduction and expression of the 400 kilobase amyloid precursor protein gene in transgenic mice. Nat Genet 5:22–30
79. Pearson BE, Choi TK (1993) Expression of the human beta-amyloid precursor protein gene from a yeast artificial chromosome in transgenic mice. Proc Natl Acad Sci USA 90:10578–10582
80. Giraldo P, Montoliu L (2002) Artificial chromosome transgenesis in pigmentary research. Pigment Cell Res 15:258–264
81. Schedl A, Beermann F, Thies E, Montoliu L, Kelsey G, Schütz G (1992) Transgenic mice generated by pronuclear injection of a yeast artificial chromosome. Nucleic Acids Res 20:3073–3077
82. Moreira PN, Pozueta J, Giraldo P, Gutiérrez-Adán A, Montoliu L (2006) Generation of yeast artificial chromosome transgenic mice by intracytoplasmic sperm injection. Methods Mol Biol 349:151–161
83. Moreira PN, Pozueta J, Pérez-Crespo M, Valdivieso F, Gutiérrez-Adán A, Montoliu L (2007) Improving the generation of genomic-type transgenic mice by ICSI. Transgenic Res 16:163–168
84. Yang XW, Model P, Heintz N (1997) Homologous recombination based modification in

Escherichia coli and germline transmission in transgenic mice of a bacterial artificial chromosome. Nat Biotechnol 15:859–865
85. Camper SA, Saunders TL (2000) Transgenic rescue of mutant phenotypes using large DNA fragments. In: Accili D (ed) Genetic manipulation of receptor expression and function. Wiley, New York, NY, pp 1–22
86. Heintz N (2000) Analysis of mammalian central nervous system gene expression and function using bacterial artificial chromosome-mediated transgenesis. Hum Mol Genet 9:937–943
87. Van Keuren ML, Gavrilina GB, Filipiak WE, Zeidler MG, Saunders TL (2009) Generating transgenic mice from bacterial artificial chromosomes: transgenesis efficiency, integration and expression outcomes. Transgenic Res 18:769–785
88. Giraldo P, Giménez E, Montoliu L (1999) The use of yeast artificial chromosomes in transgenic animals: expression studies of the tyrosinase gene in transgenic mice. Genet Anal 15:175–178
89. Testa G, Vintersten K, Zhang Y, Benes V, Muyrers JP, Stewart AF (2004) BAC engineering for the generation of ES cell-targeting constructs and mouse transgenes. Methods Mol Biol 256:123–139
90. Ohtsuka M, Kimura M, Tanaka M, Inoko H (2009) Recombinant DNA technologies for construction of precisely designed transgene constructs. Curr Pharm Biotechnol 10:244–251
91. Heintz N (2001) BAC to the future: the use of bac transgenic mice for neuroscience research. Nat Rev Neurosci 2:861–870
92. Peterson KR (2003) Transgenic mice carrying yeast artificial chromosomes. Expert Rev Mol Med 5:1–25
93. Yang XW, Gong S (2005) An overview on the generation of BAC transgenic mice for neuroscience research. Curr Protoc Neurosci. Chapter 5:Unit 5.20
94. Deal KK, Cantrell VA, Chandler RL, Saunders TL, Mortlock DP, Southard-Smith EM (2006) Distant regulatory elements in a Sox10-beta GEO BAC transgene are required for expression of Sox10 in the enteric nervous system and other neural crest-derived tissues. Dev Dyn 235:1413–1432
95. Sparwasser T, Eberl G (2007) BAC to immunology–bacterial artificial chromosome-mediated transgenesis for targeting of immune cells. Immunology 121:308–313
96. Regales L, Giraldo P, García-Díaz A, Lavado A, Montoliu L (2003) Identification and functional validation of a 5′ upstream regulatory sequence in the human tyrosinase gene homologous to the locus control region of the mouse tyrosinase gene. Pigment Cell Res 16:685–692
97. Millot B, Montoliu L, Fontaine ML, Mata T, Devinoy E (2003) Hormone-induced modifications of the chromatin structure surrounding upstream regulatory regions conserved between the mouse and rabbit whey acidic protein genes. Biochem J 372 (Pt 1):41–52
98. Montoliu L, Roy R, Regales L, García-Díaz A (2009) Design of vectors for transgene expression: the use of genomic comparative approaches. Comp Immunol Microbiol Infect Dis 32:81–90

Chapter 4

Gene Targeting Vector Design for Embryonic Stem Cell Modifications

Thomas L. Saunders

Abstract

The use of genetically engineered mice to understand gene function is widespread. Changes to the mouse genome can be introduced with gene targeting vectors or with transgenes. Targeting vectors are usually used to ablate gene expression while transgenes are designed to express proteins that are normally absent from the organism. For example, gene targeting in mouse embryonic stem cells can be used to generate a mutant mouse model that fails to express a physiologically important protein. Transgenes that express the missing protein or a substitute for the missing protein can be used to assess possible gene therapies for the mutant mouse. Both gene targeting and transgene approaches can be used to study regulatory elements that control gene function. Putative control elements can be added to or removed from the chromosome with targeting vectors. Transgenes carrying long DNA sequences that include different combinations of potential control elements can be introduced into the genome to assess their effects on gene expression. The exploration of how genes interact to control development, homeostasis, and pathophysiological conditions can be dissected by introducing carefully designed genetic constructs into model organisms.

Abbreviations

BAC	Bacterial artificial chromosome
ES cell	Embryonic stem cell
IKMC	International Mouse Knockout Consortium
Kb	Kilo base pairs of DNA
PGKneo	DNA cassette featuring neomycin phosphotransferase II expression controlled by the phosphoglycerol kinase 1 promoter/enhancer
ZFN	Zinc finger nuclease

S. Pease and T.L. Saunders (eds.), *Advanced Protocols for Animal Transgenesis*, Springer Protocols,
DOI 10.1007/978-3-642-20792-1_4, © Springer-Verlag Berlin Heidelberg 2011

4.1 Introduction to Gene Targeting Design

Genetically engineered mice are used to understand gene function in cancer, development, immunology, metabolism, neurology, reproduction, and stem cells. Gene-targeted mice are used to model disease and to pilot gene therapies for disease amelioration. After it is decided to establish a new mouse model, the investigator needs to carefully consider the final structure of the targeted chromosome. Multiple plans for gene targeting vectors and screens to detect homologous recombination can be plotted on paper. Once the process of cloning the vector for ES cell electroporation is underway, changes to the project are less likely as the investment in cloning, ES cells, screening for recombination, and mouse breeding proceed.

This chapter outlines the essential steps in gene targeting vector design after a gene has been selected for modification. These include the preparation of in silico wild type and gene-targeted chromosome sequences, the identification of arms of homology, and methods to screen for homologous recombination; design elements that will maximize the probability of a desired outcome. The ultimate goal is to provide an animal model that will enable the investigator to address fundamental hypotheses about mechanisms that contribute to human health and disease.

4.1.1. Gene Targeting Procedure Outline

- Identify the gene of interest (GOI)
- Examine GOI in Vista Genome Browser
- Obtain Genomic DNA Sequence for GOI
- Obtain DNA Sequence of Targeting Vector Plasmid
- Prepare in silico representations of wild type gene, targeting vector, and targeted chromosome
- Prepare strategy to identify gene-targeted ES cells
- Prepare strategy to genotype gene-targeted mice

4.1.2. Principles and Applications

4.1.2.1. Gene Targeting Background

Gene function is studied in gene-targeted mice (gene "knockout" mice). Chromosomes in these mouse models are modified by homologous recombination with gene targeting vectors in ES cells. After confirmation of the desired genetic change, ES cell-mouse chimeras produce gametes with the changed chromosomes and transmit the modified chromosome to their offspring. Mice that inherit the genetically engineered chromosome are then tested to determine whether they differ from normal wild type mice.

Initially, the strategy for gene targeting in mice was to replace critical exons with drug selection cassettes to produce null alleles of genes [1, 2, 82, 84]. These designs effectively eliminate gene function and produce phenotypes. Some mutations resulted in embryonic or perinatal lethality that made it impossible to study pathophysiology in adult animals. This problem was resolved with the application of the Cre/loxP system to gene targeting [3]. In combination with Cre recombinase expression driven by cell-specific promoters [4] it was possible to produce cell-specific gene deletions to bypass embryonic lethality. The next refinement was the introduction of inducible Cre recombinase transgenes that provide temporal control of Cre recombinase expression, in addition to tissue-specific expression [5–7]. Databases of Cre transgenic mouse strains are available (http://www.creportal.org, http://nagy.mshri.on.ca/cre_new/).

In addition to producing null alleles of genes it is possible to introduce new genetic sequence on the chromosome (gene "knockin" mice). The choice of what to "knock in" can include a point mutation to change an amino acid in a protein [8], or to introduce a reporter molecule such as lacZ [9] or fluorescent proteins [10, 11], or sequences from other species [12–14]. The FLEX switch cassette can be used to inactivate the targeted gene by substituting a lacZ cassette flanked by DNA sequences, recognized by Cre or FLPe recombinases. Expression of recombinases inverts the FLEX switch to inactivate the reporter and activate a critical exon to restore a normal or modified exon in the gene [15].

Two additional recombinases ϕC31 and Dre [16, 17] promise to increase the sophistication of genetically engineered mouse models. Reagents designed for recombinase mediated cassette exchange with existing gene-targeted ES cell lines that carry loxP and FRT recognition sites and heterotypic sites have been developed [18, 19]. The efficiency of modifying targeted ES cells with these cassettes is much higher than the initial targeting of the gene in ES cells. Cre and FLP recombinases are active when introduced into fertilized mouse eggs by pronuclear microinjection [20, 21]. This suggests that it may be possible to co-inject recombinases with their cassettes to directly alter targeted genes in eggs and bypass the need to subject ES cells to another round of DNA electroporation with all of the attendant consequences of screening for recombination and time to expand targeted ES cell clones.

4.1.2.2. Gene Targeting Principles

Several factors are known to affect the efficiency of homologous recombination between gene-targeting vectors and embryonic stem cells. The use of genomic DNA in the targeting vector arms of homology that is isogenic with the ES cell line in question improves recombination efficiencies [22–24]. The total length of genomic DNA in the arms should be 5–10 kb, but do not need to be perfectly divided between the 5′ and 3′ arms [25]. Other considerations include whether the gene is expressed in ES cells. In this case a promoterless drug selection cassette can be used in the gene-targeting vector. Homologous recombination will use the endogenous promoter to confer resistance to selection [26]. The consequences of including repetitive DNA sequences in targeting vectors can adversely affect homologous recombination frequencies. Repeat sequences should be omitted from the arms of homology when possible.

The mechanism of homologous recombination between cloned DNA constructs and chromosomes in ES cells is not completely understood. One factor under experimental control is the source of the genomic DNA in the 5′ and 3′ arms of the targeting vector. The DNA sequences should be an exact match to the chromosomal DNA in the ES cells. The source of genomic DNA can be a genomic library of the mouse strain used to generate the ES cells or from inbred mice of the same strain used to produce the ES cells. It is inadvisable to use ES cells in culture as a source of genomic DNA because they may be contaminated by genomic DNA from feeder cells or because the ES cells are produced in a mixed genetic background ([27–29]). Intuition suggests that as the arms of homology in targeting vectors increase in length, the frequency of homologous recombination between targeting vectors and chromosomes in ES cells should increase. Experimental evidence does not support this suggestion. For example, Valenzuela et al. [30] observed an average targeting efficiency of 3.8% with bacterial artificial chromosome (BAC) targeting vectors that averaged 112 kb in the arms of homology. This is similar to results obtained with arms in the 5–10 kb range.

An unanticipated side effect of gene targeting is the observation that commonly used drug selection cassettes (PGKneo [1], MC1neo [31]) interfere with gene expression when they remain on the chromosome. For example, homozygous $Fgf8^{tm1.1Mrt\ tm1.1Mrt}$ mice that retain the FRT flanked (flrted) PGKneo cassette die the day after birth [32]. After FLP recombinase is used to remove the flrted cassette, homozygous $Fgf8^{tm1.3Mrt/\ tm1.3Mrt}$ mice survive normally. The mechanism of interference was determined to be cryptic splice sites and donors within the PGKneo and MC1neo cassettes [8, 32, 33] that cause reduced transcript levels of the targeted gene. Phenotypes that arise from the undesired transcript trapping activity of the PGKneo cassette can confound the phenotypes caused by changes in target gene expression. It is advisable to design gene-targeting vectors so that the neo cassette can be removed from the chromosome.

A second mechanism by which the PGKneo cassette can disrupt gene function results from the bidirectional promoter activity of the PGK promoter. This may have played a role in the embryonic lethality of normal blastocysts microinjected with *Men1* gene-targeted ES cells [34]. Simple disruption of the *Men1* gene could not explain the observation, since ES cells engineered to carry a conditional allele of *Men1* produced germline chimeras [35]. Although there are numerous examples of neo cassette interference with gene expression, cassette insertion in a gene does not always interfere with expression [36, 37]. Because the effect of the PGKneo drug resistance cassette on gene expression is not predictable, it is not advisable to rely on PGKneo cassettes for production of hypomorphic or null alleles for the purpose of activating gene expression by removal of the cassette with a recombinase enzyme such as FLPe or Cre. A strategy to eliminate the drug selection cassette from the targeted locus is essential.

A third factor that affects the frequency of targeted homologous recombination is the presence of repetitive DNA sequences in the arms of homology in the targeting vector. One pathway cells use to repair double stranded breaks in chromosomes is homologous recombination with sister chromatids (reviewed in [38]). The same pathway can use nonallelic homologous DNA sequences for recombination to repair double-strand breaks. The DNA repeat sequences that make up at least 40% of the mouse genome sequence provide numerous substrates for homologous recombination. Classes of repeats include dinucleotide repeats, long interspersed repeat 1 elements (LINE1), short interspersed sequences (SINEs), early transposons (ETns), endogenous retrovirus-like elements (LRVs), and other elements with long terminal repeats (LTR) such as intracisternal A-particles (IAP) [39, 40]. There is abundant evidence that nonallelic repeat sequences are recombinogenic with each other [38, 41–43]. A systematic comparison of gene targeting vectors with different repeat sequence compositions is not available. However, anecdotal evidence from our laboratory and others supports the conclusion that repetitive sequences in arms of homology suppress correct homologous recombination. Attempts to target DNA sequences such as an exon surrounded by several kb of repeats on both the 5′ and 3′ sides resulted in one instance of homologous recombination in 2,304 ES cell clones (unpublished observations, T.S.). A targeting vector that included 1 kb of repeats at the 5′ end of an arm of homology failed to give correct targeting in 480 clones. After the arm was shortened by 1 kb to remove the repeats, correctly targeted clones were identified (unpublished observations, T.S). Thus, when a targeting vector that contains a high proportion of repetitive sequence is introduced into ES cells by electroporation, the number of possible targets for nonallelic homologous recombination will greatly exceed the desired gene target. As a result, there is a higher probability that recombination will occur between repetitive elements scattered through the genome instead of the desired homologous recombination event involving the targeted gene.

4.1.2.3. Gene Targeting Vector Design

The essential features of a gene targeting vector design are use of a plasmid that includes a drug selection cassette such as PGKneo that is flanked by recombinase recognition sites such as FRT. LoxP sites are reserved for the modification of exons. Once a conditional allele has been produced by flanking an exon with loxP sites it can be combined with a library of Cre transgenic mice. Compared to other recombinases, many more cell-specific Cre recombinase mice are available. The target sequence is most often an exon that will cause an mRNA reading frame shift during splicing that results in nonsense mediated decay of mRNA, although the target may be a microRNA that regulates gene expression or other nonprotein coding sequence. A common, but not necessary feature of targeting vectors is the inclusion of a negative selection cassette designed to express the diphtheria toxin A chain [44] or herpes simplex virus thymidine kinase [1]. The effectiveness of negative selection varies between targeting experiments and may or may not dramatically improve the efficiency of homologous recombination with the targeting vector from a few percent to a few fold.

Variations on this model include the use of splice acceptors for targeted trapping that make use of the endogenous promoter of the GOI to express protein fusions between the enzyme that confers resistance to drug selection with a reporter, such as beta-galactosidase or green fluorescent protein. In this case, the cells that express drug resistance are marked by the reporter. This approach is generally limited to genes expressed in ES cells, estimated to be about 60% of all genes in the genome [26, 45]. A second variation with a splice acceptor design is the FLEX system that makes it possible to inactivate the targeted gene and simultaneously express a different gene. FLP recombinase can be used to re-activate the targeted gene and produce a floxed conditional allele that can be inactivated by independent expression of Cre recombinase. With the demonstration that phiC31-attP/attB [46] and Dre-*rox* [16, 47] recombinase systems function in ES cells and mice it is possible to design experimental models that are only limited by the imagination of the scientist.

The international knockout mouse project (IKMC) uses a "knockout first" targeting vector design. The targeting vector includes a splice acceptor designed to substitute the expression of a lacZ cassette for the expression of the endogenous gene 5′ of a floxed exon. Expression of FLPe removes the splice acceptor and drug resistance cassette and the remaining floxed exon produces a conditional allele that can be inactivated upon the expression of Cre recombinase. One consideration that applies to targeted genes and gene trapped alleles that rely on a splice acceptor to reduce gene expression is that measurable levels of normal gene expression may occur as a result of RNA splicing around the splice acceptor. Thus, the use of a splice acceptor may result in a hypomorphic allele instead of a gene knockout.

When multiple targeted alleles of the same gene need to be generated (for example, an allelic series of reporter molecules or point mutations knocked in to the same allele) it is advantageous to strategically position an FRT-loxP sequence on the chromosome so that dual recombinase-mediated cassette exchange (RMCE) can be used. The rate-limiting step in gene targeting is often the identification of correctly targeted ES cells. After the initial identification of a targeted clone, the generation of an allelic series can be more efficiently achieved by dual RMCE instead of retargeting the wild type allele. Dual RMCE is achieved simply by co-electroporating existing targeted ES cells with both Cre and FLP and a cassette that will be introduced into existing recombinase recognition sites [18, 19, 81].

Targeting designs that do not use a splice acceptor to modify the GOI are used to introduce point mutations, to knock in heterologous genes to be expressed under the control of the regulatory elements of the GOI, or to generate conditional alleles. Designs that omit the splice acceptor-reporter/drug selection cassette used in the original knockout design will not mark cells that express the GOI, but may be simpler to clone in the molecular biology laboratory.

4.2 Gene Targeting Vector Materials

4.2.1. Software and Internet Resources

Tools for gene targeting vector design include (1) DNA sequence manipulation software packages such as DNAStar or VectorNT, (2) access to the public genomic sequence through ENSEMBL, NCBI, or UCSC genome browsers, (3) access to an image-based sequence browser such as the Vista genome browser that includes data on conserved noncoding sequences. These tools are used to design a customized gene nucleotide by nucleotide. Another useful resource is the Custom Design Tool from the High Throughput Gene Targeting group at the Wellcome Trust Sanger Institute (http://www.sanger.ac.uk/htgt). The investigator enters the ENSEMBL identification of the gene in question, identifies the exon to be used for conditional mutagenesis, and the software produces a solution with PCR primers for the cloning of the targeting vector and screening for recombination (Fig. 4.1). Access to the software requires the user to provide a username and password.

4.2.2. Online Mouse Genome Browsers

Vista Genome Browser
http://pipeline.lbl.gov/cgi-bin/gateway2?bg=mm7&selector=vista

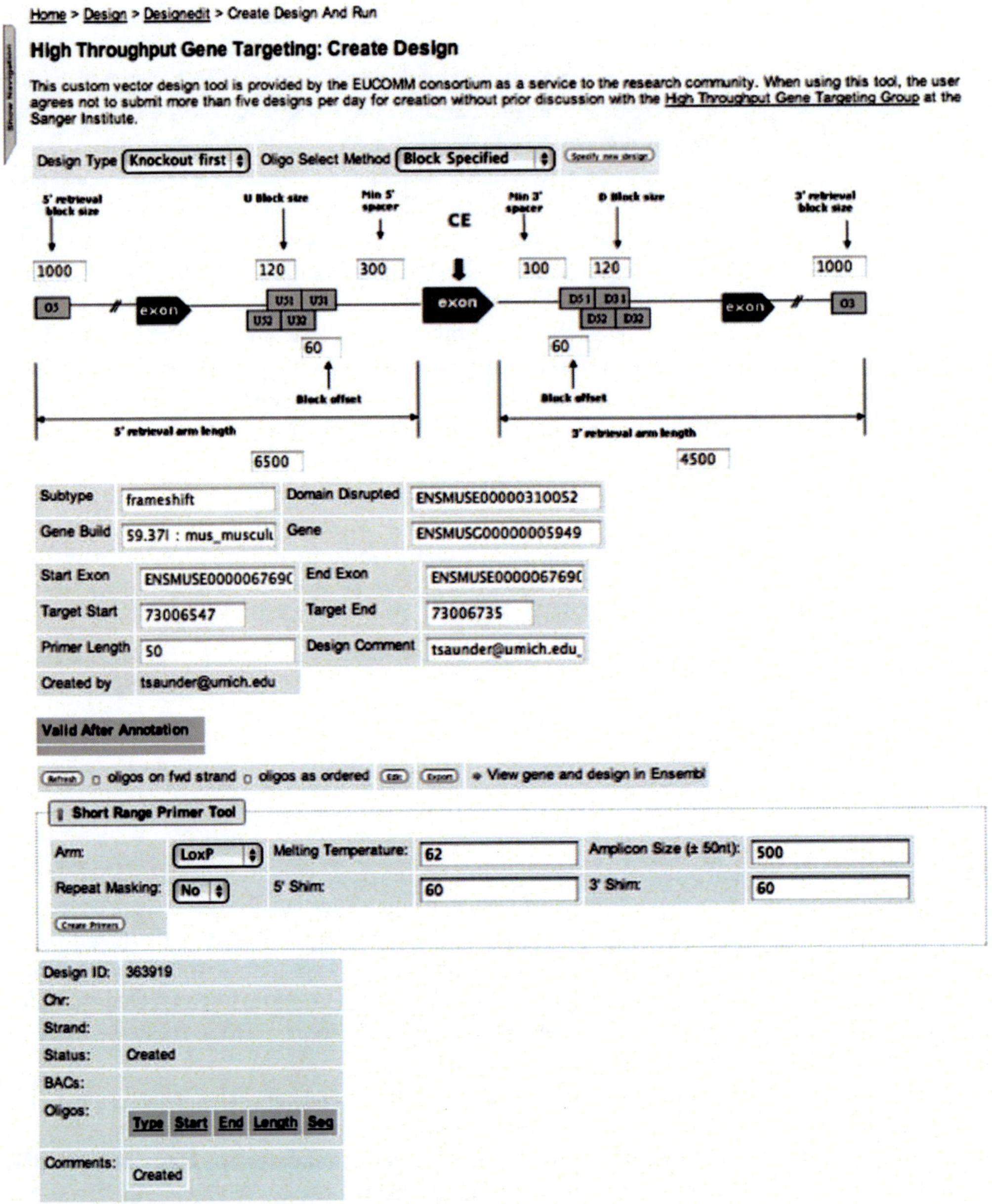

Fig. 4.1 Output from Sanger Institute custom design tool showing the solution for the *Ctns* mouse gene. The software is located at http://www.sanger.ac.uk/htgt.

ENSEMBL Genome Browser
http://ensembl.org/Mus_musculus/Info/Index
NCBI Genome Browser
http://www.ncbi.nlm.nih.gov/mapview/map_search.cgi?taxid=10090
UCSC Genome Browser
http://genome.ucsc.edu/cgi-bin/hgGateway?db=mm2

4.2.3. Gene Targeting Vector Plasmids and Internet Resources

Plasmids for gene targeting vector construction are available from many sources (*see* Subheading 4.2.4). A plasmid that leaves behind a loxP and an adjacent FRT site is essential for the production of an allelic series by RMCE strategy ([18, 19, 48]. Because isogenic DNA can profoundly influence the efficiency of homologous recombination, it is essential to know the precise source of genomic DNA used to construct the targeting vector. Genomic DNA for use in targeting vectors can be obtained in the form of a BAC clone from libraries such as those maintained at the BACPAC Resources Center (http://bacpac.chori.org). Purified genomic DNA can be obtained from the Jackson Laboratory DNA resource or other sources (Table 4.1). Although DNA can be

Table 4.1
Mouse ES cell lines and isogenic DNA sources

ES cell	Genetic background	Genomic DNA source
AB1 & AB2 [55] & sublines	129S7/SvEvBrd-*Hprt*$^{b-m2}$	
Bruce4 [Hughes et al. 2007]	C57BL/6-Thyl.1(congenic)	Stock 000664 JAX[a]
CJ7 [71]	129S1/SvImJ	Stock 002448 JAX[a]
D3 [72]	129S2/SvPas	Strain Code 476 CRL[b]
E14 [73] & sublines	129P2/OlaHsd	129P2/OlaHsd Harlan[c]
G4 [27]	(129S6/SvEvTac X C57BL/6NCr)F_1	Taconic[d]
J1 [74]	129S4/SvJae	Stock 002448 JAX[a]
JM8 [75] & sublines	C57BL/6N	Stock 000664 JAX[a]
GSI-1	129X1/SvJ	Stock 000691 JAX[a]
Pat5 [76]	129X1/SvJ	Stock 000691 JAX[a]
R1 [Nagy et al. 1993]	(129X1/SvJ X 129S1/SvImJ)F_1	Stock 000691 JAX[a]
RW4	129X1/SvJ	Stock 000691 JAX[a]
TC1 [77]	129S6/SvEvTac	Taconic[d]
TBV2 [78]	129S2/SvPas	Strain Code 476 CRL[b]
VGB6 [29] lines	C57BL/6NTac	Taconic[c]
V6.5 [79]	(129S4/SvJae X C57BL/6)F_1	Stock 000664 JAX[a]
W4 [80]	129S6/SvEvTac	Taconic[d]
W9.5 [16]	129S1/SvImJ	Stock 002448 JAX[a]

Numbers in square brackets refer to citations describing ES cell line genetic background. *See* also Simpson et al. (1997). RW4 and GSI-1 ES cell lines were derived by Genome Sciences, Inc.
[a]Isogenic genomic DNA available from the Jackson Laboratory DNA Resource http://www.jax.org/dnares/index.html
[b]Mice available from Charles River Laboratory http://www.criver.com
[c]Mice available from Harlan http://www.harlan.com
[d]Isogenic genomic DNA available from Taconic Farms http://www.taconic.com

purified directly from ES cells grown on gelatin, this is not advisable for cells grown on feeder cells or for hybrid F1 ES cells since the 5′ and 3′ arms of homology for the targeting vector might be cloned from two different genetic backgrounds.

4.2.4. Sources of Plasmids for Gene Targeting Vector Construction

http://egtc.jp/action/main/system?name=trapvector#content
http://genetrap.helmholtz-muenchen.de/ggtc/info/protocols/vectors.php
http://www.addgene.com
http://www.cmhd.ca/genetrap/vectors.html
http://www.cmhd.ca/genetrap/vectors.html
http://www.eucomm.org/htgt/backbones
http://www.eucomm.org/htgt/cassettes
http://www.eucomm.org/information/targeting/
http://www.med.umich.edu/tamc/mta.html
http://www.sanger.ac.uk/PostGenomics/genetrap/vectors/

4.3 Gene Targeting Vector Transgene Protocol

4.3.1. DNA Cloning

A detailed description of DNA cloning equipment and reagents is beyond the scope of this chapter. The reader is referred to many excellent manuals on this topic such as Surzycki [49]. Experimental details on BAC recombineering can be found in Johansson et al. [50] and in Gong et al. [51].

4.3.2. Genomic Structure Discovery for Gene Targeting Vectors

After the GOI is identified, the next step is to evaluate its genomic structure in the Vista genome browser (*see* Subheading 4.2.2 and [52]). For example, analysis of the *Ctns* gene shows a predicted structure of 11 exons. Exon one is a short noncoding exon and exon 11 has a longer untranslated region (Fig. 4.2). Repetitive DNA sequences are present in introns one, two, four, and nine. Assuming the design uses 3 kb arms of homology, use of exons in the 3′ half of the gene would be preferable, in order to avoid the presence of repetitive sequences that might reduce the efficiency of homologous recombination.

Selection of an exon for a conditional allele is accomplished by examining the reading frames of the exons in the ENSEMBL genome browser (*see* Subheading 4.2.2). Navigation to the *Ctns* transcript shows the exon–intron structure at the nucleotide level (Fig. 4.3a). The list of exons indicates that elimination of exons 3, 4, 5, 7, and 10 will result in out-of-frame splicing likely to result in nonsense mediated RNA decay [53]. Nonsense mediated decay is dependent on the presence of a premature stop codon in an exon. The predicted mRNA sequence should be analyzed in silico to ensure that a premature stop codon will occur in the mRNA. Removal of exons 6, 8, and 9 will produce an in-frame splice

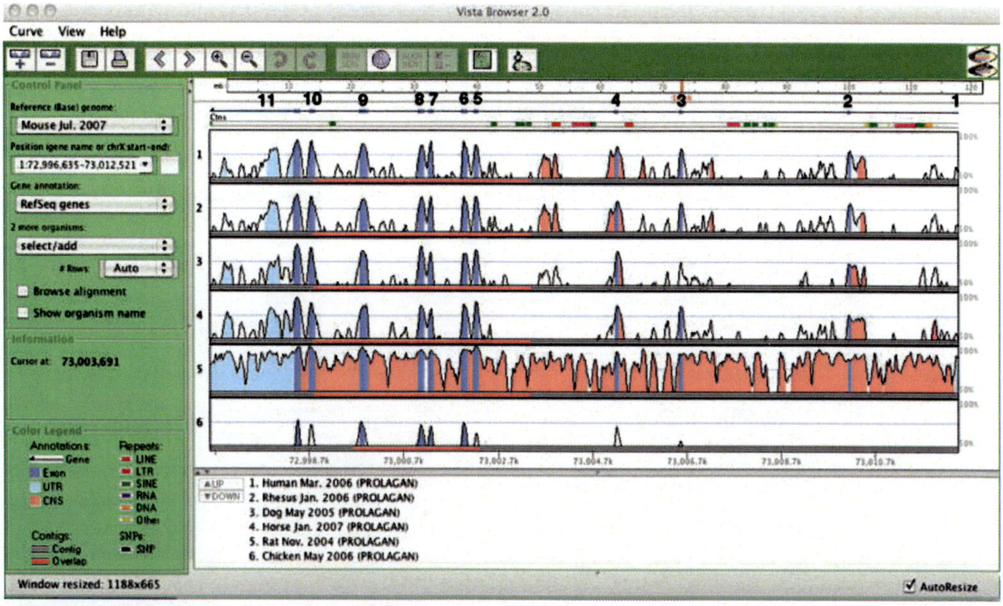

Fig. 4.2 Mouse *Ctns* genomic structure as viewed in Vista Genome Browser. Exons are numbered at the *top* of the figure. *Blue boxes* beneath *numbers* indicate exon locations. *Light blue* indicated untranslated mRNA sequences. Repeat sequences are noted as *red*, *green*, and *pink boxes*. *Lines* 1–6 indicate conserved sequences between mouse and other species. Significant conservation occurs in the noncoding mouse and human in introns 1, 2, 3, and 4. Note that the direction of transcription is from 3′ to 5′ on the reverse strand of chromosome 11.

that might produce a truncated protein with residual function (Fig. 4.3b). Ideally the out-of-frame exon nearest to the 5′ end of the gene should be targeted for a conditional allele since its elimination is most likely to result in a nonsense mRNA and loss of transcript translation.

In the Vista browser, the selection of exon 3 as the exon to flank with loxP sites suggests an upstream arm ending short of exon 2 so that the hybridization probe can include exon 2 sequence to increase probe specificity. This will result in 5′ arm of homology of approximately 3 kb, including 541 bp of repetitive sequences. On the downstream side of exon 3, the repetitive sequences are more prevalent. A short arm of approximately 1 kb free of repeats could be used. This would facilitate a PCR-based screen across a short arm of homology. Alternatively, a 4.5-kb arm could be used. It would include 1,162 bp of repeat sequences and end 400 bp downstream of the last repeat in intron 5 (between exons 5 and 6). A hybridization probe would be based on exon 6 and downstream sequence.

4.3.3. Genomic DNA Sequence Discovery for Gene Targeting Vectors

The next step in the process is to obtain genomic sequence 20 kb upstream and downstream of *Ctns*. This can be downloaded as an EMBL or GenBank file from ENSEMBL. The EMBL format is preferred because the repeat sequences are annotated. The sequence of the plasmid to be used for cloning the targeting

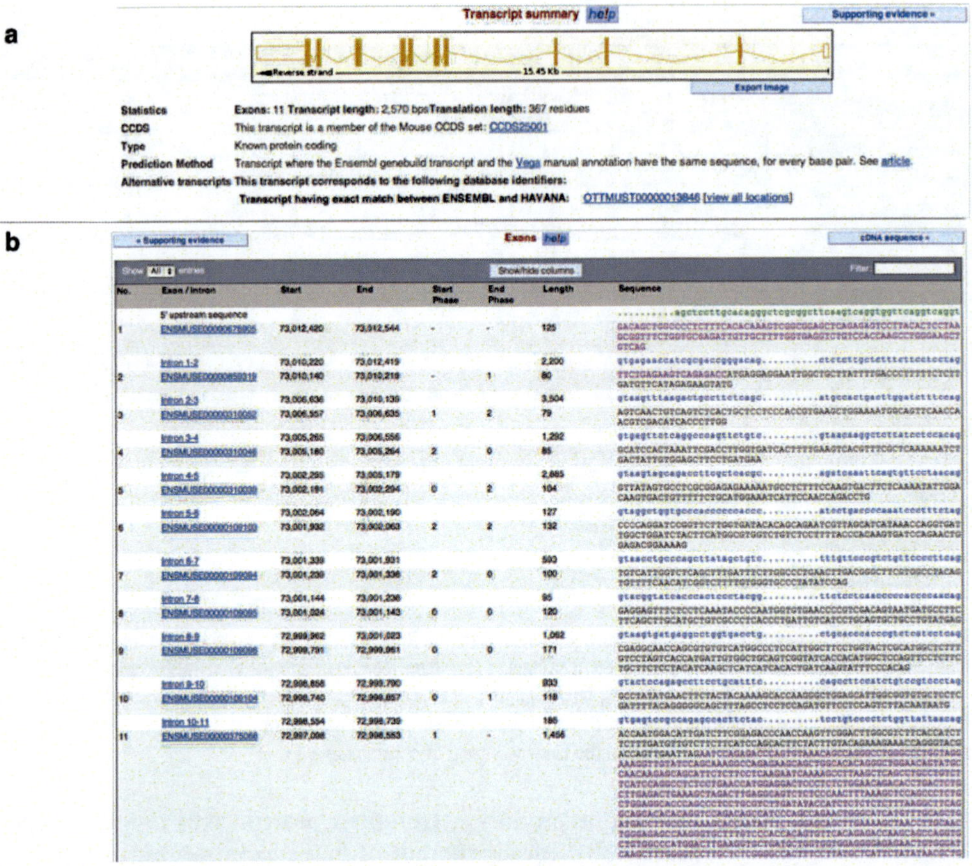

Fig. 4.3 Mouse *Ctns* transcript, exon and intron sequences obtained from ENSEMBL. (**a**) Transcript model. (**b**) Exon and intron sequences indicate exact location of exons and intron lengths. The *reading frame* of each exon is also shown.

vector should be available for analysis. The genomic sequence is decorated with restriction enzyme sites that are useful for detecting homologous recombination (Table 4.2). An in silico version of the gene targeting vector is constructed with the plasmid sequence. Exon 3 with 100 bp upstream and downstream will be PCR amplified from an isogenic DNA source and cloned into the Bam HI site of ploxPFLPneo (Fig. 4.4). After the in silico sequence of the wild type and the targeted alleles are prepared, their restriction maps can be compared (Fig. 4.5). An examination of the restriction enzymes mapped to the wild type and targeted genes shows that the Sph I restriction enzyme cuts outside of the arms of homology to produce a 14 kb fragment. After introduction of the ploxPFLPneo vector, the restriction map changes so that Sph I will produce 9 kb 5′ and 6 kb 3′ fragments that can be used to identify homologous recombination of the targeting vector.

Table 4.2
Useful restriction enzymes for the detection of homologous recombination in mouse ES cells by Southern blot analysis of restriction fragment length patterns

Good performance	Variable performance	Poor performance
Age I	Eco RI	BspD I
Asp718	Kpn I	Cla I
Bam HI	Sac I	Hind II
BbvC I		Not I
Bgl I		Pvu I
Bgl II		Sac II
Eco RV		Sal I
Hind III		Sma I
Nco I		Xba I
Nsi I		Xho I
Pst I		Xmn I
Pvu II		
Sac I		
Sca I		
Spe I		
Sph I		
Stu I		
Sst I		

The final step prior to cloning the targeting vector is to establish that the restriction enzyme and hybridization probe will detect the expected size DNA fragments when ES cell DNA is analyzed. Since the public sequence is based on C57BL/6, it is expected that C57BL/6 ES cell DNA will produce the expected size fragments. If DNA from 129 mouse ES cells is used, it is possible that Southern analysis will not produce the expected fragments because of polymorphisms in the DNA sequence. When a PCR screen is planned a positive control DNA construct that includes sequences inside the drug selection cassette and outside the short arm of homology is cloned. This construct is used to prepare copy standards by spiking ES cell DNA. The primers should detect 0.01 copies of the control construct for a pool of ES cell clones to be genotyped [54]. Ideally, the size of the PCR product will be different from the size of the product

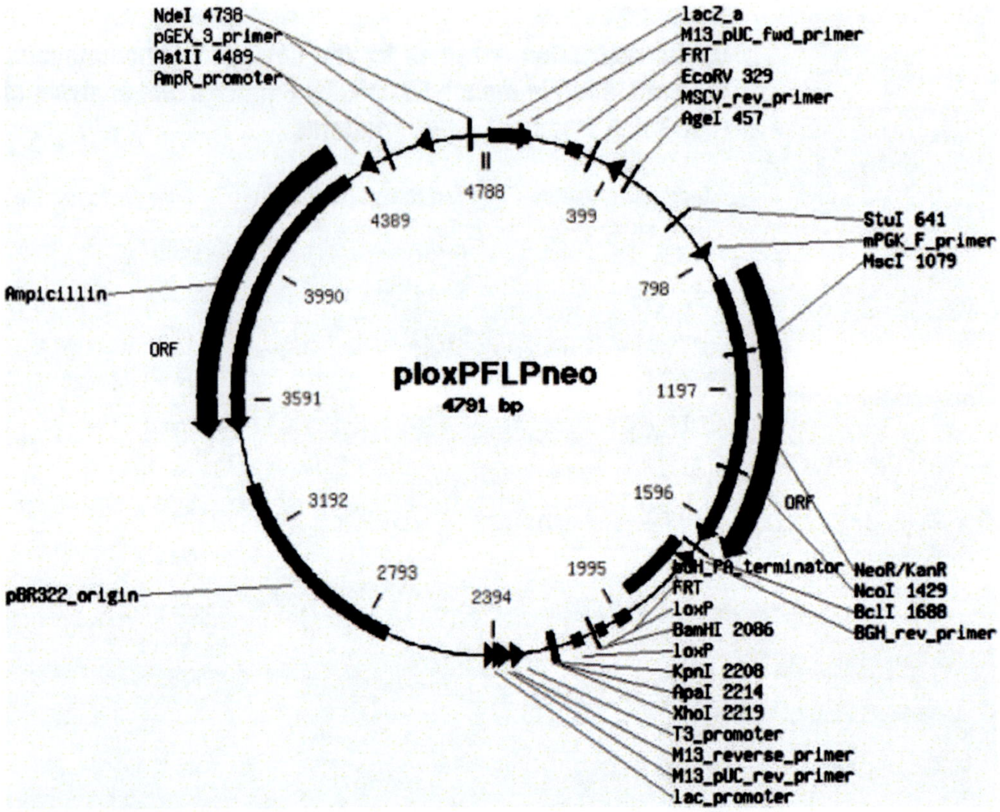

Fig. 4.4 ploxPFLPneo plasmid map. The essential features of the plasmid are the FRT flanked PGKneo cassette and the adjacent loxP- Bam HI -loxP cassette. To construct a targeting vector that will generate a conditional allele, the critical exon is cloned into the Bam HI site. Arms of genomic homology are cloned into the unique restriction sites 5′ of the flirted PGKneo sequence and 3′ of the floxed Bam HI site. DNA sequence for this plasmid is located at http://www.med.umich.edu/tamc.

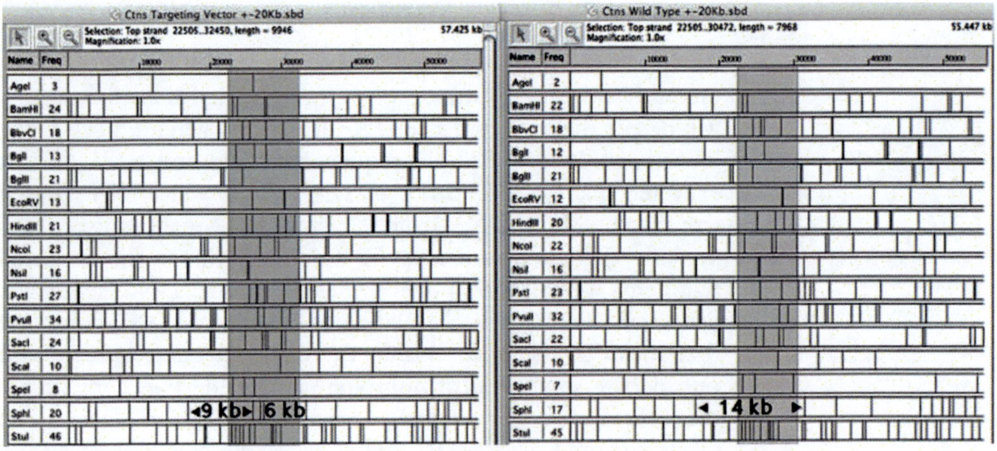

Fig. 4.5 Comparison of restriction enzyme maps between wild type *Ctns* and gene targeted *Ctns* chromosomes. DNA sequences 20 kb 5′ and 3′ of *Ctns* were obtained from ENSEMBL and a restriction map was generated with the enzymes

generated by the targeted gene [55]. A third approach to identifying homologous recombination is to use quantitative PCR to measure the number of wild type target sequences in the ES cell. In this case a TaqMan® probe for the GOI is prepared so that it is interrupted by plasmid sequences. The probe will not detect random targeting-vector integrations. Heterozygous ES cells cross the cycle threshold after homozygous ES cells, typically by a factor of one ΔCt [30, 56, 57].

4.3.4. Targeted Trapping Designs for Gene-Targeting Vectors

Gene targeting plasmids such as ploxPFLPneo depend on the mouse *Pgk1* promoter to drive expression of neomycin phosphotransferase. After electroporation with a targeting vector that includes PGKneo, the majority of the G418 resistant ES cells carry random integrations of the targeting vector. Correctly gene targeted ES cells comprise only a few percent of drug resistant colonies. It is possible to use a promoterless drug resistance cassette instead of *Pgk1*. In this type of experiment, a splice acceptor is substituted for the promoter and expression of the neo cassette depends on the activity of the targeted gene's promoter. This strategy is effective for genes that are expressed in ES cells. A simple method to identify such genes is to check the genetrap database (http://www.genetrap.org). If multiple genetrap clones are present then it is likely that the gene expression levels in ES cells are high enough to confer G418 resistance. Use of a targeted trapping approach can dramatically enrich the number of correctly targeted ES cell clones [26]. A number of gene designs in the EUCOMM pipeline are based on targeted trapping and splice-acceptor/drug resistance cassettes are available from EUCOMM [58].

4.4 Results

4.4.1. Gene Targeting Vector Results

A completed gene-targeting vector project will include the targeting vector plasmid, which can be assembled by standard plasmid cloning or by recombineering in BACs [30, 59, 60]. A screen for homologous recombination either by Southern blot, PCR, or copy number counting is established. A plan to detect the (1) wild type, (2) targeted, (3) conditional, and (4) null alleles is in place (Fig. 4.6). DNA sequencing of the clone targeting vector is used

Fig. 4.5 (Continued) listed in Table 4.2. The *gray columns* on each side of figure correspond to sequences contained within the cloned targeting vector. Restriction enzymes are listed on the *left*. *Vertical lines* represent restriction sites relative to the DNA sequence at the *top* of the figure. Changes in the Sph I restriction fragment sizes after homologous recombination are indicated.

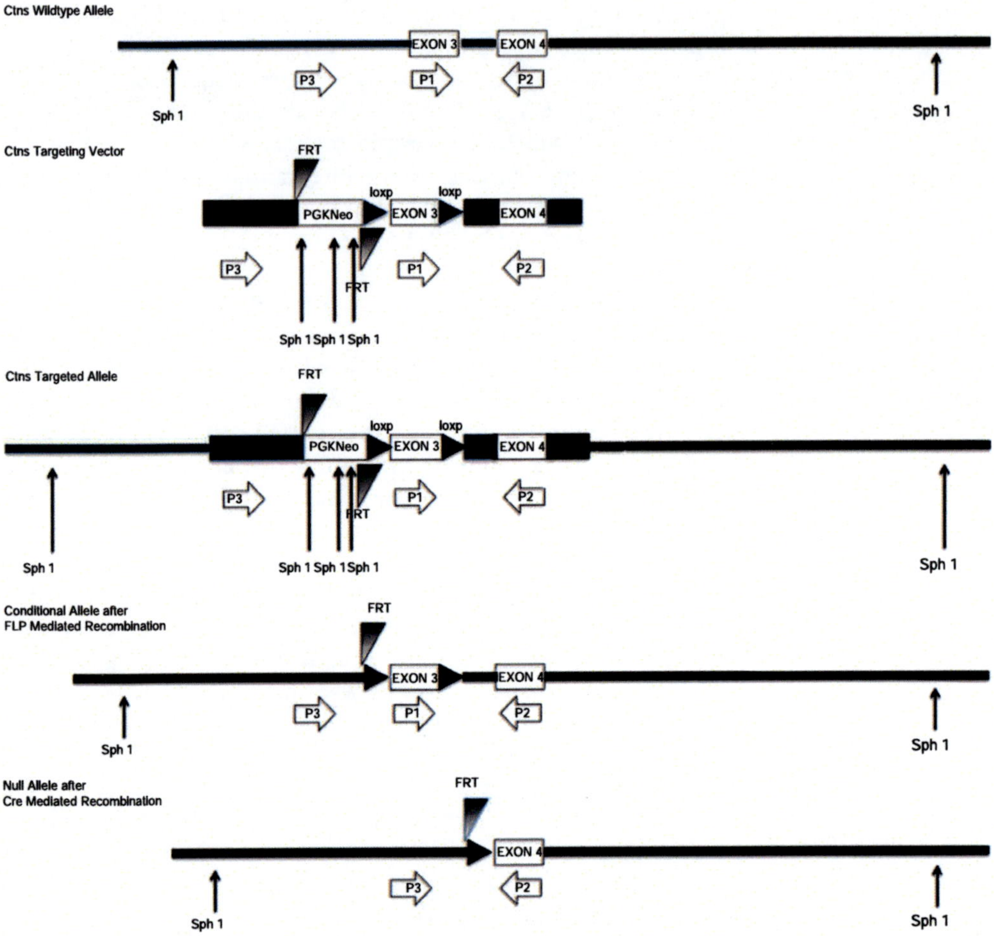

Fig. 4.6 *Ctns* Gene Targeting Diagram. Construction of a conditional allele for *Ctns*. *Top line* shows a diagram of the genomic locus showing exons three and four. *Second line* shows the structure of a targeting vector designed with a FRT flanked PGKneo cassette and loxP flanked exon four. Note the several additional Sph I restriction enzyme sites in the PGKneo cassette. *Third line* shows the *Ctns* targeted allele after homologous recombination with the targeting vector. *Fourth line* shows the Ctns conditional allele after FLPe or FLPo is used to remove the PGKneo cassette and leave behind a single FRT site. *Fifth line* shows the Ctns null allele after Cre mediated recombination. P1, P2, and P3 are primers that can be used to distinguish between the wild type allele, the floxed allele, and the null allele. Note that since the primers are internal to the targeting vector that they cannot be used to discriminate between gene targeted ES cells and ES cells carrying a random insertion of the targeting vector. In principle any new genetic element, such as a fluorescent reporter protein, a cDNA designed to express a mutant form of the endogenous protein, a cDNA for a recombinase, or genetic elements to confer inducibility of gene expression can be inserted between the loxP sites flanking exon three.

to show that the upstream and downstream arms of homology are in the same 5′–3′ orientation with respect to the floxed exon; to show that both loxP sites are present and in the same orientation, that both FRT sites are present and in the same orientation. When the targeting vector and the screen are ready the next step is to move the project into the ES cell laboratory and electroporate isogenic ES cells with the targeting vector (*see* Chap. 14).

4.5 Troubleshooting

4.5.1. DNA Cloning Roadblocks

DNA cloning may present a challenge to laboratories that infrequently use DNA cloning in their research. Or the sequence complexity may interfere with DNA cloning. In this case, it may be cost effective to employ a commercial company to synthesize and or clone the targeting vector or hybridization probes for a gene-targeting project.

4.5.2. Screening for Homologous Recombination by Southern Blot

Failure to identify ES cell clones that have undergone homologous recombination with the targeting vector is the most common set back in gene targeting. Southern blot probes can be optimized by shifting the sequences selected for the probe to a different area that does not contain repetitive sequences.

4.5.3. Screening for Homologous Recombination by PCR

PCR screens often present both false positive and false negative results. Measuring the limits of detection of DNA input in screens is essential. An insensitive screen will not detect recombination when too little DNA is available for the screening assay.

4.5.4. Failure to Detect Homologous Recombination

It is not unusual for homologous recombination to occur in 1–3% of clones with promoter-based drug selection cassettes (PGKneo). The author observed one experiment in which recombination was detected once in twenty-four 96-well plates or once in 2,300 clones. It may be necessary to screen additional clones to find the clone with the desired recombination event.

4.5.5. Failure to Obtain Germline Transmission

When targeted ES cell clones are identified, there is no guarantee that any individual clone will form a germline chimera. If the ES cell line used in the experiment is genetically unstable then it is possible that all of the targeted ES cell clones will be unusable. It may be possible to obtain a germline competent ES cell clone by subcloning or from a repository or to obtain an existing mouse model from a repository to answer the research question (Table 4.3) [61, 83].

4.6 Conclusion and Outlook

The international consortium to knockout every gene in the mouse (IKMC) will deliver thousands of ES cell lines and hundreds of genetically engineered mouse strains. The distribution of ES cell lines to research groups for the independent derivation of gene-targeted mouse strains will establish hundreds of additional mouse strains. Despite this effort, the need to

Table 4.3
Databases of genetically modified mice and ES cells

Canadian Mouse Mutant Repository (CMMR) http://www.cmmr.ca
Center for Animal Resources and Development Database (CARD) http://cardb.cc.kumamoto-u.ac.jp/transgenic/index.jsp
Cre Mouse Portal at The Jackson Laboratory http://www.creportal.org
Cre-X-Mice: A Database of Cre Transgenic Lines http://www.nagy.mshri.on.ca/cre
Deltagen Mouse Knockout Database http://www.deltagen.com
European Conditional Mouse Mutagenesis Program (EUCOMM) http://www.eucomm.org
European Mouse Mutant Archive (EMMA) http://www.emmanet.org
Federation of International Mouse Resources (FIMRe) http://www.fimre.org/
German Gene Trap Consortium (GGTC) http://genetrap.helmholtz-muenchen.de
International Gene Trap Consortium (IGTC) http://www.igtc.org or http://www.genetrap.org
International Knockout Mouse Consortium (IKMC) http://www.knockoutmouse.org
International Mouse Strain Resource (IMSR) http://www.findmice.org/
Japan Mouse/Rat Strain Resources Database http://www.shigen.nig.ac.jp/mouse/jmsr/top.jsp
Knockout Mouse Project (KOMP) Repository http://www.komp.org
Mutant Mouse Regional Resource Centers (MMRRC) http://www.mmrrc.org
RIKEN BioResource Center http://www2.brc.riken.jp/lab/animal/search.jsp
Taconic Knockout Repository http://kodatabase.taconic.com/database.php
Texas A&M Institute of Genomic Medicine (TIGM) http://www.tigm.org
Unitrap Database of Gene Trapped ES Cell Clones http://www.unitrap.cbm.fvg.it

generate custom designer genes in ES cells will continue. For example, the derivation of mouse strains that include point mutations in genes to model diseases caused by mutations found in human patients will continue [62, 63, 85]. The discovery of the role of microRNAs in disease has inspired the development of disease models based on gene targeting technology [64, 86]. The deletion of noncoding regulatory elements is another area where custom gene-targeted mouse lines are of value and are unlikely to be prioritized in the IKMC [65]. Chromosome engineering to model human diseases caused by aneuploidy and chromosome duplications/deletion can be most precisely reproduced by chromosome engineering in ES cells [66, 67]. The IKMC resources will be invaluable to numerous research groups and accelerate the pace of progress. Demand will continue for custom gene modifications in addition to the genetrap, conditional alleles, knockout first alleles, definitive null alleles, targeted nonconditional alleles, and retargetable alleles available from the IKMC.

The advent of zinc finger nuclease (ZFN) technology makes it possible to generate gene knockouts directly by microinjecting fertilized mouse eggs [68]. This approach takes advantage of an established technique and bypasses the need for ES cells. Besides producing simple gene knockouts it is possible to introduce new information on the mouse chromosome with zinc fingers. In this variation, DNA molecules that incorporate arms of homology 5′ and 3′ to the target sequence of the ZFN are co-introduced into the zygote with ZFN mRNA. The integration of the target sequence is then verified by genotyping and expression analyses [69, 70]. Whether ZFN will supplant the use of ES cells for the manipulation of the mouse genome is not clear. The use of ZFN technology promises access to the genomes of other model organisms for genetic manipulation and to advance scientific understanding.

References

1. Tybulewicz VLJ, Crawford CE, Jackson PK, Bronson PT, Mulligan RC (1991) Neonatal lethality and lymphopenia in mice with a homozygous disruption of the c-abl proto-oncogene. Cell 65:1153–1163
2. Zijlstra M, Li E, Sajjadi F, Subramani S, Jaenisch R (1989) Germ-line transmission of a disrupted beta 2-microglobulin gene produced by homologous recombination in embryonic stem cells. Nature 342(6248):435–438
3. Gu H, Marth JD, Orban PC, Mossmann H, Rajewsky K (1994) Deletion of a DNA polymerase beta gene segment in T cells using cell type-specific gene targeting. Science 265 (5168):103–106
4. Orban PC, Chui D, Marth JD (1992) Tissue- and site-specific DNA recombination in transgenic mice. Proc Natl Acad Sci USA 89 (15):6861–6865
5. Kühn R, Schwenk F, Aguet M, Rajewsky K (1995) Inducible gene targeting in mice. Science 269:1427–1429
6. Feil R, Brocard J, Mascrez B, LeMeur M, Metzger D, Chambon P (1996) Ligand-activated site-specific recombination in mice. Proc Natl Acad Sci USA 93:10887–10890

7. Furth PA, St Onge L, Böger H, Gruss P, Gossen M, Kistner A, Bujard H, Hennighausen L (1994) Temporal control of gene expression in transgenic mice by a tetracycline-responsive promoter. Proc Natl Acad Sci USA 91:9302–9306
8. Nagy A, Moens C, Ivanyi E, Pawling J, Gertsenstein M, Hadjantonakis AK, Pirity M, Rossant J (1998) Dissecting the role of N-myc in development using a single targeting vector to generate a series of alleles. Curr Biol 8:661–664
9. Le Mouellic H, Lallemand Y, Brûlet P (1992) Homeosis in the mouse induced by a null mutation in the Hox-3.1 gene. Cell 69: 251–264
10. Ikawa M, Kominami K, Yoshimura Y, Tanaka K, Nishimune Y, Okabe M (1995) A rapid and non-invasive selection of transgenic embryos before implantation using green fluorescent protein (GFP). FEBS Lett 375:125–128
11. Livet J, Weissman TA, Kang H, Draft RW, Lu J, Bennis RA, Sanes JR, Lichtman JW (2007) Transgenic strategies for combinatorial expression of fluorescent proteins in the nervous system. Nature 450(7166):56–62
12. Cretekos CJ, Wang Y, Green ED, Martin JF, Rasweiler JJ 4th, Behringer RR (2008) Regulatory divergence modifies limb length between mammals. Genes Dev 22:141–151
13. Hanks MC, Loomis CA, Harris E, Tong CX, Anson-Cartwright L, Auerbach A, Joyner A (1998) Drosophila engrailed can substitute for mouse Engrailed1 function in mid-hindbrain, but not limb development. Development 125:4521–4530
14. Zou YR, Gu H, Rajewsky K (1993) Generation of a mouse strain that produces immunoglobulin kappa chains with human constant regions. Science 262:1271–1274
15. Schnütgen F, De-Zolt S, Van Sloun P, Hollatz M, Floss T, Hansen J, Altschmied J, Seisenberger C, Ghyselinck NB, Ruiz P, Chambon P, Wurst W, von Melchner H (2005) Genomewide production of multipurpose alleles for the functional analysis of the mouse genome. Proc Natl Acad Sci USA 102:7221–7226
16. Anastassiadis K, Fu J, Patsch C, Hu S, Weidlich S, Duerschke K, Buchholz F, Edenhofer F, Stewart AF (2009) Dre recombinase, like Cre, is a highly efficient site-specific recombinase in E. coli, mammalian cells and mice. Dis Model Mech 2:508–515
17. Raymond CS, Soriano P (2007) High-efficiency FLP and PhiC31 site-specific recombination in mammalian cells. PLoS ONE 2:e162
18. Osterwalder M, Galli A, Rosen B, Skarnes WC, Zeller R, Lopez-Rios J (2010) Dual RMCE for efficient re-engineering of mouse mutant alleles. Nat Methods. doi:10.1038/nmeth.1521
19. Singla V, Hunkapiller J, Santos N, Seol AD, Norman AR, Wakenight P, Skarnes WC, Reiter JF (2010) Floxin, a resource for genetically engineering mouse ESCs. Nat Methods 7:50–52
20. Araki K, Araki M, Miyazaki J-I, Vassalli P (1995) Site-specific recombination of a transgene in fertilized eggs by transient expression of Cre recombinase. Proc Natl Acad Sci USA 92:160–164
21. Schaft J, Ashery Padan R, van der Hoeven F, Gruss P, Stewart AF (2001) Efficient FLP recombination in mouse ES cells and oocytes. Genesis 31:6–10
22. te Riele H, Maandag ER, Berns A (1992) Highly efficient gene targeting in embryonic stem cells through homologous recombination with isogenic DNA constructs. Proc Natl Acad Sci USA 89:5128–5132
23. van Deursen J, Wieringa B (1992) Targeting of the creatine kinase M gene in embryonic stem cells using isogenic and nonisogenic vectors. Nucleic Acids Res 20:3815–3820
24. Zhou L, Rowley DL, Mi QS, Sefcovic N, Matthes HW, Kieffer BL, Donovan DM (2001) Murine inter-strain polymorphisms alter gene targeting frequencies at the mu opioid receptor locus in embryonic stem cells. Mamm Genome 12:772–778
25. Hasty P, Abuin A, Bradley A (2000) Gene targeting, principles, and practice in mammalian cells. In: Joyner AL (ed) Gene targeting: a practical approach, 2nd edn. Oxford University Press, London, pp 1–35
26. Friedel RH, Plump A, Lu X, Spilker K, Jolicoeur C, Wong K, Venkatesh TR, Yaron A, Hynes M, Chen B, Okada A, McConnell SK, Rayburn H, Tessier-Lavigne M (2005) Gene targeting using a promoterless gene trap vector ("targeted trapping") is an efficient method to mutate a large fraction of genes. Proc Natl Acad Sci USA 102:13188–13193
27. George SH, Gertsenstein M, Vintersten K, Korets-Smith E, Murphy J, Stevens ME, Haigh JJ, Nagy A (2007) Developmental and adult phenotyping directly from mutant embryonic stem cells. Proc Natl Acad Sci USA 104:4455–4460
28. Ioffe E, Liu Y, Bhaumik M, Poirier F, Factor SM, Stanley P (1995) WW6: an embryonic stem cell line with an inert genetic marker that can be traced in chimeras. Proc Natl Acad Sci USA 92:7357–7361

29. Poueymirou WT, Auerbach W, Frendewey D, Hickey JF, Escaravage JM, Esau L, Dore' AT, Stevens S, Adams NC, Dominguez MG, Gale NW, Yancopoulos GD, DeChiara TM, Valenzuela DM (2007) Nat Biotechnol 25:91–99
30. Valenzuela DM, Murphy AJ, Frendewey D, Gale NW, Economides AN, Auerbach W, Poueymirou WT, Adams NC, Rojas J, Yasenchak J, Chernomorsky R, Boucher M, Elsasser AL, Esau L, Zheng J, Griffiths JA, Wang X, Su H, Xue Y, Dominguez MG, Noguera I, Torres R, Macdonald LE, Stewart AF, DeChiara TM, Yancopoulos GD (2003) High-throughput engineering of the mouse genome coupled with high-resolution expression analysis. Nat Biotechnol 21:652–659
31. Thomas KR, Capecchi MR (1987) Site-directed mutagenesis by gene targeting in mouse embryo-derived stem cells. Cell 51:503–512
32. Meyers EN, Lewandoski M, Martin GR (1998) An *Fgf8* mutant allelic series generated by Cre- and Flp- mediated recombination. Nat Genet 18:136–141
33. Jacks T, Shih TS, Scmitt EA, Bronson RT, Bernards A, Weinberg RA (1994) Tumour predisposition in mice heterozygous for a targeted mutation in *Nf1*. Nat Genet 7:353–361
34. Scacheri PC, Crabtree JS, Novotny EA, Garrett-Beal L, Chen A, Edgemon KA, Marx SJ, Spiegel AM, Chandrasekharappa SC, Collins FS (2001) Bidirectional transcriptional activity of PGK-neomycin and unexpected embryonic lethality in heterozygote chimeric knockout mice. Genesis 30:259–263
35. Crabtree JS, Scacheri PC, Ward JM, Garrett-Beal L, Emmert-Buck MR, Edgemon KA, Lorang D, Libutti SK, Chandrasekharappa SC, Marx SJ, Spiegel AM, Collins FS (2001) A mouse model of multiple endocrine neoplasia, type1, develops multiple endocrine tumors. Proc Natl Acad Sci USA 98:1118–1123
36. Zhu Y, Romero MI, Ghosh P, Ye Z, Charnay P, Rushing EJ, Marth JD, Parada LF (2001) Ablation of NF1 function in neurons induces abnormal development of cerebral cortex and reactive gliosis in the brain. Genes Dev 15:859–876
37. Shavit JA, Motohashi H, Onodera K, Akasaka J, Yamamoto M, Engel JD (1998) Impaired megakaryopoiesis and behavioral defects in *mafG*-null mutant mice. Genes Dev 12:2164–2174
38. Hedges SJ, Deininger PL (2007) Inviting instability: transposable elements, double-strand breaks, and the maintenance of genome integrity. Mutat Res 616:46–59
39. Horie K, Saito ES, Keng VW, Ikeda R, Ishihara H, Takeda J (2007) Retrotransposons influence the mouse transcriptome: implication for the divergence of genetic traits. Genetics 176:815–827
40. Richard GF, Kerrest A, Dujon B (2008) Comparative genomics and molecular dynamics of DNA repeats in eukaryotes. Microbiol Mol Biol Rev 72:686–727
41. Kang YK, Park JS, Lee CS, Yeom YI, Chung AS, Lee KK (1999) Efficient integration of short interspersed element-flanked foreign DNA via homologous recombination. J Biol Chem 274:36585–36591
42. Konkel MK, Batzer MA (2010) A mobile threat to genome stability: the impact of non-LTR retrotransposons upon the human genome. Semin Cancer Biol 20:211–221
43. McVean G (2010) What drives recombination hotspots to repeat DNA in humans? Philos Trans R Soc Lond B Biol Sci 365: 1213–1218
44. Yagi T, Ikawa Y, Yoshida K, Shigetani Y, Takeda N, Mabuchi I, Yamamoto T, Aizawa S (1990) Homologous recombination at c-*fyn* locus of mouse embryonic stem cells with use of diphtheria toxin A-fragment gene in negative selection. Proc Natl Acad Sci USA 87:9918–9922
45. Skarnes WC (2005) Two ways to trap a gene in mice. Proc Natl Acad Sci USA 89:5128–5132
46. Belteki G, Gertsenstein M, Ow DW, Nagy A (2003) Site specific cassette exchange and germline transmission with mouse ES cells expressing phiC31 integrase. Nat Biotechnol 21:321–324
47. Kranz A, Fu J, Duerschke K, Weidlich S, Naumann R, Stewart AF, Anastassiadis K (2010) An improved Flp deleter mouse in C57BL/6 based on Flpo recombinase. Genesis 48:512–520
48. Ohtsuka M, Ogiwara S, Miura H, Mizutani A, Warita T, Sato M, Imai K, Hozumi K, Sato T, Tanaka M, Kimura M, Inoko H (2010) Pronuclear injection-based mouse targeted transgenesis for reproducible and highly efficient transgene expression. Nucleic Acids Res 38: e198
49. Surzycki S (2000) Basic techniques in molecular biology. Springer, Berlin
50. Johansson T, Broll I, Frenz T, Hemmers S, Becher B, Zeilhofer HU, Buch T (2010) Building a zoo of mice for genetic analyses: a comprehensive protocol for the rapid generation of BAC transgenic mice. Genesis 48: 264–280
51. Gong S, Kus L, Heintz N (2010) Rapid bacterial artificial chromosome modification for

large-scale mouse transgenesis. Nat Protoc 10:1678–1696
52. Boffelli D, Nobrega MA, Rubin EM (2004) Comparative genomics at the vertebrate extremes. Nat Rev Genet 5:456–465
53. Baker KE, Parker R (2004) Nonsense-mediated mRNA decay: terminating erroneous gene expression. Curr Opin Cell Biol 16:293–299
54. Joyner AL, Skarnes WC, Rossant J (1989) Production of a mutation in mouse En-2 gene by homologous recombination in embryonic stem cells. Nature 338:153–156
55. McMahon AP, Bradley A (1990) The Wnt-1 (int-1) proto-oncogene is required for development of a large region of the mouse brain. Cell 62:1073–1085
56. Ginzinger DG (2002) Gene quantification using real-time quantitative PCR: an emerging technology hits the mainstream. Exp Hematol 30:503–512
57. Soliman GA, Ishida-Takahashi R, Gong Y, Jones JC, Leshan RL, Saunders TL, Fingar DC, Myers MG Jr (2007) A simple qPCR-based method to detect correct insertion of homologous targeting vectors in murine ES cells. Transgenic Res 16:665–670
58. Friedel RH, Seisenberger C, Kaloff C, Wurst W (2007) EUCOMM–the European conditional mouse mutagenesis program. Brief Funct Genomic Proteomic 6:180–185
59. Chan W, Costantino N, Li R, Lee SC, Su Q, Melvin D, Court DL, Liu P (2007) A recombineering based approach for high-throughput conditional knockout targeting vector construction. Nucleic Acids Res 35:e64
60. Testa G, Zhang Y, Vintersten K, Benes V, Pijnappel WW, Chambers I, Smith AJ, Smith AG, Stewart AF (2003) Engineering the mouse genome with bacterial artificial chromosomes to create multipurpose alleles. Nat Biotechnol 21:443–447
61. Saunders TL (2010) A survey of internet resources for mouse development. Meth Enzymol 476:3–21
62. Eitzman DT, Westrick RJ, Shen Y, Bodary PF, Gu S, Manning SL, Dobies SL, Ginsburg D (2005) Homozygosity for factor V Leiden leads to enhanced thrombosis and atherosclerosis in mice. Circulation 111:1822–1825
63. Huang X, Fu Y, Charbeneau RA, Neubig RR (2009) GNAI2 and regulators of G protein signaling as a potential Noonan syndrome mechanism. Med Hypotheses 73:56–59
64. Elia L, Quintavalle M, Zhang J, Contu R, Cossu L, Latronico MV, Peterson KL, Indolfi C, Catalucci D, Chen J, Courtneidge SA, Condorelli G (2009) The knockout of miR-143 and -145 alters smooth muscle cell maintenance and vascular homeostasis in mice: correlates with human disease. Cell Death Differ 16:1590–1598
65. Visel A, Zhu Y, May D, Afzal V, Gong E, Attanasio C, Blow MJ, Cohen JC, Rubin EM, Pennacchio LA (2010) Targeted deletion of the 9p21 non-coding coronary artery disease risk interval in mice. Nature 464:409–412
66. Olson LE, Richtsmeier JT, Leszl J, Reeves RH (2004) A chromosome 21 critical region does not cause specific Down syndrome phenotypes. Science 306:687–690
67. Tybulewicz VL, Fisher EM (2006) New techniques to understand chromosome dosage: mouse models of aneuploidy. Hum Mol Genet 15 Spec No 2:R103–R109
68. Carbery ID, Ji D, Harrington A, Brown V, Weinstein EJ, Liaw L, Cui X (2010) Targeted genome modification in mice using zinc-finger nucleases. Genetics 186:451–459
69. Cui X, Ji D, Fisher DA, Wu Y, Briner DM, Weinstein EJ (2011) Targeted integration in rat and mouse embryos with zinc-finger nucleases. Nat Biotechnol 29:64–67
70. Meyer M, de Angelis MH, Wurst W, Kühn R (2010) Gene targeting by homologous recombination in mouse zygotes mediated by zinc-finger nucleases. Proc Natl Acad Sci USA 107:15022–15026
71. Swiatek PJ, Gridley T (1993) Perinatal lethality and defects in hindbrain development in mice homozygous for a targeted mutation of the zinc finger gene Krox20. Genes Dev 7:2071–2084
72. Doetschman TC, Eistetter H, Katz M, Schmidt W, Kemler R (1985) The in vitro development of blastocyst-derived embryonic stem cell lines: formation of visceral yolk sac, blood island and myocardium. J Embryol Exp Morph 87:27–45
73. Hooper M, Hardy K, Handyside A, Hunter S, Monk M (1987) HPRT-deficient (Lesch-Nyhan) mouse embryos derived from germline colonization by cultured cells. Nature 326:292–295
74. Li E, Bestor TH, Jaenisch R (1992) Targeted mutation of the DNA methyltransferase gene results in embryonic lethality. Cell 69:915–926
75. Pettitt SJ, Liang Qu, Rairdan XY, Moran JL, Prosser HM, Beier DR, Lloyd KC, Bradley A, Skarnes WC (2009) Agouti C57BL/6N embryonic stem cells for mouse genetic resources. Nat Methods 6:493–495
76. Domino SE, Zhang L, Gillespie PJ, Saunders TL, Lowe JB (2001) Deficiency of reproductive tract (1,2)fucosylated glycans and normal fertility in mice with targeted deletions of the

FUT1 or FUT2 (1,2)fucosyltransferase locus. Mol Cell Biol 21:8336–8345

77. Deng C, Wynshaw-Boris A, Zhou F, Kuo A, Leder P (1996) Fibroblast growth factor receptor 3 is a negative regulator of bone growth. Cell 84:911–921
78. Vetter K (2000) Die Funktion von FGF2 und FGF8 während der Entwicklung des Nervensystems in der Maus. Doctoral dissertation, University of Technology, Munich
79. Rideout WM 3rd, Wakayama T, Wutz A, Eggan K, Jackson-Grusby L, Dausman J, Yanagimachi R, Jaenisch R (2000) Generation of mice from wild-type and targeted ES cells by nuclear cloning. Nat Genet 24:109–110
80. Auerbach W, Dunmore JH, Fairchild-Huntress V, Fang Q, Auerbach AB, Huszar D, Joyner AL (2000) Establishment and chimera analysis of 129/ SvEv- and C57BL/6-derived mouse embryonic stem cell lines. Biotechniques 29:1024–1032
81. Araki K, Imaizumi T, Sekimoto T, Yoshinobu K, Yoshimuta J, Akizuki M, Miura K, Araki M, Yamamura K (1999) Exchangeable gene trap using the Cre/mutated lox system. Cell Mol Biol 45:737–750
82. Cheah SS, Behringer RR (2001) Contemporary gene targeting strategies for the novice. Mol Biotechnol 19:297–304
83. Hughes ED, Qu YY, Genilk SJ, Lyons RH, Pacheco CD, Lieberman AP, Samuelson LC, Nasonkin IO, Camper SA, Van Keuren ML, Saunders TL (2007) Genetic variation in C57BL/6 ES cell line. Mamm Genome 18:549–558
84. Kendall SK, Samuelson LC, Saunders TL, Wood RI, Camper SA (1995) Targeted disruption of the pituitary glycoprotein hormone alpha-subunit produces hypogonadal and hypothyroid mice. Genes Dev 9:2007–2019
85. Martin MS, Dutt K, Papale LA, Dubé CM, Dutton SB, de Haan G, Shankar A, Tufik S, Meisler MH, Baram TZ, Goldin AL, Escayg A (2010) Altered function of the SCN1A voltage-gated sodium channel leads to gamma-aminobutyric acid-ergic (GABAergic) interneuron abnormalities. J Biol Chem 285:9823–9834
86. Miyaki S, Sato T, Inoue A, Otsuki S, Ito Y, Yokoyama S, Kato Y, Takemoto F, Nakasa T, Yamashita S, Takada S, Lotz MK, Ueno-Kudo H, Asahara H (2010) MicroRNA-140 plays dual roles in both cartilage development and homeostasis. Genes Dev 24:1173–1185
87. Nagy A, Rossant J, Nagy R, Abramow-Newerly W, Roder JC (1993) Derivation of completely cell culture-derived mice from early-passage stem cells. Proc. Natl. Acad. Sci. U.S.A. 90:8424–8428
88. Simpson EM, Linder CC, Sargent EE, Davisson MT, Mobraaten LE, Sharp JJ. 1997. Genetic variation among 129 substrains and its importance for targeted mutageneeis in mice. Nat. Genet 16:19–27
89. Szabó P, Mann JR (1994) Expression and methylation of imprinted genes during in vitro differentiation of mouse parthenogenetic and androgenetic embryonic stem cell lines. Development 120:1651–1660

Chapter 5

Transgenic Production Benchmarks

Thomas J. Fielder and Lluis Montoliu

Abstract

The efficiency with which transgenic mice can be produced via pronuclear injection of DNA constructs is subject to a large number of variables ranging from human to mechanical to biological. Transgenic core facilities, which are often run like small businesses that must attract and satisfy clients, would benefit from knowing how their efficiency compares to that of other facilities, and whether significant improvements in any phase of the process can be realistically achieved. Communicating knowledgeably with clients about the amount of variation to expect at each step is also important in maintaining good working relationships. We describe a world-wide survey of transgenic core facilities designed to estimate the average yields that can be expected at each step in the process, as well as the inherent variability at each step. The survey has been conducted with the support of the International Society for Transgenic Technologies (ISTT). Some descriptive statistics, calculated from a subset of the data collected so far, are presented for different strains of mice and types of DNA constructs.

5.1 Introduction

In the early 1980s, the first reports of transgenic mice produced by the microinjection of DNA into the pronuclei of fertilized ova appeared in the scientific literature [1–3]. Currently, this technique is used on a daily basis to produce transgenic mice in hundreds of academic and commercial institutions throughout the world, and transgenic mice remain one of the most commonly used in vivo research models. A PubMed search using the keywords "transgenic" and "mouse" reveals an average of well over 5,000 articles published per year for the last 5 years.

The methods by which transgenic mice are used to test hypotheses in biomedical research have become increasingly sophisticated. However, the most commonly used procedure for the production of transgenic mice has remained largely unchanged since the 1980s [4, 5]. This procedure can be divided into the following five steps:

S. Pease and T.L. Saunders (eds.), *Advanced Protocols for Animal Transgenesis*, Springer Protocols,
DOI 10.1007/978-3-642-20792-1_5, © Springer-Verlag Berlin Heidelberg 2011

1. Cloning, testing, and purification of the transgene DNA.
2. Superovulation and mating of mice to produce fertilized ova.
3. Microinjection of DNA solution into pronuclei.
4. Transfer of injected ova into oviducts of pseudopregnant females.
5. Genotyping of offspring to identify founders.

Usually, founders are then bred to produce enough offspring from each founder line to verify transgene expression patterns and carry out the desired experiments.

Other procedures used to generate transgenic mice include exposing early embryos to retroviruses [6], injecting lentiviral vectors into the cytoplasm or underneath the zona pellucida [7], or inserting the transgene into a permissive locus such as Hprt via homologous recombination in embryonic stem cells [8]. These procedures will not be considered in this chapter.

The equipment used to carry out the pronuclear injection of DNA can easily cost more than US $50,000, and the number of founders produced from a given microinjection session is highly dependent on the skill and experience of the person performing the microinjections and embryo transfer surgeries. For these reasons, most transgenic mice are produced by core facilities, established by universities and companies to perform microinjections and embryo transfers for their researchers.

5.2 Core Facility Considerations

Most transgenic mouse core facilities are required to recover some fraction of their operating costs by charging fees for their services. This results in core facilities being subject to many of the same forces governing the operation of a small business. Expenses must be tracked. Customer relationships must be established and nurtured. Even competition from other core facilities, for both customers and trained employees, may be an important consideration. Above all, the performance of the facility must be evaluated and improved or maintained.

Until now, the community of transgenic core facilities has lacked objective performance standards for DNA microinjection. While some published data exists as to what is achievable in terms of embryo yields, pup yields, and transgenic pup yields [5, 9–16], day-to-day variations are an inherent feature of the process that makes it difficult to know what to expect in terms of average yields. The typical facility lacks the resources to make a systematic study of most variables, and often must rely on personal communication with other facilities to decide whether or not its yields are acceptable. This process is greatly complicated by many small variations in techniques, mouse strains, reagents, materials, equipment, husbandry practices, and so forth.

From the customer's viewpoint, the critical measures of a core facility's performance are the number of transgenic mice with the desired expression patterns that are produced, the fees charged to produce them, and the time needed to produce them. Costs, yields, and timelines are particularly important considerations for prospective clients striving to make their grant proposals as competitive as possible. Thus, it is helpful if core facilities are able to predict yields with some accuracy. Of equal, if not greater, importance is a reliable estimate of the variation to be expected in these yields. Clients who are unfamiliar with the process of making transgenic mice need to be aware of this variation to avoid conflating averages with guarantees. Even experienced clients can suffer from a selective memory, causing the occasional outlier, good or bad, to improperly color their perception of a facility's overall performance.

Good communication with clients can be helpful in dispelling misperceptions about performance, but without objective performance standards, the only recourse is to compare a given outcome with an average for that facility. This makes it difficult to respond to clients who claim that other facilities have better yields. Furthermore, newly established facilities may not have enough data to make a reasonable calculation of their average yield.

In early 2008, a proposal was drafted by one of us (TJF) for the ISTT to carry out a large-scale survey and analysis of DNA microinjection data, collected from as many core facilities as possible, with the goal of establishing objective performance standards for average yields, expected variation, and principle sources of variation.

5.3 Performance Factors

Before we describe the details of this survey, let us examine the principle factors that determine the yield of transgenic mice from a given microinjection session, which can be stated as follows:

- Number of ova harvested
- Percentage of ova that are fertilized
- Percentage of fertilized ova that survive the injection process
- Percentage of injected ova that are successfully transferred to foster mothers
- Percentage of transferred ova resulting in newborn pups
- Percentage of pups that survive to be genotyped

- Percentage of genotyped pups that are positive for the transgene
- Percentage of transgenic offspring that express the transgene

The number of ova harvested for each microinjection session depends on the number of females used, and the efficiency of the superovulation process. While simply using more females per session will increase yields, there are practical limits imposed by the desirability of finishing the entire process within a standard workday, the need to harvest embryos while they are still at the pronuclear stage, and the mandates of government and institutional policies to minimize the number of animals used. The expense of purchasing and housing the mice is a large fraction of the overall cost of a microinjection session, so there is always a trade-off between maximizing yields and minimizing costs. Furthermore, the response of individual mice to superovulation can vary greatly, even for age-matched inbred mice, and the response is strain-dependent and age-dependent [10, 15]. (See Nagy et al. [5] for details of the superovulation process.)

The percentage of harvested ova that are fertilized is a function of the fraction of successful matings and the fecundity of the stud males. Again, strain-to-strain variation is an important factor. Males should not be mated either too frequently or too infrequently. Not only does superovulation increase the number of oocytes released by each female, but it also forces the females into estrus, so they will be receptive to mating. Nevertheless, even under ideal conditions, some fraction of females will not mate successfully, and the number of fertilized ova will vary significantly from mouse to mouse.

The fraction of ova that survive the injection process is dependent on the skill and experience of the person performing the injections. A survival rate of 50% is not unusual for beginners, but rates of around 90% can be expected with sufficient practice. Mechanical trauma is undoubtedly the major cause of cell death, with stickiness of the needle and flow rate of the DNA solution being important factors. Operators must learn to recognize when the injection needle needs to be changed, as evidenced by cell contents sticking to it. Needle characteristics (diameter of tip, length of taper) are also important factors. Generally, a new needle must have its tip broken against the holding pipet in order to achieve a suitable inside diameter that results in an appropriate flow rate of DNA, and the flow rate must also be adjusted at the pressure regulator. Each needle will have slightly different characteristics, and operators must learn from experience how to recognize and adjust to these differences.

Cell lysis due to mechanical trauma is easy to recognize and is generally apparent less than an hour after injection, allowing those embryos to be excluded from the transfer process. More subtle forms of trauma, for example from contaminants in the DNA solution, will not be evident if the embryos are transferred to foster mothers the same day that they are injected [17]. Embryos can be incubated overnight and only those developing to the two-cell stage transferred, but this carries its own risks, such as suboptimal culture conditions.

The percentage of embryos that are successfully transferred to foster mothers is generally determined almost entirely by the experience of the operator. When the classic method of transfer through the infundibulum is used [5], blood from the torn bursa can obscure the infundibulum and may have negative effects if it enters the oviduct. This problem can be avoided by entering the oviduct through a hole in the side of the oviduct, produced with the tip of a 27g needle.

In this case, the hole should be positioned at a point on the oviduct outside of the bursa, between the bursa and the ampulla, and the transfer pipet inserted so that it is pointing toward the ampulla. With experience, transfer rates of 100% should be routine using either methods.

The percentage of embryos that become newborn pups depends on a number of factors. Subtle mechanical trauma or DNA contaminants may prevent the embryo from developing, or the transgene product itself may have deleterious effects. Insertional mutagenesis, caused by the transgene interrupting a gene at a random location, whose encoded function is essential for embryo development, can also account for up to 5–10% of embryo and fetal deaths in utero [5, 18].

Transferring too many or too few embryos to a given foster mother may reduce yields. Yields are highly strain dependent, which is one reason why hybrid strains are so often used as embryo donors for DNA microinjection, in spite of the disadvantage that using hybrids results in pups with mixed and variable genetic backgrounds. The strain of the foster mothers may also be a factor (the vast majority of core facilities use either outbred or hybrid mothers). Opinions vary as to whether it is better to transfer embryos to both oviducts or a single oviduct of each foster mother, even though it has been clearly established that transuterine migration does not occur [19].

All of the above yields may also be affected by the media used during ova harvesting, injection, and transfer, as well as other factors, such as the hyaluronidase treatment used to remove cumulus cells from ova, the mineral oil used for microdrop culture, the temperature at which the ova are held before, during, and after injection, and the elapsed time between harvest and transfer to foster mothers.

Survival rates of live pups can be affected by environmental conditions, husbandry practices, nutrition, litter sizes, the quality of maternal care, and deleterious effects of the transgene product.

The rate of transgenesis (defined as the percentage of pups that carry the transgene) has been shown to be affected by the concentration of DNA being injected, and linear molecules result in higher yields of transgenic mice than do circular molecules [20]. If the concentration is too low, pup yields may be good but very few will be transgenic. If the concentration is too high, the total pup yield will probably be lower than normal, but a higher percentage of the pups may be transgenic. It is essential that the method used to genotype the founders be tested for sensitivity and specificity, to avoid false negative and false positive results.

The percentage of transgenic founders that express the transgene is unlikely to be influenced by any factors under the control of the core facility and will be directly influenced by the transcriptional regulatory DNA elements that have been included in the construct [21, 22]. Of course, proper design of the transgene and its promoter is critical and clients should be encouraged to sequence the final product and perform functional testing (e.g., in vitro transfection).

Transgenes are integrated into the genome in a random fashion, usually as head-to-tail concatemers consisting of multiple copies [23]. The site of integration is most likely a major determinant of transgene activity, with copy number also playing a role. On occasion transgene integrations interrupt genes and cause mutations [24]. For these reasons, each founder from a given microinjection must be treated as a separate line and multiple founder lines must be analyzed to rule out phenotypic effects caused by other genes whose expression may be affected by the integration of the transgene. Transgenes may integrate after the first cell division, producing a mosaic founder, a situation that appears to occur in most transgenic founder mice produced by standard pronuclear injection [25]. Also, multiple integration sites may be present in some founders, which can result in differential expression patterns in offspring from a single founder, due to segregation of the different integration sites.

5.4 Survey Design Considerations

In May of 2008, we began to receive input from a small group of advisors as to the design of our survey of DNA microinjection data. It was immediately apparent that the process would be far too complicated if we tried to capture data on every possible source of variation. Furthermore, we agreed that participation would be inversely proportional to the length of the survey.

Some sources of variation have already been well documented, such as the effect of background strain on embryo yields [9, 10, 15], so it was deemed essential to gather information about donor strains in order to better estimate the inherent variations caused by differences between individual mice, constructs, and personnel. For the same reason, we chose to distinguish between relatively small, typically cDNA-based constructs and those employing the genomic DNA-based bacterial artificial chromosome (BAC), P1-derived artificial chromosome (PAC), and yeast artificial chromosome (YAC) cloning vectors [21].

Most of the time spent designing the survey was devoted to the definition of terms. We decided that we should not gather data for which a precise definition could not be agreed upon. Thus, for example, we decided not to gather data on transgene expression because this is determined in different ways by different clients, and is not necessarily determined for every founder line. Similarly, we elected not to distinguish between injections performed by inexperienced and experienced operators because there is no generally accepted definition of what constitutes an experienced injectionist. We also decided not to request the number of superovulated females that were found to have a vaginal plug the day after mating, after learning that some facilities process all superovulated females, regardless of whether a vaginal plug is found or not. We also wanted to use definitions that were as inclusive as possible. For example, in deciding how to define the yield of pups from a given injection session, we chose to include all pups born or delivered by caesarian section, regardless of whether they were alive or dead at birth, since some facilities only count live pups, while others count both live and dead pups.

In technical terms, this survey is nonrandom, incidental, and voluntary. We chose to gather the data in the form of an Excel spreadsheet, with each column constituting a separate category of data, and each row corresponding to a single injection session, i.e., 1 day of injection. (During the data-gathering phase, we were somewhat surprised to learn that some facilities combine the results from all injection sessions for a given construct, which prevented them from submitting their data in the required format, and also prevented the analysis of their performance per microinjection day.) The 12 categories we eventually settled on were defined as follows:

1. Date – the date of the microinjection session
2. Construct – participants were asked to submit either a unique name or number for each different construct
3. DNA type – "1" for plasmid-based constructs, or "2" for BAC/PAC/YAC constructs
4. Embryo strain – standard nomenclature, including vendor designations, was requested, with embryos from hybrid crosses designated as F2

5. Donor females superovulated – the number of embryo donors from which oviducts were harvested, irrespective of whether they were plugged or not by stud males
6. Embryos harvested – the total number of ova collected, excluding those that were visibly abnormal, but irrespective of whether they were fertilized or injected
7. Embryos injected – the total number of fertilized ova that were injected, regardless of whether they survived the injection process
8. Embryos transferred – the total number of injected ova that were eventually transferred to pseudopregnant foster mothers, regardless of when they were transferred
9. Pseudopregnant females used – the total number of females to whom microinjected embryos were transferred, irrespective of whether they became pregnant or had pups
10. Strain of pseudopregnant females used – again, standard nomenclature was requested
11. Total pups – the total number of pups going to term, regardless of whether they survived or not, including full-term fetuses delivered by caesarian section, but excluding embryos tested at mid-gestation
12. Transgenic pups – total number of pups which tested positive by PCR or Southern blot for the transgene

We included a 13th column where participants could record any incident that may have adversely affected the outcome of the session, or influenced it in some defined way. Some examples of suggested entries for this column included such factors as a new microinjectionist being trained, construction noise present in the animal facility, embryos lost due to technician error or equipment malfunction, etc. This column could also be used to distinguish between different operators in the same facility. We felt that allowing facilities to qualify their data in this way would encourage greater participation and reduce the temptation to omit data from "bad" injection sessions.

In addition to collecting data for individual injection sessions, we also posed a series of questions whose answers were designed to classify facilities based on the following characteristics:

- Size (number of people performing microinjections)
- How many years they have been in existence
- The type of clients they usually serve (academic or commercial, internal or external)
- Whether the microinjection duties are performed by one person for each session

- Who performs the final purification of the transgene DNA
- Whether toxicity testing is routinely performed on the transgene DNA by culturing injected embryos to see if they develop normally
- Whether embryo donors are checked for the presence of a vaginal plug, and whether unplugged females are used anyway
- Whether injected embryos are routinely incubated overnight before being transferred to foster mothers
- Whether embryos are routinely transferred to one or both oviducts

This section of the survey also asked the submitter to describe restrictions their facility places on various aspects of their microinjection service, including:

- If DNA purified outside the facility is ever accepted for injection
- If egg donors are restricted to certain specified strains
- Whether the facility accepts BAC-, PAC-, or YAC-based transgenes
- Does the facility record the total number of oocytes collected

A free text field is included in this section to allow the submitter to describe any other significant restrictions or rules in effect.

How might these characteristics and policies influence the data from individual injection sessions?

One can reasonably expect that facilities with a larger number of microinjection technicians and a wider client base would benefit from more experience, which may be reflected in higher average yields. The size of a facility may also be an indication of higher than average institutional support, which could result in the hiring of more experienced staff and the acquisition of better equipment.

Purification of the transgene fragment is a critical step that can have a huge effect on the yield of transgenic mice. Accurate determination of the final concentration is also extremely important. Facilities that purify and quantify all of the transgenes they inject may thus have more consistent yields, compared to those who inject client-purified DNA.

Injecting a small number of ova and culturing them overnight or several days to see what fraction of them develop to two-cell embryos or later stages should allow the facility to reject transgene preparations that exhibit toxicity. This in turn could be expected to result in higher average yields of transgenic mice.

Facilities that only use embryo donors that mated successfully, as judged by the presence of a vaginal plug, could be expected to have a higher percentage of fertilized ova, on average.

Facilities that culture their injected embryos overnight before implanting them into foster mothers could be expected to have a higher percentage of embryos develop to term, assuming culture conditions are optimal.

Opinions vary as to whether one-sided or two-sided embryo transfers result in better yields, or if they produce equivalent yields. One-sided embryo transfer can be accomplished more quickly, resulting in less surgical stress that may be reflected in better yields. However, since transuterine migration does not occur [19], transferring the same number of embryos to both sides could result in a higher rate of successful implantation of the transferred embryos. By comparing data from facilities that state a definite preference for one method over the other, it may be possible to definitively answer this question.

Since we were aware, while designing the survey, that facilities differ in these practices, we wanted to be able to take these factors into account when performing the final analysis. Not only did we wish to see if these practices had the expected effect on yields, we also wanted to estimate the variability in yields that is not due to these practices. Furthermore, for facilities that have changed significantly over their lifetimes in terms of size, client base, and the practices mentioned above, our design allows them to submit more than one facility description, each linked to a separate data file. This allows individual facilities to see how these changes have influenced their yields.

The confidentiality of participants and their data was one of the most important considerations during the design phase. We considered methods of gathering data that would allow complete anonymity, but ultimately rejected this notion for two reasons. One is that, despite our best efforts to precisely define the data fields and explain them to participants, we anticipated (rightly so, in retrospect) that problematic entries would require human intervention and communication with the participants about their data. The other reason is that we wanted to offer each participant a personalized analysis of their data, compared with the group averages and variabilities, as an incentive for participation in the survey. It seemed that the most reliable way to do this was to maintain knowledge of which participant had submitted a particular data file, but at the same time confine this knowledge to a very limited number of individuals, namely, the authors of this chapter.

5.5 Results

In order to gauge the effectiveness of the data for identifying the major sources of variation in yields, we conducted a pilot study using data from the first 12 facilities to submit descriptions and data files [26]. Briefly, the statistical technique known as principal component analysis was used to identify, from the very large number of potential correlations, a small subset that accounted for 88% of the variation in yields across injection days and across facilities. While this analysis did not produce any surprises (the number of transgenic pups obtained will obviously be highly correlated with the number of ova harvested, injected, and implanted), it did demonstrate that we would be able to achieve several major goals of the survey, i.e., to identify the major sources of variability, to allow individual facilities to see where they rank in terms of average yields, and to pinpoint the specific areas in which a given facility could reasonably hope to improve their yields.

We received facility descriptions and data files from 67 different institutions, comprising more than 17,000 individual days of injection, nearly 2.7 million ova harvested, and more than 250,000 offspring produced.

A thorough analysis of this data will not be a trivial task. About half of the reported injection days lack complete data (e.g., some facilities do not record the total number of ova harvested), which means that each statistic, such as the fraction of harvested ova that were injected or the fraction of injected ova that survived to be transferred into foster mothers, must be calculated from a different set of injection days.

Tables 5.1–5.3 present descriptive statistics of a subset of the data. For the purpose of this analysis, we included only those injection days for which complete data was reported. Additionally, we limited this analysis to the five most commonly used donor strains of this data subset, with two hybrid strains, B6SJLF2 and B6CBAF2, and three inbred lines, C57BL/6J, C57BL/6N, and FVB/N. Note that this analysis does not distinguish between different vendors of any of these lines, although the full data set does distinguish between them to the extent that participants provided this information. In the case of different vendors of C57BL/6N and FVB/N, there is no evidence of significant genetic differences among the various substrains. However, a number of mutations and SNPs distinguish C57BL/6J from C57BL/6N [27, 28] so the exact origins of the two hybrid lines could be significant. Results were calculated separately for injection days using small plasmid-based transgenes (DNA type 1) and for those using BAC-, PAC-, or YAC-based transgenes (DNA type 2).

As shown in Table 5.1, a total of 6,341 injection days with plasmid-based transgenes were included in this analysis. For BAC/PAC/YAC transgenes, a total of 786 injection days were included.

In Table 5.2, the average yields per day of injection are presented for the same DNA types and donor strains.

Table 5.1
Totals for each data category, grouped by embryo strain and DNA type (1 = small plasmid-based constructs, 2 = BAC/PAC/YAC constructs)

DNA type	Embryo strain[a]	Injection days	Donor females superovulated	Embryos harvested	Embryos injected	Embryos transferred	Pseudopregnant females used	Total pups	Transgenic pups
1	C57BL/6J	950	12,723	265,296	133,073	96,748	3,782	10,188	1,314
1	C57BL/6N	926	13,889	271,338	151,394	110,185	4,546	10,781	1,398
1	FVB/N	2,671	37,478	578,599	399,699	311,307	14,708	41,475	5,760
1	B6SJLF2	1,087	9,237	344,232	215,596	160,529	6,944	48,423	5,947
1	B6CBAF2	707	5,208	144,331	92,809	60,820	2,208	11,525	1,358
2	C57BL/6J	86	1,061	29,064	13,982	11,278	405	869	105
2	C57BL/6N	100	1,895	39,043	25,824	18,853	761	1,151	117
2	FVB/N	255	3,539	52,677	38,408	28,782	1,299	3,711	405
2	B6SJLF2	311	2,347	93,122	60,067	45,711	1,976	12,161	904
2	B6CBAF2	34	284	6,082	4,594	3,067	110	571	27

[a]The exact origins of B6SJLF2 and B6CBAF2 mice were not determined, so parental line could be C57BL/6J, C57BL/6NTac, or other substrain. For the data presented here, FVB/N or C57BL/6N mice from different vendors were grouped together

Table 5.2
Average values per injection day

DNA type	Embryo strain	Donor females superovulated	Embryos harvested	Embryos injected	Embryos transferred	Pseudopregnant females used	Total pups	Transgenic pups
1	C57BL/6J	13	279	140	102	4	11	1.4
1	C57BL/6N	15	293	163	119	5	12	1.5
1	FVB/N	14	217	150	117	6	16	2.2
1	B6SJLF2	8	317	198	148	6	45	5.5
1	B6CBAF2	7	204	131	86	3	16	1.9
2	C57BL/6J	12	338	163	131	5	10	1.2
2	C57BL/6N	19	390	258	189	8	12	1.2
2	FVB/N	14	207	151	113	5	15	1.6
2	B6SJLF2	8	299	193	147	6	39	2.9
2	B6CBAF2	8	179	135	90	3	17	0.8

Table 5.3
Average ratios and standard deviations

DNA type	Embryo strain	Embryos harvested per donor	Fraction of harvested ova that were injected	Fraction of injected ova that were transferred	Injected ova transferred per foster mother	Fraction of transferred ova resulting in pups	Fraction of pups that were transgenic	Transgenic pups per 100 injected ova
1	C57BL/6J	24 ± 14	0.52 ± 0.24	0.74 ± 0.19	25 ± 9	0.11 ± 0.10	0.13 ± 0.16	1.0 ± 1.7
1	C57BL/6N	20 ± 8	0.57 ± 0.19	0.73 ± 0.14	25 ± 6	0.11 ± 0.09	0.12 ± 0.17	0.9 ± 1.4
1	FVB/N	17 ± 6	0.68 ± 0.19	0.76 ± 0.18	22 ± 8	0.13 ± 0.12	0.16 ± 0.18	1.4 ± 2.0
1	B6SJLF2	38 ± 12	0.66 ± 0.18	0.75 ± 0.17	24 ± 6	0.29 ± 0.17	0.14 ± 0.14	2.8 ± 3.1
1	B6CBAF2	29 ± 12	0.69 ± 0.19	0.67 ± 0.18	28 ± 7	0.20 ± 0.14	0.12 ± 0.14	1.5 ± 1.9
2	C57BL/6J	29 ± 12	0.48 ± 0.16	0.80 ± 0.15	27 ± 8	0.08 ± 0.09	0.16 ± 0.19	0.9 ± 1.4
2	C57BL/6N	21 ± 6	0.59 ± 0.21	0.68 ± 0.16	23 ± 6	0.08 ± 0.08	0.10 ± 0.15	0.6 ± 1.4
2	FVB/N	15 ± 6	0.70 ± 0.21	0.72 ± 0.21	22 ± 6	0.13 ± 0.12	0.11 ± 0.16	1.0 ± 1.5
2	B6SJLF2	41 ± 11	0.65 ± 0.16	0.77 ± 0.15	23 ± 4	0.26 ± 0.15	0.08 ± 0.08	1.5 ± 1.7
2	B6CBAF2	22 ± 10	0.81 ± 0.23	0.68 ± 0.19	29 ± 6	0.20 ± 0.13	0.08 ± 0.14	0.6 ± 0.8

Table 5.3 lists the averages and standard deviations for seven ratios, calculated per injection day: harvested ova/donor, injected ova/harvested ova, transferred ova/injected ova, number of transferred ova per foster mother, transgenic pups/total pups, and number of transgenic pups per 100 injected ova. These ratios constitute the most important statistics for judging the success of a facility.

We would like to emphasize that the statistics in Tables 5.1–5.3 describe a subset of the total data collected and illustrate only the most basic analyses that can and will be performed on the entire data set. We are therefore purposely avoiding drawing any conclusions from the statistics presented here, preferring to base those on the full set of analyses. However, it is interesting to note similarities and differences between this subset of our data and data presented in other smaller studies. For example, Auerbach et al. [9] reported that the FVB/N strain produced more transgenic founders per 100 injected eggs than their hybrid strain, B6D2F1/NTac. By contrast, in this subset of our data, the yield of founders per 100 injected B6SJLF2 hybrid eggs was twofold greater, on average, than that for FVB/N (Table 5.3), while the FVB/N yield was nearly equal to the yield for the B6CBAF2 strain.

Brinster et al. [20] reported founder yields that were similar to ours, while Van Keuren et al. [29] reported a rate of about two founders per 100 injected B6SJLF2 eggs with BAC transgenes, slightly higher than our yield of 1.5 founders per 100 injected eggs.

It is important to note that each of the studies cited above are based on data collected in a single transgenic core facility, while our data come from a wide variety of facilities with different levels of expertise and many other differences in procedures and reagents. In the case of Auerbach et al. [9], injected eggs were almost always incubated overnight before implanting them into foster mothers. Some facilities in our study almost always implanted eggs the same day they were injected, some almost always implanted the next day, and some used both techniques frequently. It is conceivable that this variable could have some strain-specific effects.

Gathering data on the basis of individual injection days (with associated dates) enables the analysis of performance trends over time (see Fielder et al. [26] for examples). Facilities were encouraged to submit as much data as possible, even for injection days where the injectionist was new and still being trained. Calculating trends over time may serve to define the typical learning curve and predict the amount of training that should be necessary for a new injectionist. While we purposely did not collect data on such factors as changes in microinjection equipment, media, personnel, caging equipment, and so forth, it is possible that the influence of such changes may be evident in the trends over time (and, to the extent that these factors are included by each participant as incidental data, we will take them into account during the analysis). It is likely, however, that the effect of such changes will be obscured by the inherent day-to-day variability resulting from stochastic fluctuations in egg yields, fertilization rates, pup yields, etc. Indeed, one of the major goals of the survey is to estimate this inherent variability. This information can then be used by individual facilities to better educate their clients as to what constitutes reasonable expectations for the outcome of a given day of injection.

Final analyses of the complete data set should be ready for dissemination to individual participants in mid-2011. Overall results will be posted on the ISTT website (http://www.transtechsociety.org).

The authors would like to thank Anna Auerbach, Kerri Kluetzman, and Aimee Stablewski for their valuable assistance in designing this survey, and Laura Barrios for professional statistical advice. We would also like to thank all of the participants in the survey.

References

1. Gordon JW, Scangos GA, Plotkin DJ, Barbosa JA, Ruddle FH (1980) Genetic transformation of mouse embryos by microinjection of purified DNA. Proc Natl Acad Sci USA 77:7380–7384
2. Gordon JW, Ruddle FH (1981) Integration and stable germline transmission of genes injected into mouse pronuclei. Science 214:1244–1246
3. Palmiter RD, Chen HY, Brinster RL (1982) Differential regulation of metallothionein-thymidine kinase fusion genes in transgenic mice and their offspring. Cell 29:701–710
4. Si-Hoe SL, Wells S, Murphy D (2001) Production of transgenic rodents by the microinjection of cloned DNA into fertilized one-cell eggs. Mol Biotechnol 17:151–182
5. Nagy A, Gertsenstein M, Vintersten K, Behringer R (2003) Manipulating the mouse embryo: a laboratory manual. Cold Spring Harbor Laboratory Press, Cold Spring Harbor, NY
6. Jaenisch R (1976) Germ line integration and Mendelian transmission of the exogenous Moloney leukemia virus. Proc Natl Acad Sci USA 73:1255–1259
7. Lois C, Hong EJ, Pease S, Brown EJ, Baltimore D (2002) Germline transmission and tissue-specific expression of transgenes delivered by lentiviral vectors. Science 295: 868–872
8. Bronson SK, Plaehn EG, Kluckman KD, Hagaman JR, Maeda N, Smithies O (1996) Single-copy transgenic mice with chosen-site integration. Proc Natl Acad Sci USA 93:9067–9072
9. Auerbach AB, Norinsky R, Ho W, Losos K, Guo Q, Chatterjee S, Joyner AL (2003) Strain-dependent differences in the efficiency of transgenic mouse production. Transgenic Res 12:59–69
10. Byers SL, Payson SJ, Taft RA (2006) Performance of ten inbred mouse strains following assisted reproductive technologies (ARTs). Theriogenology 65:1716–1726
11. Gates AH, Bozarth JL (1978) Ovulation in the PMSG-treated immature mouse: effect of dose, age, weight, puberty, season and strain (BALB/c, 129 and C129F1 hybrid). Biol Reprod 18:497–505
12. Johnson LW, Moffatt RJ, Bartol FF, Pinkert CA (1996) Optimization of embryo transfer protocols for mice. Theriogenology 46:267–276
13. Shirley B, Condon-Mahony M, Wortham JW Jr (1985) Effects of season, environmental temperature, size of dams, and age of breeder males on numbers of embryos obtainable from superovulated mice. Exp Biol 44:101–108
14. Shirley B, Wortham JW, Condon-Mahony M (1986) Mating and embryo yield of mice

injected with gonadotropins on specific days of the estrous cycle and in acyclic periods. Exp Biol 46:83–88
15. Spearow JL (1988) Major genes control hormone-induced ovulation rate in mice. J Reprod Fertil 82:787–797
16. Watson JG, Wright RW Jr, Chaykin S (1977) Collection and transfer of preimplantation mouse embryos. Biol Reprod 17:453–458
17. Wall RJ, Paleyanda RK, Foster JA, Powell A, Rexroad C, Lubon H (2000) DNA preparation method can influence outcome of transgenic animal experiments. Anim Biotechnol 11:19–32
18. Jaenisch R, Breindl M, Harbers K, Jähner D, Löhler J (1985) Retroviruses and insertional mutagenesis. Cold Spring Harb Symp Quant Biol 50:439–445
19. Rülicke T, Haenggli A, Rappold K, Moehrlen U, Stallmach T (2006) No transuterine migration of fertilised ova after unilateral embryo transfer in mice. Reprod Fertil Dev 18:885–891
20. Brinster RL, Chen HY, Trumbauer ME, Yagle MK, Palmiter RD (1985) Factors affecting the efficiency of introducing foreign DNA into mice by microinjecting eggs. Proc Natl Acad Sci USA 82:4438–4442
21. Giraldo P, Montoliu L (2001) Size matters: use of YACs, BACs and PACs in transgenic animals. Transgenic Res 10:83–103
22. Montoliu L, Roy R, Regales L, García-Díaz A (2009) Design of vectors for transgene expression: the use of genomic comparative approaches. Comp Immunol Microbiol Infect Dis 32:81–90
23. Palmiter RD, Brinster RL (1985) Transgenic mice. Cell 41:343–345
24. Meisler MH (1992) Insertional mutation of 'classical' and novel genes in transgenic mice. Trends Genet 8:341–344
25. Whitelaw CB, Springbett AJ, Webster J, Clark J (1993) The majority of G0 transgenic mice are derived from mosaic embryos. Transgenic Res 2:29–32
26. Fielder TJ, Barrios L, Montoliu L (2010) A survey to establish performance standards for the production of transgenic mice. Transgenic Res 19(4):675–681
27. Zurita E, Chagoyen M, Cantero M, Alonso R, Gonzalez-Neira A, Lopez-Jimenez A, Lopez-Moreno JA, Landel CP, Benitez J, Pazos F, Montoliu L (2010) Genetic polymorphisms among C57BL/6 mouse inbred strains. Transgenic Res 20:481–489
28. Freeman HC, Hugill A, Dear NT, Ashcroft FM, Cox RD (2006) Deletion of nicotinamide nucleotide transhydrogenase: a new quantitative trait locus accounting for glucose intolerance in C57BL/6J mice. Diabetes 55:2153–2156
29. Van Keuren ML, Gavrilina GB, Filipiak WE, Zeidler MG, Saunders TL (2009) Generating transgenic mice from bacterial artificial chromosomes: transgenesis efficiency, integration and expression outcomes. Transgenic Res 18:769–785

Chapter 6

Generation of Transgenic Mice by Pronuclear Microinjection

Katja Becker and Boris Jerchow

Abstract

The introduction of a transgene into a fertilized oocyte by pronuclear microinjection of a solution containing the construct of choice is probably the most straightforward method to generate a genetically modified organism. This technique has been adapted to a number of vertebrate and invertebrate species and readily yields founder animals carrying the transgene if performed correctly. We will here describe the generation of transgenic mice from small transgenes and bacterial artificial chromosome (BAC) type transgenes by random integration. Moreover, we provide a method for the transposase-mediated integration of a transgene that is flanked by transposase recognition sequences. While not all species require the use of a sophisticated setup, the generation of transgenic mice is technically challenging mainly because of the small size of the oocyte and the need for well-defined buffer and media conditions. Moreover, manipulated embryos have to be put back into an environment where they can develop to term, and this environment can only be the oviduct of a recipient mouse that has been prepared to allow pregnancy.

6.1 Introduction

In principle, transgenic mice are quite easily generated: fertilized oocytes are isolated from female mice that have been mated to stud males, a DNA solution is injected into one of the pronuclei, and the manipulated oocytes are transferred into the oviduct of a pseudopregnant recipient mouse. Less than 3 weeks later, transgenic offspring are born and can be identified usually by PCR genotyping. Unfortunately, there are a number of potential pitfalls along this way and it is of paramount importance to have adequate equipment for oocyte manipulation and embryo transfer as well as controlled environmental settings for the mouse colony at one's disposal. We will start out by describing the instrumental setup needed followed by materials and chemicals and will continue with a detailed description of all steps on the way to the successful generation of a transgenic mouse line. In our description, we will name the suppliers and manufacturers of the products we currently and successfully use. However, other labs obtain the

S. Pease and T.L. Saunders (eds.), *Advanced Protocols for Animal Transgenesis*, Springer Protocols,
DOI 10.1007/978-3-642-20792-1_6, © Springer-Verlag Berlin Heidelberg 2011

same or possibly better results with different equipment and/or reagents. We have no financial interest in any of the companies mentioned below.

6.2 Setup

6.2.1. Instrumentation

To visualize the pronuclei of fertilized oocytes, an inverted microscope with adequate magnification is needed. Two manipulators that can be equipped with capillaries for holding and the injection of the DNA solution into the oocyte have to be connected to this microscope. These manipulators can be electrically or mechanically controlled. The holding capillary in turn is connected via oil or air-filled tubing to a device that is used to grasp and release the oocytes by adjusting the pressure to under or over pressure, respectively. The injection capillary finally is connected to an electronic device that will inject defined amounts of the DNA solution contained in the capillary into the pronucleus of the oocyte.

Due to the extreme precision that is needed to direct the injection capillary into the pronucleus, it has to be made sure that all vibrations from the surrounding environment that might be caused by building ventilation systems, heavy equipment, staff working in the vicinity, external traffic or other sources, are not transmitted to the embryo manipulation setup.

This is our setup (Fig. 6.1):

- Vibration isolation table
- Leica DM IRB injection microscope, 10× eyepieces, 5× and 40× lenses

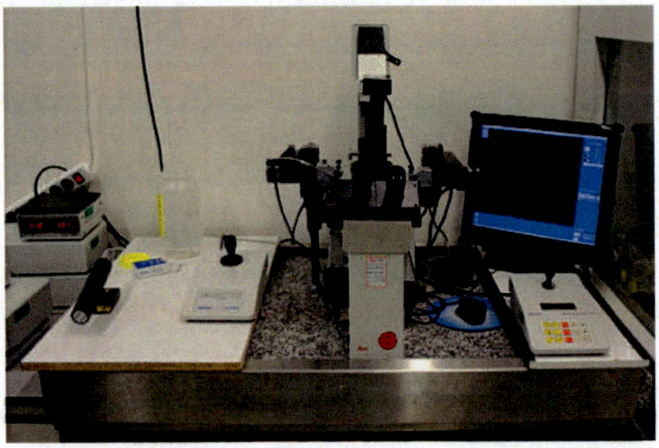

Fig. 6.1 Microinjection setup.

Fig. 6.2 Schematic representation of the injection chamber. The chamber has been milled from aluminum plate with 5 mm thickness. The central opening is approximately 45 mm wide and 20 mm high. Vacuum grease is applied to the rim of the hole on the lower side and a coverslip is fastened to produce the actual chamber for injection.

- Eppendorf micromanipulators and injection control (holding: PatchMan Micromanipulator, injection: Micromanipulator 5171 and Transjector 5246)
- Frame for injection chamber, custom made, Fig. 6.2
- Leica MZ 16, Plan 1.0×, 0.71–11.5 for oocyte collection and embryo transfer surgery equipped with two Schott KL 1500 "cold light" illumination devices with flexible light guides and a Leica light base
- Needle puller with FB255B 2.5 mm square box filament, Sutter Instruments Model P-97; settings: $P = 500$, heat 580, pull 80, vel 40, time 150, ramp 560
- Heated microscope stage HT 200, Minitüb Abfüll- und Labortechnik GmbH & Co. KG

6.2.2. Surgical Instruments

- Two pairs of forceps Inox 2
- Two pairs of forceps FST by Dumont, No. 3
- Scissors, Asculap BC060R
- Autoclip applier, Becton Dickinson, Clay Adams MikRon 9 mm
- Wound clips, Becton Dickinson, Clay Adams Auto Clip 9 mm
- Surgical clip, Fine Science Tools, Dietrich Bulldog Clamp 28 mm, serrefine serrated straight, No. 18050-28

6.2.3. Buffers, Media, and Other Chemicals

- Hormones for Superovulation:
  - Pregnant mare's serum gonadotropin (PMSG) Intergonan, Intervet, 1000 I.E.
  - Human chorion gonadotropin (hCG) Ovogest, Intervet, 1500 I.E.

  Hormones are dissolved according to the manufacturer's recommendations and frozen in 1.5-ml reaction tubes in aliquots containing 50 I.U. at −20°C. Before injection, tubes are filled to 1 ml with cold sterile PBS and thoroughly mixed after thawing.

- Anesthesia:
  - Mixture of 2% Xylazin (Ceva Sante Animale) and 10% Ketamin (WDT) in sterile PBS
- Analgesia:
  - Tamgesic (Buprenorphinhydrochloride), Schering-Plough, diluted to 50 μg/ml in sterile PBS (1 μl or 0.05 μg are applied per g bodyweight)
- Media and Chemicals:

  All media, chemicals, and enzymes that come into contact with embryos should be of "mouse embryo tested" quality, if available. Otherwise, they should at least be cell culture tested.
  - M2, Sigma M7167
  - KSOM embryo culture medium (for best results, we set up our own KSOM; the composition can be found here: http://www.mshri.on.ca/nagy/KSOM.HTML)
  - Embryo-tested water is used to prepare the medium (Sigma W1503)
  - Mineral oil, Sigma M8410
  - Hyaluronidase, Sigma H4272 working solution at 300 μg/ml in M2
- Buffers
  - BAC DNA microinjection buffer

  10 mM Tris–HCl, pH 7.5, 0.1 mM EDTA, 100 mM NaCl in embryo-tested water, Sigma W1503

  The buffer is sterile filtered through 20-μm bottle top filters. To exclude contaminations from the filter material, the first third of the filtrate is discarded.

  Add polyamines from 1,000× stock stored at −20°C to the following final concentrations:

  30 μM Spermine, Sigma S1141, 70 μM Spermidine, Sigma S2501
  - Standard microinjection buffer: We use the same buffer as for BAC DNA microinjection without polyamines

6.2.4. Other Consumables

- 1.5-ml polypropylene reaction cups with lid, Josef Peske GmbH & Co. KG, 421-800
- 3.5-cm petri dishes, Nunc 24045
- 10-cm petri dishes, generic
- Capillaries:
  - Capillaries for embryo handling and transfer, Retransfer Capillaries, BioMedical Instruments, Blaubrand, 2 mm, fire polished, ID = 108–114 μm

  - Holding capillaries, BioMedical Instruments, ends fire polished, holding capillary standard, ID= 20–25 μm, angle 15°
  - Injection capillaries, pulled from capillaries purchased from Science Products, GB120F-10, OD 1.2 mm, ID 0.69 mm ID, standard wall with inner filament
- Microloader, Eppendorf No. 5242 956.003
- Mouth aspirator tube assembly, Sigma A5177 5EA
- Syringes, B. Braun Omnifix-F 1 ml, ref. 9161406V
- Cover slips, Menzel-Gläser, 100 pieces, 24 × 50 mm, BB024050A1
- Silicon vacuum grease, Merk 1.07922, 100 g

6.3 Procedures

6.3.1. Preparation of the Transgene

A basic transgene contains a promoter and the coding sequence of a gene that will be expressed after integration of the construct into the genome followed by a polyadenylation signal. Plasmid-type transgenes should contain suitable endonuclease restriction sites to separate the actual transgene from the vector backbone. Failure to do so can result in epigenetic silencing of the transgene and the absence of protein expression. Moreover, the addition of an intron to the transgene helps to counteract gene silencing. Standard DNA mini, midi or maxi preparations can be used. After restriction, the DNA is separated on a standard agarose gel stained with ethidium bromide. The band of interest is cut from the gel and purified using a spin column according to the manufacturer's protocol. However, contrary to this protocol, DNA is eluted with microinjection buffer and the concentration is adjusted to 2–3 ng/μl with microinjection buffer. To obtain optimal results, before dilution with microinjection buffer the DNA concentration should be greater than 20 ng/μl. In this way residual contaminations of the preparation can be adequately diluted and not harm zygotes upon injection.

The preparation of high-quality DNA suitable for pronuclear microinjection from bacterial artificial chromosomes (BACs) is much more challenging than plasmid DNA preparation. However, the fidelity of transgenes expressed from a BAC recapitulating the endogenous expression pattern of the chosen promoter is much higher than with plasmid-type transgenes. This is mainly due to the fact that the much greater size of the BAC minimizes the effect of endogenous regulatory elements at the site of transgene integration into the host genome (positioning effects; for more details refer to Chap. 9 and to [1]). There are several ways to

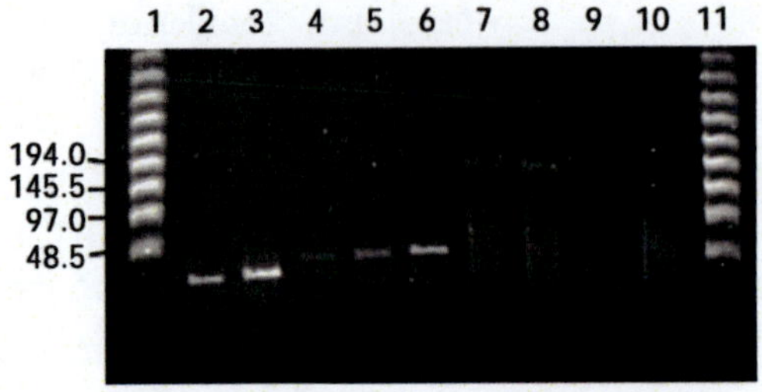

Fig. 6.3 BAC DNA after pulsed field gel electrophoresis (PFGE). Lanes 1 and 11: molecular weight standards; relevant bands are marked in kb to the *left*. Lanes 2, 3, and 7–10: different volumes of solutions of two linearized BAC DNA constructs have been loaded. Lanes 4–6: λ-DNA as mass standards (7, 21, 35 ng). PFGE of BAC DNA together with mass standards is necessary to determine concentration of the BAC DNA solution and to confirm sufficient integrity of the preparation to be used in pronuclear microinjection.

prepare high-quality BAC DNA. In any case, the quality and quantity of the DNA has to be checked by pulsed field gel electrophoresis (PFGE) before attempting to generate transgenic mice from a construct. Moreover, BAC DNA has to be stored in microinjection buffer containing polyamines to prevent degradation. A band of the expected molecular weight has to be clearly visible before attempting to generate BAC transgenic mice from a given preparation (Fig. 6.3). For microinjection, BAC DNA is diluted with BAC microinjection buffer to a final concentration of 2 ng/μl.

Recently, we have developed a method to enhance the integration rate of small transgenes into the host genome [2]. In short, specific inverted repeats are added on either side of the transgene to make it a transposable element. DNA of the transposable element is prepared by standard procedures and dissolved in microinjection buffer at a concentration of only 0.4 ng/μl. RNA of a hyperactive variant of the Sleeping Beauty transposase SB100 is added to a final concentration of 5 ng/μl (see Chap. 11 for more details). This mixture is injected in the same way as a standard DNA solution. Great care has to be taken to prevent RNase contamination. Using the transposase will yield a rate of about 50% transgenic offspring as opposed to 10% with standard techniques. Only single copies of the transgene will integrate at a given location, while standard transgenes always integrate as concatamers of several head to head or head to tail copies. The latter can be problematic, when sites are introduced that are meant to be used for further manipulations of the genome, such as loxP or frt sites. Moreover, concatamers are more prone to gene silencing than single insertions. However, the size of the transgene flanked by the inverted repeats is limited to a maximum of about 7 kb.

6.3.2. Superovulation

To obtain a maximum number of fertilized oocytes from a minimum number of animals, a well-working superovulation protocol is

Fig. 6.4 Oviducts lined up in droplets of medium before preparation of oocytes. Bigger droplets are in place for collection and washing of oocytes after hyaluronidase treatment.

needed. In principle, two possible routes can be followed to yield best results depending on the strain of donor females. Either young mice are used before the first natural estrus at an age of 3–4 weeks or adult mice that are 2–3 months old. In both cases 5 I.U. of PMSG in 100 μl PBS are injected intraperitoneally between 11 a.m. and 1 p. m. on day –2. After 46–48 h, 5 I.U. of hCG in 100 μl PBS are applied via the same route and females are mated to stud males. The next morning, oocytes are harvested (day 0.5). Mice are kept under a 12/12 h light cycle with lights on at 7 a.m. and off at 7 p.m.

6.3.3. Harvesting of Oocytes

Donor females are sacrificed by cervical dislocation the morning after mating between 8 and 10 a.m. We do not recommend utilizing CO_2 or anesthetics to sacrifice animals since this may adversely affect the zygotes. The abdomen is opened with a pair of fine sharp scissors. The oviducts are removed and put into a reaction cup containing M2 medium preheated to 37°C. Oviducts can now be transferred to the injection lab. Oviducts are put into single drops of M2 (50 μl each, Fig. 6.4). The number of oviducts that can be handled at a time depends on the experience of the operator. Always keep in mind that oocyte quality will decrease more quickly if kept below 37°C. While an experienced person can handle up to 30 oviducts the beginner should start out with two to a maximum of four mice at a time.

Under a dissecting microscope at around 12.5× magnification oviducts are torn open with fine forceps (Fig. 6.5) to release the cumulus complex (a cluster of oocytes embedded in cumulus cells). 80 μl of hyaluronidase solution are added to each drop to disperse the individual complexes. Dispersion will take about 2–5 min at room temperature and should be closely monitored under the microscope (20× magnification) since prolonged exposure

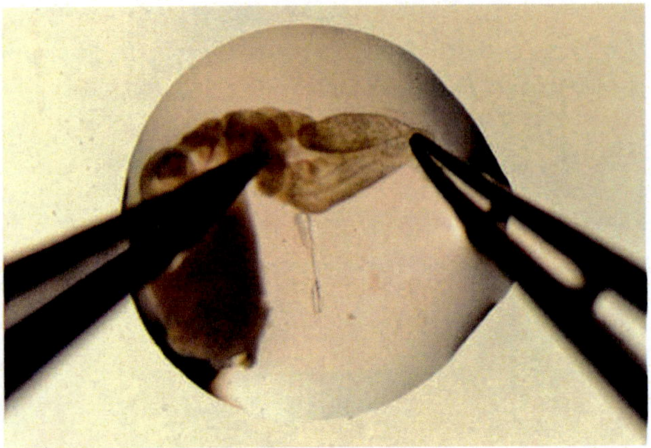

Fig. 6.5 Two fine forceps are used to rip open the individual oviducts. This will release the cumulus complexes containing the oocytes. Hyaluronidase solution is then added to separate oocytes from cumulus cells.

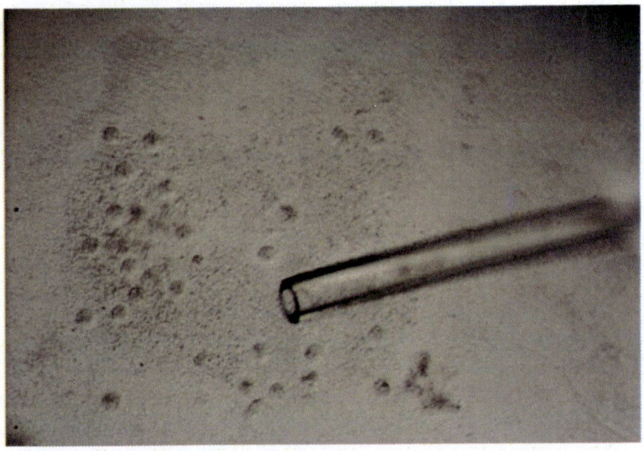

Fig. 6.6 Individual oocytes are clearly visible after hyarulonidase treatment and can be picked up with a mouth-controlled retransfer capillary. Cumulus mass cells can be seen to have fallen from around the oocytes and are settled on the dish floor.

times will adversely affect the zygotes. As soon as the oocytes are separated from cumulus cells by the enzyme, they are picked up with a finely drawn mouth-controlled capillary (Fig. 6.6), washed through three 200 μl drops of M2 (Fig. 6.4, front right), deposited into a 200 μl drop of KSOM under mineral oil and put into a cell culture incubator (3.5 cm petri dish).

6.3.4. Preparation of Injection Needles

We recommend having the manufacturer's service personnel set up the needle puller. You should make sure to always use the same quality and make of glass capillaries in order to avoid the need to make a change in the puller's settings. For further details refer to your needle puller's manual or contact the manufacturer's representative.

After pulling a number of injection needles, these are filled from the rear end using a microloader tip on a pipette with approximately 1 μl of microinjection buffer containing the transgene. Needles are kept at 4°C with the tip down to allow tiny air bubbles to go to the rear end of the needles.

6.3.5. Preparation of the Injection Chamber and Injection Setup

A thin film of vacuum grease is applied to the bottom side of the frame and this is pressed on a glass slide to make the injection chamber. The chamber is filled with 100 μl M2 medium that is then overlaid with mineral oil. It is important that the entire drop of M2 medium is covered by oil.

The injection chamber is mounted on the microscope's heating stage that has been preheated to 37°C. Now the holding capillary is lowered until it passes the oil film and enters the medium. This is controlled visually at 50× magnification. The holding capillary should not touch the bottom of the injection chamber.

Now the first injection needle is secured in its fitting. Before lowering the injection needle into the medium a compensation pressure has to be applied to counteract capillary forces that would otherwise draw up medium from the injection chamber into the injection needle. This is set on the electronic injection device (around 90 hPa). The angle of the needle is set to the lowest possible since this will reduce the amount of damage inflicted on the oocytes. Optimally, this angle should lie between 5° and 10°.

6.3.6. Injection of DNA

Transfer a number of zygotes by mouth-pipetting from the incubator into the injection chamber. As before the number that should be transferred simultaneously depends on the skill of the operator. We recommend a maximum of ten for beginners. Experienced injectionists will inject batches of up to 40 embryos and more. We recommend defining three distinct locations inside the injection chamber: (1) uninjected oocytes, (2) injected oocytes, and (3) discarded oocytes. For injection capillaries with narrow opening the injection pressure is set to "clean pressure". Upon applying injection pressure it will therefore start at 1,000 hPa and rather quickly rise to 7,000 hPa. One zygote is fixed with the holding capillary and the magnification is set to 400×. By focusing through the zygote, two separate pronuclei should be discernable (Fig. 6.7). In case these pronuclei cannot be found the oocyte should be discarded at location 3 (see example in Fig. 6.8). Sometimes the zygote will have to be released and refixed with the holding capillary to get one of the pronuclei in a proper position for injection. Penetration of the entire zygote in order to inject into a pronucleus on the far side of the cell should be avoided. Figures 6.9a through c show, how the tip of the injection needle is inserted through the *zona pellucida*, the cell wall and the nuclear membrane into one of the pronuclei. Then the injection pressure is applied until a swelling of the pronucleus is clearly visible

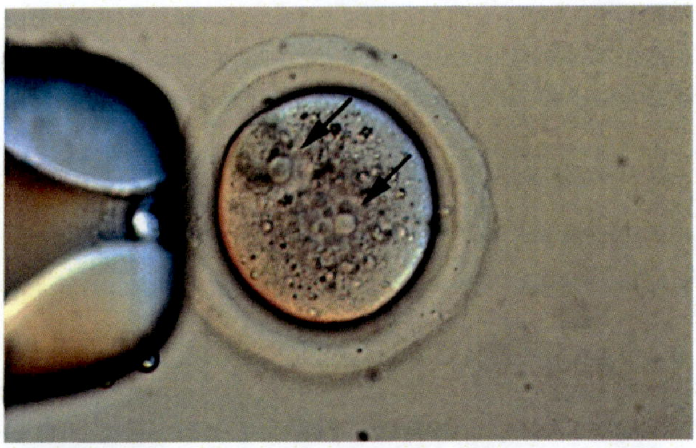

Fig. 6.7 Fertilized oocyte attached to holding capillary. The male and female pronuclei are clearly visible (*arrows*).

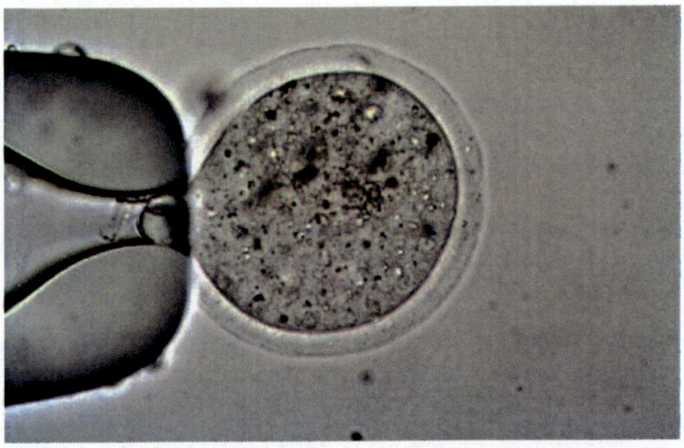

Fig. 6.8 Example of an oocyte that does not show pronuclei. Any zygote that does not appear perfectly intact is discarded.

(compare distance of arrowheads in Fig. 6.9a to distance in Fig. 6.9c). The nuclear membrane is quite elastic and not easily penetrated. If a swelling of the pronucleus cannot be observed and a little bubble forms at the tip of the needle instead (arrow in Fig. 6.10), the injection needle has to be repositioned and properly inserted into the pronucleus. It may be necessary to deeply penetrate the pronucleus, taking the tip of the needle to the far side of the membrane and then to slightly retract the needle. In the event that neither a swelling of the pronucleus nor a bubble is visible, the opening of the capillary might be too small. In this case it is possible to enlarge it by carefully hitting the needle against the holding capillary to break off the very end of the tip and leave a bigger opening. In contrast, when a swelling of the pronucleus is observed after inserting the injection capillary without applying injection pressure, the opening of the injection capillary is too big and it should be exchanged for another. A good balance between a

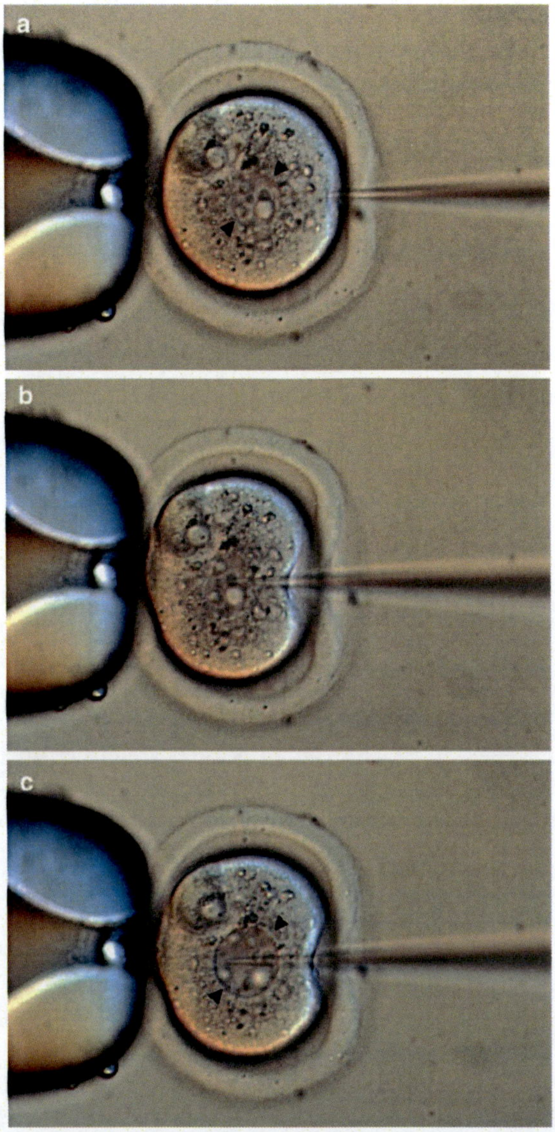

Fig. 6.9 Pronuclear microinjection. When a zygote is fixed to the holding capillary, both the injection needle filled with DNA solution as well as one of the pronuclei are brought into focus. Then the *zona pellucida* is penetrated with the injection capillary aiming at one of the pronuclei (**a**). The injection capillary is moved further to the far side of the pronucleus and then slightly pulled back to place the capillary's tip in its middle (**b**). When the injection pressure is applied a swelling of the injected pronucleus has to be clearly visible (compare *arrowheads* in **a** to those in **c**).

small and a large opening has to be found. If the opening is too big, oocytes will be damaged and the lysis rate will increase. If the opening is too small especially big constructs like BACs might be sheared. Many injectionists prefer to inject at lower pressure (e.g. injection pressure set to 1,000 hPa). The wider opening of

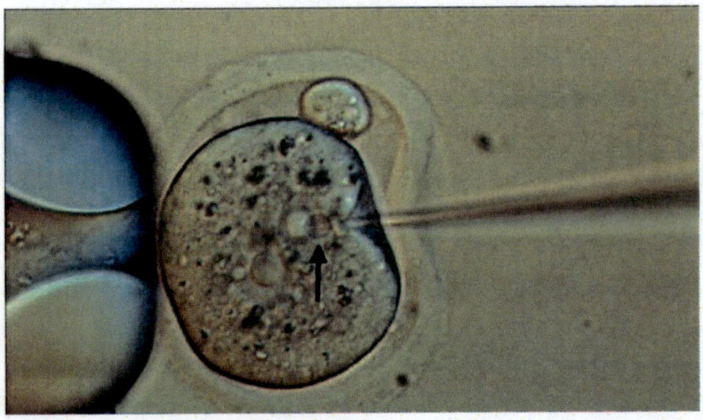

Fig. 6.10 Improper injection of a pronucleus. If the nuclear membrane has not been properly penetrated before the injection pressure is applied, a small bubble forms at the tip of the injection capillary (*arrow*) and no swelling of the pronucleus is observed. The injection capillary has to be correctly inserted into the pronucleus in a new attempt.

the injection capillary that will become necessary when attempting to inject at reduced pressure can be achieved by either breaking the tip of the capillary on the holding capillary or by applying different settings on the needle puller and/or using different capillaries. The latter should in any case contain a filament. After successful injection the zygote is deposited at location 2 and the remaining oocytes are either injected or discarded, depending on their integrity. Since damaged oocytes will quite rapidly lyse, they can be sorted out right after finishing the injection of one batch of zygotes. Injected zygotes will then either be collected in KSOM under mineral oil and put into an incubator or directly transferred into the oviducts of a recipient female (see below). Beginners might want to incubate zygotes overnight to determine the rate of oocytes that develop to the two-cell stage. Development in vitro strongly depends on optimal culture conditions. The next morning, at least 90% of the injected embryos should have developed to the two-cell stage. Uninjected embryos should be cultured in parallel to control culture conditions. It should be noted that with some strains, especially under suboptimal culture conditions, two-cell block is an issue. In this case, development in vitro will stop at the two-cell stage, eliminating the possibility of blastocyst development and transfer of the embryos into the uterus (as opposed to the oviduct) of recipient females. In our lab, we obtain best results when zygotes are directly transferred on the day of injection without further incubation in vitro.

6.3.7. Embryo Transfer

For transfer of embryos on the day of injection, recipient females (foster mice) are mated to vasectomized or otherwise sterile males on the day before the actual microinjection. Usually Swiss-derived outbred strains are preferred and ICR or CD-1 mice are widely used since they are able to give birth to big litters. We use NMRI

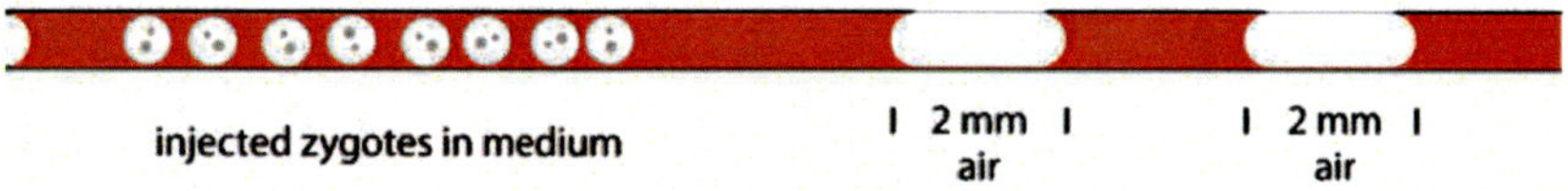

Fig. 6.11 Schematic of transfer capillary loaded with medium, air bubble, medium, second air bubble, and zygotes in medium.

females as surrogate mothers. Only those females that show a clear vaginal plug on the morning of the injection day are used. If enough foster females are available, we make a practice of using only those that still show the plug at the time surgery is started, since we have found this gives us best results.

The recipient female is anesthetized by intraperitoneal injection of 100 μl ketamin/xylazin-solution per 10 g body weight. Injected oocytes for one transfer (16–24) are mouth-pipetted from the KSOM drop from the incubator into a drop of M2. To more easily control the pipette, first a very small amount of medium is drawn up followed by a 2-mm air bubble, some medium and a second 2-mm air bubble. Then about 8–12 injected zygotes are loaded into the capillary in as little medium as possible (Fig. 6.11). Remaining zygotes are left in the incubator. The area of surgery is carefully disinfected according to local procedures. Using too much volatile disinfectant can cause a considerable drop of body temperature and lead to the death of the recipient. With a pair of forceps the fur is parted along the vertebral column. Then a single incision into the skin of approximately 1.5 cm length is made on the back of the mouse along the midline. The incision is positioned about half way between the base of the tail and a fictitious line drawn between the two ears and should start just below the last costal arch (Fig. 6.12). From this point on, all surgery is done under the surgical microscope in order to clearly visualize very small structures. The skin can be moved to the left and to the right, thereby exposing a red dot on either side of the backbone that is clearly visible under the peritoneum at around 7× magnification: the ovaries. Starting on one side, a cut of approximately 1 cm length is made through the peritoneum at one of the ovaries. The cut is held open with two pairs of fine forceps and carefully moved around until a characteristically white shiny fat pad is located. The fat pad – and only the fat pad – is carefully but firmly gripped with one pair of fine forceps and pulled out of the peritoneal cavity together with the ovary. A surgical clip is fixed to the fat pad – and only the fat pad – in a way that clearly exposes the oviduct (Fig. 6.13). The oviduct should appear swollen in a bona fide surrogate female. If not, another female should be used if available. The opening of the oviduct will always point into the direction of the mouse's tail. After increasing the magnification to around 12.5× the membrane between ovary and oviduct is opened with two pairs of

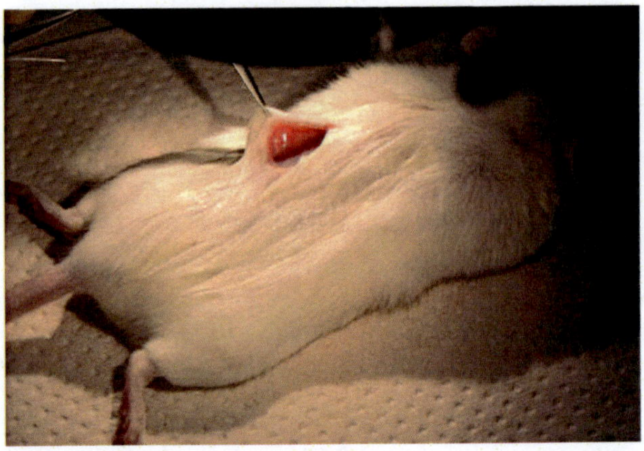

Fig. 6.12 After disinfection of the surgical area a cut into the skin of approximately 1.5 cm length is made along the midline of the foster mouse's back.

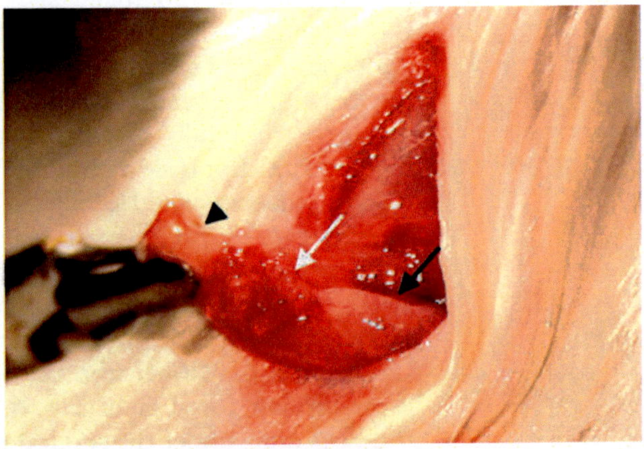

Fig. 6.13 Positioning of the oviduct for embryo transfer. The fat pad (*arrowhead*) attached to the oviduct (*white arrow*) is grasped with fine forceps to pull the oviduct out of the peritoneal cavity. The oviduct is held in place by attaching a surgical clip to the fat pad. The *black arrow points* the uterus.

very fine forceps. Care has to be taken in this and the following steps that no blood vessels are injured and that the ovary is not touched, since increased trauma inflicted on the reproductive organs will decrease the probability of a pregnancy occurring. Now the opening of the oviduct should become visible. If this is not the case the forceps can be used to carefully enlarge the gap between the ovary and oviduct. The capillary is inserted into the opening of the oviduct and the medium containing the embryos is expelled until air bubbles are visible inside the ampulla (Fig. 6.14). After the transfer into the first oviduct the surgical clip is removed and the oviduct is carefully moved back into the abdominal cavity. This is best accomplished by holding the peritoneum on both

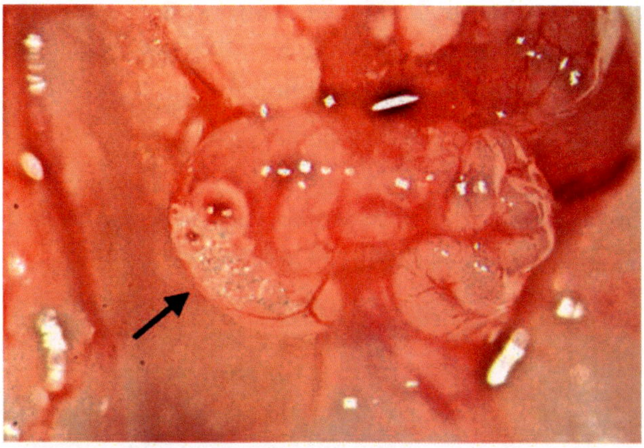

Fig. 6.14 Manipulated embryos are expelled from the retransfer capillary into the oviduct until air bubbles are clearly visible inside the oviduct (*arrow*). This is a clear indication that all content has been expelled from the capillary and that the oviduct has not been missed.

sides of the incision and pulling it up. This will pull the cut open at the same time and the oviduct will gently slide back into place. As before the skin is now moved to the other side to expose the second ovary and the transfer procedure is repeated as described. After bilateral embryo transfer, 0.05 μg per g bodyweight Bubrenorphinehydrochloride is applied intraperitoneally as an analgesic and the cut in the skin is closed with a wound clip using an autoclip wound clip system. Now the next foster mouse is anesthetized and more manipulated zygotes are transferred. Experienced injectionists might want to anesthetize two mice at a time to speed up the process. Mice will awake approximately 30 min after injection of the ketamin/xylazin mixture.

Following surgery, surrogate mothers are preferentially housed in pairs in a dedicated animal room. After about 19 days, offspring are born. Biopsies are taken either at day 3–4 after birth and animals are marked by tattooing or at 3–4 weeks of age when pups can be marked with ear punches. Biopsies come from the tail or ear, respectively, and are checked for the presence of the transgene by PCR (refer to Chap. 21).

6.4 Troubleshooting

Donor females have no oocytes

- Check superovulation protocol. The protocol we provide here works well for C57BL/6, DBA/2 and FVB/N mice. It will also work for many other strains. Some strains are very poor oocyte donors and using them for the generation of transgenic lines should be avoided. One example is the BALB/c strain. Other strains will not readily superovulate before puberty. Instead, use them at 2–3 months of age.

- Replace hormones.
- Check light cycle.

Oocytes do not show two pronuclei: Oocytes are either not fertilized or they have progressed in development beyond the point where pronuclei fuse.

- Check light cycle. If the light cycle differs from the required settings or even varies from day to day, mice will not mate or not mate at the required time.
- Check stud males. They should not be used on consecutive days and not more than three times per week. Moreover, they should be between 3 and 8 months old.

High lysis rate

- Check angle of injection capillary. If the angle is too big, the zygote will not tolerate the trauma of injection. The angle should be no greater than 10°.
- Check size of injection capillary for the same reason as the above.
- Too much DNA-solution injected. Only a slight swelling of the pronucleus should be visible upon injection.
- Bad batch of oocytes. Take other batch. If the problem persists, check superovulation protocol.

Zygotes do not develop to two-cell stage

- Check embryo culture medium. The medium is composed of six stock solutions, BSA, and water. Once you have succeeded in making your own medium (which we recommend), try to renew only single stock solutions at any one time. This way it is much easier to narrow down the cause when embryo culture fails.
- Impure DNA preparation. Repeat preparation more carefully. The higher the DNA concentration is after preparation, the greater the dilution factor for injection and therefore the cleaner the solution will be.
- Check DNA concentration. Zygotes will not develop if the DNA concentration injected is too high.
- Too much DNA solution injected. Only a slight swelling of the pronucleus should be visible upon injection.

No offspring

- The foster's reproductive tract has been damaged.
- The zygotes have been damaged, either mechanically during the injection or chemically due to impure DNA solution. Check

in vitro development beyond two-cell stage after injection with buffer only.

- Holding conditions are not optimal. Keep foster mice in a dedicated room that is especially quiet with no change in staff.

No transgenic offspring

- The gene product of the transgene might be toxic or lead to severe defects in development causing death of the embryo.
- Too little DNA solution injected. Swelling of the pronucleus should be clearly visible upon injection.
- DNA concentration too low. Check DNA concentration.

References

1. Van Keuren ML, Gavrilina GB, Filipiak WE, Zeidler MG, Saunders TL (2009) Generating transgenic mice from bacterial artificial chromosomes: transgenesis efficiency, integration and expression outcomes. Transgenic Res 18(5):769–785
2. Mátés L, Chuah MK, Belay E, Jerchow B, Manoj N, Acosta-Sanchez A, Grzela DP, Schmitt A, Becker K, Matrai J, Ma L, Samara-Kuko E, Gysemans C, Pryputniewicz D, Miskey C, Fletcher B, Vandendriessche T, Ivics Z, Izsvák Z (2009) Molecular evolution of a novel hyperactive Sleeping Beauty transposase enables robust stable gene transfer in vertebrates. Nat Genet 41(6):753–761

Chapter 7

Generation of Transgenic Rats Using Microinjection of Plasmid DNA or Lentiviral Vectors

Séverine Ménoret, Séverine Remy, Laurent Tesson, Claire Usal, Anne-Laure Iscache, and Ignacio Anegon

Abstract

The rat is an important system for modeling of human disease. The use of transgenesis is relatively uncommon in rats. In this chapter, we focus on describing efficient techniques for the generation of transgenic rats by microinjection of plasmid DNA into pronuclei and the injection of human immunodeficiency virus-1 (HIV-1)-derived lentiviral vectors into the perivitelline space of one-cell fertilized eggs. We have specifically developed optimal conditions for superovulation of prepubescent female Sprague-Dawley (CD) strains and optimal conditions for microinjection and embryo transfer into foster mothers.

Abbreviations

CD® IGS RATS	Sprague-Dawley rat strain
FSH	Purified Follicle-stimulating hormone
hCG	Human Chorionic Gonadotropin
LHRH	Luteinizing hormone releasing hormone
M16	Embryo culture medium supplemented
PBS	Dulbecco's Phosphate Buffer Solution
PMSG	Pregnant Mare's Serum Gonadotropin

7.1 Introduction

The rat is an excellent model for physiology, pharmacology, toxicology, cardiovascular disease, and immunology. Compared to mouse, the size of the rat lends itself more readily to performance of surgical procedures, multiple blood sampling (in larger volumes), tissue and organ sampling (e.g., central nervous system), and

Séverine Ménoret and Séverine Remy contributed equally and are corresponding authors.

S. Pease and T.L. Saunders (eds.), *Advanced Protocols for Animal Transgenesis*, Springer Protocols, DOI 10.1007/978-3-642-20792-1_7, analysis of organ function in vitro (e.g., heart perfusion) and is the species of choice for many experimental models [1].

The first transgenic mice were generated in 1976 by R. Jaenisch [2] after the infection of mouse embryos with Moloney leukemia retroviruses. However, these mice did not express the transgenes due to chromatin epigenetic modifications [3]. Subsequently, alternative methods were developed for generation of transgenic mice, in particular pronuclear microinjection of DNA [4].

Since the 1990s, with the development of the first transgenic rat lines [5, 6], ~250 different transgenic rat lines have been produced. A regularly updated list of all transgenic and genetically modified rat lines published since 1990 is available at:

(http://www.ifr26.univ-nantes.fr/ITERT/transgenese-rat/index.php?option=com_content&view=article&id=6&Itemid=6).

In this chapter, we describe the generation of transgenic rats by microinjection of plasmid DNA or of lentiviral vectors into rat one-cell embryos. The generation of transgenic rats remains more difficult compared to transgenic mice since rat one-cell embryos have more flexible plasma and pronuclear membranes, making injection more difficult and increasing embryo lysis [7, 8].

The methods used to induce superovulation and to manipulate rat embryos for generation of transgenic rats show considerable variation in the choice of strain [9], superovulation protocols [10], media for the culture of embryos [9] and the reimplantation of embryos directly after microinjection or after a short period of in vitro culture (1 day) [9]. With plasmid DNA microinjection [11], we obtain between 0.2 and 3% of transgenic rats/number of microinjected zygotes. To obtain 4–6 founders, we routinely perform microinjections over a period of 3–6 weeks with four microinjection sessions per week, thus around 250 embryos/week.

This chapter describes also the generation of transgenic rats using lentiviral vectors. The idea of using lentiviral vectors to generate transgenic animals emerged several years ago, as an attractive alternative method for delivering exogenous genes into cells because human immunodeficiency virus (HIV)-derived gene delivery vehicles can efficiently mediate the integration of their cargo into zygotes or other cell types, even during a quiescent phase [12]. C. Lois [13] and A. Pfeifer [14] were the first to generate transgenic mice using lentiviral vectors with a very high efficiency (80% of founder mice) and no epigenetic silencing as had been previously observed with oncoretroviral delivery of transgenes [3]. These encouraging data subsequently led the scientific community to focus on the use of lentiviral vectors as a promising tool to generate other transgenic animal species (chickens [15], pigs [16, 17], cattle [18], monkeys [19], and rats [13, 20–22]). In this chapter, we will describe the technique of microinjection of lentiviral vectors into the perivitelline space, as well as other techniques of vector delivery that have been described as applied to other species [23] but not rats.

Although the efficiency of generating transgenic rats using lentiviral transgenesis is higher (in our hands: 6–18% transgenic rat/number of transplanted zygotes) than that using DNA microinjection, the generation of lentiviral vectors, with the need for high lentiviral titers and good quality, is technically diffucult. The size of the expression cassette needs to be <9 kb, whereas the size of DNA used in microinjection can be as high as several hundred thousand kb. In the majority of cases when using lentiviral vectors, the transgene integrates at different locations within the genome [13], whereas when using DNA microinjection the transgene usually integrates as multiple copies at a single locus [24]. Consequently, lentiviral transgenesis is more demanding in terms of analyzing transgene integration in founders and their offspring, due to independent transgene segregation. However, these conditions perhaps increase the chances of obtaining founders with good transgene expression. Additionally, the exact genomic integration site can be easily determined using linker-mediated PCR (LM-PCR) techniques where single transgene copies exist in transgenic animals generated using lentiviral vectors, but not where plasmid DNA has been used.

The protocols we describe here were mostly used with the Sprague-Dawley (CD® IGS rats, Charles River) strains but also with the Brown Norway (BN/Crl) rat and Lewis (LEW/Crl) rat strains. However, they may demand further optimization for use in conjunction with other rat strains.

7.1.1. Outline of the Procedures

- Male vasectomy
- Superovulation and pseudopregnants females production
- Zygote preparation
- Microinjection of zygotes with plasmid DNA
- Microinjection of lentiviral vector to one-cell embryos
- Embryo reimplantation
- Genotyping analysis of founders by PCR, southern blot, and additionally LM-PCR for lentiviral founder animals

7.2 Materials

7.2.1. Equipment

- A basic anesthetic delivery system with an isoflurane anesthesia setup.
- Work station including:
  - Inverted microinjection microscope (NIKON DIAPHOT TMD, Narishige) equipped with 10× eyepieces,

5×, 10×, and 40× differential interference contrast objectives and with a heated stage (LINKAM MS100, Linkam Scientific Instruments, UK).

  - Microinjector – N2 pressurized microinjection system (Nikon, IM 300, Narishige).
  - Two micromanipulators for both holding and microinjection pipettes (Nikon, Narishige).
  - Video system.

- Micropipette puller (PN-30, Nikon, Narishige).
- Holding pipette (Glass capillary, Nikon, Narishige; cat. no. G-1).
- Microinjection pipette (Glass capillary with filament, Nikon, Narishige; cat. no. GD-1).
- Stereomicroscope with transmitted light base and fiber optic light source.
- Humidified incubator at 37°C and 5% CO_2.
- Egg transfer pipette (Pasteur Pipette) assembled into a mouth-operated system made up of a mouthpiece, rubber tube (~40 cm), and a pipette holder. For embryo transfer after lentiviral injection, embryo transfer pipette assembled with a membrane filter (0.02 μm, Whatman, cat. no. 6809-1102, Anotop 10) for biosafety.
- Injection chamber. Microscope slides for injection chamber (VWR international).
- Injection chamber. Coverslips (VWR international).
- 35 mm (35 × 10 mm) culture dishes (Nunc).
- 4-well plates (Nunc).
- Tips for loading capillaries with DNA plasmid or lentiviral vectors (Microloader, Eppendorf).
- Bunsen burner.
- Electric shaver.
- Surgical tools: one surgical forceps, 2× Dumont # 5 forceps, one dissection scissors. All instruments should be cleaned and sterilized with 70% ethanol.
- Surgical suture n°3/0.
- Surgical gauze.
- 2 ml syringes.
- Needles.

7.2.2. Reagents

- Pregnant Mare's Serum Gonadotropin PMSG (Folligon®, Intervet, Angers, France) at 25 IU/female/0.2 ml. Dissolve 2,500 IU in 20 ml 0.9% NaCl to give 125 IU/ml. Store in 1.8 ml aliquots (for 8× females) at −20°C for several months.

- Human Chorionic Gonadotropin hCG (Chorulon®, Intervet, Angers, France) at 30 IU/female/0.2 ml. Dissolve 1,500 IU in 10 ml 0.9% NaCl with 0.1% BSA to give 150 IU/ml. Store in 1.8 ml aliquots (for 8× females) at 4°C for 2 weeks.
- Embryo culture medium: M16 medium (M 7292, Sigma) supplemented with 10% fetal bovine serum, 2 mM L-glutamine, 100 IU/ml penicillin, and 100 mg/ml streptomycin.
- Embryo-tested bovine testis hyaluronidase (H 4272, Sigma) stock: Dissolve 30 mg in 3 ml of 0.9% NaCl to give 10 mg/ml. Store in 50 μl aliquots at −20°C for several months. For use, resuspend the aliquots in 950 μl of Dulbecco's Phosphate Buffer Solution and place it in 1 well of a 4-well plate.
- Purified DNA plasmid at 2–5 ng/ml.
- Dulbecco's Phosphate Buffer Solution (21600-069, GIBCO, Invitrogen): Dissolve one bottle (95.5 g) in 10 l H_2O. Aliquot in 500 ml and sterilize.
- Epinephrine (1 mg/ml). Amphastar Pharmaceuticals, Inc, cat no NDC 0548-9061-00.
- Embryo-tested mineral oil (M 8410, Sigma).
- 70% ethanol.
- 0.9% NaCl.
- Isopropanol.
- TE: 5 mM Tris–HCl pH 7.4 and 0.1mM EDTA.
- Tissue digestion buffer: 100 mM Tris–HCl pH 8.3, 5 mM EDTA, 0.2% (w/v) SDS, 200 mM NaCl, 100 μg/ml Proteinase K (740 506, Macherey Nagel). Add Proteinase K immediately before use from a freshly made stock solution.
- Restriction enzyme (EcoRV, PvuII, SspI).
- Herculase II Fusion DNA polymerase (Stratagene).
- TOPO TA cloning (Invitrogen).
- Gene-specific primers:

5′ gDNA-lentivirus	RU5-GSP1 5′-GCTCCTCTGGTTTCCCTTTCGCTTT-3′
	RU5-GSP2 5′-CGCCACTGCTAGAGATTTTCCACAC-3′
3′lentivirus-gDNA	3U5-GSP1 5′-GTGCTTCAAGTAGTGTGTGCCCGTC-3′
	3U5-GSP2 5′-TTAGTCAGTGTGGAAAATCTCTAGC-3′

7.2.3. Animals

- 8 × 26–30 days old mated prepubescent females.
- Fertile males (2 months old) for mating with the immature females. They should be replaced every 8 months to a year.

- Recipient females (8–16 weeks of age) that have successfully had at least 1 litter.
- Vasectomized males are needed to engender pseudo-pregnancy in the recipient females. They need to be replaced every 6–8 months.

7.2.4. Suppliers

Amphastar Pharmaceuticals, Inc., 11570 6th Street Rancho Cucamonga, California 91730 USA. Tel :(800) 423-4136 Fax : (909) 980-8296 E-mail :info@amphastar.com

Charles River, Domaine des oncins, BP 0109, 69592 L'ARBRESLE Cedex, France; phone: (33) (0)4 74 01 69 69, fax: (33) (0)4 74 01 69 99; http://www.criver.com/

Eppendorf France S.A.R.L, 60, route de Sartrouville, 78230 Le Pecq, France; phone (33) (0)1 30 15 67 40, fax: (33) (0)1 30 15 67 45, http://www.eppendorf.com/fr

Intervert S.A, rue Olivier de Serres, BP 17144, 49071 Beaucouzé, France, phone: (33) (0)2 41 22 83 83, fax: (33) (0)2 41 22 83 00

Invitrogen, P.O. Box 9418, Gaithersburg, MD 20898, USA; phone (301) 840-8000, fax (301) 670-8539; http://www.invitrogen.com/

Macherey Nagel, 1 r Gutenberg, BP 135, 67722 Hoerdt, France, phone: (33) (0)3 88 68 22 68, fax: (33) (0)3 88 51 76 88; http://www.mn-net.com/

NIKON France S.AS, 191, Rue du Marché Rollay, 94504 Champigny-sur Marne cedex, France; phone: (33) (0)1 45 16 45 16, fax: (33) (0)1 45 16 00 33; http://www.nikon.fr

Nunc distributor, VWR International S.A.S., "Le périgare" – Bâtiment B., 4ème ètage, 201, rue Carnot, 94126 Fontenay sous Bois cedex, France; phone: (33) (0)1 45 14 85 00, fax: (33) (0)1 47 23 50 56; http://www.nuncbrand.com/

Stratagene distributor, Agilent Technologies, 1 rue Galvani, 91745 Massy Cedex France; Phone: (33) 0810 446 446, Fax: 33(0)1 49 93 90 68; http://www.stratagene.com

Sigma, P.O. Box 14508, St. Louis, MO 63178, USA; phone (800) 325-3010, fax (800)325-5052; http://www.sigmaaldrich.com/

VWR International S.A.S, Le Périgares – Bâtiment B, 201, rue Carnot, 94126 Fontenay-sous-Bois, France, Phone: (33) 0 825 02 30 30, Fax: (33) 825 02 30 35, http://fr.vwr.com/app/Home

Whatman, North America, phone: 1-800-WHATMAN or 1-800-526-3593, Fax: 1-732-885-6529 or 1-732-885-6530 or 1-732-885-6531; http://www.whatman.com

7.3 Protocol

7.3.1. Male Vasectomy

1. Anesthetize male CD® IGS rats of at least 70 days of age by isoflurane (rats should weigh at least 250–400 g).
2. Place the rats on their backs to expose the abdomen.
3. Clean the animals thoroughly with 70% ethanol. All surgical procedures should be performed with sterilized instruments.
4. To open the abdominal cavity, make a transverse incision of 2.5 cm of the abdominal skin and of the abdominal wall with the dissection scissors at a point level with the top of the hind legs.
5. Apply moderate pressure to push the testes from the scrotal sac into the abdomen. With blunt forceps, grasp the left (or right) testicular fat pad and move it upwards into the abdominal cavity. The testis, vas deferens, and epidydimis will come with it and be exposed; they should be placed on a sterilized gauze compress soaked in 0.9% NaCl.
6. Visualize the vas deferens. (The vas deferens has a tube-like structure and is situated below the testis and has a blood vessel running along one side.)
7. The vas deferens can be interrupted by cauterization, which is rapid and convenient. Dissect out a short portion of the vas deferens by separating it from the surrounding tissue using a pair of forceps and blunt dissection. Hold the vas deferens in a loop with one pair of forceps and cauterize the vas deferens in two places with the tip of another heated pair of forceps. The section of vas deferens between the two cauterization points is removed. Separated and cauterization prevents the vas deferens from healing and forming a functional vas deferens.
8. Repeat this procedure on the other testis.
9. Stitch up the abdominal wall first, followed by the skin.
10. Do not mate the animals for 15 days. After this period they should be tested for effective vasectomy by two successive matings which should not give pups.

7.3.2. Superovulation and Pseudopregnant Female Production

To produce zygotes for one session of pronuclear injection, superovulate eight prepubescent female rats (26–30 days old) and mate with fertile males. This technique yields around 50% mated females. Each mated female yields around 30–40 eggs and thus a total of around 150 eggs.

1. Superovulate eight prepubescent female rats (26–30 days old) with an i.p injection of 25 I.U. of PMSG between 12 a.m. and 1 p.m. on day-2. (Light cycle 12 h each light/dark. Dark period starts at 7 p.m., mid point is 1 a.m.)

2. At day 0, between 3 and 4 p.m., inject each female i.p with 30 I.U. of hCG to induce ovulation 10–12 h later.
3. Following the hCG injection, mate females with mature fertile males (70 days old) at a 1:1 ratio in individual cages. (Vaginal plugs are not retained by rats for as long a period as in mice and may be lost before the morning. To allow identification of mated females, we recommend the use of cages with a wire grid floor and placed over trays lined with white absorbent paper, so that expelled vaginal plugs may be identified. We have a cycle of light and night of 12 h: 7 a.m./7 p.m.).
4. To obtain pseudopregnant females, put one adult female CD® IGS rats (>70 days old) with one vasectomized CD® IGS rats male at day 0. We recommend a minimum of ten couples to obtain around five mated females.
5. On the morning of day 1, identify the females with vaginal plugs. Put the females without plugs back into stock. The pseudopregnant females, identified by the presence of copulation plugs, are used later in the afternoon for reimplantation of manipulated embryos.

7.3.3. Zygote Preparation

1. Prepare a 4-well plate containing PBS in two wells and a solution of hyaluronidase at 500 μg/ml in PBS in a third well and place it for at least 1 h prior to embryo collection in an incubator at 37°C with 5% CO_2.
2. Prepare another 4-well plate with 4 wells of M16 embryo culture medium and place it for at least 1 h prior to embryo collection in an incubator at 37°C with 5% CO_2.
3. Prepare a small 35-mm Petri dish with PBS (in which to place the oviducts of mated prepubescent females).
4. Between 10 and 11 a.m., sacrifice the hormone primed and plugged females one by one by cervical dislocation after anesthetizing with isoflurane.
5. Place all sacrificed females on their backs on absorbent paper and clean thoroughly with 70% ethanol.
6. Pinch the skin at the midline and open up the abdominal cavity by a transverse incision.
7. Push up the intestine and take hold of one of the uterine horns with forceps. With small/fine scissors, make a cut between the oviduct and ovary and then through the uterus near the oviduct.
8. Transfer the oviduct (with some adjacent uterine and ovarian tissue) into a 35-mm dish with PBS.
9. Under a stereomicroscope, locate one-cell stage eggs in the upper part of the oviduct, which is swollen at this time (the

ampulla). In a Petri dish containing PBS, tear the oviduct close to where the eggs are located using fine forceps, this releases the clutch of eggs surrounded by cumulus cells.

10. The clutch of eggs is placed in one of the wells with PBS to eliminate cellular debris.
11. The eggs together with cumulus cells are then transferred to the well containing the hyaluronidase and incubated at 37°C for 3–5 min. Eggs should be pipetted up and down to eliminate cumulus cells as soon as possible (hyaluronidase can be toxic for the eggs). We carry out the removal of cumulus cells for all embryos at the same time, which is easier to manage and allows us to control the action of hyaluronidase. Embryos are then placed together in the incubator.
12. Transfer the eggs to the second well containing PBS to wash away the hyaluronidase and then place them in the well containing M16 medium. After that place the dish in the 37°C incubator.

Note: We recommend the time between euthanizing of females to embryos into the incubator does not exceed 30 min.

7.3.4. Microinjecting Zygotes with Plasmid DNA

Before microinjection, culture the embryos for 1–2 h in M16 medium at 37°C and check that the rat oocytes show visible male pronuclei.

1. Prepare a hanging drop microinjection chamber as follows: Place a drop of M16 medium (prewarmed at 37°C) in the center of a clean microscope slide and fix two small plastic coverslips (5–6 mm wide, 2 mm thick, and 2 cm length) spaced by 2 cm with silicone grease. Place a drop of medium on a small glass coverslip (5 mm wide) (*see* Fig. 7.1). Turn over the coverslip on the drop of medium and it is maintained by two small plastics coverslips in the microscope slides. With a Pasteur pipette, place two small drops of mineral oil on the left and right sides of the medium drop to surround the M16 drop, thus preventing evaporation and maintaining osmolarity.

Note: We elect to use M16 for this procedure because the embryos are on the microinjection slide for only a short period of time (between 20 and 30 min).

2. Place the slide under a stereomicroscope and transfer 20–40 zygotes, depending on the experience of the manipulator, into the upper part of the field of vision. Place the slide onto the stage of a microinjection microscope with the stage heated to 37°C.

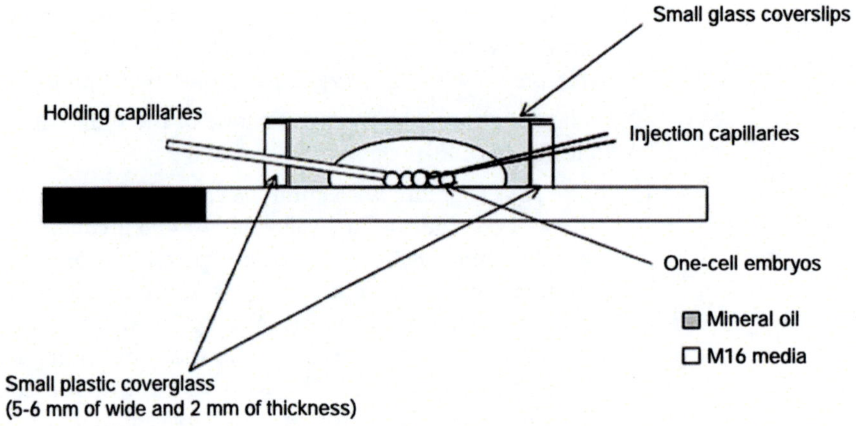

Fig. 7.1 Chamber of microinjection.

3. Place the microcapillary for holding oocytes into the connector piece.
4. Fill the filamented injection capillary with 1–2 μl of plasmid DNA, using Microloader Tips. The loaded micropipette is mounted on a micromanipulator connected to a N_2 gas-operated pressure injector.
5. Adjust the mounting angle of the holding and injection capillary to 35°.
6. Microinject each new batch of plasmid DNA at 1, 2, 4, and 6 μg/ml to 10–20 zygotes for each concentration and determine egg viability by observing development to 2-cell stage embryos after overnight culture in M16. To generate transgenic rats, use the highest concentration that will preserve embryo viability at a reasonable rate (>50%). Some decrease in viability is expected when using concentrations that result in the generation of the highest proportion of transgenic animals.
7. To inject a zygote, move the holding pipette to the collected zygotes and apply a minimum of negative pressure enough to adhere one embryo to the end of the holding pipette.
8. Move one zygote to the center of the slide and position it in the same plane of focus as the opening of the injection pipette. Examine the oocytes under high power of magnification (×400) and focus on the male pronuclei.
9. To inject, align the injection pipette in a horizontal line of the aligned pronuclei.
10. Penetrate the male pronucleus with the injection pipette and inject DNA. The male pronucleus swells when the injection is successful (*see* Fig. 7.2a).

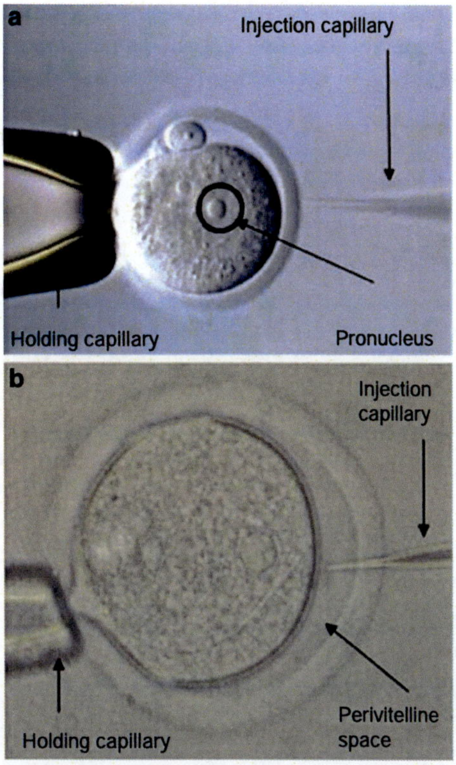

Fig. 7.2 (a) Plasmid DNA injection. Microinjection into the male pronucleus of rat embryos. (b) Lentivirus injection. Microinjection of lentiviral vector into the perivitelline space of rat embryos.

11. The equipment and procedure for microinjection of rat eggs is basically the same as those used for mouse eggs, although injection of rat eggs is more difficult and time consuming. Rat egg pronuclei are difficult to see, they are less regular and uniform than those of mouse eggs. Furthermore, the plasma and pronuclear membranes of rat eggs are more elastic than those of mouse eggs, making them more difficult to penetrate.
12. Place the injected zygotes in the lower part of the field of vision.
13. Start again with the other zygotes present on the slide. Once all the embryos have been injected, take the zygotes from the drop and place them in a well of the 4-well Petri dish containing M16 medium. Place the dish in the incubator at 37°C with 5% CO_2.
14. Take new zygotes for injection.
15. Between 31.5 and 65% of rat eggs should be viable after microinjection and are ready to be reimplanted into pseudopregnant females or cultured for 24 h in M16 medium before transfer.

7.3.5. Delivery of Lentiviral Vectors to One-Cell Embryos

Lentiviruses, as for DNA microinjection, are delivered to fertilized oocytes on the same day of collection. The use of only one-cell embryos will help to minimize mosaicism.

7.3.5.1. Use of Lentiviral Vector Stock

1. Rapidly thaw the frozen (−80°C) vector stock (<10 μl/tube) at 37°C.
2. Centrifuge at low speed (2795g) for 2 min in a bench-top centrifuge to pellet aggregates of cellular debris.
3. In a level 2 laboratory and under a biosafety laminar flow hood, transfer the supernatant to a fresh 1.5 ml microcentrifuge tube. Keep the tube on ice.
4. Vector stock is ready to be used and can be kept up to 5 days at 4°C. Minimize freeze/thaw cycles of virus (<2). A virus will lose approximately 50% of its activity with each freeze/thaw cycle.

7.3.5.2. Microinjection of Lentiviral Vectors into the Perivitelline Space of Single-Cell Embryos

Note: During the time of microinjection, the person who is doing microinjection wears two pairs of gloves and a mask.

1. In a level 2 laboratory and under a biosafety laminar flow hood, load approximately 3 μl of the viral solution into the microinjection pipette, with a microloader tip.
2. Transfer one-cell embryos (in batches of ~30) to a microdrop of embryo culture medium (M16) in the injection chamber and cover with mineral oil to prevent evaporation and maintain osmolarity.
3. Mount the chamber on the stage of an inverted microscope, and monitor the injection procedure under 400× magnification. Subzonal injections of lentivirus are performed at 37°C.
4. Hold fertilized embryos (pronuclei visible) in place against the holding pipette using gentle negative pressure. Mount the micropipette loaded with the virus onto a micromanipulator. Prior to injection, break the tip of the injection pipette to open it. Make the break twice as big as you would for injection of plasmid DNA.
5. Using the micromanipulator to guide the pipette, push the tip through the zona pellucida into the perivitelline space (region between the zona pellucida and the oocyte cell membrane). Using gentle positive pressure, the viral solution flows continuously from the pipette (When the injection pipette is broken, a continuous flow is produced due to the fact that we use an N_2 pressurized microinjection system. If the viral solution stops flowing, apply some more positive pressure). The embryo should be strongly displaced within the zona

pellucida by the pressure of the liquid. At the end of the injection, the embryo should bounce back into place. Leave the micropipette in the perivitelline space for 10–20 s before withdrawing it from the zygote. This allows time for the lentiviral particles to settle on the zona pellucida and thus will not be lost when the pipette is withdrawn (*see* Fig. 7.2b).

6. After the injection, wash the surviving embryos twice in a microdrop of embryo culture medium covered with mineral oil (to prevent evaporation) equilibrated at 37°C, 5% CO_2, and then keep them in a 37°C humidified incubator under 5% CO_2 until implantation. For embryo washing, we use a mouth pipette to which we have added a filter to prevent contamination by viral particles (Anotop 10, pore size 0.02 μm, Whatman). Clean all materials contaminated with lentiviral solution with bleach or detergent solutions.

Note: Lois's group reported on an alternative approach to generating transgenic mice, which consists of co-incubating zona-free embryos with a lentiviral suspension until they reach the morula/ blastocyst stage [13]. However, they found that the development of denuded embryos was delayed and that the rate of implantation was reduced. Unfortunately, attempts to culture denuded rat embryos have failed due to the loss of embryo integrity (blastomere separation) following cell division (unpublished data from our lab and [25]).

7.3.6. Embryo Reimplantation

Bilateral embryo transfer is used to place eggs into the reproductive tracts of pseudopregnant recipients.

1. Anesthetize a pseudopregnant female with isoflurane. Shave a broad area on each side of the rat and clean thoroughly with 70% ethanol. Be sure to perform the surgical procedure with sterile instruments. Place the rat on its right side.
2. Make a 2 cm long incision in the skin, in the middle of the lower abdomen, towards the back of the animal, and the body wall can be penetrated directly over the ovarian fat pad, which is very large in adult rats.
3. Grasp the fat pad with a forceps and gently pull the ovary out until the attached oviduct and uterus are clearly visible. Place all on a gauze compress moistened with a solution of 0.9% NaCl.
4. Correct the positioning of these organs under a stereoscopic microscope. The coiled oviduct within the transparent bursa should be clearly visible.
5. Since the bursa (a transparent membrane over the oviduct and the ovary) is more vascular in the rat than in the mouse, place two or three drops of epinephrin (acting as a

vasoconstrictor) on the top of the ovarian bursa immediately before reimplanting the eggs.

6. Make an embryo transfer pipette from a flame-pulled Pasteur pipette and load it with 10–12 zygotes aspirated in a small quantity of medium and preceded by two or three air bubbles. It is easier to insert the transfer pipette into the infundibulum if about 4 mm of the tip is bent at a 30° angle.
7. The infundibulum and the swollen ampulla are located underneath the bursa. Use two pairs of n° 3 forceps to tear the bursa covering the gap between the ovary and the oviduct.
8. The infundibulum of the rat oviduct is usually oriented horizontally, buried between the oviduct and ovary. Remove any traces of blood or other liquid with a compress.
9. Carefully widen the cleft in which the infundibulum lies with fine n°5 forceps, without causing any bleeding, and insert the reimplantation pipette containing the zygotes into the opening of the infundibulum. Slowly blow the zygotes with medium down the implantation pipette and into the oviduct, until 2–3 air bubbles become visible in the ampulla, indicating a successful transfer. Wait 30 s before removing the transfer pipette (during this time the embryos are deposited in the infundibulum).
10. With blunt forceps, pick up the fat pad and place the uterus, oviduct, and ovary back inside the abdominal cavity.
11. Stitch up the abdominal wall and the skin. Repeat the procedure using the other oviduct. Place the reimplanted rats in a cage and transfer them back to the animal facility. Pups should be born 21 days after transfer. The birth rate is around 30%.

Note: Bilateral reimplantation results in higher rates of pregnancy in transferred females (65–85%) as compared to unilateral transfer (25–50%). The size of litters after bilateral eggs transfer varies but it is usually between five and eight newborns.

7.3.7. Genotyping Analysis of Founders

This section describes the use of linked-mediated PCR to identify the insertion site of the transgene in the host genome. The genotypic screening of resulting offspring begins by first identifying the transgenic animals and second by determining the integrity of the transgene as well as whether the transgene has integrated in one or several sites within the genome. The answer to the first is obtained by PCR and the second one by Southern blot analysis. Transgene copy number can be determined by both Southern blot analysis and quantitative PCR. All of these techniques may be applied to transgenic animals generated both by microinjection of plasmid DNA and by lentiviral vectors. The description of these techniques is beyond the scope of this chapter and has already been documented [8]. The exact integration site of the transgene is determined by linked-mediated PCR [26, 27]. Due to the fact that this technique can only be applied to transgenes integrated as a single copy, it can be used to identify and analyze all single copy transgenic animals generated by use of either lentiviral vectors or plasmid DNA.

7.3.7.1. PCR Analysis

1. Extract genomic DNA from tail biopsies of 10-day-old rats by Proteinase K digestion. Briefly, incubate short tail biopsies (~6 mm) in 500 μl of Proteinase K lysis buffer overnight in a water bath at 56°C.
2. Next, centrifuge the tubes at 16099g for 10 min. Perform a phenol–chloroform DNA extraction on the supernatant for PCR genotyping.

In contrast to DNA pronuclear injection, which often leads to the insertion of multiple transgene copies in only one site in the host genome, transgenic animals generated by lentiviral perivitelline injection may carry several transgene copies in multiple integration sites.

7.3.7.2. Determination of Insertion Sites by Linker-Mediated PCR

This technique is adapted from Siebert et al. (1995).

1. 2.5 μg of transgenic rat genomic DNA is digested by blunt-ended restriction enzyme (EcoRV, PvuII, SspI). Each batch of digested genomic DNA was linked to an adaptor, created by annealing two primers (5′-GTAATACGACTCACTATAG GGCACGCGTGGTCGACGGCCCGGGCTGGT-3′ and 5′-PO4-ACCAGCCC-NH2-3′) to build separate libraries.
2. The protocol consists of two PCR amplifications using long-distance proofreading Taq DNA polymerase (Herculase II Fusion DNA polymerase, Stratagene) with the following two-step cycle parameters: 1′ at 94°C, N cycles of 2″ at 94°C – 3′ at 72°C, N′ cycles of 2″ at 94°C – 3′ at 72°C, 4′ at 67°C. The primary PCR ($N = 7$ and $N' = 32$) is made with the outer adaptor primer (AP1 5′-GTAATACGACT-CACTATAGGGC-3′) and a gene-specific primer (GSP1) placed on the lentiviral transgene (5′ gDNA-lentivirus or 3′ lentivirus-gDNA junction sites).
3. The reactions are diluted (1:50) and a secondary PCR or nested PCR ($N = 5$ and $N' = 20$) is performed with the nested adaptor primer (AP2 5′-ACTATAGGGCACGC-GTGGT-3′) and a nested gene-specific primer (GSP2).
4. The PCR products are separated on a 1.2% agarose gel (Fig. 7.3). The major PCR products can then be directly sequenced or cloned using pCRII TOPO plasmid (TOPO

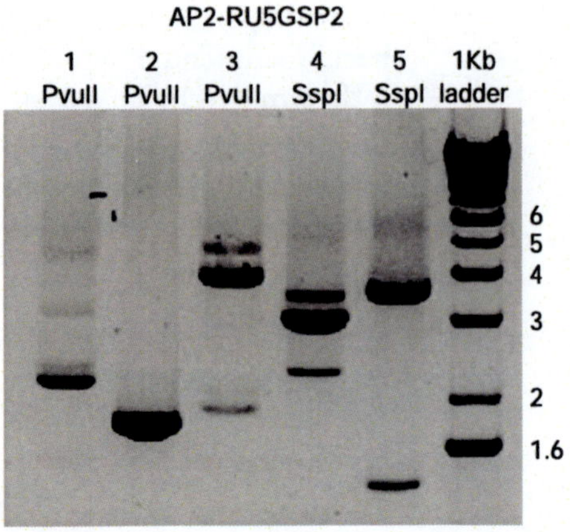

Fig. 7.3 Representative secondary PCR with AP2 primer and lentivirus specific-primer RU5GSP2 using mix and the cycling parameters described in the protocol. *Lane* 1, 2, and 3 are PvuII libraries of three transgenic rats. *Lane* 4 and 5 are SspI libraries of two transgenic rats.

TA cloning, Invitrogen), sequenced using M13 forward and M13 reverse, and further analyzed on BLAST Rat Sequences (www.ncbi.nlm.nih.gov).

7.4 Troubleshooting

7.4.1. Male Vasectomy

- Problem: My recipient females are giving birth to pups fathered by my vasectomized males.
- Solution: Vasectomies have to be performed at least 10–14 days before the first mating (to allow the recovery of males after the surgical procedure). For testing the sterility of males, we breed them two or three times before use.

7.4.2. Superovulation and Pseudopregnant Female Production

- Problem: My embryo yield is very low.
- Solution: The timing of injections is critical for collection of a sufficient number of oocytes with clearly visible pronuclei. Other techniques can be used for superovulating prepubescent females [7]: osmotic micropumps (model 1003D, Alzet, Palo Alto, CA, USA) containing partially purified Follicle-Stimulating Hormone (FSH, Vetrepharm, London, Ontario, Canada) are implanted subcutaneously or intraperitoneally, using aseptic surgical techniques in prepubescent rats between 8:00 and 10:00 a.m. on day 2. These micropumps deliver 1 μl/h, which corresponds to 0.4 IU of LHRH/day. Rats are given an i.p. injection of 1 μg of LHRH on the afternoon

of day 0 and are mated with fertile male SD rats in individual cages.

- Use dedicated mating cages, to check females that loose vaginal plugs after copulation.

7.4.3. Injecting Zygotes with DNA Plasmid

- Problem: I have no Tg animals amongst pups born.
- Solution: When using short DNA fragments (<4 Kb) with repeated sequences the yield of transgenic animals can be very low. Heating the DNA fragment prior to each microinjection to 95°C for 3 min followed by incubation in ice allows a dramatic increase in transgenic yield. (Unpublished data, – no heating: 0 transgenic rats/58 newborns; with heating 3 min at 95°C: 3 transgenics rats/67 newborns).
- The injection buffer used to dilute the plasmid for microinjection should be filtered (0.20 μm) to remove all particles.
- Problem: Few embryos are surviving pronuclear injection.
- Solution: Low survival rates are frequently caused by injection pipettes that are too large, nucleus damage, or high concentration of DNA.

7.4.4. Microinjection of Lentiviral Vectors to One-Cell Embryos

- Problem: I have no Tg animals among pups born from injected embryos.
- Solution: The expected number of transgenic rats obtained by lentiviral perivitelline microinjection depends on two crucial parameters: the lentiviral vector purity and its titer. The description of the production of lentiviral vectors for transgenesis is beyond the scope of this chapter and has been previously described in detail [28] and in Chapter 10. In general, we obtain a good yield of transgenic rats using lentiviral vectors purified by ultrafiltration or affinity chromatography, but not if the vectors were inadequately concentrated. Titers need to be $>10^8$ viral particles/ml.
- Problem: My injection pipette keeps blocking.
- Solution: If the viral solution stops flowing, increase positive pressure. If the flow does not resume then it is likely that the pipette has clogged with some large cellular debris. Apply positive pressure until the debris is expelled from the micropipette. If the debris cannot be moved, discard the pipette in a flask containing bleach and prepare another for use.

7.4.5. Embryo Reimplantation

- Problem: No pups born.
- Solution: There could be several reasons for this: (a) one-cell embryo reimplantation should be performed 1 h–2 h after zygote injection or (b) the time between loading zygotes into

the reimplantation pipette and embryo transfer must be kept short (<5 min).

7.4.6. Genotyping Analysis of Founders

- Problem: I have no Tg animals among animals born from microinjected embryos.
- Solution: Before generating transgenic animals, it is necessary to have a robust PCR with transgene specific primers tested for good efficacy and specificity for the transgene. When setting up the PCR it is important to dilute the plasmid in rat genomic DNA. In the case of transgenes of rat origin, primers should be designated in inter-exonic manner to avoid amplification of endogenous gene.
- For each sample, it is important to have a positive control by performing a supplementary PCR with a small amount of transgene (10^3 copies) or gDNA from a positive rat, mixed in each reaction to avoid false negative results due to inhibitors.
- PCR can be performed, without gDNA purification, directly on a small piece of ear or tail by using DNA release and Phire Taq DNA polymerase (Finnzymes, Espoo, Finland).

7.5 Conclusion

We describe here two techniques to generate transgenic rats by microinjection of one-cell embryos either using short DNA fragments microinjected into the male pronucleus or by microinjection of lentiviral vectors into the perivitelline space. Using these parameters, our laboratory has successfully produced 22 lines of transgenic rats by microinjection of plasmid DNA (0.2–3% of transgenic rats/number of microinjected zygotes) and six lines of transgenic rats by microinjection of lentiviral vectors (6–18% transgenic rat/number of transplanted zygotes) over several years.

References

1. Jacob HJ (2010) The rat: a model used in biomedical research. Methods Mol Biol 597:1–11, Rat Genomics: gene identification, functional genomics and model applications
2. Jaenisch R (1976) Germ line integration and Mendelian transmission of the exogenous Moloney leukemia virus. Proc Natl Acad Sci USA 73:1260–1264
3. Breindl M, Bacheler L, Fan HA, Jaenisch R (1980) Chromatin conformation of integrated Moloney leukemia virus DNA sequences in tissues of BALB/Mo mice and in virus-infected cell lines. J Virol 34:373–382
4. Gordon JW, Scangos GA, Plotkin DJ, Barbosa JA, Ruddle FH (1980) Genetic transformation of mouse embryos by microinjection of purified DNA. Proc Natl Acad Sci USA 77:7380–7384
5. Mullins JJ, Peters J, Ganten D (1990) Fulminant hypertension in transgenic rats harbouring the mouse Ren-2 gene. Nature 344:541–544
6. Hammer RE, Maika SD, Richardson JA, Tang JP, Taurog JD (1990) Spontaneous inflammatory disease in transgenic rats expressing HLA-B27 and human beta2m: an animal

model of HLA-B27 associated human disorders. Cell 63(5):1099–1112
7. Charreau B, Tesson L, Soulillou JP, Pourcel C, Anegon I (1996) Transgenic rats: technical aspects and models. Transgenic Res 5:223–234
8. Tesson L, Cozzi J, Ménoret S, Rémy S, Usal C, Fraichard A, Anegon I (2005) Transgenic modifications of the rat genome. Transgenic Res 14:531–546
9. Popova E, Bader M, Krivokharchenko A (2005) Strain differences in superovulatory response, embryo development and efficiency of transgenic rat production. Transgenic Res 14:729–738
10. Filipiak WE, Saunders TL (2006) Advances in transgenic rat production. Transgenic Res 15 (6):673–686
11. Menoret S, Rémy S, Usal C, Tesson L, Anegon I (2010) Generation of transgenic rats by microinjection of short DNA fragments. Methods Mol Biol 597:81–92, Rat Genomics: gene identification, functional genomics and model applications
12. Naldini L, Blomer U, Gallay P, Ory D, Mulligan R, Gage FH, Verma IM, Trono D (1996) In vivo gene delivery and stable transduction of nondividing cells by a lentiviral vector. Science 272:263–267
13. Lois C, Hong EJ, Pease S, Brown EJ, Baltimore D (2002) Germline transmission and tissue-specific expression of transgenes delivered by lentiviral vectors. Science 295:868–872
14. Pfeifer A, Ikawa M, Dayn Y, Verma IM (2002) Transgenesis by lentiviral vectors: lack of gene silencing in mammalian embryonic stem cells and preimplantation embryos. Proc Natl Acad Sci USA 99:2140–2145
15. McGrew MJ, Sherman A, Ellard FM, Lillico SG, Gilhooley HJ, Kingsman AJ, Mitrophanous KA, Sang H (2004) Efficient production of germline transgenic chickens using lentiviral vectors. EMBO Rep 5:728–733
16. Hofmann A, Kessler B, Ewerling S, Weppert M, Vogg B, Ludwig H, Stojkovic M, Boelhauve M, Brem G, Wolf E, Pfeifer A (2003) Efficient transgenesis in farm animals by lentiviral vectors. EMBO Rep 4:1054–1060
17. Whitelaw CB, Radcliffe PA et al (2004) Efficient generation of transgenic pigs using equine infectious anaemia virus (EIAV) derived vector. FEBS Lett 571(1–3):233–236
18. Hofmann A, Zakhartchenko V et al (2004) Generation of transgenic cattle by lentiviral gene transfer into oocytes. Biol Reprod 71 (2):405–409
19. Chan AW, Chong KY et al (2001) Transgenic monkeys produced by retroviral gene transfer into mature oocytes. Science 291(5502): 309–312
20. van den Brandt J, Wang D, Kwon SH, Heinkelein M, Reichardt HM (2004) Lentivirally generated eGFP-transgenic rats allow efficient cell tracking in vivo. Genesis 39:94–99
21. Michalkiewicz M, Michalkiewicz T, Geurts AM, Roman RJ, Slocum GR, Singer O, Weihrauch D, Greene AS, Kaldunski M, Verma IM, Jacob HJ, Cowley AW Jr (2007) Efficient transgenic rat production by a lentiviral vector. Am J Physiol Heart Circ Physiol 293:H881–H894
22. Remy S, Tesson L, Usal C, Menoret S, Bonnamain V, Nerriere-Daguin V, Rossignol J, Boyer C, Nguyen TH, Naveilhan P, Lescaudron L, Anegon I (2010) New lines of GFP transgenic rats relevant for regenerative medicine and gene therapy. Transgenic Res 19(5):745–763
23. Pfeifer A (2006) Lentiviral transgenesis – a versatile tool for basic research and gene therapy. Curr Gene Ther 6:535–542
24. Park F (2007) Lentiviral vectors: are they the future of animal transgenesis? Physiol Genomics 31:159–173
25. Dann CT (2007) New technology for an old favorite: lentiviral transgenesis and RNAi in rats. Transgenic Res 16:571–580
26. Bryda EC, Bauer BA (2010) A restriction enzyme-PCR-based technique to determine transgene insertion sites. Methods Mol Biol 597:287–299, Rat Genomics: gene identification, functional genomics and model applications
27. Siebert PD, Chenchik A, Kellogg DE, Lukyanov KA, Lukyanov SA (1995) An improved PCR method for walking in uncloned genomic DNA. Nucleic Acids Res 23(6):1087–1088
28. Rémy S, NGuyen T, Ménoret S, Tesson L, Usal C, Anegon I (2010) The use of lentiviral vectors to obtain transgenic rats. Methods Mol Biol 597:109–125, Rat Genomics: gene identification, functional genomics and model applications

Chapter 8

Generation of Transgenic Animals by Use of YACs

Almudena Fernández, Diego Muñoz, and Lluis Montoliu

Abstract

The use of genomic-type DNA constructs ensures optimal transgene expression, once inserted into the host genome, because their large size includes most if not all the regulatory elements that are needed for correct gene expression. Large heterologous DNA molecules can be easily manipulated in bacterial or yeast cells, through the use of bacterial artificial chromosomes (BACs) or yeast artificial chromosomes (YACs), respectively. YACs are the vectors that allow the manipulation of larger DNA molecules, in excess of 1 Mb (1,000 kb). Some mammalian loci (i.e., the APP locus, ~450 kb) greatly exceed the maximum size for inserts that can be accommodated by BACs. Therefore YACs are currently the only available robust and reliable solution for working with these large genes. In this chapter, we will describe several procedures directed towards the preparation of YAC DNA of a quality suitable for microinjection into mouse embryos, for the successful generation of transgenic mice.

Abbreviations

AHM	Acid hydrolyzed casein
BAC	Bacterial artificial chromosome
ddH_2O	Double distilled water
DNA	Deoxyribonucleic acid
EDTA	Ethylenediaminetetraacetic acid
ES	Embryonic stem cell
ICSI	Intracytoplasmic sperm injection
LMP	Low-melting point
Mb	Megabase (= 1,000 kilobases)
PCR	Polymerase chain reaction
PFGE	Pulsed-field gel electrophoresis
RFLP	Restriction fragment length polymorphism
SDS	Sodium dodecyl sulfate
SNP	Single nucleotide polymorphism
TAE	Buffer solution containing a mixture of Tris base, acetic acid and EDTA
TBE	Buffer solution containing a mixture of Tris base, boric acid and EDTA
YAC	Yeast artificial chromosome

S. Pease and T.L. Saunders (eds.), *Advanced Protocols for Animal Transgenesis*, Springer Protocols,
DOI 10.1007/978-3-642-20792-1_8, © Springer-Verlag Berlin Heidelberg 2011

8.1 Introduction

The use of very large genomic pieces of DNA, in the form of YACs or BACs, has become increasingly popular for the generation of transgenic animals, because their size usually includes all regulatory elements needed for specific expression domains to function correctly in an ectopic genomic location [1]. YACs exist mostly as linear molecules, although circular versions can be prepared, and carry specific auxotrophic gene markers for selection purposes in haploid yeast cells. The size of heterologous DNA cloned into YACs may vary from a few kilobases (i.e., 35 kb [2]) to more than 2 Mb (i.e., 2.4 Mb [3]). YACs are maintained as the 17th chromosome within 16 chromosome yeast cells, due to the functional elements included in their vector arms, such as a centromere, autonomous replicating sequences and telomeres. All these elements do not function within a mammalian cell and, hence, YACs need to be integrated into the host genome to be expressed adequately. However, the expression of genes within YACs does not appear to be influenced by vector sequences [4–6] and normally reach an expression level similar to that of their corresponding endogenous loci [1]. In particular, transgene expression obtained from a YAC-based approach would normally be better than plasmid-based strategies [7]. Transgenic animals carrying YAC transgenes with optimal expression levels have been obtained in several mammalian species, including mice ([4, 5]; reviewed in [1, 8]), rats [9–11], rabbits [12] and pigs [13]. Preliminary attempts have also been reported in goats [14]. Transgenic mice generated with YAC constructs have been instrumental for the production of new animal models of Alzheimer disease, overexpressing the large human APP locus [15, 16], or for producing human antibodies in mice, through the manipulation of large human chromosomal fragments harboring the immunoglobulin loci [17–19].

YACs can sometimes be of a similar size as any of the endogenous 16 yeast chromosomes, thereby co-migrating or closely migrating in pulsed-field gel electrophoresis (PFGE) and making impossible the isolation of the YAC DNA molecule without co-isolation of a given yeast chromosome. Although some studies have shown that the co-integration of contaminating endogenous yeast chromosomes does not seem to impair the expression of YAC transgenes (i.e., transfer of YAC DNA to mouse embryonic stem (ES) cells by spheroplast cell fusion: [17], [40], [20]), it is always preferable to microinject YAC DNA samples free of contaminating yeast chromosomes [1]. The problem of YACs co-migrating with other yeast chromosomes can be easily solved by the use of yeast window strains, which carry defined alterations in their karyotype and result in areas of the PFGE free of endogenous yeast chromosomes [41]. Purified YAC DNA samples can also be transfected into mammalian cells (i.e., mouse ES cells) by the use of lipofection procedures [21–23].

The addition of polyamines (spermidine and spermine) to the microinjection buffer, in the presence of high salt (100 mM NaCl) appears to be essential for keeping the DNA intact, for the prevention from shearing, and hence, for the successful generation of both YAC and BAC transgenic mice, carrying intact constructs [24–26]. The combined presence of polyamines and high salt leads to compaction of YAC DNA molecules, which are stabilized and protected, in the form of globular structures, which can be directly observed by electron microscopy [25]. In the presence of high-salt buffers, polyamines and DNA molecules form intramolecular aggregates which help to keep these large constructs intact. Without polyamines, complex intermolecular aggregates may be formed. This may result in the breakage of DNA molecules when microinjection of these suboptimal transgene samples takes place [25].

The efficiency of transgenic mouse production using YAC DNA may be lower (1–5% of newborn pups) compared to plasmid-based DNA fragments (5–10%). However, compared to plasmid-based DNA, the variability between individual YAC constructs and between YAC DNA preparations for the same construct is increased. These variations are mostly attributable to the quality of the YAC DNA preparation, which may retain a higher level of co-purified contaminants. This results in poor embryo survival after microinjection and hence, poor transgenic results, including less pups born and a lower percentage of these pups being transgenic [1].

Higher efficiencies can be achieved by intracytoplasmic sperm injection (ICSI)-mediated YAC DNA transgenesis [27, 28], where 10–35% YAC transgenic mice have been reported among newborns. The ICSI adapted procedure for delivering YAC DNA in association with sperm heads into unfertilized oocytes has been described in great detail [29]. However, ICSI-mediated procedures may result in a higher proportion of fragmented YACs integrated into the host genome, compared to standard microinjection methods. In contrast, nearly all YAC transgenic mice generated by ICSI-mediated microinjection appear to be non-mosaic and transmit the transgene to the progeny at Mendelian ratios. By comparison, a majority of transgenic mouse founders generated through standard pronuclear microinjection methods, including YAC DNA microinjection, are mosaics [5, 30, 31].

Due to the large size of the YAC DNA molecules and the molarity of the YAC DNA samples, microinjection will result in the introduction of a small number of DNA molecules into the pronucleus of fertilized mouse eggs. Using the appropriate conversion factors and Avogadro's number, about 400 DNA molecules would be microinjected for a standard small DNA construct, 5 kb in size, at a DNA concentration of 1 ng/μl, with a delivered volume of 2 pl into the mouse oocyte pronucleus. Equivalent calculations for a solution of YAC DNA, 500 kb in size, at the same concentration and with the same delivered volume, will result in only about four molecules being microinjected. Surprisingly, although the overall transgenesis efficiency is lower with YACs, compared to standard smaller DNA molecules, clearly it is not 100 times lower, as we might expect, given the calculation above. It is tempting to speculate that the presence of repeated sequences (yeast telomeres) at either end of the YAC DNA molecule might render them more "attractive" to the endogenous integration machinery.

Usually, YAC transgenic mice carry one or very few copies integrated (1–5) [1]. Rarely, they carry more than five copies integrated into one single locus. The integrity of the YAC DNA transgene should be addressed and confirmed using a variety of methods, including systematic genotyping across the entire YAC DNA molecule (with several PCR reactions) and/or the use of known genetic polymorphisms (RFLPs, SNPs). Standard Southern-blot [4] and fingerprint analyses [17] should also be carried out, to verify the integrity of the YAC transgene. Eventually, the co-segregation of several independent markers across the YAC DNA molecule among the F1 progeny will be an additional proof of the linked co-integration of the YAC transgene.

YAC DNA solutions can be maintained at 4°C and should never be frozen, because the YACs will shear into smaller DNA molecules. YAC DNA solutions should be freshly prepared as close to the microinjection session as possible, and microinjected within the following 2 months. In order to avoid breaking these large DNA molecules, pipetting of YAC DNA solutions should be always carried out with utmost care, proceeding slowly and using cut-off tips, to enlarge the opening of the tip and thus avoid breaking these large DNA molecules. Vortex or intense agitation should never be applied to YAC DNA solutions. Wear gloves all the time to avoid contamination of the YAC DNA samples with nucleases, which would quickly digest and degrade the YAC DNA sample.

YAC DNA solutions cannot be concentrated by standard DNA precipitation methods and resuspended into smaller volumes, as would be the case for smaller DNA plasmid-type molecules. Attempting to precipitate YAC DNA solutions will result, unavoidably, in the destruction of the YAC DNA molecules. Therefore, in order to keep the molecules intact, YAC DNA must be obtained and stored, always, in agarose plugs, or in solution, preserved with high salt and polyamines.

During microinjection of YAC DNA samples into the pronucleus of fertilized mouse oocytes, the inner diameter of the microinjection needle tip should be slightly larger (i.e., tip can be carefully broken by touching the holding needle) than for plasmid-based DNA fragments. The injection pressure should be kept as low and, most importantly, as constant as possible, to avoid shearing of these large DNA molecules upon exposure to extreme pressure changes during the procedure. This is best achieved by the use of automated microinjection devices (i.e., Femtojet, Eppendorf).

Purification of YAC DNA begins with the large-scale preparation of yeast agarose plugs carrying the YAC DNA molecules, followed by a PFGE step and a subsequent concentration step, which can be achieved through a second standard agarose electrophoresis gel or by the use of ultrafiltration units. The essential protocol has been outlined before [32–35] and should produce enough YAC DNA of a quality suitable for microinjection purposes. Next, we will describe and comment upon the main procedure itself and alternative methods for preparation of YAC DNA of a quality suitable for the generation of transgenic mice.

8.2 Equipment

Agarose block formers (plug molds, GE Healthcare Life Sciences).

Medium-size centrifuge (for 50-ml plastic tubes).

Centrifuge for 1.5-ml Eppendorf tubes.

Flask, 1-l.

Hemocytometer.

Water bath, 37–40°C.

Eppendorf tube, 1.5 ml.

Falcon tubes, 50 ml.

Pipette tips, cut-off, yellow tips (200 μl) and blue tips (1,000 μl).

PFGE comb with preparative slot in center.

Millipore dialysis filter (Millipore VMWP02500, pore size 0.05 μm).

Millipore flat forceps (for handling dialysis filters).

Millipore ultrafiltration unit (Millipore Ultrafree MC 30,000 NMWL UFC3 TTK 00).

Petri dishes, 10 cm diameter.

Pulsed-field gel electrophoresis (PFGE) apparatus (i.e., Gene Navigator [GE Healthcare Life Sciences] or CHEF DR [Bio-Rad]).

Standard horizontal gel electrophoresis apparatus (for DNA).

Plastic ruler.

Scalpel (sterile, single use).

UV light transilluminator.

Water bath or hot block, 40–65°C.

Nanodrop spectrophotometer (Thermo Scientific) or fluorometer.

8.3 Reagents and Solutions

Overnight yeast culture carrying the YAC of interest

Selection medium for YACs (i.e., AHC or drop-out medium, depending on the yeast strain)

ddH_2O sterile, MilliQ grade (embryo-tested)

EDTA 500 mM (pH 8.0) and also 50 mM (pH 8.0)

NaCl 5 M

Tris–HCl 1 M (pH 7.5)

Agarose SeaPlaque LMP GTG (Lonza)

Agarose NuSieve LMP GTG (Lonza)

Gelase (Epicentre) or Agarase (New England Biologicals)

Lambda (λ) DNA multimers (New England Biologicals or Roche)

PFGE yeast chromosome markers (Roche)

Solution I
~1 M Sorbitol (Merck, autoclaved)

20 mM EDTA (pH 8.0) (autoclaved)

14 mM β-mercaptoethanol (Merck)

2 mg/ml Zymolyase-20 T (MP Biomedicals catalog number 320921)

Sterile water

Prepare fresh, do not store

Solution II
~1 M Sorbitol (Merck, autoclaved)

20 mM EDTA (pH 8.0) (autoclaved)

2% SeaPlaque GTG LMP agarose (Lonza catalog number 50111)

14 mM β-mercaptoethanol (Merck)

Sterile water

Prepare fresh. Melt the Sorbitol plus EDTA and agarose in the microwave

Avoid boiling. Equilibrate in a water bath at 37–40°C

Add the β-mercaptoethanol and keep the solution at 37–40°C until use

Solution III
~1 M Sorbitol (Merck, autoclaved)

20 mM EDTA (pH 8.0) (autoclaved)

10 mM Tris–HCl (pH 7.5) (autoclaved)

14 mM β-mercaptoethanol (Merck)

2 mg/ml Zymolyase-20 T (MP Biomedicals catalog number 320921)

Sterile water

Prepare fresh, do not store

Solution IV
1% lithium dodecyl sulfate (Sigma L4632)

100 mM EDTA (pH 8.0) (autoclaved)

10 mM Tris–HCl (pH 8.0) (autoclaved)

Sterile water

Filter-sterilize (0.22 μm). Store at room temperature

100% NDS buffer
500 mM EDTA

10 mM Tris base

34 mM N-laurylsarcosine

Mix 350 ml of water with 93 g of EDTA and 0.6 g of Tris base. Equilibrate to pH >8.0 with solid NaOH pellets. Add 5 g of N-laurylsarcosine (Sigma) pre-dissolved in 50 ml of water. Equilibrate to pH 9.0 with 10 M NaOH and bring the final volume to 500 ml with water. Filter-sterilize (0.22 μm). Store at 4°C. The final concentration of NDS buffer is 20%.

YAC Equilibration buffer
1× TAE buffer

100 mM NaCl

0.030 mM Spermine (Sigma, tetrachloride, S1141)

0.070 mM Spermidine (Sigma, trihydrochloride, S2501)

Ethidium bromide (EtBr) staining solution
Add 50 μl of 10 mg/ml EtBr stock solution per 1 l of 0.5× TBE PFGE running buffer

TBE buffer (used at 0.5×; stock solution is 10×)
10× stock solution: 890 mM Tris base, 890 mM Sodium Borate, 20 mM EDTA (pH 8.0)

TAE buffer (used at 0.25×, 0.5× and 1.0×; stock solution is prepared as 50×)
50× stock solution: 2,000 mM Tris acetate, 50 mM EDTA (pH 8.0)

1,000× Polyamine mix
30 mM Spermine (Sigma, tetrahydrochloride, S-1141)

70 mM Spermidine (Sigma, trihydrochloride, S-2501)

Working concentrations for 1× polyamines are 30 μM spermine and 70 μM spermidine, also known and called as 100 μM polyamines mix. Both reagents are dissolved together in sterile H_2O, filter sterilized (0.22 μm), stored in aliquots at −20°C and can be used with confidence for long periods of time (>1 year). The two polyamines are highly hygroscopic. Therefore, we recommend you to avoid weighing these compounds and, instead, order a defined amount (i.e., 1 g), do the required calculations and prepare the whole volume at once and combine in one single tube.

YAC Microinjection buffer
10 mM Tris–HCl (pH 7.5)

0.1 mM EDTA (pH 8.0)

100 mM NaCl

1× Polyamine mix

Aliquot, filter-sterilize (0.22 μm), and store the mixture (without polyamines) at 4°C for several months. The ready-to-use YAC microinjection buffer (polyamines added) should be prepared fresh for each experiment and should not be stored.

8.4 Preparation of YAC DNA for Microinjection Purposes

8.4.1. Large-Scale Preparation of Yeast Agarose Plugs for Isolation of YAC DNA

1. Inoculate 200 ml of appropriate selection medium for YACs with 1 ml of the corresponding yeast overnight culture in 1-l flask. Let the culture grow at 30°C with vigorous shaking (250 rpm) until saturation (1–2 days).
2. Count the number of yeast cells with a hemocytometer. A saturated yeast cell culture should contain between 0.75 and 1×10^8 yeast cells/ml.

3. Spin down the yeast cells in a 50-ml tube at 600 g for 5 min at room temperature. Discard the medium.
4. Resuspend and wash the cells in 50 mM EDTA (pH 8.0). Use 40 ml of this solution per 100 ml of original medium. Spin down the cells as in **step 3**. Repeat this washing step one more time with 20 ml of 50 mM EDTA (pH 8.0). Discard the medium. Additional washing steps can be included at will, in order to obtain cleaner yeast cell preparations. Failure to remove debris adequately will also impact in the quality of the final YAC DNA preparation. After each centrifugation step all remaining liquid should be removed. Last drops should be carefully aspirated. The procedure can be halted at this step and the yeast cells kept in 50 mM EDTA (pH 8.0) solution at 4°C until they can be processed. However, a drop in YAC DNA quality may occur if the unprocessed yeast cells are maintained for longer periods (>1 week) in this solution.
5. Weigh the yeast cell pellet (assuming a density of 1 g/ml and subtracting the weight of the plastic tube). The yeast cell pellet will weigh between 2.0 and 2.5 g.
6. Warm the yeast cell pellet at 37–40°C for 30 s in a water bath. Immediately add enough pre-warmed *Solution I* to give a final concentration of, approximately, 8×10^9 yeast cells/ml. Resuspend the cells by careful swirling. The volume of liquid added should be kept as small as possible, with a maximum being equal or similar to the volume of cells. For optimal results, add half of the volume of the yeast cell pellet (1–1.25 ml). However, irrespective of the volume used, the yeast cells should be adequately dispersed into this limited amount of liquid to achieve a homogeneous solution before proceeding to the next step.
7. Immediately add an equal volume of pre-warmed (37–40°C) *Solution II* and keep the tube in a water bath. Mix quickly, but gently, and pipette (use cut-off yellow tips) 80-μl aliquots into agarose block formers (plug mold, Fig. 8.1a) previously

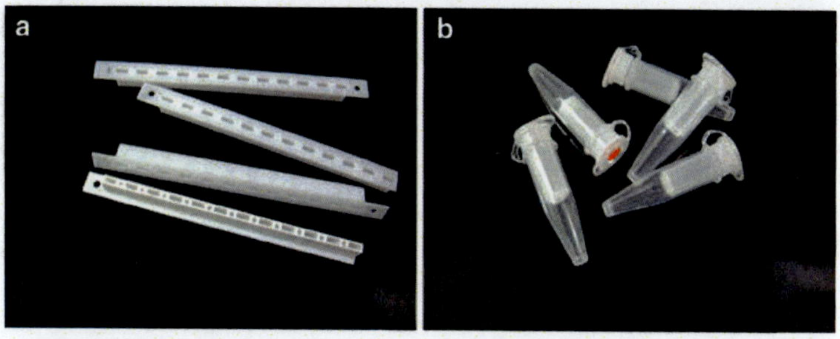

Fig. 8.1 **(a)** Agarose block formers (plug molds) [Gene Navigator System, GE Heathcare Life Sciences). Picture by Lluis Montoliu. **(b)** Ultrafiltration units (Millipore). Picture by Lluis Montoliu.

bottom-sealed with tape and placed on ice. Proceed as quickly as possible and avoid trapping air bubbles in the plugs. Gently shake the tube with cells and agarose every 20 s in the water bath (37–40°C) to prevent too early solidification of the agarose, before pipetting is finished. This will make a final concentration of 4×10^9 yeast cells/ml of agarose plug.

8. Chill on ice for 10 min until the agarose plugs solidify.
9. Transfer the agarose plugs into *Solution III* for spheroplasting, using 8 ml of solution per ml of plug (according to the total volume obtained in **step** 7). Incubate at 37°C for 2–3 h with gentle agitation, in a water bath.
10. Decant *Solution III* and replace it with an identical volume of *Solution IV* (8 ml of solution per ml of plug). Continue the incubation at 37°C with gentle agitation for at least 1 h. Replace the buffer with fresh Solution IV and continue incubating overnight (>12 h, fine up to 24 h) at 37°C with gentle agitation.
11. On the next day, decant the buffer and wash the agarose plugs with *20% NDS buffer* using the same volume as before (8 ml of buffer per ml of plug). Proceed for 2 h with gentle agitation at room temperature. Repeat this washing step two times. Agarose plugs can be loaded directly onto PFGE gels or stored indefinitely in this buffer at 4°C. Alternatively, agarose plugs can also be stored in 50 mM EDTA (pH 8) solution for longer periods of time.

8.4.2. Purification of YAC DNA with Two Gel Electrophoresis Steps

1. Cast a 1% agarose SeaPlaque GTG (Lonza) PFGE gel using 0.5× TBE as a buffer and a comb with a preparative slot in the center (~150 ml for the Gene Navigator casting gel set-up). The central preparative slot should be 4–5 cm wide, thus enabling loading 8–10 agarose plugs (Subheading 8.4.1).
2. Load the agarose plugs vertically and consecutively into the preparative slot (8–10 blocks, depending on the size of the slot), which should be centrally located (Fig. 8.2a–c). Include marker lanes on both sides with a very small slice (1/4 or 1/8) of the same batch of agarose plugs, and additional marker lanes with λ DNA multimers and/or known yeast chromosome PFGE markers.
3. Cover/seal all slots with 1% SeaPlaque GTG (FMC) LMP, let it solidify, and start the gel with appropriate running conditions to ensure optimal resolution in the desired chromosomal size range. The running buffer must be cooled to 10°C with the help of an external cooling device, to avoid aberrant DNA separation.

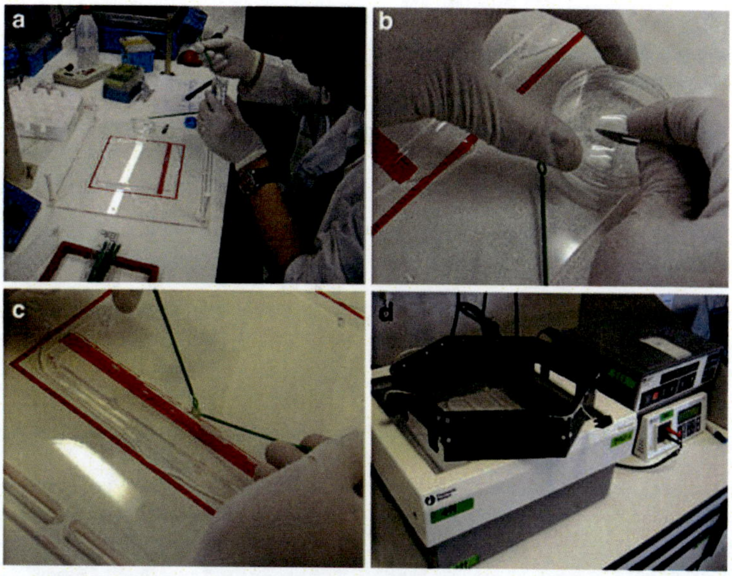

Fig. 8.2 (**a**) Retrieval of YAC DNA-containing agarose plugs from equilibrating buffer. The polymerized PFGE gel, wet with running buffer, on the table, is ready to be loaded. (**b**) Trimming the agarose plugs to prepare 1/4 and 1/8 slices for marker lanes, on a Petri dish, with a sterile scalpel. (**c**) Loading YAC DNA-containing agarose plugs into the PFGE gel using single-use sterile plastic inoculating loops. (**d**) Gene Navigator PFGE system (GE Healthcare Life Sciences). All pictures by Lluis Montoliu.

4. After the gel has run, cut-off marker lanes plus a small part of the preparative lane on either side and stain these two external parts with *EtBr staining solution* during 30 min with gentle shaking (Fig. 8.3a). The central part of the gel containing most of the preparative slot remains unstained in cold running buffer.
5. Using a UV transilluminator, mark the locations of the desired YAC chromosomal bands and two additional bands (usually the endogenous yeast chromosomes located above and below the YAC of interest) on the gel slices by cutting a nick with a scalpel.
6. Re-assemble the EtBr-stained and marked parts of the PFGE gel next to the preparative central lane and carefully cut out the YAC-containing agarose slice and the two additional slices containing endogenous yeast chromosomes using the marked nicks as a reference and a ruler as a guide. Aim to produce agarose slices no thicker than 5–6 mm. Then, remove the YAC DNA-containing and the endogenous yeast chromosome-containing slices and transfer them to different tubes with 1× TAE buffer. The rest of the PFGE gel can be now stained with *EtBr staining solution* to confirm that the desired YAC DNA and neighboring endogenous

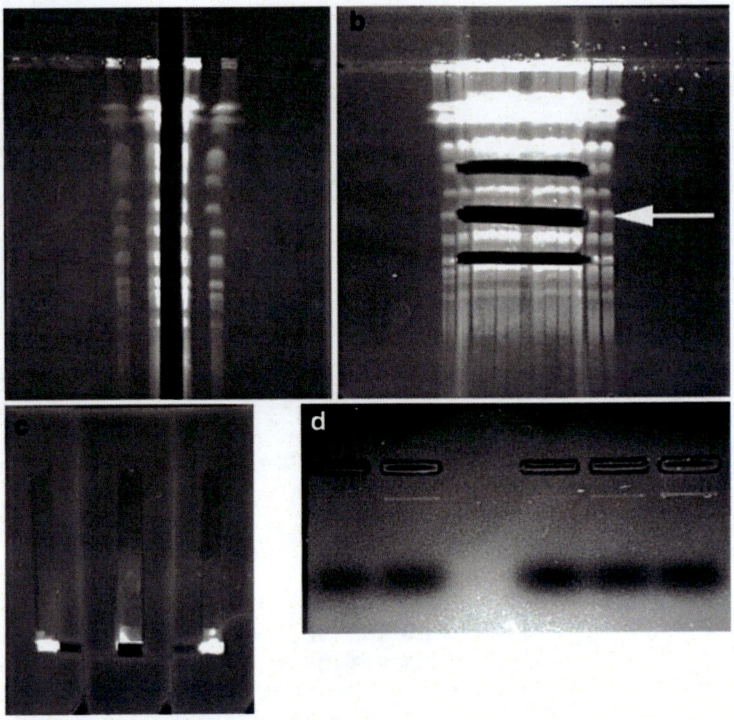

Fig. 8.3 **(a)** Side lanes of the first PFGE gel stained with EtBr solution. The central part, containing the preparatory slot and the desired YAC DNA, has been removed. **(b)** PFGE re-assembled, after staining also the central part and after having removed the agarose slices containing the YAC (indicated by an *arrow*) and neighboring endogenous chromosomal bands. **(c)** Second standard gel electrophoresis, stained with EtBr solution, both the central and side lanes. Please note a cube of agarose missing in the central part, corresponding to the location of the concentrated YAC DNA, after the second gel run. **(d)** Assessing the concentration of the YAC DNA stock solution by comparison to EtBr-stained YAC DNA bands of known concentration. *Left lanes*, 1 and 2 μl of a new YAC DNA stock solution. *Right lanes*, 5, 10 and 20 ng of a control YAC DNA, previously obtained and quantified, as a reference.

chromosomal bands have been adequately cut out (Fig. 8.3b).

7. Equilibrate the three gel slices, including the YAC DNA slice, in 1× TAE buffer three times for 30 min each. At this step the agarose slices can be kept at 4°C and the protocol can be resumed within the next few days (but within 1 week).

8. For the second electrophoresis gel step, carefully position the gel slices on a minigel DNA electrophoresis tray (at 90° angle, in relation to the previous PFGE gel run, Fig. 8.4). Place the YAC-containing agarose slice in the middle surrounded by the two marker slices. The three agarose slices should be placed and aligned at the same level. Remove all drops of buffer with a tissue paper.

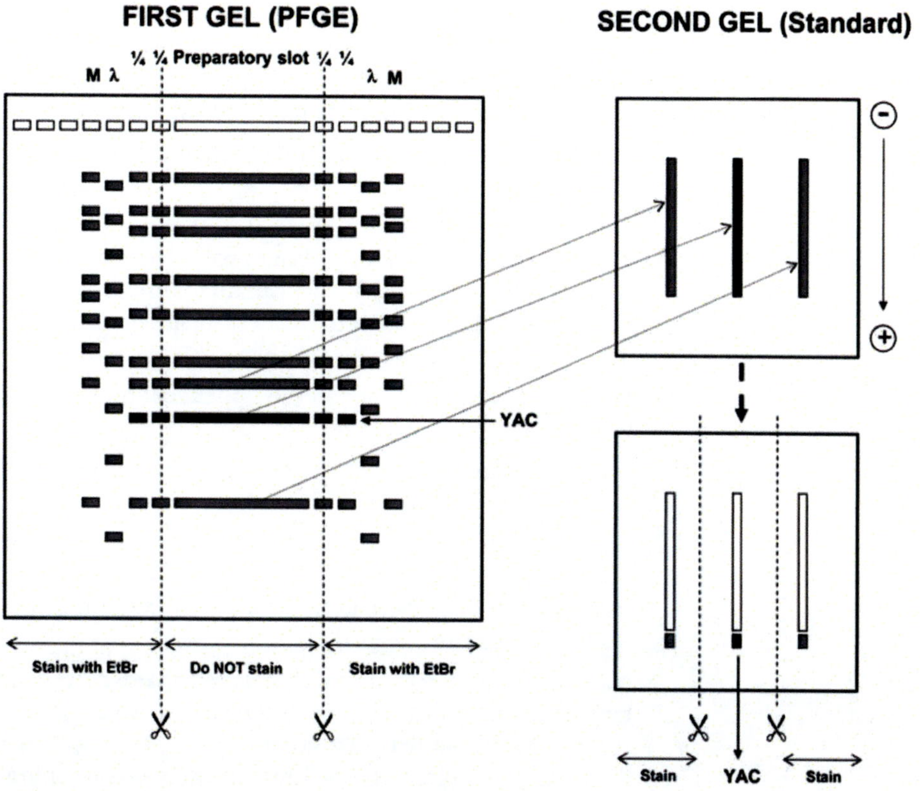

Fig. 8.4 Scheme illustrating the purification and concentration of YAC DNA using the two gels approach. First (*left*) the PFGE gel, then (*right*) the standard horizontal electrophoresis gel, shown before (*above*) and after the run (*below*), with the concentrated YAC DNA shown as a *black square*, indicated by an *arrow*.

9. Embed the gel slices in a 4% NuSieve GTG LMP (FMC) agarose gel in 1× TAE (approx. 50 ml) and run the gel for 6–9 h at 60–90 V.
10. After the gel run, cut-off one side of the gel and stain with *EtBr staining solution* for 30 min, as described in **step 4**. If the run is complete you should see all DNA from the long agarose slice being concentrated at the border and within the first 3–4 mm of the 4% NuSieve agarose gel. However, if DNA is still visible within the preparative PFGE slice, then the electrophoresis must proceed for a longer time. Then, you should use the other side for staining and visualizing.
11. Using the nicks of the marker lanes as a reference and a ruler remove the corresponding part of the central YAC-containing lane. Carefully transfer this gel slice (it should be a cube-shaped agarose portion) into a 15 ml plastic tube with 1× TAE buffer and keep it at 4°C. The rest of the gel can be stained now in EtBr staining solution to verify that all DNA has been cut out (Fig. 8.3c).

12. Wash and equilibrate the YAC DNA-containing gel slice, in fresh 1× TAE buffer three times for 30 min each. At this step the agarose slice can be kept at 4°C and the protocol can be resumed within the next few days (but within 1 week).
13. Equilibrate the agarose gel cube containing YAC DNA in excess of freshly prepared *YAC equilibration buffer* (minimum 15 ml) for at least 2 h.
14. Transfer the YAC DNA gel slice onto a sterile surface (i.e., a sterile plastic Petri dish), and carefully remove all drops of equilibration buffer with the help of tissue paper. Be careful not to dry out the agarose gel cube.
15. Weigh the gel slice using a sterile Eppendorf tube (previously tared to zero).
16. Melt the agarose cube by placing the tube in a hot block or water bath for 10 min at 65°C.
17. Spin the tube in a microcentrifuge at top speed (10,600–20,800 g) for 5 s and place it back in a water bath set at 40°C for temperature equilibration (5–10 min).
18. Add 4–8 units of gelase/agarase per 100 mg of agarose gel cube to the tube containing the melted agarose. It is important to pre-warm the aliquot of the enzyme solution, kept at −20°C, to be added, first to room temperature and then to 40°C (i.e., place the yellow cut-off tip containing the gelase solution inside an open empty Eppendorf tube floating in a 40°C water bath for 15–30 s) to avoid immediate re-polymerization of the agarose gel upon contact.
19. Place the tube back in the 40°C water bath and let the gelase solution settle. After 5 min, gently pipette up and down two or three times with a cut-off blue-tipped pipette to start carefully mixing the gelase solution. Proceed with the digestion for 2–3 h at 40°C, mixing the sample gently every hour (pipetting up and down carefully and slowly with a cut-off blue tip).
20. Chill the tube on ice for 5–10 min and check for the completeness of the agarose-gel digest.
21. Centrifuge the digest at maximum speed for 15–20 min.
22. Prepare a Petri dish with 40 ml of *YAC microinjection buffer*. Carefully place (i.e., use Millipore flat forceps) a Millipore dialysis filter floating on the buffer surface with the glossy side up.
23. Carefully spot the digested agarose with YAC DNA liquid solution from Subheading 8.4.2, **step 21** (<200 μl) onto the center of the Millipore dialysis filter. Allow the dialysis to proceed quietly, without any shaking or movement, for 2–3 h.

24. Carefully, pipette off the solution (using a cut-off yellow tip), and transfer the YAC DNA solution into a sterile Eppendorf tube. This forms the stock YAC DNA solution. Recoveries between 50 and 70% of the original volume spotted onto the dialysis filter are normal. Store the YAC DNA stock solution at 4°C, do not freeze, and do not centrifuge more than very brief spins (1–2 min, full speed).
25. Measure the concentration of the YAC DNA stock solution using a spectrophotometer (i.e., a NanoDrop), or better, a fluorometer, and then confirm the obtained concentration on a standard horizontal agarose gel stained with EtBr, where previously obtained YAC DNA preparations can be added as a reference. Do not allow the samples to enter more than 2–3 cm into the gel matrix (Fig. 8.3d). The entire procedure should yield YAC DNA preparations at concentrations ranging between 5 and 20 ng/μl.
26. The integrity of the YAC DNA molecules can be checked by loading an aliquot of the preparation into a PFGE or, when available, by analyzing the presence of intact globular DNA molecules at the level of electron microscopy, as described [25].
27. YAC DNA solutions should be microinjected at around 1–2 ng/μl. Use *YAC microinjection buffer* to dilute out the original YAC DNA stock solution. Additional dilution (in ½ steps) might be required if embryo toxicity is observed (>50% embryos failing to divide and develop to 2-cell eggs after microinjection). YAC DNA samples can be efficiently microinjected down to 0.5 ng/μl of DNA. Lower YAC DNA concentrations may result in zero transgenic animal births.

8.4.3. Purification of YAC DNA with Ultrafiltration Units

This is a method alternative to the two gel steps procedure, described in Subheading 8.4.2.

1. Follow Subheading 8.4.2 exactly as described up to Subheading 8.4.2, **step 6**.
2. Then, jump to Subheading 8.4.2, **step 12**, thereby avoiding the steps describing the second gel run (Subheading 8.4.2, **steps 7–11**), and proceed normally until Subheading 8.4.2, **step 21**. In this case, the agarose slice containing the YAC DNA will be larger, coming directly from the PFGE gel run, and hence, we recommend cutting it down to 4–5 pieces and distribution into separate Eppendorf tubes, for better melting and digestion processes.
3. After checking for completeness of the agarose digestion and finishing the centrifugation step (Subheading 8.4.2, **step 21**) you can combine all YAC DNA liquid solutions into one single tube for further processing. Reserve an aliquot for DNA

quantification purposes and for calculating the enrichment factor after the concentration achieved by the ultrafiltration units.

4. Transfer up to 400 μl of the digested agarose solution containing YAC DNA into each upper reservoir of a Millipore ultrafiltration unit (Fig. 8.1b), and centrifuge for 2 min at 3,800 g. Check the amount of liquid that has gone through the membrane (lower reservoir). This liquid should not contain YAC DNA, since that will remain in the upper reservoir of the unit.
5. Continue with additional centrifugation steps (in rounds of 2 min, at 6,000 rpm) until about 320 μl (out of the initial 400 μl) have passed through the membrane (there should be about 80 μl left in the upper reservoir).
6. Incubate the tubes at 4°C for a few hours (i.e., overnight). Resuspend the YAC DNA (possibly attached to the surface of the membrane) by pipetting up and down with a cut-off yellow tip (maximum 2–3 times) very carefully and slowly.
7. Collect all concentrated volumes and combine the resulting volume of concentrated YAC DNA solution into one single tube.
8. Proceed with the dialysis step on floating Millipore filters as described in Subheading 8.4.2, **steps 22–27**, including the measurement of YAC DNA concentration (compare it with the value obtained in Subheading 8.4.3, **step 3**) and the assessment of YAC DNA integrity, as indicated before. A concentration factor of 5–7 times is expected.

8.5 Discussion

We have described a method for obtaining agarose plugs carrying YAC DNA (Subheading 8.4.1) and two methods for isolating the YAC DNA from the agarose plugs, using a two consecutive gel steps (Subheading 8.4.2) or ultrafiltration units (Subheading 8.4.3).

Regarding the yield of the procedures, you should expect to produce 75 to 100 × 80-μl agarose plugs from 200 ml of saturated yeast cell culture (Subheading 8.4.1). This should be plenty of plugs for subsequent YAC DNA preparation and for any associated tests.

Subsequently, each YAC DNA preparation experiment will require about 8–10 agarose plugs and will produce (Subheading 8.4.2) 150–200 μl of a 5–20 ng/μl YAC DNA solution. Therefore, the expected total yield of the entire process can vary between 1 and 4 μg of YAC DNA/200 ml of saturated yeast culture. Using ultrafiltration units (Subheading 8.4.3), this alternative procedure can yield YAC DNA preparations with higher concentrations (10–60 ng/μl). However, getting the YAC DNA back into solution can be more cumbersome if it gets attached to the membrane of the ultrafiltration units, and overly aggressive attempts for resuspension may lead to shearing of the YAC DNA sample. On the positive side, the ultrafiltration method can be easily applied to the concentration of BAC DNA solutions from PFGE agarose slices, after separation from vector sequences.

Alternative methods have been developed to increase the yield of YAC DNA during the amplification and purification steps [2, 36]. Targeted genetic modification of YAC DNA vector sequences by homologous recombination in yeast cells may allow the accumulation of several copies of the YAC DNA molecules per yeast cell (normally there should be only one copy of YAC DNA per yeast cell), thereby resulting in a potential higher yield. This introduces the opportunity to apply higher dilution factors, resulting in cleaner YAC DNA preparations and therefore eventually translating into higher efficiencies of transgenesis (i.e., up to 10% of transgenic pups born, [36]). However, the presence of several copies of YAC DNA molecules per yeast cell will also impact on the ease by which they can be further modified by homologous recombination (i.e., [37]).

8.5.1. Notes to Subheading 8.4.1: Preparation of Agarose Plugs Carrying YAC DNA

The right amount of yeast cell culture must be used, as recommended in Subheading 8.4.1, **step 1** and Subheading 8.4.1, **step 2**. Overloading the process with larger quantities of yeast culture, will increase the risk of producing a YAC DNA preparation with larger quantities of co-purified contaminants that will be difficult to remove and will impact in the survival ratio of the microinjected mouse embryos.

In the spheroplasting step (Subheading 8.4.1, **step 9**), the enzymes will degrade the yeast cell wall and convert the yeast cells into spheres, hence the name "spheroplasting". This is one of the most important steps determining the quality and yield of the DNA. An inefficient spheroplasting step will result in very poor yield (not enough yeast cells will be exposed to the cell lysis steps to follow). Correspondingly, overdigesting the sample or using enzymatic batches of lower quality, might result in DNA degradation and/or bad electrophoretic mobilities. Using an enzyme of the highest quality is strongly recommended. Zymolyase (MP Biologicals) enzymatic preparations can be very expensive, but it is worth investing the money and purchasing the best enzymes for this fundamental step in YAC DNA preparation.

8.5.2. Notes to Subheading 8.4.2: Isolation of YAC DNA Using Two Gels

There are two common methods devised for extracting YAC DNA from the PFGE agarose plugs, using two consecutive gels or ultrafiltration units. The first method, described in Subheading 8.4.2, involves the use of two gel electrophoresis steps [4, 33–35, 38, 39]. This is the most reliable and robust of the two methods for routinely obtaining YAC DNA of the highest quality and concentration. The first gel electrophoresis step is the PFGE. Thereafter, the cut slice of agarose containing the YAC DNA of interest is run on a standard electrophoresis at a 90° angle to the PFGE run. The YAC DNA is forced to move out of the agarose slice and to penetrate into a thicker low-melting agarose gel, thereby promoting concentration of YAC DNA molecules into a much smaller volume. The purpose of this method is to convert a "slice" of agarose into a "cube" of agarose, of a smaller volume. The volume of this cube will determine the final concentration of the YAC DNA precipitation.

Various companies make PFGE apparatus that are suitable for YAC DNA preparation. We have always used and recommend the Gene Navigator system (Fig. 8.2d, GE Healthcare Life Sciences, originally made by LKB) because it is a robust, simple and reliable apparatus. Alternatively, the various CHEF DR systems (Bio-Rad) are more sophisticated and modern but also suitable for PFGE purposes.

For better resolution and PFGE running conditions, we recommend equilibration of the YAC DNA-containing agarose plugs (from Subheading 8.4.1) with PFGE running buffer (TBE 0.5×) or TE (pH 8.0) before loading the gel. Equilibration is performed with at least four consecutive washes of 30 min each in excess of buffer. Unused and pre-equilibrated agarose plugs can be kept at 4°C for a short period of time (<1 week) if they are destined to be reloaded into another PFGE gel. Otherwise, agarose plugs should be returned to 20% NDS buffer solution or, better still, 50 mM EDTA (pH 8) solution, for long-term storage at 4°C.

Both TBE and TAE buffers are adequate and can be used for PFGE at 0.5× and 0.25×–0.50× dilutions, respectively, thereby providing the sufficient low ionic strength required to avoid overheating of the buffer during the run, while maintaining an increased DNA mobility. We have used both over the years and obtained good results with both. However, bear in mind that PFGE running conditions for TBE 0.5× will be different from TAE 0.5× buffer, since they do not share the same ionic strength.

When loading the agarose plugs into the PFGE gel (Subheading 8.4.2, **step 2**, Fig. 8.2c) make sure the preparative slot is centrally located and not wider than 4–5 cm, to avoid smiling of the chromosomal bands, which will later interfere with the exact determination of DNA band position and might result in decreased yield and lower DNA concentrations. The PFGE gel must be kept wet, with excess of PFGE running buffer during the entire loading procedure, to avoid drying out the agarose gel matrix (Fig. 8.2a). Agarose plugs are best retrieved from solution and loaded into the preparative slot of the PFGE handling with single-use sterile plastic inoculating loops (Fig. 8.2a–c). The agarose plugs should occupy the entire height of the gel, therefore it may be necessary to trim off (with a sterile single-use scalpel) some of the agarose plug to prevent DNA from coming off the plug and running over the gel, instead of through the gel, resulting in aberrant electrophoretic images (Fig. 8.2b). When pouring and casting the PFGE gel and when running the PFGE gel inside the PFGE chamber, make sure all are absolutely level (you can use a spirit level to confirm exact positioning), in order to avoid any loss of DNA during the gel run. All agarose used for PFGE purposes must be of the highest quality (GTG) and have a low-melting point (LMP). Excellent results are obtained with SeaPlaque (Lonza) agarose.

There are many PFGE running conditions, according to the equipment used and the YAC chromosomal band size to be resolved. A wide range of chromosomal sizes (50–2,000 kb) can be recovered at the Gene Navigator system by using 0.5× TAE buffer at 10°C, running at 180 V with a pulse of 30 s for 12 h followed by a pulse of 60 s for 15 h. Other suitable PFGE programs for the Gene Navigator system are as follows: (a) for optimally separating 250 kb YACs you can use 0.5× TBE at 250 V and 10°C with 6 h at 9 s pulses followed by 12 h at 15 s pulses, or (b) to purify larger YACs, in the range of 500–600 kb, you can choose between 0.5× TBE at 200 V during 24 h at a single pulse of 60 s and a two-pulses program of 12 h pulsing at 50 s followed by 14 h pulsing at 100 s. Avoid using too much running buffer within the PFGE tank (not more than 2 mm over the surface of the gel) to improve the quality of the chromosomal separation.

The second electrophoresis gel (Subheading 8.4.2, **step 9**) run can be left overnight at 40–60 V. However, it is essential to use a pump to assist the circulation of buffer. To prevent overheating, this second gel can be run inside the cold room. The agarose for the second gel must also be of the highest quality (GTG), LMP and should allow preparation of high (4%) percentage gels. Excellent results are obtained with NuSieve (Lonza) agarose.

Both sides (with marker lanes) of the second gel should be stained with EtBr to precisely identify the position of the central YAC DNA band that will have run to a similar position (Subheading 8.4.2, **step 10**). There are no differences in mobility in standard linear electrophoresis between chromosomal DNA bands of different size. Therefore, with a razor-blade it is possible to mark the position of the DNA in the marker lanes, which then can be used to localize the YAC DNA band. It is very important to precisely locate the position of YAC DNA. It is always better to lose some YAC DNA by cutting closer to the concentrated core, rather than try to cover a greater area (and hence, more volume) since this will result in a more diluted YAC DNA preparation. Bear in mind that the actual volume of this agarose cube will approximate to the final volume of the liquid YAC DNA sample you will have at the end of the procedure. Usually the concentration factor achieved with this second gel run is X4–X6.

The digestion of the YAC DNA containing agarose slice with gelase is an important step, crucial for the success of the entire procedure (Subheading 8.4.2, **step 18**). If the enzyme is added directly from −20°C freezer it will promote the re-polymerization of the agarose and the digestion will not progress. We suggest it may be an improvement to add half of the calculated amount of units of the enzyme at the beginning of the digestion step and, after 1 h and 30 min, add the remaining half of the calculated enzyme volume, thereby facilitating the digestion of the agarose sample. Carefully pipette up and down with cut-off blue tips to allow adequate mixing of the enzyme and the melted agarose. Do not vortex the preparation nor keep it outside of the water bath for more than 5–10 s at each step, in order to avoid the re-polymerization of the agarose gel sample. The Gelase or Agarase buffers commercially provided with the enzymes are not recommended for digestion of YAC DNA containing agarose slices. Instead, straight 1× TAE buffer should be used.

Checking for complete agarose gel digest (Subheading 8.4.2, **step 20**). This is again a very important step. The appearance of a pale brown or white opaque cloud within the tube clearly indicates that the digestion has not been complete. Sometimes you do not see white opaque tube contents, but you cannot pipette out the entire solution. This is because there is a plug at the end of the tube made of re-polymerized remaining undigested agarose. In this case, go back to Subheading 8.4.2, **step 18**, add more Gelase and perform a second incubation with additional enzyme.

You should aim to obtain a YAC DNA preparation of the highest possible concentration. The higher the dilution factor applied to the original YAC DNA stock solution, the better the quality of the YAC DNA microinjection sample will be, because any trace of co-purified contaminants and toxic reagents that could impair the mouse embryo development will be titrated out more effectively. For this reason, observation of embryo survival tests after microinjection are strongly recommended (Subheading 8.4.2, **step 27**).

8.5.3. Notes to Subheading 8.4.3: Isolation of YAC DNA Using Ultrafiltration Units

The use of ultrafiltration units is an alternative purification protocol for concentrating YAC DNA preparations without the use of a second gel (Subheading 8.4.2), described earlier. The agarose slice containing the YAC DNA of interest is melted and digested, directly as it comes from the PFGE run. The resulting YAC DNA-containing liquid solution is then concentrated by subsequent centrifugation steps, using ultrafiltration units with a membrane pore cut-off of 30,000 MW (Fig. 8.1b). Water and small molecules can pass through the filter, whereas DNA will be retained in the upper container, in a smaller volume, therefore concentration factors of up to 6–7 times can be rapidly achieved. Ideally, YAC DNA preparations of higher concentration should be obtained (10–60 ng/μl). However, great care should be taken to try to recover the YAC DNA molecules that can sometimes be irreversibly attached to the filter membrane, thereby yielding YAC DNA preparations of a much lower concentration. This procedure can also be applied to BAC DNA preparations, processing excised BAC DNA-containing agarose slices from a PFGE exactly as described for YACs.

References

1. Giraldo P, Montoliu L (2001) Size matters: use of YACs, BACs and PACs in transgenic animals. Transgenic Res 10:83–103
2. Schedl A, Beermann F, Thies E, Montoliu L, Kelsey G, Schütz G (1992) Transgenic mice generated by pronuclear injection of a yeast artificial chromosome. Nucleic Acids Res 20:3073–3077
3. Den Dunnen JT, Grootscholten PM, Dauwerse JG, Walker AP, Monaco AP, Butler R, Anand R, Coffey AJ, Bentley DR, Steensma HY et al (1992) Reconstruction of the 2.4 Mb human DMD-gene by homologous YAC recombination. Hum Mol Genet 1(1):19–28
4. Schedl A, Larin Z, Montoliu L, Thies E, Kelsey G, Lehrach H, Schütz G (1993) A method for the generation of YAC transgenic mice by pronuclear microinjection. Nucleic Acids Res 21:4783–4787
5. Montoliu L, Umland T, Schütz G (1996) A locus control region at −12 kb of the tyrosinase gene. EMBO J 15:6026–6034
6. Peterson KR (2003) Transgenic mice carrying yeast artificial chromosomes. Expert Rev Mol Med 5:1–25
7. Forget BG (1993) YAC transgenes: bigger is probably better. Proc Natl Acad Sci USA 90:7909–7911
8. Giraldo P, Montoliu L (2002) Artificial chromosome transgenesis in pigmentary research. Pigment Cell Res 15:258–264
9. Fujiwara Y, Miwa M, Takahashi R, Hirabayashi M, Suzuki T, Ueda M (1997) Position-independent and high-level expression of human alpha-lactalbumin in the milk of transgenic rats carrying a 210-kb YAC DNA. Mol Reprod Dev 47:157–163
10. Takahashi R, Ito K, Fujiwara Y, Kodaira K, Kodaira K, Hirabayashi M, Ueda M (2000) Generation of transgenic rats with YACs and BACs: preparation procedures and integrity of microinjected DNA. Exp Anim 49:229–233
11. Takahashi R, Ueda M (2010) Generation of transgenic rats using YAC and BAC DNA constructs. Methods Mol Biol 597:93–108
12. Brem G, Besenfelder U, Aigner B, Müller M, Liebl I, Schütz G, Montoliu L (1996) YAC transgenesis in farm animals: rescue of albinism in rabbits. Mol Reprod Dev 44:56–62
13. Langford GA, Cozzi E, Yannoutsos N, Lancaster R, Elsome K, Chen P, White DJ (1996) Production of pigs transgenic for human regulators of complement activation using YAC technology. Transplant Proc 28:862–863
14. Zhang XF, Wu GX, Chen JQ, Zhang AM, Liu SG, Jiao BH, Cheng GX (2005) Transfer of an expression YAC into goat fetal fibroblasts by cell fusion for mammary gland bioreactor. Biochem Biophys Res Commun 333:58–63
15. Lamb BT, Sisodia SS, Lawler AM, Slunt HH, Kitt CA, Kearns WG, Pearson PL, Price DL, Gearhart JD (1993) Introduction and expression of the 400 kilobase amyloid precursor protein gene in transgenic mice. Nat Genet 5:22–30
16. Pearson BE, Choi TK (1993) Expression of the human beta-amyloid precursor protein gene from a yeast artificial chromosome in transgenic mice. Proc Natl Acad Sci USA 90:10578–10582
17. Jakobovits A, Moore AL, Green LL, Vergara GJ, Maynard-Currie CE, Austin HA, Klapholz S (1993) Germ-line transmission

and expression of a human-derived yeast artificial chromosome. Nature 362:255–258
18. Zou X, Xian J, Davies NP, Popov AV, Brüggemann M (1996) Dominant expression of a 1.3 Mb human Ig kappa locus replacing mouse light chain production. FASEB J 10:1227–1232
19. Popov AV, Zou X, Xian J, Nicholson IC, Brüggemann M (1999) A human immunoglobulin lambda locus is similarly well expressed in mice and humans. J Exp Med 189:1611–1620
20. Davies NP, Huxley C (1996) YAC transfer into mammalian cells by cell fusion. Methods Mol Biol 54:281–292
21. Strauss WM, Jaenisch R (1992) Molecular complementation of a collagen mutation in mammalian cells using yeast artificial chromosomes. EMBO J 11:417–422
22. Strauss WM, Dausman J, Beard C, Johnson C, Lawrence JB, Jaenisch R (1993) Germ line transmission of a yeast artificial chromosome spanning the murine alpha 1(I) collagen locus. Science 259:1904–1907
23. Strauss WM (1996) Transfection of mammalian cells via lipofection. Methods Mol Biol 54:307–327
24. Schedl A, Montoliu L, Kelsey G, Schütz G (1993) A yeast artificial chromosome covering the tyrosinase gene confers copy number-dependent expression in transgenic mice. Nature 362:258–261
25. Montoliu L, Bock CT, Schütz G, Zentgraf H (1995) Visualization of large DNA molecules by electron microscopy with polyamines: application to the analysis of yeast endogenous and artificial chromosomes. J Mol Biol 246:486–492
26. Van Keuren ML, Gavrilina GB, Filipiak WE, Zeidler MG, Saunders TL (2009) Generating transgenic mice from bacterial artificial chromosomes: transgenesis efficiency, integration and expression outcomes. Transgenic Res 18:769–785
27. Moreira PN, Giraldo P, Cozar P, Pozueta J, Jiménez A, Montoliu L, Gutiérrez-Adán A (2004) Efficient generation of transgenic mice with intact yeast artificial chromosomes by intracytoplasmic sperm injection. Biol Reprod 71:1943–1947
28. Moreira PN, Pérez-Crespo M, Ramírez MA, Pozueta J, Montoliu L, Gutiérrez-Adán A (2007) Effect of transgene concentration, flanking matrix attachment regions, and RecA-coating on the efficiency of mouse transgenesis mediated by intracytoplasmic sperm injection. Biol Reprod 76:336–343
29. Moreira PN, Pozueta J, Giraldo P, Gutiérrez-Adán A, Montoliu L (2006) Generation of yeast artificial chromosome transgenic mice by intracytoplasmic sperm injection. Methods Mol Biol 349:151–161
30. Whitelaw CB, Springbett AJ, Webster J, Clark J (1993) The majority of G0 transgenic mice are derived from mosaic embryos. Transgenic Res 2:29–32
31. Moreira PN, Pozueta J, Pérez-Crespo M, Valdivieso F, Gutiérrez-Adán A, Montoliu L (2007) Improving the generation of genomic-type transgenic mice by ICSI. Transgenic Res 16:163–168
32. Huxley C, Hagino Y, Schlessinger D, Olson MV (1991) The human HPRT gene on a yeast artificial chromosome is functional when transferred to mouse cells by cell fusion. Genomics 9:742–750
33. Schedl A, Grimes B, Montoliu L (1996) YAC transfer by microinjection. Methods Mol Biol 54:293–306
34. Umland T, Montoliu L, Schütz G (1997) The use of yeast artificial chromosome for transgenesis. In: Houdebine LM (ed) Transgenic animals. Generation and use. Harwood Academic Publishers, Amsterdam, pp 289–298
35. Hiemisch H, Umland T, Montoliu L, Schütz G (1998) The generation of transgenic mice with yeast artificial chromosomes. In: Cid-Arregui A, Garcia-Carranca A (eds) Microinjection and transgenesis, strategies and protocols. Springer, Berlin, pp 297–308
36. Shen S, Harmar A, Hastie N (2006) Modification and amplification of yeast artificial chromosomes. Methods Mol Biol 349:67–74
37. Giraldo P, Giménez E, Montoliu L (1999) The use of yeast artificial chromosomes in transgenic animals: expression studies of the tyrosinase gene in transgenic mice. Genet Anal 15:175–178
38. MacKenzie A (2006) Production of yeast artificial chromosome transgenic mice by pronuclear injection of one-cell embryos. Methods Mol Biol 349:139–150
39. Peterson KR (2007) Preparation of intact yeast artificial chromosome DNA for transgenesis of mice. Nat Protoc 2: 3009–3015
40. Jakobovits A, Lamb BT, Peterson KR (2000) Production of transgenic mice with yeast artificial chromosomes. Methods Mol Biol 136: 435–453
41. Hamer L, Johnston M, Green ED (1995) Isolation of yeast artificial chromosomes free of endogenous yeast chromosomes: construction of alternate hosts with defined karyotypic alterations. Proc Natl Acad Sci USA 92: 11706–11710

Chapter 9

BAC Transgenes, DNA Purification, and Transgenic Mouse Production

Michael G. Zeidler, Margaret L. Van Keuren, and Thomas L. Saunders

Abstract

Transgenic mouse models open new avenues of research to understand gene function, to mark cells, and to model human genetic diseases. The use of large DNA transgenes provides more information to cells and tissues in the mouse so that gene expression occurs at physiological levels in the appropriate cell types while recapitulating normal developmental time frames. Genomic libraries prepared in bacterial artificial chromosomes (BACs) for the mouse and human genome sequencing projects are a ready source of large DNA transgenes. The use of recombineering technology to modify BAC clones for (1) expression of proteins resulting from point mutations, (2) marking of specific cell populations with fluorescent protein reporters, or (3) cell-specific expression of exogenous proteins that can be used to control inducible gene expression and enhance the utility of BAC transgene-based experimental models. Large BAC transgenes and small plasmid transgenes are used to produce transgenic founder mice by the microinjection of purified DNA into the pronuclei of fertilized mouse eggs in both cases. Despite this similarity, the application of small DNA injection methods to large DNA transgenes leads to the integration of DNA fragments instead of intact BAC transgenes. Important technical differences between small and large DNA transgenesis include differences in DNA purification, microinjection buffers, and genotyping strategies to detect genomic integration. Careful attention to detail results in rewarding mouse models that can be used to identify disease-causing mutations in spontaneous mouse mutants, to purify rare cells that express fluorescent markers, and to explore genetic elements that control gene expression.

Abbreviations

BAC	Bacterial Artificial Chromosome
bp	Base pair
C.H.O.R.I	Children's Hospital Oakland Research Institute
NCBI	National Center for Biotechnology Information
NEB	New England Biolabs
PA	Polyamine
PFGE	Pulsed Field Gel Electrophoresis
pg	Picogram
SSAHA	Sequence Search and Alignment by Hashing Algorithm
UCSC	University of California Santa Cruz

S. Pease and T.L. Saunders (eds.), *Advanced Protocols for Animal Transgenesis*, Springer Protocols,
DOI 10.1007/978-3-642-20792-1_9,

9.1 Introduction

The advent of transgenic technology provided scientists with a powerful tool to investigate gene function in a small mammal whose physiology resembles that of humans in many ways. Advances in genomic library construction facilitated the progress of genome sequencing projects. These two technologies are combined in the production of transgenic mice with large genomic fragments contained in bacterial artificial chromosomes (BACs). Cloned genomes in BAC libraries provide immediate access to genes and the regulatory elements required for tissue- and cell-specific gene expression. Unlike small plasmid transgenes, BAC transgenes often show endogenous gene expression levels and developmental patterns of expression. When BAC transgenes contain all necessary regulatory sequences the result is integration-site-independent, copy-number-dependent, gene expression. BAC transgenes can be modified by recombineering to drive fluorescent protein reporters, Cre recombinase enzyme, or point mutations according to regulatory elements in the BAC. Numerous biological questions such as modeling human diseases in mice, accelerating discovery in gene mapping projects, and the generation of "humanized mice" to improve understanding of human gene regulation have been studied with BAC transgenic mice.

Procedures to identify BACs of interest in the public genome sequence and download of the corresponding DNA sequences are discussed. Methods for the acquisition of BAC clones, the preparation of BAC clone DNA, and verification of BAC clone identity are provided. Purification of BAC clone DNA for pronuclear microinjection and the identification of transgenic founders are considered.

9.1.1. Outline of the Procedure

- Identify BAC clone for the gene of interest and obtain BAC DNA sequence.
- Acquire the BAC clone from a resource such as C.H.O.R.I.
- Purify and quantitate BAC DNA for restriction mapping.
- Verify identity of BAC DNA by PFGE restriction enzyme mapping.
- Prepare BAC DNA for pronuclear microinjection.
- Identification of transgenic founder animals.

9.1.2. Principles and Applications

The value of BAC and other large transgenes is that they include all of the genetic information necessary for reproduction of endogenous expression patterns for genes contained within them [1]. The average size of genomic inserts in human and mouse BAC libraries is less that 200 kb [2, 3]. BAC clones used to generate transgenic mice contain distant regulatory elements required for expression simply because they include more DNA sequence than phage or cosmid library clones [4, 5]. Numerous studies have demonstrated that BAC transgenes can control cell specific and temporal regulation of genes [4, 6–8]. On those occasions when the gene of interest is so large that it cannot be contained within a single BAC, alternative strategies include the use of YAC transgenes from libraries with average sizes of 820 kb [9] to produce transgenic mice with 520 kb transgenes [10] or linking BAC clones to produce a larger 413 kb genomic clone for transgenesis [32].

The use of pronuclear microinjection to produce mouse models is direct and rapid. When the goal is to study the interaction of genes carried by the BAC with existing mutant genes or other genetic modifications in mouse strains, progress can be accelerated by preparing fertilized eggs for microinjection from mice with the mutant or modified gene [11, 12]. The earliest uses of BACs in transgenic mice were complementation studies to map genes in spontaneous mouse mutants. BAC transgenic mice successfully rescued circadian rhythm and deafness mutations and accelerated the discovery of the genes responsible for the phenotypes [11, 12]. Methods to modify BACs by recombineering techniques were subsequently established [13–15]. Prior to the use of BAC transgenes, cell-specific transgene expression in the central nervous system was problematic [16]. The use of BAC transgenes engineered to express green fluorescent protein is the foundation of the Gene Expression Nervous System Atlas [17] that is designed to document gene expression patterns in the central nervous system. In the arena of functional genomics, BACs are the basis of gene targeting vectors for the ablation of thousands of genes in the Knock Out Mouse Project [18, 19]. A powerful application of BAC transgenesis is the combination of knockout mice with human BAC transgenes that permit the study and function of human gene expression in animal models that lack the endogenous gene. These are "humanized" transgenic mice in which human gene expression replaces and compensates for the lack of mouse gene expression. [20–22]. BAC transgenesis amplifies the power of transgenic technology for the identification of disease genes and generation of animal models of disease for biomedical research. Key parameters for successful transgenic production are BAC DNA purity and integrity, optimal microinjection DNA concentration, and robust detection of transgene integration.

9.2 Materials

9.2.1. Equipment

- Computer with internet connection
- DNASTAR Lasergene or other DNA sequence analysis software
- Refrigerator, Fisher (4°C)
- Freezer, Fisher (−80°C)
- Autoclave
- Balance, Fisher
- Bacteria incubator, Fisher (set to 37°C)
- Cell culture incubator (humidified, 37°C, 5% CO_2)
- Nanodrop spectrophotometer, Thermo Scientific
- Shaking incubator (set to 37°C)
- Refrigerated floor centrifuge (4°C)
- Refrigerated tabletop centrifuge (4°C)
- CHEF-DR III PFGE system, BioRad (cat. no. 170-3700)
- Micromanipulation workstation for pronuclear microinjection (*see* Chap. 6)
- Materials to transfer microinjected eggs to recipients (*see* Chap. 6)

9.2.2. Reagents

- Anotop 10 Syringe Filters, sterile 0.02 μm pore size (Whatman cat. no. 6809-1102)
- BAC purification kit (Nucleobond BAC 100, Macherey-Nagel, Clontech Cat. No. 740 579)
- BAC buffer kit (optional: Buffers S1, S2, S3 Clontech cat. nos. 740516.1, 740517.1, 740518.1)
- Bromphenol blue (Sigma, cat. no. B3269)
- Chloramphenicol (Roche, cat. no. 10 634 433 001. 2,000× stock solution: 30 mg/ml dissolved in EtOH. Working solution: 15 μg/ml dissolved in ethanol)
- Embryo tested water (Sigma cat. no. W1503)
- EDTA, 0.5 M solution (Sigma cat. no. E7889)
- Ethanol, anhydrous (Fisher cat. no. AC61510-0010)
- Ethidium bromide (Sigma, cat. no. E7637)
- Glycerol (Roche, cat. no. 03 117 499 001)
- Isopropanol (Fisher cat. no. AC14932-0010)
- NaCl, 5 M solution (Sigma cat. no. S5150)
- M16 mouse embryo medium (Sigma cat. no. M7292)

- Mineral oil, embryo tested (Sigma cat. no. M5310)
- Sterile H_2O (Invitrogen, cat. no. 15230-001)
- Spermine (Sigma, tetrahydrochloride, cat. no. S-1141)
- Spermidine (Sigma, trihydrocholoride, cat. no. S-2501)
- Sucrose (Sigma, cat. no. S0389)
- Tris–HCl, pH 7.5, 1 M solution (Sigma cat. no. T2319)
- 10× TBE (Invitrogen, cat. no. 1558-044)
- MidRangePFG Marker II (NEB, cat. no. N3552S)
- LB broth base (Invitrogen, cat. no. 12780-052)
- Pipet tips (Fisher, cat. nos.)
- 50 ml polypropylene conical tubes (Fisher, cat. no. 14-959-49A)
- 14 ml polypropylene round bottom tube (Fisher, cat. no. 14-959-10B)
- 1.5 ml microcentrifuge tube (Fisher, cat. no. 02-681-8)
- Whatman filter paper (Fisher, cat. no. 09-790-14G)
- Cryogenic vials (Dot Scientific, cat. no. T311-2)
- Thin Wall Glass Capillaries (World Precision Instruments cat. no. TW100F-4)

9.2.3. Buffers and Other Solutions

- 0.5× TBE: add 100 ml 10× TBE to 1,900 ml H_2O.
- 10% glycerol: mix 10 ml glycerol and 90 ml H_2O. Autoclave. Store at room temperature.
- 6× DNA loading buffer (40% sucrose, 0.25% bromphenol blue), dissolve 25 mg bromphenol blue and 4 g sucrose in 8 ml water, adjust final volume to 10 ml with water.
- 70% EtOH: mix 30 ml H_2O and 70 ml ethanol. Store at room temperature.
- Ethidium bromide (10,000×), dissolve 10 mg/ml in H_2O, store at room temperature, in glass bottle protected from light.
- LB-broth: add 20 g LB broth base to 100 ml H_2O. Autoclave, store at room temperature.
- PA Stock solution, 1,000×. Since the polyamines are very hygroscopic, it is suggested that small quantities (1 g) should be ordered and then all of it should be prepared at once to produce a 1,000× stock solution. Prepare 2,000× spermine by dissolving 1 g spermine in 47.9 ml embryo tested water (60 μM). Prepare 2,000× spermidine by dissolving 1 g spermidine in 28.1 ml embryo tested water (140 μM). Mix together 25 ml of spermine and 25 ml of spermidine to produce 50 ml of 1,000× PA Stock, 30 μM spermine, 70 μM spermidine. Filter sterilize (0.2 μM filters) and store 1 ml aliquots in sterile tubes at −80°C.

- PA Microinjection Buffer: 10 mM Tris–HCl (pH 7.5), 0.1 mM EDTA (pH 8.0), 100 mM NaCl, 30 mM spermine, 70 mM spermidine. To prepare 10 ml of polyamine microinjection buffer mix together 0.1 ml of 1 M Tris–HCl, pH. 7.5, 0.02 ml of 0.5 M EDTA, 0.2 ml of 5 M NaCl, and 0.01 ml of 1,000× PA stock. Filter sterilize and store at 4°C.

9.2.4. Suppliers

- BioRad Laboratories, 1000 Alfred Nobel Drive, Hercules, CA 94547, USA; phone: 510-724-7000; http://www.biorad.com
- Decon Labs, Inc., 460 Glennie Circle, King of Prussia, PA 19406, USA; phone: 800-332-4647 or 610-755-0800; http://www.deconlabs.com
- DNASTAR, 3801 Regent Street, Madison, WI 53705, USA; phone: 608-258-7420, fax: 608-258-7439; http://www.dnastar.com
- Dot Scientific, 4165 Lippincott, Burton, MI 48519, USA; phone: 810-744-1478; http://www.dotscientific.com
- Fisher Scientific, 2000 Park Lane Drive, Pittsburgh, PA 15275, USA; phone: 800-766-7000; http://www.fishersci.com
- Invitrogen, 5791 Van Allen Way, Carlsbad, CA 92008, USA; phone: 760-603-7200; http://www.invitrogen.com
- Macherey-Nagel, Inc., 2580 Emrick Boulevard, Bethlehem, PA 18020, USA; phone: 484-821-0984, toll free phone 888-321-6224, fax: 484-821-1272; http://www.macherey-nagel.com
- New England Biolabs, 240 County Road, Ipswich, MA 01938-2723, USA; phone: 978-927-5054; http://www.neb.com
- Roche Applied Science, 9115 Hague Road, P.O. Box 50414, Indianapolis, IN 46250-0414, USA; phone: 800-262-1640, fax: 800-428-288; http://www.roche-applied-science.com
- Sigma-Aldrich, P.O. Box 14508, St. Louis, MO 63178, USA; phone: 314-771-5765, fax: 314-771-5757; http://www.sigmaaldrich.com
- Thermo Scientific NanoDrop Products, 3411 Silverside Rd., Bancroft Building, Wilmington, DE 19810, USA; phone: 302-479-7707, fax: 302-792-7155; http://www.nanodrop.com
- Whatman Inc., Building 1, 800 Centennial Avenue, Piscataway, NJ 0885, USA; phone: 800-942-8626, fax: 732-885-6529; http://www.whatman.com
- World Precision Instruments, 75 Sarasota Center Boulevard, Sarasota, FL 34240-9258, USA; phone: 941-371-1003, fax: 941-377-5428; http://www.wpiinc.com

9.3 Protocol

9.3.1. Identify a BAC with the Gene of Interest

There are multiple public online tools to find a BAC clone that contains the gene of interest, including NCBI Map Viewer, Ensembl Genome Browser, and UCSC Genome Browser. The name of a BAC is composed of an abbreviation of the library and its coordinates on a microtiter plate. For instance, RP24-338R12 is a BAC clone of library RPCI-24 and located on microtiter plate 338, row R, column 12. The predominant mouse BAC libraries are RPCI-23, average insert size 197 kb [2] and RPCI-24, average insert size 155 Kb. The RPCI-23 library was derived by C.H.O.R.I. from brain and kidney tissue of 5-week-old female C57BL/6J mice using pBACe3.6 as cloning vector. The RPCI-24 library has been derived by C.H.O.R.I. from male C57BL/6J mice using pTARBAC1 as cloning vector [23]. See more detailed information at http://bacpac.chori.org/libraries.php?disp=t

It is often the case that genetically engineered BACs are used to generate transgenic mice. If the BAC in question has been modified, then the laboratory that recombineered the BAC should provide the DNA sequence of the BAC. With the DNA sequence in hand the next step is to purify BAC DNA and match the restriction map of the BAC with the in silico map (*see* Subheading 9.3.4 and beyond).

1. This protocol reflects the 2009 Ensembl browser update; it may not faithfully reproduce later updates. Open the Ensembl Genome Browser: http://www.ensembl.org/index.html in your web browser. Set browser preferences to allow popup windows.
2. In the pull down menu of the search box, choose the desired genome database, e.g., "mouse." To search for your gene of interest enter the name of your gene in the search box and press "go."
3. If your gene is not found by Ensembl, it may be that the search term is an unrecognized synonym for the gene of interest. If you are not certain of the correct gene nomenclature, use the Mouse Genome Informatics website (http://www.informatics.jax.org) to identify the correct name of the gene of interest and repeat the Ensembl search.
4. Verify you chose the right database (e.g., mouse). Click on the "Region in detail" link on the highlighted line giving the name of the gene of interest. A figure of the genomic contig centered on your gene of interest will appear.
5. On the left panel, select "Configure this page." In the pop-up window under the "Main panel," choose "External data,"

click on "DAS BAC map" and "(force) Labels." Click on the "Save and close" button.

6. After the page refreshes BACs will be listed beneath the "Region in detail." It may be necessary to zoom out for a display of BACs from 5′ to 3′ ends in relation to the contig. Depending on the size of the gene of interest, it may be completely contained within a BAC.
7. Gold rectangles with black outlines indicate BACs positioned with read end pairs. Gold rectangles without outline indicate BACs with available fingerprinting maps. Black rectangles represent BACs with single SSAHA mapped read ends.
8. Choose a BAC for use as a transgene. Ideally, the BAC will contain the gene of interest in the center with as much 5′ and 3′ genomic sequence as possible. This will increase the probability that the gene's endogenous expression pattern and developmental timing will be reproduced in transgenic mice.

9.3.2. Importing BAC DNA Sequence Data

Ensembl exports BAC DNA sequences in various formats that include data sets such as gene information, repeat features, marker features, etc. DNA analysis software such as SeqBuilder interprets data sets and displays them as the sequence is viewed and manipulated.

1. Continued from Subheading 9.3.1, **step 8**. Note the chromosome number that carries the gene of interest. Click on the BAC of interest. The name, genome coordinates of the BAC, and the strain that codes for the gene (+ or −) will show up in a popup window (e.g., RP23-214G8, 1000637203 100929106, Strand -).
2. Enter the start and end coordinates in the location boxes on the "Region in Detail" window and click the "Go" button. When the window refreshes the display will correspond to the DNA sequence in the BAC.
3. Select "export data" on the left side of the screen.
4. In the popup window, decide on a format and on which features to include. To work with SeqBuilder, choose EMBL. Use the "Output:" pull down menu to select EMBL format then click the "Next" button. The next window will allow you to choose HTML, text, or compressed text output files. Click on the "Text" hyperlink.
5. A new window opens with the exported file in text format. Use the browser to "Select all," paste the selection into a new document in a text editor, and save the document as an unformatted text file.

6. In SeqBuilder, use the "Open" command in the "File" menu to import the text file. The BAC can be decorated with restriction enzyme sites for mapping purposes.
7. Verify the presence of the gene of interest in the sequence to confirm that it is contained in the selected BAC.
8. Add the sequence of the pBACe3.6 vector (NCBI Locus CVU80929) to RPCI-23 library clones to complete the sequence file. In the RPCI-23 library, genomic DNA was cloned between the Eco RI sites at positions 10 and 2,801.
9. Add the sequence of the pTARBAC1 vector (NCBI Locus AY487252) to RPCI-24 library clones to complete the sequence file. In the RPCI-24 library, genomic DNA was cloned between the Bam HI sites at position 1 and 2,810.
10. To preserve changes or edits, save the file as a SeqBuilder file.

9.3.3. Order the BAC Clone

Clones can be ordered from C.H.O.R.I., imaGenes GmbH (Germany), Geneservice Limited (United Kingdom), or other companies. C.H.O.R.I.'s service will be described in detail. Note that C.H.O.R.I. does not guarantee that clones will match predicted DNA sequence downloaded genome browsers. It is therefore important to map BACs after they are received with restriction enzymes to validate each clone.

1. Go to http://bacpac.chori.org/order_clones.php.
2. Enter the name of the BAC of choice into the query box. Multiple BACs can be listed. Specify whether a LB Stab or a FISH Confirmed Clone is preferred. Click the "Clone(s) Verify" button.
3. If the BAC clone identity is verified the "Clone Ordering Page" will open. Follow the instructions to pay for and receive the clone containing the gene of interest.
4. The BAC arrives in DH10B *E. coli* as a stab in LB-agar, which can be stored for a few weeks at 4°C.
5. The first thing to do after the LB stab arrives is to freeze down an aliquot of an overnight culture in 10% glycerol at −80°C.

9.3.4. BAC DNA Purification

BACs are large circular DNA molecules (average sizes exceed 150 Kb) and are prone to fragmentation if handled improperly. BAC cloning vector backbones confer chloramphenicol resistance. They replicate as single copies in bacteria, thus BAC DNA yields are always lower than yields of high copy plasmids from equal volumes of bacterial culture. Always store BAC DNA in PA buffer at 4°C. To prevent sheared DNA never freeze, boil, or vortex BAC DNA. Use only wide bore pipet tips and pipet gently. Kits for BAC DNA purification are available from Qiagen,

Macherey-Nagel (distributed by Clontech) and Marligen. Each of the kits follows a similar protocol and gives similar results. We obtain better DNA yields from Nucleobond kits with the modified protocol below.

1. Grow BAC over night in 250 ml LB containing chloramphenicol (15 μg/ml) at 37°C in a shaking incubator (250 rpm).
2. Cool centrifuges and rotors to 4°C.
3. Add RNAse (kit reagent) to bacterial Cell Suspension Buffer (kit reagent).
4. Place Whatman filter into small funnel. Insert small funnel in Nucleonond BAC 100 column. Equilibrate filter and column by adding 10 ml equilibration buffer. Let it drain by gravity flow. The filter is used to remove precipitates from the bacterial lysate and prevent columns from becoming clogged.
5. Take a 1 ml aliquot of bacteria culture, spin it down in a tabletop microcentrifuge (4°C, 5,000 × *g*), resuspend in 1 ml 10% glycerol, and store it at −80°C.
6. Spin down the remaining bacteria for 10 min at 5,000 × *g* rpm at 4°C. Discard the supernatant.
7. Resuspend cell pellet in 30 ml cell suspension buffer per 250 ml of cells grown, by gently swirling the centrifuge bottle. If larger volumes of bacterial culture have been used, increase buffer proportionately.
8. Be generous with kit buffer volumes. Additional buffers can be purchased from Macherey-Nagel. Marligen, Qiagen, and Nucleobond buffers are interchangeable. If necessary, matching buffers from the Qiagen Plasmid Maxi Kit can be substituted for Macherey-Nagel buffers.
9. Add an equal amount of cell lysis solution. Mix gently. Do not vortex. Keep at room temperature for 5 min.
10. Add an equal amount of neutralization solution. Do not vortex. Mix by swirling. A white precipitate will form. It should be viscous. Spin down precipitate at 15,000 × *g* for 10 min.
11. Pour the neutralized cell lysate into the funnel. Let the column drain by gravity flow. Discard the flow through. Discard the filters.
12. Add 30 ml wash buffer to the columns and let it drain by gravity flow.
13. Add 15 ml elution buffer to the column and let it drain by gravity flow.

14. Add 11 ml isopropanol to eluate.
15. Spin 40 min at 10,000 × *g* at 4°C (50 ml Falcon tubes break above 10,000 × *g*).
16. A fragile white pellet will form. It will be small, sometimes hard to see. Be careful when handling the tube, since the BAC detaches easily from the tube wall. If you see a big pellet, something went wrong with the purification (e.g., RNAse was not added to the cell suspension buffer or has been degraded).
17. Resuspend the pellet in 3 ml 70% EtOH and divide it in two microcentrifuge tubes. Spin 10 min at high speed at 4°C.
18. Dry the pellets 3 min. Do not let them dry completely or shearing will occur.
19. Prefilter 1 ml of PA buffer with a 0.02 μM Anotop syringe filter. Do not use these filters with resuspended BAC DNA, the BACs will be trapped in the filters and removed from solution.
20. Resuspend the BAC DNA in 300 μl prefiltered PA buffer. BACs go easily in solution with PA buffer. If some precipitate will not dissolve, it most likely is not DNA but waste material carried over in an improperly executed BAC DNA purification procedure.
21. Quantitate BAC DNA.

9.3.5. BAC DNA Quantitation

Measure BAC DNA concentration with a NanoDrop spectrophotometer. BAC DNA concentrations are typically concentration 100 ng–200 ng/μl. Significantly higher concentrations typically result from co-purification of RNA when RNAse is omitted from the cell suspension buffer or has lost activity.

1. Open NanoDrop software and choose application.
2. Raise the lever arm and pipet 1 μl water onto the pedestal in response to the dialog box. Click "OK" to initiate calibration.
3. Clean both optical surfaces of the pedestal with a KimWipe.
4. Pipet 1 μl of PA buffer onto the pedestal and click "Blank" to zero the instrument.
5. Clean the pedestal and add 1 μl of BAC DNA solution and click "Measure."
6. Clean the pedestal with a KimWipe between samples and pipet 1 μl water as the last sample.
7. Save data in computer, clean the pedestal, and quit the software program.

9.3.6. BAC DNA Restriction Mapping

Before working with a BAC it is mandatory to check its integrity on a pulsed field gel. When a BAC shears, its products will still be tens of thousands of base pairs long. Standard agarose gel electrophoresis conditions do not have the necessary resolving power to evaluate BAC DNA integrity.

1. Use the *in silico* BAC sequence file assembled in Subheading 9.3.2 to choose enzymes that cut two to four times in the BAC. The cuts should result in fragments of different lengths no smaller than 8 kb.
2. Set up restriction enzyme digests with 250–500 ng BAC DNA.
3. Set up the PFGE apparatus in a cold room or attach a circulating cooling bath.
4. Pour a 0.8% agarose gel. Make sure no bubbles disturb the gel. If imperfections are present in the gel, do not use it. Every bubble will disturb the migration of BAC DNA. Do not add ethidium bromide to the gel. Ethidium bromide will diffuse into the buffer and need to be disposed of as hazardous waste.
5. Fill the gel apparatus with 2,000 ml 0.5× TBE.
6. The MidRange PFG Marker II comes as agarose block. Use a razor blade to cut a thin slice off the marker and load it onto the gel.
7. There is no need to prepare agarose blocks to run your sample. Load samples with 6× DNA loading dye as you would for a standard horizontal agarose gel electrophoresis. Loading liquid BAC DNA samples in 6× DNA loading buffer is as effective as loading BAC DNA in agarose blocks.
8. Load between 250 and 500 ng restriction enzyme digested BAC DNA per lane into the gel wells.
9. Load one lane with undigested BAC DNA as a control.
10. Enter settings for the pulsed field gel: initial switch time: 1 s, final switch time: 25 s, run time: 17 h, 6 V, angle: 120°.
11. The actual current should be around 135–150 A when the gel run begins. It will go up overnight. If it is much higher, check the TBE buffer concentration. If the current is too high smeared bands will result.
12. After the run, soak gel in ethidium bromide bath for 30 min. Visualize bands with UV light. Soaking the gel in an ethidium bromide bath after electrophoresis reduces waste. The bath can be reused, which leaves the gel as the only hazardous waste product of the procedure. Ethidium bromide is a known mutagen. Observe safe handling and disposal procedures of ethidium bromide waste.

13. The markers should produce discrete bands of predicted molecular weight. Overloaded marker DNA will produce smeared bands in marker lanes.
14. Load different amounts of BAC DNA to make sure that bands will be clear. For instance, in case of RNA contamination the NanoDrop reading will show a high DNA concentration. In this case not all of the sample consists of BAC DNA and DNA band intensities will appear low. Contaminating RNA will migrate off the gel and will not interfere with visualization of BAC DNA fragments.
15. Compare the expected restriction enzyme pattern to the in silico restriction analysis. If they do not match, either the wrong BAC was received or the Ensembl BAC sequence is incorrect. C.H.O.R.I. will send the same BAC again for free if the restriction map is wrong. It is recommended that a different BAC be ordered if the in silico map does not match the actual map.
16. BAC DNA samples can be stored for 2 years and longer at 4°C in PA buffer without degradation.

9.3.7. BAC DNA Transgenesis

BAC DNA samples used for microinjection should be resuspended in PA buffer to improve the chance of producing transgenic founders with intact, unfragmented DNA molecules [33]. Transgenesis efficiency is similar whether circular or linearized BAC DNA molecules are microinjected. Always test the integrity of purified BAC DNA by PFGE. PCR analysis or standard gel electrophoresis will not reveal that BAC DNA molecules were degraded during purification. The range of effective microinjection BAC DNA concentrations is more narrow than the range used for microinjection of smaller molecules. The typical starting BAC DNA microinjection concentration is 0.5 ng/μl. If necessary the concentration can be reduced to increase egg survival and improve birth rates.

Sensitive genotyping assays are necessary for the detection of BAC transgene DNA within genomic DNA. In the absence of sensitive assays, false negative results (failure to detect transgenic founder animals) are likely to occur. Positive control assays are essential to rule out technical problems such as improper DNA concentration or the presence of PCR inhibitors that can cause false negative results. Artificial transgene copy standards can be prepared by mixing together known quantities of transgene DNA and genomic DNA.

9.3.7.1. BAC DNA Pronuclear Microinjection

1. Dilute BAC DNA to 0.5 ng/μl with 0.02 μm filtered PA buffer by 10 and 100-fold dilutions for best accuracy. Store DNA at 4°C, never store frozen.

2. Aliquot 50 μl of diluted DNA into four to six sterile microtubes.
3. Pull microinjection needles from thin wall glass capillaries with internal filaments on a pipet puller (WPI cat. no. TW100-4). *See* Chap. 6 for information on pulling microinjection needles.
4. Place rear end of pulled needles in resuspended BAC DNA and wait 1–2 min for solution to wick to needle tip.
5. After the meniscus is visible in the needle tip, fit the needle to the microinjector and position it in the microinjection chamber filled with fertilized eggs.
6. Perform pronuclear microinjection as usual (Chap. 6). The microinjection of properly prepared BAC DNA will proceed as smoothly and as easily as plasmid transgene DNA solutions.
7. It is not necessary to change microinjection needle geometry for BAC DNA microinjection.
8. Microinjected eggs are removed from the injection chamber and washed through four drops of 75 μl of a mouse embryo medium (M16) under mineral oil in a 35-mm cell culture Petri dish pre-equilibrated in a humidified 37°C, 5% CO_2 incubator.
9. Egg survival rates after BAC DNA microinjection will be similar to survival after plasmid DNA transgene microinjection.
10. One-cell eggs are transferred to 0.5 pseudopregnant females the day of injection (20–24 eggs per recipient, divided between the two oviducts). If necessary the eggs can be cultured overnight to the two-cell stage (18–20 eggs transferred per recipient). Chapter 6 has information on preparation of pseudopregnant recipients and transfer of eggs.
11. After transferring eggs to pseudopregnant females, birth rates should be similar to that seen with plasmid transgene microinjections.

9.3.7.2. Genotyping Potential Transgenic Founders

1. PCR genotyping assays use primers that are specific for the BAC that will not amplify mouse genomic DNA. The BAC end sequences on either side of the cloning vector can be used with primers in the vector to generate transgene specific primers. Primers for cassettes such as eGFP or Cre that are introduced into the BAC can also be the basis of transgene specific primers. Depending on the genomic DNA sources, it may be possible to use SNP assays to place multiple markers along the BAC.
2. Copy standards are prepared for 10, 1, 0.1, and 0.01 copy equivalents. Assuming that one diploid genome consists of

6×10^9 bp and that a BAC transgene is 10^5 bp then the ratio between the two is $6 \times 10^9 : 10^5$ or $1 : 1.67 \times 10^{-5}$. To prepare 1 μg of a one copy standard the mass ratio would be 1 μg genomic DNA : 16.7 pg BAC transgene. To prepare a 0.1 copy standard, combine 1 μg genomic DNA with 1.67 pg BAC DNA, etc. Transgene specific PCR products are analyzed on agarose gels.

3. Primers for that amplify endogenous mouse DNA, such as marker D5Mit425, are used as a positive internal control. D5Mit425 primers: 5′ TCGCCTTTCTTTCCCTCC 3′ and 5′ AAAATTACATTTGCATCTGGGG 3′; annealing temperature 56°C. Expected PCR product size is 114–136 bp, which varies according to the genetic background of the mouse genomic DNA.

9.4 Results

9.4.1. BAC DNA Quality Control

The identity of BAC clones is verified by examining restriction enzyme digested BAC DNA on pulsed field gels. Purified BAC DNA is quantitated, digested, and mapped on gels. The pattern of DNA bands is compared to the predicted map based on the downloaded DNA sequence. Occasionally the predicted and actual band patterns do not match each other (Fig. 9.1). In this case, the BAC clone should not be used for subsequent

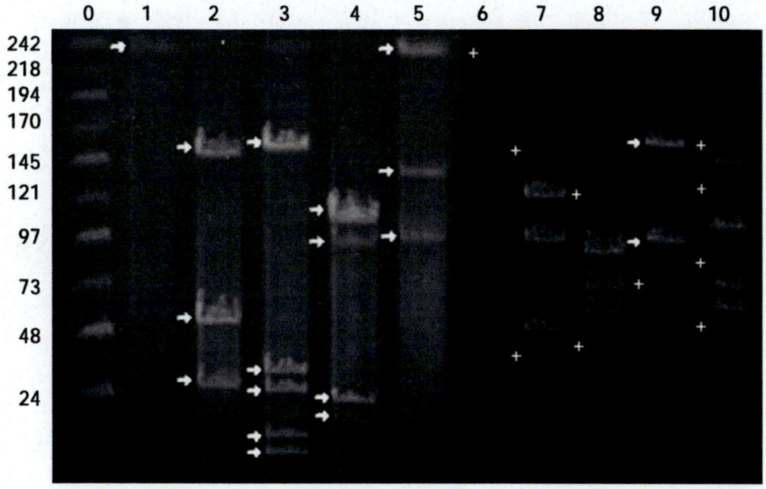

Fig. 9.1 Analysis of BAC DNA by pulsed field gel restriction mapping. *Lane* 0: MidRange Marker II, NEB. *Lane* 1: BAC 1 unrestricted. *Lanes* 2–5: BAC 1 digested with different restriction enzymes. *Lane* 6: BAC 2 unrestricted. *Lanes* 7–10 BAC 2 digested with different restriction enzymes. *White arrows* indicate locations of bands predicted by Lasergene software and verified by restriction analysis. *White plus* marks indicate locations of bands predicted by Lasergene software and not verified by restriction analysis.

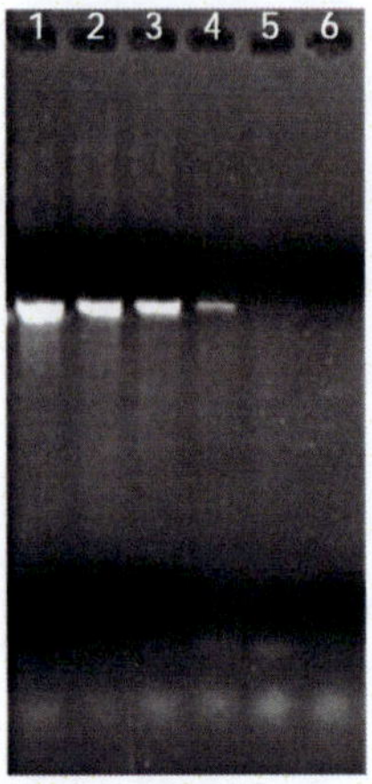

Fig. 9.2 Copy standard sensitivity of transgene specific genotyping primers. *Lane* 1: 10 copy standard; *Lane* 2: 1 copy standard; *Lane* 3: 0.1 copy standard, *Lane* 4: 0.01 copy standard; *Lane* 5: genomic DNA only (0 copy standard); *Lane* 6: water only, no DNA template. PCR amplification products were analyzed by electrophoretic separation on 0.8% agarose in TBE.

experiments. Ideally there will be other BAC clones available that can be ordered and analyzed.

9.4.2. BAC Transgene Genotyping

Genomic copy standards are prepared by spiking BAC transgene DNA into mouse genomic DNA as described earlier (Subheading 9.3.7.2). Primers specific for the transgene that do not amplify genomic DNA are tested on the copy standards to establish copy sensitivity. It is essential to have a robust and sensitive assay or else transgenic founders that carry the BAC transgene may be scored as false negatives. An example of copy standard titration is shown in Fig. 9.2. Note that the PCR reaction produces a robust signal with 10, 1, and 0.1 genomic copies of the BAC on the agarose gel. Although weak, a signal is also present at the 0.01 copy standard concentration.

A necessary control for transgene PCR is demonstration that DNA from potential transgenic founders is suitable for amplification, free of PCR inhibitors, and at an appropriate concentration for PCR. Results of amplifying mouse genomic DNA with D5Mit425 primers that recognize an endogenous DNA sequence in any mouse genome are shown in Fig. 9.3. Note that primers were labeled with 6-FAM so that the PCR fragments could be sized with a fluorescent DNA sequencing instrument. Transgene specific primers may also be fluorescently labeled so that PCR reactions can be analyzed with DNA sequencers. Mice that return a positive genotype for the endogenous sequence and a positive genotype for the BAC are scored as transgene positive and selected for further study.

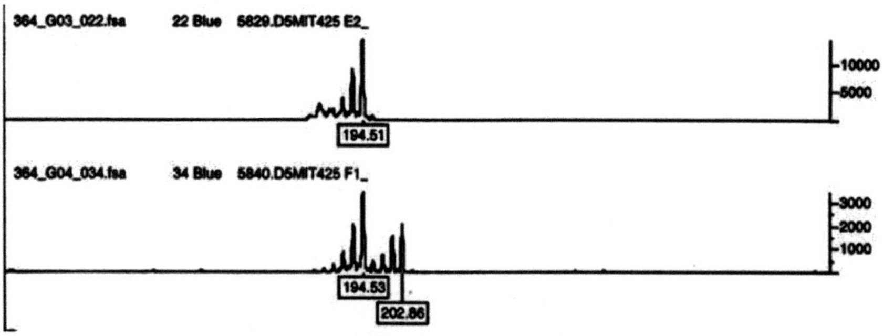

Fig. 9.3 Analysis of PCR amplification of mouse genomic DNA with D5Mit425 primers. Primers carried fluorescent tags for detection with an ABI 3730LX DNA Analyzer. In the *upper panel,* mouse genomic DNA is homozygous for SJL/J (195 bp). In the *bottom panel,* mouse genomic DNA is heterozygous for SJL/J and C57BL/6J (203 bp).

9.5 Troubleshooting

9.5.1. BAC DNA Purification

9.5.1.1. Inconsistent Yields of BAC DNA

Some BACs grow more slowly than others. It is normal for BAC DNA yields to vary between BAC clones. If insufficient DNA is purified from 250 ml of bacterial culture, purify DNA from a higher culture volume.

9.5.1.2. Low Yields of BAC DNA

- Store all buffers at the manufacturer's recommended temperatures. All buffers are stored at room temperature except Cell Suspension Buffer with RNAse (which is stored at 4°C).
- Completely remove all growth medium after centrifugation of the overnight culture. Leftover medium can change salt concentrations or pH of the DNA purification buffers.
- Do not let your pellet dry completely.
- Be careful during isopropanol DNA precipitation and the DNA pellet ethanol wash. The pellet can easily be lost by accident during both of these steps.

9.5.1.3. Bacterial Chromosome DNA Contamination

Be careful not to vortex or shake the tube during lysis. Otherwise, sheared genomic DNA will end up in your preparation. Chromosomal DNA will appear on your PFGE gel as an intensely staining smear from top to bottom of the gel lane.

9.5.1.4. RNA Contamination

- Remember to add RNAse to the Cell Suspension Buffer and to use a sufficient amount of buffer.
- Make sure all of sample enters the column. Use wash buffer to clear the column walls from clinging drops and avoid delays in lysate processing.

9.5.1.5. Viscous BAC DNA

- Sometimes the final prep is viscous. This is of no concern; the BAC should not be compromised. However, be careful when pipetting the BAC (e.g., for a restriction cut), because viscous DNA easily gets sucked out of the pipet tip by spontaneous capillary action.
- If the BAC DNA is not evenly distributed after resuspension because of its viscosity, the NanoDrop reading may not accurately reflect the DNA concentration. It is advisable to allow the BAC to come to equilibrium at 4°C for a few days before measuring the DNA concentration.
- Dilutions of viscous BAC DNA may result in microinjection DNA concentrations that are too high or too low if the BAC DNA is not evenly resuspended in microinjection, as described earlier, for quantitation.

9.5.2. Pulsed Field Gel Electrophoresis

9.5.2.1. PFGE Marker Smears

- Store the marker at −20°C to prevent degradation.
- Decrease the amount of marker you load on the gel. Use only a very thin slice, approximately 0.5 mm wide.
- Make sure to use fresh running buffer. Buffer can be reused four to five times.
- Make sure the settings of the pulsed field apparatus are correct.
- Make sure the manufacturer's recommended settings for the marker are suitable for your application. Every marker comes with a protocol that should be studied before use.

9.5.2.2. BAC DNA Smears

- If you cannot make out a distinct band and only see a smear, your BAC is degraded and needs to be repurified.
- It is not uncommon for undigested BAC DNA samples to show slight smearing in addition to sharp bands.
- Make sure to use fresh running buffer. Buffer can be reused four to five times.
- Make sure the settings of the pulsed field apparatus are correct.

9.5.2.3. Unexpected High Molecular Weight Band

Highly purified BAC DNA that has not been digested by restriction enzymes will show a high molecular weight band corresponding to closed circular supercoiled BAC DNA and a second band of the expected size that corresponds to open relaxed circular DNA molecules.

9.5.2.4. PFGE and In Silico Restriction Maps Are Different

- The incorrect BAC clone may have been received from C.H. O.R.I. Notify C.H.O.R.I. and they will send a new BAC with the same accession number at no additional cost.

- Ensembl has the wrong sequence attached to the BAC accession number. Order a different BAC covering the same region.

9.5.3. BAC Microinjection

9.5.3.1. Microinjection Needles Clog Easily

- Use 0.02 μM syringe filters to prefilter PA buffer immediately before diluting BAC DNA to microinjection concentrations.
- Needle geometry can be adjusted so that pipet tapers to the tip more rapidly. This will result in tips with larger diameter. Lower pressures need to be applied to the DNA to inject eggs. With larger tips (above 1 μm in diameter) egg lysis will increase noticeably.

9.5.3.2. BAC DNA Is Viscous

- BAC DNA at high concentrations will have the consistency of a gel instead of a fluid. Microinjection of high concentration BAC DNA will bind to the nucleolus inside the pronucleus and when the needle is removed the nucleolus will be pulled out.
- Discard the BAC DNA injection solutions, re-quantitate the BAC DNA and prepare a fresh dilution of DNA. The BAC DNA may not be evenly resuspended in PA buffer. The solution can be incubated at 65°C for an hour and then allowed to rest at 4°C overnight to achieve a uniform distribution of DNA. Do not tap, pipet, or vortex the DNA to make it go into the solution.

9.5.3.3. In Vitro and In Vivo Development Failure

- Correctly prepared PA buffer is not toxic to mouse eggs [24]. Survival and development of eggs microinjected with PA buffer and conventional microinjection buffer (10 mM Tris–HCl, pH 7.5, 0.25 mM EDTA) should be similar.
- BAC DNA concentration is too high. Birth rates of microinjected eggs below 10% occurs when large DNA molecules are microinjected at 2 ng/μl [24], 1.6–3 ng/μl [25], 5 ng/μl [26, 27]. Similarly, the microinjection of plasmid DNA transgenes at concentrations of 10 ng/μl significantly reduces birth rates [28].
- BAC DNA can be reinjected at lower concentrations (0.25 ng/μl or 0.1 ng/μl) to improve birth rates.

9.5.3.4. Transgenic Founders Are Not Identified

- Failure to identify transgenic founders may simply be a result of technical errors in the genotyping assay. Re-evaluate the sensitivity of the transgene specific genotyping assay(s) and ensure that a PCR primer pair for an endogenous mouse DNA successfully amplified all DNA samples.
- High birthrates accompanied by low transgenic rates typically occurs when BAC DNA is microinjected at too dilute

concentrations. In this case, re-quantitate the BAC DNA and prepare new dilutions of microinjection DNA as described in Subheading 9.5.3.2.

- If these steps do not remedy the situation then it is best to discard the BAC DNA sample and purify new BAC DNA for microinjection from bacterial culture.

References

1. Giraldo P, Montoliu L (2001) Size matters: use of YACs. BACs and PACs in transgenic animals. Transgenic Res 2001(10):83–103
2. Osoegawa K, Tateno M, Woon PY, Frengen E, Mammoser AG, Catanese JJ, Hayashizaki Y, de Jong PJ (2000) Bacterial artificial chromosome libraries for mouse sequencing and functional analysis. Genome Res 10:116–128
3. Osoegawa K, Mammoser AG, Wu C, Frengen E, Zeng C, Catanese JJ, de Jong PJ (2001) A bacterial artificial chromosome library for sequencing the complete human genome. Genome Res 11:483–496
4. Corpening JC, Cantrell VA, Deal KK, Southard-Smith EM (2008) A Histone2B-Cerulean BAC transgene identifies differential expression of Phox2b in migrating enteric neural crest derivatives and enteric glia. Dev Dyn 237:1119–1132
5. Nielsen LB, McCormick SP, Pierotti V, Tam C, Gunn MD, Shizuya H, Young SG (1997) Human apolipoprotein B transgenic mice generated with 207- and 145-kilobase pair bacterial artificial chromosomes. Evidence that a distant 5′-element confers appropriate transgene expression in the intestine. J Biol Chem 272:29752–29758
6. Deal KK, Cantrell VA, Chandler RL, Saunders TL, Mortlock DP, Southard-Smith EM (2006) Distant regulatory elements in a Sox10-beta GEO BAC transgene are required for expression of Sox10 in the enteric nervous system and other neural crest-derived tissues. Dev Dyn 235:1413–1432
7. Dunnick WA, Shi J, Graves KA, Collins JT (2005) The 3′ end of the heavy chain constant region locus enhances germline transcription and switch recombination of the four gamma genes. J Exp Med 201:1459–1466
8. Xing L, Salas M, Lin CS, Zigman W, Silverman W, Subramaniyam S, Murty VV, Tycko B (2007) Faithful tissue-specific expression of the human chromosome 21-linked COL6A1 gene in BAC-transgenic mice. Mamm Genome 18:113–122
9. Haldi ML, Strickland C, Lim P, VanBerkel V, Chen X, Noya D, Korenberg JR, Husain Z, Miller J, Lander ES (1996) A comprehensive large-insert yeast artificial chromosome library for physical mapping of the mouse genome. Mamm Genome 7:767–769
10. Moreira PN, Pérez-Crespo M, Ramírez MA, Pozueta J, Montoliu L, Gutiérrez-Adán A (2007) Effect of transgene concentration, flanking matrix attachment regions, and RecA-coating on the efficiency of mouse transgenesis mediated by intracytoplasmic sperm injection. Biol Reprod 76:336–343
11. Antoch MP, Song EJ, Chang AM, Vitaterna MH, Zhao Y, Wilsbacher LD, Songoram AM, King DP, Pinto LH, Takahashi JS (1997) Functional identification of the Mouse circadian *clock* gene by transgenic BAC rescue. Cell 89:655–667
12. Probst FJ, Fridell RA, Raphael Y, Saunders TL, Wang A, Liang Y, Morell RJ, Touchman JW, Lyons RH, Noben-Trauth K, Friedman TB, Camper SA (1998) Correction of deafness in shaker-2 mice by an unconventional myosin in a BAC transgene. Science 280:1444–1447
13. Yang XW, Model P, Heintz N (1997) Homologous recombination based modification in Escherichia coli and germline transmission in transgenic mice of a bacterial artificial chromosome. Nat Biotechnol 15:859–865
14. Zhang Y, Buchholz F, Muyrers JP, Stewart AF (1998) A new logic for DNA engineering using recombination in Escherichia coli. Nat Genet 20:123–128
15. Copeland NG, Jenkins NA, Court DL (2001) Recombineering: a powerful new tool for mouse functional genomics. Nat Rev Genet 2:769–779
16. Valjent E, Bertran-Gonzalez J, Herve D, Fisone G, Girault JA (2009) Looking BAC at striatal signaling: cell-specific analysis in new transgenic mice. Trends Neurosci 32:538–547
17. Gong S, Zheng C, Doughty ML, Losos K, Didkovsky N, Schambra UB, Nowak NJ, Joyner A, Leblanc G, Hatten ME, Heintz N (2003) A gene expression atlas of the central

nervous system based on bacterial artificial chromosomes. Nature 425:917–925
18. Collins FS, Rossant J, Wurst W (2007) A mouse for all reasons. Cell 128:9–13
19. Valenzuela DM, Murphy AJ, Frendewey D, Gale NW, Economides AN, Auerbach W, Poueymirou WT, Adams NC, Rojas J, Yasenchak J, Chernomorsky R, Boucher M, Elsasser AL, Esau L, Zheng J, Griffiths JA, Wang X, Su H, Xue Y, Dominguez MG, Noguera I, Torres R, Macdonald LE, Stewart AF, DeChiara TM, Yancopoulos GD (2003) High-throughput engineering of the mouse genome coupled with high-resolution expression analysis. Nat Biotechnol 21:652–659
20. Mensah-Osman E, Labut E, Zavros Y, El-Zaatari M, Law DJ, Merchant JL (2008) Regulated expression of the human gastrin gene in mice. Regul Pept 151:115–122
21. Ranatunga D, Hedrich CM, Wang F, McVicar DW, Nowak N, Joshi T, Feigenbaum L, Grant LR, Stäger S, Bream JH (2009) A human IL10 BAC transgene reveals tissue-specific control of IL-10 expression and alters disease outcome. Proc Natl Acad Sci USA 106:17123–17128
22. Sarsero JP, Li L, Holloway TP, Voullaire L, Gazeas S, Fowler KJ, Kirby DM, Thorburn DR, Galle A, Cheema S, Koenig M, Williamson R, Ioannou PA (2004) Human BAC-mediated rescue of the Friedreich ataxia knockout mutation in transgenic mice. Mamm Genome 15:370–382
23. Zeng C, Kouprina N, Zhu B, Cairo A, Hoek M, Cross G, Osoegawa K, Larionov V, de Jong P (2001) Large-insert BAC/YAC libraries for selective re-isolation of genomic regions by homologous recombination in yeast. Genomics 77:27–34
24. Van Keuren ML, Gavrilina GB, Filipiak WE, Zeidler MG, Saunders TL (2009) Generating transgenic mice from bacterial artificial chromosomes: transgenesis efficiency, integration and expression outcomes. Transgenic Res 18:769–785
25. Zeilhofer HU, Studler B, Arabadzisz D, Schweizer C, Ahmadi S, Layh B, Bosl MR, Fritschy JM (2005) Glycinergic neurons expressing enhanced green fluorescent protein in bacterial artificial chromosome transgenic mice. J Comp Neurol 482:123–141
26. Abe K, Hazama M, Katoh H, Yamamura K, Suzuki M (2004) Establishment of an efficient BAC transgenesis protocol and its application to functional characterization of the mouse Brachyury locus. Exp Anim 53:311–320
27. Schedl A, Larin Z, Montoliu L, Thies E, Kelsey G, Lehrach H, Schutz G (1993) A method for the generation of YAC transgenic mice by pronuclear microinjection. Nucleic Acids Res 21:4783–4787
28. Brinster RL, Chen HY, Trumbauer ME, Yagle MK, Palmiter RD (1985) Factors affecting the efficiency of introducing foreign DNA into mice by microinjecting eggs. Proc Natl Acad Sci USA 82:4438–4442
29. Jones JM, Datta P, Srinivasula SM, Ji W, Gupta S, Zhang Z, Davies E, Hajnóczky G, Saunders TL, Van Keuren ML, Fernandes-Alnemri T, Meisler MH, Alnemri ES (2003) Loss of Omi mitochondrial protease activity causes the neuromuscular disorder of mnd2 mutant mice. Nature 425:721–727
30. Montoliu L, Bock CT, Schütz G, Zentgraf H (1995) Visualization of large DNA molecules by electron microscopy with polyamines: application to the analysis of yeast endogenous and artificial chromosomes. J Mol Biol 246:486–492
31. Oliver ER, Saunders TL, Tarle SA, Glaser T (2004) Ribosomal protein L24 defect in belly spot and tail (Bst), a mouse Minute. Development 131:3907–3920
32. Brandt W, Khandekar M, Suzuki N, Yamamoto M, Lim K-C, Douglas Engel J (2008) Defining the Functional Boundaries of the *Gata2* Locus by Rescue with a Linked Bacterial Artificial Chromosome Transgene. J Biol Chem. 283(14):8976–8983. doi: 10.1074/jbc.M709364200
33. Margaret L. Van Keuren, Galina B. Gavrilina, Wanda E. Filipiak, Michael G. Zeidler, Thomas L. Saunders (2009) Generating Transgenic Mice from Bacterial Artificial Chromosomes: Transgenesis Efficiency, Integration and Expression Outcomes. Transgenic Res. Author manuscript; available in PMC 2011 January 5. Published in final edited form as: Transgenic Res. October; 18 (5):769–785. Published online 2009 April 26. doi: 10.1007/s11248-009-9271-2

Chapter 10

Generation of Transgenic Animals with Lentiviral Vectors

Carlos Lois

Abstract

In this chapter, we will discuss the use of lentiviral vectors to generate transgenic animals. Specifically, we will review the technology as applied to the generation of rodents and birds. But many of the procedures can be applied in other species. We review the production of vectors, their design and construction, promoters and reporters. We also discuss techniques for the introduction of vectors to rodent zygotes and bird eggs.

Abbreviations

AP	Auto fluorescence
BGH	Bovine growth hormone
CFP	Cyan fluorescent protein
CMV	Cytomegalovirus
ELISA	Enzyme-linked immunosorbent assay
GFP	Green fluorescent protein
HSVtk	Herpes simplex virus thymidine kinase
ICC	Immunocytochemistry
IRES	Internal ribosome entry site
LTR	Long terminal repeat
PGK	Phosphoglycerate kinase 1
PLAP	Placental alkaline phosphatase
siRNA	Short interfering RNAs
SV40	Simian virus 40
UbiC	Human polyubiquitin C promoter
VSVG	Vesicular stomatitis virus glycoprotein

S. Pease and T.L. Saunders (eds.), *Advanced Protocols for Animal Transgenesis*, Springer Protocols,
DOI 10.1007/978-3-642-20792-1_10, © Springer-Verlag Berlin Heidelberg 2011

10.1 Introduction

The production of transgenic animals has become one of the most powerful technical advances in the understanding of genetics [1]. Transgenic animals have been instrumental in improving our understanding of human genetic diseases and will have increasing importance in the improvement of livestock, xenotransplantation, and the production of biologically active pharmaceuticals.

The first transgenic animals were generated by infecting mouse blastocyst cells with a Moloney Leukemia virus-based retroviral vector [2]. These animals were able to transmit the integrated proviral transgene to their progeny. However, neither the founders nor the subsequent offspring expressed the exogenously introduced gene at sufficient levels to be practical for most experimental applications. Subsequent studies showed that the silencing of the proviral transgene was correlated to developmentally regulated methylation of the flanking host sequences [3].

To circumvent the developmental silencing observed with retroviral transgenics, a non-viral approach to transgenesis, called pronuclear injection, was developed [4]. In pronuclear injection, a linearized DNA sequence is introduced directly into the pronucleus of single-cell embryos by microinjection. The transgene integrates in head-to-tail tandem arrays of up to 1,000 copies at a single chromosomal site, and multiple copies of the transgene are usually required to drive the expression of the exogenous sequence to detectable levels [5] and mice with 1,000 copies of the transgene have been published [6]. Generally, transgenic animals generated from pronuclear injection transmit and reliably express the exogenous sequences. This technique has become the standard approach for generating transgenic mice.

Although pronuclear injection is widely used to generate transgenic mice, the method suffers from several limitations. First, because the technique requires visualization of the male pronucleus, discernible only in the zygotes of certain mouse strains, pronuclear injection is not practical for transgenesis in other animals. For example, the generation of transgenic birds has long been desirable because of the potential for the production of large amounts of protein in their eggs. Unfortunately, pronuclear injection is not applicable to the generation of transgenic birds, because the pronucleus is not visible in the avian embryo due to its high content in yolk [7]. Second, pronuclear injection requires penetration of both the cell membrane and pronuclear envelope for DNA delivery. As a result, it is technically challenging, requires sophisticated micromanipulation equipment and is often damaging to the embryos. Finally, under optimal conditions, superovulated mammals can yield a maximum of

25–35 embryos/female. From these starting numbers, only 3.2% of the microinjected mouse embryos [5] and 0.1% of cattle, sheep and pig embryos develop into transgenic animals. Therefore, one must inject hundreds of embryos to ensure the derivation of one transgenic line. While obtaining large numbers of embryos is possible in mice, this requirement constitutes a significant cost barrier for the generation of many different transgenic lines in other animals. This constraint seriously reduces the utility of pronuclear injection in larger mammals that may be more appropriate models for human diseases and more suitable for studying organ systems or behavior. Taken together, these limits suggest that in order for transgenesis outside of rodents to be practical, alternative methods to pronuclear injection are necessary.

The technique of transgenesis by lentiviral gene delivery to one-cell embryos overcomes many of the limitations of other techniques. In contrast to other retroviruses that proved problematic in the past, lentiviral vectors are not developmentally silenced [8]. In addition, this technique provides for reproducible tissue-specific expression from promoters with restricted specificity.

Lentiviral gene delivery is more efficient than pronuclear injection for transgenesis. Because the lentiviral delivery technique does not require visualization of the pronucleus, it can be extended to diverse mouse strains, as well as to other animal species. Using this technique we have generated transgenic rats, a species into which the introduction of exogenous genes has so far been difficult and inefficient. In addition, the method is less invasive to the embryos, more cost effective and technically less demanding. Delivering lentiviruses by co-incubation with denuded embryos (as described below) removes the need for micromanipulation. With less technical difficulty than pronuclear injection, transgenesis via lentiviral vectors will be more accessible to more laboratories. After infecting single cell embryos with lentiviral vectors, approximately 80% of surviving embryos transmit the transgene to their progeny and over 90% of transgenic founders express the transgene. Furthermore the technique works equally well with all mammals because the VSVG protein that mediates viral entry finds receptors on the cells of most species, including cattle and primates [9–11]. The combination of a high efficiency of transgenesis, low cost and scalability for high throughput makes lentiviral gene delivery an attractive alternative for the creation of transgenic animals.

The range of possible transgenes is quite extensive. Three general types of transgene are available for use with the lentiviral system. Genes of interest can be ectopically expressed in the whole animal or selected tissues, based on the choice of promoter as described below. These might be endogenous genes, overexpressed to examine the effects of dosage on their normal protein function. Alternatively, exogenous genes coding for visible markers or biopharmaceuticals may be expressed. Second, dominant negative constructs can be delivered which interfere with the function of endogenous genes [12]. Finally, non-coding RNAs, in particular short interfering RNAs (siRNAs), can also be introduced using lentiviral vectors. siRNAs have been shown to be effective in the knockdown of endogenous genes in transgenic mice generated using lentiviral vectors [13].

Beyond basic transgenesis, this technique allows for other applications such as fine-scale mutagenesis of regulatory sequences, large-scale insertional mutagenesis screens or gene trapping. It also allows for the germline introduction of exogenous genes into birds, a class of animals for which no satisfactory method existed for creating transgenics until now. In this chapter, we will discuss the details of viral production and usage in order to create transgenic rodents and birds.

10.2 Production of Lentiviral Vectors

10.2.1. Design and Construction of the Lentiviral Transfer Vector

The basic design of the transfer vector consists of a cassette in which a promoter (P) drives the expression of a gene of interest (G), and this P + G cassette is inserted in a lentiviral vector flanked by LTR sequences [8]. In most cases the engineering of a lentiviral transgene construct is a straightforward procedure by standard molecular cloning techniques. In some situations, however, some combinations of promoters and genes might be problematic, and some modifications might be necessary.

There are a few considerations that have to be taken into account before starting to engineer a lentiviral transfer vector. The first consideration is size. The capsid of the retrovirus has a maximum capacity of approximately 12 kb. Because the transfer vector contains LTRs and some sequences necessary for reverse transcription and packaging, the theoretical maximum size of the P + G cassette is 10 kb.

Second, lentiviruses have an RNA genome. Therefore, RNA processing mechanisms such as splicing and polyadenylation can impact the production of recombinant viruses. The presence of polyadenylation addition signals (such as those from SV40, BGH or HSVtk, or those present in the untranslated regions of endogenous genes) interfere with the production of the full RNA lentiviral genome and should be removed from the inserts; the 3′LTR of the virus provides the polyadenylation signal for the integrated transgene. In addition, the presence of introns in the insert can result in aberrant splicing of the lentiviral genome with subsequent loss of insert fragments from the integrated transgene. Whenever possible, it is advisable to use cDNAs because the splicing signals have been eliminated in these sequences. However, the presence of introns in the promoter may not be a problem. We have generated transgenic animals with promoters such as UbiC, or CMV/β actin (see below) that contain introns, and the expression from these transgenes is appropriate.

It is important to realize that splicing and polyadenylation signals in some cases are easily identified, but cryptic signals might also be present. If the viral titer is low, or the expression from the lentivirus is inadequate, and all other steps in the production of the virus (transfection, concentration) are correct, it is worthwhile to check whether mRNA processing has interfered with viral production. If introns or polyA addition signals are known to exist in the gene but are not possible to remove, it might be necessary to generate a transfer vector in which the P + G cassette is inserted in the opposite orientation with respect to the transcription of the lentiviral genome: the transcription of the viral genome will be transcribed 5′LTR to 3′LTR, and the P + G form the transgene will be transcribed from the 3′ to the 5′ LTR. Note, however, that some polyA addition signals (such as the one from SV40 virus) are bidirectional, and could interfere with virus production even when inserted in the reverse orientation.

10.2.2. Promoters

For ubiquitous expression the human polyubiquitinC promoter (UbiC) produces high and reliable expression in most tissues studied in both mice and rats [14]. Other promoters such as human PGK, chicken β actin, CMV enhancer/promoter or CMV enhancer/chicken β actin promoter do not provide reliable expression across all cell types. In addition, we have successfully generated transgenic animals with the myogenin promoter (specific for skeletal muscle) and lck proximal promoter (specific for T lymphocyte) [15, 16]. It is advisable to use compact promoters (<5 kb) because this will reduce the chances that splicing or polyadenylation signals might be present in those sequences. For instance, the 500 base pairs form of the human synapsin promoter provides excellent levels of expression and high specificity for expression in neurons [17].

10.2.3. Reporters

It is very convenient to include a reporter gene in the construct to be able to quickly assess the titer of the virus during production. If no reporter gene is present in the construct, viral titer should be assessed with immunocytochemistry with antibodies raised against the gene of interest to be overexpressed. In addition, once the transgenic animals are produced, the presence or absence of the reporter can be used to screen for the positive transgenic animals. We routinely include an internal ribosome entry site (IRES) reporter gene cassette in our lentivirus constructs so that we can track expression of the gene of interest simply by looking for reporter gene expression [18]. Currently, three reporter genes, green fluorescent protein (GFP) (and its variants such a Mcherry or TdtTomato or CFP [19]), lacZ and alkaline phosphatase, are widely used.

10.2.3.1. Green Fluorescent Protein

GFP is a small protein (~27 kDa) that in its native state distributes evenly throughout the cytoplasm. It offers several advantages as a reporter gene. First, it can be visualized in live cells either in vivo (by two-photon laser microscopy) or in culture. Second, it can be identified directly by its intrinsic fluorescence or through signal amplification using immunostaining with highly specific polyclonal antibodies (unfortunately, monoclonal antibodies against GFP do not produce effective staining). In addition, by adding specific amino acid sequences, GFP variants have been constructed that limit the expression of GFP to several localized subcellular compartments, such as the nucleus, mitochondria or membrane.

There are some disadvantages that one must consider when using GFP. Its intrinsic fluorescent fades significantly in the first 24 or 48 h post-cellular fixation with aldehydes. In these situations it is advisable to detect GFP expressing cells by immunostaining with antibodies raised against GFP. GFP has a significant threshold for detection. Cells that express low levels of GFP might be missed when examined for GFP intrinsic signal. Again, staining with antibodies raised against GFP will often reveal cells that express GFP below the level of intrinsic fluorescence detection.

10.2.3.2. LacZ

The main advantage of lacZ as a reporter gene is its high sensitivity. Cells expressing lacZ can be incubated for days in the presence of lacZ substrates, thus obtaining a very high amplification of the signal. It is important to remember that some tissues (chondrocytes, choroid plexus, macrophages, etc.) have some level of endogenous galactosidase activity that can be mistaken for lacZ staining. LacZ background staining can be in most cases eliminated or reduced with appropriate fixation [20].

10.2.3.3. Alkaline Phosphatase

Human placental alkaline phosphatase (PLAP), like lacZ, is also extremely sensitive and can be amplified for many hours. Some tissues exhibit some endogenous AP background activity that can be reduced with the chemical, levamisol [21].

10.2.4. Plasmid Preparation

The most common problem that researchers face when producing lentiviruses is recombination of lentiviral vectors when growing the plasmids. DNA from lentiviral backbones does not grow well in some bacterial strains. It is advisable to transform all constructs into Sure or Top10 cells. These strains have been engineered to propagate plasmids that are prone to recombination. To prevent problems associated with abnormal plasmid growth, it is also important to always grow them from a fresh streak. Plasmid yield may be low when grown from cells kept at 4°C on a plate. In addition, it is advisable to grow first a small culture (1 ml) of bacteria for 10 h, then transfer the bacteria to a large-scale culture (200 or 500 ml). Finally, to ensure optimal transfection it is important that the DNA is very clean; use Qiagen maxiprep columns or similar. To store DNA, it is convenient to dilute it at 0.5 μg/μl in water and to freeze the aliquots at −20°C.

10.2.5. Preparation of 293T Cells

The best cells for transfection are 293T cells, which can be obtained from ATCC, (http://www.atcc.org/ATCCAdvancedCatalogSearch/ProductDetails/tabid/452/Default.aspx?ATCCNum=CRL-11268&Template=cellBiology)

10.2.5.1. Cell Thawing

Take a frozen vial from a liquid nitrogen container, and immediately, incubate the vial in a 37°C water bath. The cells should thaw in less than 2 min. Without delay, transfer the cells into a 15 ml tube, and very gently, add 10 ml of pre-warmed Dulbecco's modified Eagle medium (DMEM) containing 10% fetal bovine serum (FBS) drop by drop.

It is important to preserve separate layers of the freezing medium that contains the cells and the serum-containing medium. The freezing medium contains 10–20% DMSO and cells will lyse from osmotic shock if mixed suddenly with the serum medium.

After layering the 10% FBS medium on top of the cells, spin for 5 min at 1,500 rpm (116 × *g*) in a tabletop centrifuge. The cell pellet should be visible at the bottom of the tube. With a Pasteur pipette attached to a vacuum line, remove as much medium from the top of the cell pellet as possible without disturbing the pellet. Add 4 ml of 10% FBS medium on top of the pellet, and resuspend the cells by gently pipetting up and down five to six times. Pipetting too harshly will cause the cells to lyse. Also if pipetting is not thorough enough, cells will remain aggregated in large clumps and will grow very slowly.

When the cells are appropriately resuspended, seed approximately 1–2 million cells on a 6 cm tissue culture plate. Place the plate in a tissue culture incubator (humidified, 37°C and 5% CO_2). Change the medium the next day to further growth. Confluence can occur within 24 h or longer. Once cells become confluent, they are ready to be split.

10.2.5.2. Cell Splitting

Upon cell confluence, use a Pasteur pipette connected to a vacuum line to remove all medium from the plate. Add 5 ml of room temperature PBS to the cells and swirl to wash the whole surface of the plate. With a new Pasteur pipette, remove the PBS and add 1 ml of cold 0.05% trypsin.

Thawed 0.05% trypsin should be kept at 4°C at all times to prevent autodigestion. Do not warm up trypsin before applying it to cells, or the trypsin will degrade, and the cells will not be appropriately dissociated for splitting.

Swirl the trypsin solution to ensure complete spreading over the surface of the plate. Return the plate to the 37°C incubator for 5 min. After 5 min, the cells should easily dislodge from the plate with tapping. If after tapping most cells remain attached to the plate, swirl again to cover the cells with trypsin, and return the plate to the incubator for five more minutes. Tap again to dislodge the cells. Then add 5 ml of pre-warmed 10% FBS medium and transfer the contents of the dish to a 15 ml Falcon tube. Triturate aggressively by applying the tip of the pipette against the bottom of the tube. At this point, the 293 cells are very hardy, and they can withstand very harsh trituration.

Count cells and dilute them to 0.5×10^6 cells/ml. Seed 12 ml of the diluted cells plus medium onto a 10 cm plate, and move the plate randomly to ensure even spreading of the cells and to prevent regional clumping. If the plates are moved in a circular motion, the cells clump in the center of the plate.

Once the cells in the 10 cm plate reach confluence, trypsinize as described above and dilute the cells to 0.5×10^6 cells/ml. At this stage, one can dilute the cells from a 10 cm plate into 36 ml and plate them into three 10 cm plates for transfection; or add 25 ml of the diluted cells to a 15 cm plate for further growth. As a rule of thumb, one confluent plate should contain enough cells to seed three plates of the same size to be transfected the next day.

It is important for cells to be close to confluent at the time of transfection because cells do not grow well after transfection. One of the most common reasons for low virus titer is inadequate confluency of the cells upon transfection. If cells are too confluent then the transfection will be inefficient. If they are too sparse, then they will die. Many genes are quite toxic to 293T cells upon transfection. In these cases it is very important to have cells close to confluence just before transfection, such that they reach confluency within 3–4 h after transfection.

Because production of virus is relatively time consuming, it is practical to prepare a large stock of virus and to keep it frozen in aliquots until further use. To prepare a stock of virus it is convenient to transfect cells in 10 or 15 cm plates.

10.2.5.3. Transfection

Sixteen to 20 hours after the final seeding, remove 2 ml (for a 10 cm plate seeded with 12 ml) or 5 ml (for a 15 cm plate seeded with 25 ml) of medium from each plate to be transfected. Leave cells in the incubator for at least 1 h after medium removal. Leaving plates out of the incubator for too long will change the pH and the transfection will not be efficient.

The quality of the transfection reagents is probably the main determinant for obtaining high viral titers. Pay special attention to the preparation of the 2× HBSS solution. This solution is composed of the following components:

0.28 M of NaCl

0.05 M of HEPES buffer

1.5 mM of Na_2HPO4

Fill up to 500 ml with deionized water. Titer to *exactly* pH 7.03–7.04 with NaOH and filter sterilize through a 0.22 μm filter.

Because the sodium phosphate component is included in a small amount (less than a gram per liter), use an accurate balance capable of measuring milligrams to weigh it. Calibrate the pH meter just before adjusting the pH of the solution. Leave the solution stirring for 10 min after the pH of 7.03–7.04 has been reached to ensure that it does not drift. Once the pH is stable, sterilize by filtering. HBSS prepared in this way can be kept at room temperature for at least a year.

Other transfection components include 2.5 M calcium chloride, water and the DNA to be transfected. We usually keep the 2.5 M calcium chloride solution frozen at −20°C in 5 ml aliquots and thaw immediately before transfection. It is also possible to transfect the 293T cells with commercial reagents such a lipofectamine, lipofectin, etc. However, these reagents are expensive and they might be toxic for the target cells to be infected with the concentrated virus.

To prepare for transfection, mix the DNA, water and calcium chloride in 15 ml polypropylene tubes, one for each plate to be transfected. Sterility is not required at this stage. For a 10 cm plate, mix 10 μg of transfer vector, 7.5 μg of delta 8.9 plasmid and 2.5 μg of VSVg plasmid. Add 100 μl of 2.5 M $CaCl_2$, and fill up to 1 ml with ddH_2O. For a 15 cm plate, mix 20 μg of transfer vector, 15 μg of delta 8.9 plasmid, plus 5 μg of VSVg plasmid. Add 200 μl of 2.5 M $CaCl_2$, and fill up to 2 ml with ddH_2O.

In a Biosafety level 2 hood, prepare 1 ml (for a 10 cm plate) or 2 ml (for a 15 cm plate) of 2× HBSS pH 7.03–7.04, in separate 15 ml polypropylene tubes, one for each plate to be transfected. It is important to perform this step in the hood to prevent contamination of the HBSS stock. All remaining steps will occur in the hood.

Take one plate at a time out of the incubator and place it in the hood. Mix the tube containing DNA + $CaCl_2$+H_2O and add the contents of this tube into the 2× HBSS. Aggressively pipette the mixture five times, up and down. Once mixed, add the DNA + $CaCl_2$ + HBSS mixture to the side of the plate while swirling, and put the plate back in the incubator. Again, it is important to minimize the time that each plate is outside of the TC incubator to prevent significant pH changes that will reduce the efficiency of the transfection. Repeat this transfection procedure for all remaining plates.

Three and a half hours after adding the DNA + $CaCl_2$ + HBSS mix to the cells, aspirate the medium from the plates, wash each plate twice with PBS, and add 4 ml (10 cm plate) or 8 ml (15 ml plate) of new DMEM + 10% FBS medium. If cells are not washed with PBS, some of the precipitate might remain attached to the plastic and many of the cells will die. Return the plates to the incubator for 60–72 h at which time the first round of supernatant collection will occur.

If the transfer vector contains GFP or other fluorescent reporter, the cells can be examined 24 h after transfection by live microscopy. This is also a convenient time to check for health status of the 293T cells under a microscope; 48 h after transfection, more than 90% of the cells should be transfected.

If the transfection efficiency is low, re-check the quality of the DNA, the HBSS or the 293T cells. There are some common reasons for poor transfection of 293T cells. First, the cells may be contaminated by *Mycoplasma*. Second, the cells may have been abused and allowed to become overconfluent during previous passages. If either of these situations has occurred, it is advisable to get a new seed of 293T cells from ATCC. Another reason for poor transfection is toxicity of 293T cells with expression of certain constructs at high levels. In these instances the construct will prevent the transfected 293T cells from becoming confluent. The best solution in this case is to remove the transfection mixture 3 h after adding the DNA + $CaCl_2$ + HBSS mix, and recover the virus 24 or 30 h after transfecting the cells.

10.2.6. Viral Concentration

You will collect viral supernatant from plates at 48 and 72 h after the initial transfection. Prior to collection on both days, be sure to equilibrate the ultra centrifuge, rotor and buckets to 4°C. To pre-cool the ultracentrifuge, turn it on, set temperature to 4°C and apply the vacuum. It will take at least 30 min before the centrifuge reaches 4°C. To pre-cool the swinging-bucket rotor and buckets, leave them in a refrigerator overnight. Six 10 cm plates will generate 48 ml of viral supernatant; so the centrifuge tubes should hold at least 9 ml.

To collect the viral supernatant begin by removing the viral plates from the incubator and pipetting the medium into 15 ml conical tubes (10 cm plate) or 50 ml conical tubes (15 cm plate). For the first day of virus collection, replace the medium removed with warm, fresh DMEM + 10% FBS in the amount per plate indicated above (4 ml for 10 cm plates and 8 ml for 25 ml plates). The next day you will collect this medium for the second round of viral concentration using the same process described below.

Spin the tubes at 1,500 rpm (400 × g) in a tabletop centrifuge for 5 min to remove large cellular debris. During this process under sterile conditions, add 5 ml of DMEM + 10% FBS medium into a syringe with a 0.8 μm filter attached to it, and let it the medium flow through the filter by gravity into ultracentrifuge tubes, prewashed with 100% ethanol. Pre-wetting the filter reduces the nonspecific binding of viruses that would otherwise reduce the viral titer. Wash the tubes with the medium and discard the filtered medium. Immediately after pre-wetting, add the virus suspension to the syringe, and use the plunger to push the suspension through the filter into the ultracentrifuge tubes.

Put sealed tubes containing the virus suspension into the ultracentrifuge buckets. Weigh buckets containing the virus suspension tubes, and make sure that they are completely balanced. It is very important that the tubes are correctly balanced due to the high speed at which the ultracentrifuge operates. If necessary, prepare tubes with water in other buckets for balance. When assured of balance between the tubes, tightly position the buckets in rotor in the specified slots and install the rotor in the centrifuge.

Adjust the centrifuge to spin for 90 min at 4°C and at 25,000 rpm (112,500 × *g*). Watch the centrifuge until the rotation speed is at least 3,000 rpm (463 × *g*). If the centrifuge produces any error signal, stop it immediately and check bucket placement or weight. When spinning is complete, disconnect the vacuum of the ultracentrifuge. Return viral suspension tubes to the TC hood. Discard the viral supernatant onto a biohazard waste collection. Invert the tubes over a paper tissue to drain remnants of medium.

It is usually very difficult to detect any pellet at the bottom of the tube. If a pellet is clearly visible, this probably indicates an excess of cellular debris in the supernatant. Excessive cellular debris usually indicates that many of the cells are dying after the transfection.

With a sterile Pasteur pipette, aspirate the supernatant exhaustively until the tube seems dry. Add 30 μl of cold, sterile PBS (no bicarbonate, + Ca, + Mg) to the bottom of the tube. Seal the tubes to prevent evaporation. Gently swirl the tubes at 4°C for at least 12 h in the cold room to dissolve the pellet. It is advisable to keep tubes rocking overnight to fully resuspend the virus. After 12–24 h at 4°C, pipette up and down gently to dissolve the pellet.

Two steps are important for achieving successful resuspension. First, make sure that all of the DMEM + 10% FBS medium is aspirated from the tube. Fetal bovine serum might contain some components that inactivate the virus when incubated at 4°C. Second, the virus is relatively frail, and is easily inactivated by aggressive pipetting. It is particularly important to prevent any kind of frothing at this step.

Aliquot the virus into 3 μl aliquots to store at −70°C, or use immediately. Label the date of harvest to check the titer.

10.2.7. Virus Titration

You will perform two titrations, one for each viral concentration. On the day of titration, resuspend 293T cells at 5×10^6/ml or HeLa cells at 5×10^5/ml. Alternatively for virus with tissue--specific expression, resuspend the appropriate cell type in which the promoter will be active. To determine viral titer, you will infect these cells with virus and measure reporter expression directly or through immunocytochemistry (ICC).

If it is necessary to perform immunostaining for the titration it is required to coat the plates with poly-lysine, or the cells will be the lost during the washes. Coat a 96-well plate with 100 µl/well of poly-D-lysine (10 ug/µl). Avoid using the perimeter wells. Microwave this solution for 40 s and then aspirate 90% of the solution from each well, leaving a thin coating. Let the plate dry in the culture hood with the cover ajar. During the drying process, proceed with virus dilution.

Aliquot 90 µl of the cell suspension into five 0.2 ml tubes. In addition, into one 3 µl viral aliquot add 97 µl of the cell suspension. Thoroughly mix the cell suspension and virus. With a new tip, remove 10 µl of the cell suspension with virus, and add to the first of the tubes containing 90 µl. Mix the suspension well. Continue this dilution with all six tubes. Upon completion, remove 90 µl from the most dilute virus and cell suspension, and add to the last coated well on the 96-well plate. Continue this procedure with each progressive virus concentration, moving up the column with each dilution.

Move the well plate to the incubator; 72 h after seeding the plate, you can look for the titer directly if GFP expression is detectable. Count the number of colonies in the most dilute well to estimate the titer. The final titer per microliter is equal to the number of colonies in that well divided by the initial virus volume multiplied by 10^6. For example, 30 colonies of GFP+ cells in the −6 dilution yields $3 \times 10^6/3$ µl $= 1 \times 10^6$ infectious units per microliter (IU/µl). A reasonable titer after concentration is $0.5–5 \times 10^6$ IU/µl.

If visible reporters such as GFP, lacZ or PLAP are not present in the virus, titrate by ICC with antibodies raised against the gene driven by the internal promoter. Make sure to pre-coat the wells with poly-lysine. Although the well plate has been coated, the cells can easily dislodge during the immunostaining process. Be gentle when adding and removing solutions so that you determine an accurate viral titer.

If there are no established cell lines in which the promoter is active, an approximate titration can be obtained by infecting primary cells obtained from the animal and maintaining them in culture for a few days. If there are no specific antibodies that can detect the expression of the transgene, viral titer can be determined by ELISA against the matrix protein of the virus, or by quantifying reverse transcriptase activity in the viral concentrate.

10.3 Production of Transgenic Mice or Rats with Lentiviral Vectors

The generation of transgenic animals with lentiviral vectors involves the use of some procedures that are common with other transgenic techniques such as pronuclear injection or blastocyst transfer. All aspects of superovulation, embryo collection and implantation are described elsewhere in this manual. *See* Chaps. 6 and 25.

10.3.1. Superovulation and Embryo Collection

Superovulation increases the number of ovulated eggs via administration of gonadotropins to female rodents prior to mating. These animals are superovulated with a combination of pregnant mare's serum (PMS) and human chorionic gonadotropin (hCG). To induce superovulation, prepubescent female mice (~25 days old) are injected intraperitoneally with 0.35 IU/g of PMS (Sigma G4527, 25 IU/ml in 0.9% NaCl) on day-2, followed by 0.35 IU/g of hCG (Sigma C8554, 25 IU/ml in 0.9% NaCl) 48 hours later on day 0. Prepubescent female rats between 28 and 30 days of age are injected intraperitoneally with 0.35 IU/g of PMS on day-2, followed by 0.06 IU/g of hCG 48 h later on day 0. For both rats and mice, hormone-treated females are caged with fertile males 2–3 months of age to mate overnight. On the morning of day 1, females are checked for copulation plugs. Embryos are collected from female mice on the morning of day 1, whereas embryos are collected from female rats on the afternoon of day 1.

Embryos are collected from mice and rats using the same procedure as described in [22] and Chaps. 6 and 25, in this book. After animals have been euthanised, the oviducts are excised and transferred to a dish containing M2 medium at room temperature. Newly ovulated embryos, enclosed by cumulus mass cells, are released from the swollen ampullae (the upper portion of the oviduct) by gently tugging and opening the walls of the ampullae with fine forceps. The embryos are then transferred to a dish containing hyaluronidase solution (Sigma H3884, 300 ug/ml in M2 medium). This solution enzymatically digests the cumulus cells thus releasing the embryos. As soon as the cumulus cells shed several minutes later, the embryos are transferred to fresh M2 medium to wash off the hyaluronidase solution and preserve the viability of the embryos.

In rats, the cumulus cells adhere tenaciously to the surface of some embryos and are difficult to remove completely, even with hyaluronidase. In general, their presence does not seem to hinder subsequent manipulations with the zygotes. The embryos are then transferred to microdrops of M16 medium [22] under mineral oil and cultured in a humidified 37°C incubator under 5% CO_2 until needed.

10.3.2. Delivery of Lentiviruses to Single-Cell Embryos

Lentiviruses are delivered to the fertilized oocytes on the same day of collection, targeting only single-cell zygotes to minimize mosaicism. Two methods are available to deliver the lentiviruses to the embryos.

10.3.2.1. Microinjection of Lentiviruses into the Perivitelline Space of Single-Cell Embryos

Micropipettes are prepared by pulling borosilicate glass capillaries (Sutter, 1 mm O.D., 0.7 mm I.D., no filaments) on a pipette puller (use a program similar to the one used to produce blastocyst injection pipettes). After pulling, use a razor blade to break the tip of the pipette. Sutter instruments provides a ceramic tile for breaking the tip of glass pipettes such that flush ends of approximately 10–15 μm outer diameter are produced. Before loading the pipette with the viral solution, inspect the tip under a 40× magnification objective in an inverted microscope equipped with a micrometer in one of the oculars. The best pipettes have an outer diameter between 10 and 20 μm. It is difficult to penetrate the zona pellucida with pipettes larger than 25 μm. It is difficult to load the viral solution into pipettes smaller than 10 μm. To ease the penetration of the zona pellucida, it is advisable to have a bevel-tipped pipette tip.

The micropipette is inserted into the pipette holder of a hydraulic injector (CellTram Oil by Eppendorf is very easy to use for this application). Check for trapped air bubbles in the line connecting the injector to the micropipette. Air bubbles will make it very difficult to control the flow of the virus when attempting to deliver it into the zygote.

Thaw a frozen aliquot of lentiviral concentrate to room temperature. Gently pipette the viral solution to release any large aggregates of cellular debris. Then spin the virus at 1,000 rpm (51 × *g*) for 1 min to separate large aggregates that will clog the injection pipette. To avoid any debris settled at the bottom, take the virus to be loaded from the top of the aliquot immediately after spinning. Four or five microliters of the viral suspension is then placed on a coverglass and placed on the stage of the inverted microscope. Place the tip of the glass pipette inside the viral drop, and load the viral solution into the micropipette from the tip using gentle negative pressure from the hydraulic device. It takes approximately 10 min until 3–4 μl is loaded into the tip.

Once under the microscope, it is normal to see debris in the viral supernatant. However, the particles should be smaller than 2 μm. To prevent clogging, the pipette can be moved into regions of the drop with a lower concentration of debris. If the tip becomes clogged, apply positive pressure until the mass is expelled, and start again by applying negative pressure. If the pipette clogs repeatedly during loading there may be two problems. First, the pipette tip might be too small. In this case, break a bigger tip in a new pipette. Second, the viral solution might be too dirty. Spin the virus again at low speed to get rid of debris. It is convenient to observe the loading procedure under the microscope to detect cellular debris that could clog the pipette tip and to correct problems before they occur.

Three microliters of viral solution should be sufficient to inject approximately 200 embryos. Try to reduce the time that the virus is at room temperature, because its half-life in these conditions is quite short. Some studies suggest that most of the virus will be inactive after 12 h at room temperature.

Single-cell embryos are transferred to a microdrop of M2 medium on a depression slide and covered with mineral oil to prevent evaporation and to maintain osmolarity. The slide is mounted on the stage of an inverted microscope, and the injection procedure is monitored under 400× magnification. Embryos are held in place against a fire-polished pipette using gentle negative pressure. The holder with the injection pipette containing the virus is loaded onto a micromanipulator, and the tip of the pipette is lowered into the microdrop containing the embryos under visual control with the microscope. Gently apply some positive pressure so that the viral concentrate oozes slowly but continuously from the pipette tip. It is easy to detect the flow of virus coming out off the pipette by observing the currents that it generates in the medium. In addition, cellular debris may be visualized.

Using the micromanipulator to guide the pipette, swiftly push the tip into the region between the zona pellucida and the oocytes cell membrane. To prevent rupturing the embryo, it is important not to push the pipette too deep inside of the perivitelline space. Under 400× magnification, some of the cell debris from the viral solution can be seen slowly flowing from the pipette tip into the perivitelline space. The tip is left in the perivitelline space for 10–15 s until the space expands, and the body of the ova forms a crescent shape as it is displaced by the discharging viral prep. At this point, withdraw the micropipette from the zygote.

Some problems can occur during the virus administration process. If the zona pellucida enlarges substantially when penetrated by the pipette, then the flow rate is too high, and the viral solution will be depleted before most of the embryos can be injected. Furthermore, injection of an excessive volume in the perivitelline space will collapse the embryo and result in its death. If the virus stops flowing during the injection procedure, apply some more positive pressure. If the flow does not resume, then it is likely that the pipette has clogged with some large cellular debris. Very slowly, apply positive pressure until the clog can be seen in the distal end of the pipette tip. Immediately before the clog is to be expelled from the tip, reduce the pressure to prevent a massive outflow of the virus into the medium. If the clog cannot be expelled with high pressure, then the best course of action is to discard the pipette and to load a new pipette with an appropriate tip to resume the injection.

After the injection, sort and discard any embryos that appear lysed or abnormal. The remaining embryos are transferred to M16 microdrops under oil and cultured in a 37°C incubator under 5% CO_2 until implantation. When the embryos are infected by injection through the zona pellucida, it is possible to implant them either on the same day of injection or 3 days later when they have reached the blastocyst stage.

10.3.2.2. Co-incubation of Denuded Single-Cell Embryos with Lentivirus

The zona pellucidae of the fertilized oocytes are removed by incubation in either an acidic Tyrode's solution [22] for less than a minute, or a 0.5% pronase solution in M2 medium at 37°C in a humidified 5% CO_2 incubator for several minutes. When the zonae appear to be dissolved, embryos are washed in excess M2 medium and then transferred into 10 μl microdrops of viral suspension under mineral oil. To prevent the loss of embryos due to adhesion to the glass walls of the transfer pipettes, it is important to pre-coat transfer pipettes with 1% albumin in PBS or, alternatively, to use plastic transfer pipettes. Individual embryos are cultured in separate microdrops to prevent them from adhering and producing chimeric animals.

The viral suspension can be diluted to various concentrations to roughly control the average number of proviral integrations expected per transgenic genome, since the number of insertions should scale down with the concentration of virus used. We usually incubate the zygotes in 10 μl microdrops of virus diluted to 2×10^4 IU/μl, 4,000 IU/μl and 800 IU/μl, which results in seven, three and one proviral copies per genome on average, respectively.

Zygotes are incubated in the viral suspension until the embryos reach the morula stage and are subsequently implanted. As indicated before, denuded embryos stick to the glass wall of the pipettes, resulting in a significant loss of embryos. Implantation of denuded embryos prior to the morula stage is inefficient (unpublished observations).

10.3.3. Transfer of Embryos into Recipient Females

Timed pseudopregnant females need to host the treated embryos. Sexually mature females in estrus mate with vasectomized, mature males the night before the intended day of implantation. Males are vasectomized by tying off the vas deferens at two separate locations approximately 5–6 mm apart, then cauterizing the intervening segment to sever the tube. Males are vasectomized at least 2 weeks prior to the mating to ensure that all remaining sperm in the genital tract are dead at the time of mating.

For mice, we maintain a large colony of recipient females, and at any given time, some proportion of those females will be receptive to mating, as determined by the presence of a copulatory plug the morning of the day of intended implantation. For rats, an alternative to maintaining a large colony of females is to order timed pregnant females that will deliver the day before that of intended implantation. Immediately after delivery, female rats are in estrus and will be receptive to mating that night. In addition, vaginal smears can be performed the night before implantation to select the females in estrus to be placed with the vasectomized males. Estrus synchronization by chemical means is also possible [23]. Mating is confirmed the next morning by checking for a copulatory plug.

Embryos infected by perivitelline injection should be transferred into the oviduct of timed pseudopregnant females as soon as possible for maximum rates of implantation (early stage embryos, 0–1 dpc). Denuded or injected embryos that have reached the morula or blastocyst stage should be transferred to the uterus of timed pseudopregnant females (2.5 dpc). In general, no more than 30 embryos are transferred bilaterally into the uterus [22]. Pregnancy and delivery of the transgenic litter are as usual.

10.4 Production of Transgenic Birds Using Lentiviral Vectors

The generation of transgenic birds using lentiviral vectors is an efficient, relatively simple procedure compared to the production of transgenic mice and other mammals. The method for avian transgenesis described here is similar to previous methods used to produce transgenic chickens by the microinjection of recombinant oncoretroviruses into the early embryo of freshly laid eggs [24]. Previous studies of transgenic chickens and quails produced with oncoretroviral vectors reported low or undetectable levels of transgene expression. In contrast, we and others have demonstrated that lentiviral vectors allow for high levels of transgene expression in quails, chickens and songbirds [17, 25, 26]. Although this system has only been tested in chickens, quails and zebra finches, we anticipate that lentiviral transgenesis will be possible in a wide variety of avian species.

10.4.1. Overview of the Strategy

In birds, after fertilization of the oocyte by the sperm, the embryo starts developing inside of the female's reproductive tract until the egg is laid a few days later. During the migration of the fertilized embryo through the reproductive tract, two key processes occur simultaneously. First, the cells in the embryo proliferate at a fast rate in such a way that by the time the egg is laid (3–5 days post-fertilization) the embryo will consists of 20,000–50,000 cells (depending on the species). Second, during this migration the egg will acquire a calcareous shell. Therefore, once the embryo has completed its intrauterine migration, the female will lay a single embryo consisting of tens of thousands of cells surrounded by an opaque shell.

These special characteristics of avian development pose some obstacles for the generation of transgenic birds. First, the embryo cannot be easily observed with a microscope because the shell prevents its visualization. Second, because the embryo consists of some 20–40,000 cells, it will not be possible to deliver transgenes to 100% of the embryonic cells. Accordingly, the strategy for transgenesis in birds is based on the delivery of transgenes to a percentage of the cells of the founder embryo. When this embryo develops, it will become a mosaic animal (F0) in which a percentage of its cells will contain the transgene. If some of the transgene-carrying cells in the F0 mosaic founder differentiate into sperm or oocytes, then when these mosaic founders breed, they will pass the transgene to a percentage of their progeny (F1). These progeny will become transgenic animals, carrying a copy of the transgene in all cells of the body.

10.4.2. Design of Lentiviral Vectors for Transgene Expression

The general construction of lentiviral vectors has been discussed previously in this chapter. Here we will discuss some specific considerations for the design of vectors used to make transgenic birds.

10.4.2.1. Promoters

Most of the well-characterized promoters have been studied only in mice or other mammals. Some promoter/enhancer combinations can have quite different transcriptional activity in mice and birds, which can lead to changes in tissue specificity or overall expression level from species to species. Often these differences in gene expression can be difficult to predict, especially when the promoter is a viral sequence which shares no homology with regulatory elements in the avian genome, such as the immediate early promoter from the human cytomegalovirus, CMV. In mice, the CMV promoter has been used to produce animals with strong transgene expression in most tissues [27]. In contrast, transgenic chickens made to carry the CMV promoter driving the green fluorescent protein (GFP) exhibited transgene expression primarily in the pancreas and skin, and very weak expression in other tissues [25]. Another promoter used to drive constitutive transgene expression in mice, the promoter from the human UbiquitinC gene, produces only low levels of transgene expression in quail [17], but high levels of expression in finches [26].

Before using a promoter/enhancer combination that has not been well characterized in birds, transcriptional activity should be tested either in cultured cells from that species or by in ovo injection. We have found two promoter/enhancer combinations that produce useful patterns of gene expression in transgenic quail. Using vectors carrying the human synapsin gene I promoter we have produced a line of transgenic quail with neuron-specific expression of GFP. Additionally, we found that the combination of the CMV enhancer and the chicken β-actin promoter produces ubiquitous expression in quail.

Further consideration must be given to differences in transcriptional activity of a promoter from one avian species to another. For instance, we have found that the CMV/β-actin promoter leads to high expression in quail, but weak expression in zebra finches. Before the generation of a transgenics in a novel species is attempted, candidate promoters should be tested to determine their activity in that species.

Another important consideration when choosing a promoter/enhancer combination is the possibility of side effects due to high levels of expression from ubiquitous promoters. Some constitutive promoters can overburden the normal transcriptional machinery of the cell by re-directing resources away from genes necessary for cell growth to the transgene. This effect slows cell proliferation, which puts infected cells at a selective disadvantage when compared to the naive, non-infected cells of the embryo and lowers the germline transmission frequency of adult mosaic.

We have observed this problem when generating transgenic quail that ubiquitously express GFP. When mosaic founder parent birds were generated with a lentiviral vector carrying the chicken β-actin promoter and CMV enhancer driving GFP, somatic mosaicism in adult founders was low (<1% of cells expression GFP). We found that only one out of ten founder quail produced transgenic offspring and germline transmission frequency in those animals was low (<5% of progeny were transgenic).

In contrast, transgenic quails generated with a neuron-specific promoter (human synapsin I) driving GFP expression showed very high levels of germline transmission. Transgenic quail with neuron-specific expression had a germline transmission frequency between 5 and 33%. Five out of six mosaic quail produced transgenic offspring. Additionally somatic mosaicism in the neuron-specific transgenics was greater (up to 10% of neurons in these animals expressed GFP) than that of quail generated with the Chicken β-actin/CMV vector (less than 1% of cells expressed GFP in the mosaic founders).

Special care should be used when choosing promoters that may express in the germline of the developing bird. The additional demands made on the cell's transcriptional machinery can create an obstacle for transgenesis. Any manipulation that decreases the growth rate of infected germ cells may ultimately lower the germline transmission rate of mosaic founders.

10.4.3. Production of F0 Mosaic Founders

Fertile eggs from many domestic species including quail, chicken, ducks and turkey are available by mail from farms or other breeders. Since bird embryos will start to develop when kept at room temperature, it is best to select a vendor that can guarantee freshly laid eggs and that can ship overnight. Vendors that specialize in avian research are preferable. We obtain quail eggs from CBT Farms (Chestertown, MD 2160). If it is not convenient to inject eggs the day they arrive, eggs can be stored for up to 1 week at 15°C. When storing eggs for any length of time, we place eggs in a dedicated refrigerator in crates and orient eggs so that the long axis of the egg is parallel to the ground.

Once the eggs arrive in the laboratory, the crates are unpacked and the eggs placed in racks (Stromberg's Chicks, Pine River, MN; catalogue # ET124) and oriented so that the long axis of the egg is parallel to the ground. Keeping the eggs in this position will ensure that the embryo will be accessible in the window that will be created later for virus delivery. The eggs are left in racks at room temperature for 1 h; this allows the embryo to settle to the top of the egg. After 1 h, we find that a majority of embryos lie just below the highest point of the shell. We mark this point with a felt tip pen to identify the presumptive location of the embryo.

For the windowing procedure, the eggs are transferred to individual soft foam cradles. Using a hand rotary tool with a small bit, a circle with a 4 mm radius is drilled in the shell around the mark indicating the top of the egg. The window should be shallow to avoid damaging the shell membrane. We have used both electric hand drills (Dremel) and air-powered dental drills (Benco Dental) to cut the window. Shell chips and dust can accumulate during the windowing. To eliminate this shell dust, dental drills can be purchased with a second channel that expels air at the head of the drill. This air stream is useful for blowing away the shell chips and dust that accumulate during the drilling. If shell chips and dust remain after the cutting process, these should be wiped away with a Kimwipe moistened with 70% ethanol.

After the circular cut is made, the remaining shell fragments are removed with sterilized forceps to expose the shell membrane. Before removing the shell membrane, a drop of sterile phosphate buffered saline (PBS) large enough to cover the exposed shell membrane is added over the window. Recently, it has been shown that the addition of PBS before the removal of the shell membrane significantly increases hatchability [28]. The shell membrane is then removed with forceps taking care not to disrupt the vitelline membrane that surrounds the embryo and yolk.

10.4.3.1. Injection

Once the shell membrane is removed, the embryo can be observed directly. However, for precise targeting of the embryo, we place the egg in a soft cradle under a dissecting microscope with 16× magnification.

Microinjection needles are fabricated from borosilicate glass capillary tubes (Sutter catalogue # B100-75-10) pulled using a Flaming/Brown micropipette puller model P-97 equipped with a 3 mm box filament and use the following parameter settings:

Heat = Ramp + 15 Pull = 30 Vel = 120
Time = 200 Pressure = 200

A ceramic tile (Sutter) is used to score and break the pulled glass to give a blunt tip with a 15–20 μm outer diameter. Once cut, the microinjection needle is inserted into the pipette holder of a CellTram Oil hydraulic injector (Eppendorf) and front-loaded with concentrated viral solution slowly (roughly 200 nl/s). To be able to visualize the viral solution as it is injected, we add 5% phenol red before we load the viral solution into the pipette. We typically load 10–15 μl of viral solution, mixed with phenol red, into a microinjection needle. A fully loaded needle should allow for up to 5 egg injections (for quail or chicken). Considerations for virus degradation at room temperature are the same as those described above for rodent embryos.

High titer viral stocks are critical for the production of transgenic birds. We have successfully used viral stocks with titers of 10^6–10^7 IU/μl when titered on 293T cells. Using viral stocks of 10^5 IU/μl or lower will result in low rates of germline transmission frequencies in mosaic birds. Previous studies using unconcentrated retroviral vectors (10^4 IU/μl) to produce transgenic birds have reported low germline transmission frequencies (<1%) in mosaic founders.

Under a dissecting microscope, the microinjection needle is positioned above the center of the embryo using a micromanipulator. The needle tip is then advanced to penetrate the embryo, at an angle of approximately 45°. A Piezo Drill (Burleigh) attached to the microinjection needle can be used to help the needle tip rupture the vitelline membrane and penetrate the embryo. After the embryo has been penetrated, 3 μl of the vector solution is slowly injected into the subgerminal cavity, below the embryo. An injection is considered successful if the viral solution is observed spreading horizontally in a circle below the embryo and if the perimeter of the viral solution reaches the borders of the area opaca. More than 90% of the injections are successful according to these criteria. Eggs with unsuccessful injections are discarded.

10.4.3.2. Postinjection

Immediately after a successful injection, the eggs are sealed in order to prevent microbial contamination and fluid loss during incubation. To seal the eggshell, a round glass cover slip (EMS, Ft. Washington, PA; catalogue # 72196-12) is placed over the shell window and adhered to the egg with a biocompatible silicone elastomer (Kwik-Cast WPI). For quails or chickens, we routinely inject 60–80 eggs per day. Assuming a hatching rate of 5–10%, this number of injected eggs will provide a sufficient number of founder animals for breeding.

Eggs are placed blunt end up, in a forced air incubator (Brinsea) with a temperature of 38°C and a relative humidity of 45% until hatching roughly 18 days later. The method of egg turning is an important factor to consider when choosing an incubator. Artificial incubator models that keep each egg (or eggs) in a single plastic cradle and change the orientation of the cradle, such as Brinsea and Alpha Genesis, are preferable to incubators that roll the eggs, such as Grumbach.

This simple method of sealing and incubating the eggs typically yields a 10% hatch rate. Other incubation methods, both in artificial culture systems and in ovo have been described with an increased hatching rate up to 30 or 60% [29, 30]. However, since germline transmission rates are high enough in mosaics produced with lentiviral vectors, only a few founders are needed to guarantee transgenic offspring.

Husbandry concerns vary from species to species. Our laboratory has experience raising quails (*coturnix coturnix*) and zebra finches (*taeniopygia guttata*). Quails are the preferred species for testing injection methods, promoter/enhancers and general troubleshooting. They develop at a rapid rate becoming sexually mature roughly 50 days after hatching, and they are easy to house and maintain. A colony of 30 breeding quails can be easily established by purchasing ready-to-assemble breeder cages (Stromberg's) that only occupy 20 sq ft. of laboratory space. For most laboratories in an academic environment, it will be difficult to establish a colony of chickens because their large size will require a special facility for their breeding. Zebra finches require more time to become sexually mature (90 days) and must be raised by foster parents. Tutorials on quail husbandry are available with an agricultural or research focus [31].

Our quails hatch in the incubator and are immediately transferred to a brooder (Brinsea) that is maintained at a temperature of 37°C. After 5 days, the hatchlings are transferred to a larger battery cage modified for hatchlings. The floors and sides have been reinforced with tighter mesh wiring (9 mm × 9 mm hole size) and the cage is equipped with an area heater (Start N Gro; Stromberg's Chicks catalogue # SNGB) set to produce an average air temperature of 35°C. Quail can also be raised in commercially available brooders, but our design can house both adults and hatchlings on different levels of a single rack of cages. Hatchlings are fed adult food that has been finely ground to a powder in a blender. Since quail hatchlings can drown in even shallow depths of standing water, we provided drinking water either in low plastic drinking fountains (Stromberg's Chicks catalogue # NDF) or in shallow tissue culture plates.

10.4.3.3. Transgenic Offspring

Once mosaic founders have reached sexual maturity (7 weeks after hatching), these animals are bred to wild type quail. Wild type quail can be purchased from breeders and are usually inexpensive (<$10 per adult quail). However, shipping live birds can be quite expensive (approx. $300). Therefore, the best strategy is to order additional eggs and let them develop alongside the injected eggs, or alternatively, to breed mosaic founder quail to each other.

Since founder quail will be mosaic in all tissues, including the germ cells, only a percentage of their offspring will be fully transgenic. If the transgene produces a visible marker, such as green fluorescent protein, the progeny can be screened for the presence of the transgene by phenotype. In general, however, screening is accomplished by either PCR or Southern blot analysis. Genomic DNA is required for both methods and can be easily extracted from the blood of birds. The alar vein of 5-day-old hatchlings is an excellent source of blood. The vein is nicked with a hypodermic needle and 70 μl blood is collected with a heparinized capillary tube (Chase Scientific Glass, Rockwood, TN). Genomic DNA can be extracted from blood by digesting overnight in proteinase K followed by phenol chloroform [32].

Lentiviral vectors integrate stably into the host chromosome and are passed through the germline to subsequent generations. Transgene expression is consistent from generation to generation. Positively identified transgenic birds can be used to establish breeding lines.

10.5 Cloning of Integration Sites in the Genomic DNA

Retroviruses contain an RNA genome. Upon viral entry into the target cell, this RNA is copied into the cell's DNA by the viral enzyme reverse transcriptase. After reverse transcription, the viral DNA integrates into the cell's genome and is called a provirus [33]. In many cases, the levels and pattern of expression of an individual provirus depend on the locus of integration in the cell's genome. There are two important consequences derived from the integration of the virus into a particular genomic location. First, the provirus can integrate into a cellular gene and cause an increase, decrease or elimination of expression of that gene. This phenomenon is called insertional mutagenesis. Insertional mutagenesis can affect cell genes in several ways. If the integration occurs in the middle of an exon, the provirus will likely inactivate the gene. If the provirus integrates upstream of the open reading frame of the cellular gene, the internal promoter of the provirus could enhance the expression of the cellular gene. If the provirus integrates in an intron, it may have no effect on the gene's expression because in many cases the provirus will not affect the splicing processing. Second, the sequences in the genomic DNA that flank the provirus may affect the level or pattern of expression of the inserted transgene, a phenomenon known as position effect. Thus, to investigate both insertional mutagenesis and position effects it is necessary to characterize the integration sites of the provirus in the cellular genome.

Next we describe a procedure to clone insertion sites of lentiviral proviruses in the genome of transgenic animals. In short, the procedure involves amplifying the proviral DNA and its flanking sequence by PCR. This is accomplished by digesting the cell's genomic DNA and ligating it to a linker primer. After this, PCR is performed with two primers, one that recognizes the ligated linker and one that recognizes the proviral sequence. Subsequent rounds of amplification are performed with nested primers to increase the specific product (provirus + flanking region). The final PCR product is sequenced, and a sequence analysis comparison is performed against the genome of interest to identify the genomic region where the provirus is located.

10.5.1. Annealing of the Linker

Prepare a 10× annealing buffer.

10× annealing buffer contains

- 100 μl 1 M Tris pH8
- 100 μl 5 M NaCl
- 20 μl 0.5 M EDTA pH 8
- 780 μl H_2O

Split this mixture into 250 μl aliquots, and store at −20°C.

The Afl3 oligos described next will amplify the 3′ end of the provirus. To amplify the 5′ end of the provirus, use the Afl5 oligos descried at the end of the chapter, under the heading "***description of oligos***".

Resuspend the 2 Afl3 oligos (see below for sequences) at a concentration of 500 μM in H_2O and anneal as follows:

- 20 μl of Afl3 upper oligo (at 500 μM)
- 20 μl of Afl3 lower oligo (at 500 μM)
- 10 μl of 10× STE annealing buffer
- 50 μl of H_2O

Incubate at 98°C for 5 min in a heat block to denature the mixture.

To anneal oligos, turn off heating block and let it cool down to room temperature (30 min).

Chill at 4°C.

This mixture has the oligos annealed at 100 μM, final concentration.

Aliquot the annealed oligos in 10 μl aliquots, and keep at −20°C.

10.5.2. Digestion of the Genomic DNA

Extract genomic DNA using standard protocols and resuspend at approximately 1 μg/μl.

Digest genomic DNA as follows:

- 1 μg of genomic DNA
- 2 μl NEB1 buffer
- 1 μl SacI
- 1 μl MesI
- 0.3 μl BSA
- 14.7 μl H_2O

Digest between 6 h and overnight at 37°C
Inactivate digestion by incubating mixture at 65°C for 20 min

10.5.3. Ligation of Digested Genomic DNA to Annealed Linker

Mix the following components in an Eppendorf tube

10 μl of DNA digest

1.4 μl of 10× T4 ligation buffer

1 μl 100 μl of linker AFL31

1 μl T4 ligase

0.6 μl H_2O

Ligate at 16°C for 10 h
Inactivate ligation at 70°C for 20 min

10.5.4. First Round of PCR Amplification

In an Eppendorf tube, mix the following components

38.5 μl H_2O

5 μl 10× Taq buffer

1.5 μl 50 mM $MgCl_2$

1 μl of Afl3 primer (from 10 μM stock)

1 μl of HIVLTR primer (from 10 μM stock)

1 μl Taq enzyme

1 μl dNTPS (from 10 mM stock)

1 μl gDNA + AFL3 ligation

Amplify with the following program [LMPCRNOG]
95°C 2 min followed by 25 cycles of the following steps:

15 s at 94°C

30 s at 55°C

1 min at 72°C

Plus a final step, 10 min at 72°C.

Run gel to check PCR: it should be quite streaky, because this step will amplify hundreds of segments of genomic DNA that do not contain the provirus. To increase the specificity of the reaction, it is necessary to perform a second round of PCR with nested primers.

10.5.5. Second Round of LMPCR

In an Eppendorf tube, mix the following components:

5 μl Taq buffer 10×

1.5 μl $MgCl_2$ (from 50 mM stock)

1 μl AFl3nest (from 10 μM stock)

1 μl Aiv3LTR nest (from 1 μM stock)

1 μl Taq enzyme

0.5 μl PCR product (from first reaction)

39 μl H_2O

Run LMPCRNOG program again

It is possible to get a clear band at this stage. If this is the case, proceed to sequencing.

If the product is still "streaky", do a third round of amplification using LMPCRIII program as follows:

95°C 2 min followed by 59 cycles of the following steps:

95°C 1 min

58°C 1 min

72°C 1 min

Plus a final step, 10 min at 72°C.

5 μl taq buffer 10×

1.5 μl $MgCl_2$ (from 50 mM stock)

1 μl AFl3 nest (from 10 μM stock)

1 μl HIVLTR nest (from 10 μM stock)

1 μl Taq enzyme

0.5 μl LMPCR product (from second reaction)

39 μl H_2O

The product should be a tight, clear band at this stage.

10.5.6. Sequencing of PCR Product

(a) If product is >100 bp, purify PCR product with Quiaquick column (Qiagen), and sequence PCR product with AFl3nest and HIVLTRnest primers.

Determine the concentration of the PCR product by running a 2% agarose gel and comparing the PCR product with mass standards, and do sequencing as usual.

(b) If PCR product is too small for direct PCR sequencing (less than 100 base pairs), PCR again the product using the HIV3'LTRnestEcorI primers and the Afl3nest XhoI primer, and subclone into BSK as EcoRI XhoI.

Grow plasmid, purify DNA with a miniprep and sequence with T3 and T7 primers.

(c) If a + b fail, try to PCR with the alternative set of oligos described below for the 5′ end.

10.5.7. Sequence of Oligos

Sequence:

Linker Afl3-us/Afl3-ls is used for cloning 3′ junction region (the ones described in the previous protocol).

Linker Afl5-us/Afl5-ls is used for cloning 5′ junction region.

Linker Afl3 upper Strand (name: Afl3-us):

Length: 36 mer

Scale: 1 μmol

Purification: PAGE

5′-GTAATACGACTCACTATAGGGCTCCG CTTAAGGGAC-3′

Linker Afl3 lower strand (name: Afl3-ls):

Length: 17 mer

Scale: 1 μmol

Special modification: 5′ phosphate

3′ amino C7

PAGE

5′-PO^4-TAGTCCCTTAAG CGGAG-NH^2-3′

Linker Afl5 upper Strand (name: Afl5-us):

Length: 35 mer

Scale: 1 μmol

Purification: PAGE

5′-GGATTTGCTGGTGCAGTACAGGC CTTAAGAGGGAC-3′

Linker Afl5 lower strand (name: Afl5-ls):

Length: 18 mer

Scale: 1 μmol

Special modification: 5′ phosphate

3′ amino C7

Purification: PAGE

5′-PO^4-TAGTCCCTCTTAAG GCCT-NH^2-3′

For 3′ junction cloning
Primer set for first PCR:
AP1:

Length: 22 mer

Scale: 250 nmol

5′-GTAATACGACTCACTATAGGGC-3′

3′LTR:

Length: 25 mer

Scale: 250 nmol

5′-GACTTGTGGTCTCGCTGTTCCTTGG-3′

Primer set for nested PCR:
Primer name: Afl3nest

Length: 19 mer

Scale: 100 nmol

5′-AGGGCTCCG CTTAAGGGAC-3′

3′LTRNest:

Length: 25 mer

Scale: 250 nmol

5′-GGTCTCCTCTGAGTGATTGACTACC-3′

For 5′ junction cloning,
Primer set for first PCR:

Primer name: Afl5
Length: 21 mer

Scale: 100 nmol

5′-GGATTTGCTGGTGCAGTACA G-3′

Primer name: MseI 5′LTR
Length: 21 mer

Scale: 100 nmol

5′-TAGCTTGCCAAACCTACAGGT-3′

Primer set for nested PCR:

Primer name: Afl5nest
Length: 20 mer

Scale: 100 nmol

5′-AGTACAGGC CTTAAGAGGGA-3′

Primer name: MseI 5′LTR nested
5′-ACCTACAGGTGGGGTCTTTCA-3′

Appendix

Safety Guidelines for Pseudotyped Retroviruses

The biosafety office at your institution must be notified prior to use of lentiviral vectors for permission and for further institution-specific instructions. At a minimum, BL2 conditions should be used at all times when handling the virus.

All decontamination steps should be performed using 70% ethanol/1% SDS. Gloves should be worn at all times when handling lentiviral preparations, transfected cells or the combined transfection reagent. Remember that although the virus has been significantly modified for biosafety, with a VSV coat human cells can be infected even if they are not dividing. That said, the following modifications have been made to prevent viral replication:

1. The packaging genes are separated into two plasmids, which lack LTRs and have no viral packaging signal.
2. The following viral genes have been deleted from the packaging vector: env, vpr, vpu, vif and nef.
3. The vector expressing the packaged viral genome has a self-inactivating LTR (TATA box deletion) and expresses no viral gene products.
4. Envelope, in this case VSVg, is expressed on a separate vector.

For more information regarding the lentiviral constructs, please refer to the following papers.

Packaging Vectors

Zufferey R, Nagy D, Mandel RJ, Naldini L, Trono D (1997) Multiply attenuated lentiviral vector achieves efficient gene delivery in vivo. Nat Biotechnol 15(9):871–875.

Naldini L, Blomer U, Gallay P, Ory D, Mulligan R, Gage FH, Verma IM, Trono D (1996) In vivo gene delivery and stable transduction of nondividing cells by a lentiviral vector. Science 272(5259):263–267.

Self Inactivating LTR

Miyoshi H, Blomer U, Takahashi M, Gage FH, Verma IM (1998) Development of a self-inactivating lentivirus vector. J Virol 72 (10):8150–8157.

As mentioned before the viruses have been modified in such a way that it is virtually impossible to accidentally generate a replication-competent virus. However, because these viruses are pseudotyped with a VSVg envelope special considerations have to be taken into account.

- Always treat the virus solutions and the cells transfected or infected with the virus as a potential biohazard. Always wear gloves. Do not open the tubes outside of the tissue culture hood.
- Remember that you know that you are working with a biohazard, but other people that are not aware of it might come in contact with your virus solutions. Do not leave virus-containing solutions unattended. Do not leave virus-containing solutions in the hood or in the centrifuge if you are not there.
- Decontaminate thoroughly after your virus usage with 70% ethanol or 1% SDS or bleach. Wipe extensively the surfaces of

the TC hood that you have used. Clean thoroughly the buckets for the ultracentrifuge. Discard the TC plates and other virus-containing plastic into dedicated biohazard bags. Aspirate some bleach through the vacuum apparatus to decontaminate the lines. Be especially careful with glass Pasteur pipettes, for they are the main risk of accidental skin breakage.

- Restrict the number of places where the virus is produced and used. Use the viruses only in TC hoods, TC incubators and centrifuges that are labeled with a "*Biohazard: pseudotyped viruses*" sign. To prevent accidental contamination of items of common use, after using viruses, change gloves before using them.
- Do not use pseudotyped viruses encoding genes that are known or potential oncogenes in a normal laboratory setting that other researchers use. Furthermore, unless it is completely established that a gene is clearly not oncogenic, use only in a P3 (BL3) grade level room.

References

1. Jaenisch R (1988) Transgenic animals. Science 240:1468
2. Jaenisch R (1976) Germ line integration and Mendelian transmission of the exogenous Moloney leukemia virus. Proc Natl Acad Sci USA 73:1260
3. Jahner D et al (1982) De novo methylation and expression of retroviral genomes during mouse embryogenesis. Nature 298:623
4. Gordon JW, Scangos GA, Plotkin DJ, Barbosa JA, Ruddle FH (1980) Genetic transformation of mouse embryos by microinjection of purified DNA. Proc Natl Acad Sci USA 77:7380
5. Brinster RL, Chen HY, Trumbauer ME, Yagle MK, Palmiter RD (1985) Factors affecting the efficiency of introducing foreign DNA into mice by microinjecting eggs. Proc Natl Acad Sci USA 82:4438
6. Lo CW (1986) Localization of low abundance DNA sequences in tissue sections by in situ hybridization. J Cell Sci 81: 143–162
7. Ivarie R (2003) Avian transgenesis: progress towards the promise. Trends Biotechnol 21:14
8. Lois C, Hong EJ, Pease S, Brown EJ, Baltimore D (2002) Germline Transmission and Tissue-Specific Expression of Transgenes Delivered by Lentiviral Vectors. Science 295:868
9. Hofmann A et al (2004) Generation of Transgenic Cattle by Lentiviral Gene Transfer into Oocytes. Biol Reprod 71:405
10. Yang SH et al (2008) Towards a transgenic model of Huntington's disease in a non-human primate. Nature 453:921
11. Sasaki E et al (2009) Generation of transgenic non-human primates with germline transmission. Nature 459:523
12. Herskowitz I (1987) Functional inactivation of genes by dominant negative mutations. Nature 329:219
13. Tiscornia G, Singer O, Ikawa M, Verma IM (2003) A general method for gene knockdown in mice by using lentiviral vectors expressing small interfering RNA. Proc Natl Acad Sci USA 100:1844
14. Schorpp M et al (1996) The Human Ubiquitin C Promoter Directs High Ubiquitous Expression of Transgenes in Mice. Nucleic Acids Res 24:1787
15. Cheng TC, Wallace MC, Merlie JP, Olson EN (1993) Separable regulatory elements governing myogenin transcription in mouse embryogenesis. Science 261:215
16. Abraham N, Veillette A (1991) The lymphocyte-specific tyrosine protein kinase p56lck. Cancer Invest 9:455
17. Scott BB, Lois C (2005) Generation of tissue-specific transgenic birds with lentiviral vectors. Proc Natl Acad Sci USA 102:16443
18. Martinez-Salas E (1999) Internal ribosome entry site biology and its use in expression vectors. Curr Opin Biotechnol 10:458
19. Shaner NC et al (2004) Improved monomeric red, orange and yellow fluorescent proteins

derived from *Discosoma* sp. red fluorescent protein. Nat Biotechnol 22:1567
20. Shimohama S et al (1989) Grafting genetically modified cells into the rat brain: characteristics of E. coli beta-galactosidueasa reporter gene. Brain Res Mol Brain Res 5:271
21. Fields-Berry SC, Halliday AL, Cepko CL (1992) A recombinant retrovirus encoding alkaline phosphatase confirms clonal boundary assignment in lineage analysis of murine retina. Proc Natl Acad Sci USA 89:693
22. Hogan B (1994) Manipulating the mouse embryo: a laboratory manual, 2nd edn. Cold Spring Harbor Laboratory Press, Plainview, NY, pp. xvii, 497 p
23. Filipiak WE, Saunders TL (2006) Advances in transgenic rat production. Transgenic Res 15:673–686
24. Bosselman RA et al (1989) Germline transmission of exogenous genes in the chicken. Science 243:533
25. McGrew MJ et al (2004) Efficient production of germline transgenic chickens using lentiviral vectors. EMBO Rep 5:728
26. Agate RJ, Scott BB, Haripal B, Lois C, Nottebohm F (2009) Transgenic songbirds offer an opportunity to develop a genetic model for vocal learning. Proc Natl Acad Sci USA 106:17963
27. Schmidt EV, Christoph G, Zeller R, Leder P (1990) The cytomegalovirus enhancer: a pan-active control element in transgenic mice. Mol Cell Biol 10:4406
28. Speksnijder G, Ivarie R (2000) A modified method of shell windowing for producing somatic or germline chimeras in fertilized chicken eggs. Poult Sci 79:1430
29. Perry MM (1988) A complete culture system for the chick embryo. Nature 331:70
30. Andacht T, Hu W, Ivarie R (2004) Rapid and improved method for windowing eggs accessing the stage X chicken embryo. Mol Reprod Dev 69:31
31. Randall M (2001) Raising Japanese quail PhD thesis. State of New South Wales, Department of Primary Industries, Agriculture
32. Sambrook J, Russell DW (2001) Molecular cloning: a laboratory manual, 3rd edn. Cold Spring Harbor Laboratory Press, Cold Spring Harbor, NY
33. Coffin JM, Hughes, SH, Varmus, HE (2002) Retroviruses. Cold Spring Harbor Laboratory Press, Cold Spring Harbor, NY

Chapter 11

Vertebrate Transgenesis by Transposition

Aron Geurts, Darius Balciunas, and Lajos Mates

Abstract

Transposable elements represent a class of DNA molecules that have the ability to move genetic material from one location to another. Since the first recognition and description of their activities by Barbara McClintock beginning in the 1950s, researchers have been harnessing their molecular machinery to deliver and mutate genes in a variety of whole animal and single-cellular systems. In this chapter, we describe the recent advances in establishing robust gene transfer applications to vertebrate laboratory model systems by introducing transposon molecular machinery into the one-cell embryo of mouse, rats, or zebrafish. The method leads to highly reproducible transgenesis in these systems, enabling a variety of biomedical research applications using genetically modified animals.

Abbreviations

CDS	Coding sequence
EGA	Embryonic genome activation
ITR	Inverted terminal repeats
PB	piggyBac
REN	Restriction endonuclease
RT	Room temperature
RNAse	Ribonuclease
SB	Sleeping Beauty
UTR	Untranslated region

11.1 Introduction

11.1.1. Transposons as Gene Delivery Vehicles

Transposons are mobile genetic elements that move within or between genomes. If introduced to a target cell, these elements integrate into the chromosomes of the host. Therefore, similar to retroviruses, they can be considered as natural gene delivery

S. Pease and T.L. Saunders (eds.), *Advanced Protocols for Animal Transgenesis*, Springer Protocols,
DOI 10.1007/978-3-642-20792-1_11, © Springer-Verlag Berlin Heidelberg 2011

vehicles. They are primarily classified as Class I (or retrotransposons) and Class II (or DNA transposons) type of elements (for review *see* [1]).

Retrotransposons move via a "copy-and-paste" mechanism, as they are first transcribed into an RNA intermediate and subsequently, this element RNA is reverse transcribed into DNA that finally integrates into the host genome. Among retrotransposons, the non-LTR-type long interspersed nulcear elements are the most promising for genome manipulations. However, their mechanism of integration significantly hinders their utilization as it is hard to control the number of new integration events. In animal models, the integrations preferably occur in somatic tissues during embryogenesis rather than in germ cells; therefore, they are not heritable [2], and the newly inserted transposon copies are frequently 5′ truncated [1].

Conversely, DNA transposons are particularly useful tools for genome manipulations. Efficiency is an essential prerequisite for successful gene delivery, and currently there are a number of DNA transposon systems capable of efficient integration of functional genes into the genome of vertebrate cells. DNA transposons move in the host genome via a simple "cut-and-paste" mechanism, which is usually a precise event integrating the transposon without truncations [1]. Figure 11.1 outlines the structure and the transposition process of a DNA transposon. Natural DNA transposons have a simple structure consisting of a transposase-coding gene flanked by the inverted terminal repeats (ITRs). The transposase proteins bind to the ITRs and catalyze the excision and subsequent integration of the element into a new sequence environment (Fig. 11.1b). The process of transposition can easily be controlled by separating the transposase source from the transposable DNA harboring the ITRs, thereby creating a nonautonomous system consisting of the transposon vector (donor) and transposase source (helper) components (Fig. 11.1a). This two-component gene delivery system is entirely plasmid based, and in practice, any sequence of interest can be positioned between the ITR elements according to the requirements of the experiment. Due to their advantageous features, a number of DNA transposon systems have already been utilized for the purpose of transgenesis in vertebrates.

11.1.2. Transposon-Mediated Transgenesis

In the fields of biomedical research and biotechnology, there is a continued interest in developing various tools and approaches to modify the genome of vertebrates to produce transgenic animals. Initially "passive" methods were developed for transgenesis, fully relying on the host's DNA repair machinery to generate the transgene integrations. At present, pronuclear microinjection of linear DNA fragments into fertilized oocytes is the most commonly accepted method for transgenesis, first developed 30 years

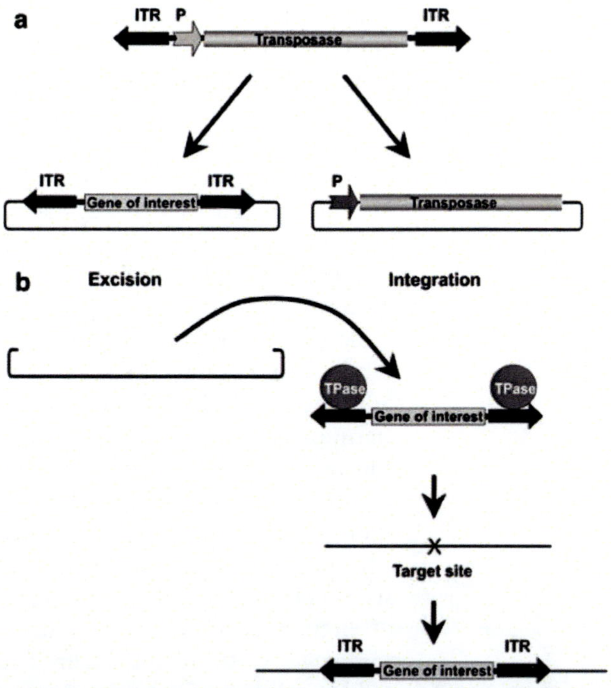

Fig. 11.1 The process of transposition. (**a**) Scheme of a class II (or DNA) transposon and that of a binary transposition system created by dissecting the transposase source from the transposon. (**b**) Outline of the mechanism of the "cut-and-paste" transposition. TPase – transposase enzyme protein. Reproduced from [34] with permission.

ago by Gordon et al. [3, 4]. The main limitations of this approach are low rates of genomic integration and integration of the injected DNA most frequently as a repetitive multicopy transgene concatemer. Such multicopy transgene concatemers may result in variegated expression and/or silencing of the transgenes [5]. In certain species, the efficiency of this protocol is limited by frequent transgene mosaicism, impairing germ line transmission of the transgenes. The reason for this is that the genomic integration of transgenes mediated by the nonhomologous end-joining repair pathway [6] generally occurs relatively late during embryonic development. For example, in zebrafish transgenesis, the mosaicism extent is high due to the low probability of transgene integration into the genome [7, 8]. In contrast, mosaicism does not seem to be a routine problem in rodents, where the same integration events predominantly happen at the one-cell stage in injected embryos.

Current "active" transgenesis techniques, where the transgene is inserted into the host genome by enzymes supplied into the embryo, involve the use of viruses such as disarmed retroviruses, especially lentiviruses, and more recently, the use of transposon systems. In comparison to standard pronuclear microinjection, both lentiviral and transposon-mediated approaches generate higher transgenic rates and mediate the insertion of single transgene units. However, lentiviruses have a strict size limitation of the transgenic DNA cargo, and high extent of transgene [9, 10]. Pronounced developmental transgene silencing is also characteristic for simple retroviruses [11] and lentiviral vectors [12]. Importantly, handling of retroviral transgenes is laborious and may bring up serious safety issues.

All these drawbacks can be circumvented by utilizing transposition-mediated gene delivery. Among transposons applicable for vertebrate transgenesis, there are systems with significantly larger cargo tolerance compared to viral vectors (*see* Subheading 11.1.2). Transgene silencing is less frequent using transposon systems than in the case of viral vectors [13]. Transposon-mediated transgenesis catalyzed by the delivery of the transposase in the form of an expression plasmid DNA also faces the transgene mosaicism problem, since oocytes are in a transcriptionally quiescent state and embryonic genome activation (EGA) starts at later stages following fertilization. However, co-injection of engineered transposons with in vitro-transcribed transposase messenger RNA (mRNA) helps overcome this limitation, because only translation of the synthesized mRNA is necessary to produce the transposase protein catalyzing the integrations. By this approach, it is possible to promote transposon-mediated integration events at early stages of development in order to reduce mosaicism and to facilitate successful transmission of the transgene to the next generation. By the same token, mRNA co-injection prevents undesired integration of the transposase-coding gene, potentially resulting in continued remobilization of the transposon. The genomic integration pattern of transposons is generally close to random, decreasing the chance for insertional mutagenesis, an undesired side effect of transgene integrations (for review *see* [14]). Importantly, the introduction of transposon systems into the embryo is based on standard pronuclear microinjection of nucleic acids. No difficulties of material preparation and no safety concerns similar to those inherent in the application of viral vectors are associated with this method.

Currently, the Sleeping Beauty (SB), piggyBac (PB), and Tol2 transposon systems are predominantly harnessed for transgenic purposes in vertebrates, mainly using the mRNA co-injection method. This method has been employed to generate transgenic zebrafish with Tol2 [15, 16] and SB [17]; transgenic Xenopus with SB [18] and Tol2 [19]; transgenic mice with Tol2 [20] and SB [21, 22]; and transgenic rats ([23] and L. Mates and A. Geurts, in preparation). We will focus primarily on the applications of the SB and Tol2 transposon systems using an mRNA transposase source for rodent and zebrafish transgenesis, but, at least for rodent transgenesis, the PB system is also applicable in a similar fashion when the transposase is supplied in trans using an expression plasmid [24].

11.1.3. Outline of the Procedure

- Subclone gene or sequence of interest into a transposon donor vector
- Prepare in vitro-transcribed transposase messenger RNA
- Prepare injection mixture
- Microinjection into vertebrate embryos
  - Zebrafish embryo production
- Screen transgenic founders
  - Develop genotyping assay

11.1.4. Principles and Applications

In case of transgenesis, a single copy insertion away from endogenous genes is desirable for most applications. Several transposon systems have been characterized for biochemical properties impacting their use as vehicles for transgenesis in certain species and under certain conditions (Table 11.1). All elements described here have a broad host range and are active in vertebrate cells, but certain elements have been shown to be more active in particular species, such as the Tol2 element, which is more active in zebrafish embryos than either SB [17, 25] or the Ac/Ds element from maize [26, 27]. In addition, unique features of each transposon

Table 11.1
Biochemical properties of the PB, SB, and Tol2 transposon systems

Transposon name	Minimal ITRs	Tolerated cargo size	Target site sequence	Overexpression Sensitivity	Integration pattern
piggyBac (PB)	235–310 bp [46]	Efficiency drops above 9.1 kb in pronucleus injected mice [24]	TTAA [47]	No	Pronounced preference for transcription units [24, 48]
Sleeping Beauty (SB)	225 bp [37]	Increased cargo size exponentially decreases the efficiency in cultured cells [37]	TA [28]	Yes [36]	No preference for genes [49, 50]
Tol2	150–200 bp [41]	11.7 kb did not reduce germ line transmission in zebrafish; 70-kb BAC insert transposed in zebrafish and mouse embryos [20, 41]	Heterogenic [15, 51]	No	Preference for transcription units [52]

system, such as the random insertion spectrum of the SB system into the genome and lack of bias toward insertion into genes; the tolerance of PB and Tol2 for large cargo sizes (10–70 kbp); and sensitivity of SB to overexpression inhibition are important to consider for optimizing your experiment (Table 11.1). Therefore, the transposon system should be selected carefully based on the actual experimental design.

Table 11.2 summarizes reported transgenic rates in rodents and zebrafish using the PB, SB, and Tol2 transposon systems as a percentage of transgenic animals per live-born founder. In the case of the PB-mediated transgenic study, plasmid DNA was used as a transposase source. However, integration and continued expression of a gene encoding the transposase could be problematic if it

Table 11.2
Transposon-mediated transgenesis in mice, rats, and zebrafish

Species	Transposon name	Transposon source	Transposase source	Transgenic efficiency[a]	Transgene copy number[b]	Source
Mouse	PB	Circular plasmid DNA	Circular plasmid DNA	35–65%	~10	[24]
	SB	Linear DNA	mRNA	14%	3	[21]
	SB100X	Circular plasmid DNA	mRNA	37%	1–2	[22]
	SB11	CpG-methylated linear DNA	mRNA	57–90%	1–11	[23]
		CpG-methylated circular DNA	mRNA	47%	1–11	[23]
	Tol2	Circular plasmid DNA	mRNA	30%[c]	1–5	[20]
Rat	SB11	CpG-methylated linear DNA	mRNA	64%	1–4	[23]
Zebrafish	SB	Circular plasmid DNA	mRNA	24–39%	ND	[44]
	Tol2	Circular plasmid DNA	mRNA	16%[d]	ND	[31]
				4%[c]	1–5	[20]

[a]As determined by PCR genotyping, live-born mouse offspring (which could be germ line mosaic) or germ line transmission to F1 zebrafish after breeding
[b]Copy number reflects the number of individual linkage groups since transposase-mediated events are single transgene insertions, usually determined by Southern blot
[c]Suster et al. used a Tol2 transposon harboring a 66-kbp insert
[d]Tol2-mediated transgenesis appears lower; however, earlier activity of the transposase results in low germ line mosaicism and, therefore, high percentage transmission to the F1 generation

led to transposon re-mobilization and reintegration. Therefore, and considering later onset of EGA in some species, we recommend the use of in vitro-synthesized transposase mRNA. In addition, variant transposases such as SB10 [28] and the hyperactive SB11[23] and SB100X [22] transposases have been used in mice in conjunction with different doses of transposon donor plasmid, modulating both the transgenic efficiency and copy number per transgenic founder (Table 11.2). A mouse codon-optimized version of the PB transposase, mPB, has recently been created [29] and is expected to be translated more efficiently, therefore reducing the amount of injected transposase required for high transgenic efficiency. However, no whole-animal transgenic studies have been reported using this new version of the PB transposase coding sequence. Finally, cis-enhancement of the transposon substrate with in vitro CpG hypermethylation dramatically increased transgenic production rates in mice compared to nonmethylated transposon substrates (Table 11.2).

In standard rodent experiments, the transposon donor plasmid can be supercoiled or linearized by restriction digest. Early studies [21] used SB10 transposase mRNA and linearized transposon donor substrate to increase the likelihood of integration into the genome, whether it is catalyzed by the transposase or mediated by spontaneous nonhomologous integration. Likewise, Carlson et al. showed that donor substrate hypermethylation and linearization led to high rates of both transposase and nonhomologous integration-mediated insertions [23]. Injecting supercoiled substrate reduces the rate of transgene insertions by nonhomologous integration [22–24], but reduces the overall transgenic frequency compared to linearized substrate [23].

Independent laboratories have generated different vectors to transcribe SB and Tol2 transposase RNA [21, 22, 25, 30, 31]. The more recent studies have employed the pT3TS specialized transcription vector containing *Xenopus laevis* beta-globin 5′ and 3′untranslated regions (UTRs) [23, 25] or the pcGlobin2 vector [32] which contains the zebrafish beta-globin UTRs [22]. Presence of these untranslated regions stabilizes mRNA and enhances translation [25, 33].

In all instances, the mixture of transposon donor plasmid and transposase source is co-injected into the pronucleus of a newly fertilized rodent embryo and transferred to a pseudopregnant surrogate, as described for mice and rats in this laboratory manual and a recent chapter on mouse transgenesis [34]. For zebrafish transgenesis experiment, transposon DNA is co-injected with transposase RNA into one- to two-cell stage embryos. In our experience, injection into one-cell embryos is highly preferred. We describe in more detail the procedure of zebrafish embryo microinjection below.

11.2 Materials

11.2.1. Equipment

- Refrigerated tabletop centrifuge capable of high speed 15,000 × g.
- FHS/LS-1B macrovisualization equipment (BLS, Ltd).
- Water bath, 37°C for REN digestion.
- NanoDrop® ND-1000 Spectrophotometer (Peqlab) or similar.
- Agarose gel running apparatus.
- Eppendorf tubes (RNase free).
- 28°C Incubator.
- Needle puller (Sutter P-87 or similar).
- Picoinjector (Harward Instruments PLI-90).
- Micromanipulator (Narishige M152).
- Stereomicroscope with epi-illumination, 10× optical magnification, and 6.5–40× zoom magnification range is suitable.
- Fluorescence microscope suitable for detecting fluorescent protein transgenes.
- Milli-Q water purification system (Millipore) or a similar source of ultrapure water.
- Manual pipette pump (Fisher Scientific 13-683°C).
- Glass pipettes (Fisher Scientific 13-678-20A).

11.2.2. Supplies and Reagents

- mMessage mMachine® T7 kit (Ambion# AM1344).
- mMessage mMachine® T3 kit (Ambion # AM1348).
- DNA prep kit (Qiagen # 27106).
- RNeasy RNA cleanup kit (Qiagen # 74204) (optional).
- Transposon donor plasmid (see below).
- Transposase source plasmid (see below).
- Borosilicate glass capillaries (World Precision Instruments #1B100F-4).
- Microcapillaries (Drummond Scientific #1-000-0010).
- Microloader tips (Eppendorf #5242 956.003).
- Agarose, QA-Agarose TM (MP Biomedicals #11AGAH0500).
- Ethidium bromide, 1% solution in water (Sigma–Aldrich #E7637).
- *Cla* I or *Xba* I, REN (New England Biolabs #R0179S or #R0145S).
- Taq DNA polymerase (New England Biolabs #M0267S).

- RiboLock™ RNAse inhibitor (Fermentas #EO0381).
- λ Eco47I DNA ladder (Fermentas SM1051).

11.2.3. Buffers and Other Solutions

- Injection buffer, EmbryoMax® (Millipore #MR-095-10 F).
- Sodium acetate 3 M pH 5.5 (RNase free) (Ambion #AM9740).
- Ethanol (RNase free) (Roth #9065.2 or similar).
- Water (RNase free) (Sigma–Aldrich #W4502) for nucleic acid precipitation.
- 70% Ethanol (RNase free) (made from the above reagents, seven parts ethanol to three parts water).
- Isopropanol (RNAse free) (Sigma–Aldrich #9516).
- Phenol/chloroform/isoamyl alcohol (Roth #A156.2 or similar).
- Chloroform/isoamyl alcohol (Roth #X984 or similar).
- 1 × TBE; 89 mM Tris–HCl, 89 mM borate, and 2 mM EDTA (Sigma–Aldrich #93306) prepared with Milli-Q water.
- TE buffer (Sigma–Aldrich #93283 or similar).
- Water (RNAse free) (Ambion #AM9938) for RNA resuspension for embryo microinjection.

11.2.4. Suppliers

Ambion (USA), Applied Biosystems, 2130 Woodward St. Austin, TX 78744-1832, USA, Phone: +1 512 651-0200, Fax: +1 512 651-0190

Aquaneering, 7960 Stromesa Court, San Diego, CA 92126, USA, Phone: 858-578-2028, Fax: 858-689-9326

BLS, Biological Laboratory Equipment, Maintenance and Service Ltd. Zselyi Aladar u. 31, H-1165 Budapest, Hungary, Phone: (36)-1- 407-2602, Fax: (36)-1-401-0925

Drummond Scientific, 500 Parkway, Box 700, Broomall, PA 19008, USA, Phone: +1 610 353-0200, Fax: +1 610 353-6204

Fermentas (Lithuania), V. Graiciuno 8, LT-02241, Vilnius, Lithuania, Phone: +370-5-2602131, Fax: +370-5-2602142

Fisher Scientific, 2000 Park Lane Drive, Pittsburgh, PA 15275, USA, Phone: +1 800 766-7000, Fax: +1 800 926-1166

Merck (Germany), Merck KGaA, Frankfurter Str. 250 64293 Darmstadt Germany, Phone: +49 6151 72-0, Fax: +49 6151 72-2000

Millipore (Germany), Millipore GmbH, Am Kronberger Hang 5, 65824 Schwalbach/Ts., Germany, Phone: 0180 5 045 645, Fax: 0180 5 045 644

NEB, New England Biolabs 240 County Road, Ipswich, MA 01938-2723 USA, Phone: 978-927-5054, Fax: 978-921-1350

Peqlab (Germany), Carl-Thiersch-Str. 2b, D-91052 Erlangen, Germany, Phone: +49(0)9131 / 610 7020, Fax: +49(0)9131/610 7099

Qiagen (UK), QIAGEN House Fleming Way, Crawley, West Sussex, RH10 9NQ, UK, Phone: 01293-422-911, Fax: 01293-422-922

Roth, Carl Roth GmbH + Co. KG, Schoemperlenstr. 3-5, 76185 Karlsruhe, Germany, Phone: +49 (0)721/5606-0, Fax +49 (0)721/5606-149

Sigma (Germany), Sigma-Aldrich Chemie Gmbh Munich, Germany, Phone: +49 89 6513 0, Fax: +49 89 6513 1169

World Precision Instruments, 175 Sarasota Center Blvd., Sarasota, FL 34240, USA, Phone: +1 941 371-1003, Fax: +1 941 377-5428

11.2.5. Sources of Plasmid Reagents:

Proper permission must be obtained for the use of SB and Tol2 transposon systems. Material Transfer Agreements (MTA) for the Sleeping Beauty reagents pT2 and pT3TS-SB11 and for the pcGlobin2-SB100X plasmid can be obtained for signature by contacting Dr. Perry Hackett and Dr. Zsuzsanna Izsvak, respectively (below). The Tol2 vector system is available from the laboratory of Dr. Stephen Ekker at Mayo Clinic by MTA. The individual plasmids described in this chapter can then be obtained from the sources below.

11.2.5.1. Sleeping Beauty Plasmid Reagents

- Transposon Donor Plasmids:
  - pT2/BH, pT2/HB, pT2/SVNeo
    Dr. Perry Hackett, Ph.D.
    University of Minnesota, Minneapolis, MN, USA
    http://www.cbs.umn.edu/labs/perry
    Dr. Zsuzsanna Izsvak, Ph.D.
    Delbrück-Centrum für Molekulare Medizin, Berlin, Germany
    http://www.mdc-berlin.de/en/research/research_teams/mobile_dna/index.html
- Transposase Source Plasmids:
  - pcGlobin2-SB100X
    Dr. Zsuzsanna Izsvak, Ph.D.
    Delbrück-Centrum für Molekulare Medizin, Berlin, Germany
    http://www.mdc-berlin.de/en/research/research_teams/mobile_dna/index.html
  - pT3TS-SB11
    Dr. Aron Geurts, Ph.D.

Medical College of Wisconsin, Milwaukee, WI, USA
http://www.mcw.edu/HMGC/Laboratories/AG.htm

11.2.5.2. Tol2 Plasmid Reagents

- Transposon Donor Plasmid:
  - pminiTol2/MCS (pDB739)
    Dr. Stephen Ekker, Ph.D.
    Mayo Clinic, Rochester, MN, USA
    http://mayoresearch.mayo.edu/mayo/research/staff/Ekker_SC.cfm
  - Additional Tol2 donor plasmids are available from Drs. Ekker and Balciunas
    Darius Balciunas
    Temple University, Philadelphia, PA, USA
    http://www.temple.edu/biology/directory/faculty/balciunas
- Transposase Source Plasmid:
  - pT3TS-Tol2 (pDB600)
    Dr. Stephen Ekker, Ph.D.
    Mayo Clinic, Rochester, MN, USA
    http://mayoresearch.mayo.edu/mayo/research/staff/Ekker_SC.cfm

11.3 Protocol for Transpositional Transgenesis

The transposon-mediated transgenesis protocol is very similar to classical transgenesis using naked DNA transgenes prepared from plasmids. The main differences lie in the materials and their preparation for injection into the fertilized oocytes and the handling of injection needles (capillaries) to avoid RNase contamination and degradation. Preparation of the transposon system components for mouse embryo microinjection and expected outcomes were also similarly described separately [34]. For technical references and protocols on pronuclear injection into the mouse or rat pronucleus, refer to the chapters by Jerchow (Chapter 6) or Anegon and Menoret (Chapter 7) in this laboratory handbook, respectively. Details on zebrafish microinjection are found here.

11.3.1. Rodent Transgenesis

Since all SB transposases demonstrate higher transposition activity with the T2-type enhanced SB ITRs [22, 35, 36], it is recommended to clone your gene of interest between the enhanced SB ITRs (*see* Subheading 11.2.5). Preferably, your transgene size should not exceed 5 kb (Table 11.1; [36, 37]); however, SB transposons of 9.1-kb have been used in mouse transgenic experiments with success [23]. Increasing the amount of transposon donor plasmid in the final injection mixture may help to increase the efficiency in case of larger transgenes. However, plasmid concentrations above 2 ng/μl can decrease the viability of the injected zygotes ([22], and unpublished observations of the authors).

It is sometimes beneficial to design the transposon transgene to allow phenotypic detection by incorporating coat color markers [38] or fluorescent proteins [22]. One copy of a CAGGS promoter [39]-driven Venus [40] marker is sufficient to detect the presence of the transgene in the genome and can be visualized using the FHS/LS-1B macrovisualization equipment (Fig. 11.2).

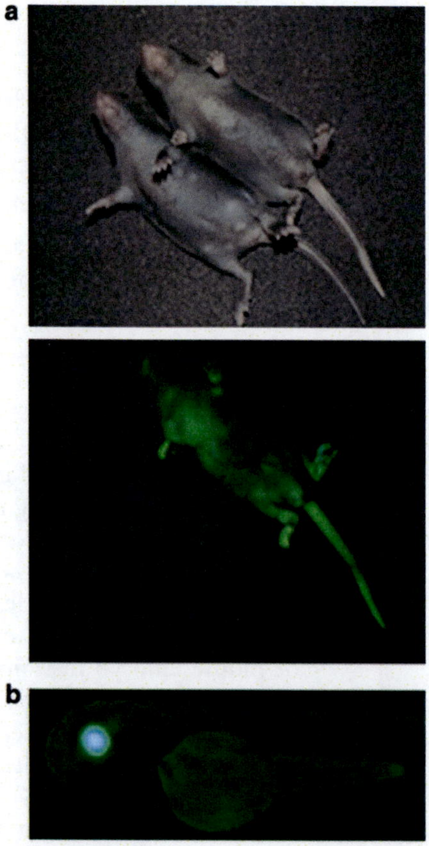

Fig. 11.2 Fluorescent protein markers of vertebrate transgenesis. (**a**) A nontransgenic and a single-copy CAGG-Venus fluorescent protein-transgenic mouse are shown side by side under brightfield and using the FHS/LS-1B light source and YFP filter (BLS Ltd). (**b**) Green fluoresecent protein in a transgenic zebrafsh under the control of lens-specific promoters under a standard fluorescence microscope equipped to detect GFP. Reproduced in part from [34] with permission.

11.3.1.1. Preparation of the Transposon Donor Plasmid

After cloning your gene of interest between the transposon ITR sequences, prepare the transposon donor plasmid using the Qiagen plasmid kit. Follow the instructions of the manufacturer. **Note:** The amount of DNA injected into a rodent or zebrafish embryo measures in the range of femtograms to nanograms, so even a few micrograms of prepared transposon donor plasmid is enough to inject hundreds or even thousands of embryos.

Prepare the plasmid DNA RNase free by phenol/chloroform extraction as described below. **Note:** All centrifugations during the following phenol/chloroform extraction are done at 15,000 × g in a tabletop centrifuge at room temperature, unless otherwise noted.

1. Increase the volume of the plasmid prepared to 400 μl with TE buffer in a 1.5-ml Eppendorf tube.
2. Add 400 μl phenol/chloroform/isoamyl alcohol.
3. Vortex for 15 s and leave on the table for 2 min. Repeat this step three times to inactivate the residual RNAse completely.
4. Centrifuge for 5 min.
5. Transfer the top layer to a new RNase-free microcentrifuge tube and add 400 ml chloroform/isoamyl alcohol (24:1, vol:vol).
6. Vortex for 15 s and centrifuge for 5 min.
7. Transfer the top layer to a new RNase-free Eppendorf tube, add 1/10 volume 3 M sodium acetate (pH 5.5, 40 μl, RNase free) and 2.5 volumes of 100% ethanol (1 ml, RNase free), and let the DNA precipitate for 30 min at −20°C.
8. Spin down for 15 min in a refrigerated centrifuge at 4°C at 15,000 × g, and discard the supernatant.
9. Wash the pellet in cold 70% ethanol (RNase free). **Note:** Keep the ethanol on the pellet for 10 min and then remove it (centrifuge if necessary as in **step 8**). Repeat this step once more to wash away completely any residual chemicals that may not be tolerated by the embryos.
10. Air-dry the pellet for 5–10 min and resuspend the pellet in 100 μl RNase-free injection buffer (EmbryoMax®, Millipore). **Note:** The buffer must be tested for the presence of RNase as the manufacturer does not guarantee it is RNase free. *See* Subheading 11.3.1.2 below. **Note**: For zebrafish embryo microinjection, we prepare plasmid DNA using Qiagen Miniprep (#27106) protocol. When using this protocol, it is essential to perform the PB buffer wash step listed as optional by Qiagen. Omission of PB wash results in RNAse contamination of DNA.
11. Measure the concentration of the plasmid using a NanoDrop® or other spectrophotometer.

11.3.1.2. Preparation of the Transposase mRNA

As described above, a variety of plasmid vectors exist for in vitro mRNA synthesis. These plasmids should be prepared using the Qiagen plasmid kit, following the manufacturer's instructions. The pcGlobin2-SB100X vector utilizes the T7 promoter for synthesis, while the pT3TS-SB11 and pT3TS-Tol2 vectors utilize the T3 promoter. For in vitro transcription, proper kit reagents must be used in each case to synthesize mRNA as described below.

1. Linearize at least 2 μg of plasmid with *Cla* I (pcGlobin2-based vectors) or *Xba* I (pT3TS-based vectors) digestion, according to the instructions of the enzyme supplier (Fig. 11.3a). 1 μg linearized plasmid will be necessary for one round of mRNA synthesis. Check for complete linearization on 1% agarose gel, comparing the predicted size to a DNA standard such as the λ Eco47I ladder.
2. Make the fully digested plasmid RNase free, by phenol/chloroform extraction as described above and modifying the volumes of the digested plasmid DNA, the phenol/chloroform/isoamyl alcohol, and the chloroform/isoamyl alcohol to 100 μl each.
3. Synthesize the mRNA using a mMessage mMachine® T7 (or T3) kit (Ambion). Follow the manufacturer's instructions. After synthesis, use the Turbo DNAse treatment and phenol/chloroform extraction options suggested in the kit

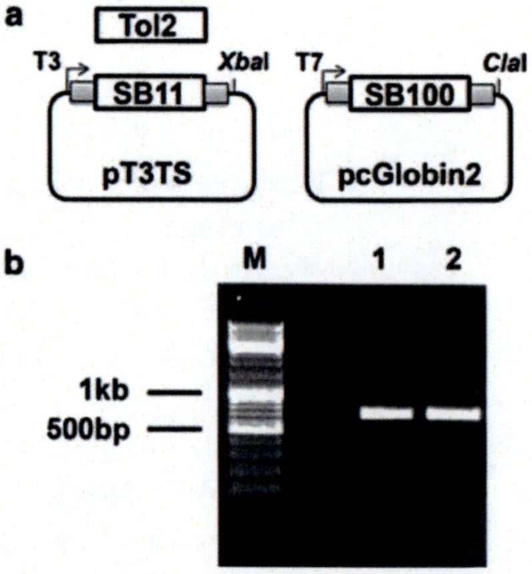

Fig. 11.3 mRNA synthesis. (**a**) Plasmids for run-off in vitro transcription of transposase mRNA are linearized with either *XbaI* or *ClaI* and used as a template in the mMessage mMachine® T3 or T7 in vitro transcription kit (Ambion). Both the pT3TS and pcGlobin2 vectors have UTR sequences flanking the transposase coding region (*gray boxes*). (**b**) mRNA quality can be determined by standard RNAse-free agarose electrophoresis and ethidium bromide staining. Shown in *lane* 2 is mRNA that was incubated for 1 h with EmbryoMax® injection buffer to detect the presence of RNAse in the buffer. Sometimes additional band(s) are visible and are a result of mRNA secondary structure. Reproduced from [34] with permission.

manual, with the modification that after the isopropanol precipitation following the phenol/chloroform extraction, wash the pellet twice in cold 70% ethanol (RNase free). Air-dry the pellet. Resuspend the mRNA in 20 μl RNase-free water from Ambion. **Note:** Do not use lithium chloride to precipitate the mRNA. **Note:** We have had success modifying the in vitro transcription reaction to include 0.5 μL of Ribolock™ RNAse inhibitor to protect the RNA from degradation. **Note:** For zebrafish transgenesis, we prefer to purify the synthesized Tol2 mRNA using Qiagen RNEasy® minicolumns following the supplementary "RNA Cleanup" protocol instead of the phenol/chloroform and precipitation methods. Resuspend the RNA in the Ambion RNAse-free water at the final step.

4. Check your synthesized mRNA on an RNase-free 1% agarose gel.

(a) Wash the running chamber, gel tray, comb, and flask for gel preparation with 70% ethanol.

(b) Incubate the running chamber, gel tray, comb, and flask for gel preparation in 0.2 N NaOH for 1 h.

(c) Rinse the running chamber, gel tray, comb, and flask for gel preparation with sterile Milli-Q water.

(d) Prepare 1× TBE gel running buffer using sterile 10× TBE buffer and sterile Milli-Q water.

(e) Prepare the 1% agarose gel using sterile 1× TBE buffer, sterile Milli-Q water, and Agarose, QA-Agarose TM, or equivalent.

(f) Load 1 μl of the final in vitro-synthesized mRNA in RNA loading buffer and a DNA size marker and run the gel. The RNA loading buffer is supplied in the mMessage mMachine® kits. A typical result is shown in Fig. 11.3b. The SB100X mRNA prepared using the T7 promoter on the *Cla* I-digested pcGlobin2-SB100X runs on a normal agarose gel between 700 and 800 bp corresponding to dsDNA size marker (Fig. 11.3b, *lane 1*). Typically, the in vitro-synthesized mRNA runs as one band on a non-denaturing gel (Fig. 11.3b, *lane 1*). Alternatively, you may see two bands (Fig. 11.3b, *lane 2*) due to secondary structures. **Note:** It is not necessary to run a Northern gel here; however, if you see more than two bands on the non-denaturing gel, a Northern gel may help to identify whether the bands are different length products or your mRNA runs aberrantly due to secondary structures (also see the troubleshooting instructions of the mMessage mMachine® T7 kit). **Note:** The new batches of the injection buffer, EmbryoMax® (Millipore), or equivalent must be tested for the presence of RNase.

Incubate 1 µl of in vitro-synthesized mRNA in 10 µl injection buffer for 1 h at 37°C and run on RNase-free 1% agarose gel as above. RNase is typically not detected. Figure 11.3b, *lane 2* shows the result of an injection buffer test without visible signs of mRNA degradation.

5. Measure the concentration of the in vitro-synthesized mRNA using a NanoDrop® or similar spectrophotometer. The typical yield is around 1 µg/µl and a total volume of 20 µl.
6. Make diluted aliquots for later use in RNase-free injection buffer (*see* Subheading 11.3.1.3 below). Store the concentrated and diluted aliquots at −80°C.

11.3.1.3. Preparation of the Microinjection Mixture

The mixture prepared for microinjection contains two components: the transposon donor plasmid and the in vitro-synthesized transposase mRNA. Both components have to be prepared RNase free in injection buffer and the appropriate volumes have to be mixed to have the final injection mixture. As described above, variant SB transposases result in different activities in the rodent embryo. In our hands the best working injection mixtures contain 15 ng/µl SB11 mRNA and 5 ng/uL transposon donor plasmid [23], or 5 ng/µl SB100X mRNA and 0.4 ng/µl transposon donor plasmid (Mates et al. 2009) [22]. In order to create the final microinjection mixture, specific amounts of transposase mRNA and transposon donor plasmid must be mixed to reach the desired final concentrations in the injection mixture. For example, to reach the final concentrations for SB100X-mediated rodent transgenesis, 10 ng/µl SB100X mRNA and 0.8 ng/µl transposon donor plasmid stock solutions should be mixed at 1:1 ratio. **Note:** In case of transgenes larger than 5 kb, it may be beneficial to increase the transposon donor plasmid concentration to 2 ng/µl final concentration for rodent embryo injections to compensate for the larger plasmid size. Prepare 10 µl aliquots of the final microinjection mixture and keep them frozen at −80°C until use. Follow the standard microinjection protocol as described in the accompanying chapters mentioned in Subheading 11.3 above for rodents or as described below for zebrafish. After each round of microinjection, discard the used aliquot.

11.3.2. Zebrafish Transgenesis

Both SB and Tol2 have been demonstrated as effective tools for zebrafish transgenesis; however, Tol2 demonstrates attributes which make it the preferred element for fish applications including increased cargo size [41] and more rapid kinetics of transposition leading to higher germline transmission rates [25]. Preparation of transposon and transposase injection mixtures for transpositional transgenesis in zebrafish is performed as described above, except that Tol2-mediated zebrafish transgenesis is optimized at 8 ng/µl transposase mRNA and 8 ng/uL transposon donor plasmid.

Zebrafish transposon constructs are easily marked by fluorescent protein reporter genes, facilitating sorting of successfully injected embryos, identification of transgenic offspring, and line maintenance. We prefer to use lens-specific fluorescent markers [17], because it leaves the possibility to reuse the same fluorescent protein for other tissue-specific labeling (excluding the eye). Other tissue-specific promoters, such as heart-specific cmlc2 [42] and TnnT2 (D. Balciunas, unpublished observations) promoters, are more suitable for scientists interested in eye development and function. One disadvantage of both lens- and heart-specific promoters is that they begin expression only as corresponding tissues form during day 1 of development. They, therefore, cannot be used to identify transgenic embryos during earlier stages of development preceding the formation of the lens and heart, respectively. Early-expressing more ubiquitous promoters have also been used, such as *X. laevis* EF1a [17] and carp b-Actin [43].

11.3.2.1. Zebrafish Embryo Production

An extensive guide to zebrafish husbandry in embryo production is available elsewhere (*The Zebrafish Book*, 5th edition, 2007, by Monte Westerfield, available from http://zebrafish.org). We will only recap the main points. Zebrafish are maintained at 10/14 (night/day)-hour cycle and typically lay embryos within 15 min after the lights come on. We typically keep males and females separated and pair them for breeding once a week. In a single mating cage (Aquaneering ZHCT100), we set up 1–3 males and 1–3 females. It is possible to delay egg laying by placing a physical divider between males and females. Once the divider is removed, fish start producing fertilized eggs. We typically collect embryos for injection 15–20 min after removing the divider. This routine allows sufficient time to load the embryos onto injection tray and inject before they begin to divide to the two-cell stage.

11.3.2.2. Preparation of Microinjection Needle

A typical zebrafish microinjection setup is shown in Fig. 11.4a. Needles are pulled using the Sutter P-87 needle puller set at following parameters according to the manufacturer's manual (arbitrary units unless specified): Heat 380, pull 100, velocity 80, time 250 (1/2 ms), and pressure 400. More detailed guidelines for needle pulling can be found elsewhere [44, 45]. The needle is backfilled with 2.5 μL of injection mixture using Eppendorf 10-μL pipette and Eppendorf Microloader tip. The backfilled needle is placed into the needle holder (Fig. 11.4b). The Picoinjector is set at 30 ms. The tip of the needle is broken off under a stereomicroscope at approximately 20× magnification using a forceps. Jagged tip of the broken needle makes it easier to pierce the chorion and is therefore preferred. However, one has to keep in mind that due to taper of the needle, diameter increases after each break.

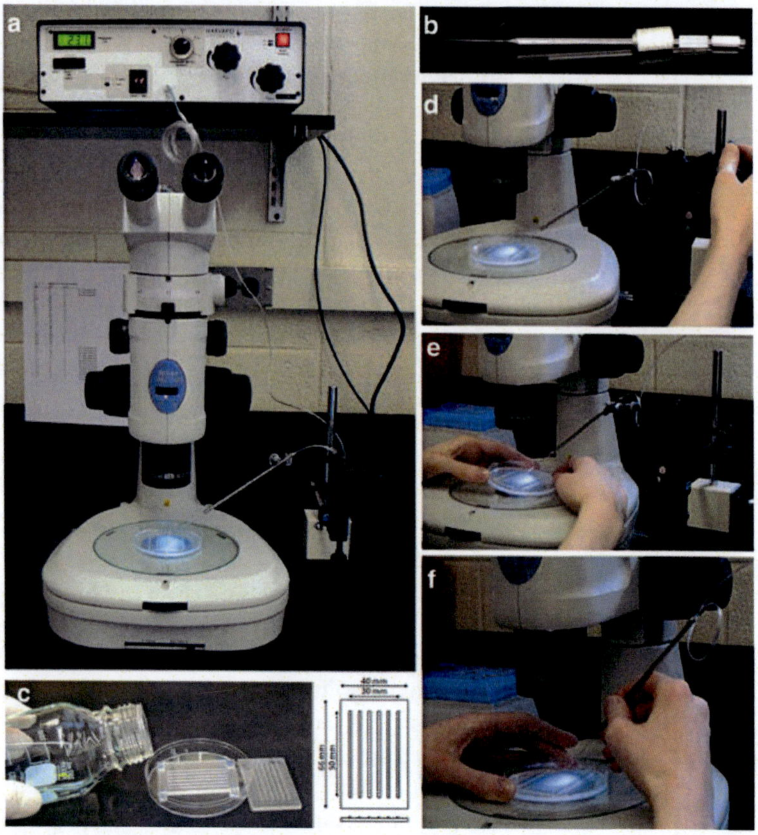

Fig. 11.4 Microinjection setup. (a) Zebrafish are microinjected in ambient conditions under a stereomicroscope. The needle is attached to a picoinjector which controls the injection time (*see* Table 11.3) and is held by a micromanipulator for fine control. (b) The needle is backfilled with 2.5 μL of injection mixture using a microloader tip. An agarose bed is poured using a custom Plexiglass mold to hold embryos for microinjection. (c) There are three basic techniques for injecting embryos: the micromanipulator technique (d), plate-moving technique (e), or pen technique (f).

To calibrate the needle, we inject ten times into a Drummond microcapillary. 1 mm length in the capillary corresponds to 30 nl volume. Calibration chart (Table 11.3) is used to adjust injection time so that each injection delivers a volume of 3 nl. If calibration results in less than 0.6 mm of solution in the microcapillary, the needle is broken off more. Conversely, if calibration results in more than 2.6 mm of solution in the microcapillary, the needle is broken off too much and a new needle must backfilled and calibrated.

11.3.2.3. Microinjection

We use an agarose plate for microinjection of zebrafish embryos ([45], also available online at zebrafish.org). A custom-manufactured plexiglass mold is used to pour the agarose plate (Fig. 11.4c). The mold has dimensions of 50 mm by 66 mm. The mold has six ridges 50 mm in length, which are 1 mm high and have one edge at a right angle to the surface of the plate.

Table 11.3
Picoinjector settings for zebrafish embryo microinjection

After ten injections into microcapillary	Setting for desired injection volume (ms)	
Length (mm)	3 nl	2 nl
0.6[a]	50	33
0.8	38	25
1	30	20
1.2	25	17
1.4	21	14
1.6	19	13
1.8	17	11
2	15	10
2.2	14	9
2.4	13	8
2.6	12	8
2.8[a]	11	7
3[a]	10	7

[a]Injection settings at or below 0.7 mm or at or above 2.7 mm are not recommended

The ridges are 1.5 mm wide, and thus the other side of the ridge is at approximately 34° angle to the surface of the mold. One percent agarose solution in egg water is prepared (egg water contains 60 μg/ml "Instant Ocean" sea salt available at most pet stores [45]). First, we pour agarose into a 60-mm Petri dish to 5 mm thickness. After agarose solidifies, four slabs are cut out and placed on the four corners of the side of the mold with the ridges (Fig. 11.4c). The mold with agarose slabs is then inverted into a standard 100-mm Petri dish so that the ridges point downward. The plate with inverted mold is placed at a slight angle, for example, by using another microinjection mold as shown in Fig. 11.4. The agarose is poured into the lower end of the microinjection plate. This prevents trapping of air bubbles under the microinjection mold. After pouring, the plate is leveled while the agarose is still liquid, and the agarose is allowed to solidify and the mold is removed.

Zebrafish embryos are loaded onto microinjection plate using a manual pipette pump with a glass pipette. Excess water is removed, leaving embryos exposed to the air. This helps maintain embryos in the groove as the needle is withdrawn without the use of plastic cover recommended elsewhere [45].

We have used three main techniques for microinjection. Each can be used to achieve high-quality injections. It is, therefore, up to an individual scientist to choose which technique works best for him/her and perfect it.

1. Micromanipulator technique (Fig. 11.4c). Micromanipulator is set at an angle so that movement along *X*-axis will result in movement of the needle down and to the left. Injection plate is placed on the microscope at about 20X magnification, and the needle is lowered to be slightly above the embryos on the vertical axis and to the right along the horizontal axis. The *X*-axis of the micromanipulator is used to move the needle to pierce the chorion and then the yolk. The left hand is used to move the plate to place the next embryo in from the needle.
2. Plate-moving technique (Fig. 11.4d). Micromanipulator is placed so that the *X*- and *Z*-axes are horizontal. The tip of the needle is placed 0.5–1 cm above the injection plate in the center of the microscope vision field, and the microscope at about 20× magnification is focused on the tip of the needle. The plate is held by both hands, with right side of the plate touching the microscope base and left side being raised for the embryos to get to the level of the needle. The plate is moved leftward for microinjection. This technique was used to make the accompanying movie. Please see the Supplementary Material in the online version for a corresponding video.
3. Pen technique (Fig. 11.4e). Microinjection plate is placed on the microscope, and the microscope at about 20× magnification is focused on the embryos. The needle holder is held like a pen for injection.

Injected embryos are kept at 28°C in Petri dishes with embryo water at a density of 100 embryos per dish or less. In a typical Tol2 transgenesis experiment, 30–50% of injected embryos survive to day 1 and are morphologically normal. It is essential to remove dead embryos daily to prevent bacterial growth. Some laboratories maintain embryos in low concentration of antibacterial agent such as methylene blue [45]; however, in our experience, methylene blue may cause an increase in autofluorescence using several fluorescent filter sets.

11.3.3. Screening Founder Animals for Transpositional Transgenesis

Setting up a genotyping PCR sensitive enough for the detection of one copy of the transgene in a diploid genome is also necessary, especially if the use of visible markers is not feasible. To reach this goal, one should optimize the genotyping PCR to stably detect 50 femtograms of transposon donor plasmid mixed with 500 ng of wild-type genomic DNA. One example of genotyping PCR

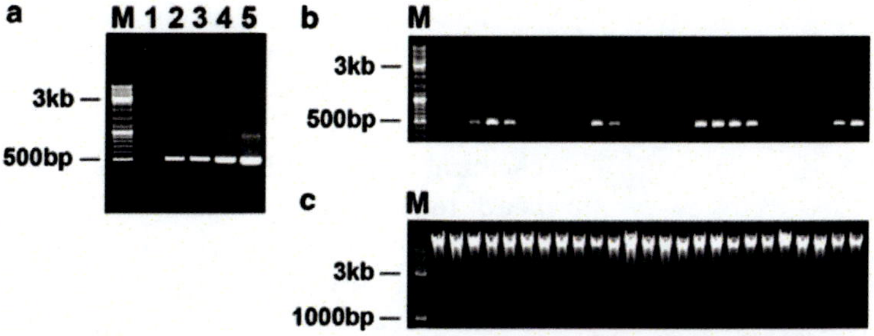

Fig. 11.5 Genotyping assay. (**a**) Different amounts of pT2/Venus plasmid mixed with 500 ng wt mouse DNA per reaction to optimize a genotyping PCR. *Lanes*: M, DNA size marker, GeneRuler DNA ladder mix (Fermentas); *lane* 1, no pT2/Venus template was added into the reaction; *lanes* 2–4, 50, 100, or 500 fg pT2/Venus template was mixed with the mouse DNA, respectively; *lane* 5, 10 ng pT2/Venus plasmid alone was used as template. (**b**) PCRs on 500 ng genomic DNA samples for the detection of the Venus transgene. (**c**) Genomic DNA samples, 500 ng each. Reproduced from [34] with permission.

optimization detecting the Venus transgene is shown in Fig. 11.5a. This sensitivity guarantees that single copy transgenes are always detected using 500 ng genomic DNA as PCR template. Genotyping PCR of a cohort of F1 generation mice, descendants of pT2/Venus [22] transgenic founders, is shown in Fig. 11.5b as an example. In zebrafish transgenesis, we always use fluorescent reporters to track both microinjection and transgenesis, making genotyping by PCR unnecessary; however, a similar screen for PCR genotyping of zebrafish founders could be designed.

11.3.4. Results

Actual transpositional transgenesis frequencies will vary depending on the transposon size and cargo. Using several transposon constructs of varying sizes, we have detected SB11- and SB100X-mediated transgenesis frequencies in mice and rats ranging from 20 to 90%, with low mosaicism. In a typical zebrafish transgenesis experiment using Tol2, we observe that about 50% of adults raised from injected embryos transmit the transgene to the next generation, sometimes with more than 80% of F1 progeny inheriting the transgene. This indicates that two or more transgenes are integrated and can be transmitted by the founder. In such cases, it may be necessary to outcross the line for multiple generations before a line with a single copy of the transgene is established.

11.4 Troubleshooting

In case of problems concerning mRNA synthesis and mRNA quality, see the troubleshooting guide of the mMessage mMachine® T7 kit (Ambion).

Decreased viability of the injected embryos may be due to residual harmful chemicals in the microinjection mixture. Therefore, careful washing of precipitated plasmid DNA and mRNA with 70% ethanol is important. Diethylpyrocarbonate (DEPC), a chemical widely used for the preparation of RNase-free solutions, is not tolerated by the embryos. Therefore, the RNase-free solutions of the above protocol are prepared without using DEPC.

Since a DNA/RNA mixture is made before the injection and maintained at room temperature in the injection needle while the injection is carried out, any contamination of prepared DNA by RNAse will result in RNA degradation and lack of transposition. We find that the optional PB wash step is absolutely essential to remove traces of RNAse H used in Qiagen resuspension buffer P1. Note that the seemingly low transgenic rate could also be due to single-copy transgene detection problems. Even the optimized genotyping PCR may fail to detect a single-copy transgene if the genomic DNA template is degraded. Good quality genomic DNA runs on agarose gel as a dominant high molecular weight band (Fig. 11.5c).

11.5 Conclusion and Outlook

Transpositional transgenesis is an easy and efficient method to generate transgenic animals carrying single- or multicopy transgenes. Single-copy transgene units may be less prone to transgene silencing and allow better control of transgene expression levels and patterns. *Sleeping Beauty* and Tol2 represent only two well-characterized elements among an ever-growing arsenal of transposons harnessed for gene transfer applications. Development of hyperactive variants of other broad host-range elements such as *piggyBac*, and methods of delivering them to embryos will add to the growing list of powerful tools for vertebrate transgenesis.

References

1. Craig NL, Craigie R, Gellert M, Lambowitz AM (2002) Mobile DNA II. ASM Press, Washington, DC
2. Kano H et al (2009) L1 retrotransposition occurs mainly in embryogenesis and creates somatic mosaicism. Genes Dev 23(11): 1303–1312
3. Gordon JW, Ruddle FH (1981) Integration and stable germ line transmission of genes injected into mouse pronuclei. Science 214 (4526):1244–1246
4. Gordon JW et al (1980) Genetic transformation of mouse embryos by microinjection of purified DNA. Proc Natl Acad Sci USA 77 (12):7380–7384
5. Wall RJ (2001) Pronuclear microinjection. Cloning Stem Cells 3(4):209–220
6. Dai J et al (2010) Non-homologous end joining plays a key role in transgene concatemer formation in transgenic zebrafish embryos. Int J Biol Sci 6(7):756–768
7. Stuart GW, McMurray JV, Westerfield M (1988) Replication, integration and stable germ-line transmission of foreign sequences injected into early zebrafish embryos. Development 103(2):403–412

8. Stuart GW et al (1990) Stable lines of transgenic zebrafish exhibit reproducible patterns of transgene expression. Development 109 (3):577–584
9. Lois C et al (2002) Germline transmission and tissue-specific expression of transgenes delivered by lentiviral vectors. Science 295 (5556):868–872
10. van den Brandt J et al (2004) Lentivirally generated eGFP-transgenic rats allow efficient cell tracking in vivo. Genesis 39(2):94–99
11. Park F (2007) Lentiviral vectors: are they the future of animal transgenesis? Physiol Genomics 31(2):159–173
12. Ellis J, Yao S (2005) Retrovirus silencing and vector design: relevance to normal and cancer stem cells? Curr Gene Ther 5(4):367–373
13. Grabundzija I et al (2010) Comparative analysis of transposable element vector systems in human cells. Mol Ther 18(6):1200–1209
14. Izsvak Z et al (2010) Translating Sleeping Beauty transposition into cellular therapies: victories and challenges. Bioessays 32(9): 756–767
15. Kawakami K, Shima A, Kawakami N (2000) Identification of a functional transposase of the Tol2 element, an Ac-like element from the Japanese medaka fish, and its transposition in the zebrafish germ lineage. Proc Natl Acad Sci USA 97(21):11403–11408
16. Korzh V (2007) Transposons as tools for enhancer trap screens in vertebrates. Genome Biol 8(Suppl 1):S8
17. Davidson AE et al (2003) Efficient gene delivery and gene expression in zebrafish using the Sleeping Beauty transposon. Dev Biol 263(2): 191–202
18. Sinzelle L et al (2006) Generation of trangenic Xenopus laevis using the Sleeping Beauty transposon system. Transgenic Res 15(6):751–760
19. Hamlet MR et al (2006) *Tol2* transposon-mediated transgenesis in *Xenopus tropicalis*. Genesis 44(9):438–445
20. Suster M, Sumiyama K, Kawakami K (2009) Transposon-mediated BAC transgenesis in zebrafish and mice. BMC Genomics 10(1): 477
21. Dupuy AJ et al (2002) Mammalian germ-line transgenesis by transposition. Proc Natl Acad Sci USA 99(7):4495–4499
22. Mates L et al (2009) Molecular evolution of a novel hyperactive Sleeping Beauty transposase enables robust stable gene transfer in vertebrates. Nat Genet 41(6):753–761
23. Carlson DF et al (2010) Efficient mammalian germline transgenesis by cis-enhanced Sleeping Beauty transposition. Transgenic Res 20(1):29–45
24. Ding S et al (2005) Efficient transposition of the piggyBac (PB) transposon in mammalian cells and mice. Cell 122(3):473–483
25. Balciunas D et al (2006) Harnessing a high cargo-capacity transposon for genetic applications in vertebrates. PLoS Genet 2(11):e169
26. Emelyanov A et al (2006) Trans-kingdom transposition of the maize dissociation element. Genetics 174(3):1095–1104
27. Emelyanov A, Parinov S (2008) Mifepristone-inducible LexPR system to drive and control gene expression in transgenic zebrafish. Dev Biol 320(1):113–121
28. Ivics Z et al (1997) Molecular reconstruction of Sleeping Beauty, a Tc1-like transposon from fish, and its transposition in human cells. Cell 91(4):501–510
29. Cadinanos J, Bradley A (2007) Generation of an inducible and optimized piggyBac transposon system. Nucleic Acids Res 35(12):e87
30. Kawakami K, Shima A (1999) Identification of the Tol2 transposase of the medaka fish Oryzias latipes that catalyzes excision of a nonautonomous Tol2 element in zebrafish Danio rerio. Gene 240(1):239–244
31. Parinov S et al (2004) Tol2 transposon-mediated enhancer trap to identify developmentally regulated zebrafish genes in vivo. Dev Dyn 231(2):449–459
32. Ro H et al (2004) Novel vector systems optimized for injecting in vitro-synthesized mRNA into zebrafish embryos. Mol Cells 17(2):373–376
33. Hyatt TM, Ekker SC (1999) Vectors and techniques for ectopic gene expression in zebrafish. Methods Cell Biol 59:117–126
34. Mates L (2011) Rodent transgenesis mediated by a novel hyperactive Sleeping Beauty transposon system. Methods Mol Biol 738:87–99
35. Cui Z et al (2002) Structure-function analysis of the inverted terminal repeats of the sleeping beauty transposon. J Mol Biol 318(5): 1221–1235
36. Geurts AM et al (2003) Gene transfer into genomes of human cells by the sleeping beauty transposon system. Mol Ther 8(1): 108–117
37. Izsvak Z, Ivics Z, Plasterk RH (2000) Sleeping Beauty, a wide host-range transposon vector for genetic transformation in vertebrates. J Mol Biol 302(1):93–102
38. Lu B et al (2007) Generation of rat mutants using a coat color-tagged Sleeping Beauty transposon system. Mamm Genome 18(5): 338–346
39. Niwa H, Yamamura K, Miyazaki J (1991) Efficient selection for high-expression

transfectants with a novel eukaryotic vector. Gene 108(2):193–199
40. Nagai T et al (2002) A variant of yellow fluorescent protein with fast and efficient maturation for cell-biological applications. Nat Biotechnol 20(1):87–90
41. Urasaki A, Morvan G, Kawakami K (2006) Functional dissection of the Tol2 transposable element identified the minimal cis-sequence and a highly repetitive sequence in the subterminal region essential for transposition. Genetics 174(2):639–649
42. Huang C-J et al (2003) Germ-line transmission of a myocardium-specific GFP transgene reveals critical regulatory elements in the cardiac myosin light chain 2 promoter of zebrafish. Dev Dyn 228(1):30–40
43. Gibbs PD, Schmale MC (2000) GFP as a Genetic Marker Scorable Throughout the Life Cycle of Transgenic Zebra Fish. Mar Biotechnol (NY) 2(2):107–125
44. Hermanson S et al (2004) Sleeping Beauty transposon for efficient gene delivery. Methods Cell Biol 77:349–362
45. Westerfield M (2007) The zebrafish book. A guide for the laboratory use of zebrafish (Danio rerio), 5th edn. University of Oregon Press, Eugene, OR
46. Li X et al (2005) piggyBac internal sequences are necessary for efficient transformation of target genomes. Insect Mol Biol 14 (1):17–30
47. Cary LC et al (1989) Transposon mutagenesis of baculoviruses: analysis of Trichoplusia ni transposon IFP2 insertions within the FP-locus of nuclear polyhedrosis viruses. Virology 172(1):156–169
48. Thibault ST et al (2004) A complementary transposon tool kit for Drosophila melanogaster using P and piggyBac. Nat Genet 36 (3):283–287
49. Vigdal TJ et al (2002) Common physical properties of DNA affecting target site selection of sleeping beauty and other Tc1/mariner transposable elements. J Mol Biol 323:441–452
50. Carlson CM et al (2003) Transposon mutagenesis of the mouse germline. Genetics 165 (1):243–256
51. Koga A (2004) Transposition mechanisms and biotechnology applications of the medaka fish Tol2 transposable element. Adv Biophys 38:161–180
52. Kondrychyn I et al (2009) Genome-wide analysis of Tol2 transposon reintegration in zebrafish. BMC Genomics 10(1):418

Chapter 12

Rat Spermatogonial Stem Cell-Mediated Gene Transfer

Karen M. Chapman, Dalia Saidley-Alsaadi, Andrew E. Syvyk, James R. Shirley, Lindsay M. Thompson, and F. Kent Hamra

Abstract

More than 20 years have passed since the advent of genetic manipulation of the mouse germline using cultures of pluripotent embryonic stem cells. Still, despite remarkable successes in the mouse, the application of stem cell cultures for transgenesis in other mammalian species has been *comparatively* nonexistent. By focusing on the laboratory rat as a widely popular model species in science, this chapter highlights several advantages of the spermatogonium as an alternative type of germline stem cell for transgenesis. Protocols for isolating, propagating, genetically modifying, and determining the germline transmission rates of spermatogonial cultures for the production of transgenic rats are introduced in detail. Although the full potential of spermatogonia has yet to be realized in animal genetics, this chapter illustrates how their application as novel germline vectors would open new doors to advance transgenic technology. Most notably, gene manipulations directly in the spermatogonium simplify production of germline founders while bypassing the intermediate production of chimeric progeny using micromanipulated embryos. Once experimental conditions for producing genetically modified animals using spermatogonial cultures are optimized, the approach holds the potential to facilitate targeted germline modifications by gene replacement, gene-insertion and/or restriction endonuclease technologies in a diversity of mammalian species.

12.1 Introduction

As germline stem cells in adult males, spermatogonial stem cells within the testes maintain continual production of spermatozoa throughout reproductive life [1, 2]. Spermatogonial stem cells from both mice and rats can now be cultured under conditions where they proliferate over multiple passages, while maintaining their ability to develop into functional spermatozoa when transferred into recipient testes [3–5]. Similar to mouse embryonic stem cells, cultures of mouse spermatogonia have been successfully modified genetically and then clonally expanded in vitro for the production of transgenic animals [6]; this includes targeted mutagenesis by gene replacement, and random mutagenesis using retroviral gene-trap vectors [7]. Despite the fact that stem

S. Pease and T.L. Saunders (eds.), *Advanced Protocols for Animal Transgenesis*, Springer Protocols,
DOI 10.1007/978-3-642-20792-1_12, © Springer-Verlag Berlin Heidelberg 2011

spermatogonia possess several inherently positive attributes for application to animal genetics, due to the relative novelty of protocols for propagating these germline stem cells in culture, they have yet to achieve the popularity of pluripotent embryonic stem cells. However, once optimized for transgenesis, a clear advantage of spermatogonial over embryonic stem cells for animal genetics will be the ability to transplant spermatogonial stem cells into testes to more directly generate germline founders without the need for any embryo manipulations. Thus, the spermatogonium's physiological state provides a streamlined strategy to manipulate mammalian genomes using germline stem cells (Fig. 12.1).

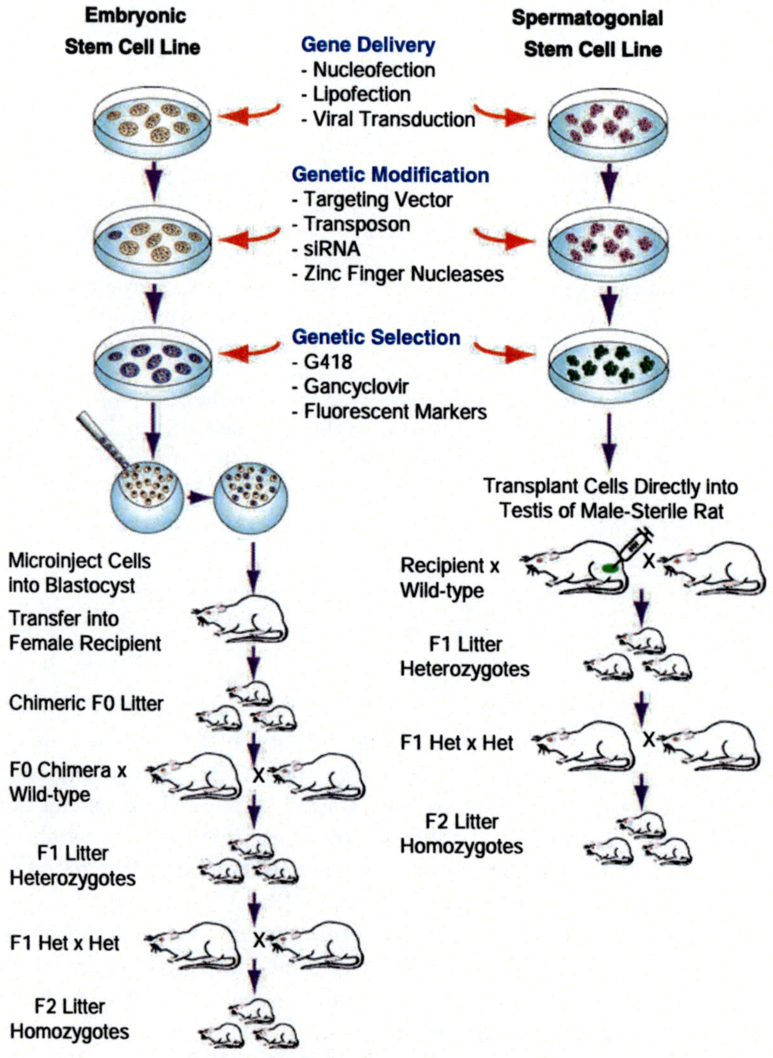

Fig. 12.1 *Stem Cell-Based Technologies for Producing Transgenic Rats*: Shown are flow diagrams comparing steps to generate new germline mutations in rats using cultures of embryonic or spermatogonial stem cells. Note the more streamlined approach to stably modify the germline in animals using stem spermatogonia.

Another positive attribute of spermatogonia as germline vectors for gene transfer in rodents (and other species) is that mutant alleles can be transmitted at maximal rates (i.e., 100%) from donor spermatogonial lines to F1 progeny by way of genetically sterile recipient-founders [8]. When used as a strategy to enrich for germline transmission of donor alleles, this mode of sterile-testis complementation with genetically modified spermatogonial stem cells in rodents is analogous to tetraploid-embryo complementation with embryonic stem cells in mice [9]. Considering these experimental factors, once culture conditions for genetic selection are further optimized, it will provide a cost-effective and technically straightforward method for transgenesis. To further advance biological research using the laboratory rat, this chapter presents experimental details for isolating, propagating, genetically modifying and transplanting fully functional rat spermatogonial stem cells into recipient-founders for the production of transgenic animals.

12.2 Isolating Rat Spermatogonial Stem Cells from Primary Testis Cell Cultures

Here, a cell culture-based protocol is presented demonstrating the isolation of spermatogonial stem cells from 23 to 24-day-old rats. This method is based on the principle that testicular somatic cells bind tightly to plastic and collagen matrices when cultured in serum-containing medium, whereas spermatogonia and spermatocytes do not bind to plastic or collagen when cultured in serum-containing medium [10]. The collagen-non-binding testis cells obtained by these procedures are thus ~97% pure spermatogenic cells. Stem spermatogonia are then easily selected from the isolated spermatogenic population during a short incubation step in culture on laminin matrix. The selected laminin-binding (Lam_B) spermatogonia are comprised of >90% undifferentiated type A spermatogonia (ZBTB16$^+$; DAZL$^+$) and are highly enriched in genetically modifiable stem cells that can develop into functional spermatozoa when transplanted into recipient rat testes. The isolated spermatogonia provide a highly potent and effective stem cell source that has been used to initiate studies on spermatogenesis [11, 12], derive proliferating lines of functional germline stem cells [3, 13], and for the production of transgenic rats [10].

12.2.1. Preparing Primary Testis Cell Culture form Neonatal Rats

12.2.1.1. Reagents

- Microdissecting scissors (cat. no. RS-5852, Roboz Surgical, Inc.).

- Operating sharp scissors (cat. no. RS-6802, Roboz Surgical, Inc.).
- 5/45° angle tweezers (cat. no. RS-5005, Roboz Surgical, Inc.).
- 23-day-old, male Sprague-Dawley rats, received at 15–16 d of age (Harlan Co. Inc.).
- *N*-2-Hydroxyethylpiperazine-*N'*-2-ethanesulfonic acid (HEPES) (cat. no. H4034-1KG, Sigma, Inc.).
- Glycine (Cat. no. 16407, USB Corp.).
- Hemacytometer (cat. no. 02-671-6, Fisher, Inc.).
- Dulbecco's modified Eagle's medium:Ham's F12 medium 1:1 (cat. no. D8437, Sigma, Inc.).
- Tissue culture-treated microplates, individually wrapped, sterile (cat. no. 3516, 3513, 3524, 3548, 3596, Corning, Inc.).
- Dulbecco's phosphate-buffered saline (PBS), 200 mg/L KCl (w/v), 200 mg/L KH2PO4 (w/v), 8 g/L NaCl (w/v), 1.15 g/L Na2HPO4 (w/v); (cat. no.D8537, Sigma, Inc.).
- Modified filtration system (MFS): use a sterile, disposable no. 10 scalpel (cat. no. 08-927-5A, Fisher, Inc.) to cut cellulose filter out of a disposable, 250-mL 0.2-μm filtration system (cat. no. 430767, Fisher, Inc.). Discard cellulose filter and replace with sterile mesh using aseptic technique.
- Sterile mesh: Sterile, 30 × 30 cm sheets of 41-μm pore, nylon spectra-mesh (cat. no. 08-670-202, Fisher, Inc.); Prepare for MFS by cutting into 5–6 cm wide squares.
- 70-μm tube top cell strainers (cat. no. 352350, Falcon, Inc.).
- Antibiotic–antimycotic solution: 10,000 U/mL penicillin G sodium (U/v), 10,000 μg/mL streptomycin sulfate (w/v), and 25 μg/mL amphotericin B (w/v) (cat. no. 15240-062, Invitrogen, Inc.).
- Glycine buffer: 1 M glycine, 20 mM HEPES, pH 7.2. To make 200 ml, add 15 g glycine and 950 mg HEPES to 180 mL deionized water. Adjust the pH of the solution to 7.2 by titrating in 0.1 N NaOH. Adjust the volume to 200 ml with deionized water. Sterilize the solution by passing it through a 0.2-μm filter into a sterile container. Store at 4°C for up to 4 months.
- Falcon brand 5, 10, and 25 ml pipettes (cat. no. 357543, 357543 and 357535, Falcon, Inc.).
- Heat-inactivated FBS: To prepare, thaw a 500-ml bottle of FBS (cat. no. S11550, Atlanta Biologicals, Inc.) received from the manufacturer (takes ~5 h at 22–24°C) and then place it into a clean water bath equilibrated to 56°C. Heat treat at 56°C for 45 min.
- Heat-inactivated horse serum [14]: The solution is received complete as a heat-inactivated solution from the manufacturer

in 500-ml bottles (cat. no. 26050-088, Invitrogen, Inc.). Thaw a 500-ml bottle of the HS (takes ~5 h at 22–24°C).

- Dispase: The solution is received complete from the manufacturer in 100-ml bottles (cat. no. CB-40235, Fisher, Inc.) that contain 50 caseinolytic units/ml.

12.2.1.2. Method

Starting Notes: Rat spermatogonial stem cell cultures are established from the testes of 23–24-day-old Sprague-Dawley rats. If the goal is to derive an isogenic spermatogonial line (*see* Subheading 12.3), testes from a single rat are used to obtain a starting source of stem spermatogonia [13]; or if larger numbers of spermatogonia are needed for experimental purposes the procedure can be scaled up by pooling testes from multiple rats [10].

Starting Notes: Prepare DHF12 medium by adding Antibiotic–antimycotic solution to 1× in Dulbecco's modified Eagle's medium:Ham's F12 medium 1:1. Use DHF12 in subsequent steps to prepare DHF12-TC (Subheading 12.2.1.2), DHF12-FBS (Subheading 12.2.1.2), and DHF12-FBS-2ME (Subheading "Matrix Selection for Laminin-Binding Spermatogonia"), SG Medium (Subheading 12.3.2.1).

The following procedure is scaled to establish a primary culture from testes pooled from five rats (ten total testes).

1. Testes are harvested from five, 23–24-day-old rats using sterile surgical technique and transferred into wells of a plastic six-well culture dish (9.6 cm^2 wells) containing 6 ml DHF12 medium at ambient conditions (22–24°C).
2. The outer, thin tunica albuginea of each testis is then carefully dissected free from the seminiferous tubules using fine forceps. The tunica albuginea is discarded, and the isolated seminiferous tubules are saved and transferred into a new well of a six-well culture dish (9.6 cm^2 wells) containing 6 ml DHF12/well.
3. The seminiferous tubules are then washed twice in fresh DHF12 by using forceps to transfer them into two consecutive wells of the six-well culture dish containing 6 ml DHF12/well.
4. The seminiferous tubules are then washed again by using forceps to transfer them into a new well of the six-well dish containing 6 ml, PBS.
5. Using forceps, the tubules are immediately transferred into a final well of the same plate containing 6 ml, 1 M glycine buffer, pH 7.2 where they are minced with scissors for ~3 min, and incubated in the same well for an additional 7 min following mincing (i.e., total of 10 min in the glycine buffer).
6. After incubation in glycine buffer, the minced tubules are transferred to a 50-ml tube, suspended to 35 ml with glycine

buffer and pelleted by centrifugation at 400 × *g* for 5 min at ambient temperature (i.e., 20–22°C).

7. The supernatant is discarded into a waste beaker (Do not aspirate cultures from this step on; pipette supernatant into waste beaker to discard wash medium). The tubular pellet is retained, suspended in 35 ml of DHF12 containing 8.5% FBS (DHF12-FBS) and then pelleted at 400 × *g* (procedure for all wash steps).
8. Repeat wash as in **step 7**.
9. After a second wash, tubules are suspended in dispase solution (200 caseinolytic units/testis) and digested for 30 min at 32.5°C with gentle rocking at 5 min intervals.
10. The digest is then adjusted to 35 ml with DHF12-FBS, positioned vertically on a G24 environmental shaker (New Brunswick Scientific) and agitated for 10 s at 220 rpm under ambient conditions (22–24°C).
11. The suspension of digested tubules is pelleted at 400 × *g*, the pellet is retained and washed twice in 35 ml DHF12-FBS/wash as described in **step 7**.
12. The pellet is suspended to 35 ml DHF12-FBS and then filtered through 41-μm mesh using a Modified Filtration System (MFS) to remove undigested tubules and genomic DNA. The top of the MFS with the mesh is unscrewed to a loose position and attached to a vacuum source. The vacuum strength is kept as low as possible, while allowing the suspension to pass through. The suspension is mixed five to ten times with a 10-ml pipette and transferred into the MFS, 10 ml at a time. If genomic DNA causes the MFS to become clogged, the mesh is removed with sterile forceps and replaced with a fresh piece.

    Note: When deriving pure spermatogonial lines from single animals, the MFS can more simply be substituted with 70-μ m, 50-ml tube-top cell strainers to clear digested tubules and genomic DNA from the testis cell suspension.

13. The filtered testis cell suspension is retained, pelleted at 400 × *g*, and then washed once following suspension to 35 ml with DHF12 supplemented with 5.5% HS and 2.4% FBS (DHF12-TC).
14. The isolated testis cells are plated into 10-cm plastic culture dishes at ~2 × 10^5 cells/cm^2 (i.e., ~10^7 cells/10 cm dish) containing 10 ml of DHF12-TC and incubated undisturbed in a humidified incubator at 32.5°C, 5.5% CO_2 for ~65 h.

    Note: From ten testes, the final cell suspension should yield about 1 × 10^8 to 1.5 × 10^8 cells, which provides enough cells to seed 10–15, 10 cm dishes, respectively.

12.2.2. Selecting Rat Spermatogonial Stem Cells from Primary Culture

Based on their ability to selectively bind to laminin matrix, cultures of ~10^6 undifferentiated rat spermatogonia, which are highly enriched in germline stem cells, can be isolated from primary testis cell cultures established from five, 23–24-day-old rats (Subheading 12.2.1.2).

12.2.2.1. Reagents

- Dulbecco's modified Eagle's medium:Ham's F12 medium 1:1 (cat. no. D8437, Sigma, Inc.).
- Antibiotic–antimycotic solution: 10,000 U/mL penicillin G sodium (U/v), 10,000 μg/mL streptomycin sulfate (w/v), and 25 μg/mL amphotericin B (w/v) (cat. no. 15240-062, Invitrogen, Inc.).
- Heat-inactivated FBS (cat. no.S11550, Atlanta Biologicals, Inc.).
- Heat-inactivated horse serum [14] (cat. no. 26050-088, Invitrogen, Inc.).
- Rat tail collagen I-coated cell culture dishes (10-cm dishes, cat. no. 08-772-75, Fisher, Inc.).
- 70-μm tube top cell strainers (cat. no. 352350, Falcon, Inc.).
- Dulbecco's phosphate-buffered saline (PBS), 200 mg/L KCl (w/v), 200 mg/L KH2PO4 (w/v), 8 g/L NaCl (w/v), 1.15 g/L Na2HPO4 (w/v); (cat. no.D8537, Sigma, Inc.).
- Laminin 1 mg/mL solution (cat. no. L2020, Sigma, Inc.); prepare laminin-coated plates a day ahead of starting experiment (*see* Subheading "Preparing Laminin-Coated Culture Dishes").
- Bovine serum albumin (BSA) (cat. no. A4503, Sigma, Inc.).
- 2-mercaptoethanol (98+%, cat. no. M3148, Sigma, Inc.).
- DHF12 medium (*see* starting notes under Subheading 12.2.1.2).

12.2.2.2. Methods

12.2.2.2.1. Matrix Selection for Laminin-Binding Spermatogonia

1. Prepare 10 cm collagen-1-coated culture dishes to process primary testis cell cultures plated in Subheading 12.2.1.2. One 10 cm collagen-1-coated dish is sufficient to process germ cells from 4 to 6, 10 cm cultures of primary testis cells. Prewash dishes from the manufacture by adding 4 ml DHF12-TC to each dish; rock dishes back and forth a couple times to wash; discard the wash medium into a waste beaker. Next, add 4 ml fresh DHF12-TC culture medium to each dish and place dishes into a humidified cell culture incubator at 32.5°C, 5.5% CO_2 until they are used in **step 5** of this subheading

   Example: Three, 10 cm collagen-1-coated cell culture dishes would be sufficient to process spermatogonia from 12 to 18

primary testis cell cultures originally plated in 10-cm dishes during Subheading 12.2.1.2.

2. Following the ~65 h primary culture step at the completion of Subheading 12.2.1.2, medium is carefully removed from sets of 4–6, 10-cm dishes of testis cell cultures, and then each culture is washed once by adding 4 ml of DHF12/dish, and gently rocking the dish by tilting back and forth one time at ~45° angles.

3. The DHF12 wash is discarded into a waste beaker (Do not aspirate supernatants from this step on; use a waste beaker to discard medium from washes), and replace with 4 ml, fresh DHF12.

4. The bound germ cells are then harvested from monolayers of adherent somatic cells by repeatedly pipetting the added 4 ml, DHF12 *gently* over the surface area of the dish (Fig. 12.2).

   Note: Somatic cell-enriched monolayers remain attached to the culture dishes (*see* Fig. 12.2), and can either be discarded or used for further experimentation if desired. Avoid pipetting too stringently during **step 4** as this will remove excessive somatic cells from the dish and contaminate the spermatogonial fraction obtained in the final step of this protocol.

5. The pooled germ cell suspensions are then pelleted at 400 × *g*, suspended in 4 ml of DHF12-TC, transferred to 10-cm^2 plastic dishes coated with collagen I (prepared in **step 1**) and incubated at 32.5°C for ~4 h.

6. Cells that do not bind to collagen I dishes (Col_{NB} cells) are harvested and pelleted at 400 × *g* (Fig. 12.2). The Col_{NB} cells are then suspended in DHF12 containing 10% FBS and 0.00024% 2-mercaptoethanol (DHF12-FBS-2ME), passed through a 70-μm mesh tube top cell strainer, and plated at ~10^6 cells/ml/well in 12-well plates precoated with ~5 μg/cm^2 mouse laminin (*see* Subheading "Preparing Laminin-Coated Culture Dishes").

   Note: Laminin-coated (~5 μg/cm^2) dishes should be prepared a day ahead of time. *See* Subheading "Preparing Laminin-Coated Culture Dishes" for preparation of laminin-coated culture dishes.

7. The plated Col_{NB} cells are incubated for 45 min at 32.5°C, after which, the unbound, laminin-non-binding cell fraction (Lam_{NB} cells) of differentiated spermatogonia and spermatocytes is separated from the undifferentiated spermatogonial cells that bind to laminin (Lam_B cells) by pipetting (Fig. 12.2). The Lam_{NB} cell fractions can be discarded.

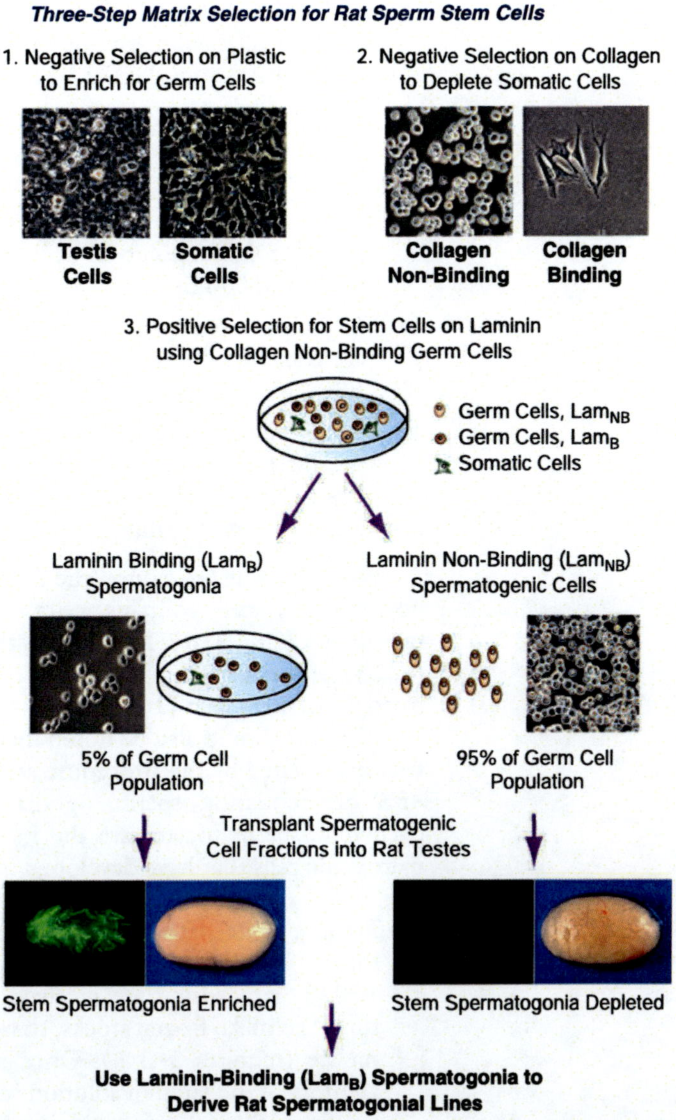

Fig. 12.2 *Isolating rat stem spermatogonia from primary culture.* A three-step method to select for rat spermatogonial stem cells in culture on (1) Plastic, (2) Collagen I, and (3) Laminin is illustrated (described in Subheading 12.2). Based on their ability to generate colonies of spermatogenesis in recipient rat testes, the resulting fraction of laminin-binding (Lam$_B$) spermatogonia obtained is highly enriched in fully functional stem spermatogonia, which can be used to derive spermatogonial lines (described in Subheading 12.3). Shown is an example in which Lam$_B$ spermatogonia were isolated from transgenic rats expressing EGFP, and then transplanted into testes of busulfan-treated rats to track fractionation of stem cell colonizing activity (*Green Fluorescence*; described in Subheading 12.8).

8. The Lam_B cells, which contain the spermatogonial stem cells, are retained bound on the plate, and then gently rinsed twice with 0.8 ml, DHF12-FBS-2ME/well/rinse.
9. Each rinse in **step 7** is discarded, and then the attached cells are incubated in 1 ml, PBS, 0.5% BSA/well at 32.5°C for 5 min to help detach them from laminin for harvesting.
10. The Lam_B cells are next harvested by gently pipetting the PBS, 0.5% BSA solution over the surface area of each well using a p1000 tip.
11. Cells harvested from up to six laminin-coated wells/plate are transferred into 15-ml tubes containing 5 ml DHF12-FBS-2ME and pelleted at 400 × *g* for 5 min.
12. The cellular pellet, now enriched in undifferentiated spermatogonia, is retained and suspended to 4 ml in fresh DHF12-FBS-2ME (or in SG Medium; Subheading 12.3.2.1) for counting.

Note: By this procedure, the freshly isolated Lam_B germ cell population typically contains >90% undifferentiated, type A spermatogonia ($ZBTB16^+$, $DAZL^+$) in the single (~88%) or paired (~12%) cell state, and is highly enriched in fully functional spermatogonial stem cells [10–12].

Note: It should also be noted that fractions of Lam_B spermatogonia isolated by this procedure contain ~4% somatic cells, and ~5% differentiating spermatogonia plus spermatocytes. Late pachytene spermatocytes are the most advanced stage of spermatogenic cells that have developed in 23 to 24-day-old rats.

12.2.2.2.2. Preparing Laminin-Coated Culture Dishes

1. Prepare frozen stocks of laminin to avoid multiple freeze thaws. Vials containing 1 mg/ml laminin solution (cat. no. L2020,Sigma, Inc.) are received frozen from the manufacturer. To make frozen stocks, thaw one vial of mouse laminin on ice (requires 1–2 h). Once thawed, make ~six 150 μl aliquots of the laminin solution in sterile microfuge tubes on ice. Store the laminin stocks at −80°C for up to 1 year.
2. The day prior to isolating spermatogonia by matrix selection, coat wells of a sterile 12-well culture dish with laminin (~5 μg/cm^2). To prepare, thaw one 150 μl aliquot from the 1 mg/ml laminin stocks on ice (requires ~30 min) and then add the entire volume of the aliquot into a sterile 50 ml tube containing 8 ml ice-cold DHF12. *Slowly* swirl or rock the tube by hand to mix the contents.
3. Add 1 ml of the diluted laminin/well into eight wells (3.8 cm^2) of a 12-well plastic culture dish. Wrap the dish with Parafilm and incubate the laminin in the wells of the culture dish overnight at 4°C.

Note: As an estimate to determine the number of wells to coat with laminin for a given experiment, four laminin-coated wells of a 12-well plate is sufficient to process spermatogonia harvested from a set of five to six 10-cm dishes of primary testis cell cultures (Subheading 12.2.1.2).

12.3 Deriving, Subculturing, and Preserving Rat Spermatogonial Stem Cell Lines

In this section, a set of protocols are presented for the derivation and routine culture of rat spermatogonial lines. Enriched fractions of undifferentiated spermatogonia isolated from primary testis cell cultures (*See* Subheading 12.2) are used as a stem cell source to derive proliferating spermatogonial lines in serum-free culture medium [13]. Established spermatogonial lines are subcultured on feeder layers of mouse embryonic fibroblasts (MEFs) without the use of proteases, and then preserved in a serum-free, spermatogonial freezing medium [13]. The generated spermatogonial stocks can be thawed after >1 year in cryostorage, re-expanded in cell number over multiple passages in culture, and then used to produce fully functional spermatozoa in recipient rat testes [8, 13]. Donor-derived spermatozoa produced in recipient rat testes yield high rates of germline transmission through natural mating [8, 13].

12.3.1. Reagents

- DMEM-high glucose: Dulbecco's modified Eagles Medium-high glucose (cat. no. D5648, Sigma, Inc.).
- Sodium Bicarbonate (cat. no. S5761, Sigma, Inc.).
- PBS: Dulbecco's phosphate-buffered saline (PBS; cat. no. D8537, Sigma, Inc.) 200 mg/L KCl (w/v), 200 mg/L KH2PO4 (w/v), 8 g/L NaCl (w/v), 1.15 g/L Na2HPO4 (w/v).
- Heat-inactivated fetal bovine serum: FBS (cat. no. 104, Tissue Culture Biologicals).
- Gelatin from Porcine Skin- Type A (cat. No. G1890, Sigma, Inc.).
- DR4 Mouse embryonic fibroblasts: MEFs (cat. no. SCRC-1045, ATCC; or cat. no. ASF 1001P, Applied Stemcell, Inc).
- Recovery cell culture freezing medium (cat. no. 12648-010, Invitrogen, Inc.).
- T225 Flasks angled neck (cat. no. 431081, Corning, Inc.).
- Costar clear TC-treated microplates, individually wrapped, sterile (cat. no. 3516, 3513, 3524, 3548, 3596, Corning, Inc.).

- Dulbecco's modified Eagle's medium: Ham's F12 medium 1:1 (cat. no. D8437, Sigma, Inc.).
- B-27 Supplement minus vitamin A (cat. no. 12587-010, Invitrogen, Inc.).
- L-glutamine (cat. no. 25030-149, Invitrogen, Inc.).
- Antibiotic–antimycotic solution: 10,000 U/mL penicillin G sodium (U/v), 10,000 μg/mL streptomycin sulfate (w/v), and 25 μg/mL amphotericin B (w/v) (cat. no. 15240-062, Invitrogen, Inc.).
- 2-Mercaptoethanol (cat. no. M3148, Sigma, Inc.).
- Recombinant human FGF2 (cat. no. F0291, Sigma, Inc.).
- Recombinant GDNF (cat. no. 512-GF, R and D Systems, Inc.).
- Dimethyl Sulfoxide: DMSO (cat no D2650, Sigma, Inc.).
- 5100 Cryo 1°C Freezing Container "Mr. Frosty" (cat. no. 15-350-50, Thermo Fisher Scientific Nalgene, Inc.).
- Cryogenic vials (cat. no. 03-337-7D, Thermo Fisher Scientific Nalgene, Inc.).
- DHF12 medium (*see* starting notes under Subheading 12.2.1.2).

Protease, nuclease and fatty acid free Bovine Serum Albumin (PNFF-BSA; cat. no. 126609, Calbiochem, Inc.).

12.3.2. Methods

12.3.2.1. Formulating Spermatogonial Culture Medium

1. Spermatogonial Culture Medium (SG Medium) is prepared by supplementing DHF12 with 20 ng/ml GDNF, 25 ng/ml FGF2, 100 μM 2-mercaptoethanol, an additional 4 mM L-glutamine (final concentration = 6 mM), and a 1× concentration of B27 Supplement Minus Vitamin A [13].

Note: Aliquots of GDNF and bFGF stocks should be prepared using PBS containing 0.1% Protease, nuclease and fatty acid free Bovine Serum Albumin (PNFF-BSA; cat. no. 126609, Calbiochem, Inc.).

12.3.2.2. Preparing Fibroblast Feeder Cell Lines

1. Primary stocks of DR4 mouse embryonic fibroblasts (MEFs) are purchased from the manufacture and expanded after plating into Dulbecco's modified Eagle's medium supplemented with 1.5 g/L sodium bicarbonate, and 15% heat-inactivated FBS (MEF medium) at 37°C/5% CO_2 for up to four passages following their thawing and initial plating from the vial received from the manufacturer.
2. Following expansion, secondary stocks of MEFs are irradiated (100 Gy) and then cryo-preserved in liquid nitrogen for future use in Recovery Cell Culture Freezing Medium (Invitrogen) according to the manufacturer's protocol.

3. Precoat tissue culture-treated dishes with a solution of sterile 0.1% gelatin for 1 h at room temperature; rinse 1× with sterile PBS before plating MEFs. To prepare 0.1% gelatin solution, dissolve gelatin in ultrapure laboratory grade water (1 g gelatin/liter) and autoclave on liquid cycle.
4. Prior to use for culture with spermatogonia, MEFs are thawed and plated into gelatin-coated dishes (4.5×10^4 cells/cm^2) in MEF medium for 16–48 h, rinsed 1× with PBS and then preincubated in SG medium for an additional 16–48 h. The SG medium used for preincubation is then discarded and spermatogonia are passaged onto the MEFs in fresh SG medium.

12.3.2.3. Deriving Rat Spermatogonial Lines

Starting Notes: The following culture steps are scaled for deriving and propagating spermatogonial lines from individual rats using SG medium. All culture steps are performed at 37°C, 5% CO_2 in a humidified incubator.

1. To derive rat spermatogonial stem cell lines, freshly isolated laminin-binding spermatogonia from individual rats are plated separately into 3.5 cm, gelatin-coated wells of a six-well culture plate in 3.5 ml SG medium/well (~0.37 ml/cm^2) at 1–5 × 10^5 cells/well.
2. After an initial selection for 40–48 h on gelatin-coated plates, spermatogonia in suspension (i.e., including loosely bound spermatogonia), are harvested free from the contaminating somatic testis cells by pipetting.
3. Harvested spermatogonia are pelleted at 400 × *g* for 4 min, the supernatant is discarded and the cellular pellet is suspended in SG medium and plated into fresh gelatin-coated wells (3.5 cm) for an additional 72–96 h.

   Note: Spermatogonia cultured on gelatin using SG medium are observed loosely bound to the culture plate, bound to residual adherent somatic testis cells, and in suspension; many of the spermatogonia in suspension adhere to each other as cellular "clusters" of variable size. In contrast, contaminating somatic testis cells attach avidly to and spread-out on the gelatin matrix.

   Note: If the somatic cell population does not look dramatically depleted after ~48 h during the second incubation on gelatin-coated plates in **step 3** (i.e., >30% confluent layer of attached somatic cells), the spermatogonia can be harvested and re-plated on fresh gelatin-coated plates in SG Medium and incubated for an additional ~48 h interval, as a third round of selection on gelatin matrix over a total of ~144 h (i.e., ~6 days).
4. After this point (i.e., after depletion of essentially all somatic testis cells), suspensions of spermatogonia derived from each

rat are harvested from the final selection step on gelatin, suspended in fresh SG medium and passaged into 1.8 cm^2 wells of a 12-well culture dish containing feeder layers of irradiated mouse embryonic fibroblasts (MEFs) (*see* Subheading 12.3.2.2 for preparing MEF feeder layers).

5. The initial passage of spermatogonial cultures after plating onto MEF feeder layers requires a 1:1 to 1:2 split into the same size wells at 14–21 days after their initial seeding onto the MEFs. In this situation, because irradiated MEF feeder layers are not as effective after 14 days in culture, fresh MEFs ($2 \times 10^4/cm^2$) are "spiked" into the on-going spermatogonial cultures on day 11–12 so to by-pass the need to passage the spermatogonia before expanding to larger numbers.
6. Once established by the second or third passage on MEFs, cultures of spermatogonia are passaged at ~1:3 dilutions onto a fresh monolayer of MEFs every 10–14 days at $1–3 \times 10^4$ cells/cm^2 for over 5 months (i.e., ~12 passages).
7. For passaging, cultures are first harvested by gently pipetting them free from the MEFs. After harvesting, the "clusters" of spermatogonia are dissociated by gentle trituration with 20–30 strokes through a P1000 pipette tip in SG Medium.
8. The dissociated cells are pelleted at $400 \times g$ for 4 min and the number of cells recovered during each passage is determined by counting on a Hemocytometer (**Note**: spermatogonial clusters are not disrupted for counting until the second passage on MEFs).
9. Spermatogonia are easily distinguished during counting as the predominant population of smaller, round cells with smooth surfaces, as compared to occasionally observed, larger and often irregular shaped irradiated MEFs (Fig. 12.3). Typically, $2–4 \times 10^6$ spermatogonia can be harvested from a single, 10-cm dish.

12.3.2.4. Cryopreserving Stocks of Spermatogonial Lines

1. Prepare Spermatogonial Freezing Medium (SG Freezing Medium) [13] by adding DMSO at a concentration of 10% (v/v) in SG Medium. Filter-sterilize and cool the prepared freezing medium on ice prior to use.

   Note: Spermatogonial Freezing Medium can be stored frozen at −20°C for up to 1 month prior to use.
2. Prepare a "Mr. Frosty" freezing container by adding 200 ml fresh isopropanol to the outer chamber. Chill the container by equilibrating it to ~4°C in a refrigerator prior to use.
3. Harvest spermatogonia as described previously in Subheading "Matrix Selection for Laminin-Binding Spermatogonia" or Subheading 12.3.2.3.

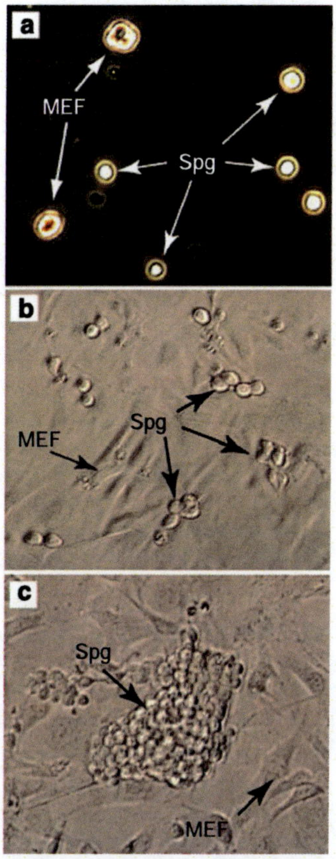

Fig. 12.3 ***Phase-contrast images of rat spermatogonial lines during subculture.*** **(a)** Appearance of rat spermatogonia (Spg) and residual irradiated mouse embryonic fibroblasts (MEFs) on a hemocytometer after harvesting from an established spermatogonial line mechanically by pipetting. **(b)** Appearance of rat spermatogonia illustrated in **(a)** ~3 days after being passaged onto fresh MEF feeder layers. **(c)** Appearance of rat spermatogonia illustrated in **(a)** ~12 days after being passaged onto fresh MEF feeder layers.

4. Suspend the harvested spermatogonial pellet in ice-cold, SG Freezing Medium at 2×10^5 to 2×10^6 cells/ml and then aliquot stocks into cryovials at 1 ml/vial. Work quickly and place filled cryovials on ice while finishing aliquots.
5. Place cryovials of spermatogonial stocks into the prechilled "Mr Frosty" and close container firmly.
6. Store the freezing container of spermatogonial stocks at −80°C for 24 h, then transfer vials into a liquid nitrogen cryostorage unit.

Note: Record each use of the "Mr. Frosty" container so that fresh isopropanol can be replenished after the fifth use (*see* Manufacturer's instructions).

12.4 Transfection of Rat Spermatogonia with Plasmid DNA by Nucleofection

In this section, we present a gene delivery protocol for rat spermatogonia, which is based on electroporation. The method consistently results in 20–40% transfection of circular plasmid DNA into freshly isolated (Subheading "Matrix Selection for Laminin-Binding Spermatogonia") or proliferating (Subheading 12.3.2.3) cultures of undifferentiated rat spermatogonia. This approach is most convenient from a user stand point, but requires a specialized Amaxa Nucleofector device from Lonza.

12.4.1. Reagents

- Undifferentiated Spermatogonia (Subheading "Matrix Selection for Laminin-Binding Spermatogonia" or Subheading 12.3.2.3).
- SG Medium (prewarmed).
- Highly purified plasmid DNA in TE buffer at 1–2 μg/μl (Qiagen, Inc. or Invitrogen, Inc. endotoxin-free maxiprep).
- Gelatin-coated plates (**step 3**, Subheading 12.3.2.2).
- Plates with fresh MEF feeder layers (Subheading 12.3.2.2).
- Nucleofector® device (cat. no. AAD-1001, Lonza).
- Cell Line Nucleofector Kit L (cat. no. VCA-1005, Lonza).

12.4.2. Methods

1. Set up Nucleofector® device according to the manufacture's instructions and enter program A-020.
2. Harvest freshly isolated spermatogonia from laminin by pipetting using PBS, 0.1% BSA (*see* Subheading "Matrix Selection for Laminin-Binding Spermatogonia"), or harvest proliferating cultures of spermatogonia from MEFs by pipetting in SG medium (*see* Subheading 12.3.2.3).

   Note: If transfecting spermatogonia from an established line maintained on MEFs, first preincubate the harvested cells on a gelatin-coated plate in SG medium for 45 min at 37°C, 5% CO_2 to deplete contaminating MEF feeder cells from the spermatogonial suspension. Then re-harvest spermatogonia for transfection.
3. After counting harvested cells, split the suspension into 15-ml tubes to obtain $2–3 \times 10^6$ cells/transfection, pellet them by centrifugation for 5 min at $400 \times g$.
4. Carefully remove the supernatant as completely as possible from the tube.
5. Suspend the spermatogonial pellet with 100 μl, Nucleofector® Solution L, add DNA (10–15 μg) and transfer the

spermatogonial suspension into a transfection cuvette provided in the kit for Solution L.

6. Insert the cuvette into the Nucleofector® holder and press the "X" button to execute program A-020.

   Note: Do not keep the cells in Nucleofector Solution longer than 15 min.

7. Promptly remove the cuvette from the Nucleofector® device and add 500 μl of prewarmed SG Medium directly to the transfected cells within the cuvette.

8. Carefully remove the cell suspension with a plastic pipette provided in the kit, avoiding visibly coagulated DNA, and plate the suspension drop-wise directly onto freshly prepared feeder layers of MEFs in SG Medium for further expansion and selection for transgenic lines (Subheading 12.3.2.2).

12.5 Transfection of Rat Spermatogonia with Plasmid DNA Using Lipofectamine 2000

Here, a second protocol is presented for transfecting rat spermatogonial stems with plasmid DNA. This protocol is based on lipofection, and therefore can be performed without special electroporation equipment.

12.5.1. Reagents

- Undifferentiated spermatogonia from protocols in Subheading "Matrix Selection for Laminin-Binding Spermatogonia" or Subheading 12.3.2.3.
- SG Medium (prewarmed).
- SG Medium without antibiotic/antimycotics (prewarmed).
- Opti-MEM (cat. no. 31985-062; Invitrogen, Inc.).
- Lipofectamine 2000 (cat. no. 11668-019; Invitrogen).
- Highly purified plasmid DNA in TE buffer at 1–2 μg/μl.
- Gelatin-coated plates (Subheading 12.3.2.2).
- Plates with fresh MEF feeder layers (Subheading 12.3.2.2).

12.5.2. Methods

1. Prepare a Transfection Mixture containing Lipofectamine 2000 (Invitrogen) and plasmid DNA in Opti-MEM, as follows:
   (a) In a 1.5-ml microfuge tube, dilute 1 μg DNA/100 μl Opti-MEM.
   (b) In a separate 1.5-ml microfuge tube, dilute 2 μl Lipofectamine 2,000/100 μl Opti-MEM.
   (c) Incubate tubes separately for 5–10 min.

(d) Combine contents of each tube together and incubate at room temperature for at least 20 min (but no longer than 6 h) to obtain the Transfection Mixture. During this incubation step, proceed to harvesting cells for transfection.

2. Harvest freshly grown Lam_B spermatogonia (Subheading "Matrix Selection for Laminin-Binding Spermatogonia") or cultures of proliferating spermatogonia grown on MEFs (Subheading 12.3.2.3) as described.

   Note: If using proliferating cultures of spermatogonia maintained on MEF feeder layers, first plate the cells onto a fresh gelatin-coated plate and incubate for 30–45 min (37°C, 5% CO_2) to deplete the number of residual MEFs present in the cell suspension.

3. Suspend spermatogonia to ~10^6 cells/ml in SG Medium (minus antibiotic/antimycotics), or DHF12-FBS + 2ME (minus antibiotic/antimycotics).

4. Add Transfection Mixture to the cell suspension at a ratio of 20% volume Transfection Mixture: 80% volume spermatogonial suspension, and incubate at 37°C, 5% CO_2 for 40–120 min (routinely 80 min) in a vented tube within a humidified incubator.

   Note: As a typical example, 40 μl volume of the Transfection Mixture prepared in **step 1** is used to transfect ~2×10^5 spermatogonia in a total transfection volume of 200 μl.

5. During transfections lasting longer than 1 h, mix the transfection by gently pipetting cells up and down two times midway through the incubation period.

6. After the transfection incubation period, wash spermatogonia by first suspending the transfection suspension to 20 times its volume using fresh culture medium (i.e., 4 ml medium/200 μl transfection reaction), and then pellet the cells for 5 min at 400 × *g*.

7. Discard the supernatant fluid, and wash the pellet(s) two additional times using fresh culture medium without antibiotic/antimycotics at an equivalent of the 20× volume/wash used in **step 6**.

8. After the third wash, suspend the cell pellet in fresh culture medium containing antibiotic/antimycotics and then plate transfected cells onto fresh MEF feeder layers for selection of transgenic spermatogonial lines (Subheading 12.7).

   Note: Take care to completely remove all supernatant after each wash in **steps 6–8** to minimize toxic effects of the transfection reaction. Do not aspirate. To do this, we highly recommend gently pouring off the supernatant after each centrifugation wash step, and then centrifuging the pellet again at 400 × *g* for ~0.5 min to re-compact the pellet before using a p200 tip to pipette off all the residual supernatant.

12.6 Lentiviral Transduction of Rat Spermatogonial Stem Cells

Over the last decade, lentiviral vectors have emerged as powerful reagents for genetically modifying the germline of rodents [15–17], including the production of transgenic rats using spermatogonial stem cells [10, 18, 19]. An advantage of this gene delivery approach is that a high percentage of spermatogonia can be stably modified genetically in culture by short periods of treatment with lentiviral particles. Following treatment with viral particles, the spermatogonial cultures can be transplanted directly into testes of recipient animals to produce germline founders without the need to subculture and further maintain spermatogonial lines. Founder male rats produced by this method can transmit the lentiviral transgene to Fl progeny at relatively high rates (i.e., up to 30% transgenic progeny/recipient founder). The following protocol describes a basic platform for concentrating spermatogonial stem cells in culture on laminin-coated plates to achieve optimal transduction with small volumes of lentiviral particles. The transduced spermatogonia can then be transplanted into recipient rat testes for production of transgenic progeny [10] (reviewed in Fig. 12.4).

12.6.1. Reagents

- VSV-G pseudotyped lentiviral particles [16].
- Laminin-coated culture dishes (Subheading "Preparing Laminin-Coated Culture Dishes").
- SG Medium (Subheading 12.3.2.1) or DHF12-10%FBS+2ME (Subheading "Matrix Selection for Laminin-Binding Spermatogonia", **step 6**).
- Laminin-Binding Spermatogonia at passage 0–14 (Subheading "Matrix Selection for Laminin-Binding Spermatogonia" or Subheading 12.3.2.3).

12.6.2. Methods

1. For viral transduction, freshly isolated laminin-binding spermatogonia (Subheading "Matrix Selection for Laminin-Binding Spermatogonia"), or spermatogonia from established primary lines (Subheading 12.3.2.3) are harvested and re-plated onto laminin-coated culture dishes (Subheading "Preparing Laminin-Coated Culture Dishes") in 96- or 48-well plates at 1–2 × 10^5 cells/cm^2 in 100 μl or 200 μl culture medium/well, respectively.

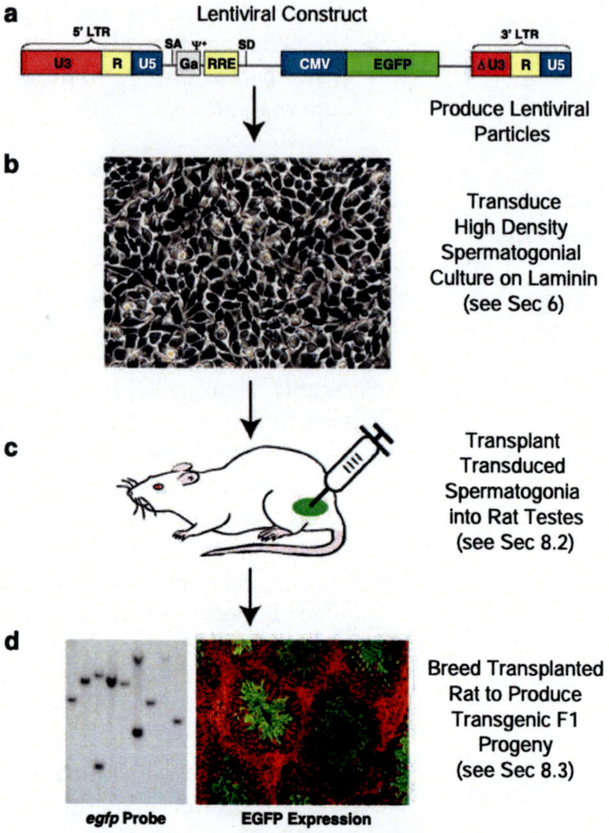

Fig. 12.4 ***Producing transgenic rats by lentiviral transduction of spermatogonial cultures.*** (**a**) Schematic of a self-inactivating lentiviral transgene construct that can be used to make lentiviral particles for stably transducing rat spermatogonial stem cells in culture. (**b**) Phase-contrast microscopy image of freshly isolated rat, Lam_B spermatogonia (described in Subheading 12.2) after plating at high density on laminin prior to overnight treatment with lentiviral particles. (**c**) Following overnight treatment with lentiviral particles, spermatogonia are harvested from laminin and transplanted into testes of busulfan-treated rats (described in Subheading 12.8). Transplanted recipient-founders can be paired with wild-type female rats at ~65 days posttransplantation to produce transgenic F1 progeny that stably express the lentiviral transgene, pHR′-CMV-EGFP-SIN_{18} [10]. (*Left*) Southern blot of F1 transgenic rat progeny produced by lentiviral transduction of spermatogonial cultures. *Lanes* show bands from transgenic rats that inherited the lentitviral, EGFP transgene. DNA standards are not shown. (**d**) (*Right*) Expression of EGFP in testes of F1 transgenic rat progeny produced by lentiviral transduction of spermatogonial cultures. Green-labeled cells in rat testis cross section show EGFP expression within elongating spermatids; *Red*-labeled cells show Cy3-conjugated anti-vimentin IgG binding in somatic cells.

Note: Freshly isolated laminin-binding spermatogonia can be re-plated either in DHF12-FBS-2ME medium, or in SG Medium at 32.5°C, 5.5% CO_2. Established spermatogonial lines are re-plated in SG Medium at 37°C, 5% CO_2 in a humidified incubator.

1. After incubation on laminin-coated dishes for 2–3 h, the attached spermatogonia are then exposed once to VSV-G-pseudotyped, lentiviral particles[1] by adding them directly into the culture at a multiplicity of infection (MOI) of ~5–10.
2. After overnight exposure of viral particles for 14–18 h, spermatogonial cultures are rinsed once with DHF12 medium, and then harvested from laminin by gentle pipetting following incubation for 4–5 min in PBS, 0.1% BSA (*see* Subheading "Matrix Selection for Laminin-Binding Spermatogonia," **steps 9–12**).
3. Harvested spermatogonia are immediately transferred into ~20× volume of the respective culture medium (i.e., either DHF12-FBS-2ME or in SG Medium), and then pelleted at 400 × *g* to remove unbound lentiviral particles.[1]
4. The pelleted spermatogonia are retained and then washed twice more by suspending them in medium to an equivalent of the 20× volume used in **step 3**, prior to pelleting at 400 × *g*.
5. After washing, spermatogonia are suspended in the desired volume of medium for either direct transplantation into recipient testes for production of transgenic animals (*see* Subheading 12.8), or for use in clonal selection protocols during culture on MEFs (*see* Subheading 12.7) prior to transplantation into recipient testes (Fig. 12.4).

Note: VSV-G pseudotyped lentiviral particles can be produced by established protocols [16], including those listed by manufacturers of lentiviral vectors (i.e., Invitrogen, Inc). Purification of lentiviral particles by centrifugation or affinity chromatography is highly recommended.

12.7 Clonal Selection for Genetically Modified Rat Spermatogonia in Culture

The ability to clonally expand rat spermatogonia following genetic manipulation with DNA constructs or mutagens will permit selection for desired genomic modifications directly in the germline to facilitate production of new mutant animals. Once spermatogonial culture conditions are optimized in the rat, such clonal selection protocols would become particularly valuable for making

[1] *Safety Information*: Lentiviral vectors are considered biohazardous reagents. Their use for studies in animals warrants an approved safety plan certified by both the parent research institution's Environmental Health and Safety program, and the Institutional Animal Care and Use Committee (IACUC). Personal protection equipment (gloves; lab coat; safety glasses) should be worn when working with lentiviral vectors to prevent exposure to the researcher. The preparation, use and disposal of lentiviral vectors should be restricted to certified biosafety cabinets to prevent exposure to the researcher and to others.

mutant rats that harbor targeted genomic modifications at specific alleles by gene replacement, gene-insertion, and/or restriction endonuclease technologies.

12.7.1. Reagents

- Established, proliferating line of rat spermatogonial stem cells (Subheading 12.3.2.3).
- Geneticin Selective Antibiotic: G418 (cat no 11811-031, Invitrogen, Inc.)
- Lentiviral or Plasmid DNA Constructs expressing a resistance gene that selects for survival in G418 containing medium (i.e., neomycin phosphotransferase gene).
- Fibroblast Feeder cell line expressing a resistance gene that selects for survival in G418 containing medium (DR4 mouse embryonic fibroblasts; cat. no. SCRC-1045, ATCC; or cat. no. ASF 1001P, Applied Stemcell, Inc.).

12.7.2. Methods

1. After transfecting spermatogonia from an established proliferating line with plasmid DNA, or following their transduction with lentiviral particles, the treated spermatogonia are plated directly into SG Medium at an equivalent of ~3×10^5 spermatogonia/well (9.5 cm^2) in a six-well plate containing freshly prepared MEFs (*see* Subheading 12.3.2.2).
2. The transfected (or virally transduced) spermatogonia are then allowed to proliferate in cell number for ~20 days after transfection with plasmid DNA (or, 8 days for virally transduced cultures). The culture medium is replenished every 2 days; and, fresh MEFs are spiked onto cultures of the transfected spermatogonia after ~10 days (*see* Subheading 12.3.2.3, **step 5**).
3. At ~20 days following gene transfer with plasmid DNA (or, after ~8 days following lentiviral transduction), cultures are harvested and then passaged onto freshly prepared MEFs in SG medium and maintained for an additional 2–3 days before initiating clonal selection in SG medium containing 50–75 μg/ml G418 (Invitrogen, Inc.).

   Note: Optimal concentrations of G418 used for selecting transgenic spermatogonia are dependent on (1) specific activity of the Geneticin lot purchased, (2) relative expression levels of the neomycin-resistance gene in spermatogonia, and (3) relative expression levels of the neomycin-resistance gene in the feeder cells.
4. After initiating selection, cultures are fed fresh SG medium containing G418 every 2 days during an 8-day selection period. Thereafter, cells are fed every 2 days using SG medium alone to expand clonally enriched lines of rat spermatogonia that can be used to produce transgenic rats, as described in the following sections.

12.7.2.1. Polyclonal Expansion

After genetic modification of spermatogonial lines using plasmid DNA or lentiviral particles, mixtures of genetically modified spermatogonial colonies can be expanded collectively and then pooled to form a library of genetically modified rat germlines for transplantation into testes of recipient-founders (*see* Subheading 12.8). This type of polyclonal expansion of transgenic spermatogonial lines using randomly integrating DNA will mobilize the production of numerous distinct transgenic rat lines with a given construct. The newly generated rat lines can then be screened to identify animals that express the transgene of interest from a locus that yields the desired levels in tissues under study. This is critical, because transgene expression in a given stem cell line will not likely predict transgene expression in other tissues. Moreover, if lentiviral vectors are used, depending on the multiplicity of infection (MOI), multiple copies of the transgene are integrated at different genomic sites in F1 progeny (Fig. 12.4). Therefore, care must be taken to follow transgene expression profiles of interest in subsequent F2 progeny in order to link it to specific lentiviral integrations. Similarly, if using transposon systems, care must be taken to follow transgene expression profiles in subsequent progeny due to the potential for multiple insertions of the respective transposon at different genomic sites.

12.7.2.2. Monoclonal Expansion

Individual G418-resistant spermatogonial colonies can also be picked from a six-well plate of transfectants using a p200 Eppendorf tip and transferred into wells of a 96-well plate to facilitate clonal expansion of mutant germlines to larger numbers upon subsequent passages into larger culture dishes. A single 10 cm dish typically yields enough spermatogonia from a given line for both preservation of frozen stocks (Subheading 12.3.2.4), and for the production of multiple recipient-founders (Subheading 12.8.2).

12.8 Germline Transmission from Genetically Modified Spermatogonia

A major attribute of spermatogonia as germline vectors is that through standard breeding, relatively high rates of germline transmission from donor rat spermatogonial lines to F1 progeny (i.e., 90–100%) can be achieved by transplanting them into testes of recipient-founders [8, 13]. We have found male-sterile, *DAZL*-deficient rats [20] to be most effective for germline transmission of mutant alleles from donor rat spermatogonial lines [8]. Maximal rates of donor germline transmission from *DAZL*-deficient rats can be attributed to their lack of endogenous sperm production, which is due to a severe, postmeiotic block in haploid germ cell development [8]. When used as a strategy to enrich for germline transmission of donor alleles, this mode of sterile-testis complementation with genetically modified spermatogonial stem cells in rats is analogous to tetraploid-embryo complementation with ES cells in mice [9]. Therefore, this approach can be used to more directly generate germline founders to rapidly produce colonies of transgenic male and female F1 breeders.

12.8.1. Reagents

- Disposable Pasteur Pipettes (cat. no. 13-678-20C, Thermo Fisher Scientific, Inc.).
- 30G Precision Glide Needles (cat. no. 305106, BD, Inc.).
- 1 ml Syringes (cat. no. 309602, BD, Inc.).
- Busulfan (cat. no. 154906, MP Biomedicals).
- Dimethyl Sulfoxide (DMSO) (cat. no. 317275, Calbiochem).
- Trypan Blue (cat. no. T6146-25G, Sigma, Inc.).
- Triadine Prep Solution, (10% povidone iodine solution, cat. no. 10-8208, Triad Disposables).
- Ethanol 200 Proof (cat. no. 111000200, Pharmco-AAPER).
- PBS: Dulbecco's phosphate-buffered saline (PBS; cat. no. D8537, Sigma, Inc.) 200 mg/L KCl (w/v), 200 mg/L KH2PO4 (w/v), 8 g/L NaCl (w/v), 1.15 g/L Na2HPO4 (w/v).
- Kimwipes (cat. no. 34155, Kimberly-Clark).
- Bead Sterilizer; Germinator 500 (Cellpoint Scientific, Inc).
- Flaming/Brown Micropipette Puller; Model P-97 (Sutter Instruments Co.).
- Glass Capillaries for needles; 100 μl micropipette (cat. no 1-000-1000, Drummond Scientific Co.).
- Heat Therapy Pump (cat. no HTP-1500, Kent Scientific Corporation or other suitable model).
- Reusable Warming Pad (cat. no. TPZ-1215EA, Kent Scientific Corporation).
- 10 ml Syringes (cat. no. 309604, BD, Inc.).
- Acepromazine (cat. no. 038ZJ03, Vedco).
- Rompun (cat. no. LA33806A, Lloyd Laboratories).
- Ketaset (cat. no. 440761, Fort Dodge Animal Health).
- Buprenex Injectable (cat. no. 12496-0757-1, Reckitt Benckiser).
- Shaving Razors – Stainless Steel Surgical Prep Blades (cat. no. 74-0001, Personna).

- Suture Thread; Spool Suture (cat. no. SUT-15-2, Roboz Surgical, Inc.).
- Suture Needles; Eye 3/8 circle (cat. no. RS-7981-4, Roboz Surgical, Inc.).
- Michel Wound Clips (cat. no. RS-9272, Roboz Surgical, Inc.).
- Michel Wound Clip Forceps (cat. no. RS-9294, Roboz Surgical, Inc.).
- Ear Puncher – 2 mm diameter (cat. no. RS-9902, Roboz Surgical, Inc.).
- Hemostat (cat. no. RS-7110, Roboz Surgical, Inc.).
- Straight Sharp Microdissecting Scissors (cat. no. RS-5882, Roboz Surgical, Inc.).
- Curved, Sharp Microdissecting Scissors (cat. no. RS-5883, Roboz Surgical, Inc.).
- Full-Serve Microdissecting Forceps (cat. no. RS-5137, Roboz Surgical, Inc.).
- Straight Tip, Dumostar Tweezers (cat. no. RS-4978, Roboz Surgical, Inc.).
- 5/45 INOX Tweezers (cat. no. RS-5005, Roboz Surgical, Inc.).
- Polyethylene capillary tubing (cat. No. 19-0040-01, GE Heatlthcare, Inc.).
- 24-day-old, busulfan-treated, male Sprague-Dawley rats (*see* Subheading 12.8.2).

Note: All reagents (solutions and tools) used for transplantation should be sterilized prior to use to ensure sterile surgical technique.

12.8.2. Methods

12.8.2.1. Generation of Recipient-Founders by Testicular Transplantation

Busulfan-treated wild-type Sprague-Dawley rats (Harlan, Inc.), or male-sterile *DAZL*-deficient, Sprague-Dawley rats at 24 days of age can be used as recipients for spermatogonial lines [8].

12.8.2.2. Recipient Preparation

1. To prepare recipients for transplanting spermatogonia, rats arrive from the supplier at 8–10 days of age, together with mother, 14–16 days prior to the transplantation procedure which is performed at 24 days of age.
2. At 12 days of age (i.e., 12 days prior to the transplantation procedure), each rat is administered a single dose of busulfan (12.5 mg/kg, i.p. for wild-type Sprague-Dawley rats; 12.0 mg/kg for *DAZL*-deficient Sprague-Dawley rats), and then housed in a quiet, clean and well-ventilated location within an approved animal facility. *Under guidelines of an*

approved safety plan[2], a 4 mg/ml working stock of busulfan in 50% DMSO is prepared by first dissolving busulfan in 100% DMSO at 8 mg/ml, and then adding and equal volume of filter-sterilized, deionized water.

Note: Recipients are pretreated with busulfan in order to deplete their endogenous spermatogonial populations prior to transplantation; this enhances engraftment by the transplanted spermatogonia.

12.8.2.3. Transplantation Procedure

1. On the day of transplantation, rat spermatogonia are harvested from culture and suspended in ice-cold, culture medium (i.e., either SG medium or DHF12-FBS-2ME) at concentrations ranging from 2 to 8 × 10^5 spermatogonia/100 μl. The cellular suspension is transferred to a sterile microfuge tube and maintained on ice until the time of transplantation.
2. Just prior to transplantation, the cell suspension is supplemented with a 20% volume of a filter-sterilized, 0.2% trypan blue solution made fresh in PBS the same day.

   Note: As an example, 100 μl of prechilled 0.2% trypan blue-solution is typically added directly to 400 μl of a cell suspension in culture medium (i.e., final trypan blue concentration = ~0.04%), and then mixed gently by pipetting up and down three to four times.
3. Once spermatogonia are harvested, the first busulfan-treated recipient rat is anesthetized by intraperitoneal (i.p.) injection of a cocktail containing 100 mg/ml ketaset, 20 mg/ml rompun, and 10 mg/ml acepromazine at 0.1 ml/100 g body weight to achieve a surgical plane of anesthesia (as demonstrated by the lack of a pedal reflex in the toe pinch test).
4. The recipient is layed on its back. The abdominal skin is then opened just rostral to the pelvis on either the left or right side of the midline, and the testis is exposed. The efferent ductules leading into the rete testis are then accessed by blunt dissection using microdissection forceps, after which an absorbent wick made of a small, ***tightly*** twisted corner of a Kimwipe is inserted directly under the efferent ductules to provide support during the injection (Fig. 12.5).

[2] *Safety Information*: Busulfan is considered a biohazardous compound. Its use for studies in animals warrants an approved safety plan certified by both the parent research institution's Environmental Health and Safety program, and the Institutional Animal Care and Use Committee (IACUC). Personal protection equipment (doubled gloves; lab coat; safety glasses; ventilation mask) should be worn when working with busulfan to prevent any type of exposure to the researcher. The preparation, use and disposal of busulfan should all be restricted to certified chemical and/or biosafety cabinets to prevent exposure to the researcher and others.

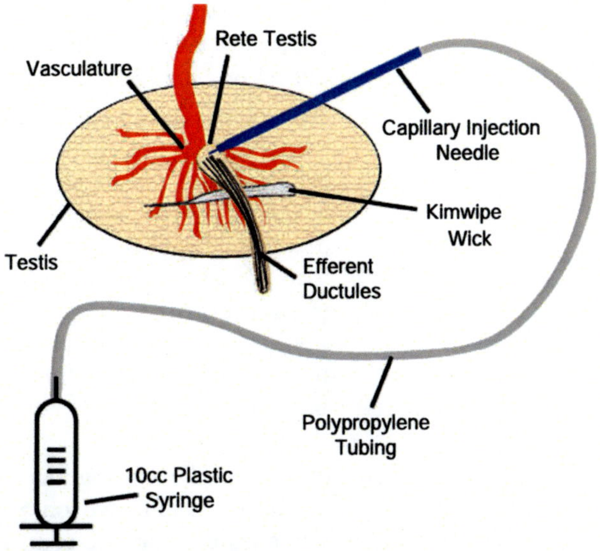

Fig. 12.5 *Transplanting spermatogonial cultures into rat testes.* Illustration summarizing key aspects of rat testicular anatomy, and surgical materials required for injecting spermatogonial suspensions into busulfan-treated rat testes. Spermatogonia are manually injected into the rete testis which resides at the base of the efferent ductules. Injecting cells into the rete testis will result in retrograde filling of the seminiferous tubules composing the testis (Subheading 12.8.2.3).

5. The ductules and wick are soaked with several drops of PBS and the ductules are further dissected up to the base of their respective testis to yield visible access to the rete, which will be the site of injection (Fig. 12.5).
6. Once the rete is exposed, the harvested spermatogonial suspension is mixed gently by pipetting up and down approximately five times with a p200 tip and then ~70–80 μl of the suspension is loaded into a 100-μl glass capillary injection needle using a flame pulled, transfer pipette (i.e., made from Pasteur pipettes) and rubber squeeze bulb. Capillary injection needles were made on settings: Heat = 305; Pull = 200; Vel = 50; Time = 150, using the Sutter Instrument Micropipette Puller.
7. The injection needle containing spermatogonia is manually inserted into the rete of the testis, and the cells are transferred into the testis by injection using a stationary 10-ml syringe (i.e., simply taped to the work bench), which is connected to the glass capillary injection needle by flexible plastic tubing (Fig. 12.5).
8. The injected testis is then carefully placed back into the abdominal cavity and the same procedure can be performed on the contra-lateral testis to achieve more optimal breeding.

9. Once injected and placed back in the abdominal cavity, the abdominal wall (sutured) and skin (wound clips) are surgically closed for each respective testis.

   Note: Care should be taken to avoid inadvertently suturing the fat pad of the epididymus to the abdominal wall to allow proper descent of the testes following surgery.
   Spermatogenesis will be disrupted if the testes do not descend.
10. The procedure can then be repeated on subsequent recipients using the same spermatogonial suspension. Spermatogonial suspensions can be maintained on ice in SG medium for up to 5 h during the transplantation of multiple recipients.

12.8.2.4. Postoperative Steps

1. After surgically closing the abdominal cavity and skin, all animals are maintained on a warming pad set to 34°C and receive postoperative care to assure their safe recovery from anesthesia and to alleviate pain and distress.
2. For recovery from anesthesia, each animal is observed with respect to its breathing rate, muscle control and external stimuli until ambulatory, prior to being housed in a quiet, well-ventilated location within the animal facility.
3. As a postoperative analgesic to alleviate pain, each rat is administered a single dose of buprenorphine hydrochloride (25 μg/kg) (Buprenex Injectable, Reckitt Benckiser) as it starts to regain consciousness. An additional dose is given every 6–12 h for the next 48 h upon signs of discomfort or pain.
4. Wound clips are removed at 21 days postsurgery.
5. The recipients are then housed together for ~65 days prior to initiating breeding studies.

12.8.2.5. Germline Transmission from Recipient-Founders

Recipient males transplanted with spermatogonial lines are paired with wild-type female Sprague-Dawley rats of similar age at 60–70 days posttransplantation. Typically, the first F1 progeny are born between 100 and 150 days posttransplantation and recipients can continue to sire litters for greater than 300 days posttransplantation due to the long-term spermatogenesis colony forming potential of laminin-binding rat spermatogonia [10, 13]. Transgenic rat progeny from recipient-founders and wild-type females are identified by genomic PCR and/or Southern Blot analysis using probes specific to the mutation of interest.

12.8.2.6. Animal Husbandry

Rats are housed in individually ventilated, Lab Products 2100 cages in a dedicated room with atmosphere controls set to 72°F, 45–50% humidity during a 12 h light/dark cycle (i.e., Light cycle = 6:00 am–6:00 pm, Central Standard Time adjusted for daylight savings time). Rats are fed Harlan Teklad Irradiated 7912, LM-485 Mouse/Rat Diet, 5% fat Diet and a continuous supply of reverse osmosis water.

Protocols for the use of rats by these procedures have been approved by the Institutional Animal Care and Use Committee (IACUC) at UT-Southwestern Medical Center in Dallas, as certified by the Association for Assessment and Accreditation of Laboratory Animal Care International (AALAC).

12.9 Conclusion

As a rapidly emerging alternative to embryonic stem cells, the use of stem spermatogonia for gene manipulation directly within the germline is technically straightforward and does not require micromanipulation of host embryos for the production of transgenic animals. Genetic manipulation at the level of the spermatogonium bypasses the need to produce, genotype and to breed intermediate colonies of chimeric progeny. Multiple F1 male and female littermates harboring the same mutant allele can be produced directly from recipient-founders transplanted with an individual spermatogonial line [8, 13]. Thus, multiple breeders of F1 progeny heterozygous for a given transgene or mutant allele can immediately be paired to rapidly establish breeding colonies that yield F2 progeny homozygous for the gene of interest (Fig. 12.1).

Acknowledgments

Studies to establish the methodology reported herein was supported by NIH grants R21RR023958 from the National Center for Research Resources and RO1HD036022 from the National Institute of Child Health and Human Development, and by the Cecil H. & Ida Green Center for Reproductive Biology Sciences at the University of Texas Southwestern Medical Center in Dallas.

References

1. Merry BJ, Holehan AM (1979) Onset of puberty and duration of fertility in rats fed a restricted diet. J Reprod Fertil 57: 253–259
2. Suzuki N, Withers HR (1978) Exponential decrease during aging and random lifetime of mouse spermatogonial stem cells. Science 202:1214–1215
3. Hamra FK, Chapman KM, Nguyen DM, Williams-Stephens AA, Hammer RE, Garbers DL (2005) Self renewal, expansion, and transfection of rat spermatogonial stem cells in culture. Proc Natl Acad Sci USA 102: 17430–17435
4. Kanatsu-Shinohara M, Ogonuki N, Inoue K, Miki H, Ogura A, Toyokuni S, Shinohara T

(2003) Long-term proliferation in culture and germline transmission of mouse male germline stem cells. Biol Reprod 69:612–616
5. Ryu BY, Kubota H, Avarbock MR, Brinster RL (2005) Conservation of spermatogonial stem cell self-renewal signaling between mouse and rat. Proc Natl Acad Sci USA 102:14302–14307
6. Kanatsu-Shinohara M, Toyokuni S, Shinohara T (2005) Genetic selection of mouse male germline stem cells in vitro: offspring from single stem cells. Biol Reprod 72: 236–240
7. Kanatsu-Shinohara M, Ikawa M, Takehashi M, Ogonuki N, Miki H, Inoue K, Kazuki Y, Lee J, Toyokuni S, Oshimura M et al (2006) Production of knockout mice by random or targeted mutagenesis in spermatogonial stem cells. Proc Natl Acad Sci USA 103: 8018–8023
8. Richardson TE, Chapman KM, Dann CT, Hammer RE, Hamra FK (2009) Sterile testis complementation with spermatogonial lines restores fertility to DAZL-deficient rats and maximizes donor germline transmission. PLoS ONE 4(7):e6308
9. Nagy A, Rossant J, Nagy R, Abramow-Newerly W, Roder JC (1993) Derivation of completely cell culture-derived mice from early-passage embryonic stem cells. Proc Natl Acad Sci USA 90:8424–8428
10. Hamra FK, Gatlin J, Chapman KM, Grellhesl DM, Garcia JV, Hammer RE, Garbers DL (2002) Production of transgenic rats by lentiviral transduction of male germ-line stem cells. Proc Natl Acad Sci USA 99:14931–14936
11. Hamra FK, Chapman KM, Nguyen D, Garbers DL (2007) Identification of neuregulin as a factor required for formation of aligned spermatogonia. J Biol Chem 282: 721–730
12. Hamra FK, Schultz N, Chapman KM, Grellhesl DM, Cronkhite JT, Hammer RE, Garbers DL (2004) Defining the spermatogonial stem cell. Dev Biol 269:393–410
13. Wu Z, Falciatori I, Molyneux LA, Richardson TE, Chapman KM, Hamra FK (2009) Spermatogonial culture medium: an effective and efficient nutrient mixture for culturing rat spermatogonial stem cells. Biol Reprod 81:77–86
14. Li P, Tong C, Mehrian-Shai R, Jia L, Wu N, Yan Y, Maxson RE, Schulze EN, Song H, Hsieh CL et al (2008) Germline competent embryonic stem cells derived from rat blastocysts. Cell 135:1299–1310
15. Dann CT (2007) New technology for an old favorite: lentiviral transgenesis and RNAi in rats. Transgenic Res 16:571–580
16. Lois C, Hong EJ, Pease S, Brown EJ, Baltimore D (2002) Germline transmission and tissue-specific expression of transgenes delivered by lentiviral vectors. Science 295: 868–872
17. Michalkiewicz M, Michalkiewicz T, Geurts AM, Roman RJ, Slocum GR, Singer O, Weihrauch D, Greene AS, Kaldunski M, Verma IM et al (2007) Efficient transgenic rat production by a lentiviral vector. Am J Physiol Heart Circ Physiol 293:H881–894
18. Kanatsu-Shinohara M, Kato M, Takehashi M, Morimoto H, Takashima S, Chuma S, Nakatsuji N, Hirabayashi M, Shinohara T (2008) Production of transgenic rats via lentiviral transduction and xenogeneic transplantation of spermatogonial stem cells. Biol Reprod 79:1121–1128
19. Ryu BY, Orwig KE, Oatley JM, Lin CC, Chang LJ, Avarbock MR, Brinster RL (2007) Efficient generation of transgenic rats through the male germline using lentiviral transduction and transplantation of spermatogonial stem cells. J Androl 28:353–360
20. Dann CT, Alvarado AL, Hammer RE, Garbers DL (2006) Heritable and stable gene knockdown in rats. Proc Natl Acad Sci USA 103:11246–11251

Chapter 13

Mouse Cloning by Nuclear Transfer

Sayaka Wakayama, Nguyen Van Thuan, and Teruhiko Wakayama

Abstract

A new mouse cloning method using nuclear injection with a piezo impact drive unit can aid in the bypass of several steps of the original cell fusion procedure. This approach has made it possible to create not only live cloned mice, but also ES cell lines from adult somatic cells via nuclear transfer. It is important to note that these techniques potentially may also be applied to the preservation of genetic material from any mouse strain instead of preserving embryos or gametes. At present, this is the only technique available for the preservation and propagation of valuable genetic resources from mutant mice that are infertile or old, or recovered from carcasses, without the use of germ cells. This technique will greatly help not only in mouse cloning, but also in other forms of micromanipulation such as intracytoplasmic sperm injection into oocytes (ICSI) or embryonic stem (ES) cell injection into blastocysts. Moreover, the piezo unit simplifies pipette preparation, as it allows one to use blunt tipped pipettes without any additional modification.

Abbreviations

EF medium	Embryonic fibroblast medium
ES cell	Embryonic stem cell
NIM	Nuclear isolation medium
NKT cell	Natural Killer T cell
NT	Nuclear transfer
ntES cell	Nuclear transfer embryonic stem cell
SCR	Scriptaid
TSA	Trichostatin A

13.1 Introduction

A new mouse cloning method using nuclear injection with a piezo impact drive unit (hereafter termed "piezo unit") [1] can aid in the bypass of several steps of the original cell fusion procedure.

S. Pease and T.L. Saunders (eds.), *Advanced Protocols for Animal Transgenesis*, Springer Protocols, DOI 10.1007/978-3-642-20792-1_13, © Springer-Verlag Berlin Heidelberg 2011

Until now, cutting the zona pellucida, inserting a donor cell into the perivitelline space, moving cells and enucleated oocytes to an electrofusion machine, applying electrofusion and later confirmation of cell fusion have been the standard approach. Utilization of the piezo method presents an alternative to many of these steps. Surprisingly, this piezo injection method can result in the production of normal mice from dead cells or even from frozen mouse bodies stored in a freezer for long periods. The challenge in this method lies in the fact that this is technically a very difficult procedure and likely to take most individuals some time to develop sufficient skill before being able to obtain useful data. Without focused practice at these procedures, production of cloned mice is impossible. However, once the piezo unit is properly set up, it will greatly help not only in mouse cloning, but also in other forms of micromanipulation such as intracytoplasmic sperm injection into oocytes (ICSI) [2, 3] or embryonic stem (ES) cell injection into blastocysts [4]. Moreover, the piezo unit simplifies pipette preparation, as it allows one to use blunt tipped pipettes without any additional modification.

This approach has made it possible to create new types of embryonic stem (ES) cell lines from adult somatic cells via nuclear transfer (ntES cell lines) [5, 6]. We have shown that such ntES cell lines have the same differentiation potential as ES cells from fertilized blastocysts [7]. Moreover, cloned mice can be obtained from these ntES cell lines using a second nuclear transfer procedure, [6, 8] which can be used as a backup for the donor cell genome and help increase the overall numbers in mouse cloning [8]. It is important to note that these techniques potentially may also be applied to the preservation of genetic material from any mouse strain instead of preserving embryos or gametes [5, 8]. At present, this is the only technique available for the preservation and propagation of valuable genetic resources from mutant mice that are infertile or old, or recovered from freeze-dried cells or carcasses, without the use of germ cells [9–12]. Here we describe our improved approaches for the production of cloned mice and establishment of ntES cell lines from somatic cells.

13.1.1. Applications

Currently, cumulus cells [1], tail tip cells (probably fibroblasts) [13], Sertoli cells [14], fetal cells [15, 16] and ES cells [17] have been used to produce cloned mice. NKT cells [18], primordial germ cells [19], hematopoietic stem cells [20], keratinocyte stem cells [21], fetal neuronal cells [22] and newborn neuronal stem cells [23, 24] have also been used (*see* review [25, 26]). The genetic background of the mouse strain used as oocyte donors is very important [16, 27]; usually cloning hybrid mouse donor cells is much easier than cloning the cells of inbred mouse strains. Therefore, each researcher must choose the donor cell type carefully according to the need. For example, ES cells are the most

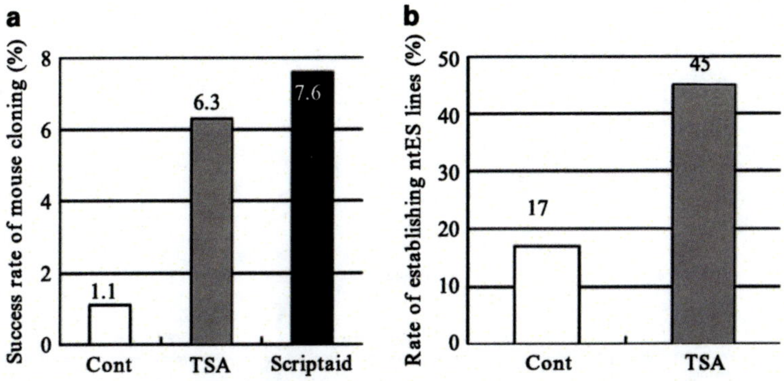

Fig. 13.1 Effects of trichostatin A (TSA) and Scriptaid on mouse nuclear transfer for full-term development and establishment of nuclear transfer embryonic cell (ntES) lines. The success rate of mouse cloning and rate of establishment of ntES cell lines increased up to sevenfold for cloning (**a**) and threefold for ntES cell derivation (**b**) by adding this histone deacetylation inhibitor into the oocyte activation medium. Key: *Cont* controls.

popular for nuclear transfer experiments because, to date, they produce the best results for production of full-term offspring. However, these are pluripotent cells, not differentiated somatic cells, so they are not appropriate for genomic reprogramming experiments. Recently, we increased the efficiency of mouse cloning and ntES cell establishment by up to sixfold by adding the histone deacetylation inhibitor (HDACi), such as trichostatin A (TSA) or scriptaid (SCR), to the oocyte activation medium (Fig. 13.1) [28–32]. This new protocol allowed us to generate cloned mice from inbred strains [28, 33]. Moreover, recently we succeeded in recovering normal live mice from frozen carcasses by using a combination of cloning and ntES cell techniques. These results suggest that even with a low success rate, mouse cloning from murine cells in any given condition is already possible.

13.1.2. Outline of the Procedure (Illustrated in Fig. 13.2)

Put donor cells into culture.

Day 3 – Hormone prime oocyte donors (PMSG).

Day 1 – Hormone prime oocyte donors (HCG). Mate foster mothers (nt pups are recovered by cesarian section).

Prepare media and microinjection needles and equipment.

Day of injection:

1. Collect oocytes
2. Enucleation
3. Prepare single donor cells
4. Nuclear injection
5. Oocyte activation and culture
6. Mate recipient females to vasectomized males

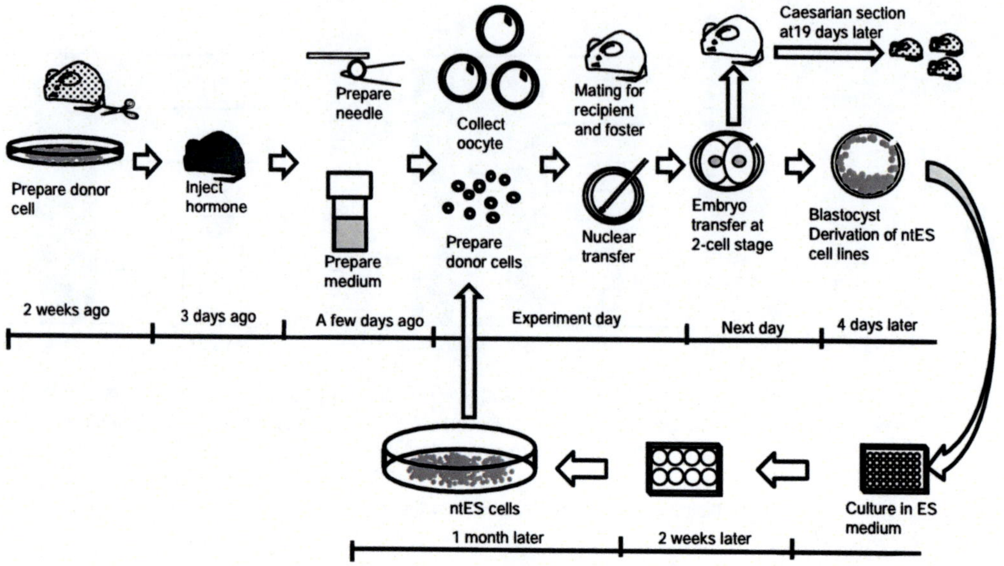

Fig. 13.2 Outline of the procedure.

Subsequent days:

7. Embryo transfer
8. Cesarian section
9. Establishment of ntES cells

13.2 Materials

13.2.1. Equipment

Inverted microscope with Hoffman optics from Olympus (Tokyo, Japan; model IX71)

Micromanipulator set from Narishige (Tokyo, Japan; model MMO-202ND, http://www.narishige.co.jp/)

Microforge (Narishige MF-900)

Warm plate (Tokai Hit, http://www.ivf.net/ivf/tokai_hit_thermo_plate-o738.html)

Piezo impact drive system from Prime Tech Ltd. (Ibaraki, Japan; model MM-150FU, http://www.primetech-jp.com/en/01products/index.html)

Pipette puller (P-97) and glass pipette (B100-75-10) from Sutter Instrument Co. (Novato, CA, USA; http://www.sutter.com)

37°C, 5% CO_2, humidified incubator

Tissue culture hood

Centrifuge

13.2.2. Mouse Strains

Normally B6D2F1 (C57BL/6 × DBA/2) mice (about 2–3 months old) are used for the provision of donor cells and oocytes. Other hybrid strains, such as BCF1 or B6129F1, can also be used as donors. Inbred stains such as C57BL/6 or C3H/He may also be used, but it should be expected that success rates for the production of cloned mice will be lower. The donor mouse strain is important for several reasons; B6D2F1 oocytes are very translucent and the metaphase spindle is easy to find; oocytes from hybrid mice are stronger for in vitro manipulation and culture and somatic cells from hybrid mice provide better donor cells for the production of cloned mice. We use the ICR (CD-1) strain of mouse for production of pseudo-pregnant surrogates, for lactating foster mothers and for vasectomized males.

13.2.3. Culture Media

13.2.3.1. Stocks

CB Stock Solution: Cytochalasin B (1 mg, Sigma-Aldrich, St Louis, MO, USA; C6762). Add 2-mL DMSO to a vial with 1-mg cytochalasin B (500 μg; 100× CB stock solution). Divide into small tubes (10–20 μL) and store at 80°C.

TSA Stock Solution: Trichostatin A (TSA; 1 mg, Sigma-Aldrich; T8552). Add 3.311-mL DMSO to a vial with 1-mg TSA (1 mM) and then take 2 μL of this stock and dilute with 198 μL of DMSO (10 μM, 200× TSA stock solution). Divide into small tubes (10–20 μL) and store at −80°C.

SRC Stock Solution: Scriptaid (SRC; 1 mg, Sigma-Aldrich; S7817). Add 3.064-mL DMSO to a vial with 1-mg SCR (1 mM) and then take 5 μL of this stock and dilute with 95 μL of DMSO (50 μM, 200× SCR stock solution). Divide into small tubes (5–10 μL) and store at −80°C.

$SrCl_2$ Stock Solution: $SrCl_2$ $6H_2O$ (Sigma-Aldrich S0390) is dissolved in distilled water (DW) at 100 mM and stored in aliquots at room temperature (10× stock solution).

EGTA Stock Solution: EGTA (Sigma-Aldrich; E8145) is dissolved in DW at 200 mM and stored in aliquots at 4°C (100× stock solution).

Equine chorionic gonadotrophin (eCG or PMSG, Sigma-Aldrich; G4527) is dissolved in normal saline at 50 IU per mL and stored in aliquots at −20°C.

Human chorionic gonadotrophin (hCG, Sigma-Aldrich; C8554) is dissolved in normal saline at 50 IU per mL and stored in aliquots at −20°C.

PMSF Stock: Phenylmethanesulfonyl fluoride (Sigma-Aldrich P7626) is dissolved in ethanol at 50 mM and stored in aliquots at 4°C (100× stock solution).

13.2.3.2. Oocyte and Embryo Culture Media

KSOM medium (Millipore, Temecula, CA, USA; MR-106-D) (http://www.millipore.com)

M2 medium (Millipore; MR-015)
M2 + hyaluronidase (Millipore; MR-051)
Acid Tyrode's solution (Millipore; MR-004-D)

13.2.3.3. Fibroblast Cell Culture Media

Tail tip or embryonic fibroblast cells are cultured in Dulbecco's Modified Eagle's Medium (DMEM) (Sigma-Aldrich D6429) supplemented with 10% fetal bovine serum (termed EF medium).

13.2.3.4. ntES Cell Establishment Medium

CultiCell Medium (Dainippon Sumitomo Pharma, Tokyo, Japan; S2211101) (http://www.ds-pharma.co.jp/english/index.html) or Stem Cell Sciences KK, Kamiarai, Tokorozawa, Saitama, Japan (http://www.scskk.com/English/product_service/Culti-Cell.html) is used for ntES cell establishment.

13.2.3.5. ntES Cell Maintenance Medium

Complete ES cell media (Specialty Media; ES-101-B) or in-house prepared mouse ES cell media.

Phosphate-buffered saline (PBS) Ca/Mg free (Millipore; BSS-1006-B).

Trypsin solution (Millipore; SM-2003-C).

13.2.3.6. Enucleation Media

Add 2 μL of CB stock solution to 198 μL of M2 medium (M2+CB medium).

The final concentration of CB is 5 μg/mL.

13.2.3.7. PVP Medium

12% PVP in M2: Add 1.2-g PVP to 8.8-mL M2 and place at 4°C overnight. Mix the solution and sterile filter it on the next day and store 1-mL aliquots at 4°C.

13.2.3.8. Oocyte Activation media

Add 10 μL of $SrCl_2$ stock solution (final 10 mM), 2 μL of EGTA stock solution (final 2 mM), 2 μL of CB stock solution (final 5 μg/mL) and 1 μL of TSA or SCR stock solution (final 50 or 250 nM, respectively) to 185 μL of KSOM medium [34].

Note: If ES cells are used as sources of donor nuclei, TSA does not work well. If G2/M phase ES cells are used as donor nuclei, cytochalasin B must be omitted from the medium.

13.2.3.9. Reprogramming Enhancement Media

For reprogramming enhancement, cloned embryos should be cultured for an additional few hours with TSA or SCR. Add 1 μL of TSA or SCR stock solution (final 50 or 250 nM, respectively) to 199 μL of KSOM medium.

13.2.3.10. Nuclear Isolation Media

The 123.0-mM KCl, 2.6-mM NaCl, 7.8-mM NaH_2PO_4, 1.4-mM KH_2PO_4, 3-mM EDTA disodium salt and 0.5-mM PMSF.

The pH is adjusted to 7.2 by addition of a small quantity of 1-M KOH.

13.2.4. Other Reagents

Distilled water (DW, Sigma-Aldrich, St Louis, MO, USA; W1503)

DMSO (Sigma-Aldrich; D2650)

Mercury (Sigma-Aldrich; 215457)

Mineral oil (Sigma-Aldrich; M5310)

Polyvinylpyrrolidone (PVP; 360 kD) (Irvine Scientific, Santa Ana, CA, USA; 99311) (http://www.irvinesci.com)

13.3 Nuclear Donor Cell Sources

13.3.1. Donor Cell Preparation 1 (from Cells Maintained in Culture)

13.3.1.1. Tail Tip Fibroblasts

Tail tip cells must be prepared at least 2 weeks before NT. Cut the tail into sections at least 2-cm long and wash in 70% ethanol. In a sterile tissue culture hood, remove the skin and cut the tail into many small pieces on a 6-cm plastic dish. Culture the fragments in 10-mL DMEM under 5% CO_2 in air until used. There is no need to passage the cells.

13.3.1.2. ES Cells

ES cell should be passaged 2 days before an experiment, or thawed at least 2 days before an experiment and cultured in six-well dishes. We do not recommend the use of cells within 1 day of passage. ES cells are the most popular cell type for NT experiments because they have been seen to give the best rate of success in the production of full-term offspring. However, each individual ES cell line, even if from the same genetic background, will lend itself to this procedure to different degrees, resulting in varying results for full-term pups produced. The number of passages of ES cell lines will also affect the rate of success. Note that these are pluripotent cells, not differentiated somatic cells, so they are not appropriate for genomic reprogramming experiments.

13.3.1.3. Method

1. Prepare cells in culture for use: remove any culture medium from the dish and wash them in PBS (Ca^{++}, Mg^{++} free). Remove the PBS, add trypsin medium and then incubate 5–20 min in the incubator. Add culture medium (including serum) and triturate the cells to produce a single-cell suspension. Spin down the cells in a centrifuge at 200 × *g* for 10 min and wash cells with PBS by centrifugation at least three times, as above. ***Note: As trypsin is very toxic at the time of nuclear injection, the donor cells must be washed thoroughly.***

2. Make a very concentrated cell suspension in EF medium. The final volume should be less than 10 μL. If the final concentration is too low, it is difficult to find an appropriate donor cell and the delay can be detrimental to the recipient oocytes awaiting NT.

13.3.2. Donor Cell Preparation 2 (Primary Cells, Directly from In-Vivo Collection)

13.3.2.1. Cumulus Cells

Cumulus cells are the easiest to prepare as nuclear donors because they can be used immediately after collection without washing, with no need to remove hyaluronidase from the medium.

13.3.2.2. Sertoli Cells

Sertoli cells are testicular sustentacular cells and the male counterparts of cumulus cells. Adult Sertoli cells are inappropriate donor cells because of their large size, but those collected from neonatal mouse testes (immature Sertoli cells) are small enough for injection and usually give better results than cumulus cells [14, 27]. Thus, the age of donors will affect the success rate. We recommend the use of newborn males at less than 6 days of age.

13.3.2.3. Method

Wash in vivo collected cells in PBS and centrifuge at 100 rpm or 3 × *g* for 10 min, at least three times, in order to remove any enzymes that may be present. Cumulus cells are an exception and may be used without washing.

Prepare concentrated cell suspension as described in Subheading 13.3.1.3.

13.3.3. Donor Cell Preparation 3 (from Frozen Mice)

13.3.3.1. Brain

Collect the brain from the frozen body and break it into small pieces (1–2 mm^3) on the dry ice. Homogenize it under 500 μL of Nuclear isolation media (NIM) in 1.5-mL tube by homogenizer pestle, then filter it by cell strainer (BD Falcon 352235) and collect the supernatant. This medium contains many naked nuclei. Introduce a few microliters of suspension into NIM on the manipulation chamber instead of PVP medium.

13.3.3.2. Blood

Collect 1–2 μL of blood cells from the frozen-thawed mouse tail and add to 500 mL of NIM. Make a condensed cell suspension by centrifugation and introduce it into NIM in the manipulation chamber instead of PVP medium.

13.4 Protocol

13.4.1. Oocyte Production

13.4.1.1. Hormone Priming for Superovulation

Inject equine chorionic gonadotrophin (eCG or PMSG; 5 IU) into the abdominal cavity 3 days before an experiment and then inject human chorionic gonadotrophin (hCG, 5 IU) 48 h later (1 day before an experiment). Usually we inject mice at 5–6 p.m.

13.4.1.2. Collection of Oocytes

1. Collect oocyte–cumulus cell complexes from the oviduct ampullae at 14–15 h after hCG injection (usually we collect oocytes at 8–9 a.m.; Fig. 13.3) and move them into M2+ Hyaluronidase medium.
2. After 5 min, pick up the good oocytes, wash them in M2 medium three times and place them in the KSOM medium prepared as above.
3. Place oocytes as drops. The number of oocytes in a drop depends on each person's skill or type of experiment. We recommend that all oocytes in one drop must be manipulated within 15 min. Therefore, place 10–30 embryos per drop, according to skill level.

13.4.2. Preparation of Enucleation and Injection Pipettes

13.4.2.1. Micropipettes

Micropipettes can be ordered from several companies (e.g., Prime Tech Ltd. http://www.primetech-jp.com/en/01products/psk.html). Alternatively, they may be made in the lab. For the holding pipette, the outside diameter (OD) should be smaller than that of the oocyte (e.g., OD 80 μm; inner diameter (ID) 10 μm). The ID

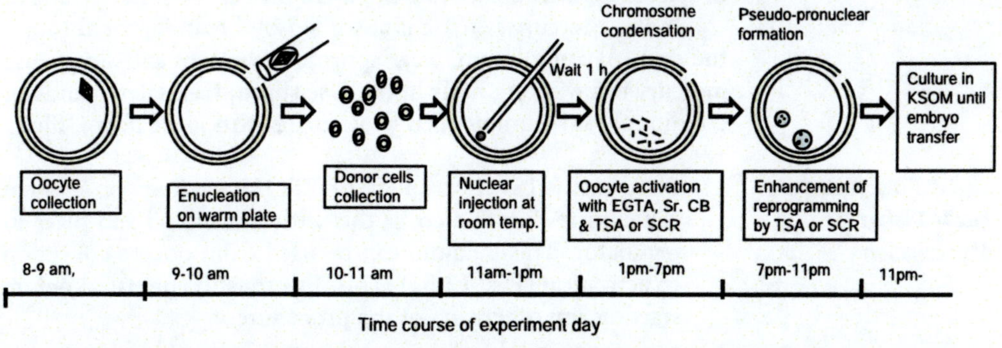

Fig. 13.3 Outline of the experimental procedure. Without diligent practice, it is very difficult to complete this procedure within the allotted time. This method is referred to as the "Honolulu method" because it was developed in a laboratory in Honolulu in 1998 [1].

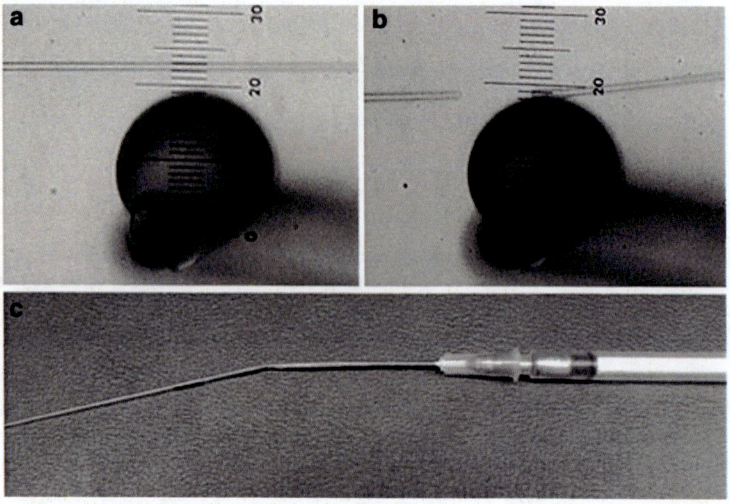

Fig. 13.4 Making pipette. The micropipette is positioned above the grass bead (a) and the filament is heated for less than 1 s. The pipette is fused to the glass bead, and when filament was cool down, pipette was broken (b) at the point of contact. The size of outer diameter must be adjusted depending on the cell size. (c) Load a small amount of mercury from back of pipette using 1-mL syringe and 30-gauge needle or 26-gauge needle with fine tubing.

of the enucleation pipette is 7–8 μm. The ID of the injection pipette depends on nuclear donor cell type: 5–6 μm for cumulus cells and 6–7 μm for fibroblasts or ES cells (Fig. 13.4a, b). If you are purchasing pipettes, ask the supplying company to bend all pipettes close to the tip (about 300 μm back) at 15–20° using a microforge.

13.4.2.2. Inserting Mercury into the Pipette

Backload a small amount (about 3-mm-long column) of mercury into the enucleation/injection pipette using a 1-mL syringe and 30-gauge needle (Fig. 13.4c). Fill the syringe part way with mercury, insert the needle into the back of the pipette and inject a mercury droplet into the needle. Store in 10-cm dish at room temperature.

Note: Mercury is toxic if absorbed by breathing or through the skin. At a minimum, wear appropriate gloves and always use mercury in a working fume hood. Use appropriate safety handling conditions, as recommended by your Institutional Safety Office.

13.4.3. Preparation of Media Dishes for Manipulation

1. Place many 15-μL droplets of KSOM medium on a 6-cm cell culture dish and cover this with mineral oil and place in incubator. This medium can be used from oocyte collection to activation (Fig. 13.5a). This dish must be prepared before starting any other part of the procedure.
2. Place ~15-μL droplets of three different media (M2, M2+CB (enucleation media) and PVP) on the lid of a 10-cm dish as shown in Fig. 13.5b and then cover this with mineral oil. This

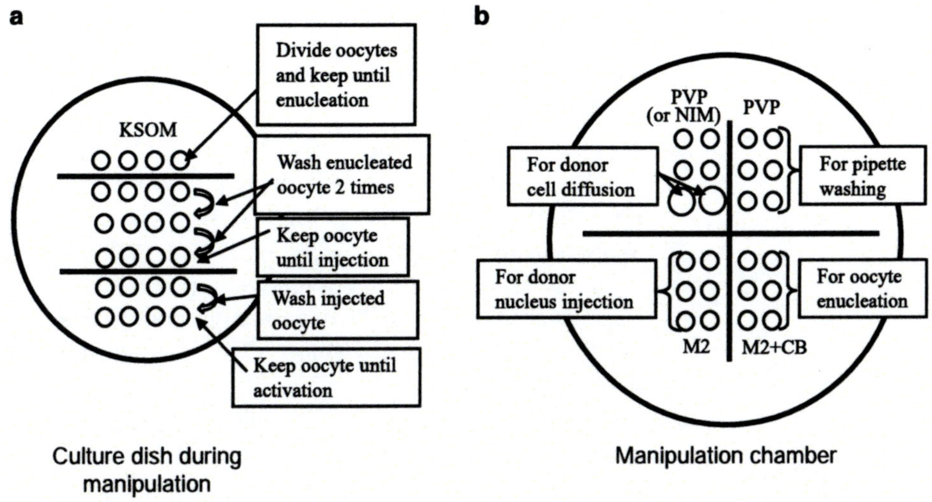

Fig. 13.5 Manipulation dish and chamber. Oocytes are kept from the time of collection to just before activation in KSOM medium in a 6-cm dish (**a**). The lid of a 10 cm dish is used as a micromanipulation chamber (**b**). Key: *PVP* polyvinylpyrrolidone, *NIM* nuclear isolation medium.

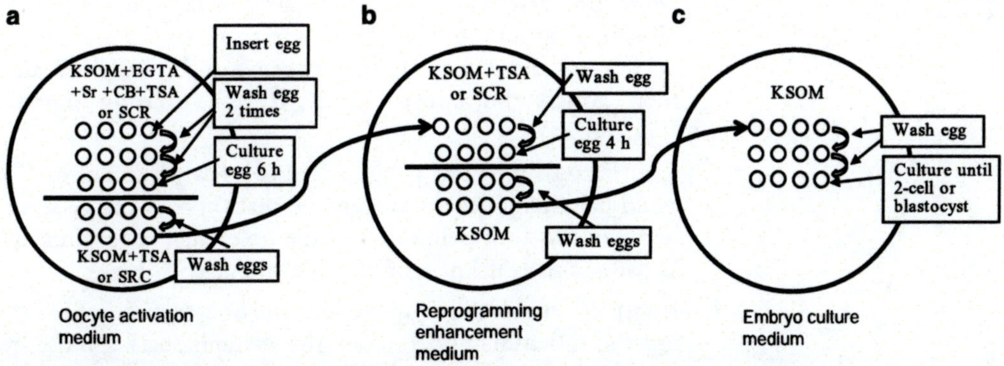

Fig. 13.6 Oocyte activation, enhancement of reprogramming and long-term culture. Reconstructed oocytes are exposed to the activation medium (**a**) for 6 h. Move embryos to the reprogramming enhancement medium (**b**) and culture for a few hours. Move embryos to the long-term culture dish (**c**) until transfer into a recipient female.

chamber can be used for both enucleation and microinjection. Draw a line on the dish to distinguish these media.

3. Place ~15-μL droplets of oocyte activation media: KSOM+EGTA+Sr+CB+TSA (use SCR when somatic cell nuclei donors are derived from inbred mice [33]), re-programming enhancement media: KSOM+TSA or SCR and embryo culture media: KSOM, on the 6-cm dish as shown in Fig. 13.6a–c and then cover this with mineral oil. Draw a line on the dish to distinguish these media.

13.4.4. Setting Up the Micromanipulator

1. Attach the enucleation pipette to the pipette holder of the piezo unit. The top of the pipette holder must be screwed in tightly. Position the piezo unit on the micromanipulator.
2. Expel any air and oil and a few drops of mercury from the enucleation pipette in the PVP medium. Wash both the inside and outside of the pipette using PVP medium.
3. Adapt and fit the pipette to the piezo unit. While expelling the air and mercury from the pipette in the PVP droplet, the piezo unit must be applied with high power (more than 10 units) and high speed (more than 10 units) for at least 1 min continuously.
4. Attach the holding pipette on the opposing side of the micromanipulator.

13.4.5. Enucleation of Oocytes

1. Place one group of oocytes in an M2+CB droplet into the micromanipulation chamber and wait at least 5 min before starting enucleation. The cytochalasin B makes the oolemma more flexible and reduces lysis.
2. Locate the metaphase II spindle inside the oocyte. It can be recognized without any staining using Nomarski or Hoffman optics. Rotate the oocyte to place the spindle at between 2 and 4 o'clock or between 8 and 10 o'clock and then attach the oocyte firmly to the holding pipette (Fig. 13.7a).
3. Cut through the zona pellucida using a few piezo pulses. To avoid damaging the oocyte, ensure that there is a large space between the zona pellucida and the oolemma, approximately as wide as the thickness of the zona pellucida.
4. Insert the enucleation pipette into the oocyte without breaking the oolemma and remove the metaphase II spindle by aspiration with a minimal volume of cytoplasm. The oocyte membrane and spindle must be pinched off slowly, draw the needle out carefully until the oolemma seals and do not apply piezo pulses to cut the membrane (Fig. 13.7b). The MII spindle is harder than the cytoplasm, so you can feel its consistency through the micromanipulator.
5. Wash the enucleated oocytes twice in KSOM to remove the cytochalasin B completely, and return to the incubator in KSOM medium under oil for at least 30 min before starting donor cell injection (Fig. 13.5a). If the CB is not completely washed out, many oocytes will lyse after injection.
6. If you feel tired at this point, take a short break before starting the next step. From the next step, there is no respite and the procedure requires intense concentration.

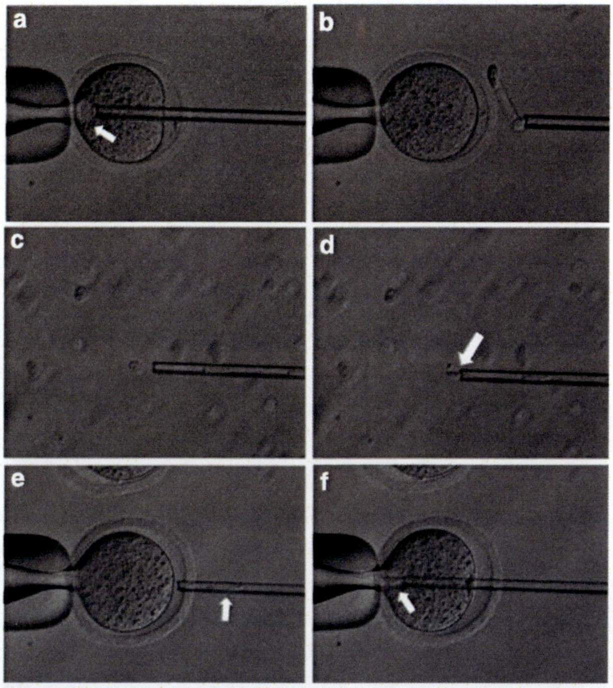

Fig. 13.7 Nuclear transfer using the piezo unit. (**a**) Rotate the oocyte, locate the metaphase II spindle and place it between the 8 o'clock and 10 o'clock position, or between the 2 o'clock and 4 o'clock position (see *arrow*). Then stabilize the oocyte on the holding pipette. (**b**) Remove the spindle by suction without breaking the plasma membrane and gently pull the pipette away from the oocyte. (**c**) and (**d**) Donor nuclei are gently aspirated in and out of the injection pipette until their nuclei are largely devoid of visible cytoplasmic material (*arrow*). (**e**) Hold the enucleated oocyte and cut the zona pellucida using piezo pulses (*arrow*). (**f**) Insert the injection pipette into the enucleated oocyte. Apply a single piezo pulse to break the membrane, and then inject the donor nucleus immediately (*arrow*). Gently withdraw the injection pipette from oocyte.

13.4.6. Donor Nucleus Injection

1. Pick up 1–3 μL of condensed donor cell suspension and put the cells in a PVP medium droplet in the micromanipulation chamber (Fig. 13.5b). Mix the donor cells with PVP medium gently but completely, using sharp forceps. Do not scratch the bottom of the chamber.
2. Place about 10–20 enucleated oocytes into M2 medium. The number of oocytes per droplet depends on each individual's skill level. Each group should be finished within 15 min.
3. Remove the donor nuclei from the cells by gently aspirating them in and out of the injection pipette until each nucleus is clearly separate from any visible cytoplasmic material (Fig. 13.7c, d). Take up a few nuclei into the injection pipette.
4. Stabilize an enucleated oocyte using a holding pipette. Cut the zona pellucida using a few piezo pulses (Fig. 13.7e).

5. Reduce the power level of the piezo unit (power levels 1–2 and speed 1). The oolemma is weaker than the zona pellucida and the survival rate of oocytes after injection will be better with this reduced power.
6. Push one nucleus forward until it is near the tip of the pipette and advance the pipette until it almost reaches the opposite side of the oocyte's cortex (Fig. 13.7f). Do not apply the piezo unit's power until the pipette reaches the opposite side. If the piezo power is applied with the tip of the pipette in the middle of the oocyte, the oocyte will die after injection.
7. Apply one weak piezo pulse to puncture the oolemma at the pipette tip. This is indicated by a rapid relaxation of the oocyte membrane (Fig. 13.7f). Expel the donor nucleus into the enucleated oocyte cytoplasm immediately with a minimal amount of PVP medium. Gently withdraw the injection pipette from the oocyte.
8. Wash the injection pipette with PVP medium by expelling some mercury and applying power from the piezo unit. This washing step is essential to prevent the pipette from getting sticky.
9. Keep the injected oocytes in this drop for at least 10 min then transfer them into KSOM medium (Fig. 13.5a) and culture for at least 30 min in the incubator before activation.

Note: The process of nuclear transfer should be performed at room temperature (25–26°C). Do not use warm plate for injection process.

13.4.7. Activation and Embryo Culture

1. Prepare the oocyte activation medium (Fig. 13.6) at least 30 min before use and equilibrate in a CO_2 incubator.
2. Transfer and culture each group of oocytes into drops of activation medium and wash twice, then culture for 6 h in a 5% CO_2 incubator at 37°C (Fig. 13.6a).
3. All embryos must be washed twice in TSA or SCR-KSOM medium to remove the cytochalasin B completely (Fig. 13.6a, below the line). Examine the rate of oocyte activation. If NT and activation are done properly, those oocytes should each possess two or three pseudo-pronuclei after 6 h in culture as above [1].
4. Move the cloned embryos to a different dish. Because some of the chemicals used for activation can diffuse to other drops through the mineral oil and are embryotoxic, all embryos should be moved to different culture dishes for long-term culture. Some batches of mineral oil are toxic. So it is advisable to test all new batches of oil by culture of embryos in drops of

KSOM, overlaid with test oil. Always set up a control group of embryos cultured under oil of known quality.

5. Additional culture of cloned embryos in TSA-KSOM for 10 h or SCR-KSOM for up to 24 h will enhance the reprogramming (Fig. 13.6b, [33]).
6. Wash the cloned embryos and move them to another dish (Fig. 13.6c); culture them to the two-cell stage (next day) or to the blastocyst stage (3 days later).

13.4.8. Embryo Transfer and Cesarian Section

1. Mate estrous ICR female mice with normal males on the same day or 1–2 days before the experiment; these will be used as foster mothers for receipt of NT pups derived by cesarian section at E19.5. Foster mothers are needed for NT pups because NT litters are too small to stimulate lactation in the surrogate mothers.
2. Mate estrous ICR female mice with vasectomized males on the same day as the experiment; these will be used as pseudo-pregnant (surrogate) mothers.
3. Transfer the two-cell (24 h after NT) or four- to eight-cell (48 h) cloned embryos into oviducts of 0.5 days post-copulation (dpc), or morulae/blastocysts (72 h) or blastocysts (96 h) into the 2.5-day post-coitum (dpc) pseudo-pregnant female mice, respectively [35].
4. A cesarian section is required to recover the cloned mice fetuses securely (*see* point 9 in Subheading 13.6). Euthanize the surrogate mother at 18.5 or 19.5 dpc. Remove the uterus from the abdomen and dissect out the cloned pups with their placentas. Wipe away the amniotic fluid from the skin, mouth and nostrils and stimulate the pups to breathe by rubbing the back or pinching them gently with blunt forceps. Warm to 37°C.
5. To transfer the cloned pups to the cage of a naturally delivered foster mother, first remove the mother from the cage. Take some soiled bedding from the cage and nestle the cloned pups in the bedding material so that they take on the odor of the bedding. Keep the pups warm while allowing them to remain in the bedding for a few minutes. Remove some pups from the foster mother's litter and then mix the cloned pups with the foster female's pups. Return the mother to the cage and leave the animal room quietly.

13.4.9. Establishing ntES Cell Lines from Cloned Embryos

1. The preparation of mouse embryonic fibroblasts for use as ES cell feeder cells is described in Chapter 14. Prepare a 96-well plate of mitotically inactivated feeder cells at least 1 day before ntES cell establishment.

2. Change the medium of the dish from DMEM to at least 200 μL of CultiCell medium before plating cloned embryos. The CultiCell medium does not contain fetal calf serum. Serum may contain potential differentiation factors. Therefore, it is important to use serum-free medium for establishment of new ntES or ES cell lines. Alternatively, you can use the new ES cell establishment "3i medium" which inhibits GSK3, MEK and FGF receptor tyrosine kinases and enhances the ES cell establishment rate significantly [36]. *See* Chapter 19, Rat and Mouse ES cell Derivation by 2i/3i Method.
3. Remove the zona pellucida from cloned blastocysts (Day E 3.5) using acidic Tyrode's solution. The zona pellucida will dissolve within 30 s and prolonged exposure to acid Tyrode's solution will decrease the quality and survival of embryos. Therefore, before dissolving the zona pellucida completely, pick up the embryos and wash them several times in M2 medium. The remaining thinned zonae are easily broken by repeated pipetting.
4. Plate each cloned blastocyst onto the feeder cell of 96-well multidishes one by one (Fig. 13.8a).
5. Culture the multidish for 10–14 days in an incubator without changing the medium. During this period, the cloned

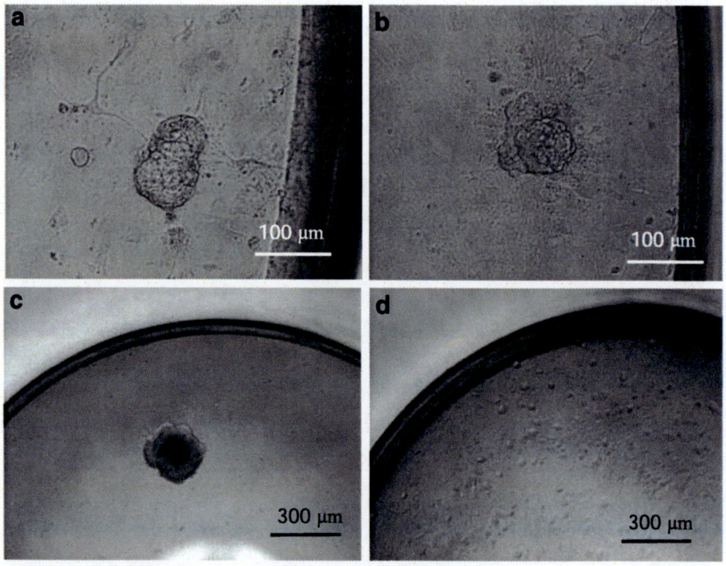

Fig. 13.8 Establishment of ntES cell lines. (**a**) Zona-free cloned blastocysts are plated on 96-well dishes. (**b**) Cloned blastocysts are attached onto the feeder cells, then the trophoblast cells spread and the inner cell mass (ICM) appears 7–9 days after plating. (**c**) The ICMs grow to almost 5–10 times as large as the original at 11–14 days after plating. (**d**) Two days after trypsinization, some wells show newly established ntES cell lines.

blastocyst will attach to the surface of the feeder layer and the inner cell mass (ICM) can be seen to grow (Fig. 13.8b, c).

6. Some of the wells should develop ICMs, which form large clumps. When those clumps appear, treat them with trypsin and disaggregate the cells using a 200 μL pipette. Then replate the suspension into another well (pre-plated with feeders) of the same multidish.
7. When ES-like cell colonies dominate the well (Fig. 13.8d), the cells should be expanded in a clonal manner, gradually to 48 well, 24 well, 12 well and then 12.5 cm^2 flask and 25 cm^2 flask by repeated passages several times. As these ES cell lines are to be used for nt at a later date, we do not use feeder cells to support ES cell culture from 48-well dishes onwards. After the cell numbers have increased, the cells should be cryopreserved in the manner usual for ES cells [37].

13.4.10. Production of Cloned Mice from ntES Cell Nuclei

Nuclear transfer may be accomplished by using ntES cells as donors (Fig. 13.2), repeating the procedure from section 13.3 onwards. Unlike somatic cells, ntES cells (like ES cells in general) [7] will divide indefinitely, so they offer the possibility for use without limitation, unlike somatic cells that have a limited life in cell culture. Moreover, the overall success rates of cloning from individuals are increased when ntES cell lines are used as intermediate nuclear donors [8].

13.5 Results

The genetically modified mouse is a powerful tool for research in medicine and biology. However, in one large-scale study on ethyl-nitrosourea (ENU) mutagenesis, infertility was listed as a phenotypic trait in more than half of the mutants described [38]. Overcoming this infertility is a challenge worth undertaking, as the ability to maintain such types of mutant mice as genetic resources would afford numerous advantages for research in human infertility and reproductive biology. Unfortunately, the success rate of somatic cell cloning is very low. Even when cloning a phenotypically sterile mouse is successful, it will still be necessary to clone all subsequent generations. This represents another significant barrier, as the success rate of repeated cloning from cloned mice decreases for each successive generation after the first NT [39]. On the other hand, the ntES cell establishment rate is nearly ten times higher than the success rate of cloned mice, even from the so-called "unclonable" mouse strains (Fig. 13.9). Therefore, we recommend the establishment of ntES cell lines at the same

Fig. 13.9 Comparison of the success rate of cloned mice and the rates of establishment of ntES cell lines from different mouse strains.

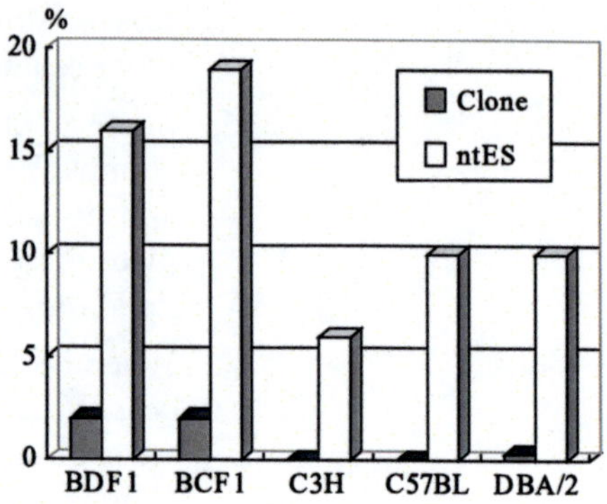

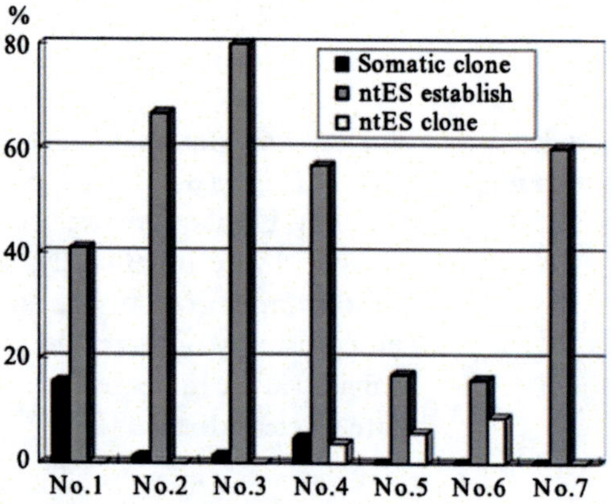

Fig. 13.10 Comparison of the success rates of producing cloned mice, establishing ntES cell lines and producing cloned mice from the ntES cell nuclei. Six out of seven donor mice could produce clones using either technique [8].

time as cloning to preserve the donor genome because these lines can then be used as an unlimited source of donor nuclei for subsequent rounds of NT. For example, because of the generally low success rate of cloning, we were able to generate cloned animals from only four out of seven somatic cell donor mice. We were ultimately able to obtain cloned mice from six out of seven individuals by using either somatic cells or ntES cells [8] (Figs. 13.10). Senescent mice are often infertile and the cloning success rate decreases with age, making it almost impossible to produce cloned progeny directly from such animals. We succeeded in establishing ntES cell lines from all aged mice, regardless of sex or strain. The ntES cells were then used to generate cloned mice by a second NT. In addition, healthy offspring were obtained from all aged donors via germline transmission of the ntES cells in chimeric mice [9]. For example, we found a mutant, hermaphrodite and sterile mouse in our ICR strain mouse-breeding colony.

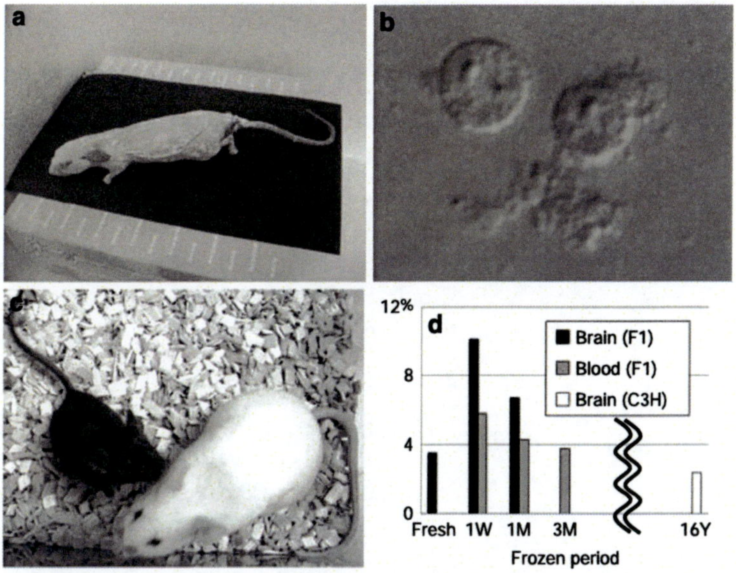

Fig. 13.11 Production of cloned mice from bodies frozen for 1 week, 1 month, 3 months or 16 years without cryoprotectant [12]. (**a**) A frozen mouse. (**b**) Donor nuclei derived from frozen brain by simple homogenization. (**c**) Cloned mouse and its foster mother. (**d**) Rate of establishing ntES cell lines. F1: B6D2F1 mice; C3H: C3H/He mice were used as donors, respectively.

Unfortunately, ICR is one of the most difficult strains to clone from [28, 40]. However, ntES cell lines from a tail tip fibroblast of this particular mouse were established successfully. Although this mouse was lost soon after tail-tip biopsy, we attempted to make cloned mice from its ntES cell nuclei, but we could not obtain full-term offspring. Finally, using tetraploid complementation [41], we obtained two cloned mice. Using the diploid chimera method, most of the mutant mouse genes were transmitted to the next generation via the ntES cells. As a second example, recently we developed new NT techniques, which allowed us to resurrect normal mice from bodies kept frozen at −20°C for up to 16 years without any cryoprotection (Fig. 13.11a) [12]. Although we could not produce cloned offspring from the somatic cell nuclei, several ntES cell lines were established from the cell nuclei of most organs. Finally, healthy cloned mice were produced from these ntES cells by a second round of nuclear transfer (Fig. 13.11c).

Thus, this technique is applicable for the propagation of a variety of animals, regardless of age or the potential for fertility. If a valuable mouse is unexpectedly lost, then it is clearly possible, now, to generate live copies of such animals from tissue that was otherwise considered lifeless.

13.6 Troubleshooting

1. *Cannot make a hole using the piezo drill?*

If you cannot cut the zona pellucida, check the connection between the pipette and pipette holder. The top of the pipette holder must be screwed in tightly. Expel all oil inside the pipette, as oil may have reduced the piezo power transmission. There should be a slight negative pressure inside the pipette to enhance the piezo power.

2. *Sticky pipettes?*

PVP will cover both the inside and outside of the pipette to keep the surface slick. Washing the pipette in PVP medium is very important and will affect not only oocyte survival rate, but also embryo development after NT. Without this step, the pipette soils rapidly and needs to be changed.

3. *Difficulty in finding the oocyte spindle?*

The room temperature is very important as the spindle microtubules will disperse and become unclear at room temperature. However, the spindle will become visible again if the oocytes are cultured at 37°C for 30 min before enucleation. Oocyte transparency also depends on the mouse strain; B6D2F1 is better than some others.

4. *Donor cells aggregate in PVP medium?*

Mix the donor cells with PVP medium using sharp tweezers for at least 30 s. If the donor cells are not mixed sufficiently in PVP medium, they will aggregate and it will be difficult to isolate single cells. ES cells are especially sensitive and fragile and it is better to make new ES cell suspension drops every 30 min.

5. *Difficulty in releasing donor nuclei from the pipette?*

Probably the pipette is too dirty. It must be washed frequently using PVP medium by expelling some mercury and applying power from the piezo unit.

6. *Oocyte lysis after nuclear injection?*

Either you used too large a pipette, the room temperature is too high or the pipette insertion was too shallow. A large pipette or warm temperatures increase the rate of oocyte lysis. The injection pipette must be inserted very deep into the oocyte before applying the piezo pulse. In addition, you need practice. If you are a beginner, all oocytes will lyse immediately after injection. One month after starting practice, about half of your oocytes might survive.

One year later, about 80% of oocytes will survive, if you continue to practice diligently. If oocytes are transferred to KSOM medium immediately after injection, 10–20% of them undergo lysis from the damage of injection. The oocyte membrane must be allowed to recover before transfer to KSOM; this takes about 10 min.

7. *Oocytes die during activation?*

The activation medium must be checked before use, using intact oocytes. During strontium treatment, up to 10% of the oocytes will die and the medium will become dirty. This is normal and the surviving oocytes are usually undamaged.

8. *No pseudo-pronuclear formation after oocyte activation?*

There are several reasons why oocytes do not form pronuclei. Usually it is because of failure to break the donor cell membrane or failure of oocyte activation. The injection pipette must be smaller than the donor cell. If the donor cell has a tough cell membrane (e.g., tail tip fibroblasts), apply piezo power to break the donor cell membrane at the time of cell pickup.

9. *Cloned neonate dies at birth?*

All cloned mice allowed to go to full term gestation to date have been born with abnormal and hypotrophic placenta and often die immediately after birth from respiratory failure. At this point, there is no way to avoid this lethal phenotype.

10. *Pups do not develop to full term?*

If you have no success in getting full-term development, change the donor cell type from somatic to ES cells, or other hybrid mouse strains. Also, try making up new embryo culture medium. However, the most important solution is to keep practicing. Technical skill is essential. Do not give up!

References

1. Wakayama T, Perry AC, Zuccotti M, Johnson KR, Yanagimachi R (1998) Full-term development of mice from enucleated oocytes injected with cumulus cell nuclei. Nature 394:369–374
2. Kimura Y, Yanagimachi R (1995) Intracytoplasmic sperm injection in the mouse. Biol Reprod 52:709–720
3. Wakayama T, Yanagimachi R (1998) Development of normal mice from oocytes injected with freeze-dried spermatozoa. Nat Biotechnol 16:639–641
4. Kawase Y, Iwata T, Watanabe M, Kamada N, Ueda O, Suzuki H (2001) Application of the piezo-micromanipulator for injection of embryonic stem cells into mouse blastocysts. Contemp Top Lab Anim Sci 40:31–34
5. Wakayama S, Ohta H, Kishigami S, Thuan NV, Hikichi T, Mizutani E, Miyake M, Wakayama T (2005) Establishment of male and female nuclear transfer embryonic stem cell lines from different mouse strains and tissues. Biol Reprod 72:932–936
6. Wakayama T, Tabar V, Rodriguez I, Perry AC, Studer L, Mombaerts P (2001) Differentiation of embryonic stem cell lines generated from adult somatic cells by nuclear transfer. Science 292:740–743
7. Wakayama S, Jakt ML, Suzuki M, Araki R, Hikichi T, Kishigami S, Ohta H, Van Thuan

N, Mizutani E, Sakaide Y, Senda S, Tanaka S, Okada M, Miyake M, Abe M, Nishikawa S, Shiota K, Wakayama T (2006) Equivalency of nuclear transfer-derived embryonic stem cells to those derived from fertilized mouse blastocysts. Stem Cells 24:2023–2033

8. Wakayama S, Mizutani E, Kishigami S, Thuan NV, Ohta H, Hikichi T, Bui HT, Miyake M, Wakayama T (2005) Mice cloned by nuclear transfer from somatic and ntES cells derived from the same individuals. J Reprod Dev 51: 765–772
9. Mizutani E, Ono T, Li C, Maki-Suetsugu R, Wakayama T (2008) Propagation of senescent mice using nuclear transfer embryonic stem cell lines. Genesis 46:478–483
10. Ono T, Mizutani E, Li C, Wakayama T (2008) Nuclear transfer preserves the nuclear genome of freeze-dried mouse cells. J Reprod Dev 54:486–491
11. Wakayama S, Kishigami S, Van Thuan N, Ohta H, Hikichi T, Mizutani E, Yanagimachi R, Wakayama T (2005) Propagation of an infertile hermaphrodite mouse lacking germ cells by using nuclear transfer and embryonic stem cell technology. Proc Natl Acad Sci USA 102:29–33
12. Wakayama S, Ohta H, Hikichi T, Mizutani E, Iwaki T, Kanagawa O, Wakayama T (2008) Production of healthy cloned mice from bodies frozen at −20 degrees C for 16 years. Proc Natl Acad Sci USA 105:17318–17322
13. Wakayama T, Yanagimachi R (1999) Cloning of male mice from adult tail-tip cells. Nat Genet 22:127–128
14. Ogura A, Inoue K, Ogonuki N, Noguchi A, Takano K, Nagano R, Suzuki O, Lee J, Ishino F, Matsuda J (2000) Production of male cloned mice from fresh, cultured, and cryopreserved immature Sertoli cells. Biol Reprod 62:1579–1584
15. Ono Y, Shimozawa N, Ito M, Kono T (2001) Cloned mice from fetal fibroblast cells arrested at metaphase by a serial nuclear transfer. Biol Reprod 64:44–50
16. Wakayama T, Yanagimachi R (2001) Mouse cloning with nucleus donor cells of different age and type. Mol Reprod Dev 58:376–383
17. Wakayama T, Rodriguez I, Perry AC, Yanagimachi R, Mombaerts P (1999) Mice cloned from embryonic stem cells. Proc Natl Acad Sci USA 96:14984–14989
18. Inoue K, Wakao H, Ogonuki N, Miki H, Seino K, Nambu-Wakao R, Noda S, Miyoshi H, Koseki H, Taniguchi M, Ogura A (2005) Generation of cloned mice by direct nuclear transfer from natural killer T cells. Curr Biol 15: 1114–1118
19. Miki H, Inoue K, Kohda T, Honda A, Ogonuki N, Yuzuriha M, Mise N, Matsui Y, Baba T, Abe K, Ishino F, Ogura A (2005) Birth of mice produced by germ cell nuclear transfer. Genesis 41:81–86
20. Inoue K, Ogonuki N, Miki H, Hirose M, Noda S, Kim JM, Aoki F, Miyoshi H, Ogura A (2006) Inefficient reprogramming of the hematopoietic stem cell genome following nuclear transfer. J Cell Sci 119:1985–1991
21. Li J, Greco V, Guasch G, Fuchs E, Mombaerts P (2007) Mice cloned from skin cells. Proc Natl Acad Sci USA 104:2738–2743
22. Yamazaki Y, Makino H, Hamaguchi-Hamada K, Hamada S, Sugino H, Kawase E, Miyata T, Ogawa M, Yanagimachi R, Yagi T (2001) Assessment of the developmental totipotency of neural cells in the cerebral cortex of mouse embryo by nuclear transfer. Proc Natl Acad Sci USA 98:14022–14026
23. Inoue K, Noda S, Ogonuki N, Miki H, Inoue S, Katayama K, Mekada K, Miyoshi H, Ogura A (2007) Differential developmental ability of embryos cloned from tissue-specific stem cells. Stem Cells 25:1279–1285
24. Mizutani E, Ohta H, Kishigami S, Van Thuan N, Hikichi T, Wakayama S, Kosaka M, Sato E, Wakayama T (2006) Developmental ability of cloned embryos from neural stem cells. Reproduction 132:849–857
25. Wakayama T (2007) Production of cloned mice and ES cells from adult somatic cells by nuclear transfer: how to improve cloning efficiency? J Reprod Dev 53:13–26
26. Thuan NV, Kishigami S, Wakayama T (2010) How to improve the success rate of mouse cloning technology. J Reprod Dev 56:20–30
27. Inoue K, Ogonuki N, Mochida K, Yamamoto Y, Takano K, Kohda T, Ishino F, Ogura A (2003) Effects of donor cell type and genotype on the efficiency of mouse somatic cell cloning. Biol Reprod 69:1394–1400
28. Kishigami S, Bui HT, Wakayama S, Tokunaga K, Van Thuan N, Hikichi T, Mizutani E, Ohta H, Suetsugu R, Sata T, Wakayama T (2007) Successful mouse cloning of an outbred strain by trichostatin A treatment after somatic nuclear transfer. J Reprod Dev 53:165–170
29. Kishigami S, Mizutani E, Ohta H, Hikichi T, Thuan NV, Wakayama S, Bui HT, Wakayama T (2006) Significant improvement of mouse cloning technique by treatment with trichostatin A after somatic nuclear transfer. Biochem Biophys Res Commun 340:183–189
30. Kishigami S, Van Thuan N, Hikichi T, Ohta H, Wakayama S, Mizutani E, Wakayama T (2006) Epigenetic abnormalities of the

mouse paternal zygotic genome associated with microinsemination of round spermatids. Dev Biol 289:195–205

31. Kishigami S, Wakayama S, Thuan NV, Ohta H, Mizutani E, Hikichi T, Bui HT, Balbach S, Ogura A, Boiani M, Wakayama T (2006) Production of cloned mice by somatic cell nuclear transfer. Nat Protoc 1:125–138
32. Rybouchkin A, Kato Y, Tsunoda Y (2006) Role of histone acetylation in reprogramming of somatic nuclei following nuclear transfer. Biol Reprod 74:1083–1089
33. Van Thuan N, Bui HT, Kim JH, Hikichi T, Wakayama S, Kishigami S, Mizutani E, Wakayama T (2009) The histone deacetylase inhibitor scriptaid enhances nascent mRNA production and rescues full-term development in cloned inbred mice. Reproduction 138:309–317
34. Kishigami S, Wakayama T (2007) Efficient strontium-induced activation of mouse oocytes in standard culture media by chelating calcium. J Reprod Dev 53:1207–1215
35. Nagy A, Gertsenstein M, Vintersten K, Behringer R (2003) Manipulating the mouse embryo. Cold Spring Harbor Laboratory Press, New York
36. Ying QL, Wray J, Nichols J, Batlle-Morera L, Doble B, Woodgett J, Cohen P, Smith A (2008) The ground state of embryonic stem cell self-renewal. Nature 453:519–523
37. Schatten G, Smith J, Navara C, Park JH, Pedersen R (2005) Culture of human embryonic stem cells. Nat Methods 2:455–463
38. Hrabe de Angelis MH, Flaswinkel H, Fuchs H, Rathkolb B, Soewarto D, Marschall S, Heffner S, Pargent W, Wuensch K, Jung M, Reis A, Richter T, Alessandrini F, Jakob T, Fuchs E, Kolb H, Kremmer E, Schaeble K, Rollinski B, Roscher A, Peters C, Meitinger T, Strom T, Steckler T, Holsboer F, Klopstock T, Gekeler F, Schindewolf C, Jung T, Avraham K, Behrendt H, Ring J, Zimmer A, Schughart K, Pfeffer K, Wolf E, Balling R (2000) Genome-wide, large-scale production of mutant mice by ENU mutagenesis. Nat Genet 25:444–447
39. Wakayama T, Shinkai Y, Tamashiro KL, Niida H, Blanchard DC, Blanchard RJ, Ogura A, Tanemura K, Tachibana M, Perry AC, Colgan DF, Mombaerts P, Yanagimachi R (2000) Cloning of mice to six generations. Nature 407:318–319
40. Saito M, Saga A, Matsuoka H (2004) Production of a cloned mouse by nuclear transfer from a fetal fibroblast cell of a mouse closed colony strain. Exp Anim 53:467–469
41. Nagy A, Rossant J, Nagy R, Abramow-Newerly W, Roder JC (1993) Derivation of completely cell culture-derived mice from early-passage embryonic stem cells. Proc Natl Acad Sci USA 90:8424–8428

Chapter 14

Gene Targeting in Embryonic Stem Cells

Elizabeth D. Hughes and Thomas L. Saunders

Abstract

Genetic modification of mouse embryonic stem cells is a powerful method to study gene function in whole animal models. The ability to re-design genes in mouse to reproduce genetic defects found in human patients gives researchers a wide open arena for biomedical research. Successful manipulation of ES cells requires culture conditions that restrain differentiation and support robust growth. It is essential to use culture conditions that are carefully calibrated to maintain ES cells in pluripotent condition because the only test for germline transmission is the time-consuming process of preparing ES cell-mouse chimeras and breeding them for several months. Fastidious adherence to culture conditions will maintain the capability of ES cells to differentiate into many cell types. Inadequate care of ES cells will degrade the ES cell quality and result in cells that fail to transmit mutant genes through the germline. Mouse ES cell culture is a well-established procedure. After targeting vector electroporation and identification of euploid gene-targeted clones, generation of chimeras from three clones is normally sufficient to transmit engineered mutations through the germline and produce a new mouse strain. Although the process of generating a new mouse model is not always routine, careful attention to ES cell culture is rewarded by the production of robust gene-targeted ES cell clones that regularly go germline.

Abbreviations

2-ME	Beta-Mercaptoethanol
DMEM	Dulbecco's Modified Eagle Medium
DMSO	Dimethyl sulfoxide
D-PBS	Dulbecco's phosphate buffered saline
ES cells	Embryonic stem cells
FBS	Fetal bovine serum
FIAU	1-(-2-Deoxy-2-fluoro-1-b-D-arabino-furanosyl)-5-iodouracil
LIF	Leukemia Inhibitory Factor (ESGRO®)
MEF	Mouse embryonic fibroblast feeder cells
MEM	Minimal Essential Medium
Pen/strep	Penicillin/streptomycin antibiotic mix for cell culture
SDS	Sodium dodecyl sulfate

S. Pease and T.L. Saunders (eds.), *Advanced Protocols for Animal Transgenesis*, Springer Protocols,
DOI 10.1007/978-3-642-20792-1_14, © Springer-Verlag Berlin Heidelberg 2011

14.1 Introduction

Genes in mouse embryonic stem (ES) cells can be customized to address research questions in biomedical science and developmental biology. When pluripotent ES cell lines are used, they will contribute to the germline of ES cell-mouse chimeras when they are returned to their original developmental stage in the mouse blastocyst. The resulting chimeras will then transmit the customized gene to its offspring and a novel mouse strain will be available to study the gene in question. Examples of genetic modifications include null alleles (inactive in all cells, [1]), conditional alleles (gene inactivated in specific cell types [2]), or subtle changes such as point mutations that introduce amino acid changes to alter protein phosphorylation states and activity [3] or deletions of functional domains in proteins [4]. Numerous other genetic modifications are possible – the principle limitation is the investigator's imagination.

Mouse ES cell lines are routinely used to generate novel mouse models. The process of genetic modification in ES cells is called gene targeting. It is performed by introducing gene targeting vectors into ES cells and identifying cells that undergo homologous recombination with the vector to produce the desired change in the gene in question. Positive selection is applied to the ES cells, such as a toxic antibiotic (G418) to eliminate ES cells that did not take up the vector. Negative selection to eliminate ES cells that take up the vector randomly can also be used [5]. For those unfamiliar with the field, the array of available mouse ES cell lines can be bewildering. For example, a survey of 6,914 mouse strains produced from genetically modified ES cells shows that 99 different ES cell lines were used to generate these animal models (this list is available for download at http://www.transtechsociety.org/members/methods.php). The ten most commonly used ES cell lines in the survey are R1, E14, J1, E14.1, D3, CJ7, RW-4, W9.5, TC-1 and E14Tg2a. Irrespective of the desired genetic modification or ES cell line, all gene-targeting projects have the same criteria for success: live mice carrying the customized gene.

14.1.1. Outline of the Procedure

- Identify the gene in question
- Design and test a screen for homologous recombination
- Construct the gene targeting vector
- Purify linearized targeting vector for electroporation
- Electroporate ES cells
- Culture ES cells under selective pressure (e.g. G418)

- Pick drug-resistant ES cell clones into 96-well plates
- Split 96-well plate cultures to cryopreserve ES cell clones
- Grow up ES clones in duplicate 96-well plates for DNA extraction
- Extract DNA from 96-well plates
- Screen DNA plates to identify gene-targeted ES cell clones
- Expand gene-targeted ES cell clones from cryopreserved 96-well plates
- Chromosome count and verify gene targeting in expanded clones
- Prepare ES cell-mouse chimeras from gene-targeted clones

14.1.2. Principles and Applications

The key to success in gene targeting is to recognize that mouse ES cells are sensitive to culture conditions and require meticulous culture techniques to maintain pluripotency. Like any cell line maintained in long-term culture, ES cells are susceptible to spontaneous genetic changes such as chromosomal duplication [6–8]. Most ES cell lines are cultured on feeder layers of mitotically inactivated mouse embryonic fibroblasts. In the typical experiment, ES cell cultures are handled daily and maintained for weeks. All methods described here require fastidious aseptic technique to avoid loss of experiments to bacterial or other contamination.

Before a gene-targeting experiment is initiated it is essential to verify that culture conditions and reagents, especially fetal bovine serum and feeder cells, will support the vigorous growth of pluripotent ES cells without inducing differentiation. When a new ES cell line is introduced to the laboratory, appropriate culture conditions need to be established and the ES cells tested for pluripotency by preparing ES cell-mouse chimeras with host blastocysts of the appropriate background and demonstrating that the chimeras will transmit the ES cell haplotype through the germline. Germline testing should also be performed every time a major expansion of an ES cell line is undertaken. Without these precautions the success of a gene-targeting project may be compromised even before the electroporation step. The ultimate goal of a mouse ES cell laboratory is to produce ES cell clones that are not only correctly gene targeted, but also germline competent.

The success of ES cell culture work depends upon many factors, one of which is the correct functioning of equipment. A quality control schedule to check that all equipment is working properly is essential (are autoclaves effective? Do biological safety cabinets provide sterile work areas?). Include a biological indicator with the load when you sterilize materials. Freezers that hold cryopreserved ES cell clones (−80°C and liquid nitrogen storage) should be alarmed so that irreplaceable samples can be rescued before complete meltdowns. An alarm/monitor system that will telephone staff when critical equipment fails will help prevent catastrophic failures. Cell-culture safe cleaning solutions such as 7× Cleaning Solution (ICN Biomedicals Inc) should be used to clean glassware that will come in contact with cells or media. Wherever possible, use sterile disposable plastics.

Not all mouse ES cell lines are cultured in the same medium formulation. Some may require sodium pyruvate and/or nonessential amino acids and/or nucleosides. Some may require growth on feeder cells while others grow on gelatin-treated plasticware. Feeder cells isolated from transgenic mouse embryos that carry drug resistance markers will survive and support ES cell growth during drug selection [9, 10]. Always verify the culture conditions with the provider of the ES cells that are in use in the laboratory. Whenever cells are imported into a dedicated cell culture facility it is essential to test for mycoplasma contamination. This includes ES cell lines from all repositories associated with the international program to knockout every gene in the mouse. Mycoplasma species contamination is common in cell cultures [11]. It spreads easily between cultures by aerosols or shared pipettes when changing growth media. It will seriously reduce the efficiency of generating germline gene-targeted ES cell clones. Every cell line imported is considered potentially contaminated until tests prove that mycoplasma is absent. The MycoAlert kit (Lonza) is a simple and sensitive assay that detects multiple mycoplasma species. Matching the media and other culture conditions to the ES cell line and testing for possible mycoplasma contamination are critical steps when a new ES cell line is established in the cell culture facility.

ES cells can acquire chromosomal abnormalities during culture that will make them incapable of generating germline chimeras [12]. These changes include loss of chromosomes, duplication of chromosomes and translocation events, as well as more subtle changes. These mutations secondary to the introduced genetic modification may also result in unexpected phenotypes of any resulting animals, and complicate interpretation of experimental results [13]. It is recommended to use more than one ES clone for blastocyst injection in a gene targeting experiment, to control for these subtle secondary mutations. Chromosome counting is a relatively simple procedure that will allow the exclusion of ES cell lines with gross chromosomal abnormalities from being used in blastocyst injection. Research effort and materials can then be more efficiently used on only those clones most likely to be germline competent – that is, those with euploid chromosome counts.

14.2 Materials

14.2.1. Equipment

- Refrigerator.
- −20°C freezer.
- −80°C freezer.
- 37°C waterbaths.
- Autoclave or sterilizer.
- Biological safety cabinet.
- 5% CO_2, 37°C, humidified incubator.
- Pipetting aid for pipettes, and pipettors for 200 μl and 1,000 μl tips.
- 8 or 12 channel multichannel pipettor, up to 200 μl.
- 8 or 12 channel multichannel aspirator.
- 8 or 12 multi-channel programmable repeating pipettor up to 1.2 ml.
- Aspirator (either house vacuum or vacuum pump).
- Liquid nitrogen storage container for cells.
- Centrifuge for 96-well plates.
- Tabletop centrifuge, with rotors for 15 and 50 ml conical centrifuge tubes.
- Inverted cell culture microscope, with 5× and 10× objectives, 10× oculars.
- Upright microscope, equipped with 10× and 100× (oil) objectives, 15× oculars.
- Luminometer for use with Mycoplasma detection kit.
- Sterile surgical instruments: scissors, forceps, scalpels.
- Cellector tissue sieve and pestle (Bellco Glass, #1985-85000).
- Trypsinizing flask (Bellco Glass, #1989-00125).
- Electronic colony counter pen (Bel-Art Products #378620000).
- Magnetic stirring plate.
- Electroporator (Biorad Gene Pulser Xcell Electroporation System).
- 55°C Oven.

14.2.2. Sterile Disposable Cell Culture Plastics

- All plasticware should be non-pyrogenic and sterile.
- Individually wrapped pipettes – 1, 5, 10 and 25 ml (Costar stripette paper-wrapped disposable polystyrene serological pipettes).

- 200 μl pipette tips (VWR Rak Pak #L1-PL-1030-290-306).
- 1,250 μl pipette tips, sterile (Thermo Scientific Matrix Pipet Tips #8042).
- 200 μl Round, MultiFlex Tip, Sterile (Dot Scientific, Inc., Burton, MI).
- 15 ml polystyrene conical centrifuge tubes (BD Falcon #352095).
- 50 ml polypropylene conical centrifuge tubes (BD Falcon #352098).
- 2 ml cryovials (free-standing vials, internal threads, Dot Scientific #T311-2).
- 0.2 μm pore size filter sterilization units, 150 ml (Thermo Scientific #156-4020).
- 0.2 μm pore size filter sterilization units, 500 ml (Thermo Scientific #156-0020).
- 6-well flat bottom cell culture dishes (Corning Costar #3516).
- 12-well flat bottom cell culture dishes (BD Falcon #353043).
- 24-well flat bottom cell culture dishes (Corning Costar #3524).
- 48-well flat bottom cell culture dishes (Corning Costar #3548).
- U-bottom 96-well cell culture dishes (BD Falcon #351177).
- Flat bottom 96-well cell culture dishes (BD Primaria #353872).
- 60 mm diameter cell culture dishes (BD Falcon #353004).
- 100 mm diameter cell culture dishes (Corning #430167).
- 150 mm diameter cell culture dishes (Corning #30599).
- Electroporation cuvettes (BioRad #165-2088).
- Non-porous 96-well plate seal (Whatman #7704-0001).

14.2.3. Cell Culture Media Components and Reagents

- ES-qualified fetal bovine serum (FBS); FBS lots from several vendors are tested. Once a lot is identified enough FBS for 12 months work is purchased and stored at −80°C. The reputation of the FBS vendor is not as important as the performance of the serum with your ES cell lines.
- DMEM (high glucose) (Invitrogen #11965).
- LIF (Millipore #ESG1107).
- 200 mM glutamine stock solution (Invitrogen #25030).
- Beta-mercaptoethanol (2-ME) (Sigma #M-7522).

- Sodium pyruvate solution (Invitrogen #11360).
- MEM-non-essential amino acids solution (Invitrogen #11140).
- Penicillin/streptomycin solution (pen/strep) (Invitrogen #15070), Pen/strep is not routinely used in cell culture. Pen/strep may mask low-grade contaminants, but in its absence contaminants will be revealed quickly. Pen/strep is omitted from media in Subheadings 14.4.1–14.4.7, and included in media when new ES cell clones are in process in Subheadings 14.4.9–14.4.12 and 14.4.14.
- 0.05% trypsin/EDTA (Invitrogen #25300).
- D-PBS (Invitrogen #14190).
- DMSO (Sigma #D2650).
- Gelatin (Sigma #G-1890).
- DNAse (Roche #10104159001).
- Sarcosyl (*N*-Lauroylsarcosine sodium salt, Sigma #L9150).
- SDS, 10% solution (Sigma #L4522).
- Sterile 5 M NaCl (Sigma #59222C).
- Colcemid (Invitrogen #15210-040).
- Gurr's Buffer (Invitrogen #10582-013).
- Giemsa Stain (Invitrogen #10092-013).

14.2.4. Mycoplasma Detection Reagents

- Mycoalert Detection Assay (Lonza #LT07-418).
- MycoAlert Control Set (Lonza #LT07-518).

14.2.5. Mouse ES Cell Lines

- Specific ES cell lines of particular interest may be obtained from the investigators who originated them.
- Rl (Mutant Mouse Regional Resource Centers, http://www.mmrrc.org).
- E14Tg2a.4 (MMRRC, http://www.mmrrc.org).
- AK7 (MMRRC, http://www.mmrrc.org).
- D3 (American Type Culture Collection #CRL-1934). The ATCC has a number of mouse ES cell lines in its catalog.
- W4 (Taconic Farms # ES W4129S6).
- Bruce4 (Millipore #CMIT-2).
- JM8A3, JM8.N4 and JM8.F6 (http://www.komp/org).
- Other suppliers include The Jackson Laboratory, Thermo (Hyclone, Open Biosystems, Primogenix).

14.2.6. Mice

- Pregnant mice at mid-gestation (embryonic day 14.5–16.5) for primary fibroblast production for use as feeder cells.

14.3 Laboratory Protocols

14.3.1. Media and Reagent Formulations

14.3.1.1. Reagents

- 100× 2-ME: dilute 0.07 ml 2-ME in 100 ml cell-culture grade water. Filter sterilize and store in 10 ml aliquots at 4°C for no more than 2 weeks.
- 0.1% (w/v) Gelatin: prepare 0.1% gelatin in cell-culture grade water. Autoclave to sterilize. Store at room temperature.
- DNAse, 13.33 mg/ml: Prepare DNAse stock by dissolving 100 mg DNAse in 7.5 ml of 0.3 M NaCl in 50% glycerol. To make 13.33 mg/ml DNAse combine 0.45 ml of 5 M NaCl sterile stock solution, 4.69 ml of 80% sterile glycerol stock solution, 2.36 ml sterile cell culture grade water to 100 mg of sterile DNAse.
- Add 2 ml D-PBS to 2mg mitomycin-C (Sigma # M4287, 2 mg in sterile vial) to prepare 1 mg/ml mitomycin-C stock.
- NaCl/Ethanol suspension (15 ml 5 M NaCl in 1 L 95% ethanol), stored at −20°C for DNA precipitation.
- Hypotonic solution for chromosome spreads: 0.075 M KCl.
- Fixative for chromosome spreads: mix together 30 ml methanol with 10 ml glacial acetic acid just before use. Methanol should be purchased in 500 ml glass bottles.
- 10% SDS.
- 25:24:1 mixture of phenol:chloroform: isoamyl alcohol.
- 24:1 mixture of chloroform:isoamyl alcohol.

14.3.1.2. ES Cell Media to Make 500 ml

- 0.05 ml LIF (10^3 units/ml final concentration).
- 5 ml 100× 2-ME reagent.
- 5 ml 200 mM glutamine.
- 5 ml pen/strep (only if required).
- 75 ml FBS (15% FBS final concentration).
- 410 ml DMEM.

14.3.1.3. MEF Isolation Media to Make 500 ml

- 5 ml 100× 2-ME reagent, addition of 2-ME increases MEF attachment to dishes and increases MEF yield when dishes are trypsinized.
- 5 ml 200 mM glutamine.
- 5 ml pen/strep.
- 50 ml FBS (10% FBS final concentration).
- 435 ml DMEM.

14.3.1.4. MEF Culture Media to Make 500 ml

- 5 ml 200 mM glutamine.
- 50 ml FBS (10% FBS final concentration).
- 445 ml DMEM.

14.3.1.5. ES Cell Cryopreservation Media to Make 10 ml

- 9 ml ES culture media.
- 1 ml DMSO (10% DMSO final concentration).

14.3.1.6. ES Cell Electroporation Media

- Same as Subheading 14.3.1.2.
- Add pen/strep.

14.3.1.7. Lysis Buffer for 96-Well ES Clone Plates

- 1 ml 1 M Tris-HCl, pH 8.
- 0.2 ml 5 M NaCl.
- 2 ml 500 mM EDTA.
- 0.5 g Sarcosyl.
- Sterile water to 100 ml final volume.
- Add 1 mg/ml Proteinase-K to Lysis Buffer immediately before use.

14.3.1.8. Mouse Embryonic Fibroblasts

- See Protocol 14.4.1. Isolation of Primary Mouse Embryonic Fibroblasts.

14.4 Protocols

14.4.1. Isolation of Primary Mouse Embryonic Fibroblasts (MEFs) for Use as Feeder Cells

The purpose of this protocol is to prepare frozen stocks of MEFs. Mouse ES cells are cultured on layers of mitotically inactivated fibroblasts. An alternative to MEFs is the use of STO cell layers, which are plated at twice the density of MEFs [14]. With an input of 10^8 fibroblasts, this protocol will bank down 400 vials at passage number three and will yield enough MEFs for 50 gene-targeting projects.

1. Warm media and D-PBS in water bath. Thaw trypsin and place on ice. Treat thirty 150 mm dishes with gelatin (cover dish surface with 8 ml gelatin solution and aspirate).
2. Place 10 ml D-PBS in six 100 mm dishes; place 5 ml D-PBS in one 100 mm dish.
3. Kill mid-gestation pregnant mice. Remove uterine horns with embryos, cutting away the mesentery, and place in the first 100 mm dish with 10 ml PBS.
4. Cut along the length of uterus with scissors to release embryos. Cut the umbilical cord to separate placenta from embryos, and place in the second 100 mm dish.

5. Cut open the amniotic sac and place the embryos one-by-one in the third 100 mm dish.
6. Open new pack of sterile instruments and wash each embryo by moving them to the fourth 100 mm dish.
7. Eviscerate and decapitate embryos. Hold embryo by head with abdomen facing up, use fine blunt forceps to puncture body wall and remove internal organs. Cut off head at neck.
8. Place carcasses in the fifth 100 mm dish. Discard heads and viscera.
9. Rinse carcasses in the sixth 100 mm dish, and place in the dish containing exactly 5 ml D-PBS. Mince carcasses with crossed scalpels.
10. Place sterile Cellector cup in bottom of empty 100 mm dish. Pour minced tissue into the cup. Rinse plate with 5 ml D-PBS, then with 10 ml trypsin and pipette into cup to collect all remaining tissue.
11. Force tissue through screen with sterile glass pestle. Continue to press tissue through screen until as much as possible has gone through. Pipette 10 ml trypsin through the screen, and bottom of screen to remove adhering tissue. Repeat with one more 10 ml aliquot of trypsin.
12. Total volume is now 40 ml (10 ml D-PBS + 30 ml trypsin). Pipette into sterile trypsinizing flask. Add a sterile stir bar and 0.4 ml DNAse reagent.
13. Place on stirring plate inside 5% CO_2, 37°C, humidified incubator for 60 min. If needed, add more DNAse to reduce viscosity.
14. Pour contents through the side arm of trypsinizing flask into a 50 ml tube containing 10 ml of MEF isolation media.
15. Centrifuge the cell suspension for 10 min at $400 \times g$.
16. Aspirate the supernatant, leaving a small amount in tube to resuspend pellet. Break up cell pellet and resuspend in 30 ml MEF isolation media.
17. Repeat **step 16** to wash cells a second time.
18. Resuspend the cells in 15 ml MEF isolation media.
19. Count the viable cells. Expect 5×10^7–10^8 cells from ten e15.5 embryos.
20. Plate the cells at 6×10^6 per gelatin-coated 150 mm dish in 25 ml MEF media. The cells are at passage one.
21. Place the dishes in 5% CO_2, 37°C, humidified incubator.
22. Replace the media on the following day with MEF isolation media.

23. Observe the dishes daily until the cells are confluent.
24. Place 10 confluent dishes in culture hood. Wash the dishes with 15 ml D-PBS. Add 10 ml trypsin. Place in 5% CO_2, 37°C, humidified incubator for 5 min. Pipette cells from 4 dishes into 50 ml tube containing 10 ml MEF isolation media.
25. Centrifuge the cell suspension for 5 min at 400 × *g*.
26. Resuspend cells in 20 ml of 10% DMSO in MEF isolation media.
27. Aliquot 1 ml of cells into each of 20 cryovials. Depending on the growth of the cells there may be 5–10 × 10^6 cells per vial. Immediately place in small Styrofoam box or freezing container and transfer to −80°C freezer. When these vials are thawed, plate each onto one 150 mm dish per vial (*see* Subheading 14.4.2). When confluent, trypsinize and re-plate onto five gelatin-coated 150 mm dishes at a 1:5 ratio for expansion. The plated MEFs are at passage two.
28. Alternatively, instead of freezing, confluent dishes may be trypsinized and re-plated onto gelatin-coated 150 mm dishes at a 1:5 ratio. The plated MEFs are at passage two.
29. Continue to culture until the cell layers are confluent.
30. Cells are frozen down as above (**steps 24–27**). MEFs are at passage three. When these vials are thawed, plate them onto one 150 mm dish (see below). The plated MEFs are at passage three.

14.4.2. MEF Test Thaw

The purpose of this protocol is to verify that MEFs frozen in vials can be recovered, to detect microbial contaminants and to estimate how many MEFs will be harvested for every vial plated onto 150 mm dishes. Once the yield of MEFs per vial is known it is a straightforward matter to plate out enough MEFs for experiments.

1. Test thaw one cryovial per batch of 20 frozen vials.
2. Place 3 ml MEF medium in 15 ml conical centrifuge tube. Use one tube per vial of thawed cells.
3. Gelatin coat one 150 mm dish per cryovial and add 23 ml MEF media per dish.
4. Thaw MEF cryovials in 37°C waterbath for 60–90 s.
5. Dry the cryovials with a paper towel, spray with 70% ethanol and place in cell culture hood.
6. Transfer thawed cells to centrifuge tubes.
7. Centrifuge the cells for 5 min at 400 × *g*.
8. Aspirate the supernatant, leaving a small volume of medium in the tube. Loosen the cell pellet by tapping the tube.

9. Resuspend the cells in 2 ml MEF media and add to 150 mm dish.
10. Place dish in 37°C, 5% CO_2, humidified incubator.
11. Observe the cells daily in incubator for 4 days for cell growth and microbial contamination.
12. Each day for 4 days trypsinize and count the cells from one 150 mm dish. Plot a growth curve and use this to estimate the MEF yield per cryovial.
13. After 4 days collect spent media for Mycoplasma test in MycoAlert kit.
14. If MEFs show microbial or Mycoplasma contamination, discard MEFs and plan a new isolation.
15. If MEFs pass test thaw, transfer cryovials to liquid nitrogen container.

14.4.3. Preparation of Mitotically Inactivated MEFs

Mitotically inactivated feeder cells are prepared every Monday and Thursday. Determine the number of dishes you need 1 week in advance. When inactivated feeders are prepared, thaw enough MEF cryovials to produce MEFs for the next preparation. The results of Subheading 14.4.2 will provide you with the number of MEFs to be expected from each vial you thaw.

1. Place 3 ml MEF medium in 15 ml conical centrifuge tube for each cryovial.
2. Gelatin coat the appropriate number of 150 mm dishes and add 23 ml MEF media per dish.
3. Thaw passage 3 MEF cryovials in 37°C waterbath for 60–90 s.
4. Dry the cryovials with a paper towel, spray with 70% ethanol and place in hood.
5. Transfer thawed cells to centrifuge tube.
6. Centrifuge the cell suspension for 5 min at 400 × *g*.
7. Aspirate the supernatant, leaving a small volume of medium in the tube. Loosen the cell pellet by tapping the tube.
8. Resuspend cells in 2 ml MEF media for each 150 mm dish to be seeded. Add 2 ml cell suspension to each 150 mm dish (final volume of MEF media is 25 ml).
9. Place dishes in 5% CO_2, 37°C, humidified incubator for 3–4 days.
10. Add 0.25 ml mitomycin-C stock to each 150 mm dish.
11. Place the dishes in 5% CO_2, 37°C, humidified incubator for 2.5–5 h.
12. Gelatin coat the necessary dishes and multiwell plates.

13. Aspirate mitomycin-C containing media and wash the dishes twice with 10 ml D-PBS.
14. Add 10 ml trypsin to each dish and place in incubator for 5 min. Pipette the cells into 15 ml tubes containing 3 ml MEF media.
15. Centrifuge the cell suspension for 5 min at 400 × *g*.
16. Aspirate the supernatant, leaving a small volume of medium in the tube. Loosen the cell pellet by tapping the tube.
17. Resuspend the cell pellet in 10 ml MEF media and plate out into dishes and wells to produce the final concentrations shown in Table 14.1. The plated inactivated MEFs are at passage four. Low passage MEFs produce the best feeder layers. It is inadvisable to use MEFs beyond passage five for feeder layers because of decreasing MEF viability.
18. Alternatively, MEFs can be mitotically inactivated by irradiation with 6,000 rads before plating the cells on the desired gelatin-coated dishes.
19. Inactivated MEFs can be cryopreserved in 10% DMSO in MEF media at 3–5 × 10^6 cells per ml for future use.

14.4.4. ES Cell Line Importation

This protocol describes how to bring new ES cell lines into the culture laboratory and expand them without placing the existing cell cultures at risk.

Table 14.1
Number of mitotically inactivated feeder cells to seed in different well and plate sizes

Plate	Diameter (cm)	Area Sq. cm	Relative size	Volume/plate (mls)	# Feeders/plate (Feeders/well)
p 150	14	150	7.5	25	7.5 × E6
p 100	9	63	3.2	10	3 × E6
p 60	5.5	20	1	5	1 × E6
p 35	3.5	9.4	1/2.1	2	5 × E5
6-well	3.5	9.4	1/2.1	2	5 × E5
12-well	2.2	3.8	1/5.22	1.5	2 × E5
24-well	1.6	1.9	1/10.6	1	1 × E5
4-well	1.6	1.9	1/10.6	0.5	5 × E4
48-well	1.1	0.9	1/22.2	0.5	5 × E4
96-well	0.64	0.32	1/62.5	0.2	2 × E4

1. Handle imported cells after all other culture work is completed.
2. Culture imported cells in designated quarantine incubators in a laboratory physically separated from the main ES cell culture facility.
3. D-PBS, trypsin and media should be aliquoted and assigned specifically to each imported cell line.
4. Do not share media between cell lines.
5. Do not work with more than one cell line in the cell culture hood.
6. Discard or re-sterilize open packages of supplies after cell culture.
7. Use cotton-plugged pipettes to passage imported clones to prevent contamination of pipette aids.
8. Wipe down the hood with disinfectant and 70% ethanol before and after cell culture.
9. Wash hands with anti-microbial soap before and after handling imported cells.
10. Observe media and feeder requirements. Do not add pen/strep to the ES media in this protocol.
11. Prepare MEF or gelatin-coated dishes. Match dish size to number of ES cells in cryovial. Use 60 mm dish for 4–5 × 10^6 ES cells.
12. Place 3 ml ES medium in 15 ml conical centrifuge tube.
13. Thaw ES cryovial in 37°C waterbath for 60–90 s.
14. Dry cryovial with paper towel, spray with 70% ethanol and place in cell culture hood.
15. Transfer thawed cells to 15 ml centrifuge tube.
16. Centrifuge the cells for 5 min at 400 × *g*.
17. Aspirate the supernatant, leaving a small volume of medium in the tube. Loosen the cell pellet by tapping the tube.
18. Resuspend cells in 5 ml ES media without antibiotics and add to 60 mm dish. Feeders should have been mitotically inactivated and plated no more than 2 days earlier.
19. Place dish in 37°C, 5% CO_2, humidified incubator.
20. Passage regularly as described in Subheading 14.4.5 until three to five 60 mm dishes are available.
21. Cryopreserve ES cells as described in Subheading 14.4.7 and conduct test thaw.
22. Transfer vials of cryopreserved imported cells to liquid nitrogen storage if they pass the test thaw, otherwise store at −80°C.

14.4.5. ES Cell Passage

ES cells are rapidly dividing cells that are typically passaged every other day. ES cells may be plated at higher densities if they are needed the following day. Care should be taken to culture ES cells to proper density (Fig. 14.1b) before passage. ES cells passaged too early may not grow quickly (Fig. 14.1a). ES cells left in culture too long before passage are at risk of differentiation (Fig. 14.1c).

1. Locate feeder dishes or gelatin treat 60 mm dishes for passage.
2. Place 3 ml ES cell media in a 15 ml conical bottom centrifuge tube.
3. Aspirate media from 60 mm dish of ES cells.
4. Wash 60 mm ES cell dish with 3 ml D-PBS and aspirate.
5. Add 2 ml trypsin and place in a 37°C, 5% CO_2, humidified incubator for 5 min.
6. Triturate the trypsinized cells with a cotton-plugged Pasteur pipette 4–6 times to form a single-cell suspension and place cells in centrifuge tube.
7. Centrifuge the cell suspension for 5 min at 400 × g.
8. Aspirate the supernatant, leaving a small volume of media (about twice the volume of the packed cell pellet).
9. Flick the tube to break up the cell pellet.
10. Resuspend cells in 10 ml ES cell media.
11. Count the cells. Expect to obtain 1–2 × 10^7 ES cells from a 60 mm dish.
12. Aspirate media from feeder dishes. Feeders should have been mitotically inactivated and plated no more than 5 days earlier.

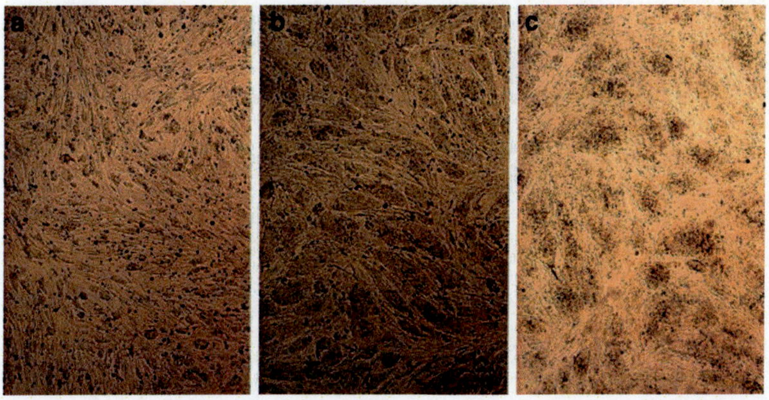

Fig. 14.1 Mouse ES cell culture. Mouse ES cells are shown in various states of confluency. (**a**) ES cells are subconfluent and will be ready to passage the following day. (**b**) ES cells ready for passage or for trypsiniation and microinjection. (**c**) ES cells that are super-confluent. Colonies have begun to fuse with each other and differentiation will soon follow.

13. Add 2 × 10^6 ES cells in 5 ml ES media to each 60 mm dish.
14. Rock the dish to evenly distribute the cells.
15. Place the dish in 37°C, 5% CO_2, humidified incubator.
16. Change the media on the following day.
17. Trypsinize and passage the cells on the second day following passage (*see* **steps 1–15**). Some ES cell lines grow more slowly, and may need another day in culture before passage.

14.4.6. ES Cell Preparation for Blastocyst Microinjection

1. ES cells growing at log phase are used to prepare a single cell suspension (*see* Subheading 14.4.5, **steps 2–11**).
2. The cell concentration is adjusted to 1 × 10^6 cells/ml in complete ES cell media.
3. Place 1 ml of cells on ice and deliver them to the microinjection laboratory.
4. Unused ES cells can be cryopreserved or re-plated.

14.4.7. ES Cell Cryopreservation

The purpose of this protocol is to freeze down ES cell lines in cryovials for long-term storage in liquid nitrogen containers. This protocol may be used for all ES cell lines and ES cell clones every time cell lines are frozen back, to ensure that the cells are viable.

1. Prepare a single cell suspension of ES cells in exponential growth (*see* Subheading 14.4.5, **steps 1–11**).
2. Place 2 × 10^6 cells in culture on a 60 mm dish and continue to passage these cells (Subheading 14.4.5) until the test thaw is complete (see below).
3. Divide the total number of ES cells by 4 × 10^6. Determine the number of cryovials that can be frozen at 4–5 × 10^6 cells per vial. Label the appropriate number of cryovials.
4. Centrifuge the cell suspension for 5 min at 400 × g.
5. Aspirate the supernatant, leaving a small volume of media (~twice the volume of the packed cell pellet).
6. Flick the tube to resuspend the cells in the remaining supernatant.
7. Add 1 ml ES Cryopreservation Medium for each vial to be frozen.
8. Resuspend ES cells in cryopreservation medium and aliquot 1 ml of cells into cryovials. Immediately place in small Styrofoam box or freezing container and transfer to −80°C freezer overnight. Record the location of the frozen ES cells in your freezer log.
9. Test thaw one cryovial of cells and plate onto 60 mm dish of feeders or gelatin, as appropriate for the ES cell line, in ES cell media without antibiotics.

10. Place the dish in 37°C, 5% CO_2, humidified incubator.
11. Observe the cells daily in incubator for 4 days and observe for ES cell growth and microbial contamination. Do not change the media.
12. After 4 days collect spent media for Mycoplasma test in MycoAlert kit.
13. If ES cells show microbial contamination repeat the freeze with the culture continued from **step 2**.
14. If ES cells are contaminated with Mycoplasma, replace the cells from a different source, if possible.
15. Transfer the frozen vials to liquid nitrogen storage once the cells have passed the test thaw. Do not allow the vials to warm up during the transfer. Record the location of the frozen ES cells in your freezer log.

14.4.8. ES Cell Line Stock Expansion

The purpose of this protocol is to generate 50 or more vials of low passage ES cells for use as the foundation for a series of gene targeting projects.

1. Thaw the ES cells (*see* Subheading 14.4.4, **steps 10–19**), observing media and feeder requirements. Do not add pen/strep to the ES media in this protocol.
2. On the next day observe the cells for attachment and contamination and change the media.
3. Change the media daily until the cells are 60–80% confluent. This may take 1–5 days. Usually the cells will be ready to passage after 2 or 3 days. Passage the cells on the fifth day after the thaw regardless of the degree of confluence, otherwise the colonies will get too large and start to differentiate.
4. Trypsinize the cells and plate them on 60 mm dishes of feeders at 2×10^6 cells per dish (*see* Subheading 14.4.5).
5. Change the media daily until the cells are 60–80% confluent.
6. Cryopreserve 8–12 vials of cells at $4\text{–}5 \times 10^6$ cells per vial (*see* Subheading 14.4.7). Plate the remaining cells on 60 mm dishes. It may be necessary to expand the cells for another round to generate enough cells to freeze back 8–12 vials.
7. Change the media daily until the cells are 60–80% confluent.
8. Trypsinize the cells and plate them on 60 mm dishes at 2×10^6 cells per dish.
9. Change the media daily until the cells are 60–80% confluent.
10. Cryopreserve the cells at $4\text{–}5 \times 10^6$ cells per vial (*see* Subheading 14.4.7). If needed, freeze the cells in batches of 20 vials to reduce the time that the cells are exposed to the cryoprotective agent. It is essential to keep the time that

elapses between the time the cells are resuspended in cryoprotective agent until the time that cell vials are placed in the −80°C to a minimum.

11. Continue until you have generated a large number of vials of frozen stocks.
12. Archive all test thaw results for future reference.

14.4.9. ES Cell Electroporation

Use ES cells in exponential growth for electroporations. Purify targeting vector plasmid DNA with a kit designed to produce endotoxin-free DNA. Linearize DNA with an appropriate restriction enzyme and use fresh phenol/chloroform to extract DNA after linearization. The following protocol will yield about 6,000 ES clones after G418 selection when the PGKl-neo selection cassette is included in the gene-targeting vector. If a different selection cassette is to be used, it may be advisable to electroporate one cuvette of cells and test different concentrations of G418 to establish the concentration that will allow colonies with integration of the vector to survive while wild type ES cells are eliminated.

1. Prepare twenty 100 mm feeder or gelatin dishes, according to the requirements of the ES cell line you are working with.
2. Trypsinize and count the ES cells (day 1) (*see* Subheading 14.4.5, **steps 2–11**). Add pen/strep to the ES media in this protocol.
3. Centrifuge the cell suspension for 5 min at 400 × g.
4. Aspirate supernatant, leaving a small volume of media (~twice the volume of the packed cell pellet).
5. Flick the tube to breakup the cell pellet.
6. Resuspend the cells in ES electroporation media at a concentration of 10^7 cells per ml.
7. Add 20 μg linearized targeting vector DNA in TE (1–2 μg/μ l) to electroporation cuvettes (BioRad #165-2088). Use three cuvettes for a single drug selection experiment (e.g. G418 treatment), and five cuvettes for a double drug selection experiment (e.g. G418 and gancyclovir treatment).
8. Add 0.8 ml of ES cells per cuvette and pipette up and down once to mix without introducing bubbles.
9. Electroporate with 250 μF and 0.3 keV, maximum capacitance, for the pulse. Record the time constant for each cuvette. The time constant should be about 4 ms.
10. Aspirate media from feeder dishes.
11. Combine all electroporated cells by pipetting into 200 ml ES cell media in a bottle. Gently mix the cells and plate out 10 ml per 100 mm dish.

12. Place the dishes in 5% CO_2, 37°C, humidified incubator.
13. On day 2, replace the media with ES cell media.
14. On day 3, replace the media with ES cell media containing 300 mg/ml G418 (or 110 μg/ml hygromycin, or 2 μg/ml puromycin as determined by the targeting vector construct).
15. On day 4, replace the media with ES cell media containing 300 mg/ml G418 (or other selection agent) and 2 μM gancyclovir if targeting vector includes HSV-TK cassette. Instead of gancyclovir, FIAU may be used as a negative selection agent [15].
16. On days 5–9, replace the media with ES cell media containing selection agents.
17. Drug-resistant ES cell colonies will appear 1 week after selection is initiated.

14.4.10. ES Cell Clone Picking

The purpose of this protocol is to transfer drug-resistant ES cell clones that have integrated targeting vector DNA to 96-well plates and to grow them up prior to cryopreservation and DNA extraction. A total of 480 clones are picked for each targeting vector.

1. Prepare five 96-well flat bottom plates with MEFs or gelatin.
2. On day 1, aspirate media from plates and add 150 μl ES culture media per well. Add pen/strep to the ES media in this protocol. Place plates in 5% CO_2, 37°C, humidified incubator.
3. Place microscope in cell culture hood for clone picking.
4. Add 25 μl trypsin/EDTA to each well of a 96-well U-bottom plate.
5. Wash one to three 100 mm dishes of ES cell clones twice with 10 ml D-PBS. Leave the last rinse on the dishes.
6. Place dish on microscope. Place a gel loading tip on a 20 μl pipettor set to 20 μl. Select a compact, undifferentiated colony in the microscope (Fig. 14.2). Scrape and draw colony into pipette tip.
7. Place colony in trypsin in well of 96-well U-bottom plate. Place one clone in every well. Change tips between colonies. Do not allow clones to remain in trypsin at room temperature for more than 30 min.
8. Remove 96-well feeder plate from incubator. Transfer 25 μl ES culture media from each well of the feeder plate to the corresponding wells of the trypsin plate with a multichannel pipettor.
9. Pipette up and down 4–6 times to disaggregate the cells with a multichannel pipettor. Transfer the cells to the 96-well feeder plate. Final volume is 195 μl per well.

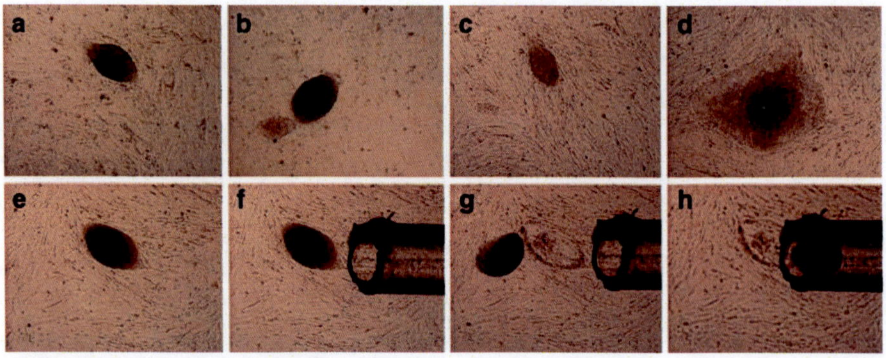

Fig. 14.2 Mouse ES cell clones and clone picking. Mouse ES cell clones are isolated with clone picking tips and transferred to trypsin in U-bottom 96-well plates for further processing. (a) ES cell clone with characteristic dome shaped morphology, notice colony has well defined edges. (b) Two ES cell clones in close proximity. Neither clone should be picked because of the likelihood of cross contamination. (c) Small ES clone with too few cells will not grow up after picking. (d) Grossly differentiated ES cell clone. Note wide skirt of flat cells around center. (e) Robust ES cell clone selected for picking. (f) Clone picking tip positioned next to clone. (g) Clone is separated from feeder layer by tip. Note scar on feeder layer where ES cells were attached. (h) ES cell clone is aspirated into tip. Colony will be transferred to trypsin in well of U-bottom 96-well plate.

10. Place the plates in 5% CO_2, 37°C, humidified incubator.
11. On day 2, replace the media with 100 μl ES cell media.
12. On day 3–6, replace the media with 100 μl ES cell media.

14.4.11. Split/Freeze of 96-Well ES Clone Plates

The purpose of this protocol is to cryopreserve ES cell clones in two 96-well plates that each replicate the original plate of clones. Cells are cryopreserved in two different −80°C freezers to guard against freezer failure. A third replica plate is continued in culture to provide cells for DNA extraction.

1. Observe 96-well plates of clones from Subheading 14.4.11. Score each well for density of cell growth. Wells without ES cells are scored as zero, wells with less than 10 colonies are scored as 1+, wells with up to 30 colonies are scored as 2+, wells with higher densities of ES cell colonies are scored as 3. When at least 60% of the wells are at least 60% confluent the cells are subjected to a Split/Freeze.
2. Prepare 2× ES cell freezing medium (20% DMSO in ES cell media).
3. On day 1, aliquot 33 μl 2× freezing medium into all wells of two flat bottom 96-well plates.
4. Rinse 96-well plate of ES clones twice with 200 ml D-PBS.
5. Add 50 μl trypsin per well and incubate at 37°C for 5 min, or up to 20 min, until the cells are rounding up.
6. Add 50 μl ES culture medium to trypsinized cells and pipette up and down to prepare a single cell suspension. Visual

observation under a microscope can be used to verify cells are disaggregated.

7. Add 33 μl of trypsinized ES cells to the 2× freezing medium in the first plate. Repeat with the second plate to produce two replica plates in freezing medium of the original plate.
8. Immediately transfer plates with ES cells in freezing medium to small Styrofoam box and place in −80°C freezer. If desired, 50 μl sterile mineral oil can be added to the wells to prevent evaporative loss. Additionally, the plates can be wrapped in Parafilm prior to placement in the −80°C freezer. In our hands we do not find these extra steps are required.
9. Add 100 μl ES cell media to the wells containing trypsinized ES cells and place plates in incubator.
10. On day 2, replace the media with 100 μl ES cell media.
11. On day 3 and 4, replace the media with 100 μl ES cell media.
12. Repeat **steps 1–9** for each additional plate of clones.
13. If possible, store replica sets of plates in separate −80°C freezers in different buildings.

14.4.12. DNA Split of 96-Well ES Clone Plates

The purpose of this protocol is to prepare two 96-well plates of ES cell clones for DNA extraction.

1. Observe 96-well plates of clones from Subheading 14.4.11, **step 11**. When at least 80% of the wells are at least 80% confluent the cells are subjected to a DNA split. This is usually at 2–5 days after the split/freeze.
2. Prepare one gelatin-treated flat bottom plate for every 96-well plate of clones.
3. On day 1, rinse 96-well plate of ES clones twice with 200 μl D-PBS.
4. Add 50 μl trypsin per well and incubate at 37°C for 5 min, or up to 20 min, until the cells are rounding up.
5. Add 50 μl ES culture medium to trypsinized cells and pipette up and down to prepare a single cell suspension. Visual observation under a microscope can be used to verify the cells are disaggregated. LIF may be omitted from the media in this protocol since the cells will be used for DNA extraction.
6. Transfer 50 μl of the cell suspension to replica gelatin-coated dishes.
7. Add 100 μl ES cell media to the wells containing trypsinized ES cells and place plates incubator.
8. On day 2, replace the media with 100 μl ES cell media.
9. On days 3–7, replace the media with 100 μl ES cell media.

14.4.13. DNA Extraction from 96-Well ES Clone Plates

The purpose of this protocol is to extract DNA from ES cell clones grown to high density in 96-well plates. The DNA preparation is crude and not all enzymes will digest the DNA with high efficiency (Table 14.2).

1. Prepare lysis buffer.
2. Gently invert plate over sink and blot on paper towels to remove media from wells.
3. Gently wash the wells twice with 100 μl D-PBS, using the multichannel pipettor.
4. Add 50 μl per well of Lysis Buffer containing Proteinase K to the empty wells.
5. Apply adhesive seals to the plates.
6. Incubate overnight at 55°C.
7. Centrifuge plates at 2,000 × *g* for 5 min to collect condensation.
8. Add 100 μl of cold NaCl/ethanol suspension and incubate at −20°C for at least 30 min, and up to several days. The presence of precipitated DNA may be visually verified with a microscope.
9. Quickly invert plate over sink to remove ethanol/lysis buffer mixture.
10. Rinse each plate three times with 100 μl/well of 70% ethanol. Quickly invert plate over sink to remove ethanol. Blot on paper towel after the last rinse.
11. Air dry 15–20 min. *Do not overdry* the plates, or you will not be able to resuspend the DNA.
12. Add 40 μl of TE or restriction enzyme buffer per well. Incubate at 65°C for 1 h to facilitate DNA resuspension. Pipette each well up and down 20–30 times to resuspend DNA. Add restriction enzyme for Southern blot digests; include BSA, 100 μg/μl and spermidine, 4 mM final concentration to improve DNA digestion. Apply adhesive seals to the plates and incubate overnight at appropriate temperature for enzyme. Samples in TE should be stored at −20°C.
13. Spin plate down 3–5 min at 2,000 × g to collect condensation.
14. Store the plates at −20°C until Southern blot gels are ready to load.
15. Add loading dye and resuspend the DNA by pipetting up and down at least 20–30 times before loading the gel. Use the contents of one well per lane of gel.

Table 14.2
Performance of restriction enzymes on 96-well plate ES cell DNA

Complete digestion	Variable digestion	Incomplete digestion
Asp718	EcoRI	BspDI
Age I	KpnI	ClaI
BamHI	SacI	HindII
BbvC I		NotI
BglII		Pvu I
Bgll		SacII
Bgll		SalI
EcoRI		SmaI
EcoRV		XbaI
EcoRV		XhoI
HindIII		XmnI
HindIII		
KpnI		
NcoI		
NcoI		
Nhe I		
Nsi I		
Pst I		
PstI		
PvuII		
Sac I		
SacI		
ScaI		
Spe I		
Sph I		
SstI		
StuI		
XbaI		
XhoI		

Enzymes in the Table 14.2 were tested by for digestion on 96-well ES cell DNA (Hughes and Saunders, unpublished [22, 23])

14.4.14. Thawing ES Cells Cryopreserved in 96-Well Plates

We typically expand no more than five clones at a time from a 96-well plate, to ensure each clone is handled optimally to maintain germline competence. In the "typical" successful gene-targeting experiment, five gene-targeted ES cell clones are enough to generate germline chimeras. Plates are thawed once to recover five clones and discarded. If additional clones are needed from the plate, the duplicate frozen plate is thawed. If more clones are needed, clones are thawed from the other four plates produced during clone picking. We mark the thawed wells and photograph the plate after the thaw to have a record of exactly which wells were expanded. Include pen/strep in the culture media throughout the expansion process.

1. Refer to the well scoring at the time of the split freeze. It will be impossible to recover viable ES cells from wells that were scored zero and wells that were scored 1+ will grow up slowly, if at all.
2. Place 120 ml water in small plastic tray (15 cm × 9 cm × 2 cm, e.g. the lid of tip box).
3. Place tray in the incubator to warm the water.
4. Replace MEF media with ES cell media in 24-well feeder plate (129 derived ES cells) or in 48-well feeder plate (C57BL/6 derived ES cells). Add pen/strep to the ES cell media in this protocol. 129 ES cells grow more vigorously than C57BL/6 ES cells and can be plated at lower densities than C57BL/6 cells when clones are thawed.
5. Remove 96-well plate from −80°C freezer.
6. Add 100 μl of 37°C ES culture media to each well identified for expansion.
7. Carefully float the plate on the water in the incubator for 2 min. Check to see if thawed completely. If not, put back for 1 min and check again.
8. Transfer entire contents of thawed well to one well in 24-well MEF plate containing 2 ml ES culture media or 48-well containing 1 ml media.
9. Visually inspect empty wells under microscope to confirm cell transfer.
10. Place MEF plate with ES cells in incubator.
11. On days 2–4, replace the media with ES cell media.
12. ES cell colonies should appear by day 3.
13. If ES cell colonies do not appear by day 5, trypsinize the well contents and add an equal volume of ES cell media. A few cells may grow up in a few more days. The trypsin treatment will break up any small colonies and spread them out. Check

lab records to determine if the well contained cells at the time of the Split/Freeze.

14. When ready on day 3 or 4, passage the ES cells to a 6-well plate or a 12-well MEF plate:
    (a) Replace media on MEF plate with ES cell media.
    (b) Aspirate media from 24- or 48-well plate.
    (c) Wash with D-PBS and aspirate.
    (d) Add 1 or 0.5 ml trypsin and place in a 37°C, 5% CO_2, humidified incubator for 5 min.
    (e) Add equal volume of ES cell media, resuspend ES cells and transfer directly to MEF plate.
15. On days 2–3, replace the media with ES cell media.
16. When ES cells are ready, passage to larger wells and 60 mm dishes until five cryovials of cells can be frozen back according to Subheading 14.4.7.

14.4.15. ES Cell Subcloning

The purpose of this protocol is to prepare subclones of ES cell clones or ES cell lines. Subclones of aneuploid targeted clones may be euploid and can be used to generate germline ES cell chimeras [6]. Occasionally ES cell clones may contain a mixture of wild type and gene-targeted ES cells. Subcloning can be used to purify gene-targeted clones. In addition, subclones of parental ES cell clones allow for the recovery of pluripotent characteristics that have been lost with increasing passage number [16].

1. Prepare three 10 cm dishes of mitotically activated feeder cells.
2. Thaw one vial ES cells frozen at 4–5 × 10^6 cells/ml onto a 60 mm dish of feeder cells and culture until ready to split (2 days after thaw).
3. Trypsinize ES cells and use serial dilutions to prepare a suspension of 3,000 ES cells/ml in culture medium. Ensure that the cells are evenly resuspended before plating.
4. Plate out the cells by dispensing 1 ml of cell suspension to each of the three 10 cm dishes of feeder cells containing 9 ml culture medium.
5. Incubate at 37°C, and renew the culture media daily until the colonies are ready to pick.
6. When the colonies are of a suitable size (usually 5–7 days later), pick colonies onto one or two 96-well plates containing feeder cells and 100 μl culture medium (*see* Subheading 14.4.10).
7. Incubate at 37°C, and renew the culture media daily.

8. When the clones grow up perform a split freeze (*see* Subheading 14.4.11).
9. Continue to incubate the DNA replicate plate at 37°C, and renew the culture media daily (*see* Subheading 14.4.12).
10. After the ES cells have grown to high density, DNA can be extracted from the cells for genotyping (*see* Subheading 14.4.13).
11. Subclones are genotyped to identify correctly targeted clones. After expansion of ES cells and chromosome counting, targeted euploid clones are used to generate ES cell-mouse chimeras for germline transmission of the targeted gene.

14.4.16. Preparing Chromosome Spreads

The purpose of this protocol is to produce chromosome spreads so that ES cells can be scored as euploid or aneuploid based on the number of chromosomes present in the cells.

1. Plate 5×10^6 ES cells in 10 ml media onto a 100 mm gelatin coated tissue culture dish.
2. Place the dish in incubator overnight.
3. Add 20 μl colcemid to cells in exponential growth phase and return to incubator for 2 h.
4. Prepare a single cell suspension by trypsin treatment:
   (a) Pipette 10 ml spent media from the dishes into 15 ml tubes, and rinse dishes with 5 ml D-PBS. Add the D-PBS to the tubes. Centrifuge 5 min at 200 × g, and aspirate supernatant leaving 4 ml in the tube.
   (b) Add 4 ml trypsin to dish and place in incubator for 5 min.
   (c) Using a Pasteur pipette, triturate and transfer the cells into 15 ml tube containing 4 ml media.
   (d) Centrifuge the cell suspension for 5 min at 200 × g.
   (e) Aspirate the supernatant, leaving a small volume of medium in the tube. Loosen the cell pellet by flicking the tube.
5. Gently resuspend cells in 10 ml of 37°C hypotonic solution (0.075 M KCl) and incubate in 37°C waterbath for 15 min.
6. Centrifuge cells at 200 × g for 5 min.
7. Aspirate supernatant, leaving a small volume of media (about twice the volume of the packed cell pellet).
8. Flick the tube to break up the cell pellet.
9. Add 1 ml of fixative dropwise with a Pasteur pipette while constantly agitating the tube to avoid cell clumping. Add additional 6 ml fixative more quickly.

10. Incubate on ice for 30–240 min. Do not leave on ice overnight.
11. Wash the cells twice with 10 ml cold fixative, centrifuging at 200 × *g* for 5 min.
12. Aspirate the supernatant, leaving a small volume of media (about twice the volume of the packed cell pellet).
13. Flick the tube to break up the cell pellet.
14. Resuspend in 1 ml cold fixative for immediate use or resuspend in 10 ml and store at −20°C. Fixed cells can be stored indefinitely at −20°C. Resuspend stored preparations in fresh fixative just before dropping slides.
15. Drop the fixed swollen cells onto humid glass microscope slides.
16. Incubate slides overnight on a stack of ten wet paper towels at room temperature, on the lab bench, uncovered.
17. Slides are stained with Giemsa Stain:
    (a) Incubate the slides in 4% Giemsa stain in Gurr's Buffer for 15 min.
    (b) Place the slides in Gurr's Buffer for 10 min.
    (c) Place the slides in distilled water for 5 min and then air dry the slides.
18. Photograph chromosome spreads at 1,000× magnification using a research grade microscope with a 100× oil objective (Fig. 14.3).
19. Use an electronic colony counting pen to count chromosomes in each spread.
20. Count 20 spreads per ES cell clone. Euploid spreads contain 40 chromosomes with normal, acrocentric or telocentric morphology. An ES cell line is suitable for microinjection if 12 or more spreads contain 40 chromosomes [6].

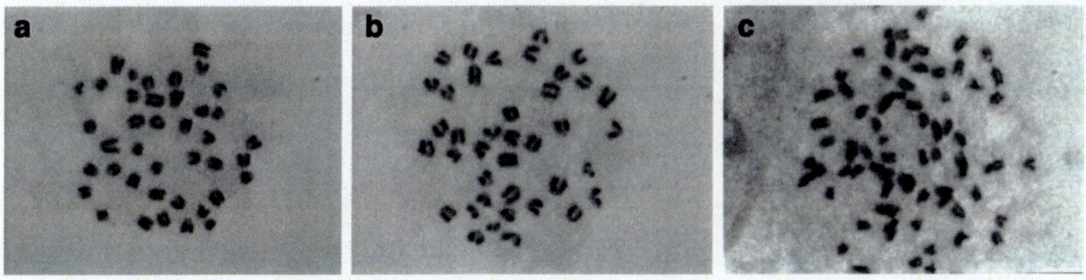

Fig. 14.3 ES cell chromosome spreads. Chromosome counts are used to determine if further work on a clone will yield results. (**a**) Euploid mouse ES cell chromosome spread that contains 40 chromosomes. Note that the chromosomes are not identified as for a karyotype, they are simply counted. (**b**) Aneuploid spread that contains only 39 chromosomes. It is possible for X:0 ES cells to contribute to female germline chimeras. If needed, the next step for this ES cell clone is PCR genotyping assay determine if it is X:Y or X:0. (**c**) Aneuploid spread that contains 2n chromosomes. Tetraploid ES cell clones should not be used to produce ES cell-mouse chimeras because they will not transmit the ES cell genome through the germline.

14.4.17. DNA Extraction for 100 mm ES Cell Dishes

The purpose of this method is to obtain enough high-quality genomic DNA from an ES cell clone to test the structure of the targeted gene with multiple restriction enzymes on a Southern blot.

1. Seed a gelatin-coated 100 mm dish with at least 5×10^6 cells.
2. Place the dishes in 5% CO_2, 37°C, humidified incubator overnight. Cells can be cultured for 1 or 2 days before collection. Change media daily.
3. Harvest cells from dish by trypsinization (*see* Subheading 14.4.5, **steps 2–7**).
4. Aspirate supernatant until almost dry.
5. Place cell pellet at −20°C overnight or until needed. Cells from one 100 mm dish should provide enough DNA for several analyses with several restriction enzymes on Southern blots.
6. Add 600 ml Lysis Buffer (10 mM Tris-HCl pH 8, 5 mM EDTA, 0.4 M NaCl).
7. Pipette up and down to obtain a uniform suspension.
8. Add SDS to 0.2%, invert gently to mix (12 µl of 10%SDS).
9. Add Proteinase K to 500 µg/ml (12.2 ml of 25 mg/ml stock).
10. Incubate overnight at 55°C with rocking.
11. Extract twice with phenol/chloroform/isoamyl alcohol (25:24:1).
12. Extract once with chloroform/isoamyl alcohol (24:1).
13. Add 2.5 volumes of 100% ethanol to aqueous phase.
14. Invert to mix.
15. Spool DNA onto flame-sealed glass Pasteur pipette.
16. Rinse with 70% ethanol.
17. Let air dry briefly and then break off tip into 1.5 ml tube and add 300 ml of TE.
18. Heat at 65°C for 30 min with gentle mixing by inversion. Store at 4°C.
19. Quantitate resuspended DNA (OD 260/280).
20. Digest 5–10 µg of DNA for each lane of a Southern blot.

14.4.18. Qualifying Fetal Bovine Serumfor ES Cell Culture

The purpose of this test is to identify lots of FBS that will support vigorous growth of ES cells without inducing differentiation and to maintain ES cells in a pluripotent state. We have observed even tissue culture tested lots of FBS can induce differentiation, or even be toxic, to ES cells (Fig. 14.4). Any ES cell that has generated

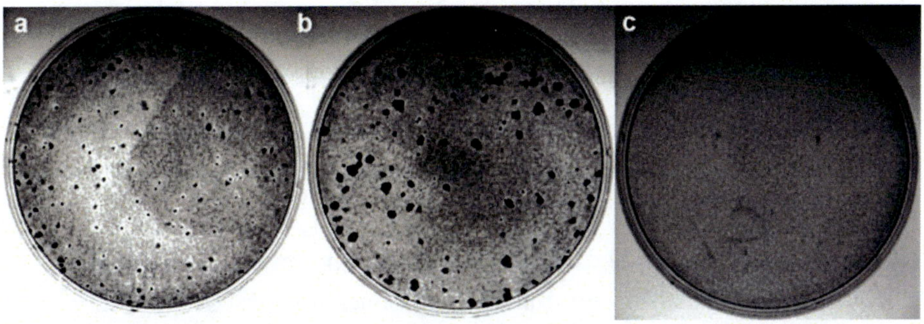

Fig. 14.4 ES cell colonies from serum test. (a) ES colonies growing robustly in serum suitable for ES culture. Most of the colonies have characteristic undifferentiated morphology. (b) ES colonies are growing, but many have differentiated in this serum. (c) Only very few, and very small colonies grew in this lot of serum. The serum used in (c) was completely unsuitable for ES cell culture.

germline gene-targeted clones, such as the R1 ES cell line [16], can be used in this test.

1. Obtain FBS samples (50–100 ml) from five or more vendors. Ask for most recent lot, storage conditions and estimated purchase price. If desired, aliquots of pre-qualified serum for ES cell culture can be sampled. However, in our experience with more than 50 lots of serum, pre-qualified serum has never been the best for ES cell culture. Reserve a 12 month supply of serum from each lot in the test. One to two 500 ml bottles of serum should be enough to target one gene.
2. Thaw one vial of R1 ES cells frozen at 4–5 $\times$ 10^6 cells/ml onto a 60 mm dish of feeder cells and culture until ready to split (2 days after thaw).
3. Prepare six 60 mm dishes of feeder cells for each serum lot to be tested and six more dishes for the control serum.
4. Prepare serum test medium for each lot of serum: 14 ml of ES media containing 30% test serum and a second aliquot of 14 ml of ES media containing 15% test serum. In addition, prepare control media with serum known to maintain ES cell pluripotency.
5. Trypsinize ES cells and resuspend at 3 $\times$ 10^3 ES cells/ml in ES cell media base that contains all ES medium ingredients except for serum.
6. Add 3 $\times$ 10^3 ES cells (1 ml) to each aliquot of 14 ml of media.
7. Aspirate media from 60 mm dishes of feeder cells and add 5 ml of serum test media to three dishes. Mix the media gently just before addition to resuspend the ES cells. At the end of this step triplicate 60 mm dishes of ES

cells for two serum concentrations will be produced for each serum lot.

8. Place the 60 mm dishes in the incubator and do not disturb them for 7–10 days.
9. At the end of the incubation period, aspirate the media from the dishes, rinse them once with PBS and fix the cells in methanol for 12 min. Stain the cells with Giemsa stain (1:10) for 10 min, rinse with water and air dry.
10. Score the ES cell clones on the dishes. Count the total number of clones on each dish and calculate the plating efficiency (clones per 1,000 ES cells) for each serum test media. Score the ES cell clones for morphology as they are counted: determine the number of ES cell clones that are too small and the number of clones that have differentiated (borders of clones are not sharp, typical dome shaped morphology is absent). Calculate the number of very small/ differentiated clones per 1,000 ES cells.
11. Serum should not be toxic at high concentrations (30%). Plating efficiency should be similar between 15% and 30% serum conditions. Serum should encourage the vigorous growth of ES cell colonies. Few ES cell clones should appear differentiated. The great majority of colonies should have classical dome-shaped appearance with sharply defined borders.
12. Determine which serum lot(s) has the best plating efficiency without demonstrated toxicity at 30% and has the least differentiation. Contact the vendor to purchase the serum. If two lots have similar biological characteristics, purchase the least expensive serum.

14.5 Troubleshooting

Gene targeting in mouse ES cells requires the successful culture and manipulation of cells in culture over a prolonged period of time. At any step, problems may arise that prevent the successful conclusion of the project or require that the gene targeting process begin anew. Most problems can be avoided by rigorous quality control. Others are less easily resolved.

14.5.1. Parental ES Cells Fail to Grow After Thaw

1. ES cells are frozen in media containing DMSO as a cryoprotective agent. Upon thawing the cells must be washed to eliminate DMSO from the culture media.
2. Be sure there are no selective agents (G418, hygromycin, puromycin, etc.) in the media used to culture the wild type parental cells.

14.5.2. ES Cells Do Not Grow Robustly

1. Be sure to use only ES-qualified materials to prepare ES culture media. ES cell culture media should be used within 2 weeks of preparation.
2. Be sure to use the correct concentration of 2-ME, and use the 100× diluted stock within 2 weeks of preparation.
3. Use feeder layers within 2 days of mitotic inactivation when thawing vials of ES cells, and within 5 days of inactivation for routine ES cell culture.
4. It is preferable to mitotically inactivate feeder cells by irradiation with 6,000 rads. Some lines of feeder cells do not support robust growth of ES cells after mitomycin-C treatment although they perform well after irradiation.

14.5.3. Contamination After Electroporation

If your ES cell cultures are contaminated with undesirable microbial agent(s) after the electroporation step and no other cells cultures are contaminated, then the targeting vector DNA is the source of contamination. To control for this possibility, always culture a sample of unmanipulated ES cells used in the electroporation to demonstrate the cells were free of contamination before exposure to the DNA.

14.5.4. Cells Do Not Die in Selection step

1. If the ES cell line in question has not been characterized for the selective agent in question (G418, hygromycin, puromycin, etc.) it may be necessary to perform a drug sensitivity curve to determine the minimum drug concentration required to kill wild type ES cells.
2. Check the label on the bottle of the selective agent – gentamicin is not the same as geneticin.

14.5.5. Too Few Colonies After Drug Selection

1. Confirm that expression of the drug selection marker (i.e. neomycin phosphotransferase) is under the control of a strongly expressed promoter in mouse ES cells [17]. Some experimental designs rely on expression from an endogenous promoter through IRES sequences, and these will fail to confer drug resistance if the endogenous gene is not expressed highly enough in ES cells.
2. On occasions we observe very few colonies after drug selection when targeting vectors include the diphtheria toxin A-chain (DT-A) negative selection cassette [18]. In these cases, ectopic or constitutive DT-A cassette expression kills the ES cells. This can be remedied by making sure that the targeting vector is completely linearized prior to electroporation or by removing the DT-A cassette from the vector.
3. Very few drug-resistant ES cell colonies will grow if the targeting vector DNA is contaminated with endotoxins or oxidized phenol. Use plasmid DNA purification kits that are

designed to produce endotoxin-free DNA. Use fresh phenol/chloroform for DNA extractions when preparing targeting vector DNA for electroporation.

4. Confirm targeting vector DNA concentration by spectrophotometer measurements and agarose gel electrophoresis. Targeting vector DNA should produce a single band of the correct molecular weight, and the band intensity should be correlated with a known DNA mass standard on the same gel.

14.5.6. Mosaic Colonies After Transient Recombinase Expression

Cells plated at clonal density before Cre and FLP recombinase can act on genomic DNA will result in mosaic ES cell colonies that are not clonal [19, 20]. Allow 48 h after electroporation with recombinase expression plasmid and then trypsinize and re-plate ES cells at a clonal density (e.g. 3,000 cells per 100 mm dish).

14.5.7. Gene-Targeted Clones Are Not Identified

1. Confirm the targeted gene is not required for survival of ES cells. XY ES cells are sensitive to gene dosage effects when genes on the X chromosome are inactivated [21].
2. Confirm the DNA used in the construction of the targeting vector is isogenic with the ES cells used in the experiment.
3. Some genetic loci are unavailable for homologous recombination.
4. Confirm the arms of homology are long enough to allow efficient homologous recombination. Total homology of 6–10 kb in the recombination arms is generally sufficient.
5. We have observed a decrease in homologous recombination in experiments where the genetic region deleted by homologous recombination with the targeting vector is more than 5 kb.
6. We recommend you consult the chapter in this book on the design of targeting vectors.

14.5.8. Mixed Populations of Cells in Picked Clones

1. Only pick ES cell colonies that are well separated from other colonies, and have a smooth, domed, circular or oval shape. Avoid irregularly shaped colonies – they may be two colonies that have grown together.
2. Mark each well on the lid as you place colonies in the wells, so you do not place more than one colony in a single well. Observe 96 well plates with a microscope after clone picking before trypsin treatment. Any wells that contain two or more colonies should be aspirated and cells discarded.
3. Use a separate tip to pick each colony.
4. Do not cross-contaminate the 96 wells during media replacement and trypsinization procedures. Do not triturate the trypsinized cells so vigorously as to generate aerosols!

14.5.9. Mixed Populations of Cells in Expanded Clones

1. Be careful to not cross-contaminate the cultures during expansion. Use separate pipettes for each clone and do not ever return a pipette to a reagent or media bottle after it has been used to dispense material to a container that has cells in it.
2. Be sure to label all tubes and dishes before you place any cells in them.

14.5.10. Failure to Recover ES Cell Clones from Frozen 96-Well Plates

1. Examine the scoring sheet and confirm there were ES cells in the particular well at the time of the split/freeze.
2. Confirm the 96-well plate has not been in storage at −80°C for more than 6 months.
3. ES cells derived from C57BL/6 mice recover more slowly from the thaw than do 129-derived cells. Use 48-well feeder plates instead of 24-well plates for C57BL/6-derived ES cell clone expansions. If the 129-derived clones have been in frozen storage for more than 6 months, use a 48-well plate of feeders for the thaw.
4. Trypsinize and replate well contents 5 days after thawing 96-well plates, even if ES cells are not visible or only very small colonies appear. The cells will often start dividing again after trypsin treatment. Small colonies growing near the plate edges are difficult to see and even very small colonies will expand, given enough time.

14.5.11. All Expanded Clones Are Aneuploid

1. Confirm your parental cells are euploid and germline competent before using them in a gene-targeting experiment.
2. Confirm your parental cells have not been in culture for an excessive number of passages before using them in a gene-targeting experiment.
3. Some parental ES cells lines are more chromosomally stable than others. For example, it is often necessary to thaw more clones from the C57BL/6 Bruce4 ES cell line than from the 129 derived R1 ES cell line [6].

14.5.12. Clones Did Not Produce Any Germline ES Cell-Chimeras Following Blastocyst Injection

1. Confirm your parental cells are euploid and germline competent before using them in a gene-targeting experiment.
2. Confirm your clones have euploid chromosome counts and chromosomes have normal morphology. Remember ES cells can harbor secondary mutations that will not be detected by chromosome counts and render them incapable of colonizing the host blastocyst or from transmitting through the germline. Microinject at least three euploid clones into blastocysts for each gene targeting experiment.
3. Confirm that the lowest available passage of cells is used for blastocyst injection, and that cells have a normal, undifferentiated morphology.

4. Confirm the cells were growing robustly when prepared for blastocyst injection. Culture ES cells in Resgro immediately before blastocyst injection. Thaw ES cell clones and culture in Resgro for 3–4 days before microinjection. Passage the clone 1–2 days before the injection, seeding it at 2–4 × 10^6 cells per 60-mm dish for injection after 1 day or 1.5–2 × 10^6 for injection after 2 days.
5. See the chapter elsewhere in this volume on combining ES cells with host blastocysts for the production of ES cell-mouse chimeras.

References

1. Kim I, Saunders TL, Morrison SJ (2007) Sox17 dependence distinguishes the transcriptional regulation of fetal from adult hematopoietic stem cells. Cell 130:470–483
2. Lapinski PE, Bauler TJ, Brown EJ, Hughes ED, Saunders TL, King PD (2007) Generation of mice with a conditional allele of the p120 Ras GTPase-activating protein. Genesis 45:762–767
3. Huang X, Fu Y, Charbeneau RA, Saunders TL, Taylor DK, Hankenson KD, Russell MW, D'Alecy LG, Neubig RR (2006) Pleiotropic phenotype of a genomic knock-in of an RGS-insensitive G184S Gnai2 allele. Mol Cell Biol 26:6870–6879
4. Wada M, Saunders TL, Morrow J, Milne GL, Walker KP, Dey SK, Brock TG, Opp MR, Aronoff DM, Smith WL (2009) Two pathways for cyclooxygenase-2 protein degradation in vivo. J Biol Chem 284:30742–30753
5. Capecchi MR (1994) Targeted gene replacement. Sci Am 270:52–59
6. Hughes ED, Qu YY, Genik SJ, Lyons RH, Pacheco CD, Lieberman AP, Samuelson LC, Nasonkin IO, Camper SA, Van Keuren ML, Saunders TL (2007) Genetic variation in C57BL/6 ES cell lines and genetic instability in the Bruce4 C57BL/6 ES cell line. Mamm Genome 18:549–558
7. Liu X, Wu H, Loring J, Hormuzdi S, Disteche CM, Bornstein P, Jaenisch R (1997) Trisomy eight in ES cells is a common potential problem in gene targeting and interferes with germ line transmission. Dev Dyn 209:85–91
8. Sugawara A, Goto K, Sotomaru Y, Sofuni T, Ito T (2006) Current status of chromosomal abnormalities in mouse embryonic stem cell lines used in Japan. Comp Med 56:31–34
9. Tucker KL, Wang Y, Dausman J, Jaenisch R (1997) A transgenic mouse strain expressing four drug-selectable marker genes. Nucleic Acids Res 25:3745–3746
10. Linnell ER, Lerner CP, Johnson KA, Leach CA, Ulrich TR, Rafferty WC, Simpson EM (2001) Transgenic mice for the preparation of puromycin-resistant primary embryonic fibroblast feederlayers for embryonic stem cell selection. Mamm Genome 12:169–171
11. O'Connell RC, Wittler RG, Faber JE (1964) Aerosols as a source of widespread mycoplasma contamination of tissue cultures. Appl Microbiol 12:337–342
12. Longo L, Bygrave A, Grosveld FG, Pandolfi PP (1997) The chromosome make-up of mouse embryonic stem cells is predictive of somatic and germ cell chimaerism. Transgenic Res 6:321–328
13. Kumar RA, Chan KL, Wong AH, Little KQ, Rajcan-Separovic E, Abrahams BS, Simpson EM (2004) Unexpected embryonic stem (ES) cell mutations represent a concern in gene targeting: lessons from "fierce" mice. Genesis 38:51–57
14. Martin GR, Evans MJ (1975) Differentiation of clonal lines of teratocarcinoma cells: formation of embryoid bodies in vitro. Proc Natl Acad Sci USA 72:1441–1445
15. Chen YT, Bradley A (2000) A new positive/negative selectable marker, puDeltatk, for use in embryonic stem cells. Genesis 28:31–35
16. Nagy A, Rossant J, Nagy R, Abramow-Newerly W, Roder JC (1993) Derivation of completely cell culture-derived mice from early passage embryonic stem cells. Proc Natl Acad Sci USA 90:8424–8428
17. Tybulewicz VLJ, Crawford CE, Jackson PK, Bronson PT, Mulligan RC (1991) Neonatal lethality and lymphopenia in Mice with a homozygous disruption of the c-abl proto-oncogene. Cell 65:1153–1163
18. Yagi T, Nada S, Watanabe N, Tamemoto H, Kohmura N, Ikawa Y, Aizawa S (1993) A novel negative selection for homologous recombinants using diphtheria toxin A fragment gene. Anal Biochem 214:77–86

19. Schaft J, Ashery-Padan R, van der Hoeven F, Gruss P, Stewart AF (2001) Efficient FLP recombination in mouse ES cells and oocytes. Genesis 31:6–10
20. Yu H, Kessler J, Shen J (2000) Heterogeneous populations of ES cells in the generation of a floxed Presenilin-1 allele. Genesis 26:5–8
21. Yu RN, Ito M, Saunders TL, Camper SA, Jameson JL (1998) Role of Ahch in gonadal development and gametogenesis. Nat Genet 20:353–357
22. Nagy A, Gertsenstein M, Vintersten K, Behringer R (2003) Manipulating the mouse embryo: a laboratory manual, 3rd edn. Cold Spring Harbor Laboratory Press, Cold Spring Harbor, NY
23. Ramirez-Solis R, Rivera-Perez J, Wallace JD, Wims M, Zheng H, Bradley A (1992) Genomic DNA microextraction: a method to screen numerous samples. Anal Biochem 201: 331–335

Chapter 15

The Importance of Mouse ES Cell Line Selection

Wojtek Auerbach and Anna B. Auerbach

Abstract

The choice of a specific parental ES cell line used to create a genetically modified mouse has critical impact on the overall success of the project – affecting costs, complexity of effort, and time to efficient project completion. Despite the importance of making a thoughtful choice, many people default to employing a familiar cell line previously used in their laboratory or they are limited to the cell lines available from the service facility at their Institution. With careful consideration of a few key experimental parameters, the investigator can easily avoid unnecessary mistakes.

Historically, substrains of 129 strain mice have been used to create the majority of parental ES cell lines now in common use. Often these cell lines have been expanded many times and shared among colleagues, rather than being obtained directly from the laboratory in which they were originally established. Although this was a common practice in the early days of ES cell technology, avenues are now in place to assure optimal cell line history, health, genetic integrity, and performance. Methods for establishing new ES cell lines have greatly improved. Well-validated ES cell lines are now commercially available from a variety of genetic backgrounds beyond the 129-strain. In this chapter, we will discuss critical factors to consider, when choosing an ES cell line for a project.

Abbreviations

ES cells	Embryonic stem cells
MEF	Mouse embryonic fibroblasts
BRL cells	Buffalo rat liver cells
LIF	Leukemia inhibitory factor
FBS	Fetal bovine serum
KOSR	Knockout serum replacement
BMP4	Bone morphogenetic protein 4
GSK3	Glycogen synthase kinase 3
MEK	MAP kinase
FGF	Fibroblast growth factor
GLT	Germline transmission
ATCC	American Type Culture Collection
BAC	Bacterial artificial chromosome

S. Pease and T.L. Saunders (eds.), *Advanced Protocols for Animal Transgenesis*, Springer Protocols,
DOI 10.1007/978-3-642-20792-1_15,

SW	Swiss Webster
DMEM	Dulbecco's Modified Eagle Medium
IKMC	International Knockout Mouse Consortium
KOMP	NIH Knockout Mouse Project
EUCOMM	European Union Conditional Mouse Mutagenesis Program
NorCOMM	North American Conditional Mouse Mutagenesis
NEAA	Non-essential amino acids
Pen-Strep	Penicillin and streptomycin
PE	Plating efficiency
TC	Tissue culture
TE	Targeting efficiency
EP	Electroporation

15.1 Introduction

The first ES cell line was derived in 1981 by researchers experienced in culturing teratocarcinoma cell lines [1]. The authors selected the 129 mouse strain, since it shows a frequent occurrence of testicular teratomas. They hypothesized that this tendency would increase the chance of establishing an ES cell line. Soon after this account, several more ES cell lines, mostly from 129 strains of mice, were obtained. [2–5] The 129 mouse strain was not widely available, therefore scientists began to maintain their own colonies of mice, often selecting for specific traits or even breeding with mice from other strains. This husbandry led to genetic variations between animals from different centers. Eventually the 129-strain was officially divided into several substrains [6, 7].

ES cell lines were also derived from mouse strains other than 129. These included cell lines from C57BL/6 [8–10] and BALB/c mice [11]. The use of these specialized cell lines for gene targeting was limited mainly to the founder laboratory.

Initially, ES cells were cultured on top of a monolayer of growth-inactivated mouse embryonic fibroblasts (MEFs) or on the immortalized fibroblast cell line STO, as feeder cells. It has been shown that the addition of media conditioned by teratocarcinoma cells [2], buffalo rat liver (BRL) cells [12] or MEFs helped to maintain ES cells in an undifferentiated state. In 1988, the critical component of the conditioned medium was identified to be the cytokine, Leukemia Inhibitory Factor (LIF) [13, 14]. It was determined that ES cells could be grown in the presence of purified LIF, on gelatin, without feeder cells, although long-term culture without feeders required a period of adaptation.

The most common way of culturing ES cells is still to use a feeder-based system with medium containing LIF and ES cell-tested fetal bovine serum (FBS). These culture conditions, although ideal for 129-strain ES cells, are less supportive for ES cell lines derived from other mouse strains, as evidenced by their accelerated loss of undifferentiated characteristics compared to 129 cell lines [15].

In order to obtain proper culture conditions for ES cells, it is important to select an FBS lot suitable for support and growth of undifferentiated ES cells. FBS provides nutritive supplementation of the basal growth medium and the attachment factors. Unfortunately, FBS may also contain components that can promote differentiation of ES cells. Since FBS is produced from a dynamic biological source, the concentration of all of these components differs from lot to lot. Each lot of FBS must be screened before it is used as a component of ES cell culture medium. Ideally, for best results, each lot of FBS should be tested in conjunction with the specific ES cells of intended use. Commercially "pretested" sera are generally screened with ES cells from 129 strain. A lot that may be good for the tested ES cells might not be ideal for other ES cell lines. The leading sera companies are now providing more characterized products. Sera from each lot are tested for concentration of many components. This allows users to select for screening sera with a characterized panel that resembles those lots used previously with success.

Two trends emerged in efforts to facilitate development and maintenance of ES cells from strains other than 129; the use of conditioned media [16, 17] and serum-free medium [18, 19]. One example of the former is a medium conditioned by rabbit fibroblasts expressing recombinant rabbit LIF. The latter is now available in the form of complete serum-free medium like ESGRO (Millipore) or N2B27-based media like GS2-M (Stem Cells) or by substituting FBS for Serum Replacement – KOSR (Invitrogen) in standard ES cell medium. The benefit of serum-free media is that it does not contain any factors that may promote differentiation. The disadvantages are lack of attachment factors and often observed slower growth of cultured ES cells.

In addition to LIF other proteins like BMP4 [18] or Wnt [20] or small molecules inhibitors of GSK3 [21], MEK [22, 23] or FGF receptor inhibitor [24] have been shown to prevent differentiation of ES cells. Some of those have synergistic effect when combined with LIF [20]. Thus, there are now more reliable methods for derivation and maintenance ES cells from practically any mouse strain. The decision to either derive de novo ES cell lines or obtain an established ES cell line from academic laboratory, or commercial supplier should be taken on a case-by-case basis. The decision will be influenced by the availability of in-house technical experience, budget, and the availability and quality of an ES cell line from a particular background as well as planned future use for the cell line.

15.2 Most Commonly Used ES Cell Lines, Their Sources, and Sources of Mice from the Same Genetic Background

Many different ES cell lines have been developed since 1981, more so in recent years. It would be impossible to describe all of them. In this section, we describe some of the most often used ES cell lines (Table 15.1). This includes a few of historic importance that may now only be available at high passage numbers and thus not recommended for further use.

15.2.1. ES Cell Lines Derived from 129-Strain of Mice

The most widely used for gene-targeting experiments are ES cell lines derived from the 129-mouse strain. This was the source of the first ES cell line ever established [1]. The main reason this strain has been the most widely used to date is because of the relative ease with which it is possible to establish new ES cell lines from this strain, as well as to maintain them undifferentiated, in culture. Such cell lines are the "work horse" of most gene-targeting Facilities. However, it has been demonstrated that numerous substrains of this strain exhibit demonstrable physiological and genetic differences [6, 25]. Thus, discrepancies due to different genetic backgrounds may arise when comparing phenotypes of genes targeted in different 129-derived ES cell lines. This mixed genetic background could complicate gene-targeting experiments, when constructs and ES cells are not derived from the same 129 substrain. In addition to some intentional inbreeding to increase the rate of testicular teratocarcinomas, or outbreeding to C3HeB strain to introduce agouti coat color, there was also undocumented breeding to an unknown strain causing the most widely available 129/SvJ (now 129X1/SvJ) to be genetically significantly different from the other 129 substrains and recently more accurately classified as a recombinant congenic strain. This led to a new classification of 129 strain, dividing different substrains into three groups: parental (P), Steel (S), and contaminated (X). Those substrains were genetically separated before the various ES cell lines were established. Some of them were kept in closed colonies in individual researchers' Institutions and were not commercially available. The complex and previously not entirely recognized situation led to erroneous breeding of some knock out mice obtained by manipulation of ES cells from one 129 substrain to mice from a different substrain of 129.

Below is a description of most commonly used ES cell lines. Also *see* Table 15.1 for more information on availability of cells

Table 15.1
Most commonly used ES cell lines. Also provided are availability of cells, and availability of the mice from the same genetic background for breeding as well as recommended embryo donor strain for chimera production

Name	Genetic background	Availability	Source of mice	Coat color	References	Host	Comments
CCE	129S5/SvEv		Taconic Farms	White-bellied agouti	[4]	C57Bl/6	
D3	129S2/SvPas	ATCC	Pasteur Institute	White-bellied agouti	[5]	C57Bl/6	Available only at high passage
E14	129P2/OlaHsd		Harlan	Pink-eyed, chinchilla	[26]	C57Bl/6	Contaminated Bovine mycoplasma; more creamy chimeras, better contribution.
E14TG2A	129P2/OlaHsd		Harlan	Pink-eyed, chinchilla	[27]	C57Bl/6	HPRT-deficient subline of E14
E14Tg2a.4	129P2/OlaHsd	Bay Genomics	Harlan	Pink-eyed, chinchilla	[43]	C57Bl/6, B6D2/F1/J	Feeders-independent subclone of E14G2a, used in gene trap project by BayGenomics
J1	129S4/SvJae	ATCC	Jackson Laboratories	White-bellied agouti	[29]	C57Bl/6	
AB1	129S7/SvEvBrd	A. Bradley, Sanger Institute	A. Bradley, Sanger Institute, Not available from commercial supplier	White-bellied agouti	[28]	C57Bl/6	
AB2.1, AB2.2	129S7/SvEvBrd	ATCC (AB2.2)		White-bellied agouti	[51]	C57Bl/6	
R1	129X1/SvJ x 129S1/Sv	A. Nagy (MSHRI), ATCC- late passage cells	Jackson Laboratories	Agouti in F0	[30]	C57Bl/6 for blast inj. and albino	Segregates in F1 agouti and chinchilla and further in F2. 129X1/SvJ is contaminated inbred strain

(continued)

Table 15.1 (continued)

Name	Genetic background	Availability	Source of mice	Coat color	References	Host	Comments
		and R1E subclone,				outbred for agg.	
AK7.1	129S4/SvJae	P. Soriano (MSH)	Jackson Laboratories	White-bellied agouti	[33]	C57Bl/6	Employed primarily in Soriano's gene trap experiments
RW4	129X1/SvJ	ATCC, Primogenix, Jax	Jackson Laboratories	agouti	[32]	C57Bl/6	129X1/SvJ is contaminated inbred strain
CJ7	129S1/SvImJ	T. Gridley (Jakson Lab.)	Jackson Laboratories	White-bellied agouti	[31]	C57Bl/6	
W4	129S6/SvEvTac	Taconic Farms	Taconic Farms	White-bellied agouti	[15]	C57Bl/6	
G4	129S6/SvEv x C57Bl/6N	A. Nagy (MSHRI	Taconic Farms (129) and CRL (B6)	Agouti	[37])	C57Bl/6 for blast. inj. and albino outbred for agg.	Heterozygous in agouti (A) locus
V6.5	C57Bl/6 x 129S4/SvJae	Open Biosystems	Jackson Laboratories	Agouti	[35]	B6D2/F2	Heterozygous in agouti (A) locus
V17.2	BALB/c x 129S4/SvJae	Open Biosystems	Jackson Laboratories	Agouti	[35]	B6D2/F2; for blast. and outbred albino for eight-cell inj.	Heterozygous in agouti (A) locus

VGF1	129S6/SvEv x C57Bl/6N	Parental line not available	Taconic Farms	Agouti	[36]	C57Bl/6 and albino outbred for 8-cell inj.	Heterozygous in agouti (A) locus
Bruce4	C57Bl/6-*Thy1.1*	Stewart C. (NCI) or Ozgene limited	C. Stewart	Black	[9]	BALB/c, B6 (Cg)-Tyr^{c-2J}/J	Congenic line (not pure inbred)
CMTI-2	C57Bl/6-*Thy1.1*	Millipore	C. Stewart	Black	http://www.millipore.com	BALB/c, B6 (Cg)-Tyr^{c-2J}/J	Bruce4 line adapted to grow in feeder-free and serum-free ESGRO complete medium
HGTC-8	C57Bl/6J	NHGRI	Jackson Laboratories	Black	[19]	BALB/c	
B6(Cg)-Tyr^{c-2J}/J-PRX-B6-albino	B6(Cg)-Tyrc-2J/J	Jackson Laboratories Millipore	Jackson Laboratories	Albino	http://jaxmice.jax.org/	C57Bl/6	Developed by R. Wesselschmidt, Primogenix Inc.; passage 10 Distributed by Millipore as B6-White
Lex3.13	C57Bl/6N	Parental line not available	TIGM closed colony; NCI	Black	[40]	C57Bl/6-Tyr^{c-Brd} or B6 (Cg)-Tyr^{c-2J}/J	TIGM gene-trapped clones; passages 18 and higher
C2	C57Bl/6N	A. Nagy (MSRI)	Taconic Farms	Black	[42]	Albino outbred for agg. and 8-cell inj.	Possible C2J and BALB/c as host blastocysts. Recommended medium RESGRO; used by NorCOMM
VGB6	C57Bl/6N	Parental line not available	Taconic Farms	Black	[17]	BALB/c, B6 (Cg)-Tyr^{c-2J}/J; outbred albino for 8-cell inj.	Regeneron-originated KOMP gene targeting

(continued)

Table 15.1 (continued)

Name	Genetic background	Availability	Source of mice	Coat color	References	Host	Comments
JM8F6	C57Bl/6N	KOMP repository	CRL	Black	[41]	C57Bl/6-Tyr^{c-Brd}	F – feeder dependent, subclone of JM8 line; used for EUCOMM and conditional gene targeting KOMP
JM8N4	C57Bl/6N	KOMP repository	CRL	Black	[41]	C57Bl/6-Tyr^{c-Brd}	N – Feeder independent subclone of JM8 line
JM8A3	C57Bl/6N	KOMP repository	CRL	Agouti	[41]	C57Bl/6-Tyr^{c-Brd} and C57Bl/6	A – agouti, restored agouti locus in JM8F6 line, heterozygous in agouti (A) locus
							heterozygous in agouti (A) locus
BALB/cByJ-PB150.18	BALB/cByJ	Jackson Laboratories	Jackson Laboratories	Albino	NA	C57Bl/6	Obtained by Predictive Biology, Inc. used to produce random integration and gene targeting

blast. inj. - blastocyst injections, agg. - aggregations, 8-cell inj. - ES cell injections into 8-cell stage embryos and mice from same genetic background and recommended embryo host strain for ES cell injections.

15.2.1.1. CCE [4]

This cell line was obtained from 129S5/SvEv mice.

ES cell lines derived from 129S substrains carry A^w/A^w (white-bellied agouti) allele on the Agouti gene A.

Recommended blastocyst donor strain for microinjection: ***a/a*** at the Agouti locus A (i.e., non-agouti coat color) and Tyr^+/Tyr^+, for example, C57BL/6J or C57BL/6N mice.

Chimeras have agouti patches over black background. The more agouti color is present, the stronger the chimera.

Germline transmission (GLT) is obtained by breeding chimeric males to black B6 mice like C57BL/6J or C57BL/6N. Transmission is made evident by the presence of agouti pups in progeny. Half of pups should be carrying targeted mutation. Heterozygous animals could be backcrossed to C57BL/6 strain of mouse to obtain congenic B6. Alternatively proven GLT chimeras can be bred to 129 mice from the same substrain as ES cell line origin, to obtain pure inbred mice, although unlike B6, 129 mice are not often used for phenotyping for variety of reasons. The same two approaches could be used for any pure agouti (A/A) ES cell lines (for example, AB1, CJ7, J1, or W4).

15.2.1.2. D3 [5]

Derived from 129S2/SvPas mice. This was the first ES cell line to be distributed by ATCC. Currently it is only available at late passage number and thus recommended for in vitro differentiation study only. ATCC now has a wider selection of ES cell lines available.

15.2.1.3. E14 [26]

Derived from 129P2/OlaHsd mice. A^w/A^w (white-bellied) on the Agouti gene A, $Oca2^p$ $Tyr^{c\text{-}ch}$/ $Oca2^p$ $Tyr^{c\text{-}ch}$ dilution gene pink-eyed (*p*) and chinchilla.

Recommended blastocyst donor strain: ***a/a*** at the Agouti locus A and Tyr^+/Tyr^+ such as C57BL/6J or C57BL/6N mice.

E14 chimeras have pigmented patches of two colors – agouti and cream over black background (Fig. 15.1). In order from weaker to strongest chimeras they might be: agouti on black, agouti and cream on black or prominent cream color over agouti.

Cream color is obtained only if both component of developing hair bud (epidermal and mesenchymal) and melanocytes (neural crest lineage) are derived from ES cells. Thus, stronger the ES cell contribution – more cream color is present.

Germline transmission is obtained by breeding chimeric males to black B6 mice like C57BL/6J or C57BL/6N and transmission is evident by the presence of agouti pups in progeny. Half of pups should be carrying targeted mutation. To create coisogenic strain

Fig. 15.1 Chimeric mice with strong contribution of manipulated E14 ES cells. Courtesy Pan Y-X; mice were described in [54]. See also comments in the text 15.2.1.3.

of mutant mouse, chimeras with highest rates of germline transmission are bred to 129P2/OlaHsd females available from Harlan. All progeny is genotyped. Alternatively, heterozygous F1 animals could be backcrossed to C57BL/6 strain of mouse from the same source as that used for breeding with chimeras.

While known to produce strong chimeras this line, as well as its derivatives, is contaminated with bovine mycoplasma and for this reason might not be accepted by some Facilities.

15.2.1.4. E14.TG2a [27]

An HPRT-deficient subline of *E14* that allows use of the HPRT gene as a selection cassette. Currently, neomycin, hygromycin, or puromycin resistance genes are more commonly used for this purpose.

15.2.1.5. AB1 & AB2.1 AB2.2 [28]

Derived from 129S7/SvEvBrd mice. A BAC genomic library was developed at Welcome Trust Sanger Institute from the same substrain of mice. Both BAC library and mice are available from the Sanger Institute. Injections and breeding should follow the same scheme as other lines derived from S substrains of 129 mice. For details see description for CCE cell line.

15.2.1.6. J1 [29]

Established in the Laboratory of R. Jaenisch from 129S4/SvJae mice. Injection and breeding recommendation is the same as for CCE line, with exception that 129S4/SvJae mice should be used to maintain mutation in 129-strain. Initially mice from this substrain were maintained only at Harvard but now are also available from Jackson Laboratory.

15.2.1.7. R1 [30]

This ES cell line was established from an intercross between two 129 substrains 129X1/SvJ x 129S1/Sv and thus represents an F1 genetic background.

It was the first ES cell line from which fully ES cell line-derived embryos were produced. This is most probably related to it's hybrid nature and heterozygosity at a large number of loci. Although ES cell contribution in chimeras is defined by agouti coat color, there is a segregation of agouti and chinchilla genes in the F1 generation and even further in F2.

R1 ES cells are A^{w}/A^{w} *Oca2*^{p} *Tyr*$^{c\text{-}ch}$/*Oca2*^{p} *Tyr*^{c} or A^{w}/A^{w} *Oca2*^{p} *Tyr*^{c}/*Oca2*^{p} *Tyr*^{c}

Suggested blastocyst donor strain: *a/a* at the Agouti locus A and *Tyr*$^{+}$/*Tyr*$^{+}$, for example, C57BL/6J or C57BL/6N mice.

Chimeras have agouti patches over black background.

Germline transmission is obtained by breeding chimeric males to black B6 mice like C57BL/6J or C57BL/6N and is evident by the presence of agouti pups in the progeny. Half of agouti pups could be carrying mutation. To create coisogenic strain of mutant mouse, heterozygous animals need to be backcrossed to either 129X1/SvJ, 129S1/Sv, or C57BL/6 strain for many generations. In the latter case, the next generation coat color genes would segregate further and agouti plus chinchilla pups would be observed.

R1 and aggregation: The R1 cell line retains its full developmental potential for an extended number of passages and is also able to contribute to the germline after aggregation with outbred morulae. Working with outbred animals brings the advantage of reduced costs, for a number of reasons. They have a relatively long lifespan, are resistant to disease, have high fecundity and are less expensive to purchase.

Commonly used morula donor strains: ICR, CD1, or Swiss Webster (SW) are *Tyr*^{c}/*Tyr*^{c} (albino). Chimeras have agouti patches over white background. Increased agouti color signifies higher ES cell contribution.

Germline transmission is obtained by breeding chimeric males to albino mice such as CD1 or SW and it is evident by the presence of agouti pups in progeny. To create coisogenic mutant mouse strain, heterozygous animals need to be backcrossed to either 129X1/SvJ or 129S1/Sv, or C57BL/6 strain of mouse for a minimum of ten generations. Alternatively, for experimentation where genetic background is of a lesser importance or where a random genetic population is desired, mutant mice can be maintained and analyzed on a mixed genetic background.

15.2.1.8. CJ7 [31]

Derived from 129S1/SvImJ mice. Injections and breeding should follow the same scheme as most other lines derived from any S substrain. For details see description for CCE cell line.

15.2.1.9. RW4 [32]

The first commercially available BAC library was made from this cell line and distributed by Stratagene. Unfortunately, it was later shown that the 129X1/SvJ substrain from which this ES cell line was derived is highly contaminated with genomic regions of non-129 origin [6, 25]. In addition, this strain is homozygous for Cdh23^{ahl}, the age-related hearing loss mutation, which on this background results in progressive hearing loss with onset prior to 3 months of age, making some behavioral studies impossible.

15.2.1.10. W4 [15]

This line was developed, from 129S6/SvEvTac mice from Taconic Farm, by the authors of this chapter, for use in a service facility. Cells were uniformly expanded and a large number of vials frozen at early passage. Later they were transferred to Taconic Farm for distribution making it convenient "one place shopping" for ES cells and mice of the same genetic background. Blastocyst injection and chimera breeding protocol are the same as for most other 129S-derived lines (see description for CCE cell line for breeding details). W4 ES cells are also able to contribute to chimeras by aggregation with outbred morulae.

15.2.1.11. AK7.1 [33]

Established from 129S4/SvJae mice (the same as J1). This cell line was employed primarily in Soriano's gene trap experiments.

15.2.2. Hybrid ES Cell Lines

It is common practice to mate chimeras produced from 129-derived ES cells injected into C57BL/6 blastocysts, to C57BL/6 mice. F1 mice harboring the introduced mutation are identified, interbred and phenotyped on this mixed genetic background. Alternatively, they are bred for several generations to C57BL/6 mice to eventually produce a new congenic strain [34]. In such a case, there may be an advantage to using a 129B6/F1 hybrid ES cell line. Not only are these cells more reliable for karyotype stability and transmission than 129-derived ES cells due to hybrid vigor [35], but would also save one generation of backcrossing to C57BL/6 mice.

Hybrid cell lines might be also an alternative where ES cells from the desired inbred strain (such as BALB/c, for example) are not efficient in gene targeting and/or producing chimeras. Researchers, who do not have experience with culturing C57BL/6 ES cell, requiring more rigorous culture conditions (*see* Subheading 15.3) may choose to use a 129B6/F1 hybrid line. There are only a handful of known hybrid ES cell lines at present as follows.

15.2.2.1. V6.5 [35]

Derived from C57BL/6 x 129S4/SvJae embryo at the laboratory of R. Jaenisch and distributed by OpenBiosystems. Shown to be able to produce live animals by tetraploid complementation. Cells are A^{w}/a (heterozygous) on the Agouti gene A, (white-bellied black Agouti).

Recommended blastocyst donor strain: C57BL/6, *a/a* at the Agouti locus A· Chimeras have agouti patches over black background.

Germline transmission is obtained by crossing chimeras with C57BL/6 mice. Presence of agouti pups in the litter is evidence of germline transmission. Pups may be either agouti or black, because ES cells are heterozygous in the Agouti locus. Progeny of both colors have an equal chance of carrying targeted mutation. All progeny need to be genotyped. For the sake of efficiency, genotyping might be limited to the progeny from litters containing agouti pups. The only exception would be when the gene of interest is located on chromosome 2, where the agouti gene is also located, in which case, depending on location of the allele carrying the mutation the heterozygous pups would be always black or agouti, respectively.

15.2.2.2. V17.2 [35]

Derived from BALB/c x 129S4/SvJae embryo are available from OpenBiosystems. BALB/c genetic background is desired in immunological studies, because of ease of obtaining hybridoma cell lines in this background, plus the body of immunological research published to date. ES cells are A^{w}/a (heterozygous) on the Agouti gene A, (white-bellied black Agouti), Tyr^{+}/Tyr^{c}.

Blastocyst donor strain: BALB/c, these are *A/A* at the Agouti locus A, and Tyr^{c}/Tyr^{c} (albino). Chimeras have agouti patches over white background.

Germline transmission is obtained by crossing chimeras with BALB/c mice. Presence of agouti pups in the litter is evidence of germline transmission. Pups are agouti or white. Progeny of both colors may carry targeted mutation, because ES cells are heterozygous at the Agouti locus. All progeny needs to be genotyped. For efficiency, only progeny from litters with agouti pups can be genotyped. The chromosome 2 exception applies here also. This cell line was shown to be able to produce live animals by tetraploid complementation [35] as well as produce entirely ES cell-derived mice by injections into the eight-cell outbred embryos (unpublished data).

15.2.2.3. VGF1[36]

Derived from 129S6/SvEvTac × C57BL/6NTac embryo at Regeneron Pharmaceuticals and used to generate hundreds of knockouts with Velocigene Technology (For complete list of publications see http://www.velocigene.com/pubs). This line is able to produce entirely ES cell-derived mice by injection into eight-cell outbred embryos [17] and produced GLT F0 mice even after ten rounds of electroporations.

F0 males should be bred with C57BL/6NTac females. Presence of agouti pups in the litter is evidence of germline transmission. Progeny of both colors might carry targeted mutation, because ES cells are heterozygous in the Agouti locus. Black F1 heterozygotes can be backcrossed to C57BL/6NTac mice in order to obtain coisogenic strain.

15.2.2.4. G4 [37]

Derived from 129S6/SvEvTac x C57BL/6NCrl embryo. Developed at and distributed by A. Nagy lab, MSHRI. This ES cell line is able to generate chimeric mice by aggregation with outbred morulae, as well as to support life of ES cell-derived mice in tetraploid complementation [37]. Chimera production by blastocyst injection or morula aggregation should follow same schema as for R1 cells. GLT chimeras can be bred to one of the parental lines (C57BL/6NCrl or 129S6/SvEvTac). Since Y chromosome in this line is from 129S6/SvEvTac strain F1 het females should be bred to C57BL/6NCrl males to eventually obtain congenic B6 mice. Either males or females can be used in the following generations.

15.2.3. ES Cell Lines Derived from C57BL/6 Strains of Mice

15.2.3.1. Bruce4 [9]

Derived from C57BL/6-*Thy1.1* mice. This is a congenic, not a pure C57BL/6 strain [38].

Blastocyst donor strain: *a/a* at the Agouti locus A and Tyr^{c}/Tyr^{c} (albino B6). Using strain B6(Cg)-$Tyr^{c\text{-}2J}$/J strain results in the production of moderate chimeras, but even <50% chimeras can transmit through germline [38].

Germline transmission is tested by breeding with B6(Cg)-$Tyr^{c\text{-}2J}$/J mice, black mice being evidence of GLT although they will be heterozygous at the tyrosinase locus $Tyr^{c\text{-}2J}$/ Tyr^{+} and Thy1 locus.

15.2.3.2. HGTC-8 [19]

Derived and maintained in serum-free medium (KO-DMEM +KOSR) and used primarily at the NIH. Injection and breeding protocol is the same as for any other black B6 lines, (*see* Bruce4).

15.2.3.3. B6(Cg)-$Tyr^{c\text{-}2J}$/J-PRX-B6-albino#1

This ES cell line was derived from a B6(Cg)-$Tyr^{c\text{-}2J}$/J embryo and is being distributed by Jackson Laboratories. Millipore distributes a similar line under the name *PluriStemB6-white*. ES cells are homozygous *a/a* on the Agouti gene and $Tyr^{c\text{-}2J}$/ $Tyr^{c\text{-}2j}$ (albinoB6) and can be injected into C57BL/6J blastocysts (*a/a* on the Agouti gene A and Tyr^{+}/Tyr^{+}). Chimeras have white patches on a black background. More white color indicates higher ES cell contribution. Chimeras are bred with B6(Cg)-$Tyr^{c\text{-}2J}$/J mice and presence of white pups in F1 generation is evidence of GLT. Half of them should be het for a mutation of interest. Those pups are pure B6(Cg)-$Tyr^{c\text{-}2J}$/J strain. Since B6(Cg)-$Tyr^{c\text{-}2J}$/J mice are poor breeders F1 pups are often bred to C57BL/6J mice rather than maintained in a B6(Cg)-$Tyr^{c\text{-}2J}$/J background.

15.2.4. C57BL/6 ES Cell Lines Used by Large-scale Knockout Programs

The members of the International Knockout Mouse Consortium (IKMC) are working together to mutate all protein-coding genes in the mouse genome using a combination of gene trapping and gene targeting in C57BL/6N mouse ES cells.

15.2.4.1. VGB6 [17]

Used by Regeneron to produce knockout ES cell clones for the NIH Knockout Mouse Project (KOMP). Targeted ES cells are available from the KOMP repository. This cell line was isolated from inbred C57BL/6NTac embryo (*a/a*, Tyr^{+}/Tyr^{+}).

Blastocyst donor strain: For chimera production recommended embryo donor strain is albino (Tyr^{c}/Tyr^{c}). Excellent results were obtained using BALB/c or B6(Cg)-$Tyr^{c\text{-}2J}$/J strains for blastocyst injections, SW for eight-cell embryo injections (Fig. 15.2) and ICR for ES-morula aggregations. Also C57BL/6-$Tyr^{c\text{-}Brd}$ mice can be used as blastocyst donors, available from Harlan Laboratories [39]. BALB/c mice respond poorly to superovulation and produce few blastocysts. Thus, they are not a first choice as blastocyst donors. Similarly, B6(Cg)-$Tyr^{c\text{-}2J}$/J mice are poor breeders and expensive to purchase. On the other hand, injections into eight-cell outbred embryos can result in production of entirely ES cell-derived F0 mice [17].

Chimeras have pigmented patches (black to agouti) over albino background with the strongest chimeras being completely black.

Fig. 15.2 Chimeric mice generated by injection of manipulated VGB6 ES cells into 8-cell stage SW embryos.

To test for GLT, chimeras can be bred with any albino females. When using albino B6 mice (*a/a*, Tyr^c/Tyr^c) evidence of GLT is the presence of black pups in the progeny. BALB/c mice are *A/A* at the agouti locus and the evidence of GLT is the presence of agouti pups in the F1 generation. Whereas SW mice, an outbred strain, are not fixed at the agouti locus, so the presence of either agouti or black pups would be evidence of GLT. To obtain the mutation in a pure inbred background, proven germline transmitting males should be bred with C57BL/6NTac females. Entirely ES cell-derived F0 mice obtained from injections into eight-cell embryos, if fertile, are 100% germline transmitters and can be directly bred with C57BL/6NTac females without previous mating with an albino female to test for GLT. All F1 progeny must be genotyped.

15.2.4.2. Lex3.13 [40]

Derived from C57BL/6N mice. This line was used by Lexicon Pharmaceuticals, Inc. to produce the gene trap library of over 350,000 cell lines that are distributed by the Texas Institute for Genomic Medicine (TIGM). Unmanipulated late passage cells are also available. It is best to follow the injection and breeding protocols provided by TIGM. On average only 50% of clones produce GLT chimeras, thus it is advisable to inject at least four clones to have >90% chance that at least one will go through germline.

15.2.4.3. JM8F6 [41]

The JM8 ES cell line was isolated from inbred C57BL/6NCrl mice. JM8.F6 is a subclone of the JM8 line selected for the higher GLT rate that the parental line exhibited and was used by The Wellcome Trust Sanger Institute to generate conditional knockouts for the European Union Conditional Mouse Mutagenesis Programme (EUCOMM) and KOMP. GLT rate is host-strain dependent and is higher when C57BL/6-$Tyr^{c\text{-}Brd}$ mice are used as blastocyst donor compared to BALB/c.

15.2.4.4. JM8A3 [41]

The non-agouti mutation in C57BL/6 inbred strains is due to an integration of a retrotransposon in the Agouti gene abolishing transcription of Agouti mRNA. The JM8A3 line was created by restoring the wild-type agouti locus in JM8.F6 ES cells. Thus the ES cells are *A/a* (heterozygous, one allele restored,) on the Agouti gene (agouti B6).

Blastocyst donor strain: The goal of this modification was to enable use of B6 embryos (*a/a*, Tyr^+/Tyr^+) as blastocyst donors for injection, this being the most widely used strain for this purpose. Chimeras should have agouti patches over black background. In our experience, the patches have edges blending in agouti with the strongest chimeras looking like darker agouti. Chimeras should be bred with C57BL/6NCrl females.

Germline transmission is confirmed by the presence of agouti pups along with black in the litter. Progeny of both colors might carry targeted mutation, because ES cells are heterozygous (*A/a*), thus all progeny need to be genotyped. If only black F1 mice would be used for further breeding the mutation will be on a pure C57BL/6N background. The only exception is when the gene of interest is located on chromosome 2 (where the Agouti locus is located), in which case, depending on the location either on the chromosome with the corrected *A* or the mutated *a*, the pups carrying the mutation would be always black or agouti, respectively. Alternatively, inject ES cells into blastocysts from an albino strain such as C57BL/6-$Tyr^{c\text{-}Brd}$ (*a/a*, Tyr^{c}/Tyr^{c}) or other albino B6 or BALB/c (*A/A*; Tyr^{c}/Tyr^{c}). GLT data for this ES cell line are predominantly from injections into albino embryos (C57BL/6-$Tyr^{c\text{-}Brd}$ BALB/c).

15.2.4.5. C2 [42]

This cell line is used by the North American Conditional Mouse Mutagenesis (NorCOMM) program in Canada. It was derived from C57BL/6NTac embryo same as VGB6, thus should follow same injection and breeding scheme. This cell line is able to contribute to GLT chimeras either by aggregation with outbred morulae or injection into eight-cell embryos, but not efficiently if cultured in regular ES cell medium containing FBS. Culture in conditioned medium (RESGRO) or serum-free medium [42] is recommended.

15.2.5. Feeder-Independent Cell Lines

It is an appealing idea to grow ES cells without a feeder layer. It eliminates the need to produce and inactivate mouse embryonic fibroblasts. When judging the morphology of ES cells, it should be noted that ES cell colonies grown on gelatin are usually more flat than when grown on feeders, unless an activator of the beta-catenin (like Wnt protein or GSK3 inhibitor) is added to the media. Usually, new ES cell lines are derived on feeders and later adapted to growth on gelatin. It was often reported that they went through some type of crisis during the adaptation period. This creates a risk of selection of a subpopulation of cells with possible chromosomal changes. This situation might be different for cell lines established without feeders from the beginning.

Below are listed 129- and B6-derived ES cell lines adapted to growth without feeders.

15.2.5.1. E14Tg2a.4 [43]

Feeder-independent subclone of E14Tg2a, used in the BayGenomics gene trap project [44] and the Sanger Institute Gene Trap Resource. This line although producing strong chimeras is unable to make fully ES cell-derived mice (Stewart F. Personal communication).

15.2.5.2. CMTI-2 http://www.millipore.com

Bruce4 line adapted to grow in feeder-free and serum-free ESGRO complete medium distributed by Millipore.

15.2.5.3. JM8.N4 [41]

A feeder-independent, subclone of JM8 line; used for some EUCOMM and KOMP conditional gene targeting. This line was later withdrawn due to the duplication of a small region on chromosome 10 [41].

15.2.5.4. JM8.A3N1 [41]

A feeder-independent, subclone of JM8A3 line; used for most of EUCOMM and KOMP conditional gene targeting. Both JM8. N4 and JM8A3.N1 show higher chromosomal instability than feeder-dependent line (W. Skarnes, personal communication).

15.2.6. ES Cell Lines from Other Mouse Strains

ES cell lines have also been developed from many other strains including the following: BALB/c, C3H, CBA, DBA, FVB (all available from Millipore), NOD, [52, 53], A/J and NZW. These cells may have different culture requirements, depending on how they were established. There is not enough literature at this time to evaluate their performance. While importing one of those lines, it is best to follow the original culture conditions, but it may also be necessary to test electroporation conditions and drug resistance levels before use for gene targeting.

15.2.7. Commercial Suppliers of ES Cell Lines

It is always recommended to use ES cells, which were obtained directly from the laboratory in which they were established, or the commercial supplier where they were either made or deposited for distribution by the creator. Although a second-hand gift from a next door laboratory might sound like a money saver, only an established organization can guarantee that the cells are of proper karyotype, mycoplasma-free, of known passage number and are able to maintain the same characteristics between different batches. Below is a list of vendors that either serve as distributors of cell lines developed in research laboratories or which have produced their own:

ATCC – http://www.atcc.org

Open Biosystem – http://www.openbiosystems.com

Jackson Laboratories – jaxmice.jax.org/cells

Millipore – http://www.millipore.com

Primogenix – http://www.primogenix.com

Predictive Biology – http://www.predictivebio.com

15.2.8. Repositories of Targeted ES Cell Clones

Although unmanipulated, usually later passage parental ES cell lines are available from most repositories, the repositorie' main goal is to distribute targeted ES cells to avoid the duplication of work and to streamline phenotypic studies. Thousands of new targeted ES cell lines have been deposited into repositories annually, in recent years, so there is a good probability that the modification one plans to make was already made and is available at a fraction of the price, time and effort it would take to make it again. So please, before attempting to make a knockout of your favorite gene, check those repositories to see if the same modification has already been made and is now available:

MMRRC – http://www.mmrrc.org – a collection of cell lines deposited by many individual researchers including BayGenomics gene trap clones.

KOMP – http://www.komp.org – collection of definitive null and conditional KO in B6 ES cells.

EuMMCR – http://www.eummcr.org – European Mouse Mutant Cell Repository distributes material arising within the EUCOMM consortium.

TIGM – http://www.tigm.org – maintains a KO gene trap library in C57BL/6N and 129/SvEvBrd ES cells.

For additional details, *see* Chapter 2, "Global Resources."

15.3 ES Cell Culture Media

The most widely used growth medium for culturing ES cells was developed years ago and optimized for 129-derived ES cells. Its main components are high glucose DMEM, FBS, and LIF. DMEM and Ham's F12 are sometimes mixed (DMEM/F12) to combine the higher concentrations of components in DMEM with the wider range of Ham's F12 ingredients. The addition of F12 provides components such as biotin, putrescine, lipoic acid, glycine, proline, copper, and zinc that are not present in DMEM.

Lately, several companies have started selling a DMEM version that is either pretested or specially adjusted for the culture of ES cells. The most widely known is KO-DMEM from Invitrogen (Cat. No. 10829-018) with lowered osmolality optimized to approximate that of mouse embryonic tissue. The basal media is then supplemented with 15% FBS, L-Glutamine (or GlutaMax), Sodium Pyruvate, Non-Essential Amino Acids (NEAA), 2-Mercaptoethanol, LIF and antibiotics (Pen-Strep). The use of antibiotics is controversial. Although it can prevent the bacterial contamination, it also can mask mycoplasma contamination.

15.3.1. Complete ES Cell Medium Preparation

Solutions

Dulbecco's Modified Eagles Medium (DMEM + 4,500 mg/L Glucose); Invitrogen/Gibco, Cat. No. 11960-044, or

D-MEM/F-12 (1×), liquid 1:1 Invitrogen/Gibco, Cat. No. 11320-082.

Or KnockOut-DMEM Invitrogen/Gibco, Cat. No. 10829-018.

200 mM L-Glutamine (100×) Invitrogen/Gibco Cat. No. 25030-081.

Penicillin/Streptomycin Invitrogen/Gibco (optional), Cat. No. 15070-063.

100 mM Sodium Pyruvate (100×) Invitrogen/Gibco, Cat. No. 11360-070.

10 mM MEM Non-Essential Amino Acids, NEAA (100×) Invitrogen/Gibco, Cat. No. 11140-050.

2-Mercaptoethanol (1,000×) Invitrogen/Gibco, Cat. No. 21985-023 or Sigma-Aldrich, Cat. No. M7522).

LIF (Millipore, Cat. No. ESG1107 or ESG1106). ES qualified FBS (it can be either company pretested ex. Gibco cat no 16141-079, or individually tested-See 15.3.2 for details on screening of FBS).

Preparation

1. 500 ml bottle of DMEM (or DMEM/F12 or KO-DMEM)
2. Add 90 ml of ES cell-qualified FBS
3. Add 12 ml of L-Glutamine
4. Add 6 ml of Pen-Strep
5. Add 6 ml of Sodium Pyruvate
6. Add 6 ml of MEM NEAA
7. Add 1.2 ml of 2-Mercaptoethanol
8. Add 60 μl of LIF

   ES cell medium stored at 4°C is good for 2 weeks.

- FBS is used for nutritive supplementation of ES cell growth medium and also contains attachment factors. For ES cell culture usually it is used at 15–20% concentration. Unfortunately, FBS also contains factors that can promote differentiation of ES cells. Thus, each lot must be screened prior to use. In addition, each lot of FBS is different from the last one, or the next one, so screening is a continuous task. Several companies sell ES cell-qualified FBS, which is a good choice for anyone growing ES cells on an occasional basis. For the best results, FBS should be tested for its ability to support the ES cell line that will be used. This approach could pose a problem for service Facilities offering multiple ES cell lines of different origin.

15.3.2. Screening of FBS for ES Cell Growth Support

15.3.2.1. Test of FBS for ES Cell Growth Support in 15% FBS (Normal Culture Condition)

Prepare six-well TC dishes with feeder layers in complete ES cell growth medium with 15% FBS. Use two wells for each sample of serum (e.g., different suppliers and/or different lot numbers) and the control serum. Use the currently used serum or serum proven to support ES cell growth and GLT as control. Seed duplicate plates for each sample. Plate 1.2×10^5 ES cells (suspended in nonsupplemented DMEM) on each well of the six-well feeder plate.

Change media every day, paying attention to the use of correct media. Passage cells every other day on a feeder layer and count the cell number to calculate multiplying factor (number of cells harvested at each passage/number of cells plated).

Re-plate cells at 3×10^5 per 6-cm plate at every passage. After three passages, taking the number of cells plated and the total number of cells 2 days later at each passage, calculate the total number of doublings during entire period, (i.e., in Table 15.2, FBS A, calculate $24.1 \times 22.7 \times 17.3 = 9{,}464$).

15.3.2.2. Plating Efficiency (PE) and Morphology Test in Different FBS Concentrations

At the third passage, plate cells on gelatin plates (500 cells/6-cm dish or 2,000 cells/10-cm dish). For each serum batch, plate cells in ES media containing 10, 15, or 30% FBS.

Change media every day. After 7 days stain the cells on all plates with trypan blue or alkaline phosphatase (AP). While staining with trypan blue allows for counting colonies (Table 15.4) and morphology observation, AP staining additionally allows for observation of the proportion of undifferentiated cells within each colony. Observe colonies under microscope. Sort out colonies into the four morphology categories (Table 15.3) based on their size and presence or lack of differentiation. Count number of colonies in each group. Use total colony number (Table 15.4) to calculate plating efficiency (number of colonies/number of cells plated) expressed as percentage.

Note: If you are planning to microinject ES cells to test for GLT (and we recommend you do), pick few colonies for each serum you plan to test from 15% concentration before staining. To limit number of injections, this step is usually limited to two best sera, based on in vitro tests (morphology, plating efficiency, growth rate), plus the control serum.

The number of colonies may fall slightly in 30% FBS DMEM, but for some samples the number still increases.

The best serum is the one that supports faster growth of ES cells while maintaining majority of colonies as undifferentiated.

Table 15.2
Example of record keeping for the evaluation of ES cell culture growth rate

	Plating	1st Passage	2nd Passage	3rd Passage
FBS A Lot #:	Thaw and plate 1.2×10^5 cells/ well on 6-well feeder plate	2.9×10^6	6.8×10^6	5.2×10^6
		(×24.1)	(×22.7)	(×17.3)
		Plate 3×10^5 on 6 cm feeder	Plate 3×10^5 on 6 cm feeder	**×9,464**
FBS B Lot #:		2.7×10^6	7.4×10^6	5.7×10^6
		(×22.5)	(×24.7)	(×19.0)
		Plate 3×10^5 on 6 cm feeder	Plate 3×10^5 on 6 cm feeder	**×10,559**
FBS C Lot #:		2.7×10^6	7.3×10^6	5.4×10^6
		(×22.5)	(×24.3)	(×18.0)
		Plate 3×10^5 on 6 cm feeder	Plate 3×10^5 on 6 cm feeder	**×9,841**
Control FBS Current FBS		2.9×10^6	7.0×10^6	5.6×10^6
		(×24.1)	(×23.3)	(×18.7)
		Plate 3×10^5 on 6 cm feeder	Plate 3×10^5 on 6 cm feeder	**×10,500**

Cells are cultured over three consecutive passages in regular culture conditions, on a 60 mm plate with a feeder layer. Media contain 15% FBS from each serum batch to be tested. In bold total number of doublings during testing period. Its higher values indicate a better ES cell growth. It is desired serum characteristics reflecting better ES cell support

Table 15.3
Morphology observation of ES cell colonies grown in different lot of serum after 7 days from plating

Category	Description	FBS A Company lot #:	FBS B Company lot #:	FBS C Company lot #:	Control FBS
1	Small colonies (too small to pick)	% Of total number of colonies	%	%	%
2	Larger, undifferentiated colonies	%	%	%	%
3	Partially differentiated colonies	%	%	%	%
4	Mostly differentiated colonies	%	%	%	%

Table 15.4
Plating efficiency on gelatin or feeder layer

		FBS A		FBS B		FBS C		Control FBS	
		Gelatin	Feeder layer	Gelatin	Feeder layer	Gelatin	Feeder layer	Gelatin	Feeder layer
10%	No. colonies/plate								
	Average								
	% PE								
15%	No. colonies/plate								
	Average								
	% PE								
30%	No. colonies/plate								
	Average								
	% PE								

Notes:
10% concentration of serum is used to test serum ability to support ES cell growth
15% concentration of serum is used for regular ES tissue culture
30% concentration of serum is used to test its potential toxicity.

15.3.2.3. Alternative FBS Screening Test (B. Wang, Personal Communication)

Different lots of sera can be tested by culturing electroporated ES cells with selection to determine the growth rate and morphology of the G418 positive clones. The colonies should grow big enough for picking in 8 days after electroporation. If not, the growth is too slow. These are more harsh condition for the cells. Most lots of ES-qualified serum fail to pass this kind of test.

Note: Unfortunately, all testing protocols need to be repeated for each parental ES cell line used in gene-targeting experiments.

15.3.3. Alternatives to Serum Use

ES cells can be cultured in more defined conditions with serum substitutes and a cocktail of hormones, growth factors, vitamins, and serum albumins or albumin substitutes.

15.3.3.1. Knockout Serum Replacement

Knockout Serum Replacement (KOSR) is chemically better defined than FBS. The exact formulation of KOSR is proprietary. Direct replacement of FBS by KOSR in ES cell media allows the growth of undifferentiated ES cells [19]. The best performance is achieved when KOSR is used with Knockout DMEM. KOSR can exhibit considerable variability between different lots [45] thus, for optimal results each lot should be tested similarly to FBS prior to use for ES cell culture. Additionally, since KOSR does not contain cell adhesion factors, or trypsin inhibitors, medium containing FBS has to be used after trypsynization and when plating ES cells on gelatinized plates without feeders.

15.3.3.2. ESGRO

Millipore Cat. No SF002-500 – a basal serum-free medium, for use with LIF and BMP4. This medium was developed for culture of ES cells in serum-free and feeder-free conditions [18]. Cells previously grown in medium containing FBS need a period of adaptation. In our hands, adapted cells grew significantly slower than the parental cells, which affects their usefulness in gene-targeting experiments.

ESGRO-Complete PLUS (Millipore Cat. No SF001-500P) – a complete formulation which does not require the addition of LIF. It contains a GSK3β inhibitor to enhance viability of mouse embryonic stem (ES) cells and increased maintenance of pluripotency in the absence of serum and feeder cells.

Another trend in culturing ES cells is the use of media conditioned by cells secreting proteins that prevent differentiation of ES cells. For example, RESGRO (Millipore Cat. No SCM002) previously called TX-WES [16] is a medium conditioned by rabbit fibroblasts expressing recombinant rabbit LIF.

According to the manufacturer RESGRO Culture Medium has the capacity to rescue established ES cell lines that have started drifting and either generate low percentage chimeras or have lost germline transmission capability. Differentiation, which is present in the ES cells but not visible with traditional medium, will become recognizable when using RESGRO Culture Medium. After two passages, a clear difference is seen between differentiated and undifferentiated ES cells, at which time undifferentiated cells can be selected for by subcloning. Rescue protocol is available at the manufacturer Web site – (http://www.millipore.com/userguides/tech1/pc1809en00).

15.3.3.3. VGB6-Medium

VGB6-Medium developed for the maintenance of undifferentiated B6 ES cells contains media conditioned by L cells expressing Wnt3a protein [17]. It was recently shown that culture in this media for few passages could rescue clones that lost their ability to produce chimeras [42].

15.4 Comparison of Gene Targeting and Germline Transmission Efficiencies in C57BL/6 vs. 129 ES Cell Lines

In the past, most gene targeting was accomplished using 129-derived ES cell lines. If for phenotyping, mice had to be in a particular genetic background (e.g., C57BL/6), heterozygous mice were backcrossed for many (10) generations until the congenic strain was produced, or the results were published before the congenic stage was obtained. It was shown [21] that it is possible to achieve enormous gain in time and resources if ES cells from the same genetic background needed for phenotyping are used for gene targeting experiments at the outset, even if targeting efficiency (TE) and GLT are lower in this particular ES cell line. In this case, the authors compared a single B6 line with a single 129 line, so the results may only be reflective of the differences between those two particular cell lines. Variation in TE and especially GLT between different cell lines of the same genetic background are common. To our best knowledge no comparison between statistically significant numbers of ES cell lines from both genetic backgrounds has ever been published.

There are many factors affecting TE and/or GLT, any of which might have more impact than the genetic background of the ES cells, although B6 lines do require more controlled culture conditions than 129 lines. In our opinion, the factor, which has the biggest effect on targeting efficiency is the targeting vector itself, the length of arms of homology and the source and quality of the DNA. The use of one short arm of homology, a commonly used approach, facilitates screening by PCR, whereas use of two long arms of homology greatly increases TE (W. Skarnes, personal communication). Even slight changes in the length of homology arms achieved by use of a different enzyme for linearization, might have effect on TE. Other factors influencing TE are culture and electroporation (EP) conditions. EP settings, amount of DNA used, number of ES cells, ratio of the amount of DNA to the number of ES cells, purity and integrity of the targeting construct; all will have effect on TE. In addition, cells from the same cell line will have a different doubling rate, when grown on feeders or on gelatin, in medium with or without serum, which would have effect on both TE and GLT. The faster the ES cells divide, the greater the chance they will contribute to the developing chimera [46]. It also has to be taken in to account that aneuploid cells (especially with trisomy 8) might have a growth advantage, but are unable to produce live animals [47, 48]. This is why even, if small percent of ES cells are trisomic, such a cell line should be subcloned before injection. Otherwise, trisomic cells will outcompete normal cells but will not allow GLT. Another factor influencing GLT is the correct combination of ES cells and host embryos with regard to their genetic backgrounds. The best example of such a good match is the combination of 129 ES cells and C57BL/6 blastocyst donor. Such an optimal combination is also different depending of the stage of embryos used for injections. For example, for eight-cell stage injections or morula aggregations, outbred embryos are a better host than inbred, although they are not useful for blastocyst injections.

In summary, if the goal is to have a mutation in a C57BL/6 genetic background it is better to use C57BL/6 ES cells rather than electroporate 129 ES cells and follow up with lengthy backcrossing, even though working with C57BL/6 ES cells will require more rigorous culture conditions and the use of different media (*see* Subheading 15.3).

15.5 Which ES Cell Line to Use for Gene Targeting: Factors to Consider

When planning to make a new modification in the mouse genome the first question should be: "Has the same or very similar modification already made and if so, is the mouse or ES cell line with such a modification available"? With the international initiative (IKMC) to create either definitive null or conditional KO for every protein-coding gene in the mouse genome, it is a good probability that either the mouse or ES cells with the desired modification, has already been produced. In such a situation, it is advisable to use existing resources instead of creating a new cell line or genetically altered mouse. Repository databases are good sources of information (*see* Subheading 15.2.7 and Chapter 2). Such resources can provide an enormous savings in time, effort and assets, as is their purpose. If the mutated ES cell line is available, the production of a mutant mouse requires knowledge and understanding of the most appropriate combination of ES cells and embryo and subsequent breeding protocols (*see* Subheading 15.5 above and Chapter 22). If based on the data from the analysis of an existing mutation, plans are undertaken to make a more elaborate modification, then use of the same ES cell line as previously used would simplify the phenotyping and interpretation of the obtained results. In the past, when different modifications for the same gene were generated in different genetic backgrounds, interpretation of the results was often complicated or the results were not reproducible [49, 50].

In general, when planning a new project involving gene targeting in ES cells, the answers to the following question should be taken into consideration before starting construction of targeting vector and choosing an ES cell line for the experiment:

- Is the genetic background important for phenotyping mice for this project?
- Does an ES cell line from the desired genetic background already exist and is it available?
- Is the available ES cell line well validated?
- Are mice from exact strain available for further breeding with chimeras?

15.5.1. Pros and Cons of Making Your Own ES Cell Line

Although recent improvement in media and protocols has made derivation of new ES cell lines easier than in the past, the following should be taken into account when deciding between use of existing lines and attempting to create a new one:

- Experience – as for many other complicated procedures, previous experience can affect the outcome of such an attempt.

- Ability to characterize multiple ES lines – since usually multiple lines are created at the same time, further analysis of karyotypes and GLT potential need to be performed on all of them, before deciding to use one particular line. Such tests should be repeated after each expansion of stocks for future use. The use of time, costs, and other resources, all have to be taken into consideration.
- Throughput – unless a specific unique genetic background is absolutely necessary (existing previous random or spontaneous modification, especially multigenic), derivation of a new ES cell line for single modification is hardly justifiable. The situation might be different for high-throughput laboratories or service Facilities, where the availability of early passage ES cell lines could provide enough advantage to justify the cost and effort necessary for testing.

15.6 Vendors of ES Cell-Related Products

Millipore (ES cell lines, media, LIF, MEFs)

Invitrogen/Gibco (media, sera, media additives, serum replacement, LIF)

ThermoFisher (Hyclone sera, media, supplements)

Stem Cell Technologies (media, sera, supplements, MEFs)

Global Stem (MEFs, media, ES cells)

Gemini Bioproducts (sera, media, supplements)

Corning (plasticware)

Nunc (plasticware)

Mice can be purchased from the following vendors:

Charles River Laboratory (http://www.crl.org)

The Jackson Laboratory (http://www.jax.org)

Harlan Laboratories (http://www.harlan.com)

Taconic Farms (http://www.taconic.com)

References

1. Evans MJ, Kaufman MH (1981) Establishment in culture of pluripotential cells from mouse embryos. Nature 292:154–156
2. Martin GR (1981) Isolation of a pluripotent cell line from early mouse embryos cultured in medium conditioned by teratocarcinoma stem cells. Proc Natl Acad Sci USA 78:7634–7638
3. Robertson EJ, Kaufman MH, Bradley A, Evans MJ (1983) Isolation, properties and karyotype analysis of pluripotential (EK) cell lines from normal and parthenogenetic embryos. In: Silver LM, Martin GR, Strickland S (eds) Teratocarcinoma stem cells. Cold Spring Harbour Conferences on Cell Proliferation 10. CSHL Press, NY

4. Bradley A, Evans M, Kaufman MH, Robertson E (1984) Formation of germ-line chimaeras from embryo-derived teratocarcinoma cell lines. Nature 309:255–256
5. Doetschman TC, Eistetter H, Katz M, Schmidt W, Kemler R (1985) The in vitro development of blastocyst-derived embryonic stem cell lines: formation of visceral yolk sac, blood island and myocardium. J Embryol Exp Morph 87:27–45
6. Simpson EM, Linder CC, Sargent EE, Davisson MT, Mobraaten LE, Sharp JJ (1997) Genetic variation among 129 substrains and its importance for targeted mutagenesis in mice. Nature Genet 16:19–27
7. Festing MF, Simpson EM, Davisson MT, Mobraaten LE (1999) Revised nomenclature for strain 129 mice. Mamm Genome 10:836
8. Ledermann & Bürki (1991) Establishment of a germ-line competent C57BL/6 embryonic stem cell line. Exp Cell Res 197:254–258
9. Kontgen F, Suss G, Stewart C, Steinmetz M, Bluethmann H (1993) Targeted disruption of the MHC class II Aa gene in C57BL/6 mice. Int Immunol 8:957–964
10. Lemckert FA, Sedgwick JD, Korner H (1997) Gene targeting in C57BL/6 ES cells Successful germ line transmission using recipient BALB/c blastocysts developmentally matured in vitro. Nucleic Acids Res 25:917–918
11. Noben-Trauth N, Köhler G, Bürki K, Ledermann B (1996) Efficient targeting of the IL-4 gene in a BALB/c embryonic stem cell line. Transgenic Res 5:487–491
12. Smith AG, Hooper ML (1987) Buffalo rat liver cells produce a diffusible activity, which inhibits the differentiation of murine embryonal carcinoma and embryonic stem cells. Dev Biol 121:1–9
13. Williams RL, Hilton DJ, Pease S, Willson TA, Stewart CL, Gearing DP, Wagner EF, Metcalf D, Nicola NA, Gough NM (1988) Myeloid leukaemia inhibitory factor maintains the developmental potential of embryonic stem cells. Nature 336:684–687
14. Smith AG, Heath JK, Donaldson DD, Wong GG, Moreau J, Stahl Mand Rogers D (1988) Inhibition of pluripotential embryonic stem cell differentiation by purified polypeptides. Nature 336:688–690
15. Auerbach W, Dunmore JD, Fairchild-Huntress V, Fang Q, Auerbach AB, Huszar D, Joyner AL (2000) Establishment and chimera analysis of 129/SvEv and C57BL/6-derived ES cell lines. Biotechniques 29:1024–1032
16. Schoonjans L, Kreemers V, Danloy S, Moreadith RW, Laroche Y, Collen D (2003) Improved generation of germline-competent embryonic stem cell lines from inbred mouse strains. Stem Cells 21:90–97
17. Poueymirou WT, Auerbach W, Frendewey D, Hickey JF, Escaravage JM, Esau L, Doré AT, Stevens S, Adams NC, Dominguez MG, Gale NW, Yancopoulos GD, DeChiara TM, Valenzuela DM (2007) F0 generation mice fully derived from gene-targeted embryonic stem cells allowing immediate phenotypic analyses. Nat Biotechnol 25:91–99
18. Ying QL, Nichols J, Chambers I, Smith A (2003) BMP induction of Id proteins suppresses differentiation and sustains embryonic stem cell self-renewal in collaboration with STAT3. Cell 115:281–292
19. Cheng J, Dutra A, Takesono A, Garrett-Beal L, Schwartzberg PL (2004) Improved generation of C57BL/6J mouse embryonic stem cells in a defined serum-free media. Genesis 39:100–104
20. Ogawa K, Nishinakamura R, Iwamatsu Y, Shimosato D, Niwa H (2006) Synergistic action of Wnt and LIF in maintaining pluripotency of mouse ES cells. Biochem Biophys Res Commun 343:159–166
21. Sato H, Amagai K, Shimizukawa R, Tamai Y (2009) Stable generation of serum- and feeder-free embryonic stem cell-derived mice with full germline-competency by using a GSK3 specific inhibitor. Genesis 47:414–22
22. Nagy A, Gertsenstein M, Vintersten K, Behringer R (2003) Manipulating the mouse embryo: a laboratory manual. CSHL Press, NY, Third Edition
23. Buehr M, Smith A (2003) Genesis of embryonic stem cells. Philos Trans R Soc Lond B Biol Sci 358:1397–402
24. Ying QL, Wray J, Nichols J, Batlle-Morera L, Doble B, Woodgett J, Cohen P, Smith A (2008) The ground state of embryonic stem cell self-renewal. Nature 453:519–23
25. Threadgill DW, Yee D, Matin A, Nadeau JH, Magnuson T (1997) Genealogy of the 129 inbred strains: 129/SvJ is a contaminated inbred strain. Mamm Genome 8:390–393
26. Handyside AH, O'Neill GT, Jones M, Hooper ML (1989) Use of BRL-conditioned medium in combination with feeder layers to isolate a diploid embryonal stem cell line. Roux's Arch Dev Biol 198(48–55):14
27. Hooper ML, Hardy K, Handyside A, Hunter S, Monk M (1987) HPRT-deficient (Lesch-Nyhan) mouse embryos derived from germline colonization by cultured cells. Nature 326:292–295
28. McMahon AP, Bradley A (1990) The Wnt-1 (int-1) protooncogeneis required for development of a large region of the mouse brain. Cell 62:1073–1085

29. Li E, Bestor TH, Jaenisch R (1992) Targeted mutation of the DNA methyltransferase gene results in embryonic lethality. Cell 69: 915–926
30. Nagy A, Rossant J, Nagy R, Abramownewerly W, Roder JC (1993) Derivation of completely cell culture-derived mice from early-passage embryonic stem cells. Proc Natl Acad Sci USA 90:8424–8428
31. Swiatek PJ, Gridley T (1993) Perinatal Lethality and defects in hindbrain development in mice homozygous for a targeted mutation of the zinc finger gene Krox20. Genes Dev 7:2071–2084
32. Shipley JM, Wesselschmidt RL, Kobayashi DK, Ley TJ, Shapiro SD (1996) Metalloelastase is required for macrophage-mediated proteolysis and matrix invasion in mice. Proc Natl Acad Sci USA 93:3942–3946
33. Chen WV, Soriano P (2003) Gene trap mutagenesis in embryonic stem cells. Meth Enzymol 365:367–86
34. Wolfer DP, Crusio WE, Lipp HP (2002) Knockout mice: simple solutions to the problems of genetic background and flanking genes. Trends Neurosci 25:336–340
35. Eggan KH, Loring AJ, Jackson-Grusby L, Klemm M, Rideout WM, Yanagimachi R, Jaenisch R (2001) Hybrid vigor, fetal overgrowth, and viability of mice derived by nuclear cloning and tetraploid embryo complementation. Proc Natl Acad Sci USA 98:6209–6214
36. Valenzuela DM, Murphy AJ, Frendewey D, Gale NW, Economides AN, Auerbach W, Poueymirou WT, Adams NC, Rojas J, Yasenchak J, Chernomorsky R, Boucher M, Elsasser AL, Esau L, Zheng J, Griffiths JA, Wang X, Su H, Xue Y, Dominguez MG, Noguera I, Torres R, Macdonald LE, Stewart AF, DeChiara TM, Yancopoulos GD (2003) High-throughput engineering of the mouse genome coupled with high-resolution expression analysis. Nat Biotechnol 21:652–659
37. George SH, Gertsenstein M, Vintersten K, Korets-Smith E, Murphy J, Stevens ME, Haigh JJ, Nagy A (2007) Developmental and adult phenotyping directly from mutant embryonic stem cells. Proc Natl Acad Sci USA 104:4455–4460
38. Hughes ED, Qu YY, Genik SJ, Lyons RH, Pacheco CD, Lieberman AP, Samuelson LC, Nasonkin IO, Camper SA, Van Keuren ML, Saunders TL (2007) Genetic variation in C57BL/6 ES cell lines and genetic instability in the Bruce4 C57BL/6 ES cell line. Mamm Genome 18:549–58
39. Seong E, Saunders TL, Stewart CL, Burmeister M (2004) To knockout in 129 or C57BL/6: that is the question. Trends Genet 20:59–62
40. Hansen GM, Markesich DC, Burnett MB, Zhu Q, Dionne KM, Richter LJ, Finnell RH, Sands AT, Zambrowicz BP, Abuin A (2008) Large-scale gene trapping in C57BL/6N mouse embryonic stem cells. Genome Res 18:1670–1679
41. Pettitt SJ, Liang Q, Rairdan XY, Moran JL, Prosser HM, Beier DR, Lloyd KC, Bradley A, Skarnes WC (2009) Agouti C57BL/6N embryonic stem cells for mouse genetic resources. Nature Meth 6:493–496
42. Gertsenstein M, Nutter L, Reid T, Pereira M, Stanford WL, Rossant J, Nagy A (2010) Efficient generation of germ line transmitting chimeras from C57BL/6N ES cells by aggregation with outbred host embryos. PLoS ONE 5:11260
43. Smith AG (1991) Culture and differentiation of embryonic stem cells. J Tissue Cult Meth 13:89–94
44. Stryke D, Kawamoto M, Huang CC, Johns SJ, King LA, Harper CA, Meng EC, Lee RE, Yee A, L'Italien L, Chuang PT, Young SG, Skarnes WC, Babbitt PC, Ferrin TE (2003) BayGenomics: a resource of insertional mutations in mouse embryonic stem cells. Nucl Acids Res 31:278–281
45. Chaudhry MA, Vitalis TZ, Bowen BD, Piret JM (2008) Basal medium composition and serum or serum replacement concentration influences on the maintenance of murine embryonic stem cells. Cytotechnology 58:173–179
46. Wang Z, Jaenisch R (2004) At most three ES cells contribute to the somatic lineages of chimeric mice and of mice produced by ES-tetraploid complementation. Dev Biol 1 275 (1):192–201
47. Liu X, Wu H, Loring J, Hormuzdi S, Disteche CM, Bornstein P, Jaenisch R (1997) Trisomy eight in ES cells is a common potential problem in gene targeting and interferes with germ line transmission. Dev Dyn 209:85–91
48. Longo L, Bygrave A, Grosveld FG, Pandolfi PP (1997) The chromosome make-up of mouse embryonic stem cells is predictive of somatic and germ cell chimaerism. Transgenic Res 6:321–328
49. Sibilia M, Wagner EF (1995) Strain-dependent epithelial defects in mice lacking the EGF receptor. Science 269:234–8
50. Threadgill DW, Dlugosz AA, Hansen LA, Tennenbaum T, Lichti U, Yee D, LaMantia C, Mourton T, Herrup K, Harris RC, Barnard JA, Yuspa SH, Coffey RJ, Magnuson T (1995) Targeted disruption of mouse EGF

receptor: effect of genetic background on mutant phenotype. Science 269:230–4

51. Soriano P, Montgomery C, Geske R, Bradley A (1991) Targeted disruption of the c-src proto-oncogene leads to osteopetrosis in mice. Cell 64:693–702
52. Hanna J, Markoulaki S, Mitalipova M, Cheng AW, Cassasy JP, Staerk J, Carey BW, Lengner CJ, Foreman R, Love J, Gao Q, Kim J, Jaenisch R (2009) Metastable pluripotent states in NOD-mouse-derived ESCs. Cell Stem Cell 4:513–524
53. Ohta H, Ohinata Y, Ikawa M, Morioka Y, Sakaide Y, Saitou M, Kanagawa O, Wakayama T (2009) Male germline and embryonic stem cell lines from NOD mice: efficient derivation of GS cells from a non-permissive strain for ES cell derivation. Biol Reprod 81:1147–1153
54. Pan Y-X, Xu J, Xu MM, Rossi G, Matulonis JE, Pasternak GW (2009) Involvement of exon 11-associated variants of the mu opioid receptor MOR-1 in heroin, but not morphine, actions. Proc Natl Acad Sci USA 106:4917–4921

Chapter 16

Tetraploid Complementation Assay

Marina Gertsenstein

Abstract

The generation of mouse chimeras by combining host embryos with genetically modified embryonic stem (ES) cells is a necessary step towards establishment of genetically modified mouse strains. This protocol describes the procedures necessary for the production of mouse chimeras by tetraploid complementation assay. This technique can be used to generate ES cell-derived embryos or animals, and to rescue extraembryonic defects. It provides a powerful tool for direct analysis of phenotype and for studies of cell fate during mouse development. This method can speed up the production of genetically modified strains directly from hybrid ES cells. The Transgenic facility at the Samuel Lunenfeld Research Institute of Mount Sinai Hospital and more recently at the Toronto Centre for Phenogenomics has been successfully using this protocol since early 1990s.

Abbreviations

AC	Alternating electric current
BSA	Bovine serum albumin
DC	Direct electric current
E1.5 etc	Embryonic stage 1.5 and further
EGFP	Enhanced green fluorescent protein
FBS	Fetal bovine serum
hCG	Human chorionic gonadotropin
lacZ	A reporter gene encoding beta galactosidase
MEF	Mouse embryonic fibroblasts
PMSG	Pregnant mare's serum gonadotropin

Selected Vendors:
Millipore: www.millipore.com
Sigma: www.sigmaaldrich.com
Invitrogen: www.invitrogen.com
Harlan: www.harlan.com
Charles River Labs: www.criver.com

S. Pease and T.L. Saunders (eds.), *Advanced Protocols for Animal Transgenesis*, Springer Protocols, DOI 10.1007/978-3-642-20792-1_16, © Springer-Verlag Berlin Heidelberg 2011

Taconic: www.taconic.com
BLS Ltd: www.bls-ltd.com
Becton Dickinson: www.BD.com
Fine Science Tools: www.finescience.com
VWR: www.vwr.com
Drummond: www.drummondsci.com

16.1 Introduction

The protocol below is based on the procedures described by Nagy et al. [1]. Several different methods can be used to induce tetraploidy (reviewed in [2]). The electrofusion of blastomeres of 2-cell stage embryos for the generation of tetraploid embryos described here was first developed by Kubiak and Tarkowski [3]. Most tetraploid embryos die shortly after implantation, but when complemented with diploid embryos, their contribution is primarily restricted to the extraembryonic tissues: the primitive endoderm of the yolk sac and the trophoblast layer of the placenta, and excluded from the primitive ectoderm lineage [4]. Aggregation of tetraploid and diploid embryos is used to segregate embryonic and extraembryonic phenotypes and rescue or bypass extraembryonic defects of the mutations [5, 6]

ES cells have limited ability to contribute to the trophoblast lineage [7]. When ES cells are complemented by tetraploid embryos they colonize the embryo proper, the amnion, the allantois as well as the mesoderm layer of the yolk sac, while tetraploid cells are excluded from these lineages and restricted to the extraembryonic tissues [8, 9] resulting in nearly completely ES cell-derived embryos. The aggregation of ES cells with tetraploid embryos provides a rapid test for the developmental potential of ES cells and allows the generation of mutant embryos directly from ES cells for analysis of their phenotype [10]. Moreover, it is possible to derive viable and fertile animals carrying mutations directly from ES cells and speed up traditional breeding using characterized F1-hybrid ES cells introduced into tetraploid embryos by blastocyst microinjection or aggregation methods [11, 12].

Occasionally, a small contribution of tetraploid cells can be observed in the embryo proper; therefore, reporters such as EGFP or lacZ are often used for the tetraploid host embryos in order to make them easily recognizable and to confirm ES cell origin of mutant embryos when they are dissected at mid-gestation [13]. For more details on the applications of the tetraploid complementation assay and chimera analysis in general, *see* [14–16].

16.2 Materials

16.2.1. Equipment, Tools, and Plasticware

- Stereomicroscope(s) with transmitted light for embryo manipulations and incident light or fiber optics with a gooseneck for embryo transfer surgery. The use of two microscopes is convenient for the fusion of 2-cell stage embryos and for embryo transfers into pseudopregnant females; a single microscope is also sufficient, however. We find that frosted instead of transparent glass in the base of the microscope with transmitted light often provides better visualization of the zonae pellucidae, which is helpful for zona removal.
- Inverted microscope with phase contrast for ES cell culture observation.
- Biosafety Cabinet for routine cell culture.
- Humidified incubator(s) at 37°C, 5% CO_2 for embryo and ES cell culture.
- Cell-fusion instrument [e.g., CF-150B pulse generator with 250 μm electrode chamber (BLS Ltd, Hungary, www.bls-ltd.com)].
- Aggregation needle DN-09 (BLS Ltd, Hungary, www.bls-ltd.com).
- Sterile Petri and tissue culture dishes (35, 60, 100 mm), organ culture dishes (Falcon 35–3037). 35 mm Easy Grip Falcon 35–3001 tissue culture plates are well suited for making depressions in the plastic for aggregates.
- Sterile Pasteur pipettes and plastic pipettes for tissue culture (1 ml, 5 ml, 10 ml).
- Sterile 1 cc syringes, 26 G 1/2 and 30 G 1/2 needles. To make a flushing needle, the sharp tip of 30 G 1/2 needle is cut off and/or polished on a sharpening stone or sand paper. The flushing needle is flushed with 70% ethanol before and after use.
- Bunsen or alcohol burner for pulling embryo-manipulating pipettes.
- Embryo-manipulating pipettes: Pasteur pipettes or glass microcapillaries (e.g., Drummond 1-000-0400 or 1-000-0500) are drawn by hand over the flame and broken flat with an inner diameter slightly larger than an embryo (>100 μm). It is very important to flame polish the tip to prevent damaging zona-free embryos. Embryo-manipulating pipettes are connected through elastic rubber tubing (e.g., VWR 62996–350) to an aspirator mouthpiece (Drummond 2-000-000 or Sigma A5177 Aspirator Tube Assembly). Pasteur pipettes fit into standard 1,000 μl pipettor tips and microcapillaries are

inserted into silicone tips (Drummond 1-000-9003 or Sigma A5177).

- Surgical instruments (e.g., Fine Scientific Tools – FST): sharp fine-pointed scissors, fine forceps (e.g., Dumont #5), straight or curved blunt forceps with serrated tips, serrefine (e.g., FST18050-28).
- AUTOCLIP Wound Clip Applier (Becton Dickinson 427630) and AUTOCLIP Wound Clips, 9 mm (Becton Dickinson 427631).

16.2.2. Mouse Stock

Outbred stocks such as ICR/CD-1 (e.g., Harlan Sprague Dawley, Charles River Laboratories or Taconic) are commonly used as donors of host embryos as well as recipients of manipulated embryos. For analysis of mutant embryos at mid-gestation, outbred females are generally mated with homozygous transgenic males expressing ubiquitously a reporter such as EGFP [17] to generate tetraploid host embryos. The presence of a reporter assists in identification of cells derived from tetraploid embryos. If fluorescent reporters are not available and where albino animals have been used for the generation of tetraploid embryos, then ES cell contribution of 129B6F1 hybrid or other pigmented strains can be determined by fetal eye color, starting at E11.5. The details of mouse colony management and procedures involved in the production of superovulated embryo donors and pseudopregnant recipients are described in [18] as well as elsewhere in this book (Chaps. 6 and 25).

16.2.3. ES Cells

Before attempting to generate completely ES cell-derived embryos or animals from genetically modified ES cell clones it is very important to establish that a nonmanipulated parental ES cell line in existing culture conditions has such developmental potential. We use aggregation with tetraploid ICR embryos for R1 and G4 ES cells derived in A. Nagy's laboratory.

R1 ES cells [9] were derived from a hybrid of two 129 substrains: 129X1 female crossed with 129S1 male. 129X1 is white-bellied, pink-eyed, light chinchilla (A^{w}/A^{w} $Oca2^{p}$ $Tyr^{c\text{-}ch}/Oca2^{p}$ $Tyr^{c\text{-}ch}$) and 129S1 is white (or light)-bellied agouti (A^{w}/A^{w}). Thus, R1 ES cells are homozygous for white-bellied agouti allele at the agouti locus and heterozygous for chinchilla ($Tyr^{c\text{-}ch}$) and pink-eyed dilution ($Oca2^{P}$): $A^{W}A^{W}$ *Cc-ch Pp*; they have light-bellied agouti coat color and black eyes. For more details on 129 substrains see http://www.informatics.jax.org/mgi-home/nomen/strain_129.shtml

G4 ES cells were established from the cross of 129S6/SvEvTac female and C57BL/6NTac male (129S6B6F1). They are heterozygous for white-bellied agouti allele at the agouti locus and homozygous for tyrosinase (Tyr^{+}) and pink-eyed dilution (Oca2[+]) loci: $A^{W}a$ *CC PP*. Nonmanipulated G4 ES cells aggregated with tetraploid ICR embryos produced completely ES cell-derived pups at a rate of ~30% per number of aggregates transferred up to passage 14. After one or two electroporations, 75% of G4 subclones produced newborn pups at ~25% rate [12, 19].

16.2.4. Reagents

16.2.4.1. Embryo Culture

During the aggregation experiments, embryos are cultured in vitro for 2 days and at least 24 h without zona pellucida. Zona removal dramatically increases the embryos' sensitivity, which makes optimal culture conditions absolutely necessary for success. Embryo culture medium can be purchased commercially (e.g., Millipore) or prepared as described in [18]; aliquots are stored at 4°C and should not be stored as such for more than 2 weeks. The quality of water is very critical for media preparation. Water should be obtained from a regularly maintained Milli-Q (Millipore) filtration system preferably pretreated by deionization or purchased commercially (e.g., Invitrogen, Sigma). Disposable plasticware is highly recommended; if glassware is used, it should never be exposed to detergents or organic solvents. All chemicals should be of highest grade, embryo tested if available and once purchased, used only for media preparation.

Embryos are cultured in organ culture dishes or in microdrops covered with embryo-tested light mineral oil (e.g., Millipore ES-005-C or Sigma M8410). As embryos are intolerant of pH and temperature fluctuations, the time between euthanizing the embryo donors and placing the embryos in the culture dish should be minimized. To test in vitro culture conditions, zygotes are cultured for 96 h. More than 80% should reach blastocyst stage. For additional details on preimplantation embryo in vitro culture, refer to [18]. The list of necessary reagents is below.

- M2 (e.g., Millipore MR-015-D) is a HEPES-buffered medium used during embryo collection and other manipulations while in room atmosphere such as electrofusion and zona removal. Aliquots are stored at 4°C and brought to room temperature prior to use. If M2 is brought to 37°C it should be done on a warm plate or in a 37°C oven. If you must use a 5% CO_2 incubator to warm M2 media, make sure the cap is tightly closed to avoid exposure of the media to CO_2 as it will alter the pH. Embryos should not be kept in M2 for a prolonged period of time and need to be rinsed through several drops of CO_2-equilibrated embryo culture medium before being placed in the incubator. Historically, we use M2 medium for all embryo manipulations outside the incubator; however, M2 can be substituted with FHM (HEPES-buffered KSOM medium) or any other HEPES-buffered media corresponding to the embryo culture medium used in subsequent procedures.

- KSOM-AA (e.g., Millipore MR-121-D) is a bicarbonate-buffered medium with nonessential amino acids used for embryo culture [20–22]. Ideally, freshly prepared medium as well as the air space in a tube should be gassed with 5% CO_2 gas mixture and re-gassed after opening because medium rapidly becomes alkaline outside the incubator. Embryo culture medium is gas-equilibrated by placing the open tube or prepared microdrop dishes in the incubator at least a few hours or better 16–20 h before use.
- Embryo-tested light mineral oil (e.g., Millipore ES-005-C or Sigma M8410) is used to overlay microdrops of embryo culture medium. Unopened containers with oil can be stored at room temperature (below 30°C) away from normal light. We prefer to aliquot oil using proper aseptic procedures, store in the fridge, and incubate with loose cap overnight in CO_2 incubator.
- Acid Tyrode's solution (e.g., Sigma T1788) is used for zona removal. Aliquots are stored at –20°C. One thawed aliquot should be kept at 4°C for no longer than 2 weeks and brought to room temperature prior to use.
- 0.3 M mannitol (Sigma M4125) is used for 2-cell stage embryo electrofusion. It is prepared in ultrapure embryo-tested water (e.g., Sigma W1503) containing 0.3% BSA (e.g., Sigma A3311), filtered and stored at –20°C. Aliquots are thawed prior to use and not re-frozen.

16.2.4.2. ES Cell Culture

ES cells are typically grown on mitotically inactivated mouse embryo fibroblasts in an ES cell medium containing FBS. Before purchasing a new batch, we routinely test FBS lots in vitro as described in several publications [18, 23] and whenever possible by tetraploid complementation assay. FBS is kept frozen at −80°C up to 2 years and must be used within 4 months after thawing. If the whole bottle is not going to be used during that time, it is best to prepare aliquots and re-freeze them. The ES cell medium is made fresh as necessary, kept at 4°C and used within 3–4 weeks.

- ES cell medium
  - Dulbecco's modified Eagle's medium (DMEM) (Invitrogen 11960)
  - 15% FBS (ES cell qualified)
  - 2 mM GlutaMAX™ (Invitrogen 35050) or L-Glutamine (Invitrogen 25030)
  - 0.1 mM 2-mercaptoethanol (Invitrogen 21985–023)

  - 0.1 mM MEM nonessential amino acids (Invitrogen 11140)
  - 1 mM Sodium pyruvate (Invitrogen 11360)
  - 1,000 U/ml LIF (Millipore ESG1107)
  - 50 U/ml Penicillin and 50 μg Streptomycin (Invitrogen 15140) – optional
- 0.05% Trypsin/EDTA (Invitrogen 25300) or 0.25% Trypsin/EDTA (Invitrogen 25200)
- 0.1% gelatin in sterile water (Millipore ES-006B)
- Ca/Mg free D-PBS (Millipore ES-1006-B or Invitrogen 14190)

16.3 Protocol

16.3.1. Collection of 2-Cell Stage Embryos

Table 16.1 describes the co-ordination of mouse and ES cell protocols necessary for the generation of embryo donors and recipients as well as the preparation of ES cells for the aggregation experiment.

Two-cell stage embryos collected at E1.5 are used for generation of tetraploid embryos by electrofusion. Fused embryos are cultured overnight and aggregated with ES cells (or diploid embryos) the following day when they reach the "4-cell stage."

1. One day before embryo collection: prepare culture plates using KSOM-AA medium (e.g., one organ culture dish

Table 16.1
Timeline and co-ordination of mouse and ES cells preparation

	Day 1	Day 2	Day 3	Day 4	Day 5	Day 6	Day 7
Mouse embryo donors	PMSG injection		hCG injection & mating	S/O plug checking	E1.5 S/O embryo donors		
Mouse embryo recipients				Mating with VAS males	VAS plug checking		E2.5 pseudo pregnant recipients
ES cells	Thawing	Media change	Passage or thawing	Media change	Passage on gelatin	Trypsinization for aggregation	
Embryo manipulations				Prep' of embryo culture dishes	E1.5 embryo collection & electrofusion Prep' aggregation plates	Zona removal & aggregation with ES cells	Uterine embryo transfer

S/O superovulation, *VAS* vasectomized males

and/or one dish with microdrops overlayed with embryo-tested mineral oil). Draw the line(s) on the bottom of the microdrop dish to distinguish two groups of embryos: those, subjected to the electric pulse but not fused yet, and those fused after the pulse. Collected embryos can be kept in an organ culture dish or in a separate drop of the same or different dish before the pulse (Fig. 16.1a, b).

2. Alternatively, place the tubes containing KSOM-AA medium and embryo-tested mineral oil with the caps loose into the incubator to equilibrate overnight or at least for a few hours before embryo collection.

3. On the day of the embryo collection and electrofusion: bring M2 medium to room temperature and prepare culture dishes using equilibrated KSOM-AA medium and oil if not prepared the day before. Pull embryo-manipulating pipettes.

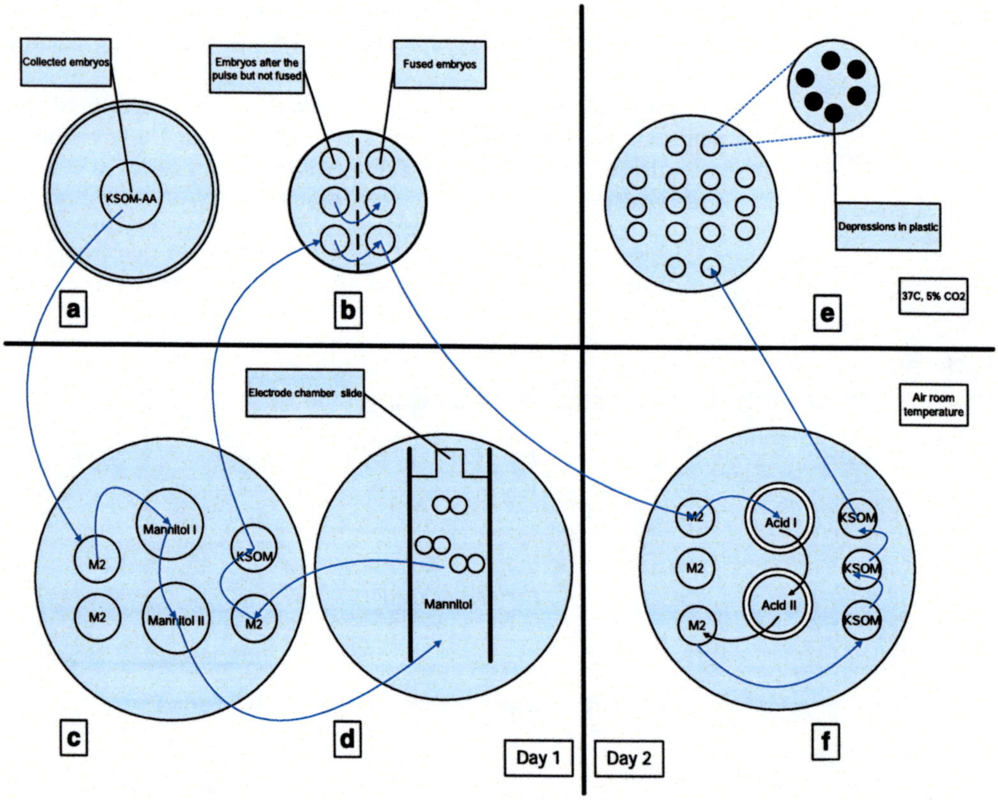

Fig. 16.1 Electrofusion and zona removal scheme. (a) Embryo culture dish; (b) Microdrop dish divided into the zones with not fused and fused embryos after the pulse; (c) Mannitol and media washes plate; (d) Electrode chamber slide with 2-cell stage embryos oriented by AC field; (e) Aggregation plate with depression wells made in the plastic; (f) Zona Pellucida removal by acid Tyrode's.

4. Dissect the oviducts and place them in a drop of M2 medium. The time between euthanizing embryo donors by cervical dislocation and placing the embryos in culture dishes should be kept to a minimum (ideally no more than 30 min). Do not dissect more donor females than can be handled in 30 min.
5. Transfer one oviduct into a small drop of M2 medium under stereomicroscope; insert the flushing needle attached to a 1 or 5 ml syringe filled with M2 into the infundibulum, and gently press the tip of the flushing needle against the bottom of the dish to hold it in place. The use of fine forceps helps to hold the needle in the right position. Flush M2 medium through the oviduct; observe its swelling. Proceed with the remaining oviducts, keeping the time of manipulations to a minimum.
6. Collect the embryos and wash them through several M2 drops to get rid of all debris and several drops of equilibrated KSOM-AA medium. Transfer the embryos into a prepared KSOM-AA embryo culture dish and place it back in the incubator. We find that leaving E1.5 embryos in the incubator for at least 15 min before applying the electric pulse improves the fusion rate.

16.3.2. Generation of Tetraploid Embryos

1. The fusion of blastomeres of 2-cell stage embryos occurs when DC electric pulse is applied perpendicular to the plane of the blastomeres' contact area. The suggested parameters for different types of electrode chambers of CF-150B BLS fusion instrument are listed in Table 16.2. The actual parameters vary depending on the instrument and the mouse strain – they need to be determined in a pilot experiment. The voltage for the square wave DC pulse applicable for electrofusion of mouse embryos is recommended to be 1–1.5 kV/cm [24]. The goal is to reach a 90% fusion rate in 30–45 min without embryo lysis. The adjustable 1 MHz AC field set up at 0.5–1 V orients the 2-cell stage embryos in the electrode chamber using nonelectrolyte solution, so that the blastomeres contact area is parallel to the electrodes and

Table 16.2
Suggested parameters for different electrode chambers of CF-150B BLS fusion instrument

Electrode chamber	DC voltage	Duration	# of pulses	AC voltage
250 μm	30 V	40 μs	1–2	1 V
500 μm	50 V	35 μs	2	2 V
1,000 μm	160 V	36 μs	2	2.2 V

enables the simultaneous fusion of a group of embryos instead of individual embryo fusion performed in electrolyte solution.

2. Thaw a frozen aliquot of 0.3 M mannitol, turn on the cell-fusion machine (we leave it on during the embryo collection). Make sure the switch on the back of the machine is on the "nonelectrolyte" or "normal" setting and set up other parameters. We routinely apply one or two pulses of 30 V and 40 μs for the fusion of ICR embryos in a 0.3 M mannitol solution using a CF-150B cell-fusion BLS instrument with a 250 μm electrode chamber.
3. Wipe the electrode chamber slide with 70% ethanol. The chamber must be cleaned thoroughly before use.

Note: Do not immerse the electrodes in ethanol for any length of time as it will lead to the electrodes' damage!

4. Connect the electrodes to the pulse generator and place the slide into a 100 mm Petri dish. Use the electrode clip or tape to secure it to the dish to prevent it from moving. The same dish can be used for all embryo washes before and after application of the electric pulse if only one microscope is available. Alternatively, set up an additional plate for washes on the second microscope; it helps to speed up the manipulations without the need to move the dish and adjust the focal plane. We find it more convenient.
5. Place several drops of the M2 medium (e.g., 50 μl) for "before" and "after" the pulse and 1–2 drops of the mannitol solution in 100 mm Petri dish (e.g., 100–200 μl) See Fig. 16.1c.
6. Pick up 2–3 groups of 25–30 embryos from embryo culture dish (Fig. 16.1a) and transfer them into "before the pulse" M2 drops (Fig. 16.1c). The number of embryos in the group is determined by the total time they can all be handled, so their manipulations outside the incubator do not exceed 15–20 min. No more than 2–3 groups of 10–20 embryos should be used for initial experiments.
7. Place one large mannitol drop over the chamber to cover both electrodes.

Note: The mannitol drop over the electrode chamber should not be used for longer than 15 min at a time. It must be replaced with fresh mannitol for every new batch of embryos.

8. Pick up the first group of 25–30 embryos in a minimal volume of M2 media and transfer it into the first mannitol drop. Quickly collect the embryos (they will be floating) and move them into the second mannitol drop (Fig. 16.1c). The embryos should be well equilibrated in mannitol before

placing them in the electrode chamber or they will float to the surface of the mannitol instead of resting between the electrodes. It is important not to transfer M2 medium into the electrode chamber otherwise embryo orientation by the AC field will not work efficiently. However, embryos should be kept in mannitol for a minimum amount of time.

9. Place the embryos between the electrodes, leaving space between them (Fig. 16.1d). An AC field set up in advance will orient the embryos so the blastomeres' contact area will be parallel to the electrodes, while they are dropped into the chamber. It is also possible to manually adjust the AC field after placing all the embryos into the electrode chamber, gradually increasing the voltage until all of them are fully aligned with the cleavage plane parallel to the electrodes (Fig. 16.2a).

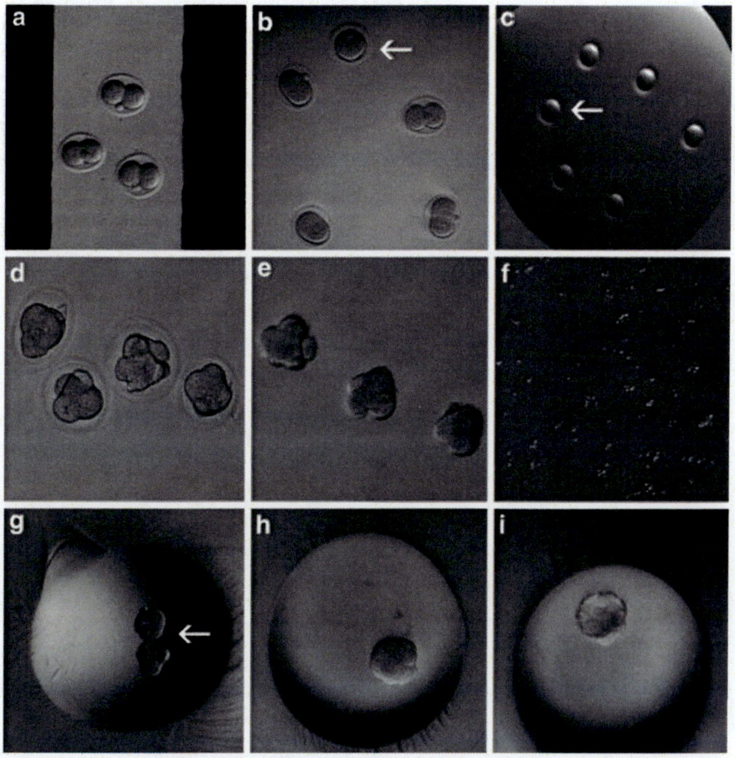

Fig. 16.2 Production of tetraploid embryos and their aggregation with ES cells. (**a**) Orientation of embryos in electrofusion chamber between electrodes under AC field; (**b**) Two-cell stage embryos undergoing fusion after application of the pulse, *arrow* indicates fused embryo; (**c**) Depression wells (one marked by the *arrow*) in embryo culture microdrop; (**d**) "Four-cell" stage tetraploid embryos after overnight culture; (**e**) Zona removal; (**f**) Lifted ES cell colonies ready for aggregation; (**g**) Aggregate of two "4-cell" stage tetraploid embryos and a clump of ES cells marked by the *arrow*; (**h**) Compacted morula after overnight culture; (**i**) Blastocyst after overnight culture.

10. Manually adjust the few embryos that are not aligned by the AC field. Push the trigger pulse and immediately transfer the embryos from the electrode chamber into an "after the pulse" M2 drop (Fig. 16.1c).
11. Proceed with the rest of the embryos until all the embryos in the first batch receive the pulse.
12. Wash the embryos through 2–3 drops of M2 medium to dispose of all traces of mannitol and then through 2–3 drops of equilibrated KSOM-AA medium. Place the embryos into the embryo culture plate in the incubator (Fig. 16.1b).
13. Wipe the electrode chamber slide with 70% Ethanol.
14. Proceed with the rest of the collected embryos as described in Subheading 16.3.2, **steps 6–12**.
15. Approximately 30 min after the pulse application, carefully assess the morphology of the fusing embryos after the second batch is placed in culture (Fig. 16.2b). Under optimal conditions, around 90% of embryos should fuse in 45–60 min. Select successfully fused embryos that look like a 1-cell stage embryo, and move them to new KSOM-AA drops that are marked as "fused" (Fig. 16.1b). Remove the embryos that cleaved before fusion, they can be used for aggregation as diploid embryos in a parallel experiment if necessary. Return the dish to the incubator.

Note: It is very important to select only completely fused embryos and transfer them into a fresh drop on the "fused" side of the dish. Since embryos are recovered at the late 2-cell stage, the second mitotic division is expected soon after the fusion. If not checked in time, fused and cleaved tetraploid embryos cannot be distinguished from nonfused diploid 2-cell stage embryos.

16. A second pulse can be applied to the embryos that did not fuse after 1–1½ h.
17. After overnight incubation, the "4-cell" stage embryos are used for aggregation in the afternoon as described in Subheading 16.3.6. The development of fused embryos to the "4-cell" stage should be at least 80% in optimal culture conditions. Embryos arrested at the "1-cell" stage are not used for aggregations. "2-cell" stage embryos are delayed and some may be aggregated later in the day after they have developed to the "4-cell" stage. Aggregation of "2-cell" stage tetraploid embryos is not efficient in our hands.

16.3.3. Preparation of the Aggregation Plates

The plates are usually prepared in the afternoon of the day before aggregation; alternatively use equilibrated KSOM-AA medium and oil if it is done on the day of aggregation.

1. Place microdrops (~3 mm in diameter or 10–15 μl) onto 35 mm Falcon 353001 Easy Grip dish using 1 cc syringe filled with KSOM-AA medium or micropipettor. We usually make two rows of four to five drops in the middle of the plate and two more rows of three drops on each side but the actual configuration of the microdrops is an individual choice (Fig. 16.1e). Immediately cover the drops with embryo-tested mineral oil.
2. Wipe the aggregation needle with 70% ethanol, if required the needle can be autoclaved. Press the needle into the plastic and make a slight circular movement. Do not press too hard or the plate will crack; however, not enough pressure will result in too shallow of a depression. The goal is to create a small cavity with a smooth and transparent surface that is deep enough to hold the aggregate safely when moving the plate to the incubator. Make six to eight depressions per microdrop, position them in the circle approximately halfway between the center and the edge of the microdrop (Fig. 16.2c). Do not make depressions too close to the edge – the embryos will be difficult to manipulate. Avoid the center, so that in the event that air bubbles are accidentally introduced to the drop, the embryos will remain visible.
3. Leave a few microdrops on the side without depressions; they will be used for the final selection of ES cell clumps as described in Subheading 16.3.6. We usually make depressions for 40–60 aggregates per plate to limit the time of embryo manipulations outside the incubator.
4. Place the aggregation plate in the incubator until it is needed.

16.3.4. Zona Removal

Acid Tyrode's solution is used to dissolve the glycoprotein membrane surrounding the embryo called the zona pellucida. If the acid is diluted with HEPES-buffered medium, it will not work efficiently; on the other hand any acid transferred into embryo culture media will damage the embryos. Hence, it is very important to transfer minimal amounts of solutions between drops and use multiple washes.

1. Place several drops of M2 media and acid Tyrode's in 100 mm Petri dish (Fig. 16.1f). The use of the tissue culture plate's lid or bacterial grade dish helps to reduce stickiness. The temperature of the acid Tyrode's should not exceed room temperature or it will act too quickly and may lead to the embryos' damage, increased stickiness, and difficulties in embryos' manipulations.
2. Transfer a few groups of embryos from the embryo culture dish and place them in M2 drops. The number of embryos manipulated at a time depends on the speed of manipulations.

With practice, it is possible to manipulate 30–50 embryos, but initially start with no more than 5–10 embryos in each group. Manipulations outside of the incubator should be limited to 20 minutes.

3. Transfer a group of embryos with a minimal volume of media into the first acid drop. Gently pipet the embryos around in acid and place them into second fresh drop of acid (Fig. 16.1f). Keep moving the embryos and observe zona dissolution that should happen within a few seconds, unless too much M2 was carried over with embryos (Fig. 16.2d, e).

Note: It is very important to fire-polish the tip of the pipette as zona-free embryos can be easily damaged by sharp edges.

4. As soon as the zona dissolves, immediately transfer the embryos with a minimal volume of acid into the drop of M2 media. Rinse them through 2–3 drops of M2 media to remove any remaining acid. Spread the embryos in the last drop of M2 and do not allow them to touch each other (Fig. 16.2e).
5. Proceed with the zona removal on the remaining embryos.
6. Wash all denuded embryos through several drops of equilibrated KSOM-AA; place them individually into the aggregation plates, directly inside or outside the depression wells (Fig. 16.3) depending on the way the aggregates will be assembled, as described in Subheading 16.3.6.
7. Keep the plates with embryos in the incubator until the ES cells are ready.

16.3.5. Preparation of ES Cells for Aggregation

It is important to have optimal ES cell culture conditions at all times, but particularly for ES cell clones introduced into mice to preserve their developmental potential. ES cells should be kept subconfluent in order to maintain their undifferentiated state, i.e. split at densities that are neither too high nor too low. Generally, ES cells are not kept for more than 2 days without passaging (1:5–1:7) and are used for experiments when growing exponentially. The medium is changed daily and never allowed to become yellow. More details on ES cell culture are provided in [18] and elsewhere in this book (Chap. 17).

1. Thaw a vial of ES cells on a plate with mitotically inactivated mouse embryonic fibroblasts (MEF) 3 days before aggregation; the surface depends on the number of ES cells frozen in the vial, 35 mm or 60 mm plates are sufficient. If the cells were frozen properly, they should be subconfluent and ready for passage in 2 days. A vial of unknown viability should always be thawed earlier, e.g., 5 or more days before aggregation, to ensure their timely recovery and provision of the quantity of cells required for experiment. Change the medium the day after thawing.

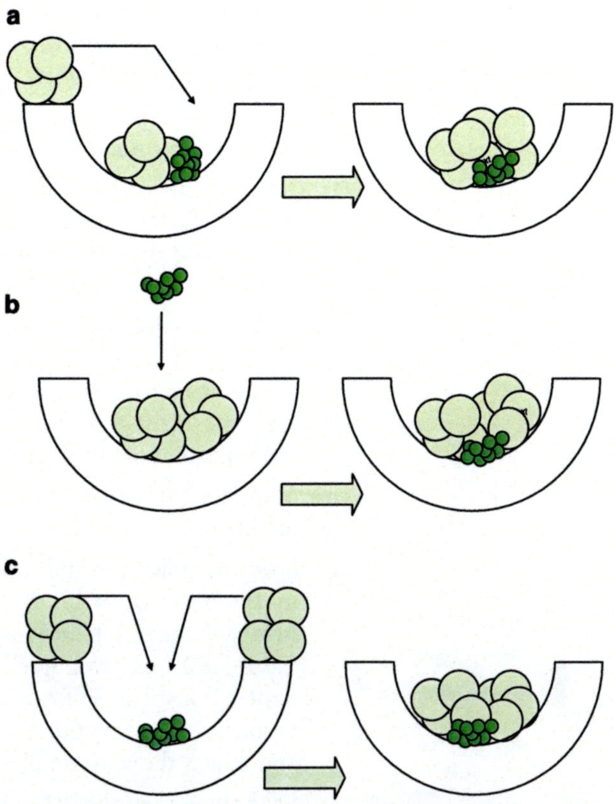

Fig. 16.3 Assembly of tetraploid embryos <–> ES cells aggregates. Right after zona removal denuded embryos are placed into microdrops of aggregation plate inside the depressions in one of the following ways: (**a**) one embryo in each depression well, (**b**) two embryos in each depression well; or (**c**) beside the depressions making sure that they do not touch each other. When the ES cells are ready the clumps are positioned next to the embryos (**a** and **b**) or distributed into the empty depressions (**c**), the embryos are then placed inside the depression next to the clumps. In all cases, special care should be taken to make sure both embryos and ES cell clumps touch each other.

2. One day prior to the aggregation, passage subconfluent ES cells on gelatinized plates as described below. For most clones 24 h is enough, while 48 h may be necessary for slower growing clones. This sparser than usual passage should produce the colonies of 8–15 cells that will be lifted by gentle trypsinization immediately before aggregation.
   (a) Add 0.1% gelatin to two to three dishes (e.g., 2 ml per 60 mm dish).
   (b) Aspirate the medium from ES cell plate and wash the cells with PBS.
   (c) Add trypsin to cover the cells (e.g., 1 ml per 60 mm dish), place it in the incubator for 5 min.

(d) Neutralize the trypsin with ES cell medium containing FBS and re-suspend the cells by gentle pipetting, ensure single cell suspension.

(e) Transfer the cell suspension to a tube and centrifuge at 200 g for 5 min.

(f) Remove the supernatant and gently re-suspend the pellet in fresh ES cell medium. The volume depends on cell density and the surface, e.g., a subconfluent 60 mm dish can be re-suspended in 5 ml of medium.

(g) Leave the tube undisturbed for 3–5 min to allow for the majority of large ES cell clumps and feeders to settle. Alternatively, place the cell suspension back into the original dish and put it back in the incubator for 10 min to allow the MEF to reattach (preplating).

(h) Aspirate the gelatin solution from the plates prepared earlier.

(i) Seed the cells from the top portion of the cell suspension in the tube or from carefully tilted original dish on new gelatinized plates using different dilutions (e.g., 1:10–1:50). For example, 0.2 ml, 0.4 ml, and 0.6 ml of 5 ml cell suspension from a subconfluent 60 mm dish can be plated on three 60 mm plates. Check the cell density and adjust the volume if necessary.

(j) Seed the rest of the cells on one or more plates, they will serve as a back-up and/or may be frozen.

(k) Incubate overnight.

3. Next day lift small colonies of 8–15 cells by gentle trypsinization immediately before the aggregation as follows.

   (a) Aspirate the medium and wash the cells first with PBS, and then with trypsin (this is optional, it helps to loosen up the cells and minimize the amount of trypsin in the next step).

   (b) Add a minimal amount of trypsin to just cover the cells, e.g., 0.3–0.5 ml per 60 mm plate, place in the incubator for 1–2 min or leave at room temperature. Watch under the microscope, gently swirl the plate to detach the colonies and tap at the microscope stage until all colonies are lifted. Do not over-trypsinize, as cells will become sticky and hard to manipulate.

   (c) As soon as the colonies are detached, quickly add ES cell medium to the plates (e.g., 4 ml per 60 mm dish). Do not pipette. However, if ES cell clumps are larger than required 8–15 cells, a very gentle pipetting can be used being careful not to break the clumps into single cell suspension.

4. Loosely connected ES cell clumps (Fig. 16.2f) are now ready for aggregation within next 1–2 h and can be kept in the original plate. Alternatively, for ease of transportation, gently transfer the suspension of clumps into 5 ml tubes (e.g., Falcon 352063 or 352058) using a pipette; be careful not to break them into single cells. Keep the cells at room temperature, as they will start attaching to the plate if placed in the incubator.

Note: If ES cells are grown on MEF up until the day of aggregation, colonies can be lifted by a very short trypsinization at room temperature leaving the majority of feeders behind. Transfer the floating ES cell clumps into new dish with medium, gently pipette if necessary to reach the clumps of the right size.

16.3.6. Assembly of Aggregates

We routinely use two tetraploid embryos at the "4-cell stage" to aggregate with ES cells in a sandwich manner as this method has proven to be more efficient in our hands. However, if it is necessary for the experimental design to have tetraploid mutant embryos of different genotypes that should not be mixed or the number of embryos is not sufficient, single tetraploid embryos may also be used. The aggregates can be assembled in either of the ways described in Subheading 16.3.6, **step 2**; they work equally well and the choice depends upon individual preference. Special care should be taken that all three components of the aggregates (two tetraploid embryos and an ES cell clump) touch each other, so that they form one embryo with integrated ES cells after overnight culture.

1. Under the dissecting microscope, collect ES cell clumps of roughly the required size and transfer them into the microdrops that do not contain depressions for the final selection and rinse from ES cell medium.
2. Select clumps of 8–15 cells and carefully transfer them individually into the depression wells using one of these three approaches, (Fig. 16.3):
   (a) Place the clump of ES cells next to one embryo, then carefully drop the second embryo into the depression to "sandwich" the ES cell clump (Fig. 16.3a).
   (b) Place the clump of ES cells next to two denuded embryos already positioned inside the well and likely fused into one embryo by that time (Fig. 16.3b).
   (c) Distribute the ES cell clumps into all empty depressions of the plate, then drop both denuded embryos into each well on top of ES cell clump (Fig. 16.3c).
3. Assemble all the aggregates in the plate (Fig. 16.2g). Check and make sure that all the embryos touch corresponding ES cell clumps.
4. Carefully put the plate in the incubator and culture overnight.
5. Proceed with the rest of the plates.

16.3.7. Embryo Transfer

The following day the majority of aggregates should reach blastocyst stage with some remaining late morulae (Fig. 16.2h, i) and are ready to be transferred into the uteri of E2.5 pseudopregnant females as described in [18] and elsewhere in this book (Chap. 17). The implantation rate of zona-free embryos is lower than for embryos with the zona intact. Two days in culture and the addition of ES cells decreases this still further. If embryos are to be dissected at different stages to assess in vivo expression we usually transfer 12–15 embryos per recipient. For the experiments left to term 18–22 embryos per recipient are transferred. In case of recipient shortage, aggregates can be transferred into the oviducts of E0.5, the uteri of E3.5 or cultured for an additional night and transferred the following day into E2.5 uteri. All these options work but we find them less efficient that the standard uterine transfer into E2.5 recipients.

16.3.8. Troubleshooting

- Low number of embryos is obtained: test different doses of hormones, time of injection, and age of donor females.
- A lot of embryos are developing to 4-cell stage before they can be fused: adjust the time of superovulation – early hormone injections promote embryo development.
- Embryos are not fusing efficiently: test different pulse parameters changing one at a time, e.g., increasing the voltage or duration and monitoring the time of fusion in different groups of embryos.
- Embryos are lysing after pulse application: likely too high AC field was used. Adjust and test with the same or different pulse parameters.
- Embryos are not aligning in the electrode chamber: too much HEPES-buffered medium was transferred with the embryos into Mannitol drop. Make sure embryos are well equilibrated in Mannitol before moving them to electrode chamber.
- Aggregates do not look viable after overnight culture: too large a clump of ES cells was used.
- Aggregates look like two embryos attached to each other after overnight culture: two embryos and ES cell clump did not touch each other at the time of aggregation to form one embryo.
- There is no pregnancy: embryo transfer surgery technique failed or recipient did not ovulate.
- There are no pups born but there are implantation sites: ES cell quality and culture conditions do not support the survival of ES cell-derived animals. Dissect at mid-gestation to determine the embryonic stage of lethality, test parental ES cell line and other subclones. Try different FBS lot for culture of ES cells.

References

1. Nagy A, Gocza E, Diaz EM, Prideaux VR, Ivanyi E, Markkula M, Rossant J (1990) Embryonic stem cells alone are able to support fetal development in the mouse. Development 110:815–821
2. Eakin GS, Behringer RR (2003) Tetraploid development in the mouse. Dev Dyn 228:751–766
3. Kubiak JZ, Tarkowski AK (1985) Electrofusion of mouse blastomeres. Exp Cell Res 157:561–566
4. Tarkowski AK, Witkowska A, Opas J (1977) Development of cytochalasin in B-induced tetraploid and diploid/tetraploid mosaic mouse embryos. J Embryol Exp Morphol 41:47–64
5. Guillemot F, Nagy A, Auerbach A, Rossant J, Joyner AL (1994) Essential role of Mash-2 in extraembryonic development. Nature 371:333–336
6. Rossant J, Guillemot F, Tanaka M, Latham K, Gertenstein M, Nagy A (1998) Mash2 is expressed in oogenesis and preimplantation development but is not required for blastocyst formation. Mech Dev 73:183–191
7. Beddington RS, Robertson EJ (1989) An assessment of the developmental potential of embryonic stem cells in the midgestation mouse embryo. Development 105:733–737
8. Eakin GS, Hadjantonakis AK, Papaioannou VE, Behringer RR (2005) Developmental potential and behavior of tetraploid cells in the mouse embryo. Dev Biol 288:150–159
9. Nagy A, Rossant J, Nagy R, Abramow-Newerly W, Roder JC (1993) Derivation of completely cell culture-derived mice from early-passage embryonic stem cells. Proc Natl Acad Sci USA 90:8424–8428
10. Carmeliet P, Ferreira V, Breier G, Pollefeyt S, Kieckens L, Gertsenstein M, Fahrig M, Vandenhoeck A, Harpal K, Eberhardt C et al (1996) Abnormal blood vessel development and lethality in embryos lacking a single VEGF allele. Nature 380:435–439
11. Eggan K, Akutsu H, Loring J, Jackson-Grusby L, Klemm M, Rideout WM 3rd, Yanagimachi R, Jaenisch R (2001) Hybrid vigor, fetal overgrowth, and viability of mice derived by nuclear cloning and tetraploid embryo complementation. Proc Natl Acad Sci USA 98:6209–6214
12. George SH, Gertsenstein M, Vintersten K, Korets-Smith E, Murphy J, Stevens ME, Haigh JJ, Nagy A (2007) Developmental and adult phenotyping directly from mutant embryonic stem cells. Proc Natl Acad Sci USA 104:4455–4460
13. Kunath T, Gish G, Lickert H, Jones N, Pawson T, Rossant J (2003) Transgenic RNA interference in ES cell-derived embryos recapitulates a genetic null phenotype. Nat Biotechnol 21:559–561
14. Nagy A, Rossant J (1996) Targeted mutagenesis: analysis of phenotype without germ line transmission. J Clin Invest 97:1360–1365
15. Nagy A, Rossant J (2001) Chimaeras and mosaics for dissecting complex mutant phenotypes. Int J Dev Biol 45:577–582
16. Tam PP, Rossant J (2003) Mouse embryonic chimeras: tools for studying mammalian development. Development 130:6155–6163
17. Hadjantonakis AK, Gertsenstein M, Ikawa M, Okabe M, Nagy A (1998) Generating green fluorescent mice by germline transmission of green fluorescent ES cells. Mech Dev 76:79–90
18. Nagy A, Gertsenstein M, Vintersten K, Behringer R (2003) Manipulating the mouse embryo: a laboratory manual, 3rd edn. Cold Spring Harbor Laboratory Press, Cold Spring Harbor, NY
19. Vintersten K, Monetti C, Gertsenstein M, Zhang P, Laszlo L, Biechele S, Nagy A (2004) Mouse in red: red fluorescent protein expression in mouse ES cells, embryos, and adult animals. Genesis 40:241–246
20. Ho Y, Wigglesworth K, Eppig JJ, Schultz RM (1995) Preimplantation development of mouse embryos in KSOM: augmentation by amino acids and analysis of gene expression. Mol Reprod Dev 41:232–238
21. Lawitts JA, Biggers JD (1992) Joint effects of sodium chloride, glutamine, and glucose in mouse preimplantation embryo culture media. Mol Reprod Dev 31:189–194
22. Lawitts JA, Biggers JD (1993) Culture of preimplantation embryos. Methods Enzymol 225:153–164
23. Joyner AL (2000) Gene targeting: a practical approach, 2nd edn. Oxford University Press, Oxford
24. McLaughlin KJ (1993) Production of tetraploid embryos by electrofusion. Methods Enzymol 225:919–930

Chapter 17

Combining ES Cells with Embryos

Elizabeth Williams, Wojtek Auerbach, Thomas M. DeChiara, and Marina Gertsenstein

Abstract

The marvel of embryonic stem (ES) cells is that after in vitro culturing and genetic modification, they still have the ability to contribute to the developing embryo, when combined with pre-implantation embryos, to produce chimeras and even completely ES cell-derived animals. In this chapter, we will describe three methods for combining ES cells with embryos: the injection of ES cells into blastocysts, the injection of ES cells into eight-cell stage embryos and aggregation of ES cells with morulae. To date, blastocyst injection is the most commonly used method, adopted by core facilities rather than individual laboratories, partially because of the high cost of equipment and long training period required, prohibitive to some labs. The Injection of eight-cell stage embryos can be performed using the same equipment, but because fewer cells are injected per embryo this method is faster and can be learned quickly by anyone trained in blastocyst injection. The procedure makes use of less expensive outbred embryo donor mice and produces completely ES cell-derived mice when good quality ES cells are used. Morula aggregation is performed under a simple dissecting stereomicroscope, thereby lowering the startup costs, and requires a shorter training period. Although the procedure utilizes less expensive outbred strains of mice as embryo donors, the savings are partially offset by the need for larger numbers of transferred embryos per female due to the lower implantation rate of the zona pellucida (ZP) free embryos. On the other hand, morula aggregations are much faster and easier to perform than microinjections and similar to eight-cell microinjections they often result in fully ES cell-derived animals.

Abbreviations

CM	Compacted Morulae
dpc	Days post coitum
DIC	Differential interference contrast optics
ES Cells	Embryonic stem cells
HMC	Hoffman modulation contrast optics
ICM	Inner cell mass
PVP	Polyvinylpyrrolidone
SW	Swiss Webster
ZP	Zona pellucida

S. Pease and T.L. Saunders (eds.), *Advanced Protocols for Animal Transgenesis*, Springer Protocols,
DOI 10.1007/978-3-642-20792-1_17, © Springer-Verlag Berlin Heidelberg 2011

17.1 Introduction

The generation of chimeric mice from gene-targeted ES cells is an essential step for the establishment of new genetically modified mouse strains. In addition, chimeric embryos or animals can be used for direct phenotype analysis in mouse development studies [1, 2]. There are various methods that have been employed to introduce initially embryonic carcinoma, inner cell mass (ICM) cells, and later ES cells, into the host mouse embryos, which are described in this chapter (1) injection into the blastocyst cavity; (2) injection into perivitelline space of morulae or eight-cell stage embryos; and (3) aggregation with denuded zona-free embryos.

The first mouse embryonic chimeras were made by Beatrice Mintz and Andrzej Tarkowski in the 1960s by the aggregation of two cleavage stage embryos resulting in mosaic animals [3, 4]. Microinjection of embryonic cells into the blastocyst was developed by Richard Gardner and required the use of five micropipettes [5]. Subsequently, Moustafa and Brinster [6] and Babinet [7] considerably simplified this technique and it is now the most commonly used classical method for the generation of ES cell chimeras.

The injection of ES cells into decompacted morula incubated in Mg^{2+}/Ca^{2+} – free PBS [8] became a standard method for the generation of chimeras in the Pasteur Institute [9]. Morulae are relatively easy to inject compared to blastocysts, firstly because they do not require penetration through the trophectoderm, and secondly because fewer cells are usually injected. Although less pups develop to term, a higher percentage of them are high degree chimeras [10] and more efficient germline transmission was reported when eight-cell embryos were used as hosts instead of blastocysts [11, 12]. The injections into eight-cell or morula can be performed exactly the same way as into blastocyst; alter natively, a slit cut in the ZP with a fine glass needle can be made [11, 12]. Modern modification of this technique employing the laser-assisted subzonal injection was developed by Regeneron Pharmaceuticals, Inc. [13] and is described in this chapter. Piezo assisted micromanipulations also can be used for ES cell injections into blastocysts [14] or four- to eight-cell stage embryos [15].

Chimeras generated between embryonic carcinoma cells and embryos using the aggregation technique were first demonstrated by Colin Stewart [16, 17]. This method was further expanded to include ES cells, diploid and tetraploid embryos [18]. The details and applications of the tetraploid complementation assay are described in Chapter 16. Aggregation of ES cells with zona-free diploid cleavage stage embryos in depression wells made in a plastic dish is the third method presented here used to generate mouse ES cell chimeras. Alternatively, zona-free embryos can be cultured on a layer of ES cells [19] or in microwell plates [20]. The aggregation method does not require sophisticated equipment and microinjection skills. A good stereomicroscope and standard embryo manipulation techniques are employed. This method can be relatively easily established in a new lab. Aggregations are performed under lower magnification and as a result ES cells cannot be selected based on their morphology, as is routine during microinjection. Typically, inexpensive readily available outbred albino stocks are used to supply embryos for aggregation, and it is relatively easy to obtain a sufficient number of embryos. This method is not dependent on the particular developmental stage of the blastocyst, which may or may not be suitable for injection. The same is also true for morula or eight-cell stage embryos injections. An additional advantage of using earlier than blastocyst stage embryos to generate chimeras by aggregation or injection is the observation that resulting chimeras are rarely partial transmitters that are most often obtained by blastocyst injections. The possible correlation between the allocation of the germline and urogenital lineages was discussed by Saburi et al. [21] who demonstrated a single ES cell injected into eight-cell stage embryo was capable of colonizing the germ cells.

Special attention should be paid to the embryo culture conditions during aggregations as zona-free embryos are more sensitive and are maintained in culture for longer periods than injected blastocysts. Some ES cell lines (e.g., D3, [22]) do not work with the aggregation method. In general, the choice about which approach to take in the generation of chimeric mice is based on individual preference, operator skills, and availability of equipment, ES cell lines, and mouse stocks.

Most ES cell lines used for gene targeting and subsequent generation of chimeras are XY. Although it is possible to obtain germline chimeras from female ES cells [23], it is not very efficient [24]. Generally, sex ratio distortion (more male than female chimeras) is expected when XY ES cells are combined with XX embryos, since conversion of the embryo from female to male may occur in a percentage of XX embryos. Such resultant chimeras transmit the ES cell-derived genome, but occasionally chimeras with high ES cell contribution prove to be hermaphrodites and sterile. The Y chromosome may be lost during XY ES cell manipulation, leading to the generation of female chimeras that can transmit through the germ-line [25, 26].

It is essential to test both the karyotype and the contribution capacity of the parental ES cell line prior to any ES cell manipulations. ES cell aneuploidy is a major cause of failure in obtaining high contribution chimeras [27]. Trisomy of chromosome 8 is very common in ES cells and this has been seen to confer a selective growth advantage [28]. Generally speaking, the time that ES cells spend in culture prior to or during manipulation, should be kept to a minimum and ideally the earliest possible passage of ES cells should be used for the generation of chimeras [29].

The choice of mouse strain for host embryo supply depends upon the strain of ES cells to be used. It is common to have two visually distinguishable cell populations based on coat and eye pigmentation; additionally, the strain combination should provide a competitive advantage to the ES cells. Until recently the majority of currently used ES cell lines were derived from different sub-strains of 129 that efficiently generate chimeras when injected into C57BL/6 but not into CD-1 or MF-1 blastocysts [30]. On the other hand, 129 and F1-hybrid ES cells can efficiently generate germline-competent chimeras when aggregated [24] or injected into outbred albino ICR/CD-1 morula and eight-cell stage embryos [11, 12].

Ongoing large-scale mouse mutagenesis projects under the umbrella of the International Knockout Mouse Consortium (IKMC) are using C57BL/6N ES cells to mutate all protein coding genes of the mouse as this the most characterized and the strain of choice for vast majorities of studies [31, 32]. In spite of being available from the early 1990s, significantly less-targeted mutations have been reported in C57BL/6 compared to 129 ES cells until their recent use in IKMC projects [33]. C57BL/6 ES cells have been shown to be less efficient in the generation of chimeras compare to 129 [34–38] and unstable in standard culture conditions [33, 34, 39]. It is not always possible to derive C57BL/6 ES cells in standard FBS medium; serum replacement [40] and conditioned medium [41] were successfully employed to overcome this problem; however, the C57BL/6 strain is often considered recalcitrant for derivation [42].

Various host embryos and methods have been used to generate germline chimeras from C57BL/6 ES cells: BALB/c blastocyst injection [43–45] and CD-1 morula aggregation [34]. The yield of good quality embryos from BALB/c is not very consistent and other options for the supply of host embryos such as (C3H/HeNCr X BALB/cAnNCr)F_1 blastocysts have been investigated [46]. Co-isogenic C57BL/6-Tyr$^{c\text{-}2J}$ blastocysts have proven to be efficient hosts [36, 47]. C57BL/6 ES cells injected into eight-cell stage C57BL/6-Tyr$^{c\text{-}2J}$ and outbred Swiss Webster (SW) embryos yielded F0 completely ES cell-derived animals [13]. Microinjection into C57BL/6-Tyr$^{c\text{-}Brd}$ blastocysts has been shown to be a favorable combination for obtaining germline transmission from C57BL/6 ES cell line JM8 [48]. This strain is now available from Harlan Laboratories (C57BL/6- BrdCrHsd-Tyr$^{c\text{-}Brd}$) and Charles River Laboratories (B6N-*Tyr*^{c}/BrdCrCrl). C57BL/6-Tyr$^{c\text{-}2J}$ is available from The Jackson laboratory (B6(Cg)-*Tyr*$^{c\text{-}2J}$/J cat # 000058) although it has a relatively high cost and a poor breeding performance.

The historically low efficiency of the germline transmission of C57BL/6 ES cells is likely related to their instability in standard FBS culture routinely used for 129 and hybrid ES cells. It was recently identified that the addition of the small molecule inhibitors of the Erk and GSK3 pathways [49] allowed the derivation of ES cells from strains previously considered to be nonpermissive, such as nonobese diabetic NOD [50]. This system of inhibitors is currently postulated to represent a generic culture condition for the maintenance of authentic pluripotency [51]. Indeed when C57BL/6 ES cells are maintained in optimal culture using inhibitors or conditioned medium, they are capable to efficiently generate germline chimeras by both aggregation and injection into eight-cell stage outbred albino embryos using conventional or laser-assisted methods [13, 52–54].

The most straightforward way to assay for ES cell contribution is the use of pigmentation markers. Many loci influence coat color. The agouti locus affects the distribution of pigment type. Commonly used 129 ES cells are derived from mice homozygous for the white-bellied agouti (A^W/A^W) allele at the agouti locus. When these cells are injected into C57BL/6 host embryos (a/a), ES cell contribution is made apparent by agouti patches on a black background. These chimeras are mated to C57BL/6 mice and ES cell-derived germline pups are detected by the presence of agouti offspring. The albino or Tyr^c recessive allele results in no pigment production in a homozygous state and "hides" all other coat color alleles. If albino non-agouti host blastocysts (e.g., C57BL/6-$Tyr^{c\text{-}Brd}$) are injected with 129 ES cells, the resulting chimeras will have pigmented patches on a white background. Outbred mice commonly used as embryo donors for aggregations are pink-eyed albino and homozygous for the albino allele (Tyr^c/Tyr^c). However, they may be heterozygous for alleles at other coat color loci. Some of the 129 ES cell lines are homozygous [e.g., E14, [55]] or heterozygous for chinchilla ($Tyr^{c\text{-}ch}$) and pink-eyed dilution ($Oca2^p$) [R1, [24]], and segregating alleles result in a variety of coat colors in intercrosses of F_1 animals.

Heterozygous gene-targeted mice derived from a test breeding for germline transmission will be F_1 animals. Essentially, these are the result of a cross between the background strain of the ES cell line and the strain of females used to mate with chimeras. This should be either an inbred strain, as was used to provide host embryos for injection, (e.g., C57BL/6) or an outbred strain. The existence of a mutation on a mixed background conveys the advantage of hybrid vigor, but it is often necessary to present a uniform genetic background for phenotype analysis, since the phenotype of the mutation can vary according to genetic background [56, 57]. Typically, C57BL/6 is the strain of choice and

F_1 animals resulting from the injection of 129 ES cells into C57BL/6 host embryos are already on the way to generate a B6 congenic strain. Another option would be to cross germline transmitting chimeras onto 129 using the same substrain as that from which the ES cells were derived. This would result in the placement of the mutant allele directly onto an inbred 129 background and produce a co-isogenic mouse strain [58].

Chimeras from C57BL/6 ES cells lines, (aBC) injected into co-isogenic albino B6 host embryos, (aBc) will have black patches on a white background. When these chimeras are mated to albino C57BL/6 mice, the production of black offspring, (aaBBcC) is an indicator of germline transmission. C57BL/6 ES cells injected into BALB/c host embryos, (Abc) will have agouti or black patches on an albino background and can be mated to BALB/c mice (germline = black agouti pups, no germline = white), or to C57BL/6 mice (germline = black pups, no germline = agouti). The advantage of this last scenario is that heterozygous animals will be immediately on a pure B6 background.

17.2 Microinjection Materials

17.2.1. Microscope Setup

For all the equipment listed below, contact the local representative of each vendor for brochures and information on their systems. Vendors will be listed in Subheading 17.5. Before purchasing, it can be worthwhile to try a demonstration microinjection rig/ microscope or to visit other laboratories with similar equipment. Factors to take into consideration are as follows: budget, vendor supply, service availability, (in-house or vendor provided) and compatibility with existing equipment.

Dissecting Microscope – with a transmitted light base is used for embryo collection (flushing of the oviduct/uterus) and loading/moving the embryos/ES cells between the culture dish and injection chamber.

Inverted Microscope – this is used for the injection of ES cells into blastocysts or eight-cell stage embryos. The ability to attach a video camera for training and taking images for publications is an advantage.

Objectives – DIC (differential interference contrast optics) also known as Nomarski, or HMC (Hoffman modulation contrast optics) (*see* FAQ 1 and 2). Magnifications of 4×, 10×, 20×, 40×.

Micromanipulators – Three types of micromanipulators are currently available, – manual (mechanical), motorized or hydraulic (*see* FAQ 3).

Microinjection controllers – 2 types are needed – one for holding of the embryos and one for injection of the ES cells.

Oil or air can be used in the tubing (*see* FAQ 4). These can be purchased as a complete system or as individual items.

Vibration free table – (Isolation table). Depending on the lab environment and degree of vibration to be overcome, a table can be made in the lab or may be purchased. To make in the lab, materials needed are as follows: a sturdy table, a steel plate, and shock-absorbing material (such as squash or tennis balls) for placement between the table and steel plate, which will form the work surface. Alternatively, a heavy stone table can be used. In this case, the weight of the table will isolate the equipment from building vibration to a great extent. If the table is big enough, a dissecting scope for loading/moving the embryos/ES cells could be placed alongside the microinjection microscope.

Optional equipment:

Cooling stage – Cools the injection chamber to a desired temperature (*see* FAQ 5).

Laser System – Computer-controlled device utilizing an infrared laser (1,480 nm) fired through a 20× objective using a software interface to control the alignment, temperature, and delivery of a laser pulse to ablate a small portion of the ZP without causing embryonic lethality.

Piezodrill – can be used to drill through the ZP to assist in eight-cell embryo or blastocyst injections.

17.2.1.1. Frequently Asked Questions

1. *What does DIC and HMC mean?*

   DIC or Nomarski (Differential Interference Contrast) is a system of prisms that essentially act as a high-pass filter that uses interference of polarized light wavefronts, resulting in virtual tridimensional images. The technique enhances the contrast in unstained, transparent samples.

   HMC (Hoffman Modulation Contrast) is designed to enhance contrast in unstained samples by accentuating the phase gradients within the sample, thereby displaying them as levels of gray against the background.

2. *What are the better objectives to use, differential interference contrast optics (Nomarski) or light (Hoffman)?*

   DIC gives a higher resolution image therefore allowing the selection of a better quality ES cell, which will exhibit a haloed appearance. Glass slides are used for DIC because the polarized light in this system interacts with the polymers in plastic dishes to create interference patterns. Plastics can be used with HMC optics.

3. *Which are the better manipulators – manual or motorized?* Manual manipulators give you complete and immediate control of the movement of the needles but will transmit user movement irregularities. Motorized manipulators can be programmed for exact movements and positioning and generally deliver smooth movement. Motorized are more expensive to purchase than manual systems. Servicing availability and costs are also important issues to consider.

4. *Oil versus air in microinjection systems, – which gives the best control for ES cell injection?* Overall, air is less messy, but the use of oil provides an improvement in fine control of pressure, producing a more constant pressure within the injection needle. But oil droplets can form in the media in the injection needle, or air bubbles can form in the oil line, both of which can interfere with control of injections.

5. *Should one use a cooling stage or inject at room temperature?* Cooling the injection chamber to 10–14°C can improve the injection process. Placing well-expanded blastocysts on the cooling stage will slow down the expansion process, giving you more time to inject all the embryos before they become unsuitable for injection. When cooled, the ZP and trophectoderm become firmer and the embryo is less likely to collapse when penetrated. Cooling also prevents the ES cells from becoming sticky, thus maintaining them in a condition suitable for microinjection – round and easier to aspirate, for longer. Condensation can be a problem with a cooling stage but blowing a gentle stream of air under the stage will help to dispel condensation. In the absence of a temperature-controlled stage, cooling the injection media in the fridge prior to setting up the injection chamber has the same effect for a short time.

17.2.2. Pipettes

17.2.2.1. Fine Glass Holding Pipettes

Fine glass holding pipettes are used to immobilize the blastocyst during the injection of the ES cells. They can be purchased or made in the laboratory. The holding pipettes made in the laboratory are generally easy and cheap to make, and readily reproducible but the quality may be variable. The ability to control the movement of the blastocyst contributes to minimizing damage during microinjection, and important for the success of the procedure.

Materials

Laboratory made	Borosilicate Glass capillary tubes (Drummond Scientific 1-000-0500 ID 0.0314 in., OD 0.430), non filamented
	Flame – Bunsen burner or alcohol lamp
	Diamond-tipped pen
	Microforge (*see* Subheading 17.5)
	Plastic Container with plasticine, for storage
Or purchased	Holding pipette (*see* Subheading 17.5)

Method

- Hold the glass capillary at each end with the midpoint over the flame tip.
- Rotate the capillary tube between your fingers/thumbs to evenly distribute the heat through the glass. Wiggle the tube backward and forward to detect when the glass is starting to melt.
- As you feel the tube becoming pliable, slightly raise the capillary out of the flame and apply slight traction. When the middle of the capillary begins to stretch, remove it completely from the flame.
- A second pull is then applied to stretch the mid section into a thinner 5 cm section. Take care when pulling that the capillary is maintained in a horizontal position. This section should have an outer diameter of ~80–120 μm.
- Break the thin midsection of the capillary, polish the ends as described in next step and keep both pieces.
- To create a flat edge, hold the capillary between the thumb and middle finger. Rest the glass tip on the middle finger. Score the capillary at the desired length with a diamond pen and break the tip off the capillary. The length of the holding pipette required will depend on the microinjection setup.
- Place the capillary tube into the holder of the microforge and move to the vertical position. Lower the tip down close to the glass bead on the filament and inspect the edge of the cut tip at ~10× magnification. If the edge is jagged or has a rough edge, i.e., not a flat, smooth surface, you can either re-cut the end or start with a new capillary. The rough ends will not produce a fine seal to hold the embryo (*see* Fig. 17.1). Also, if the cut is not perpendicular to the bore of the pipette, then it may be difficult to (a) visualize the embryo clearly and/or (b) hold the embryo securely on the holding pipette during the injection procedure.
- The outer diameter of the capillary should be no larger than the embryo (100–120 mm) and no less than 25% of the embryo's diameter (15–100 μm).

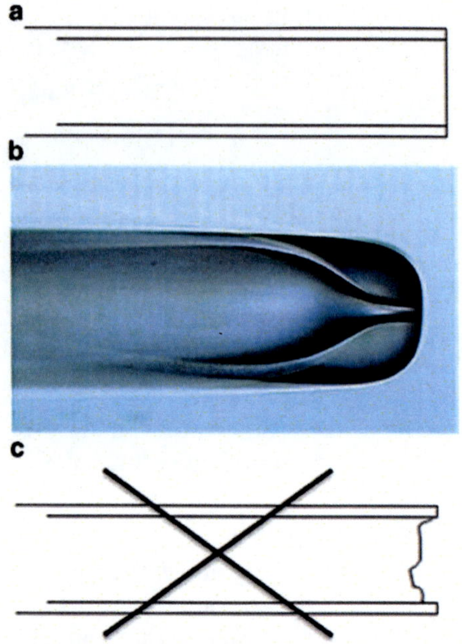

Fig. 17.1 (a) A holding pipette cut with a diamond pen to create a straight cut edge. This is suitable to use for microinjection. (b) The pipette can then be polished on the microforge to create a smooth rounded end for the embryo to attach. (c) A holding pipette with a jagged edge unsuitable for use in microinjection.

- Turn on the microforge heat so that the bead becomes orange and starts to expand. Slowly, lower the end of the capillary toward the heated bead.
- Once the end of the capillary is starting to melt, pull back the tip slightly. Watch the edges of the pipette tip smooth over and the inside edges thicken but not close. Looking through the micrometer in the eyepiece, keep applying the heat until the inner diameter is ~15 μm. Too large will cause the embryo to be drawn into the shaft of the capillary. Too small and you will not get a secure hold of the embryo.
- Once the end is polished, prepare to bend the pipette by rotating the pipette holder axis to position the capillary parallel to the bead and move the tip forward, so the filament is under the pulled section of the pipette.
- The point at which you bend and angle the needle will depend on your injection chamber and microinjection setup. At first try different angles and once you have made your optimal needle, keep it as a guide for future needle preparation. The need to bend the holding pipette will depend on the

microinjection setup. If your holding pipette comes into the injection chamber at a 10–15° angle, it is possible to use a straight holding pipette.

- Once the bend point is identified, heat the bead and lower the capillary slowly, until the capillary starts to bend slightly. Once the capillary has bent to the required angle, turn off the heat.
- Place the holding pipette in a Petri dish or plastic box, securing it in place with sticky plasticine and cover with a lid.
- Holding pipettes can be prepared prior to the injection session

17.2.2.2. Fine Glass Injection Pipette

Fine injection pipettes are used to pick up and inject the ES cells into the embryo. A sharp tip is necessary for penetration of the trophectoderm of the blastocyst with minimal damage. Injection pipettes can be purchased or made in the laboratory. Purchased injection pipettes are standardized and their dimensions are reproducible, whereas the dimensions of those made in the laboratory can vary between each microinjection session and take time to make.

Materials

Laboratory made	Glass capillary tubes (Harvard Apparatus GT100-10. 1.0 mm OD and 0.58 mm ID, no filament)
	Needle puller (*see* Subheading 17.5)
	Disposable scalpel
	Beveller/grinder (*see* Subheading 17.5)
Or Purchased	Injection needles (*see* Subheading 17.5 for the list of Vendors)

Method

- Pull the needle in the pipette puller. The best settings to produce the required needle size will need to be determined by pulling needles while varying the parameters. See pipette puller instruction manual for altering parameters. Essentially, a needle with a long tapered tip is required.
- Place a square of Parafilm™ on the dissecting scope and anchor the nontapered end of the injection pipette to the parafilm with a small piece of plasticine.
- Using a fresh scalpel blade, lightly draw the blade across the thin end of the taper to cut the glass. This should produce a needle with a concave-shaped fracture. Select only those needles which have the correct diameter.
- Assess the size and shape of the aperture under a high-power microscope under 20 or 40× objective.
- For best results, the pipette tip should be beveled to a 45–50° angle. Lower the needle tip to the stone. Wet the tip and while grinding, drop additional water on the stone. The capillary action draws the water into the pipette. The continual

movement of the water up and down and an attached vacuum system rinses out the glass debris. Clean pipette interior with acetone and air dry.

- Internal diameter should be 15–20 μm. External diameter, 20–22 μm, with bevel length ~20 μm
- The spike can be produced/pulled on a microforge
- An injection needle can be reused as long as the spike at the tip is not compromised. Cleaning the needle will remove the debris and stickiness within the needle. Do so by (1) Washing with PVP – 6–10% w/v (Polyvinylpyrrolidone MW 360,000; Sigma, catalogue# PVP360) in 1XPBS through the needle prior to and after each session, and (2) rinsing the needle in 0.25% trypsin-EDTA after each use, (Invitrogen, cat no. 25200). Store the needles in an airtight container.
- Bend needles using a microforge, as described in Subheading 17.2.2.1

17.2.2.3. Embryo-Handling Pipette

Handling pipettes (also known as transfer pipettes) are used in the collection and moving of embryos between the flushing/culture dishes and the injection chamber. These are made in the laboratory.

Materials

Glass Pasteur pipettes.

Bunsen burner or alcohol lamp.

Container.

Mouth Pipetting Device (Fig. 17.2)

Glass Pasteur pipettes

Tubing (ID 1/8″ ID; 3/16″ OD. Fisher Scientific, Cat #14-169-ID)

Flat Mouthpieces (HPI Hospital Products 1506PK)

1,000 μl filtered pipette tip

Method

- Pull the pipette over a flame as described in Subheading 17.2.2.1. Hold the glass pipette at each end with the thin end of the taper over the flame tip.
- Break at the thin midsection of the pulled glass and check the width of the opening to make sure that it is the correct size (180–200 μm). Too large (>250 μm) and you will lose control of the spontaneous capillary action. Too small and you will not be able to pick up the embryos. You can score the glass pipette tip at the required length with a diamond pen and break off the tip to create a clean end.

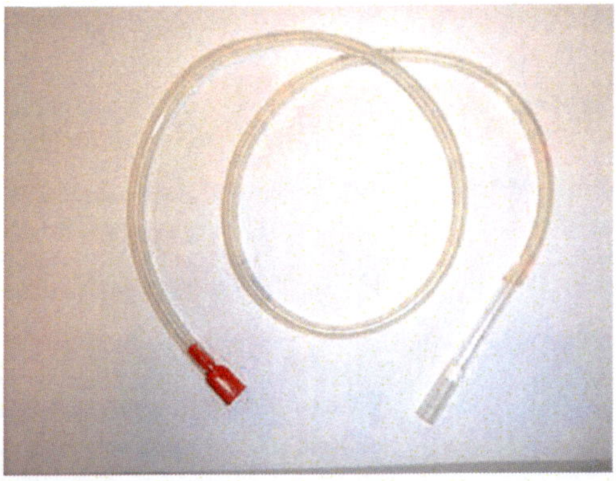

Fig. 17.2 Embryo-handling pipette. A mouth pipetting device made from tubing, a flat mouthpiece and a filtered 1,000 μl pipette tip. This device, after the addition of a pulled glass Pasteur pipette, is used for the collection and movement of embryos between wash and culture dishes and the injection chamber.

To Assemble:

- Cut a piece of tubing to the desired length (between 80 and 95 cm)
- Insert firmly the 1,000 μl filtered pipette tip into one end of the tubing
- Into the other end insert the mouth piece
- Prepare glass Pasteur pipette as above and insert into the pipette tip

17.2.3. Media

For the collection of the embryos and manipulations outside of the incubator, a HEPES-buffered embryo media, such as M2 or FHM, are used. For the culture of embryos, phosphate-buffered culture media, such as M16, KSOM +AA, or ES cell medium without LIF are used in 5–7.5% CO_2 incubator. For the injection chamber, use HEPES-buffered ES cell media without LIF or HEPES-buffered embryo culture media (*see* Subheading 17.5).

17.2.4. Injection Chambers

The injection chamber is the vessel in which you will inject the ES cells into the embryo. There is a variety of setups that can be employed. Most systems consist of a media drop covered or surrounded with mineral oil, in which the embryos and ES cells are placed for manipulation. This media drop must not be too high as excessive height can cause interference with the DIC optics, producing a suboptimal image. Care must be taken when covering the drop with oil as excess can potentially spill onto the objectives. With some methods there might be a need to bend the needles to suit the chamber (*see* Subheading 17.2.2.1).

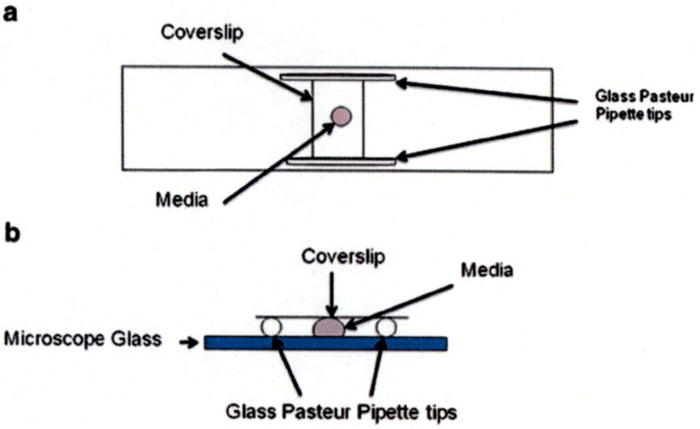

Fig 17.3 Hanging drop injection chamber. This microinjection chamber is made by using a microscope slide, seen (**a**) from above and (**b**) from the side. The two glass pasteur pipettes tips (cut to ~1 cm) are coated with vacuum grease and placed along the long edge of the slide. A small 1 cm piece of coverslip is cut. A 10 μl media drop is placed at both the center of the coverslip, and the pipette tips. The coverslip is then placed onto the pipette tips so that the two droplets join to create a media column to house the ES cells and embryos. The area around the drop and under the coverslip is filled with mineral oil.

*17.2.4.1. Hanging Drop Picture (**See** Fig. 17.3)*

Materials

Standard Microscope slide with frosted end.

Standard Coverslip.

Dow Corning® high-vacuum silicone grease (*see* Subheading 17.5).

Media – HEPES buffered (*see* Subheading 17.2.3 for details and Subheading 17.5 for Vendors).

Sigmacote (*see* Subheading 17.5).

Glass Pasteur pipette.

200 μl pipetter and tips.

Methods

- Holding the frosted end up of the slide, place a few drops of Sigmacote on the middle of the slide then rock to cover the slide with the Sigmacote. The Sigmacote enables the drop of media to bead up and not spread over the slide.
- Allow to air dry. (At this time you can prepare the entire box of slides and place them all back into the box once dried)
- Liberally spray slide with 70% ethanol and then wipe the slide with a tissue to remove any grime or dirt.
- Break off 2 × 1.5 cm pieces/rods from the tip of a glass Pasteur pipette. Using a plastic pipette tip scoop a small amount of vacuum grease and wipe along one side of the each of the glass rods. Alternatively, you can contact a machine shop to prepare 2 × 2 × 10 mm plexiglass blocks.

- Place the two glass pieces so they are running parallel to the long edges of the slide. Position them in the middle section of the slide.
- Score a line across a coverslip, 0.5 cm from the edge (enough to cover the gap between the two glass beams) and break off the end section.
- Place a 10 μl drop in the middle, between the glass rods and also in the middle of the coverslip.
- Place the coverslip on the glass rods and join the two drops of media together, forming a column between the two surfaces.
- With a plastic pipette fill the void with mineral oil. If desired, additional media can be added to the column with pipettor.

Advantages – Cheap and quick to produce, minimal amount of oil means less likelihood of spillage.

Disadvantages – The coverslip lid of chamber can obstruct and hinder the movement of the holding and injection needles. Make sure that the slide is compatible with your cooling chamber.

17.2.4.2. Petri Dish (**See** *Fig. 17.4a*)

Materials

35 mm Petri dish (WillCo-dish® with glass bottom http://www.willcowells.com/GWSB5040).

Media – HEPES buffered (*see* Subheading 17.2.3 for details and Subheading 17.5 for Vendors).

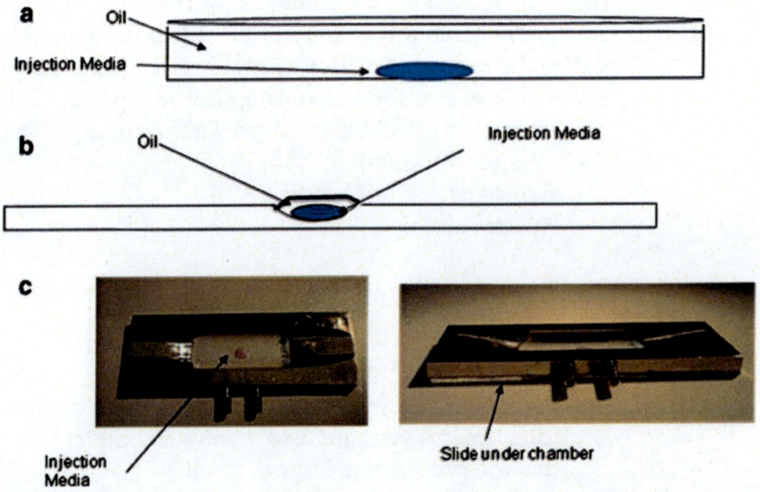

Fig. 17.4 Microinjection chambers. (**a**) 35-mm Petri Dish – A 20 μl media drop is placed in the middle of the Petri dish and then surrounded with enough mineral oil to cover the droplet. (**b**) Depression slide – A 10 μl media droplet is placed into the depression cavity then only enough oil to cover the cavity is poured over the droplet. (**c**) Metal chamber – Vacuum grease is applied to cover the entire underside of the metal chamber. Place the chamber onto the slide with firm pressure. Place a 20 μl buffered media drop onto the middle of the slide. Then aliquot enough oil into the cavity to cover the droplet.

Light Mineral oil (*see* Subheading 17.5).
Pipettes and tips.

Methods

- Place media drop 20–30 μl onto the petri dish.
- Cover with mineral oil.

Advantages – easy to use, can be prepared quickly, are cheap enough to use once and throw away, and provide easy access for maneuvering pipettes in and out of the drop. Embryos/ES cells can be added or removed while the dish is still on the microscope stage.

Disadvantages – Plastic dish on DIC microscopes do not produce the best visuals and glass-bottom dish is recommended. Oil volume has potential for spillage. Make sure that the dish is compatible with your cooling chamber.

Alternatively, whole dish can be filled with HEPES-buffered medium without the use of oil.

*17.2.4.3. Sliver Plate (***See** *Fig. 17.4c)*

Metal microscope chamber (*see* Note below).

Materials

Standard Microscope slide or coverslip.

Dow Corning® high-vacuum silicone grease (*see* Subheading 17.5).

Note: The chamber shown in Fig. 17.4c is designed to deliver cooling to a microinjection stage, in conjunction with a temperature controller (*see* Subheading 17.5). The chamber itself is hollow and cooled by chilled water, circulated via the two tube connectors that can be seen on the long side of the plate. A simple metal or plexiglass chamber that will not provide heating or cooling can be cut to the same size, with a central hole cut long enough to allow free movement of the micromanipulation pipettes.

Media – HEPES buffered (*see* Subheadings 17.2.3 for details and 17.5 for Vendors).

Sigmacote (*see* Subheading 17.5)

Pipettes and tips.

Method

- Sigmacote slide (or coverslip) as for hanging drop.
- Using a plastic pipette tip cover the metal frame's underside with vacuum grease.
- Put the microscope slide (or coverslip) and the metal chamber together, pressing hard to ensure a complete seal is formed, this will prevent the leakage of oil.
- Place a media drop in the middle of the microscope slide.
- Cover drop with mineral oil.

Advantages – as for petri dish.

Disadvantages – Initial cost of chamber.

17.2.4.4. Depression Slide (Fig. 17.4b)

Materials

Depression slide (e.g., Fisher Scientific cat # S175201 or Globe Scientific http://www.globescientific.com cat. #1341-72 or #1344-72).

HEPES-Buffered Medium (*see* Subheading 17.2.3 for details and Subheading 17.5 for Vendors).

Mineral Oil (*see* Subheading 17.5).

Pipettes and tips.

Method

- Clean Microscope slide by wiping with 70% ethanol
- Place a 10 μl media drop into the depression
- Cover with mineral oil (~70–100 μl)

Advantages –as for petri dish.
Disadvantages – Small volume chamber

17.2.4.5. Chamber Slides

Materials

Chamber slide (Fisher #12-565-16, Nunc #177380) (Lab-Tek™ Chamber Slide™ System http://www.nuncbrand.com/).

HEPES-Buffered media (*see* Subheading 17.2.3 for details and Subheading 17.5 for Vendors).

Light Mineral oil.

Method

- Detach the chamber from the slide. The remaining silicone gasket creates a well into which a drop of media is placed.
- Place a 20 μl media drop into the well.
- Cover with just enough mineral oil to surround the media drop but not to over flow the chamber.

Advantages –as for Petri dish.
Disadvantages – Full well of oil has potential for spillage and uneven well wall can interfere with movement of needles

17.2.5. Embryos

The superovulation of female mice for the production of embryos is discussed in Chapters 6 and 25. Naturally mated or superovulated females can be used (*see* FAQ 6).

17.2.5.1. Materials

Positive plugged 2.5 dpc or 3.5 dpc mice (*see* FAQ 8).

HEPES-buffered media and embryo culture media (M16 or KSOM).

30 g needle tip blunted by filing with nail file.

1 ml syringe.

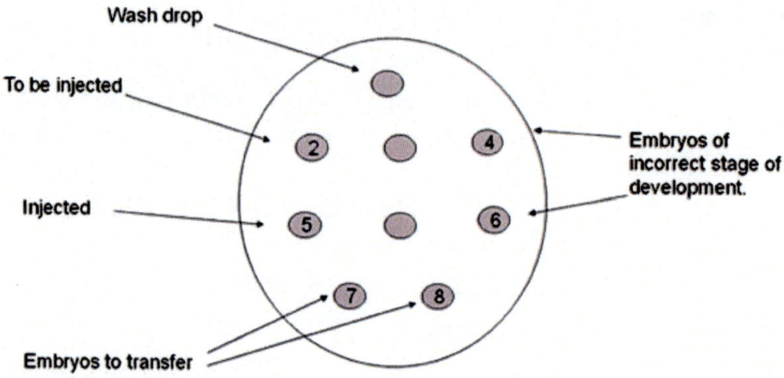

Fig. 17.5 35 mm Petri dish for culture of embryos prior, during and post ES cell injection. 20 μl M16 or KSOM droplets are placed across three rows in the dish. The top droplet is for the first placement of the embryos from which to sort, following flushing and washing. All those embryos that are at eight cell/compacted morulae or blastocyst (depending upon collection day) are placed in droplet 2. The nonviable embryos, i.e., those not at the correct stage of development into droplet 4. As the embryos are injected and returned to the culture dish they are placed into droplet 5. Once all embryos are injected, those to be embryo transferred are moved to droplets 7 and 8.

35 mm Petri dish (*see* Fig. 17.5) (GREINER Bio-one Cat # 627102).

Organ culture dishes (Falcon 35–3037).

Surgical instruments – fine scissors, blunt forceps, watchmaker forceps.

Handling pipette.

Dissecting Microscope.

17.2.5.2. Flushing Oviduct of 2.5 dpc Mice to Recover Eight-Cell Embryos

Method

1. Euthanize embryo donor mice and place on back on absorbent toweling. Spray abdomen with ethanol. This will stop the spread of hair throughout the abdomen once an incision is made in the skin.
2. Pick up the skin with the blunt forceps and cut through the lower abdominal wall and internal fascia. Then cut along the lateral walls, retracting the skin to expose the gut (*see* Fig. 17.6a).
3. Displace the abdominal cavity contents, by moving it aside over the thorax.
4. Expose the cervix, uterine horns, oviducts, and ovaries (*see* Fig. 17.6b).
5. With blunt forceps pick up the uterus just distal to the utero-tubal junction and pull it laterally toward the side of the mouse. This will stretch the uterus so that you are able to

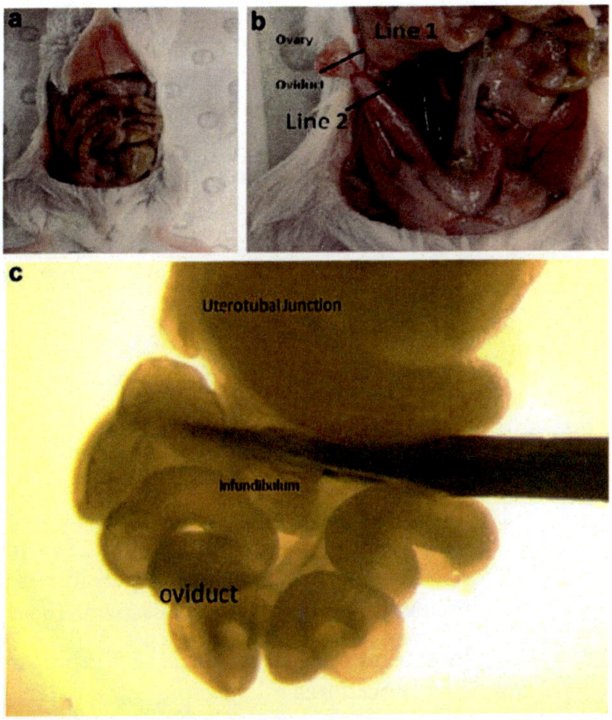

Fig. 17.6 Flushing the infundibulum. (**a**) Pick up the skin with the blunt forceps and cut through the lower abdominal wall and internal fascia. Then cut along the lateral walls, retracting the skin to expose the gut. (**b**) Once the fat is removed pull with slight traction the uterus/oviduct away from the ovary and cut in between the oviduct and ovary. Reposition the forceps and cut the uterus ~5 mm away from the oviduct. (**c**) Fill a 1-ml syringe with ~500 μl of M2 media, then affix the 30 g needle. The oviduct is anchored with one pair of fine forceps, so that the infundibulum can be identified. Rotate the oviduct with the other pair of fine forceps so that the infundibulum is in a horizontal position, with the infundibulum lying flat. Hold slightly away from the fimbrillae with forceps, then pick up needle/syringe, inserting the tip needle into the infundibulum so that the bevel is totally within. Move the forceps to surround the needle tip, holding it in place within the infundibulum. Flush through ~100–200 μl of media, watching for the embryos coming out of the other end. Repeat with all oviducts collected.

identify and then remove the fatty mesometrium adhering to the oviduct/uterus.

6. Once the fat is removed pull with slight traction the uterus/oviduct away from the ovary and cut in between the oviduct and ovary. Reposition the forceps and cut the uterus ~5 mm away from the oviduct (*see* Fig. 17.6b). Place the pieces containing oviduct and short fragment of the uterus in a dish of HEPES-buffered media. Repeat for the other side. Repeat the process for all donor mice until you have collected all oviducts to be flushed. Ideally oviducts should be flushed and embryos placed in the incubator within 30 min from the start of dissection.
7. Fill a 1-ml syringe with HEPES buffered media and then attach blunt 30 g needle.

8. The remaining procedure has to be done under a dissecting microscope equipped with transmitted light. Using fine forceps hold the infundibulum of the oviduct as you insert the needle into the opening (*see* Fig. 17.6c) and then clamp the forceps over the needle. Flush ~0.2 ml media until you see the media expelled from the end of the uterus. Repeat until all oviducts are flushed.
9. With the handling pipette, pick up all embryos in the dish and wash through 3–4 × 20 μl drops of HEPES-buffered media and then 2–3 × 20 μl drops of M16 or KSOM, prior to incubation in 20 μl drops of KSOM overlaid with oil, or in an organ culture dish with no oil, at 37°C in a 5–7.5% CO_2 humidified atmosphere.
10. As the embryos are washed, sort them into groups according to their state of expansion. Those that are ready for injection can be placed directly onto the injection chamber, in a 20 μl drop of ES cell media (without LIF) overlaid with mineral oil. Those that need more expansion can go back into the incubator until needed, in a separate dish

Note: Embryos collected at 2.5 dpc can be used for the following procedures

- Injections with ES cells
- Aggregation with ES cells
- Overnight culture for ES cell blastocyst injection next day
- Cryopreservation for future use as eight-cell/morula or blastocyst hosts

17.2.5.3. Flushing Uterus of 3.5 dpc Mice to Recover Blastocysts

1. Follow **steps 1–5** above (Fig. 17.7a).

6. After the fat is removed pull with slight traction the uterus/oviduct away from the ovary and cut at the utero-tubal junction. Reposition the forceps and cut through the cervix (*see* Fig. 17.7b). Place the uterine horn in a dish of M2 medium. Repeat on the other side and for all donor mice.
7. Fill a 1 ml syringe with buffered media and attach the 30 g needle.
8. Hold the uterus with fine forceps (at the oviduct end), and insert the needle at the utero-tubal junction and flush ~0.2 ml media through the uterine horn, in the direction of the cervix (*see* Fig. 17.7c).
9. With the handling pipette, collect, wash and sort embryos as in **steps 9** and **10** above. As you are washing embryos, remove all those embryos not at the compacted morulae (CM) or blastocyst stage, they can be discarded or in case of suitable

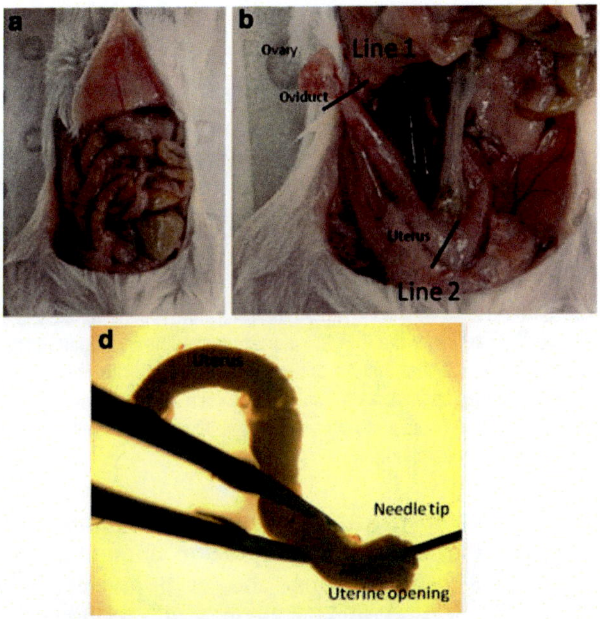

Fig. 17.7 Flushing of Uterus for Blastocysts. (**a**) Pick up the skin with the blunt forceps and cut through the lower abdominal wall and internal fascia. Then cut along the lateral walls, retracting the skin to expose the gut. (**b**) After the fat is removed pull with slight traction the uterus/oviduct away from the ovary and cut at the utero-tubal junction. Reposition the forceps and cut through the cervix- Line 1 and Line 2. (**c**) Fill approximately a 1 ml syringe with ~500 μl of M2 media and add the 30 g needle to the syringe. The uterus is then held one-third along its length with the fine forceps. Pick up needle/syringe inserting the tip of the needle into the uterus opening, so that the bevel is totally within. Move the forceps to surround the needle tip, holding it in place within the uterus. Flush through ~100–200 μl of media, watching for embryos flushed through at the other end of the horn. Repeat with all uteri collected.

developmental stage, cultured overnight or cryopreserved for future use. Place all the CM and blastocysts into a 35-mm petri dish containing a 300 μl drop of M2.

10. Separate the embryos into different oil covered drops of M16 or KSOM in a new 35 mm petri dish according to their stage of blastocoel cavity expansion – (1) no or little expansion, (2) optimal expansion, and (3) overexpanded (*see* Fig. 17.8; FAQ 9 & 10).
11. Group 2 embryos will be used first for the injections.
12. Group 1 embryos should be left to culture. Once all embryos from group 2 have been injected, look through group 1 and select those that are now expanded enough for injection (*see* FAQ 11)

17.2.5.4. Frequently Asked Questions

6. ***Is it better to use, superovulated or naturally mated females?***
Naturally mated females produce fewer embryos per female than superovulated females, but embryos may be more consistently "injectable". Superovulation means using less mice, but can result in the production of lower quality embryos

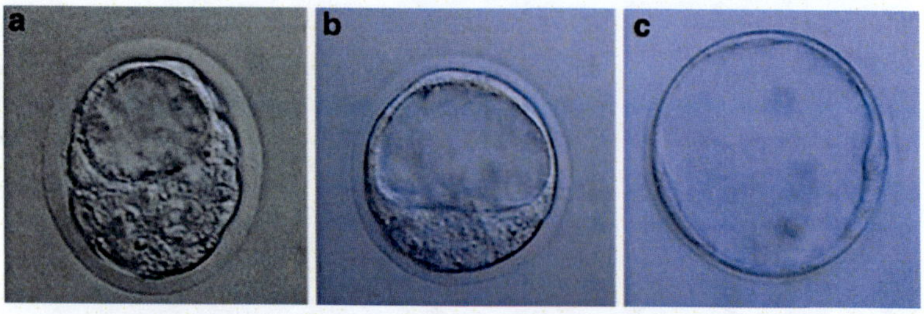

Fig. 17.8 Different stages of Blastocyst Development. It is important that the correct stage of blastocyst embryo is used for injection. (a) Early stage – cavity is forming and the Inner Cell Mass (ICM) is still over 50% of the whole embryo. At this stage, with a large ICM, the needle can injure the ICM therefore reducing the likelihood of the embryo developing any further. (b) Correct stage. Approximately 30% of embryo is made up of the ICM. This is the best stage at which to inject. (c) Overexpanded embryo. Here the ICM appears as a thin layer of cells within the embryo. At this stage, the embryo is too difficult to inject due to the release of pressure from within the blastocoel's cavity. The embryo will then collapse as the needle punctures the zona and trophectoderm.

compared to collection from naturally mated females. There may be a higher proportion among them that are zona free at 3.5 dpc flushing, and/or be overexpanded at the time of injection, making them harder to inject.

7. *What is the best time to collect the embryos prior to blastocyst injection?*
   Embryos may be collected as blastocysts on the morning of injection or as morulae on the day prior to microinjection. Collecting at the morula stage will allow you to know the day before roughly how many embryos are available for injection. If embryos numbers are insufficient, eight-cell stage embryos can be thawed from frozen stocks the day before blastocysts are needed. Keep in mind that embryos may develop a little more slowly in vitro than in vivo.

8. *If "in vitro is better than in vivo", will collecting embryos too early cause a delay, such that embryos are not at an optimal stage by the time of injection?*
   By collecting the day before and culturing overnight blastocoels may form more slowly, but not significantly so. Embryos cultured from one-cell stage to blastocyst stage may be a little more delayed by day 3.5. The development of the embryo will be severely compromised if culture conditions are suboptimal.

9. *What to do with embryos that are too expanded for injection?*
   The embryos too expanded for injection can be used to supplement the numbers of embryos at the time of embryo transfer.

10. *What is the optimal stage of blastocyst development for microinjection.*
Blastocysts should be well expanded, with the ICM taking up approximately only 30% of the embryo. Overexpanded embryos will be harder to penetrate and the embryo will collapse more readily, creating a barrier toward depositing the ES cells in the blastocoel cavity. Injecting into blastocysts with very small cavities may increase the likelihood of physical damage to the ICM by the injection needle, but the trophectoderm is easier to penetrate than that of very expanded blastocysts.

11. *At the time of injection none of the blastocysts are expanded enough to inject.*
Continue the culture of ES cells or split them for use the next day. Put the embryos in a HEPES-buffered media drop covered in oil, and place on a cooling stage or in the fridge at +4°C for a few hours. Culture overnight in a drop of KSOM covered with oil at 37°C. The exposure of embryos to cold temperatures is expected to delay their development and as a result, they may be suitable for injection the next day.

12. *At the time of injection the blastocysts are too expanded.*
Change the timing of the hormone injection and light cycle so that mating can occur later, or begin injecting earlier in the day. Try collecting 8-cell embryos the morning before and culturing overnight, to slow their development just a little.

13. *How to minimize the risk of low embryo numbers on the day of injection?*
To safeguard against this scenario, facilities with the skill and capacity for cryostorage can freeze down donor strain 8-cell stage embryos or vitrify blastocysts. This will enable the injectionist to always have an optimal number of embryos on injection days [59]. For blastocyst injection, eight-cell stage embryos can be thawed the day before and cultured overnight. Blastocysts can be thawed on the morning of the injection. Frozen eight-cell stage embryos can be thawed and used for injection of ES cell embryos on the same day (*see* Subheading 17.3.2 for injection of eight-cell stage embryos). Embryo yields will always fluctuate and if

cryopreservation is not a possibility, then using more donor females is the only other option.

14. *What if the ES cells are not optimal on the morning of injection?*
    If it is possible to plate a new vial of ES cells and culture them for injection the following day then the embryos can be placed in M2 medium under oil and kept at 4°C overnight. Alternatively, place the embryos on the cooling stage in M2 in the injection chamber during day and transfer them to a dish containing M16 or KSOM in the incubator for culture overnight. These two procedures should delay development such that the blastocysts may still be suitable for injection the next day. If there is no possibility of plating new ES cells, you can vitrify the blastocysts for use another day (*see* Chapter 23, on the Cryopreservation of Embryos)

17.3 Injection Procedures

17.3.1. Injection of ES Cells into the Blastocyst

17.3.1.1. Materials

Blastocysts.

Injection chamber.

ES cells (see preparation of cells in Subheading 17.3.2.1).

Glass Micropipettes – injection, holding, and handling.

17.3.1.2. Method

- Place the injection and holding pipettes into micropipette holders. Attach the holders to the micromanipulators. Maneuver the manipulators so that the needles are in position over the media drop in the injection chamber.
- Set the cooling stage to 12–14°C.
- Prepare the injection chamber
- Use a dissecting microscope with a transmitted light base, load the ES cells into the injection chamber. Place them across the midline of the drop. Do not overfill the chamber with cells as cellular debris can cause sticky needles.
- Pick up approximately 20 embryos and place them at the top of the injection chamber (*see* FAQ No 15).
- Transfer the loaded chamber to the microscope stage.

- Place the media drop in the center of the microscope field of view and focus on the embryos with the 4× objective.
- Lower the tips of the holding pipette and injection pipette into the injection chamber and position them near the embryos.
- Move to a higher magnification (40× objective) (*see* FAQ No 16).
- Place the tip of the injection pipette in the center of the field of view. Rotate the holder until the edges of the beveled tip line up with each other, facing down toward the bottom of the field of view.
- Pick up an embryo by moving the holding pipette to a position next to a blastocyst and applying negative pressure to the pipette. This will draw the embryo onto the end of the holding pipette. Apply enough pressure so that the embryo sits tightly on the pipette. Too much pressure will cause the embryo to be sucked up into the holding pipette. Too little and the embryo will move off the holding pipette as you try to inject (*see* Fig. 17.9)
- The embryo's orientation can be changed by slightly releasing the holding pressure and using the injection needle to roll the embryo into an optimal position for injection (*see* FAQ No 17).
- After the blastocyst is in the desired position, raise it slightly off the bottom of the slide. Some injectionists like to rest the blastocyst on the bottom of the slide to reduce the embryo's mobility during microinjection.

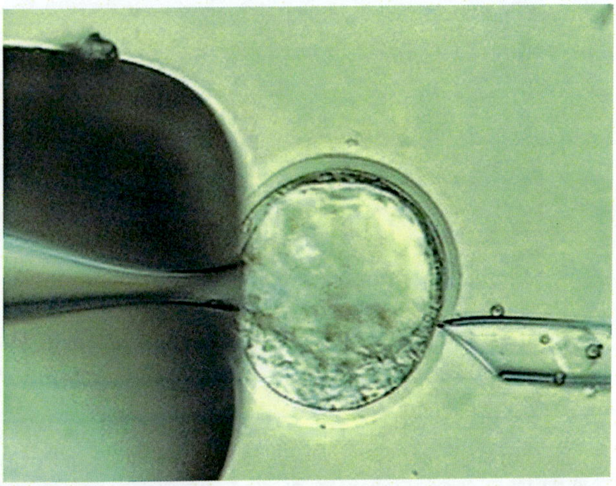

Fig. 17.9 The correct positioning to Hold Blastocyst for injection. The Blastocyst is held so that the ICM is at the 6 or 12 O'clock position. Injecting in this position will (1) reduce the damage to the ICM as the needle penetrates through the embryo; (2) make it easier to identify the junctions between the trophectoderm cells; (3) prevent the embryo moving from its position at the time of injection.

- Move the 20× objective into place (*see* FAQ No 16). Raise the injection pipette above the bottom of slide as you move the chamber around searching for the best ES cells (*see* FAQ 18). Lower the pipette in line with the selected ES cells trying not to contact other ES cells as you maneuver the pipette. Damaged ES cells create floating debris in the media, which can hamper the injection by making the injection pipette sticky and clogging the pipette.
- Apply negative pressure to the injection pipette to bring ES cells into the pipette (*see* FAQ No 19).
- Move the 40× objective into place. Bring the tip of the injection pipette into the same focal plane as the equator of the blastocyst (Fig. 17.10b).
- Draw the injection pipette away from the blastocyst and with a quick flip of the wrist, apply a fast forward thrust to break through the zona and pierce the blastocyst without hitting the holding pipette with the tip of the injection pipette (*see* FAQ No 20).
- There will be a slight flow of fluid out of the injection site (*see* FAQ 21). As the blastocoel cavity begins to collapse, slowly expel 10–16 ES cells (*see* FAQ No 22 &23) into the embryo.
- Slowly withdraw the injection needle. A fast withdrawal can cause ES cells to rush out, following the track of the needle (*see* FAQ No 24).

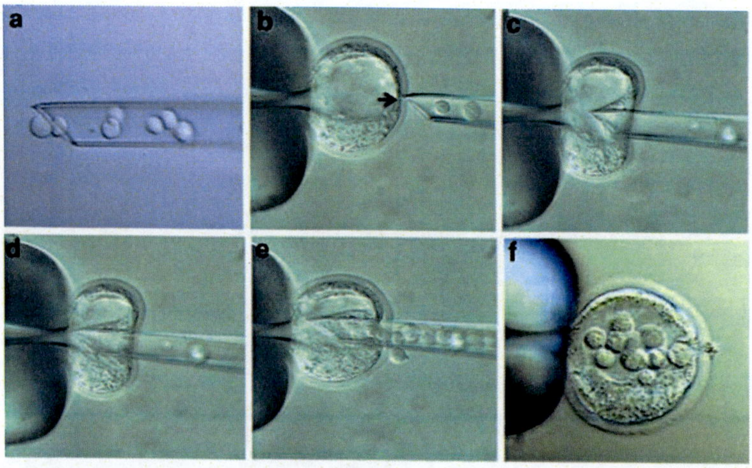

Fig. 17.10 Injection of ES cells into Blastocyst. (**a**) ES cells are drawn up into the injection needle; (**b**) using positive pressure, move the ES cells to the tip of the needle. Ensure that the tip of the needle is within the same focal plane as the blastocyst's equator, then align the tip in between two trophectoderm cells, as shown by *arrow*; (**c**, **d**) smoothly and swiftly penetrate the zona and trophectoderm in one quick motion. Be careful not to push/puncture too far through the embryo as this can cause damage to the embryo/ICM and the pipette tip; (**e**) release the ES cells over the ICM and then remove the pipette slowly out of the embryo. If ES cells start to follow, stop the tip at the trophectoderm and (**f**) let the ES cells float back onto the ICM. Culture embryo until all have been injected and ready to transfer.

- Place the injected blastocyst at the bottom of the drop.
- Repeat this procedure with all the embryos in the chamber.
- After all the embryos are injected, raise the injection and holding pipettes out of the drop. Be careful not to bump or break the fragile pipettes. Return the chamber to the dissecting/viewing microscope and remove the embryos from the chamber using the handling pipette and place them in a 35 mm petri dish containing four 75 μl drops of M16 or KSOM. Wash the embryos through the four drops to remove the M2 medium.
- At this point, you can remove some of the ES cells and media and replace with new media and ES cells. If there is a lot of debris, make a new slide.
- Collect the next batch of embryos that have reached the optimal stage and add them to the injection chamber.
- Continue until all viable embryos have been injected
- Once finished, the injected blastocysts can be examined to see how many have collapsed and how many are beginning to reform blastocoel cavities.

For implantation of blastocysts into recipient female mice, refer to Subheading 17.4.2

17.3.1.3. Frequently Asked Questions

15. *How many embryos should I place into the injection chamber at one time?*
    This largely depends on the skills of the operator and quality of ES cells. Twenty to thirty is a reasonable number to inject at one time or as many blastocysts as can be microinjected in 1 h.

16. *What is the best magnification for microinjection?*
    This is a subjective decision but it is important that you are able to identify the junction between trophectoderm cells so that you can guide the tip of the microinjection pipette between the junction. Lower magnifications (10× objective) allow for greater ease in maneuvering around the injection chamber and setting up the pipettes. 20× is suitable for the selection and loading of ES cells and positioning pipettes next to embryos and also for injection. 40× will allow the technician to distinguish the trophectoderm cell junctions and identify a precise point of entry but can present a narrow field of view for some. If an option, 35× is perfect.

17. *What is the best position to hold the embryo at the time of injection?*
The embryo should be held with the ICM either at the top or the bottom of the view (*see* Fig. 17.9). This will allow you not only to get the best view of where the ICM is and how well expanded the embryo is but it gives you the most support for the embryo when injecting. Holding the embryo at the equator gives the most stability and anchors the embryo against the force necessary for injection. This also reduces the likelihood of poking the inner cell mass with the injection needle. Ideally, the tip of the holding pipette, the equator of the blastocyst and the tip of the injection needle should be placed in the same focal plane.

18. *How can I determine which are the best ES cells to pick up?*
If your microscope has DIC (Nomarski) objectives, the ES cells will look like ball bearings, smooth, spherical, and symmetrical. Do not pick up cells that are not spherical and smooth, have a fuzzy appearance, or are too big to easily fit into the injection pipette.

19. *How many ES cells should I load into the injection pipette?*
There is no limit to the number of ES cells that can be loaded into the injection pipette at any given time. It should be noted though that an increased volume of fluid in the injection pipette reduces control of the flow of cells and media into and out of the needle, especially when needles are homemade and vary in internal diameter. The inability to precisely control the flow out of the injection needle can result in too many ES cells or too much media injected into the embryo. It is best to load up enough ES cells for a few embryos at a time.

20. *What is the best point of entry into the embryo?*
The optimal point of entry is the junction between two trophectoderm cells adjacent to the ICM (*see* Figs. 17.9 and 17.10). This junction can be readily seen by adjusting the focal plane up and down while looking at the side of the blastocyst. Look for a junction at the widest point of the embryo. Injection at another position may cause the embryo to be rolled away from the holding pipette. If the embryo is held in position as in FAQ 17, the turgidity of the embryo will also help to prevent damage. Puncturing at this point minimizes the damage to the embryo.

21. ***What if the blastocyst is collapsing after the pipette penetrates the blastocyst prior to releasing the ES cells?***
Blastocysts begin to collapse as soon as the blastocyst's integrity is compromised. If the blastocyst collapses completely as soon as you penetrate through the trophectoderm, then they may be already too expanded. As you proceed with ES cell injection you can observe the blastocyst inflate as you inject ES cells and medium. The goal is to avoid blastocyst collapse while penetrating the trophectoderm cells.

22. ***What is the best position to deposit the embryos inside the cavity?***
Ideally the ES cells are deposited on top of the ICM, the original anatomical location of the ES cells. Whether to inject with the ICM at the top or the bottom of the field of view is really an operator's decision. The embryos will collapse with the rupture of the blastocoel cavity allowing the ES cells to come in contact with the ICM. The important point is not to touch the ICM with your injection needle as the damage will reduce the likelihood of the embryos surviving the injection.

23. ***What is the best number of ES cells to inject into the blastocyst?***
The quality of the ES cells will greatly influence the strength and frequency of chimeras produced. The ES cell strain of origin also has a strong influence, as more "robust" ES cell lines will require the injection of fewer cells than less robust lines. Titration experiments show that the best number of ES cells to microinject is 10–15 ES cells per blastocyst [60] although up to 25 cells have been injected in case of B6 ES cells [48]. On the other hand, it has been shown that one to three ES cells injected into eight-cell stage embryo or blastocyst could be enough to contribute to germ cells or colonize host embryo [21, 61]. In any case, the recommended number of cells varies between parental lines, particular clones, and specifics of ES cell culture, it should be determined empirically.

24. ***What to do when the ES cells are rushing out during injection or as the pipette is withdrawn?***
This is the result of a sudden increase in the pressure inside the blastocoel cavity relative to the injection chamber, caused by the addition of excess media with the injected ES cells. If there is a sudden rush of media/ES cells into the

blastocoel cavity, slowly withdraw the injection pipette, keeping the needle tip at the point of entry. This will allow the pressure to re-equilibrate with excess media flowing slowly out and the ES cells to stay inside the blastocoel. Once the blastocyst seems to contract a little, complete withdrawal of the needle or continue injecting the required amount of ES cells.

17.3.2. Injection of ES Cells into Eight-Cell Stage Embryos

When ES cells are injected into blastocyst stage embryos they have to compete with host ICM cells to contribute to the epiblast of the developing embryo. At the eight-cell stage, each uncommitted blastomere has the potential to develop as part of either the embryo proper or the extra-embryonic tissue. When ES cells, which are derived from primitive ectoderm cells of ICM, are injected into pre-compaction stage embryos and then cultured overnight, they are observed to migrate to the embryo center to form the ICM, instead of the host blastomeres. When good quality ES cells are used for injection this method can give frequently rise to fully ES cell-derived F0 generation mice. Inbred or hybrid ES cells can be injected into either inbred or less expensive outbred host embryos. When injecting inbred ES cells into outbred strain host embryos, fully ES cell-derived mice are obtained in high yield, but because this is a more challenging environment for ES cell contribution, chimeras may also be obtained in the F0 litters. Although eight-cell embryo injections can be performed with the same basic equipment as blastocyst injections, by a trained injector, the use of a laser or piezo-electric drill, to make a hole in the ZP, facilitates the injection of ES cells by minimizing damage to the host embryo and shortening training time.

17.3.2.1. Equipment

Same as for blastocyst injection.

Optionally, microscope may be equipped with XY Clone Laser system (Hamilton-Thorne Biosciences) or piezo-electric drill (*see* Subheading 17.5)

17.3.2.2. Materials

Uncompacted eight-cell stage embryos.

Injection chamber.

ES cells trypsinized to single cell suspension.

Glass Micropipettes – injection, holding, and handling.

Beveled ES cell injection needle (for straight or laser-assisted injections).

Blunt ES cell injection needle (for Piezo drill-assisted injections).

ES cell medium without LIF.

KSOM.

Primetech piezo micromanipulator – refer to manufacturer's operation instructions.

Fluorinert – for piezo-electric drill only.

Laser ablation system – refer to manufacturer's operating instructions.

17.3.2.3. Methods

Preparation of ES Cells for Injection into Eight-Cell Stage Embryos

- ES cells are grown in 24-well tissue culture plates containing a monolayer of mitotically inactivated embryonic fibroblasts (*see* Chapter 14, *Gene Targeting in ES cells*) to achieve 30–50% confluence on the day of injection.
- Change medium 3–4 h prior to scheduled injection time.
- To harvest the ES cells, remove the medium, wash cells with 1 ml Ca, Mg-free phosphate-buffered saline, replace with 0.5 ml 0.25% Trypsin-EDTA and incubate for 10 min at 37°C
- Using a sterile 1 ml tip, pipette the ES cells gently up and down to form a single cell suspension.
- Add 1 ml of ES cell medium without LIF to inhibit the trypsin and spin the cells at 2,000 × g for 5 min.
- Remove supernatant and resuspend the cells in 1 ml of ES cell medium without LIF.
- Place on ice and incubate for 30 min prior to injections
- During the injection period gently mix the cells occasionally.
- For best results ES cells should be used within 2 h.

Beveled Injection Pipette Only Injections

- All injections are performed at room temperature in 35-mm dish in ES cell medium without LIF, overlaid with filtered mineral oil. Alternatively, the whole dish can be filled with HEPES-buffered medium without the use of oil.
- Load 10–20 eight-cell stage embryos into the injection dish.
- Add ES cells into the injection dish.
- Focus the microscope on the embryos with the 10× objective.
- Lower the needles into and place in position near embryos.
- Move to a higher magnification (20× objective).
- Pick up an embryo with the holding pipette by applying gentle negative pressure.
- Orient the embryo so that a space between two blastomeres is presented at the 3 o'clock position.
- Rest the embryo gently on the bottom of the slide.
- Using suction pick 7–9 round and shiny ES cells with the injection pipette.

- Bring the injection pipette into the same focal plane as injection site.
- Using a firm movement, insert the pipette through the ZP about half way into the embryo.
- Deposit 7–9 ES cells.
- Slowly withdraw the injection pipette.
- Place the injected embryo at the bottom of the dish.
- Pick up the next embryo and repeat the procedure.
- Once all embryos are injected, raise the injection and holding pipettes above the dish and move the dish to the dissecting microscope. With the handling pipette, wash the embryos in KSOM and transfer embryos to a new dish with KSOM or ES cell medium without LIF for overnight culture.

Laser-Assisted Injections with a Beveled Injection Pipette

- Follow steps for the preparation of cells as described in 17.3.2.2.
- Using suction pick 7–9 round and shiny ES cells into the injection needle.
- Using gentle suction through the holding pipette, rotate an eight-cell embryo to identify a region of the embryo suitable for the laser perforation of the zona pellucida (zp) such that the distance between the nearest blastomere and target site in the zp site is maximized
- Pick up the embryo with the holding pipette by applying gentle negative pressure. The drilling site is oriented at the one o'clock position of the embryo. (Fig. 17.11)
- Center the innermost isotherm ring (red) over the zp. (Fig. 17.11a)
- Make a hole in the zp with a single laser pulse. (Fig. 17.11b)
- Introduce the injection pipette into the zp opening and insert it under and along the zp to minimize damage to the blastomeres. (Fig. 17.11c)
- Deposit seven to nine ES cells into the embryo at the ten o'clock position, the greatest distance from the perforation site, to help prevent "backflow" of the ES cells out of the laser hole in the zp.

Piezo Drill-Assisted Injections with a Blunt Injection Pipette

- Back-fill the blunt-ended injection pipette with fluorinet.
- Insert needle into pipette holder.
- Set a piezo speed at 6 and intensity at 5.
- Prepare cells for injection, as described in Subheading 17.3.2.2.

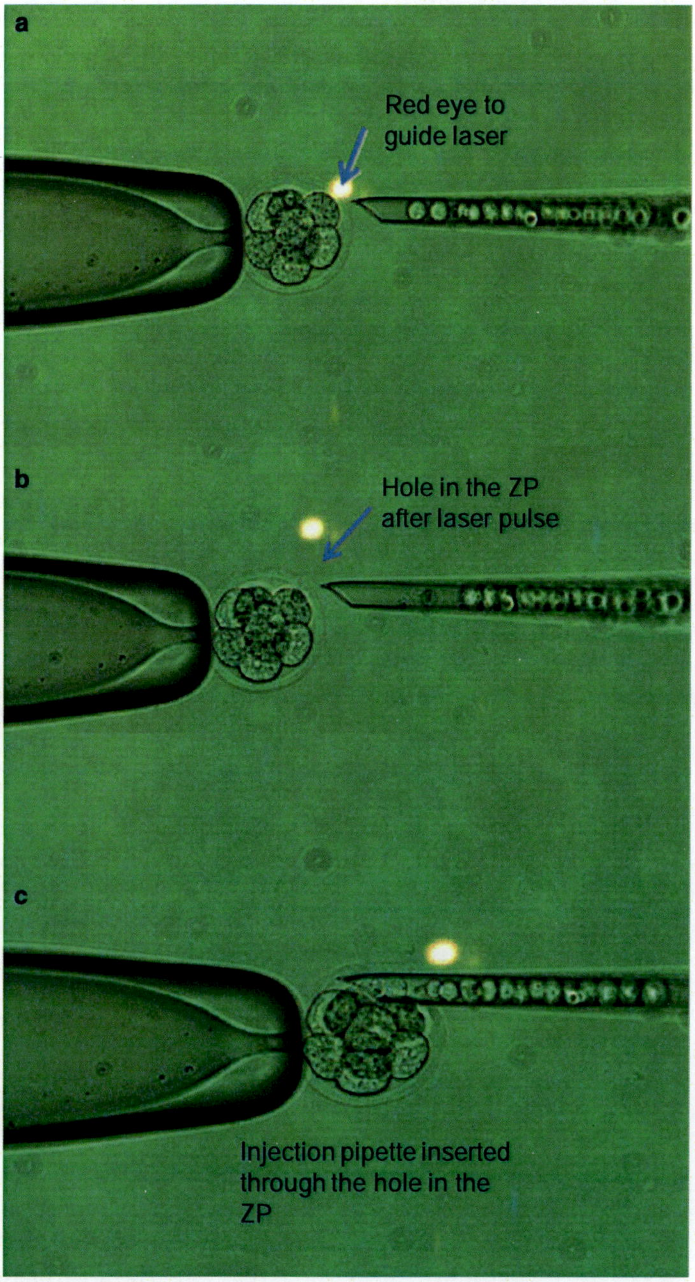

Fig. 17.11 Laser assisted 8-cell injection.

- Using suction pick up 7–9 round and shiny ES cells with the injection pipette.
- With the holding pipette position the embryo so that a space between blastomeres is placed at the three o'clock position.

- With a single piezo pulse insert the injection pipette through the zp about half way into the embryo.
- Deposit 7–9 cells and slowly withdraw the pipette.

Culturing Embryos After ES Cell Injection

Note: This is critical for obtaining the highest yield of fully ES cell-derived mice.

- Transfer eight-cell embryos injected with ES cells into a 35 mm dish containing a 100–200 μl drop of KSOM or ES cell medium without LIF overlaid with mineral oil. Alternatively, embryos can be cultured in an organ culture dish without oil. Some clones contribute better in one or the other medium so normally half the embryos are cultured in each medium and then transferred to the uterine horns of separate recipient females.
- Culture overnight in the 5–7.5% CO_2 incubator at 37°C.
- Following the postinjection culture period, observe embryos under the dissecting scope. Those that progressed to the morula or blastocyst stage are deemed to be viable and ready to be transferred into recipient females.

17.3.2.4. Frequently Asked Questions

25. *How many cells should be injected per embryo?*
    Although completely ES cell-derived mice can be produced by injection of a single ES cell the optimal number of cells for most of lines is 6–9. Similar to the results obtained with blastocyst injections, more robust ES cells lines will require fewer injected cells. In general, we have found that the injection of fewer than six ES cells more often than not results in production of chimeric F0 mice. The injection of more than nine cells has no added benefit because the excess cells tend to leak out through the opening in the zp, especially with the laser-assisted method.

26. *What is the main advantage of injecting eight-cell embryos instead of blastocysts?*
    Two main advantages: ES cells introduced into uncompacted eight-cell embryos have a competitive advantage over host blastomeres for production of fully ES cell-derived F0 mice. Since these mice are 100% germline transmitters fewer breeders need to be set up to generate F1 mice. This injection procedure is about twice as fast as blastocyst injection because about half as many ES cells are injected per embryo leading to higher productivity. Finally with no trophectoderm to penetrate eight-cell stage embryos do not collapse and are easier to inject than blastocysts.

27. *What strain of mice should be used for production of eight-cell stage embryos for injections?*
Both inbred and outbred eight-cell stage embryos can be used. The choice depends on the research objective. If the goal is to obtain F0 cohorts for direct phenotyping from targeted ES cells, it can be more efficient to use eight-cell stage host embryos from inbred mouse strains. If cost is an issue, outbred strains can be used. But when injecting inbred ES cells into the more challenging environment of outbred host embryos, chimeras may be found among the fully ES cell-derived mice in the F0 litters. This is less of an issue when injecting more robust hybrid ES cell lines.

28. *Does it work with every ES cell line?*
If the ES cell clone is able to contribute to chimeras from blastocyst injections, it will also produce chimeras or fully ES cell-derived mice by eight-cell embryo injections. Not every ES cell clone is able to produce viable mice from the injection of inbred eight-cell embryos. These clones can, however, produce fully ES cell-derived mice and or chimeras by injection of outbred embryos.

29. *Is it necessary to use a laser to inject eight-cell stage embryos?*
The injection of these embryos can be done with good quality injection needles without additional equipment. Use of a laser or piezo drill usually facilitates the process for less-experienced injectors.

30. *Should eight-cell embryos be injected when they are already fully compacted (morula)?*
Injections of compacted embryos are more difficult technically since there is less periviteline space to deposit ES cells. In addition, the internal blastomeres of a compacted embryo are already committed to the ICM lineage and such injections would lead to production of chimeras, not fully ES cell-derived animals. However, it has been demonstrated that these animals will have higher degree of chimerism than those obtained by blastocyst injections in same clone comparisons [11, 12].

31. *Can earlier stage embryos be injected?*
Fully ES cell-derived mice have been produced by injection of four-cell stage embryos [15]. One disadvantage of this approach is that with standard animal facility light cycles, this

stage is usually achieved in the middle of the night. Chimeras were produced by injection into periviteline space of one-cell embryos but the ES cells incorporate into developing embryos only at the eight-cell stage [62].

32. *Will the embryo form a chimera or completely ES cell-derived F0 mouse if the ES cells are still present between the blastomeres and the ZP after overnight culture?*
We have not observed this particular situation. Using eGFP-labeled ES cells we observe labeled cells in the morula's center after overnight culture. It is known that it only takes 1–3 ES cells in this location to form the ICM lineage.

33. *Can embryo transfer be done immediately after injections?*
The ability to produce fully ES cell-derived mice from eight-cell injection is enhanced by overnight culture. Unlike the host embryos, ES cells are adapted to a culture environment that gives them a competitive advantage to form the ICM lineage during the culture period. Transferring embryos immediately after injection is not recommended unless chimeras instead of F0 mice are the goal.

34. *What medium should be used for overnight culture of injected embryos?*
Embryo culture medium such as KSOM can be used for all ES cell lines. We found that for more challenging ES cell<->host embryo strain combinations, overnight culture of injected embryos in the presence of the medium without LIF used to culture those ES cells results in greater ES cell contribution to the developing F0 mice. We routinely culture half the injected embryos in KSOM and half in their ES cell culture medium because some clones have a postinjection culture medium preference.

35. *Can frozen embryos be used for injections?*
Frozen un-compacted eight-cell embryos can be used for injections soon after thawing. This approach allows for better planning of the injections, since variability in embryo yields obtained from superovulated females can be eliminated.

17.3.3. Aggregation of ES Cells with Morula Stage Embryos

17.3.3.1. Equipment

Any stereomicroscope with transmitted light and magnification of 16× and 40–50× can be used for aggregations as long as it allows good assessment of embryo quality and visualization of zona pellucida.

17.3.3.2. Pipettes

Standard drawn capillary embryo manipulating pipettes connected to an aspirator mouthpiece described in Subheading 17.2.2.3 are used for aggregations. However, it is absolutely essential to polish the tip of the pipettes on the flame or with the filament or glass anvil on the microforge to prevent damage to zona-free embryos.

17.3.3.3. Media

KSOM-AA medium – Millipore MR-121D

M2 medium – Millipore MR-015D

Embryo-tested light mineral oil – Millipore ES-005-C or Sigma M8410

Acid Tyrode's solution – Sigma T1788

Aggregation needles-BLS Ltd, Hungary, DN-09

ES cell culture media and reagents as described in Chapter 14.

For more details on the culture of zona-free embryos *see* Chapter 16

17.3.3.4. Embryos

Outbred CD-1/ICR stock is commonly used for provision of host embryos donors for aggregation. Embryos are collected at 2.5 dpc as described in Subheading 17.2.5.1

17.3.3.5. Method

The protocol below is based on the procedures originally described by Nagy et al. [18] and in more detail by Nagy [63]. Additional schemes and pictures can be found in the Tetraploid Complementation Assay (*see* Chapter 16). Aggregation plates, ZP removal and ES cells are prepared exactly the same way as described in this chapter.

Preparation of the Aggregation Plate

- Ideally, the plates should be prepared in the afternoon of the day before aggregation. If that is not possible, the media and oil are equilibrated for at least few hours or overnight before aggregations.
- Using a 1-ml syringe or micropipettor place ~3 mm diameter or 10–15 μl microdrops of KSOM onto a 35-mm tissue culture dish (we find Falcon 35–3001 Easy Grip dishes suitable for this purpose) and cover them with embryo-tested mineral oil. Larger drops are more affected by movement that can lead to displacement of the aggregates.

- Wipe the aggregation needle with 70% ethanol. Press the needle into the plastic and make a slight circular movement. Make 6–8 depressions per microdrop, positioning them in the circle approximately halfway between the center and the edge. Leave a few microdrops on the side without depressions; they will be used for the final selection of ES cell clumps. We usually make depressions for 40–60 aggregates per plate to limit the time of embryo manipulations outside the incubator.
- Place the aggregation plate in the incubator until it is needed.

Removal of Zona Pellucida

- Morula stage embryos are harvested as described in Subheading 17.2.5.1 and kept in an incubator until needed. The timing of zona removal is usually co-ordinated with and performed just prior to the preparation of the ES cells, so that the embryos can be aggregated immediately after the cells are ready.
- Place two ~100–150 μl drops each of M2 media and acid Tyrode's solution in a 100-mm Petri dish.
- Transfer a few groups of embryos from the embryo culture dish and place them in M2 drops. The number of embryos manipulated at a time depends on the speed of manipulations. With practice, it is possible to remove the zona from 30–50 embryos at once but if new to the technique, do not start with more than 5–10 embryos.
- Transfer a group of embryos with a minimal volume of media into the first acid drop. Rinse the embryos in it and tranfer them into the second drop of acid spreading them around. Observe under the microscope.
- As soon as the zona is dissolved (Fig. 17.12a), immediately transfer the embryos with a minimal volume of acid into the drop of M2 media. Spread the embryos and do not allow them to touch each other. It usually takes only few seconds for the zona to dissolve.
- Wash all denuded embryos through several drops of M2 media to remove any remaining acid and then through equilibrated KSOM-AA drops.
- Place the embryos into the aggregation plates either directly inside or outside the depression wells, depending on the way the aggregates will be assembled.
- Keep the plates with embryos in the incubator until the ES cells are ready.

Assembly of Aggregates

Initially two diploid eight-cell or morula stage embryos were used for aggregation with a clump of cells [16, 18] based on the hypothesis that a single cleavage stage embryo may be unable to develop properly with a group of integrated foreign cells, leading

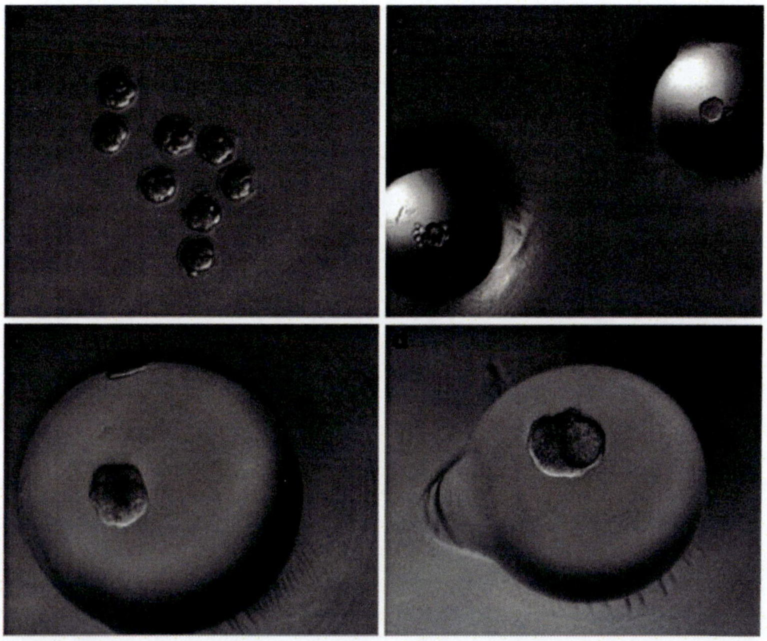

Fig. 17.12 Aggregation of ES cells with diploid embryos **(a)** Zona removal by Acid Tyrode's solution; **(b)** Depression wells in embryo culture microdrops containing embryo <->ES cells aggregates; **(c)** Compacted morula after overnight culture; **(d)** Blastocyst after overnight culture.

to abortive embryogenesis. Later experiments showed that efficient generation of chimeras can be achieved by the aggregation of ES cells with single diploid embryos [19], and single diploid embryos are now routinely used for aggregations. As for injections, it is beneficial to aggregate eight-cell embryos before compaction leading to the production of fully ES cell-derived F0 animals, but all stages of good quality embryos collected in the morning of 2.5 dpc, including compacted morulae may be used. The aggregates can be assembled in either of two ways described below, determined by individual preference; they work equally well as long as the embryo and ES cell clump touch each other.

- Collect the clumps of ES cells and transfer them into a microdrop of the aggregation plate that does not contain depressions for final selection and rinse from ES cell medium.
- Select several clumps of 8–15 ES cells and carefully transfer them individually into the depression wells using one of the following ways:

1. Place the clump of ES cells next to a zona-free embryo already positioned inside the depression well.
2. Distribute the ES cell clumps into empty depressions of the plate, then drop one denuded embryo into each well on top of ES cell clump.

- Assemble all the aggregates in the plate in this manner (Fig. 17.12b).

  Check the plate and make sure that all the embryos touch ES cell clumps.
- Carefully put the plate in the incubator and culture overnight.

The following day, the majority of the aggregates should reach blastocyst stage with some remaining late morulae (Fig. 17.12c and d) and will be ready to be transferred into the uteri of 2.5 dpc pseudopregnant females as described in Subheading 17.4.4. The implantation rate of zona-free cultured embryos is lower than for the embryos with the zona intact. We usually transfer 18–22 embryos per recipient. The number of pups and chimeras among them largely depends on the quality of ES cells. On average ~30% of aggregates will reach term and produce pups, of which at least 50% is expected to be chimeric to different degrees including some fully ES cell-derived as judged by coat color. Aggregation chimeras are most often either full or nontransmitters and rarely partial transmitters as it tends to be the case with blastocyst injection chimeras.

17.3.3.6. Frequently Asked Questions

36. *Can aggregates be cultured in ES cell medium?*
ES cells and cleavage stage embryos have different nutritive requirements. Normally, we rinse ES cell clumps in KSOM media before placing them next to the embryos and then culture the aggregates in KSOM overnight. Although the addition of FBS containing ES cell medium to M16 was found beneficial in at least one report [64] and eight-cell stage embryos injected with ES cells are routinely cultured overnight in ES cell medium as described above, in zona-free aggregates there is a danger of ES cells taking over or even destroying the embryos when ES cell medium is used for overnight culture. If you decide to experiment with the overnight culture conditions of aggregates, use smaller than regular ES cell clump (3–5 cells) and titrate the amount of ES cell medium added to the embryo culture.

37. *What is the best cell number in the clump of ES cells?*
It is hard if not impossible to accurately count number of cells in the clump. We routinely use a variety of sizes within 8–15 range for 129 and F1-hybrid ES cells. For C57BL/6 ES cells cultured in 2i media we tend to use smaller clumps of 7–10 cells [53].

38. *Why aggregates did not develop to blastocysts after overnight culture?*

Check culture conditions including medium, oil, CO_2, and water level in the incubator. Test culture conditions using wild-type zygotes, ~80% of them should reach blastocyst stage in 96 h after collection. There is also a possibility that the number of ES cells in the clump was too high slowing embryo development.

17.4 Uterine Embryo Transfer

The embryo transfer is in reality the most vital part of the genetically altered mouse production process and should be mastered before proceeding with the generation of chimeras. If you are not able to produce a viable pregnancy, all the work in creating the embryos has been in vain. There are some variations in this technique, each one with its own rationale but in the end the controlling factor will be the mouse resources available on the day, embryo numbers, skill level and personal preference. These factors will be discussed in FAQ. In many transgenic labs, the animal facility is separate from the embryo manipulation lab. A safe reliable transport vessel is vital to move the embryos to the surgery suite. A small Petri dish with buffered media drops covered in mineral oil containing the embryos is sufficient. This dish can then be placed in a larger Petri dish and placed along side plasticine to anchor the dish during transport. The dish should then be placed into a larger transport container like a small styrofoam insulated box. If the surgical suite has an incubator, the embryos can be kept there in the original culture dish withdrawing a group of embryos into a drop of HEPES-buffered media enough for one recipient at a time. Depending on the local regulations embryo transfer surgeries may need to be done in laminar flow hood or biosafety cabinet.

17.4.1. Materials

Dissecting scope(s) with transmitted light base and fiber optic illuminator.

Warming pad.

Pseudopregnant female at 2.5 dpc (*see* FAQ 36 and FAQ 37).

Surgery instruments – small scissors, blunt and fine forceps, serrefine clamp.

Wound stapler (Becton Dickinson #427630) or suture material and clamp.

Anesthesia/analgesia.

Opthalmic eye ointment.

1 ml syringe and 27 g needle.

27 or 30 g needle.

Mouth pipette and transfer pipette (*see* Subheading 17.2.2.3).

Embryos in buffered media under mineral oil, 30-mm Petri dish.

17.4.2. Method

- Anesthetize the mouse and place on a tissue.
- Place a few small drops of spare buffered media onto the underside lid of the Petri dish. Set aside.
- Take up a small amount of buffered media into the transfer pipette, then load the embryos from the transport dish and expel into the loading dish droplets. This will exclude any oil from the mouth pipette assembly. The presence of oil may interfere with the smooth movement of media in the pipette and could affect the handler's ability to load the pipette for transfer.
- Take up a small amount of buffered media into the transfer pipette, then a small air bubble, then load up the embryos in a minimal volume of media (*see* FAQ 40; Fig. 17.13).
- Place the transfer pipette aside, well out of the way from being knocked or dropped.
- Check that the mouse is fully anesthetized by the absence of pedal reflexes and place a drop of eye gel/ointment on each eye to prevent the eyes from drying out during surgery. Then place the mouse on its belly.
- Spray 70% ethanol along the midline of the back.
- Make a small incision ~1 cm vertically along the midline of the mouse's back in line with the last rib (*see* Fig. 17.14)
- Wipe incision with a tissue sprayed with 70% ethanol, this will remove any loose fur.
- Slide the skin laterally to expose the fascia overlaying the left ovary and oviduct. The ovary can be seen alongside the fat pad (*see* Fig. 17.14b).

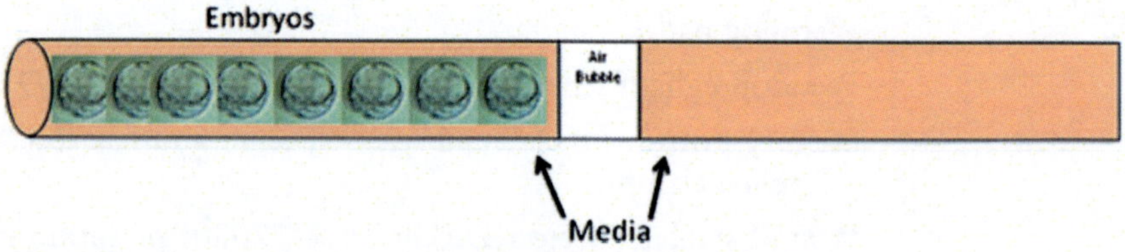

Fig. 17.13 Loading Embryos into the transfer pipette. Embryos to transfer are placed together within the one culture droplet. Insert a polished handling pipette into the mouth pipetting device. Insert pipette tip into a clear media drop and let about 2–3 cm of media move up the pipette. Then create an air bubble by putting slight negative pressure on the mouth piece. Then return to the media drop with the embryos in it and let the capillary action take all the embryos into the pipette. Use some suction if necessary. Make sure the embryos are in as little media as possible. Draw up a little bit of media after the embryos so that they do not fall out when first touching the uterus.

- Using fine forceps pick up the body wall overlaying the oviduct and cut through with the small scissors. The incision must be comparable in size with the ovary so that after the transfer of the embryos the ovary can be placed back into the cavity without damaging it (*see* Fig. 17.14c).
- Be careful not to cut any blood vessels. If there are vessels in the body wall overlaying the oviducts, pull the fascia away until a clear section is now over the oviduct.
- Using fine forceps grasp the fat pad attached to the oviduct and pull out the oviduct and upper part of the uterus. Clamp fat pad with serrafine forceps and lay over the middle of the back, across the spine (*see* Fig. 17.14d). Check the ovary for the signs of recent ovulation (*see* FAQ 38): the presence of corpora lutea (yellow bodies) forming from corpora hemorrhagica (bloody bodies) at the sites of oocytes extrusion from Graafian follicles; the lack of corpora lutea and hemorrhagica

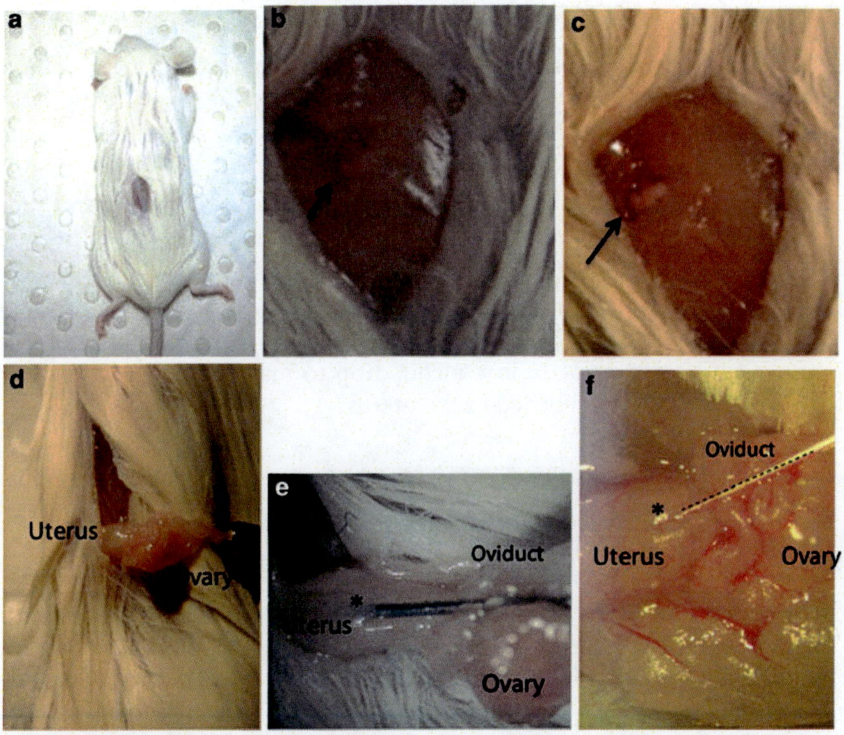

Fig. 17.14 Uterine Transfer. (**a**) The mouse is anesthetized, then a 1 cm incision is made along the midline of the back along the level of last rib. Wipe away any loose hair. (**b**) Slide the skin sideways, so that the incision is now over the ovary. You can see the ovary and fat pad through the fascia. The arrow is pointing to the ovary. (**c**) Cut through the fascia so that the ovary and fat pad are exposed. (**d**) Using a serafine clip, take hold of the fat pad and pull out the ovary, oviduct and first 1–2 cm of the uterine horn. Lay this horizontally across the back of the mouse. (**e**) Pick up the needle and embryo transfer pipette and make a hole with the needle through the uterine wall a few millimeters below the utero-tubal junction, at *Asterisk*. (**f**) Remove the *needle* and insert the tip of *pipette* with embryos loaded into the small hole. Slowly expel the embryos into the uterus and slowly withdraw the *pipette*.

means the recipient did not ovulate and should not be used for embryo transfer.

- Place the mouse with tissue on the scope platform and maneuver mouse so that the oviduct/uterus is in view through the eyepieces. Maneuver the light source so that this area is brightly illuminated.
- Pick up the transfer pipette and 27 or 30 g needle in the one hand (*see* Fig. 17.15). It is best to practice this holding position prior to embryo transfer. While practicing, and until comfortable and confident holding both, it may be best to make the hole in the uterus first and then pick up the transfer pipette
- Look for a site in the uterine wall a few millimeters down from the utero-tubal junction that is clear of blood vessels. Hold the top of the uterus with blunt forceps and make a hole with the 27 or 30 g needle (bevel up). Slide the needle slightly forward and back a few times to ensure that the needle has pierced the uterine lumen and is not in the uterine wall (*see* Fig. 17.14e).
- Withdraw the needle and insert the transfer pipette into the hole, until the air bubble is at the edge of the hole in the uterus (*see* Fig. 17.14f).
- Slowly expel the embryos into the uterus. Watch the air bubble move down the pipette and slowly retract the pipette so as to monitor the air bubble until it reaches the end of the pipette.
- Withdraw the pipette slowly so as to prevent a sudden backflow of the media out of the hole. Blow out the transfer pipette in a media drop to make sure that all embryos were deposited in the uterus.

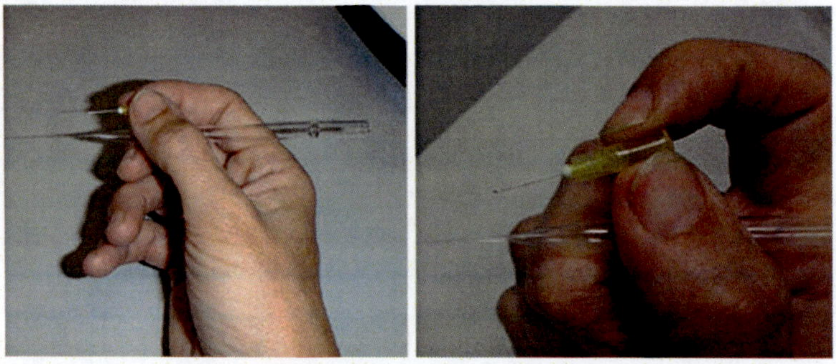

Fig. 17.15 Holding the needle tip and embryo transfer handling pipette. Place mouse on microscope stage ready for embryo transfer. Remove 27–30 g needle from the packaging and put on microscope stage near your mouse. Pick up your embryo-handling device as you would for embryo transfer. With your free hand pick up the needle tip and then place between thumb and first finger over the pipette. Looking down the microscope, you can now make a hole through the uterus wall with the needle, then drop it beside the mouse and insert the pipette tip into the hole made by the needle tip without moving your hands or eyes from the microscope.

- Remove the serrafine forceps and with the fine forceps pick up the edge of the incision in the fascia. With the blunt forceps tuck the uterus and oviduct back into the abdominal cavity.
- Reposition the skin incision over the oviduct on the right side (*see* FAQ 39) cut through the fascia over the ovary and transfer embryos to the other uterine horn.
- Close the skin using sutures or wound clips.
- Loosely wrap the mouse in a tissue and place in a fresh cage on a heating pad. Monitor at intervals until the mouse recovers from the anesthetic.

  Note: Even the best transfer may not result in pregnancy and live births. There are many external factors that can adversely influence the success of embryo transfers. Due care must be taken to minimize these factors.
- Check to see if the females has signs of recent ovulation (*see* FAQ 38)
- Pseudopregnant strain choice – outbred vs. hybrid embryo recipients (*see* FAQ 37)
- Pseudopregnant mouse characteristics – a young female may be too small to carry a pregnancy, while an overweight female will have excess fat and capillary network creating a more difficult embryo transfer. It may also be more difficult to fully anesthetize heavier females.
- Number of embryos transferred – too few embryos can result in a low implantation rate which can create oversized fetuses and therefore birth difficulties. Too few pups born can also cause cannibalism. If possible, supplement manipulated embryos with wild-type embryos that will produce pups with a different coat color from the manipulated embryos. Too many embryos transferred can result in a high implantation rate creating resorption of fetuses, runted fetuses/pups, and birthing difficulties.
- Donor embryo Strain: Outbred and hybrid mice create more robust embryos than inbred strains, which increases pregnancy rates.
- Environmental: A change in the cage environment can cause the pregnancy to abort, embryos to resorb or mothers to cannibalize their young. It is important therefore not to change their diet or bedding material while they are pregnant; minimize staff changes in the room and carefully monitor the light cycle. External weather/seasonal changes, e.g., high or low humidity can also stress expectant mothers.
- Skill level of technician: Practice with dye or WT embryos prior to transferring manipulated embryos.

- Number of embryo recipients per cage: Place the number of females per cage to correspond to cage size. Overcrowding can cause cannibalism or trampling/underfeeding of newborns. Typically, we house two recipient females per cage.

  Other mice in the cage: If there are skittish or nonpregnant females alongside pregnant recipients, these may stress the expectant female or they in turn can cause cannibalism of newborns.
- Mouse husbandry: Technicians should take extra care of the recipients. Give them lots of bedding, and material suitable for them to create a nest. Change the cage a few days before due date. Check for births but take care not to disturb the recipients and pups for a few days following birth.
- Animal Rooms: Excessive or unexpected noise or construction work can cause failed implantation.
- Diet: Follow the recommended breeding diet for the mouse strain in question. For example, a diet with 6% crude fat content will provide better support for pregnancy and lactation than a 4% crude fat diet.

17.4.3. Frequently Asked Questions

36. *Which age of pseudo, 0.5 dpc or 2.5 dpc, should one use to embryo transfer blastocysts?*
    The due date of the pregnancy is always calculated from the day that the pseudopregnant female showed a copulation plug and not the age of the embryos. At 2.5 dpc, the embryos must be put into the uterus NOT the oviduct. Advantages of transferring blastocysts into a 2.5 dpc uterus are that the embryos are put directly into the environment where they will implant and do not have to travel down the oviduct. The surgery is easier to perform and not as invasive to the mouse. As pseudopregnant recipient females must be produced prior to blastocyst injection, the number of plugs will be known 2 days prior to injection. Therefore if more recipients are needed, more can be set up for a 0.5 dpc transfer on the day of injection. There is no hard evidence supporting one pseudopregnant age over the other, so in the end the decision to use 0.5 dpc or 2.5 dpc will come down to personal preference, embryo numbers and numbers of available recipients.

37. *What is the best strain to be used for pseudopregnant recipients?*
    Outbred or hybrid mouse strains can be used as recipients. Examples of outbred stocks include NIH Swiss, SW, ICR, CD-1. F1 hybrid animals obtained by crossing two inbred strains express hybrid vigor; one of the most commonly used and readily available F1 hybrids – B6D2F1 – is generated by mating C57BL/6 females with DBA/2 males.

The advantages of using the outbred strains are: easy to check for estrus, they plug readily and are able to carry larger litters; are cheaper to purchase or breed in-house and are more readily available; are less likely to cannibalize newborns and routinely make good mothers. The main concern is that they have a tendency to gain weight quickly so that there is a higher turnover of mice when selecting for estrus. Hybrids are more prone to split vulva; harder to check for estrus but plug readily; they carry smaller litters but are good mothers; commercially less available, more expensive; harder to produce in-house and leaner in bodyweight, which results in a lower turnover and therefore colony size can be kept to a minimum. The final decision is usually a personal preference and what the lab's budget will allow.

38. *How can one check a mouse is pseudopregnant?*
The presence of a vaginal plug formed from the male's coagulated ejaculate indicates that a female can be used as pseudopregnant surrogate for embryo transfer. However, the presence of a plug does not always mean that the female is pseudopregnant. If there is an excess of pseudopregnant females on the day of transfer, it can be beneficial to check for signs that the female has released her own eggs and ready to accept embryos for implantation. At the time of uterine transfer look at the ovary for the presence of corpora lutea. If they are absent, it is unlikely that the recipient will become pregnant after transfer.

39. *Should embryos be transferred unilaterally or bilaterally?*
The literature indicates that transuterine migration does not occur in the mouse [65]. There are many factors to consider when deciding on bilateral vs. unilateral implants.
For bilateral transfers, advantages are as follows: more embryos are transferred per mouse, which does increase the surgery time taken/mouse but reduces surgery time per experiment as less mice are used. This can reduce the number of mice needed for the pseudopregnant colony. Fetuses are more spread out throughout the horns and litter sizes can be higher. If there is an error or disaster with the transfer pipette during the implant procedure, fewer embryos are lost. Disadvantages are that the surgery takes longer and is more invasive to the mouse.
For unilateral transfers, the advantages are that it takes less time per mouse and is less invasive for each recipient mouse. The disadvantages are that for a given number of embryos, more pseudopregnant recipients are needed, therefore surgery

time is longer. In order to produce more recipients, it may be necessary to increase colony size. Also, fetuses will implant in one uterine horn, which may cause birthing difficulties.
In the end it is personal preference, the number of embryos to transfer, and the number of available pseudopregnant recipients on any given day that dictate which approach to take when implanting embryos.

40. ***What is the best number of embryos to transfer to a pseudopregnant recipient?***
When determining number of embryos to transfer, factors to consider are as follows:

- What pseudopregnant mouse strain will be used – outbred or hybrid (*see* FAQ 37). Outbred recipients can carry larger litters.
- The embryos are manipulated so there will be some attrition between transfer and implantation.
- Bilateral or unilateral embryo transfers – more embryos will be transferred per recipient in a bilateral transfer
- If there are only a few pseudopregnant recipients and many embryos to transfer.
- The skill of technician

On average, 8–15 manipulated blastocysts are transferred per recipient in case of unilateral transfer and up to 20 per mouse in case of bilateral uterine transfer for the production of ESC chimeras.

17.5 Vendors

Vendors in this section are listed as suggestions only. Please contact each company to obtain product information. Web sites are listed below which contain contact details for each specific region.

Microscopes – dissecting and inverted
Leica – http://www.leica-microsystems.com/

Nikon – http://www.nikoninstruments.com/Products

Olympus – http://www.olympusamerica.com/seg_section/seg_home.asp

Micromanipulators
Narishige – http://www.narishige-group.com.

Leica – http://www.leica-microsystems.com/

Eppendorf – http://www.eppendorf.com

Cooling stage

Contact individual microscope vendors for temperature-controlled stages

20–20 Technologies Bionomic System. http://20-20tech.com/bionomic.html

Linkham Scientific Instruments. http://www.linkam.co.uk/

Brook Industries. http://www.kaker.com/mvd/data/Brook_Industries.html

Laser Systems

XYClone Laser, Hamilton-Thorne Biosciences, Beverly, MA. http://www.hamiltonthorne.com/

Octax Laser, MTG, Bruckberg, Germany. http://mtg-de.com/

Saturn Laser, Research Instruments, Falmouth, United Kingdom. http://www.research-instruments.com/

Microforge

Narishige Microforge, Narishige International, East Meadow, NY. http://www.narishige-group.com.

deFonbrune Microforge, Geneq. Montreal, Canada. http://www.geneq.com

TPI Microforge, Glassworx. http://www.theglassworx.com

Replacement Microforge Filaments: http://www.theglassworx.com/; www.fishersci.com

Bevellers

EG-44 Microgrinder, Narshige International, East Meadow, NY. http://www.narishige-group.com.

BV10 Microelectrode Grinder, Sutter Instrument, Novato CA. http://www.sutter.com/

Pipette Pullers

Sutter. http://www.sutter.com/

Borosilicate Glass capillary tubes

World Precision Instruments. http://www.wpiinc.com/

Sutter. http://www.sutter.com/

Drummond Scientific Company (http://www.drummondsci.com/)

Ready Made Holding pipette

BioMedical Instruments. http://www.pipettes.de/index.php?catid=5

Eppendorf. http://www.eppendorf.com

Humagen Pipettes from Origio. http://www.origio.com/

Cook Medical. http://www.cookmedical.com/wh/home.do

Ready Made Injection Pipettes

Humagen Injection Pipettes from Origio.

BioMedical Instruments. http://www.pipettes.de/index.php?catid=5

Eppendorf. http://www.eppendorf.com

The Pipette Company. http://www.pipetteco.com/

Cook Medical. http://www.cookmedical.com/wh/home.do

Chamber slides

Fisher # 12-565-16

Nunc # 177380

Aggregation needle

BLS Ltd: www.bls-ltd.com

Media

M2 Sigma: www.sigmaaldrich.com, Millipore: http://www.millipore.com/

M16 Sigma: www.sigmaaldrich.com

KSOM + AA, Millipore- http://www.millipore.com/stemcell/stma/embryoculture&type=memb

Light mineral oil for embryo culture – Sigma or Millipore

Invitrogen: www.invitrogen.com

Other reagents
Dow Corning® high-vacuum silicone grease – Sigma
Fluorinet – FC-77, Sigma
Sigmacote: Sigma: www.sigmaaldrich.com

Animal suppliers
Harlan: www.harlan.com
Charles River Labs: www.criver.com
Taconic: www.taconic.com

References

1. Nagy A, Rossant J (2001) Chimaeras and mosaics for dissecting complex mutant phenotypes. Int J Dev Biol 45(3):577–582
2. Tam PP, Rossant J (2003) Mouse embryonic chimeras: tools for studying mammalian development. Development 130(25): 6155–6163
3. Tarkowski AK (1961) Mouse chimaeras developed from fused eggs. Nature 190: 857–860
4. Mintz B (1962) Experimental study of the developing mammalian egg: removal of the Zona Pellucida. Science (New York, NY) 138(3540):594–595
5. Gardner RL (1968) Mouse chimeras obtained by the injection of cells into the blastocyst. Nature 220(5167):596–597
6. Moustafa LA, Brinster RL (1972) Induced chimaerism by transplanting embryonic cells into mouse blastocysts. J Exp Zool 181 (2):193–201
7. Babinet C (1980) A simplified method for mouse blastocyst injection. Exp Cell Res 130(1):15–19
8. Lallemand Y, Brulet P (1990) An in situ assessment of the routes and extents of colonisation of the mouse embryo by embryonic stem cells and their descendants. Development 110(4):1241–1248
9. Tajbakhsh S, Bober E, Babinet C, Pournin S, Arnold H, Buckingham M (1996) Gene targeting the myf-5 locus with nlacZ reveals expression of this myogenic factor in mature skeletal muscle fibres as well as early embryonic muscle. Dev Dyn 206(3):291–300
10. Stewart CL (1993) Production of chimeras between embryonic stem cells and embryos. Meth Enzymol 225:823–855
11. Tokunaga T, Tsunoda Y (1992) Efficacious production of viable germ-Line chimeras between embryonic stem (ES) cells and 8-cell stage embryos. Dev Growth Differ 34(5):561–566
12. Yagi T, Tokunaga T, Furuta Y, Nada S, Yoshida M, Tsukada T, Saga Y, Takeda N, Ikawa Y, Aizawa S (1993) A novel ES cell line, TT2, with high germline-differentiating potency. Anal Biochem 214(1):70–76
13. Poueymirou WT, Auerbach W, Frendewey D, Hickey JF, Escaravage JM, Esau L, Dore AT, Stevens S, Adams NC, Dominguez MG, Gale NW, Yancopoulos GD, DeChiara TM, Valenzuela DM (2007) F0 generation mice fully derived from gene-targeted embryonic stem cells allowing immediate phenotypic analyses. Nat Biotechnol 25(1):91–99
14. Eggan K, Akutsu H, Loring J, Jackson-Grusby L, Klemm M, Rideout WM 3rd, Yanagimachi R, Jaenisch R (2001) Hybrid vigor, fetal overgrowth, and viability of mice derived by nuclear cloning and tetraploid embryo complementation. Proc Natl Acad Sci USA 98(11):6209–6214
15. Huang J, Deng K, Wu H, Liu Z, Chen Z, Cao S, Zhou L, Ye X, Keefe DL, Liu L (2008) Efficient production of mice from embryonic stem cells injected into four- or eight-cell embryos by piezo micromanipulation. Stem Cells Dayton OH 26(7):1883–1890
16. Stewart C (1980) Aggregation between teratocarcinoma cells and preimplantation mouse embryos. J Embryol Exp Morphol 58:289–302
17. Stewart CL (1982) Formation of viable chimaeras by aggregation between teratocarcinomas and preimplantation mouse embryos. J Embryol Exp Morphol 67:167–179
18. Nagy A, Gocza E, Diaz EM, Prideaux VR, Ivanyi E, Markkula M, Rossant J (1990) Embryonic stem cells alone are able to support fetal development in the mouse. Development 110(3):815–821

19. Wood SA, Allen ND, Rossant J, Auerbach A, Nagy A (1993) Non-injection methods for the production of embryonic stem cell-embryo chimaeras. Nature 365(6441):87–89
20. Khillan JS, Bao Y (1997) Preparation of animals with a high degree of chimerism by one-step coculture of embryonic stem cells and preimplantation embryos. Biotechniques 22(3):544–549
21. Saburi S, Azuma S, Sato E, Toyoda Y, Tachi C (1997) Developmental fate of single embryonic stem cells microinjected into 8-cell-stage mouse embryos. Differ Res Biol Divers 62(1): 1–11
22. Doetschman TC, Eistetter H, Katz M, Schmidt W, Kemler R (1985) The in vitro development of blastocyst-derived embryonic stem cell lines: formation of visceral yolk sac, blood islands and myocardium. J Embryol Exp Morphol 87:27–45
23. Voss AK, Thomas T, Gruss P (1997) Germ line chimeras from female ES cells. Exp Cell Res 230(1):45–49
24. Nagy A, Rossant J, Nagy R, Abramow-Newerly W, Roder JC (1993) Derivation of completely cell culture-derived mice from early-passage embryonic stem cells. Proc Natl Acad Sci USA 90(18):8424–8428
25. Stewart CL, Kaspar P, Brunet LJ, Bhatt H, Gadi I, Kontgen F, Abbondanzo SJ (1992) Blastocyst implantation depends on maternal expression of leukaemia inhibitory factor. Nature 359(6390):76–79
26. Eggan K, Rode A, Jentsch I, Samuel C, Hennek T, Tintrup H, Zevnik B, Erwin J, Loring J, Jackson-Grusby L, Speicher MR, Kuehn R, Jaenisch R (2002) Male and female mice derived from the same embryonic stem cell clone by tetraploid embryo complementation. Nat Biotechnol 20(5):455–459
27. Longo L, Bygrave A, Grosveld FG, Pandolfi PP (1997) The chromosome make-up of mouse embryonic stem cells is predictive of somatic and germ cell chimaerism. Transgen Res 6(5):321–328
28. Liu X, Wu H, Loring J, Hormuzdi S, Disteche CM, Bornstein P, Jaenisch R (1997) Trisomy eight in ES cells is a common potential problem in gene targeting and interferes with germ line transmission. Dev Dyn 209(1): 85–91
29. Fedorov LM, Haegel-Kronenberger H, Hirchenhain J (1997) A comparison of the germline potential of differently aged ES cell lines and their transfected descendants. Transgen Res 6(3):223–231
30. Schwartzberg PL, Goff SP, Robertson EJ (1989) Germ-line transmission of a c-abl mutation produced by targeted gene disruption in ES cells. Sci NY NY 246(4931): 799–803
31. Austin CP, Battey JF, Bradley A, Bucan M, Capecchi M, Collins FS, Dove WF, Duyk G, Dymecki S, Eppig JT, Grieder FB, Heintz N, Hicks G, Insel TR, Joyner A, Koller BH, Lloyd KC, Magnuson T, Moore MW, Nagy A, Pollock JD, Roses AD, Sands AT, Seed B, Skarnes WC, Snoddy J, Soriano P, Stewart DJ, Stewart F, Stillman B, Varmus H, Varticovski L, Verma IM, Vogt TF, von Melchner H, Witkowski J, Woychik RP, Wurst W, Yancopoulos GD, Young SG, Zambrowicz B (2004) The knockout mouse project. Nat Genet 36(9):921–924
32. Collins FS, Finnell RH, Rossant J, Wurst W (2007) A new partner for the international knockout mouse consortium. Cell 129(2): 235
33. Hughes ED, Qu YY, Genik SJ, Lyons RH, Pacheco CD, Lieberman AP, Samuelson LC, Nasonkin IO, Camper SA, Van Keuren ML, Saunders TL (2007) Genetic variation in C57BL/6 ES cell lines and genetic instability in the Bruce4 C57BL/6 ES cell line. Mamm Genome 18(8):549–558
34. Auerbach W, Dunmore JH, Fairchild-Huntress V, Fang Q, Auerbach AB, Huszar D, Joyner AL (2000) Establishment and chimera analysis of 129/SvEv- and C57BL/6-derived mouse embryonic stem cell lines. Biotechniques 29(5):1024–1028, 1030, 1032
35. Ware CB, Siverts LA, Nelson AM, Morton JF, Ladiges WC (2003) Utility of a C57BL/6 ES line versus 129 ES lines for targeted mutations in mice. Transgen Res 12(6):743–746
36. Seong E, Saunders TL, Stewart CL, Burmeister M (2004) To knockout in 129 or in C57BL/6: that is the question. Trends Genet 20(2):59–62
37. Ward CM, Barrow KM, Stern PL (2004) Significant variations in differentiation properties between independent mouse ES cell lines cultured under defined conditions. Exp Cell Res 293(2):229–238
38. Hansen GM, Markesich DC, Burnett MB, Zhu Q, Dionne KM, Richter LJ, Finnell RH, Sands AT, Zambrowicz BP, Abuin A (2008) Large-scale gene trapping in C57BL/6N mouse embryonic stem cells. Genome Res 18(10):1670–1679
39. Brook FA, Gardner RL (1997) The origin and efficient derivation of embryonic stem cells in the mouse. Proc Natl Acad Sci USA 94 (11):5709–5712
40. Cheng J, Dutra A, Takesono A, Garrett-Beal L, Schwartzberg PL (2004) Improved

generation of C57BL/6J mouse embryonic stem cells in a defined serum-free media. Genesis 39(2):100–104
41. Schoonjans L, Kreemers V, Danloy S, Moreadith RW, Laroche Y, Collen D (2003) Improved generation of germline-competent embryonic stem cell lines from inbred mouse strains. Stem Cells Dayton OH 21((1):90–97
42. Batlle-Morera L, Smith A, Nichols J (2008) Parameters influencing derivation of embryonic stem cells from murine embryos. Genesis 46(12):758–767
43. Ledermann B, Burki K (1991) Establishment of a germ-line competent C57BL/6 embryonic stem cell line. Exp Cell Res 197(2): 254–258
44. Kontgen F, Suss G, Stewart C, Steinmetz M, Bluethmann H (1993) Targeted disruption of the MHC class II Aa gene in C57BL/6 mice. Int Immunol 5(8):957–964
45. Lemckert FA, Sedgwick JD, Korner H (1997) Gene targeting in C57BL/6 ES cells. Successful germ line transmission using recipient BALB/c blastocysts developmentally matured in vitro. Nucl Acids Res 25(4): 917–918
46. Pacholczyk G, Suhag R, Mazurek M, Dederscheck SM, Koni PA (2008) Generation of C57BL/6 knockout mice using C3H x BALB/c blastocysts. Biotechniques 44(3): 413–416
47. Schuster-Gossler K, Lee AW, Lerner CP, Parker HJ, Dyer VW, Scott VE, Gossler A, Conover JC (2001) Use of coisogenic host blastocysts for efficient establishment of germline chimeras with C57BL/6J ES cell lines. Biotechniques 31(5):1022–1024
48. Pettitt SJ, Liang Q, Rairdan XY, Moran JL, Prosser HM, Beier DR, Lloyd KC, Bradley A, Skarnes WC (2009) Agouti C57BL/6N embryonic stem cells for mouse genetic resources. Nat Meth 6(7):493–495
49. Ying QL, Wray J, Nichols J, Batlle-Morera L, Doble B, Woodgett J, Cohen P, Smith A (2008) The ground state of embryonic stem cell self-renewal. Nature 453(7194):519–523
50. Nichols J, Jones K, Phillips JM, Newland SA, Roode M, Mansfield W, Smith A, Cooke A (2009) Validated germline-competent embryonic stem cell lines from no obese diabetic mice. Nat Med 15(7):814–818
51. Silva J, Smith A (2008) Capturing pluripotency. Cell 132(4):532–536
52. Sato H, Amagai K, Shimizukawa R, Tamai Y (2009) Stable generation of serum- and feeder-free embryonic stem cell-derived mice with full germline-competency by using a GSK3 specific inhibitor. Genesis 47(6):414–422
53. Gertsenstein M, Nutter LM, Reid T, Pereira M, Stanford WL, Rossant J, Nagy A (2010) Efficient generation of germ line transmitting chimeras from C57BL/6N ES cells by aggregation with outbred host embryos. PLoS ONE 5(6):e11260
54. Kiyonari H, Kaneko M, Abe S, Aizawa S (2010) Three inhibitors of FGF receptor, ERK, and GSK3 establishes germline-competent embryonic stem cells of C57BL/6N mouse strain with high efficiency and stability. Genesis 48(5):317–327
55. Hooper M, Hardy K, Handyside A, Hunter S, Monk M (1987) HPRT-deficient (Lesch-Nyhan) mouse embryos derived from germline colonization by cultured cells. Nature 326(6110):292–295
56. Sibilia M, Wagner EF (1995) Strain-dependent epithelial defects in mice lacking the EGF receptor. Sci NY NY 269(5221): 234–238
57. Threadgill DW, Dlugosz AA, Hansen LA, Tennenbaum T, Lichti U, Yee D, LaMantia C, Mourton T, Herrup K, Harris RC et al (1995) Targeted disruption of mouse EGF receptor: effect of genetic background on mutant phenotype. Sci NY NY 269(5221): 230–234
58. Silva AJ, Simpson EM, Takahashi JS, Lipp H-P, Nakanishi S, Wehner JM, Giese KP, Tully T, Abel T, Chapman PF, Fox K, Grant S, Itohara S, Lathe R, Mayford M, McNamara JO, Morris RJ, Picciotto M, Roder J, Shin H-S, Slesinger PA, Storm DR, Stryker MP, Tonegawa S, Wang Y, Wolfer DP (1997) Mutant mice and neuroscience: recommendations concerning genetic background. Neuron 19(4):755–759
59. Parker-Thornburg JV, Alana JL, Smith CN, Detry M, Rojas ML, Baskin KK (2005) Cryopreserved morulae can be used to efficiently generate germline-transmitting chimeras by blastocyst injection. Transgen Res 14(5): 685–690
60. Rossant J, Merentes-Diaz E, Gocza E, Ivanyi E, Nagy A (1991) Developmental potential of mouse embryonic stem cells. In: Bavister BD (ed) Serono symposium on preimplantation embryo development. Springer, NY, pp 157–165
61. Wang Z, Jaenisch R (2004) At most three ES cells contribute to the somatic lineages of chimeric mice and of mice produced by ES-tetraploid complementation. Dev Biol 275(1): 192–201
62. De Repentigny Y, Kothary R (2010) Production of mouse chimeras by injection of embryonic stem cells into the perivitelline space

of one-cell stage embryos. Transgen Res 19(6):1137–1144
63. Nagy A (2003) Manipulating the mouse embryo: a laboratory manual. Cold Spring Harbor Laboratory Press, Cold Spring Harbor, NY
64. Kondoh G, Yamamoto Y, Yoshida K, Suzuki Y, Osuka S, Nakano Y, Morita T, Takeda J (1999) Easy assessment of ES cell clone potency for chimeric development and germ-line competency by an optimized aggregation method. J Biochem Biophys Meth 39(3): 137–142
65. Rulicke T, Haenggli A, Rappold K, Moehrlen U, Stallmach T (2006) No transuterine migration of fertilised ova after unilateral embryo transfer in mice. Reprod Fertil Dev 18(8):885–891

Chapter 18

Derivation of Murine ES Cell Lines

Kristina Nagy and Jennifer Nichols

Abstract

Embryonic stem (ES) cells have had a tremendous impact on the field of genetics and are widely used as a means for precise genetic modification of the mouse genome. This chapter will give a background to the use of these versatile cells and provide practical protocols for their derivation and culture. These techniques critically depend on attention to detail and quality control. We will thoroughly discuss potential pitfalls and provide additional techniques that can be used in cases where derivation has previously failed.

Abbreviations

2i	Two inhibitors
ES	Embryonic Stem (cells)
FBS	Fetal Bovine Serum
hiPS	Human induced Pluripotent Stem (cells)
ICM	Inner Cell Mass
iPS	Induced Pluripotent Stem (cells)
KO-DMEM	Knock-out Dulbecco's Modified Eagles Medium
LIF	Leukemia Inhibitory Factor
MEFs	Mouse Embryonic Fibroblast (cells)
NEAA	Nonessential Amino Acids
PBS	Phosphate Buffered Saline
SR	Serum Replacement
TVP	Trypsin Versene Phosphate

18.1 Introduction

Mouse embryonic stem (ES) cells have served as invaluable tools for studies in developmental biology and genetics for a very long time. There is no doubt that these and many other research areas

S. Pease and T.L. Saunders (eds.), *Advanced Protocols for Animal Transgenesis*, Springer Protocols,
DOI 10.1007/978-3-642-20792-1_18, © Springer-Verlag Berlin Heidelberg 2011

would not have advanced to their current state without the use of ES cells. ES cells are continuous cell lines that – if cultured under optimal conditions (we will get back to this point later) – will maintain a pluripotent, undifferentiated phenotype. These cells can be induced to differentiate into any and all of the cell types found in the body of a mouse, including germ cells. Since the cells can easily be propagated in culture in very large numbers, it is possible to introduce and screen for very rare genomic events. One of these, homologous recombination, is generally the purpose for working with ES cells, since they offer the possibility of producing ES cells with specific genetic modifications. If ES cells are placed back into an in vivo environment (either by injection into a host blastocyst or by aggregation with a host morula, as outlined in Chapter 17, the resulting animal will be chimeric. That is, to say it will be made up of cells with two different genotypes – those of wild type host embryo origin and those carrying the desired genomic alteration, originating from the ES cells. If some of the germ cells in the chimera arise from the ES cell contribution, then the genetic alteration can be passed along to the next generation and thus a new genetically modified mouse line is generated. Chimeric animals are also used sometimes to study more complex phenotypes as described in Chapters 16 and 17.

Introduction of transgenes or a reduction of a particular gene product can be accomplished either by pro-nuclear injection (*see* Chapter 6), or by the use of ES cells. Finally, ES cells are also widely used in vitro, for example, to study the differentiation of particular cell types or toxicology.

For a long time, the only genetic background that allowed for the establishment of germline ES cells from mice was 129. It is still not entirely clear why this particular family of inbred strains is superior in this regard, but perhaps one explanation could be the genetic makeup of 129 mouse strains, specifically in their signaling pathways. Interestingly, hybrid vigor was observed in mouse ES cells when R. Jaenisch's laboratory established F1 hybrid ES cells from a cross between 129 and C57BL/6, and showed that these cells have an even higher developmental potential than the classic 129 derived lines. Indeed, hybrid lines are the choice for performing tetraploid complementation (Chapter 16). When using some F1 hybrid lines, not only will the resulting completely ES cell-derived embryos develop to term, but live adult animals can easily be obtained.

A few years ago, the Yamanaka laboratory made an incredible discovery. They established a method for obtaining induced pluripotent stem cells (iPS cells) by reprogramming somatic cells simply by expressing four transcription factors: Oct4, Sox2, Kfl4, and c-Myc. It did not take long before there was speculation about whether ES cells would be needed at all in the future, or whether we could now completely turn to iPS cells for all the purposes for which ES cells had traditionally been used. In looking at the field in which human applications for ES cells are explored, there is no doubt that iPS cells have become a major player, as many of the ethical issues surrounding the process of creating hES cells have been eliminated by the introduction of hiPS cells. In the mouse however, the situation is different. Far from all ES or iPS cell lines are of the same developmental capacity. When deriving new lines of either sort, one should not expect to find more than 10% of newly established cell lines to be of a quality that is adequate for making highly germline competent chimeric mice. For this reason, it is wise to work with an ES cell line that has been extensively tested and proven. As of today, iPS cell lines are still relatively "new" and few of them have been used extensively. There is an additional advantage to working with ES cells rather than iPS cells – something of which iPS cells will never be capable. Consider this scenario; a combination of two genetic modifications, when present in a homozygous form, gives a very early embryonic phenotype. Part of the problem, we speculate, may be extra-embryonic, since this is often the case in an early developmental abnormality. One way to examine this situation and to access other phenotypes that might be present is through tetraploid embryo complementation (Chapter 16). For this however, one needs ES (or iPS) cells with the double-homozygous genotype. Assuming that the lines were created by homologous recombination in ES or iPS cells, one possibility would be to create homozygosity in the original cell line by gene targeting, then target the cells for the second mutation, and create a second homozygous mutation. In reality, this is hardly feasible, and even if one would succeed, there is a high risk that the ES/iPS cells would lose their pluripotency through the very extensive and repeated culture, cloning, and selection process. A more straightforward alternative is to create double-heterozygous animals, cross these, collect preimplantation stage embryos, and establish new ES cell lines from the double-homozygous embryos. These can then be used in tetraploid complementation assays.

18.1.1. Outline of the Procedure

The procedure that will be detailed here below is relatively straightforward. Anyone with experience in ES cell culture and with preimplantation stage mouse embryos should be able to follow this with ease, although it may be time-consuming and at times, tedious.

Briefly, blastocyst stage embryos are placed on a regular feeder layer of mouse embryonic fibroblasts in ES cell media. The embryos are allowed to attach to the feeders and the inner cell masses start growing out. Once this initial outgrowth has reached a certain size, it is dissociated and placed on a fresh layer of feeder cells. New colonies will appear and these will be propagated as regular ES cell cultures.

The key points to remember for success are the following:

1. Pay extreme attention to culture conditions. Inspect the cultures at the intervals listed in the protocol. Do not leave cultures unattended over the weekend.
2. Make careful records of the progress of each outgrowth, including imaging.
3. Pay utmost attention to the timing of the dissociation of the initial outgrowth.
4. Take extreme care during the first and second trypsinizations of outgrowths. Re-picking may be needed.
5. Freeze down an early passage as soon as possible.

18.2 Materials and Equipment

18.2.1. Equipment

- Inverted stereomicroscope with transmitted illumination.
- Humidified incubator (37°C, 5% CO_2).
- Vertical flow hood – equipped for regular tissue culture.
- Horizontal flow hood (optional, but highly recommended).
- Surgical instruments (e.g., Fine Scientific Tools – FST): sharp fine-pointed scissors, fine forceps (e.g., Dumont #5 or ss/mc), straight or curved blunt forceps with serrated tips, serrafine (e.g., FST #18050-28).
- Refrigerator 4°C.
- Freezer −20°C.
- Freezer −80°C.
- LN_2 storage for cryogenic vials.

18.2.1.1. Notes on Equipment

1. The incubator should be reserved for ES cell derivation purposes. If at all possible, try not to place primary cultures in the same incubator as established cell lines. Even though good practice should prevent cross-contaminations, there is always a risk that newly established cell lines could introduce pathogens in the other cultures.
2. It is advisable to have a horizontal flow hood available for placing embryos in culture, picking of outgrowths, and potential subcloning procedures. Some have succeeded in not introducing any contamination when performing these steps on the bench, or by placing a microscope in a vertical flow hood, but this is suboptimal practice.

3. Good inverted microscopes equipped with an image capture device are essential. You will be checking the cultures on a daily basis and it is important for future experiments to have documented at which stage the particular steps of the protocol were executed.

18.2.2. Materials

- Tissue culture plastic: four- and six-well plates, 60 and 100 mm dishes.
- Sterile conical 15 and 50 ml tubes.
- Plastic disposable or tissue-culture dedicated glass 1, 5, 10, and 25 ml pipettes.
- Cryogenic vials.
- Sterile 5 cc syringes.
- 26G ½″ needles.
- Bunsen burner.
- Embryo handling pipettes (see below).
- Mouth aspiration piece [round: Sigma #5177 or flat: Biotec Inc #MP-001-Y (http://biodiseno.com)].
- Elastic silicon or latex tubing with 1/8 in. inner diameter and 1/32 in. wall (VWR International #62996-350).
- Slow-rate cooling box ("Mr Frosty" Sigma-Aldrich #C1562) or a small Styrofoam box with slots for cryogenic vials.
- Pasteur pipettes.

18.2.2.1. Making a Mouth Pipetting Device

To make an embryo-handling pipette, hold the thin part of a long Pasteur pipette over a Bunsen burner flame until the glass slightly melts. At this point, make a quick pull to produce a thinly drawn-out section. Break the glass about 2–3 cm from the widening of the neck by simply snapping it. Polish the end in the flame just enough to eliminate any sharp edges but not so much that the glass melts. Cut an approximately 50 cm length of silicon tubing, attach a P1000 (blue) tip to one end (place the narrow end in the tubing), and a mouth aspiration piece in the other. If a mouth aspiration piece is not available, it is possible to use a P200 (yellow) tip instead, however, this is much less comfortable in the long run. Place the wide end of the drawn-out Pasteur pipette in the blue tip.

Note: Mouth pipetting may not be allowed at all institutions. Always follow the local guidelines, and never use mouth pipetting for hazardous, toxic, or infectious materials.

18.2.3. Reagents

- PBS Ca^{2+} Mg^{2+} free (Invitrogen Life Technologies 14190-144).
- Trypsin, 0.25%, 1 mM EDTA (Invitrogen Life Technologies 25200-072).
- DMEM high glucose (Invitrogen Life Technologies 11960-044).

- Fetal Bovine Serum (FBS) (HyClone, Invitrogen).
- Leukemia Inhibitory Factor (LIF) (Millipore – Chemicon LIF2010).
- Sodium Pyruvate 100× stock (Invitrogen Life Technologies 11360-070).
- Nonessential Amino Acids 100× stock (NEAA) (Invitrogen Life Technologies 11140-050).
- GlutaMax 100× stock (Invitrogen Life Technologies 35050-079).
- Beta-Mercaptoethanol (Sigma M7522): Dilute 70 μl 14.3 M BME in 100 ml PBS to make 100× stock solution.
- Penicillin/Streptomycin 100× stock (Invitrogen Life Technologies 15140-148).
- DMSO (Sigma D5879).
- MitomycinC (Sigma M4287).
- Embryo-tested light mineral oil (Sigma M8410).
- Acid Tyrode's solution (Sigma T1788).
- M2 embryo culture media (Millipore MR-015-D).
- KSOM embryo culture media with amino acids and glucose (Millipore MR-121-D).
- Porcine gelatin (Sigma G1890).

18.2.3.1. Notes on Reagents

1. It is immensely important to ensure the highest possible quality of all reagents and media components. Always keep in mind that it is not enough to keep ES cells proliferating. If pathogens enter the culture, the cells can no longer be used for producing chimeras. Culture conditions that fail to provide optimal support for an undifferentiated state of the cells is equally detrimental. Suboptimal conditions will quickly lead to the accumulation of chromosomal and epigenetic changes that prevent the successful transmission of the ES cell genome through the germline. To visualize this scenario, imagine the ES cell culture in your hands as a soup of millions of cells. The majority of these cells are normal and just what you expect to find. A small portion however will acquire genetic abnormalities. Point mutations, chromosomal translocations, and epigenetic changes occur at random – this is the driving force of evolution. If the culture conditions are ideal, that is, specifically suited to be optimal for the "normal" population, then the proportion of abnormal cells will stay low and increase only slowly over extended passages. However, if the culture conditions are not optimal, abnormal cells may be favored, and over time, these will take over the cultures.

2. The quality of the FBS is perhaps the most crucial for successful ES cell culture. Each new batch should be rigorously tested against a known lot to support ES cell maintenance and germline transmission of chimeras. FBS is stable for several years when frozen at −80°C. Once thawed, it should be stored in the dark at 4°C and used within 4 months. Ready-made media should also be stored at 4°C in the dark and used within 4 weeks.

18.2.4. Suppliers

- Invitrogen Life Technologies (media and media supplements).
- SIGMA (reagents).
- Falcon (disposable tissue culture consumables).
- Fisher (general laboratory equipment).
- VWR (general laboratory equipment).
- Specialty Media (mouse embryo culture media).
- High-Clone, Wisent, Invitrogen (fetal bovine serum).
- Millipore (LIF, reagents).
- HPI Hospital Products Med. Tech, SIGMA or Biotec Inc (mouth aspiration piece).

18.3 Protocols

18.3.1. Preparation of Culture Media

Culture media should always be prepared fresh when possible, and stored no longer than 4 weeks at 4°C in the dark. Some protocols suggest warming the media to 37°C before use. We discourage this practice for two reasons: (1) there is no advantage to warming the media to 37°C. ES cells will be just fine if the media is at room temperature. (2) Warming often takes place in a heated water bath. This piece of equipment is a perfect sanctuary for unwanted microorganisms, unless adamantly cared for, including regular and frequent disinfection.

1. To a bottle of 500 ml DMEM High glucose, add the following:
2. 90 ml FBS
3. 500,000 U LIF
4. 6 ml NEAA
5. 6 ml Sodium Pyruvate
6. 6 ml Glutamax
7. 6 ml Pen/Strep
8. 6 ml Betamercapto ethanol

Mix well by carefully swirling the bottle. Do not shake, as this will cause extensive foaming.

18.3.2. Gelatinizing Tissue Culture Plates

ES cells are generally cultured on top of a layer of mitotically inactivated mouse embryonic fibroblasts (MEFs). In some cases however, the presence of MEFs is less desirable: isolation of genomic DNA from the ES cells, testing FBS batches, etc. For these occasions, treating the tissue culture dishes with gelatin will allow the ES cells to adhere to the plastic. One should keep in mind though that for most ES cell lines, the morphology will change from a three-dimensional dome-shape with sharp edges to a more flat and less well-defined appearance. Provided that the culture conditions are optimal, the ES cells will regain their original appearance once they are placed back on MEFs.

1. Dissolve porcine gelatin (Sigma G1890) in ddH_2O by heating and stirring to prepare a 0.1% solution.
2. Autoclave and let cool down.
3. Cover the entire surface of the tissue culture dish with a thin layer of gelatin solution (e.g., 2 ml in a 60 mm diameter dish).
4. Let the dish stand at room temperature in the vertical flow hood for 30 min.
5. Remove the gelatin.
6. The dish is now ready for use right away or can be stored in the humidified sterile tissue culture incubator for up to 3 days.

Note: It is critically important to autoclave the gelatin before use to avoid pathogen contaminations in the culture.

18.3.3. Testing FBS Batches

Each lot of FBS – whether advertised as "ES cell tested" by the supplier or not – varies in its ability to support the undifferentiated state of ES cells during culture. Testing a number of batches against a known lot that has proven to produce a good rate of germline competent chimeras after gene targeting, subcloning, and extended culture is the only way to find a source of FBS that is worth working with. This process is tedious and costly, and so is FBS in itself. For these reasons, it is worth purchasing a large number of bottles of a favorable lot after undertaking this rigorous testing protocol. FBS can be stored at −80°C for at least 2 years.

1. Contact a number of FBS suppliers and ask them for 50 ml aliquots of as many FBS lots as available. At this time, ask how many bottles they have of each lot and request that they hold the quantity that you plan to purchase. This process may require some negotiation skills, as most suppliers are not openly inviting the testing of their serum lots.

2. Acquire FBS of a known "good" lot as well as ES cells of high quality (proven good germline competence) as a positive control.
3. Make up four different ES media aliquots with each FBS lot: 10% + LIF, 15% + LIF, 15% − LIF, and 30% + LIF FBS. All other ingredients should be kept as the regular ES media.
4. Thaw the ES cells using the positive control FBS batch.
5. Culture the cells until you have enough cells for the following step. Calculations will vary depending on the size of tissue culture plates used.
6. Trypsinize the cells and divide them into as many tubes as FBS batches to be tested multiplied by 4 (if you have ten batches to test plus a positive control batch, that means 11 × 4 = 44 tubes).
7. Centrifuge the cells and remove the supernatant.
8. Resuspend each tube with media prepared with the different FBS lots and concentrations.
9. Plate the cells onto gelatinized dishes at regular splitting ratio (day 0).
10. Feed the cells the following day (day 1).
11. Record the colony morphology the next day (day 2). Discard those batches in which the cells are not growing well at 30% FBS (sign of toxicity). Trypsinize the cells, count them, and plate all regular density.
12. Feed the cells the following day and record colony morphology again (day 3). You may see some differences between batches by now. Those with obvious decreased cell proliferation rate at 30% (FBS toxicity) and those with differentiated morphology can be discarded.
13. Record the colony morphology the next day (day 4). Repeat the above discarding selection in **step 12**. Trypsinize the cells, count them, and plate all at the same density.
14. Record the colony morphology the next day (day 5). This is usually the best time point for making a final assessment. However, if the batches still seem to be very similar, continue to passage them once or twice.
15. An ideal FBS batch should show no growth rate decrease when used at 30% concentration, and at the same time result in colony morphology as similar to the positive control as possible. Pay special attention to the 10% FBS condition. A good FBS lot can support ES cells also at this concentration. The LIF minus group is the most stringent test of all. Sometimes this is the only condition in which a significant

difference can be detected between batches of FBS. This does not mean that the difference is unimportant.

16. Once two to three lots have been identified as the "best," an order should be placed with the supplier for one 500 ml bottle. The remaining bottles should be kept on hold.
17. The same ES cells that were used for the tests should then be split at limited dilution, and plated on MEFs. Colonies should be picked and clones expanded three to four passages. This should all be done using the selected new FBS lots, and the positive control should be included.
18. Three clones from each of the best batches (plus the control) should be used for morula/blastocyst injection or morula aggregation to create chimeric animals.
19. The chimeras should be mated to detect germline competence of the ES cells.

Those lots that produce comparable results to the positive control can be deemed as an ES cell-qualified batch.

Note 1: The ES cell line used as positive control should be given some consideration. We routinely use proven germline competence R1 cells, as these seem to be sensitive and keen indicators of FBS quality. However, although very likely, there is no ultimate guarantee that the batch that is good for one ES cell line will be equally adequate for another. For this reason, it is best to test FBS batches on the ES cell line that will be mostly used in the facility for future experiments.

Note 2: ES cells cultured on gelatinized plastic always take on a more flat and less well-defined appearance than those cultured on MEFs. This is normal. The reason for removal of MEFs from culture for testing of serum lots is that MEFs can mask the inadequacy of an FBS lot ability to support ES cells in their undifferentiated state.

18.3.4. Preparation of MEFs

Mouse embryonic fibroblast feeder layers should be prepared 1 day before an ES cell derivation experiment is scheduled to take place. It is important to use a low passage number (no higher than 3 or 4) of high-quality MEFs that have previously been shown to support undifferentiated ES cell propagation. The MEFs should be mitotically inactivated the day before use.

1. Prepare a 10 μg/ml MitomycinC in regular MEF media.
2. Aspirate the media from MEFs that have been grown to confluency, and add the MitC-containing media.
3. Incubate for 3 h at 37°C. Take extreme care not to spill media containing MitC, as it is toxic!
4. Wash very carefully three times with PBS.

5. Trypsinize the MEFs and seed at a density of 1.5×10^5 cells per well in four-well dishes.
6. Use the MEF-coated plates within 2–3 days.

Note 1: It is extremely important to carefully wash the MEFs with PBS to remove any residual MitomycinC from the media.

Note 2: It is possible to use gamma irradiation of MEFs instead to achieve a mitotic arrest. However, this requires specialized equipment that is not readily available everywhere. If irradiation is used, it is very important to take care to keep the cultures sterile during the procedure, which usually will take place outside the regular tissue culture facility.

18.3.5. Collection of Blastocyst Stage Embryos

Blastocyst stage embryos are collected from the uterus of 3.5 dpc (days post coitus) pregnant female mice. If young females are available, these can be superovulated using standard hormone treatment protocols [1]. If however the females to be used are over the optimal age for superovulation for that particular genetic background, natural mating should be used. Both options are good as long as the blastocysts produced are of good quality.

Not all blastocysts will result in an outgrowth with a morphology worth picking and only a portion of those that do will result in ES-like colony growth. Many of those that do look promising after the initial picking will loose their encouraging morphology during the initial passages, and of those that retain their good morphology, some will prove to have an abnormal karyotype or have lost their pluripotency to more subtle genomic or epigenetic changes. For this reason, it is important to start each derivation experiment with an adequate number of blastocysts. For a permissive strain such as 129, 25–50 embryos are sufficient, but where other strains are concerned, much higher numbers for the production of 2–3 good ES cell lines may be required.

1. On the day before embryo collection, prepare 30 mm tissue culture dishes with microdrops of KSOM-AA embryo culture media. Cover the microdrops with light paraffin oil and place in the 37°C, 5% CO_2 incubator overnight.
2. The next morning, euthanize the embryo donor mice by the method your Institute officials recommend, whether that be Veterinarian, IACUC, or Senior animal care staff (*see* **Notes**).
3. Wash the abdomen with 70% EtOH.
4. Cut the abdominal skin and then the abdominal wall with fine scissors in a manner that allows the internal organs to remain as sterile as possible (*see* **Notes**).
5. Grasp one ovary with fine forceps and cut across the fat pad above it. Cut the adherent viscera along the uterus. Cut the

cervix and then along the other uterine horn. Finally cut the fat pad above the second ovary.

6. Lift out the whole uterus into an empty sterile 60 mm Petri dish.
7. Fill a 5 cc syringe with M2 mouse embryo culture media and attach it to a 26G short needle.
8. Insert the needle at the very top of one uterine horn and grasp the uterus and needles with fine forceps to prevent the needle from sliding out. Press the plunger. The corresponding uterine horn should swell up and the media readily come out through the cervix. At least 0.5 ml of media should be flushed through each horn to ensure that all embryos have been recovered.
9. Proceed the same way with the other uterine horn.
10. Discard the uterus and move the Petri dish to a dissecting stereomicroscope with illumination from below.
11. Using a mouth pipetting device (as described in the Subheading 18.2.2), collect the blastocysts, and place them in a small drop of M2 media.
12. Wash, count, and sort the embryos.
13. Wash them through three drops of KSOM-AA medium and place them in the preincubated KSOM-AA dish.

Note 1: Work swiftly, and do not euthanize more mice then can be handled in such a way that the embryos are all collected and placed in the incubator within 30 min.

Note 2: When lifting out the uterus from the abdominal cavity, make sure not to let it touch any of the outer body surfaces or the dissection pad. Likewise, use separate scissors and forceps to cut the skin and the body wall. These precautions should be taken to minimize the risk of contaminating the embryos with organisms present on the skin of the mouse.

18.3.6. Plating and Initial Culture of Blastocysts

Once the blastocysts have been collected, washed, sorted and counted, it is time for plating them on the MEFs that have been prepared the day before. Only fully expanded blastocysts (Fig. 18.1c) should be plated. Embryos that are at the morula or early blastocyst stage (Fig. 18.1a, b) should be allowed to develop further by continued culture in KSOM-AA drops under oil until they are fully expanded.

1. Remove the embryos from the incubator.
2. Place one embryo into one well of a four-well plate.
3. Try to place the embryos at the middle of the well. If this is not successful, try swirling the dish gently and then let it stand until the embryo settles in the middle.

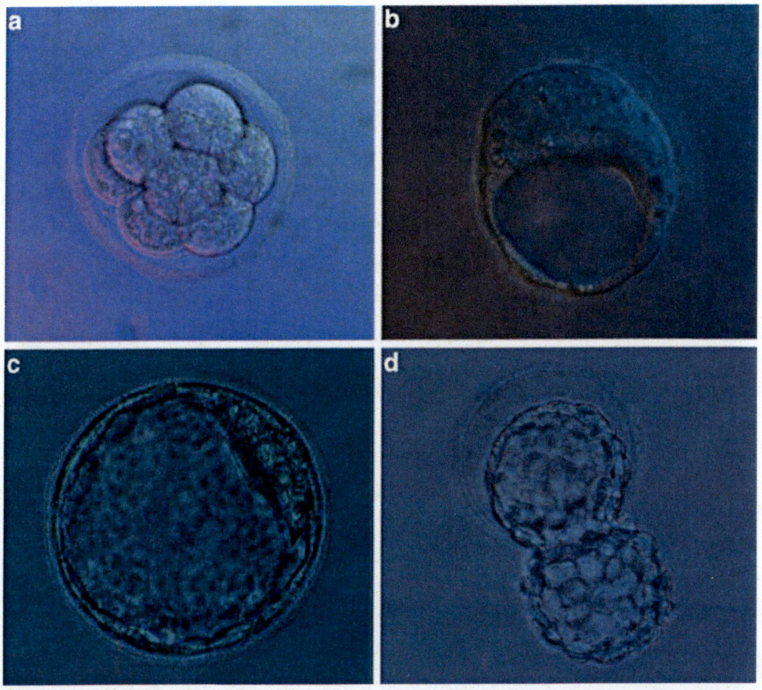

Fig. 18.1 Late preimplantation mouse embryos. (**a**) Morula containing eight well-separated blastomeres prior to compaction. (**b**) Early blastocyst, note that the blastocoel occupies almost half the volume of the embryo. (**c**) Fully expanded blastocyst, note that the blastocoel accounts for most of the volume of the embryo. (**d**) Hatching blastocyst, the embryo is emerging from the zona pellucida prior to implantation.

4. Move the dishes back to the incubator very carefully. Shaking or knocking will result in the embryo moving to the edge of the well where it will be very difficult to inspect the outgrowth.
5. Leave the dishes alone for 48 h – no touching, no moving, no looking and ideally, do not even open the incubator door.

18.3.7. Dissociation of the Primary Outgrowth

Forty-eight hours after plating, it is time to do the first inspection of the plates. There is variation between different genetic backgrounds with regard to how fast the embryos attach and start to grow. In most cases however, the blastocysts will have hatched (Fig. 18.1d) and attached (Fig. 18.2a) after 2 days, i.e., by the time of first inspection. If this is the case, half of the media should be changed and the plates placed back in the incubator again. If the embryo is not firmly attached, gently put the plate back into the incubator until the next day. From this point onward, the cultures should be inspected every day, and half the media changed every other day as soon as attachment of the embryo to the feeder layer will permit. Soon after attachment, cells from the embryo will start to proliferate (Fig. 18.2b). Often, the trophoblast cells will form a typical large flat monolayer of cells that spread out around the attachment site. In the

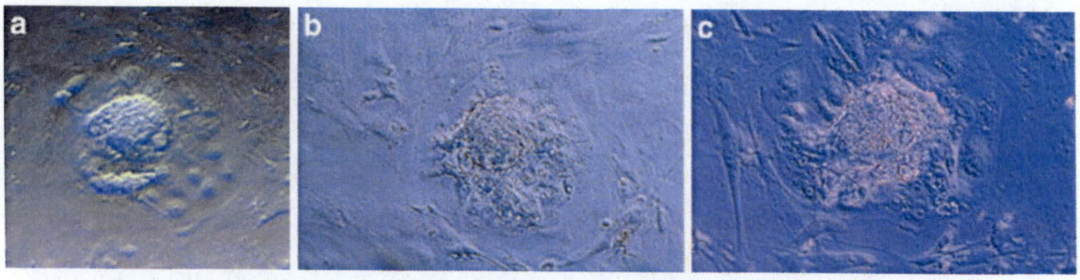

Fig. 18.2 Blastocyst outgrowths. (a) Hatched blastocyst attached to MEF layer. (b, c) Postattachment proliferation. The inner cell mass forms a central mound on flat trophectoderm cells. The outgrowth of small compact cells will give rise to ES cells.

middle, the primitive endoderm and ectoderm cells will form a multilayered, dense structure (Fig. 18.2c). The cells in this tightly packed area contain the cells that eventually will become ES cells. When referring to "picking the outgrowth", we mean the separation of this area from the rest of the cells present in the culture. This "clump" will grow during the coming days and then start to differentiate – that is, loose its distinct morphology. Picking has to be done at the optimal time point where the outgrowth is at its largest, but before it starts to differentiate. This timing is crucial for success. If picking is done too early, the cells will die off; if it is done too late, the cells have differentiated and will never become ES cells.

1. Move the culture to a horizontal flow hood equipped with a stereo dissecting microscope.
2. With a freshly drawn and carefully fire-polished Pasteur-pipette attached to a mouth pipetting device, carefully pick the well-rounded tight clump of cells described above. The inner diameter of the picking pipette should be slightly larger than what is used for embryo handing. This will allow for picking the outgrowth without squishing it too much.
3. Place the clump in a 25 μl drop of Trypsin/EDTA placed on a tissue culture dish.
4. Place the dish in the incubator for 2 min.
5. With a narrow Pasteur-pipette, pipette the clump in and out until it breaks up into five to six smaller clumps. Do not break it up to single cells! The size of this pipette should be about half of those used for embryo handling.
6. Add the clumps to one well of a four-well dish with MEFs.
7. Place the dish back in the incubator.

18.3.8. Expansion

During the following week, it will become evident if the cells will be viable and proliferate, and have the potential to become an ES cell line. The cultures should now be examined every day for the appearance of ES-like colonies. Three scenarios might occur:

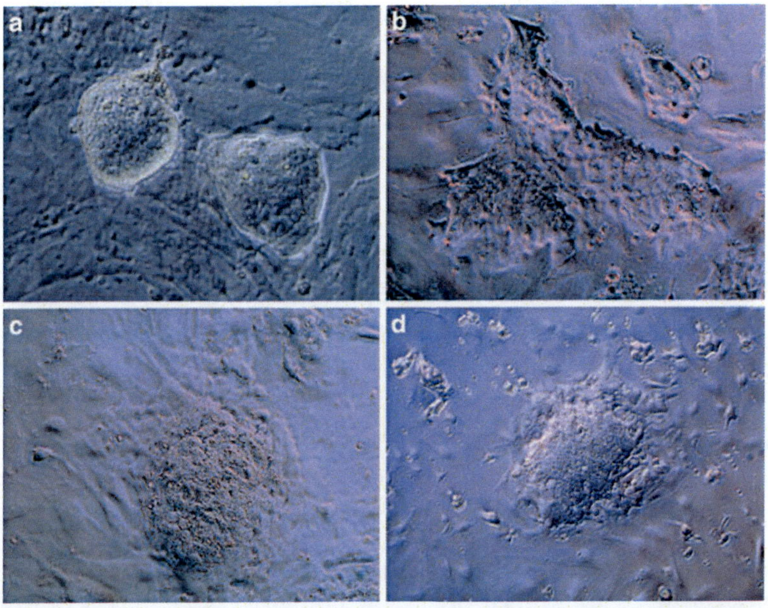

Fig. 18.3 Types of cells that can arise after passaging blastocyst outgrowths. (a) Dome-shaped colonies of ES cells. (b) Flat cells with epithelial-like morphology. (c, d) Colonies with irregular borders that lack the characteristic dome-shaped ES cell colony morphology.

(a) No growth. Change the media on the culture every other day and wait 8–10 days before "giving up" on a culture. If by the 10th day no colonies are visible, the culture can be discarded.

(b) Colonies with a clear ES-like morphology (Fig. 18.3a) appear after 3–6 days. This is the ideal scenario. Proceed by passaging the culture as described in the protocol below.

(c) Colonies appear with a mixed morphology; some resembling ES cells and others with flat, epithelial-like morphology (Fig. 18.3b), or colonies with ill-defined borders (Fig. 18.3c, d). In this case, it is important to separate the ES-like colonies from those with other characteristics. This is best done by manually picking the colonies with the desired morphology into a new well with fresh MEFs.

The newly established ES cell lines should be treated with great care. The passaging ratio should be kept to 1:3 for the first few passages. Once the cells show a robust growth, it is increased to 1:4–1:8 depending on the growth rate of the cells. The aim should be to passage the cells every 2–3 days, without ever letting the culture become overgrown.

18.3.8.1. Passaging of ES Cells

1. Aspirate the medium.
2. Wash once with PBS.

3. Add a small volume (1/5 of the media volume) of Trypsin/EDTA.
4. Incubate at 37°C for 2–5 min.
5. Examine the culture under a dissecting microscope. When the majority of the ES colonies have detached, it is time to proceed to the next step. Otherwise, place the culture back in the incubator for another 1–2 min.
6. Add an equal volume of media to the cells and pipette up-and-down vigorously five to ten times until a single cell suspension is obtained. Check for remaining clumps under the microscope.
7. Add the cell suspension to a conical tube and centrifuge at 1,000 rpm (200 × *g*) for 3 min.
8. Discard the supernatant; resuspend the cells in fresh media and plate on fresh MEFs.

18.3.9. Freezing, Thawing and Archiving

ES cells will by virtue of their nature always strive to differentiate – just as their origin, the primitive ectoderm cells of the inner cells mass would have done in vivo. No matter how much effort is put into finding and maintaining the best possible culture conditions for preserving pluripotency, with time, more and more of the cells will acquire characteristics that render them unable to contribute to all tissues of a developing embryo. Since this is at best a slow process, early passages will have the highest potential for pluripotency. It follows, then, that cryopreservation of newly established ES cell lines at an early stage is of utmost importance. A good practice is to freeze down a smaller number of vials from each passage until the quantity is large enough to cryopreserve a large pool for future use. Should this last passage prove to be of insufficient quality, one can go back to an earlier passage and attempt re-expansion with the goal of generating a pool of ES cells with a higher developmental potential. A protocol for freezing ES cells in given in Chapter 14.

18.3.10. Pathogen Testing

Although the initial source of ES cells is embryos that originate from the sterile internal environment of the female reproductive tract, cell lines resulting from these experiments are still regarded as primary cultures. Even if great care is taken to maintain good sterile technique, it is possible for microorganisms to invade the cultures. For this reason, we recommend supplementing the culture media with Penicillin/Streptomycin and to culture the cells in a dedicated incubator until lines have been established and screened for pathogens.

Bacterial contaminations are usually detected by the rapid change of pH in the cultures (the media turns yellow in a much shorter time than expected, based on cell density). Yeast infections are readily visible to the eye in a standard microscope when the cultures are inspected. In a mold contamination, thin thread-like structures (hyphae) expand from one or more focal points in the plate. If these cultures are not discarded, and the hyphae are allowed to reach the surface of the media, they will form spores and within a very short amount of time, these spores will quite readily infect not just other nearby cultures but the whole incubator.

If a contamination is detected, the cultures should be discarded immediately, and the incubator sterilized either by washing the walls and autoclaving the shelves or by running a decontamination program.

18.3.11. Screening

Unfortunately, far from all ES cell lines that show a good morphology and growth rate in vitro actually prove to be "good" ES cell lines. Assuming that the end goal is to produce chimeric animals in which the ES cell contribute to a large portion of the tissues, including the germ cells, newly established ES cell lines need to be screened to determine their suitability for such an application. The first line of action should be to determine the sex of the lines. This is most easily accomplished by PCR reaction, using primers for the SRY gene. XY ES cell lines are usually preferred for the generation of chimeras, as it is possible to produce larger numbers of offspring in a shorter time from male chimeras. Karyotyping of new ES cell lines is the next step and should be performed to determine the proper diploid state of all chromosomes. To save on resources, a simple chromosome count can be performed [1]. Cell lines that prove to be at least 80% diploid based on this simple method are worth screening in more detail by G-banding or spectral karyotyping (SKY) techniques, for detection of translocations that cannot be seen by simple chromosome counting. If the ES cell line is intended to be used for several gene targeting experiments, it should first be expanded for the cryopreservation of a large number of cells, divided between multiple vials. One vial of the passage intended for future experiments should then undergo the ultimate test of pluripotency: aggregation/injection with/into mouse embryos and proven germline transmission. Although costly and time-consuming, such an approach will pay off greatly in the end, as proven germline competent cells can be used for experimentation.

18.4 Troubleshooting

Attempts to derive ES cells using conventional culture conditions will not always result in the successful establishment of new lines. Usually, roadblocks are related to suboptimal culture conditions, the wrong timing of initial dissociation, too harsh or long trypsin treatment in subsequent passages, pathogen contamination, or factors related to the specific genetic background of the embryos.

18.4.1. Culture Conditions

The most common reason for ES derivation to fail can be tracked down to suboptimal culture conditions. To determine if the conditions are adequate to support preimplantation stage embryo development, 0.5 dpc embryos (from a strain that does not exhibit the two-cell block) should be collected and cultured in vitro. If 80% of the embryos reach the blastocyst stage by day 4.5, the culture conditions can be regarded as satisfactory (a slight delay from in vivo development speed should be expected and is not a cause for worry). If the results are unsatisfactory, the first thing to do should be to replace culture media, trypsin, and PBS with fresh batches. Do not forget to use fresh batches of all media components as well. Next, the incubator should be checked with an external thermometer and CO_2 monitoring device as internal monitors can fail. Care should also be taken not to keep the embryos out of the incubator for longer periods unless it is absolutely necessary.

18.4.2. Picking and Passaging

If the embryos develop well in culture, the blastocysts hatch, attach to the MEFs and develop into nice outgrowths but few or no colonies form after picking, the problem may lie at the critical picking step. There are three important things to keep in mind (a) wait until the outgrowth is of optimal size and morphology before picking. There should be a clear, tight, three-dimensional area in the middle of the large flat cells that spread out. (b) Pick the outgrowth carefully with a drawn-out and fire-polished Pasteur pipette and (c) keep the time in trypsin short and make sure to leave rather larger clumps instead of making a single cell suspension.

Usually, it is possible to use trypsin to passage the cells after the initial picking. However, it is also possible to passage the cells by mechanical dissagregation for the initial two to three passages where numbers of cells surviving is likely to be low. This is done with a drawn-out Pasteur pipette that has *not* been fire-polished. The sharp edge is used to cut the colonies into five to ten small clumps that in turn are transferred to culture on fresh MEFs.

18.4.3. Pathogen Contamination

What to do if a contamination has been detected? The short answer is: discard the cells. In the vast majority of cases, this indeed is the wisest thing to do. However, there might be instances where repeating the derivation process would be impossible or extremely tedious. In these cases, one might attempt to rescue the cultures by treating them with antimicrobial agents. Keep in mind, though, that any chemicals that are added to the culture media might have an impact on the final quality of the ES cells. Great caution should be taken to (a) limit the spread of the contamination, (b) thoroughly test cells again after treatment has been discontinued to make sure that all contamination has been eliminated, and (c) carefully assess that the ES cells have retained their developmental potential despite the treatment.

18.4.4. Genetic Background

Some genetic backgrounds, such as for example 129 or certain hybrid combinations are well recognized as being permissive for the derivation of ES cells. Other inbred strains, mixed or outbred backgrounds, can make the derivation much more difficult. In these situations, several different approaches have been attempted. These approaches can be used individually or in combination, and when adopted have the potential to increase the efficiency with which ES cell lines can be derived.

18.4.5. Basal Media and Serum Variations

Some groups have reported success when using KO-DMEM (Invitrogen Life Technologies 10829-018) or DMEM low glucose (Invitrogen Life Technologies 10567-014) instead of the standard high-glucose DMEM presented here. Knockout serum replacement (SR) from Invitrogen Life Technologies (10828-028) can be used if there is any indication that your FBS batch is suboptimal for ES cell culture or the genetic background of the mouse strain to be used. Keep in mind though that MEFs generally do not like SR, and will not adhere to the tissue culture plate if cultured in media without FBS. Under these circumstances, two separate medias need to be on hand; one for culturing MEFs until they are adherent, and the other used for culture of the ES cells. Also, because of this inability for feeders to attach in SR, MEF plates must be prepared in advance to ES cell passage, by a minimum of 12 h. If using regular FBS-containing media as described in our protocol, it is possible (this should however be reserved to the exceptional instances of MEF plate shortage) to trypsinize and plate MEFs and ES cells at the same time. The MEFs will adhere to the plastic much before the ES cells and will so form an adequate layer in time for the ES cells to attach.

18.4.6. Morula Stage Embryos as a Starting Material

In instances when the number of blastocysts is limited and this promises to reduce chances of success, you can try to collect the embryos 1 day earlier, at 2.5 dpc, when they are still at the morula stage and reside in the oviducts. The total number of embryos that can successfully be collected this way will be higher since none will be lost in the uterine crevices. Morula stage embryos are then cultured in vitro until they reach the expanded blastocyst stage.

18.4.7. Removal of Zona Pellucida

If the problem seems to stem from the inability of the blastocysts to hatch from the zona pellucida, this can be remedied relatively easily.

1. Prepare a lid of a tissue culture dish by placing two large (500 μl) drops of M2 embryo culture medium in the middle. Place three smaller (250 μl) drops of Acid Tyrode's between the large media drops.
2. Place the blastocysts in the media drop. From here, move them a few at a time to the first acid drop.
3. Without delay, move them to the second acid drop, and then to the third.

4. Once the blastocysts are in the third acid drop, most culture media that could inhibit the action of the Tyrode's solution has been removed. The embryos should now be monitored very carefully. As soon as the zona disappears, they should be moved to the second large drop of M2 media.
5. Once all embryos have been processed this way, they should be washed through three drops of ES culture media and then plated onto MEFs as described in the main protocol.
6. Note that embryos without zona pellucida are extremely fragile and sticky! They should be handled with utmost care, and when pipetted kept in the very lowest portion of the pipette tip, to prevent them from sticking to the glass and thereby be lost.

18.4.8. Immunosurgery

The blastocyst stage embryo consists of a single cell layer of trophoblast cell surrounding the whole embryo, and the so-called inner cell mass, a cluster of cells located in one distinct area. The inner cell mass in turn consists of a layer of primitive endoderm and a small group of primitive ectoderm cells right at the core of the inner cell mass. It is these core cells that eventually will give rise to ES cells. If the problem generating ES cell lines is due to the difficulty isolating cells with a pure ES-like morphology, one can try to isolate inner cell mass cells by selectively destroying the trophoblast cells before plating [2, 3].

1. Remove the zona pellucida as described above. From now on, handle the embryos with extreme care, as they are fragile and easily stick to the inside of the pipette.
2. Incubate the blastocysts in a 1:10 dilution of rabbit antibody raised against mouse erythrocytes (Rabbit anti-RBC) (e.g., Organon Teknika, Cedarlane) for 20 min at 37°C.
3. Very carefully wash the embryos three times in guinea pig complement (e.g., Sigma) to make sure that all antisera have been removed.
4. Incubate the embryos in 1:10 dilution of guinea pig complement for 10 min at room temperature.
5. Carefully watch the embryos under high-power dissecting microscope and remove them as soon as the trophoblast cells start to lyse. This is seen as a clear swelling of cells and a "bubbling appearance" of the trophectoderm. Too short incubation will result in remaining trophoblast cells, while too long incubation may damage the ICM cells.
6. Gently pipette the embryos through a narrow Pasteur pipette that has been fire-polished until all trophoblast cells have been removed. This step is critical and may require some practice.
7. Proceed to plating the ICM clumps on fresh layers of MEFs as outlined in the main protocol.

18.4.9. Delayed Blastocysts

Historically, some ES cell lines were derived by the use of "delayed" blastocysts, that is, blastocysts placed in an artificially induced in vivo diapause. Although we have not noticed any significant advantage using this technique, it is possible this approach may be useful for some difficult genetic backgrounds.

1. Mate embryo donors as usual and note copulation plugs.
2. Ovariectomize the embryo donors in the afternoon 2 days later (2.5 dpc) by surgically removing both ovaries. To avoid this surgical procedure, the effect of ovariectomy can be mimicked by utilizing an estrogen antagonist such as tamoxifen (Sigma T5648). Inject 0.1 ml of a 100 μg/ml solution in corn oil into the peritoneum at 2.5 dpc.
3. Inject the donors subcutaneously with 1 mg Depo Provera (Depot Medroxyprogesterone Acetate) (Sigma M1629) in 0.1 ml sterile saline.
4. Collect blastocysts 4–6 days after ovariectomy (6.5 dpc).
5. Proceed with plating the blastocysts as described in the main protocol.

Note that delayed blastocysts usually grow somewhat slower than those that develop normally. The timing of picking the outgrowth might need to be adjusted accordingly.

18.5 Protocol for the Use of 2i Media

Until recently, the most efficient derivation of ES cells reported was achieved by isolating epiblasts of peri-implantation embryos by removal of their surrounding trophectoderm and primitive endoderm [4]. This procedure resulted in generation of ES cell lines from all of the 129 embryos used in the study. In addition, it allowed ES cells to be derived from at least half of the CBA embryos used; this was the first time ES cells could be derived from this strain. This microsurgery has not been widely adopted, however, since it is extremely painstaking and technically demanding. Fortunately, an alternative to removal of the primitive endoderm and its differentiation-inducing properties has been provided by the availability of small molecule inhibitors that block the FGF/Erk signaling pathway. By incubating embryos from the eight-cell stage in the presence of such inhibitors, the entire ICM will be diverted into the epiblast lineage [5, 6]. Propagation of the "epiblast" thus formed in the presence of both FGF/Erk and GSK3 inhibitors enables efficient derivation of ES cells from all the strains of mice tested so far. This combination of inhibitors is used in serum-free medium without feeder cells. These conditions are termed "2i," and they tend to maintain ES cells in a less heterogeneous state compared with conventional ES cell culture conditions [7]. This efficient technology has thus opened up the possibility to derive ES cells from rare compound transgenic genotypes and mouse disease models such as the non-obese diabetic (NOD) mouse.

For the derivation of ES cells using 2i, embryos are preferentially isolated at the eight-cell stage of development from the oviduct and cultured in the presence of FGF/Erk and GSK3 pathway inhibitors for a few days to prevent formation of the primitive endoderm and promote expansion of the epiblast. The trophectoderm is then removed by the simple process of immunosurgery (*see* Subheading 18.5.5.1) and the isolated epiblast allowed to grow for a further few days. ES cell lines are established by disaggregation of the epiblast into single cells, which will then form colonies that can be further disaggregated to expand the cell line. Because the cells in the epiblast are maintained in a fairly homogeneous pluripotent state, there is more flexibility for the timing of disaggregation than with the method described above. ES cells can be maintained in serum-free medium in the presence of 2i, but their clonogenicity can be further improved by the addition of LIF.

18.5.1. Materials

Please refer to Section 18.2.2

18.5.2. Reagents

Please refer to Section 18.2.3

In addition:

Apo-transferrin (Sigma-Aldrich T1147)

- B27 (Invitrogen 17504-044)
- BSA (Invitrogen 15260-037)
- Chicken serum (Sigma-Aldrich C5405)
- Complement sera from guinea pig, lyophilized (Calbiochem (Merck) 234395)
- DMEM/F12 (Invitrogen 42400-010)
- EDTA (Sigma-Aldrich E6758)
- GSK3 inhibitor, Chir99021 (Stemgent 04-0004)
- Insulin (Sigma-Aldrich I1882)
- MEK inhibitor, PDO325901 (Stemgent 04-0006)
- N2B27 medium (NDiff N2B27, Stem Cell Science SCS-SF-NB-02 or Millipore SF002-500) (or see recipes for home-made N2B27)
- Neurolbasal™ Medium (Invitrogen 21103-049)
- Progesterone (Sigma-Aldrich P8783)
- Putrescine (Sigma-Aldrich P5780)

- Rabbit anti-mouse antiserum (Sigma-Aldrich M5774)
- Sodium Selenite (Sigma-Aldrich S5261)
- Trypsin (Invitrogen 15090-046)
- TVP

18.5.3. Suppliers

18.5.4. Media

18.5.4.1. Guinea Pig Complement

Please see Section 18.2.4

Reconstitute with 5 ml sterilized milliQ water, filter, divide into 100 μL aliquots on ice and store at −80°C. Thaw immediately before use.

18.5.4.2. Trypsin Versene Phosphate

To 500 ml PBS add 0.186 g EDTA, 5 ml chicken serum, and 5 ml of 2.5% trypsin. Filter, aliquot and freeze at −20°C and store frozen until required. Once thawed, do not re-freeze, but store at 4°C

18.5.4.3. N2B27 Medium Preparation

18.5.4.3.1. N2 100× Stock Solution

For 10 ml, mix 1 ml insulin (final concentration 2.5 mg/ml) with 1 ml apo-transferrin (final concentration 10 mg/ml), 0.67 ml BSA (final concentration 5 mg/ml), 33 μl progesterone (final concentration 2 μg/ml), 100 μl putrescine (final concentration 1.6 mg/ml), 10 μl sodium selenite (final concentration 3 μM), and 7.187 ml DMEM/F12. Store at 4°C

18.5.4.4. DMEM/F12-N2 Medium

To 100 ml of DMEM/F12, add 1 ml of N2 100× stock solution. The final concentration of each component of N2 in the DMEM/F12 medium is: insulin, 25 μg/ml; apo-transferrin, 100 μg/ml; progesterone, 6 ng/ml; putrescine, 16 μg/ml; sodium selenite, 30 nM; BSA 50 μg/ml. Store at 4°C

18.5.4.5. Neurolbasal/B27 Medium

To 100 ml of Neurolbasal™ medium, add 2 ml of B27 and 0.5–1 ml of 200 mM L-glutamine. Store at 4°C

18.5.4.6. N2B27 Medium

Mix DMEM/F12-N2 medium with Neurolbasal/B27 medium in the ratio of 1:1. Add β-mercaptoethanol to a final concentration of 0.1 mM from the 0.1 M stock. Store at 4°C

18.5.5. Method

The derivation and handling of ES cells using 2i is different in many respects, specifically because of the lack of serum in the medium, which causes the cells to grow in tight balls or domes that have a tendency to detach from the dish. The presence of the inhibitors also reduces differentiation and broadens the expression of pluripotency markers. The procedure for ES cell derivation in 2i is outlined here.

1. Prepare 2i embryo culture stock by adding 10 μM PD0325901 and 30 μM CHIR99021 to 1 ml KSOM. Add 10 μL of this to 990 μL KSOM in the central well of an organ culture dish. Preequilibrate in a humidified incubator with 5% CO_2 at 37°C. Evaporation can be minimized by putting about 5 ml PBS into the outer well.
2. Flush embryos from oviducts at eight-cell stage (E2.5) using M2 medium (*see* Chapter 17, Combining ES cells With Embryos). Collect embryos, rinse and place into a preequilibrated organ culture dish and culture for 1 day.
3. Prepare and preequilibrate a fresh organ culture dish containing N2B27+2i+ LIF in the central well and PBS in the outer well. Transfer embryos (now at the blastocyst stage, but lacking primitive endoderm) to the new dish. Incubate for two more days.

18.5.5.1. Immunosurgery and Expansion of ES Cell Lines

4. If the embryos have not hatched, place into a drop of Tyrode's solution, acidic to remove the zona pellucida. Monitor dissolution of the zona under the dissecting microscope. This procedure is described in Subheading 18.4.7.
5. Prepare and preequilibrate an organ culture dish containing N2B27+20% anti-mouse serum. Transfer the denuded embryos into this and incubate for about 1 h.
6. Rinse three times in preequilibrated N2B27.
7. Preequilibrate an organ culture dish of N2B27 and add freshly thawed guinea pig complement at 20% immediately before transfer of embryos. Incubate for about 10 min.
8. Transfer embryos into an organ culture dish of N2B27. It is possible to genotype embryos in advance of expanding the ES cell lines using the trophectoderm lysate. If this facility is required, transfer embryos to individual small drops of medium under oil. Incubate for about 1 h.
9. Gelatinise a 96-well plate and preequilibrate with N2B27+2i +LIF, about 0.2 ml per well.
10. Remove trophectoderm lysate using a finely drawn Pasteur pipette with a diameter fractionally larger than the ICM. Transfer the lysate to a PCR tube if genotyping is required.
11. Place each isolated "epiblast" into a separate well of the 96-well plate.
12. Incubate for 3–7 days. The "epiblasts" should increase several fold in size and assume a spherical shape. This step is quite flexible, since the "epiblast" will not differentiate in this

medium, but it may become necrotic in the centre if allowed to grow too big.

13. Disaggregate each "epiblast" by placing it into a small drop of trypsin using a mouth-controlled finely drawn Pasteur pipette. It should decompact and assume a raspberry-like appearance. At this stage, transfer it into a fresh well of the same medium in a 96-well plate. Be sure to carry over as little trypsin as possible. Multiple ES colonies should appear and expand over the next few days.
14. Gradually expand the ES cells by trypsinizing the whole well to progressively bigger wells until the line is established, when it can be cryopreserved, as described for conventionally derived ES cells.
15. Cells derived in 2i can be transferred on to MEFs with serum-containing medium if this is the preferred culture regime.

References

1. Nagy A, Gertsenstein M, Vintersten K, Behringer R (2003) Manipulating the mouse embryo: a laboratory manual, 3rd edn. Cold Spring Harbor Press, New York
2. Solter D, Knowles BB (1975) Immunosurgery of mouse blastocyst. Proc Natl Acad Sci USA 72:5099–5102
3. Cruz YP, Treichel RS, Harsay E, Chi KD (1993) Mouse blastocyst immunosurgery with commercial antiserum to mouse erythrocytes. In Vitro Cell Dev Biol 29A:671–675
4. Brook FA, Gardner RL (1997) The origin and efficient derivation of embryonic stem cells in the mouse. Proc Natl Acad Sci USA 94: 5709–5712
5. Nichols J, Silva J, Roode M, Smith A (2009) Suppression of Erk signalling promotes ground state pluripotency in the mouse embryo. Development 136:3215–3222
6. Yamanaka Y, Lanner F, Rossant J (2010) FGF signal-dependent segregation of primitive endoderm and epiblast in the mouse blastocyst. Development 137:715–724
7. Ying QL, Wray J, Nichols J, Batlle-Morera L, Doble B, Woodgett J, Cohen P, Smith A (2008) The ground state of embryonic stem cell self-renewal. Nature 453:519–523

Chapter 19

Rat Embryonic Stem Cell Derivation and Propagation

Ping Li, Eric N. Schulze, Chang Tong, and Qi-Long Ying

Abstract

Embryonic stem (ES) cells have been routinely used to create loss-of-function mutations or gene replacement by homologous recombination in mice, providing an invaluable tool to address fundamental biological questions. Although mouse ES cells have been available for the past 29 years, authentic rat ES cells have only recently been established. The efficient derivation of multiple rat ES cell lines by independent investigators will accelerate the development of novel laboratory tools for biomedical research. Here we provide detailed protocols for the derivation and propagation of rat ES cells and for the production of rat chimeras using rat ES cells. The availability of rat ES cells provides the opportunity to adapt the technology developed in the mouse to the rat.

Abbreviations

BMP	Bone morphogenetic protein
DA rat	Dark Agouti rat
Dpc	Day post-coitum
ES cells	Embryonic stem cells
F344 rat	Fischer 344 rat
GSK3	Glycogen synthase kinase 3
ICM	Inner cell mass
IP	Intraperitoneal
LIF	Leukemia inhibitory factor
MAPK	Mitogen-activated protein kinase
MEFs	Mouse embryonic fibroblasts
PBS	Phosphate-buffered saline
SD rat	Sprague Dawley rat

S. Pease and T.L. Saunders (eds.), *Advanced Protocols for Animal Transgenesis*, Springer Protocols,
DOI 10.1007/978-3-642-20792-1_19, © Springer-Verlag Berlin Heidelberg 2011

19.1 Introduction

Rats are a proven model organism for the study of human disease. Researchers have taken advantage of their physiological and pharmacological similarities to humans for over 150 years [1]. The demonstration of robust authentic pluripotent germline competent rat embryonic stem (ES) cells is one of the most highly anticipated developments in the field. Access to such cells will allow investigators to interrogate gene function in highly characterized rat models of disease.

Although ES cells have been routinely derived from mice since 1981 [2, 3], authentic rat ES cells have only recently been established [4, 5]. Derivation of mouse ES cells has relied on co-cultivation with feeder cells, usually mitotically inactivated mouse embryonic fibroblasts (MEFs), and the presence of fetal calf serum. Later it was shown that leukemia inhibitory factor (LIF) is the key cytokine secreted by feeder cells in supporting mouse ES cell self-renewal [6, 7]. Bone morphogenetic proteins (BMPs) can replace serum and act together with LIF to maintain an undifferentiated state of mouse ES cells in defined conditions [8]. However, these culture conditions developed for mouse ES cells do not yield ES cells from the rat.

Recently we have found that, contrary to dogma based on many years of research, both the LIF and BMP pathways are dispensable for mouse ES cell self-renewal. In fact, our findings indicate that mouse ES cell self-renewal does not require activating signals, but only that ES cells be shielded from inductive differentiation cues [9]. Based on these findings, we have developed culture media that can support efficient derivation and maintenance of ES cells from different strains of mice. The media contain three inhibitors (3i): CHIR99021, PD184352, and SU5402. CHIR99021 is a well-characterized, highly selective small molecule inhibitor of glycogen synthase kinase 3 (GSK3). PD184352 and SU5402 are selective pharmacological inhibitors used to inhibit mitogen-activated protein kinase (MAPK) and the tyrosine kinase domain of the fibroblast growth factor receptor (FGFR), respectively. Importantly, we found that a more potent MAPK inhibitor PD0325901 can be used to replace both PD184352 and SU5402.

Germline competent ES cells have been successfully derived from rat blastocysts using the 3i (CHIR99021, PD184352, and SU5402) or 2i (CHIR99021 and PD0325901) culture conditions [4, 5]. These rat ES cells can be genetically modified and robustly propagated in culture, while retaining the ability to contribute to germline competent chimeras. The availability of robust and germline competent rat ES cells will open the door to application of gene targeting and related genome engineering technologies in the species of choice for many areas of biomedical research. Here we describe detailed protocols for the derivation and propagation of rat ES cells and for the production of rat chimeras from rat ES cells.

19.1.1. Outline of the Procedure

- Collect rat blastocysts at 4.5 dpc, remove the zona pellucida and culture in rat ES cell media with feeders
- Disaggregate the outgrowths of rat blastocysts
- Propagate rat ES cells
- Freeze and thaw rat ES cells
- Introduce a transgene into rat ES cells
- Pick and expand rat ES cell colonies after gene transfection
- Prepare rat ES cells for blastocyst injection
- Inject rat ES cells into rat blastocysts
- Transfer injected blastocysts into pseudo-pregnant rat recipients

19.1.2. Principles and Applications

ES cells are derived from the inner cell mass (ICM) of a pre-implantation blastocyst [10]. ES cells can be maintained in culture indefinitely while retaining the capacity to generate nearly any type of cell in the body [11]. The pluripotency of ES cells, combined with their ease of genetic manipulation and selection, has provided a powerful means to elucidate gene function and create disease models.

ES cells have been routinely used to create loss-of-function mutations (knock-out) or gene replacement (knock-in) by homologous recombination in mice since 1989 [12]. In general, rats are a much more useful model system than mice for the study of human disease because of the following reasons: (1) Rats are approximately ten times larger than mice, which makes it easier to perform procedures such as nerve recordings, collection of tissue from small structures, and serial blood sampling. (2) Rat physiology more closely approximates that of humans than does mouse physiology. This will allow us to create rat models whose phenotypes are similar to human diseases. (3) Rats are also the preferred animal model for drug development assays. Establishing rat ES cells will allow us to apply the ES cell-based technologies developed for the mouse to the rat to create superior animal models for the study of human health and disease.

19.2 Materials

19.2.1. Equipment

- Humidified tissue culture incubator at 37°C, 5% CO_2
- Laminar flow tissue culture hood
- Inverted microscope
- Dissecting microscope (OLYMPUS S2X10)

- Centrifuge (Eppendorf, Centrifuge 5702)
- Nucleofector device (Amaxa GmbH)
- Gene Pulser XCell™ (Bio-Rad)
- Gene Pulser Cuvette (Bio-Rad)
- 37°C water bath
- Embryo-handling pipettes
- Hemocytometer (Hausser Scientific)
- Conical tubes, plastic, 15- and 50-ml sterile screw-cap
- Serological pipettes, 5-, 10-, and 25 ml sterile
- 9-inch glass pipettes
- Surgical blades, forceps, and scissors for dissection
- Flame
- Pipette pump
- Center-well organ culture dishes (BD Falcon, 353653)
- Tissue culture plates, 4-well, 12-well, and 6-well
- Tissue culture dishes, 60 mm, 100 mm, and 150 mm
- 1.5-ml Eppendorf tubes
- NALGENE™ Cryo 1°C freezing container (Cat No. 5100-0001)
- 4°C fridge, −20°C freezer, and −80°C freezer
- Liquid nitrogen tank

19.2.2. Animals

Mice. CF-1 (Charles River Laboratory Stain Code 023) and Tg (DR4)1Jae/J (Jackson Laboratory Stock number 003208) (DR-4) strains of mice are used to prepare MEFs. MEFs prepared from the DR-4 mouse strain are resistant to G418, 6-thioguanine, puromycin, and hygromycin.

Rats. Dark Agouti inbred rats (DA/OlaHsd), Fischer 344 inbred rats (F344/NHsd), and Sprague Dawley outbred rats (Hsd:Sprague Dawley SD) are available from Harlan Laboratories.

19.2.3. Reagents

- GMEM (Sigma, G5154)
- DMEM/F12 (Sigma, D6421)
- Neurobasal™ medium (Invitrogen, 21103-049)
- B-27 Supplement (50×; Invitrogen, 17504-044)
- Fetal bovine serum (FBS), heat-inactivated (Hyclone)
- L-Glutamine (200 mM, Invitrogen, 25030-081)
- Gelatin (Sigma, G1890). Dissolve 5 g in 500 ml dH_2O to give a 1% gelatin stock. Autoclave and store in 50 ml aliquots at 4°C for up to three months.

- Penicillin/Streptomycin (Invitrogen, 15140-122)
- Trypsin (2.5%; Invitrogen, 15090-046)
- Phosphate Buffered Saline (PBS, PH 7.2) (Invitrogen, 20012-050)
- Dulbecco's Phosphate Buffered Saline (D-PBS) with Ca^{2+} and Mg^{2+} (Invitrogen, 14040-133)
- Anti-rat whole serum (Sigma, R5256)
- Non-Essential amino acids solution (10 mM, Invitrogen, 11140-050)
- Sodium pyruvate solution (100 mM, Invitrogen, 11360-070)
- β-mercaptoethanol (Sigma, M7522): Prepare 0.1 M stock solution by diluting 100 μl β-mercaptoethanol with 14.1 ml H_2O. Sterilize through 0.2 μm filter and store at 4°C for up to 1 month.
- Insulin (Sigma, I1882): Dissolve in sterile 0.01 M HCl overnight at 4°C to give a 10 mg/ml stock solution. Store in 1 ml aliquots at −20°C. Insulin does not dissolve well, so ensure the suspension is mixed well before aliquoting.
- Apo-transferrin (Sigma, T1147): Dissolve in sterile H_2O to give a 100 mg/ml stock solution. Store in 1 ml aliquots at −20°C.
- Progesterone (Sigma, P8783): Dissolve 6 mg in 10 ml ethanol to give a 0.6 mg/ml stock. Sterilize through 0.2 μm filter and store at −20°C.
- Putrescine (Sigma, P5780): Dissolve 1.6 g in 10 ml H_2O to give a 160 mg/ml stock and filter sterilize. Store at −20°C.
- Sodium selenite (Sigma, S5261): Dissolve 2.59 mg in 5 ml H_2O to give a 3 mM stock, filter through a 0.2 μm filter and store at −20°C.
- Bovine Serum Albumin (BSA), Fraction V (Invitrogen, 15260-037). 75 mg/ml in PBS
- EDTA (Invitrogen, 15575-020)
- Dimethyl Sulfoxide (DMSO) (Sigma, D2438)
- Chicken serum (Sigma, C5405)
- M2 medium (Sigma, M7167)
- M16 medium (Sigma, M7292)
- Tyrode's Solution (Sigma, T1788)
- Mitomycin C (Sigma, M4287)
- CHIR99021 (Axon Medchem BV, Axon 1386). Dissolve 4 mg in 860 μl DMSO to give a 10 mM stock.

- PD0325901 (Axon Medchem BV, Axon 1408). Dissolve 4 mg in 830 μl DMSO to give a 10 mM stock.
- SU5402 (Calbiochem, 572630). Dissolve 1 mg in 675 μl DMSO to give a 5 mM stock.
- PD184352 (Selleck, S1020). Dissolve 4 mg in 836 μl DMSO to give a 10 mM stock.

19.2.4. Medium and Solutions

- N2 100× stock (10 ml): In 7.187 ml DMEM/F12 medium, add 0.67 ml of 75 mg/ml BSA, 33 μl of 0.6 mg/ml progesterone solution, 100 μl of 160 mg/ml putrescine solution, 10 μl of 3 mM sodium selenite solution, 1 ml of 100 mg/ml apo-transferrin and 1 ml of 10 mg/ml insulin. Mix well by pipetting and store in 1 ml aliquots at −20°C.
- DMEM/F12-N2 medium: To 100 ml of DMEM/F12, add 1 ml of N2 100× stock solution. The final concentration of each component of N2 in the DMEM/F12-N2 medium is: Insulin 10 μg/ml, transferring 100 μg/ml, progesterone 20 ng/ml, putrescine 16 μg/ml, sodium selenite 30 nM, BSA 50 μg/ml.
- Neurobasal/B27 medium: To 100 ml of Neurobasal™ medium, add 2 ml of B27 and 0.5 ml of 200 mM L-glutamine solution.
- N2B27 medium: Mix DMEM/F12-N2 medium with Neurobasal/B27 medium at the ratio of 1:1. Store at 4°C in dark for up to 1 month.
- PB1 medium: Add the following components to D-PBS with Ca^{2+} and Mg^{2+}: BSA (3 g/L), glucose (1 g/L), sodium pyruvate (0.036 g/L), phenol red (0.005 g/L), penicillin (100 units/ml), streptomycin (100 μg/ml).
- Rat ES medium: Rat ES cells can be derived and propagated using either 3i or 2i medium. 3i medium: N2B27 medium supplemented with 3 μM CHIR99021, 0.8 μM PD184352, and 2 μM SU5402. 2i medium: N2B27 medium supplemented with 3 μM CHIR99021 and 1 μM PD0325901 (**Note**: Rat ES cells grow better in 2i medium than in 3i medium).
- 0.025% trypsin/EDTA solution: In 500 ml sterile PBS, add 5 ml of 2.5% trypsin, 5 ml of chicken serum and 0.5 ml of 0.5 M EDTA. Mix well and store in 30 ml aliquots at −20°C.
- MEF medium: In 500 ml GMEM medium, add 50 ml of heat-inactivated fetal bovine serum, 5 ml of 200 mM L-glutamine solution and 5 ml of penicillin/streptomycin solution. Store at 4°C for up to 1 month.
- Freezing medium: 10% DMSO in MEF medium.
- 0.1% gelatin: Add 50 ml of 1% gelatin to 450 ml PBS, store at 4°C for up to 1 month.

19.2.5. Suppliers

Charles River Laboratories International, Inc. 251 Ballardvale Street, Wilmington, MA 01887, USA; Phone: 781-222-6000; http://www.criver.com/.

The Jackson Laboratory, 600 Main Street, Bar Harbor, ME 04609, USA; Phone: 207-288-6000; http://www.jax.org/.

Harlan Laboratories, Inc. 8520 Allison Pointe Blvd., Suite 400, Indianapolis, IN 46250, USA; Phone: 800 793-7287; http://www.harlan.com/.

Calbiochem, a Brand of EMD Chemicals, Inc. 480 S. Democrat Road, Gibbstown, NJ 08027, U.S.A; Phone: 856 423-6300; http://www.calbiochem.com/.

Selleck Chemicals Co.,Ltd. Suite204, 543-mornington Ave, London ON, Canada; Phone: +1-519-852-5693; http://www.selleckchem.com.

Axon Medchem BV, Postbus 770, 9700 AT Groningen, The Netherlands; Phone: 31-50-3118007; http://www.axon-medchem.com/.

Invitrogen, 5791 Van Allen Way, Carlsbad, CA 92008, USA; phone: 800 955-6288; http://www.invitrogen.com/.

Sigma, P.O. Box 14508, St. Louis, MO 63178, USA; Phone: 800 325-3010; http://www.sigmaaldrich.com/.

VWR Scientific, 1310 Goshen Pkwy., West Chester, PA 19380, USA; Phone: 800 932-5000; http://www.vwrsp.com/.

Bio-Rad Laboratories, 1000 Alfred Nobel Drive, Hercules, CA 94547, USA; Phone: 510 724-7000; http://www.bio-rad.com/.

19.3 Protocols

19.3.1. Preparation of MEFs from Mouse Embryos

Rat ES cells are routinely cultured on feeders. MEFs prepared from CF-1 and DR-4 mice both work well for the culture of rat ES cells.

1. Set up matings of mice.
2. 13.5–14.5 days after the plug is observed, sacrifice the pregnant mouse by cervical dislocation or CO_2 asphyxiation. The morning when the plug is observed is defined as day 0.5.
3. Sterilize the abdominal skin and fur with 70% alcohol, make an incision down the midsection using scissors, and expose the uterine horns.
4. Dissect out individual embryos. Place the embryos in a 100-mm dish of PBS.

5. Decapitate and eviscerate the embryos using sterile forceps, transfer the carcasses to a 100-mm dish of PBS.
6. Finely mince the carcasses with a sterile scissor or scalpel blade, pipette up and down 8–10 times with a 10-ml pipette, then transfer to a 50-ml conical tube.
7. Centrifuge at 200 × *g* for 3 min, aspirate the supernatant, resuspend the cell pellet using 10 ml 0.025% trypsin/EDTA solution and place the tube in 37°C water bath for 20–30 min. Shake the tube every 5 min to resuspend the cells.
8. Centrifuge at 200 × *g* for 3 min, aspirate off the supernatant.
9. Wash the cell pellet twice with 30 ml MEF medium by resuspension and centrifugation.
10. Resuspend the cells in 10–15 ml MEF medium and count the cell number with a hemocytometer.
11. Plate the cells onto tissue culture dishes at a density of 5×10^6 cells/150-mm dish in 25 ml MEF medium, change the medium 24 h after plating.
12. When the cultures become confluent, rinse once with 20 ml PBS, aspirate the PBS, add 10 ml 0.025%trypsin/EDTA solution and incubate at 37°C for 5 min.
13. Add 10 ml MEF medium to neutralize the trypsin, pipette up and down to dissociate the cells and transfer to a 50-ml conical tube.
14. Centrifuge at 200 × *g* for 3 min, aspirate off the supernatant.
15. Resuspend in MEF medium and split the cells 1:5 onto 150-mm dishes.
16. After 3–4 days, the culture will become confluent. Harvest the cells for either cryopreservation in liquid nitrogen or continuous expansion. MEFs should not be used past six passages due to senescence, but will have to be determined empirically by each investigator.

19.3.2. Preparation of Feeder Layers

Prior to use as feeders for the culture of undifferentiated rat ES cells, MEFs must be mitotically inactivated either by γ-irradiation or by mitomycin C treatment.

19.3.2.1. Mitotic Inactivation by γ-Irradiation

1. Harvest MEFs and resuspend them with 10–30 ml MEF medium in a 50-ml conical tube.
2. Expose the cells to 6,000 rads from a γ-irradiation source.
3. Count the cell number with a hemocytometer. The cells can then be cryopreserved or used as feeders by plating at a density of $2–3 \times 10^4$ cells/cm^2 (**Note**: Coat the dish with 0.1% gelatin before plating the mitotically inactivated MEFs).

19.3.2.2. Mitotic Inactivation by Mitomycin C

1. Grow MEFs in a 150-mm culture dish to confluent.
2. Aspirate the medium off the dish.
3. Add 12–15 ml of MEF medium with 10 μg/ml mitomycin C to the dish.
4. Incubate for 2–3 h at 37°C.
5. Aspirate off the medium and wash three times with 15 ml PBS.
6. Harvest the cells by trypsinization and resuspend them in MEF medium.
7. Count the cell number with a hemocytometer. The cells can then be cryopreserved or used as feeders by plating at a density of 2–3 × 10^4 cells/cm^2 (**Note**: Coat the dish with 0.1% gelatin before plating the mitotically inactivated MEFs).

19.3.3. Derivation of ES Cells from Rat Blastocysts

19.3.3.1. Derivation Step I: Recovery of Rat Blastocysts

1. Euthanize timed-pregnant rats at 4.5 dpc by CO_2 inhalation and cervical dislocation (Please obey all local regulations regarding animal euthanasia).
2. Lay the animal on its back and cleanse the abdominal skin and fur with 70% alcohol.
3. Make an incision down the midsection using scissors, expose the uterine horns.
4. Grasp uterus just above cervix with fine forceps and cut with fine scissors. Pull the uterus upward and use fine scissors to trim the mesometrium away close to the wall of the uterine horns. Then cut between the oviduct and the ovary.
5. Transfer the uterus into a 6-cm dish of PBS. Cut each horn near the cervix and flush the blastocysts from each horn toward the cervix with 0.5–1 ml N2B27 medium.
6. Collect blastocysts using embryo pipette and wash them through several drops of N2B27 medium.
7. Transfer blastocysts through two drops of acidic Tyrode's solution to wash out the N2B27 medium, and then transfer the blastocysts to a fresh drop of acidic Tyrode's solution.
8. Check blastocysts under microscope during the treatment in acidic Tyrode's solution, transfer the embryos into a fresh drop of N2B27 medium immediately after the zona pellucida has dissolved.

19.3.3.2. Derivation Step II (Optional): Isolation of the ICM by Immunosurgery

1. Mix 100 μl *anti*-rat whole serum with 400 μl PB1 medium in a center-well organ culture dish.
2. Transfer rat blastocysts with the zona pellucida removed to the above dish and incubate for 1–3 h.

3. Wash the blastocysts three times in drops of PB1/10% FBS solution.
4. Dilute rat serum (source of complement) with PB1 solution at a 1:5–10 ratio (**Note**: Rat serum is best when collected "in house". Fill 1.5 ml Eppendorf tubes with it and freeze immediately at −80°C. Thaw the rat serum immediately before use).
5. Add the blastocysts to the complement and incubate in a humidified incubator with 5% CO_2 at 37°C. The outer trophectoderm cells should begin to lyse after a few minutes incubation.
6. If lysed cells can be seen around the perimeter of the blastocyst, draw the blastocyst up into a narrow pipette to remove cell fragment and place isolated ICMs in culture as described below.

19.3.3.3. Derivation Step III: Disaggregation of Outgrowths

1. One day before the rat ES cell derivation, coat 4-well plates with 0.1% gelatin, then mitotically inactivated MEFs cultured in MEF medium.
2. On the day of the rat ES cell derivation, replace the MEF medium with 500 μl rat ES cell medium in each well of the 4-well plate pre-seeded with feeders.
3. Transfer blastocysts or the isolated ICMs by immunosurgery into the prepared 4-well plate. Up to ten blastocysts or ICMs can be placed in one well unless the nature of the experiment dictates otherwise.
4. Four to five days after culture, disaggregate the outgrowth of each individual embryo by gently detaching the outgrowth using a mouth-controlled, finely drawn, plugged Pasteur pipette with a tip diameter that is just bigger than the outgrowth. Each outgrowth is handled separately in all subsequent manipulations.
5. Transfer each outgrowth into a sterile 1.5 ml Eppendorf tube with 200 μl 0.025% trypsin/EDTA solution, incubate at room temperature for 3 min.
6. Add 1 ml MEF medium to the tube to neutralize the trypsin, gently pipette up and down 4–6 times with a 1 ml pipette tip to break up the outgrowth into small clumps.
7. Centrifuge at 200 × *g* for 3 min.
8. Aspirate off the supernatant and resuspend in 1 ml PBS.
9. Centrifuge again at 200 × *g* for 3 min.
10. Aspirate off the supernatant and resuspend the cell pellet in 500 μl rat ES cell medium.
11. Transfer the cells into one well of the 4-well plate pre-seeded with MEFs.

12. Repeat the procedure with each outgrowth. Rat ES cell colonies will emerge 3–5 days after plating.

19.3.3.4. Derivation Step IV: Expansion of Rat ES Cell Colonies

1. Three to five days after the first disaggregation, aspirate off the medium, add 200 μl 0.025% trypsin/EDTA solution to each well and incubate at 37°C for 2–3 min (**Note**: If the colonies are floating, collect the colonies by centrifuge at 200 × *g* for 3 min before trypsinization).
2. Add 600 μl MEF medium to neutralize trypsin, pipette up and down 4–6 times to dissociate the colonies into single cells or small, 2–6 cell clumps.
3. Transfer the cell suspension to a sterile 1.5 ml Eppendorf tube, centrifuge at 200 × *g* for 3 min.
4. Aspirate off the supernatant and resuspend the cell pellet with 500 μl rat ES cell medium.
5. Transfer the cell suspension to a well of the 4-well plate pre-seeded with feeders.
6. Two to three days after plating, split rat ES cells at 1:1 ratio (i.e., transfer cells from one well of the 4-well plate into another well of the 4-well plate pre-seeded with feeders) by repeating above procedures.
7. Split the cells every 2–3 days at 1:1 ratio for the first 3–5 passages.
8. Split the cells at 1:2 or 1:3 when the culture reaches 60–80% confluence.

19.3.4. Freezing Rat ES Cells

1. Grow rat ES cells in a 6-well plate on feeders with rat ES cell medium until the cells become 60–80% confluent (approximately 1–2 million cells/well).
2. Aspirate off the medium and add 0.5 ml pre-warmed 0.025% trypsin/EDTA solution. Washing with PBS before trypsinization is not necessary (**Note**: If cells are floating, collect them by centrifuging at 200 × *g* for 3 min before trypsinization).
3. Incubate at 37°C in a humidified 5% CO_2 incubator for 2–3 min.
4. Add 2 ml MEF medium, pipette up and down with a 5 ml pipette 4–6 times to dissociate the cells into single cells.
5. Transfer cell suspension into a sterile 15 ml conical tube, centrifuge at 200 × *g* for 3 min.
6. Aspirate the supernatant, then resuspend the cell pellet in 1 ml of freshly prepared freezing medium.
7. Aliquot 0.5 ml of cells into two cryotubes.
8. Freeze the vials at −80°C overnight before transfer to liquid nitrogen for long-term storage (**Note**: The vials of cells can be

directly put inside a –80°C freezer, or alternatively, put the vials into a NALGENE™ Cryo 1°C freezing container and transfer the container to a −80°C freezer. The latter will result in better cell recovery).

9. Transfer vials to liquid nitrogen the following day for long-term storage.

19.3.5. Thawing Rat ES Cells

1. Remove one vial of rat ES cells (approximately 0.5–1 million cells) from liquid nitrogen.
2. Thaw vial rapidly at 37°C water bath, rinse outside of the cryotube with 70% ethanol.
3. Transfer cells to a sterile 15-ml conical tube with 10 ml MEF medium.
4. Centrifuge at 200 × *g* for 3 min.
5. Aspirate off the supernatant and resuspend the cell pellet in 2 ml rat ES cell medium.
6. Transfer the cell suspension into one well of the 6-well plate pre-seeded with feeders.
7. Split the cells 1:2 or 1:3 every 2–3 days. (It is not necessary to change the medium the following day after thawing).

19.3.6. Gene Transfection of Rat ES Cells

Stable gene transfection in rat ES cells can be achieved using Nucleofection or conventional electroporation methods. Nucleofection method will result in higher transfection efficiency compared with conventional electroporation method. However, we found that the conventional electroporation method works better for gene targeting through homologous recombination.

19.3.6.1. By Nucleofection

1. Grow rat ES cells in a 6-well plate on feeders with rat ES cell medium until the cells become 60–80% confluent (approximately 1–2 million cells/well).
2. Trypsinize the cells and resuspend in MEF medium
3. Count the cell number with a hemocytometer.
4. Transfer 2×10^6 rat ES cells into a 15-ml conical tube, centrifuge at 200 × *g* for 3 min.
5. Aspirate off the supernatant and wash the cell pellet once with PBS.
6. Resuspend rat ES cells with 95 μl mouse ES cell nucleofector solution and transfer to a 1.5-ml Eppendorf tube.
7. Dissolve 5 μg linearized plasmid DNA in 5 μl H_2O.
8. Add the linearized plasmid DNA to the cells and mix by gently flicking the tube.
9. Transfer sample into an Amaxa-certified cuvette and select program **A-23** to perform the electroporation.

10. After transfection, add 500 μl pre-warmed rat ES cell medium and transfer the cells to a 60-mm tissue culture dish pre-seeded with feeders and 4.5 ml pre-equilibrated rat ES cell medium.
11. One or two days after transfection, replace the medium with fresh rat ES cell medium containing the appropriate drug to select positively transfected cells.
12. Pick and expand colonies 7–9 days after transfection.

19.3.6.2. By Electroporation

1. Grow and harvest rat ES cells as described above. 5×10^6 rat ES cells are required for each transfection.
2. Wash once with PBS, and resuspend 5×10^6 rat ES cells in 0.7 ml PBS.
3. Dissolve 20–50 μg linearized plasmid DNA in 0.1 ml PBS.
4. Add the linearized plasmid DNA to the cells and mix by gently flicking the tube. The total volume should be 0.8 ml.
5. Transfer DNA/rat ES cells mixture to a Gene Pulser Cuvette (4 mm gap).
6. Set the Bio-Rad Gene Pulser XCell™ at 200 V, 500 μF.
7. Place the cuvette in the electroporation holder, press the red button to electroporate.
8. Transfer the electroporated cells to two or three 100-mm tissue culture dishes pre-seeded with feeders and culture in rat ES cell medium.
9. One or two days after transfection, change the medium with fresh rat ES cell medium containing the appropriate drug to select positively transfected cells.

19.3.7. Production of Rat Chimeras

19.3.7.1. Preparation of Vasectomized Male Rats

1. Anesthetize the male rat by IP injection of ketamine (50 mg / kg) and xylazine (10 mg/kg).
2. Make a 5-mm incision in the testes membrane close to the left side of the midline wall.
3. Carefully push the testis to the left and pull the vas deferens out and cut with fine scissors such that a portion of the vas deferens in the loop is removed.
4. Repeat the step on the other testis.
5. Sew up the skin, place the rat in a clean cage and keep it warm until the rat recovers from the anesthesia.
6. Three to four weeks after surgery, set up matings between the vasectomized male with two females to confirm the success of the vasectomy. Pregnancy is normally apparent within 2 weeks, and if not visible after this time the male can be used.

19.3.7.2. Preparation of Pseudo-Pregnant Female Rats

1. Set up matings between female rats with vasectomized males 4 days before the planned injection date.
2. Check the females for copulation plugs on the following morning.
3. Separate the plugged females from males, set them aside for uterine transfer on the injection date.

19.3.7.3. Preparation of Rat ES Cells for Injection

1. One day before the blastocyst injection, plate 0.5–1 $\times$ 10^6 rat ES cells into one well of a 6-well plate, pre-seeded with feeders, and culture in rat ES cell medium at 37°C in a humidified 5% CO_2 incubator.
2. On the day of injection, aspirate off the medium and add 0.5 ml of pre-warmed 0.025% trypsin/EDTA solution.
3. Incubate at 37°C in a humidified 5% CO_2 incubator for 3–5 min.
4. Add 2 ml MEF medium, pipette up and down with a 5-ml pipette 6–8 times to dissociate the cells into single cells.
5. Transfer cell suspension into a sterile 15-ml conical tube, centrifuge at 200 $\times$ *g* for 3 min.
6. Aspirate off the supernatant, then resuspend the cell pellet in 0.5–1 ml N2B27 medium.
7. Transfer the cell suspension into a sterile 1.5-ml Eppendorf tube and put on ice while preparing for microinjection.

19.3.7.4. Preparation of Blastocysts

1. Collect blastocysts from timed-pregnant rats at 4.5 dpc as described in Subheading 19.3.3.1.
2. Wash the blastocysts through several drops of M2 medium to rinse off the debris.
3. Transfer the blastocysts to microdrops of M16 medium in a 35-mm plastic tissue culture dish.
4. Incubate for 1–4 h at 37°C in a humidified 5% CO_2 incubator to allow the expansion of the blastocoel cavity.

19.3.7.5. Injection of Rat ES Cells

1. Assemble the microinjection setup as described [13].
2. Pick up well-expanded blastocysts and immobilize it with holding pipette.
3. Inject 12–15 ES cells into the blastocyst cavity.
4. Place injected blastocysts back into incubator, where they are kept until they are transferred to pseudo-pregnant foster mothers as described below.

19.3.7.6. Transfer of Blastocysts to Recipients

1. Anesthetize the pseudo-pregnant rat at 3.5 dpc by IP injection of ketamine (50 mg/kg) and xylazine (10 mg/kg).
2. Wipe the back of the rat with 70% ethanol.

3. Make an incision in the skin and the peritoneal wall over the left side oviduct with fine dissection scissors.
4. Apply a Serrafine clamp to the ovarian fat pad to exteriorize the ovary, the oviduct, and the upper part of the uterus.
5. Hold the top of the uterus gently with blunt forceps, make a hole in the uterus a few millimeters down from the utero-tubal junction with a 26-gauge ½-inch needle.
6. Transfer 8–10 injected blastocysts into the uterus, using traditional mouth pipetting technique, via the hole made by the needle.
7. Place the uterus, oviduct, and ovary back inside the body cavity.
8. Sew up the muscle wall and close the skin.
9. At the end of procedure, place the rat in a clean cage and keep warm until it recovers from anesthesia.

The pups will be born 18–19 days after the transfer of blastocysts into the uterus of the surrogate female rat. Assess the extent of chimerism in the pups either by the appearance of the coat color or by genotyping. Breed rats exhibiting chimerism to generate germline transmission of the ES cell genome.

Confirm the germline transmission by Southern blot analysis and PCR of tail DNA. Breed rats that are heterozygous for the mutation to generate homozygous mutant rats.

19.4 Results

19.4.1. Derivation, Propagation, and Genetic Modification of Rat ES Cells

Rat ES cell lines are established from pre-implantation blastocysts. Rat blastocysts used for ES cell derivation are obtained from timed-pregnant rats at 4.5 dpc. The number of blastocysts harvested from each timed-pregnant rat varies depending on the strains of rats. On average, 6–10 blastocysts can be harvested from each timed-pregnant DA or Fischer 344 rat. For SD rats, the average number of blastocysts per timed-pregnant female is around 10–15.

The rat blastocyst (Fig. 19.1a) will attach to the feeder layers after 3–5 days of culture in rat ES cell medium. The outgrowth of the rat blastocyst mainly contains two types of cells: large, flat trophectoderm-derived cells in the outlayer and ICM-derived cells in the center of the outgrowth (Fig. 19.1b). ES cells are originated from these ICM-derived cells. Small colonies with a typical ES cell morphology emerge 3–5 days after the first disaggregation of the outgrowth and culture in rat ES cell medium (Fig. 19.1c). These colonies can be further expanded to establish ES cell lines (Fig. 19.1d). We routinely passage rat ES cells every 3–4 days. Some rat ES cell lines have been cultured over 60 passages without overt differentiation. Like mouse ES cells, rat ES cells can also be frozen and thawed using conventional methods. Around 90% of rat ES cells will recover after freezing and thawing. So far, we have established stable rat ES cell lines from five strains of rats: SD, DA, F344, Long Evans and Brown Norway. In our hands, stable ES cell lines can be established from around 40% of the rat blastocysts plated. This derivation efficiency is similar among the five strains of rats. We have not yet tried to derive ES cell lines from other strains of rats, and so the technique presented here requires validation for other strains.

For rat ES cells to have a broad application, it is critical that they can be genetically modified while retaining the ability to contribute to different types of cells both in vitro and in vivo. We have successfully introduced transgenes into rat ES cells by nucleofection, electroporation, and lipofectamine. Rat ES cell lines with transgenes stably integrated have been established using these three transfection methods. The initial transfection efficiency in rat ES cells is comparable to that in mouse ES cells. However, following drug selection, far fewer colonies emerge in transfected rat ES cells than in mouse ES cells. This may be due to the fact that rat ES cells are exceptionally sensitive to drug selection.

19.4.2. Production of Chimeric Rats

After blastocyst injection, ES cells have the ability to colonize the embryos, form chimeric animals, and transit through the germline. Chimeric rats have been produced at high efficiency by

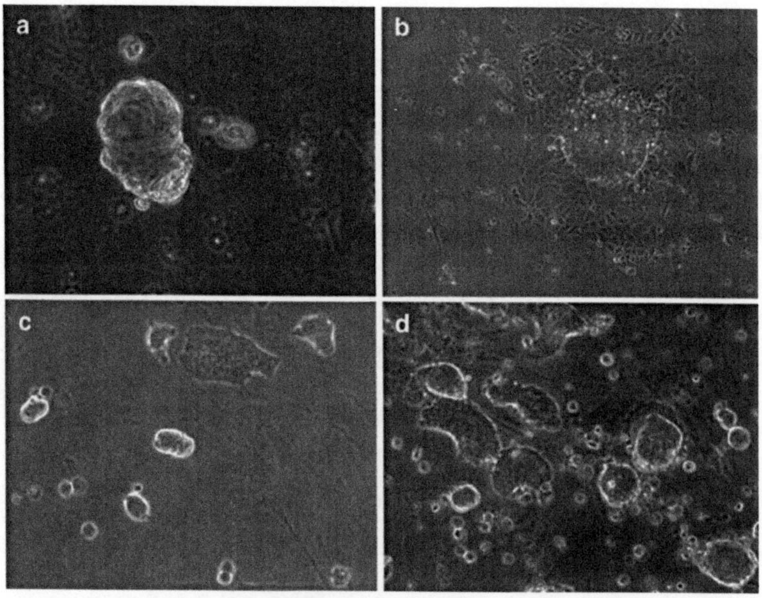

Fig. 19.1 Derivation and culture of rat ES cells. (**a**) Blastocyst flushed from uterus of a DA rat at the day of E 4.5. (**b**) Outgrowth of rat embryo cultured in 3i rat ES cell medium on feeder cells. (**c**) Primary colonies of rat ES cells after culture of disaggregated outgrowth for 3 days. (**d**) Established DA rat ES cell line (passage 10).

injection of ES cells derived from DA and SD rats into blastocysts of DA, SD, or F344 rats. The genetic background of the host embryo chosen for the production of ES cell-mouse chimeras has a dramatic effect on the successful production of chimeric mice that transmit the ES cell haplotype [14, 15]. The same is likely to be true for rat ES cells. We have generated high rates of chimerism by injection of DA ES cells into both SD and F344 rat blastocysts. However, only the DA ES cell-F344 rat chimeras have produced offspring with DA ES cell genome transmitted through the germline (Fig. 19.2), suggesting that the strain combination of ES cells and host embryos is also important for the efficient germline transmission of rat ES cells [5].

19.5 Troubleshooting

19.5.1. Culture Medium

The quality of N2B27 medium may vary from batch to batch. A poor-quality batch of N2B27 medium will adversely affect rat ES cell culture. Proper quality-control is essential to maintaining high-quality rat ES stocks. Always test each batch of N2B27 medium before use in critical experiments. Do not use commercially available N2 supplement to make N2B27 medium, as we have found that the commercially supplied N2 supplements are far inferior to in-house made N2.

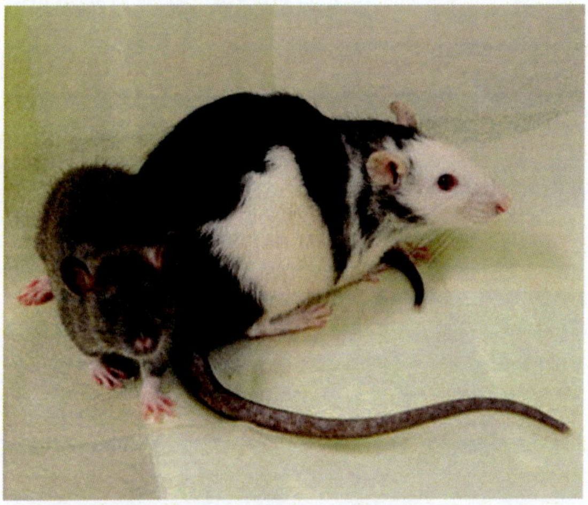

Fig. 19.2 The rat chimera and her germline offspring generated using rat ES cells. The chimeric rat was produced by injection of DA rat ES cells into a Fischer 344 rat blastocyst and subsequent transfer to a recipient pseudo-pregnant Sprague Dawley rat. The pigmented coat color denotes the presence of DA rat ES cells. The germline offspring was generated by mating the DA/Fischer 344 chimeric rat with a Sprague Dawley rat and the agouti coat color indicates the germline transmission of the DA rat ES cell genome.

19.5.2. Cell Passaging

Rat ES cells are easily detached from feeders and washed away. To passage rat ES cells, add trypsin directly to the culture after the culture medium is aspirated off. Since N2B27 is a defined, serum-free medium, it is unnecessary to wash the well in PBS. Some rat ES cells will detach and grow as floating small aggregates. In this case, collect these aggregates by centrifuge before trypsinization.

19.5.3. Cell Growth/Death

Rat ES cells proliferate slowly compared to mouse ES cells and undergo significant cell death especially immediately after passaging. This is normal and to be expected. Plate rat ES cells at a density so that they will reach 60–80% confluent after 2–3 days in culture. Passage the cells every 2–3 days. Use freshly prepared feeders. Do not add serum to the culture, as this will serve to only cause rat ES cells to die or differentiate.

19.5.4. Gene Transfection in Rat ES Cells

It is difficult to isolate gene-transfected rat ES cell colonies and establish stable cell lines, because:

1. Once small colonies are formed, rat ES cells will easily detach from feeders and float in the medium.

   Solution: use DR4 feeders to culture rat ES cells after gene transfection. Rat ES cells adhere to DR4 feeders better than CF-1 feeders.
2. Rat ES cells are very sensitive to drug selection, therefore it is important to determine each drug's killing curve for the rat ES cell lines used for gene transfection. This will help to optimize the concentration for drug selection. For drug selection in rat ES cells after gene transfection, most drugs are added at half to quarter of the concentration used for mouse ES cells. Do not add the drug continuously. This will kill all the rat ES cells even after they are transfected. You must apply a "pulsed" drug regimen: The general scheme is to expose the cells to the drug for 24–48 h, then remove the drug completely for 24–48 h. Repeat this cycle 2–3 more times. This "pulsed selection" strategy allows for the transfected GFP-positive colonies to emerge at high efficiency. We have applied the same strategy to Puromycin, G418, and Zeocin selections and achieved similar results. Puromycin, G418, and Zeocin are added at 0.4 μg/ml, 150 μg/ml, and 10 μg/ml, respectively.

19.5.5. Low Efficiency of Rat Chimera Formation and Germline Transmission

The quality of rat ES cells is the single-most important factor affecting rat chimera formation and germline transmission. Use the lowest passage number of rat ES cells available. Always maintain rat ES cells at optimal culture condition. Karyotype rat ES cells and choose the lines with highest proportion of karyotypically normal cells for blastocyst injection.

19.6 Conclusions and Outlook

The 3i/2i rat ES cell medium allows efficient derivation and robust propagation of rat ES cells. The availability of rat ES cells will lead to the development of technologies for the efficient generation of transgenic and gene-targeted models in the rat. Due to the physiological and pharmacological similarities between rats and humans, these rat models are anticipated to more closely mimic human conditions than mouse models. It is likely that rat ES cell-based technologies will provide a powerful platform for the study of human health, disease, and drug screening, within the biomedical research field.

References

1. Aitman TJ, Critser JK, Cuppen E, Dominiczak A, Fernandez-Suarez XM, Flint J, Gauguier D, Geurts AM, Gould M, Harris PC et al (2008) Progress and prospects in rat genetics: a community view. Nat Genet 40:516–522
2. Evans MJ, Kaufman MH (1981) Establishment in culture of pluripotential cells from mouse embryos. Nature 292:154–156
3. Martin GR (1981) Isolation of a pluripotent cell line from early mouse embryos cultured in medium conditioned by teratocarcinoma stem cells. Proc Natl Acad Sci USA 78: 7634–7638
4. Buehr M, Meek S, Blair K, Yang J, Ure J, Silva J, McLay R, Hall J, Ying QL, Smith A (2008) Capture of authentic embryonic stem cells from rat blastocysts. Cell 135:1287–1298
5. Li P, Tong C, Mehrian-Shai R, Jia L, Wu N, Yan Y, Maxson RE, Schulze EN, Song H, Hsieh CL et al (2008) Germline competent embryonic stem cells derived from rat blastocysts. Cell 135:1299–1310
6. Smith AG, Heath JK, Donaldson DD, Wong GG, Moreau J, Stahl M, Rogers D (1988) Inhibition of pluripotential embryonic stem cell differentiation by purified polypeptides. Nature 336:688–690
7. Williams RL, Hilton DJ, Pease S, Willson TA, Stewart CL, Gearing DP, Wagner EF, Metcalf D, Nicola NA, Gough NM (1988) Myeloid leukaemia inhibitory factor maintains the developmental potential of embryonic stem cells. Nature 336: 684–687
8. Ying QL, Nichols J, Chambers I, Smith A (2003) BMP induction of Id proteins suppresses differentiation and sustains embryonic stem cell self-renewal in collaboration with STAT3. Cell 115:281–292
9. Ying QL, Wray J, Nichols J, Batlle-Morera L, Doble B, Woodgett J, Cohen P, Smith A (2008) The ground state of embryonic stem cell self-renewal. Nature 453:519–523
10. Brook FA, Gardner RL (1997) The origin and efficient derivation of embryonic stem cells in the mouse. Proc Natl Acad Sci USA 94: 5709–5712
11. Keller G (2005) Embryonic stem cell differentiation: emergence of a new era in biology and medicine. Genes Dev 19:1129–1155
12. Capecchi MR (2005) Gene targeting in mice: functional analysis of the mammalian genome for the twenty-first century. Nat Rev Genet 6:507–512
13. Nagy A, Gertsenstein M, Vintersten K, Behringer R (2003) Manipulating the mouse embryo, a laboratory manual, 3rd edn. Cold Spring Harbor Laboratory Press, New York
14. Schwartzberg PL, Goff SP, Robertson EJ (1989) Germ-line transmission of a c-abl mutation produced by targeted gene disruption in ES cells. Science 246:799–803
15. Seong E, Saunders TL, Stewart CL, Burmeister M (2004) To knockout in 129 or in C57BL/6: that is the question. Trends Genet 20:59–62

Chapter 20

Induced Pluripotency: Generation of iPS Cells from Mouse Embryonic Fibroblasts

Han Li, Katerina Strati, Verónica Domínguez, Javier Martín, María Blasco, Manuel Serrano, and Sagrario Ortega

Abstract

The ability to directly reprogram mammalian adult somatic cells to an undifferentiated pluripotent stage similar to that of embryonic stem cells by introduction of a reduced number of transcription factors has opened new venues in many fields of Biology and Medicine. These reprogrammed cells called iPS cells (induced pluripotent stem cells) represent a powerful tool for the study of cell differentiation and pluripotency and a promise for regenerative therapy. Here, we describe a basic procedure for reprogramming of mouse embryonic fibroblasts (MEFs) to iPS cells by expression of three transcription factors: Oct3/4, Sox2, and Klf4.

Abbreviations

DMEM	Dulbecco-Modified Eagles Medium
DMSO	Dimethyl-Sulfoxide
E12.5	Embryonic day 12.5. The day in which a vaginal plug is detected is considered day 0.5 of embryonic development
E13.5	Embryonic day 13.5
EGFP	Enhanced Green Fluorescent Protein from the jellyfish *Aequorea victoria*. Enhanced means optimized by mutagenesis for its use in mammalian cells
ES cell	Embryonic Stem cell
FACS	Fluorescence Activated Cell Sorting
FCS	Fetal calf Serum
iPS cell	Induced Pluripotent Stem cell
KSR	Knockout-Serum Replacement
LB	Luria Broth
LTR	Long Terminal Repeat
MEF	Mouse Embryonic Fibroblast
N_2	Nitrogen
PBS	Phosphate Buffer Saline
PCR	Polymerase Chain Reaction

S. Pease and T.L. Saunders (eds.), *Advanced Protocols for Animal Transgenesis*, Springer Protocols,
DOI 10.1007/978-3-642-20792-1_20, © Springer-Verlag Berlin Heidelberg 2011

Pen/Strep	Penicillin/Streptomycin
RIPA	Radio-immuno-precipitation assay
Rpm	Revolutions per minute
RT	Room temperature
SCNT	Somatic Cell Nuclear Transfer
SKY	Spectral Karyotyping
SSEA-1	Stage-Specific Mouse Embryonic Antigen-1
WT	Wild Type

20.1 Introduction

20.1.1. Pluripotent Cells

Cell potency is the term used to define the ability of a particular cell type to give rise to one or more different cell lineages. During mammalian development, cellular potency decreases as the variety of cell types in the organism increases. Thus cell potency ranges from the totipotent single cell embryo, the zygote that gives rise to the whole organism, to terminally differentiated cells and adult committed progenitors that generate a single cell lineage (unipotent cells) [1]. Pluripotent cells are part of this developmental transition. These cells have the ability to give rise to multiple cell lineages. The paradigm of pluripotent cells are embryonic stem (ES) cells derived from the inner cell mass of the mammalian blastocyst and which generate all the cell lineages in the embryo, endoderm, mesoderm and ectoderm, as well as some extraembryonic tissues. Pluripotent ES cells have been the focus of attention in the last decades from various perspectives. Mouse ES cells, first established in 1981 from mouse blastocysts [2, 3] have been widely used to introduce gene targeted mutations into the germ line of ES cell-mouse chimeras that can transmit the mutation to their offspring. Moreover, ES cells provide a good experimental model for investigation of the genetic mechanisms responsible for maintaining cell pluripotency. In fact, there has been striking progress in the understanding of such mechanisms in the last few years [4–6]. But perhaps the major interest in ES cells is because of their potential for therapeutic applications in regenerative medicine since they can be induced to differentiate into multiple cell types in vitro [7]. However, while ES cells are relatively easy to establish from mouse embryos, it has proven to be very difficult to establish ES cells from other species including human. For example, pluripotent ES cells capable of germ line contribution were not successfully established from rats until 2008 [8, 9]. Human pluripotent ES cell lines that share some properties with mouse ES cells (although they are not completely equivalent) were first isolated from human embryos in 1998 [10]. The therapeutic potential of established human ES cell lines is compromised by the immune response they may elicit in the patient, leading to tissue rejection. Their potential is also limited by the ethical issues inherent in the destruction of human embryos, as is required for ES cell derivation.

20.1.2. From Adult Somatic Cells to Pluripotent Cells

For many years it was thought that the developmental transition from the pluripotent cells of the inner cell mass or their derivative, ES cells, to the unipotent adult differentiated cells was irreversible. However, in 1997, Wilmut and coworkers demonstrated for the first time that the cytosol of the mammalian oocyte contained trans-acting factors that could reprogram the nucleus of a terminally differentiated adult somatic cell to a state equivalent to the nucleus of the totipotent zygote, generating a live mammal (the sheep Dolly) from the nucleus of an adult somatic cell [11]. Thus, mammalian cloning by somatic cell nuclear transfer (SCNT) from an adult terminally differentiated cell was proven possible and demonstrated for the first time that cell differentiation was a fully reversible process (Fig. 20.1). This finding implied that the genomic modifications that take place in the nucleus to impose

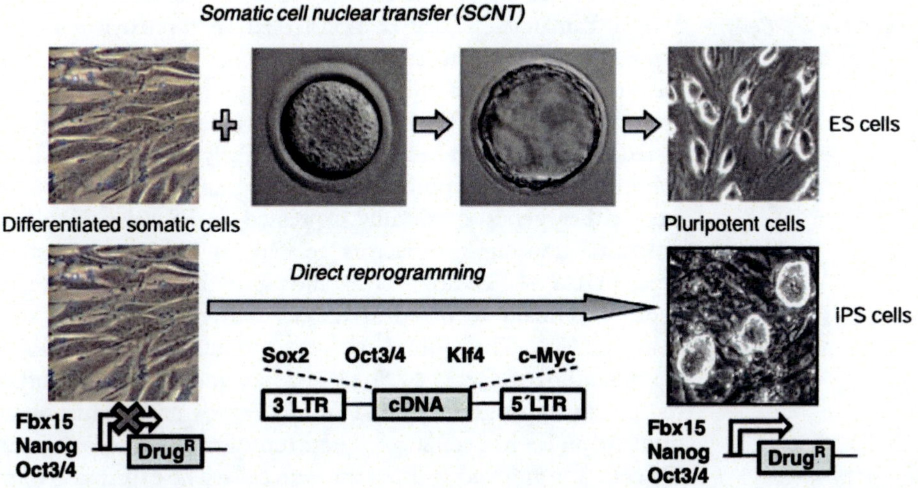

Fig. 20.1 Somatic cell reprogramming versus nuclear transfer. Somatic cell nuclear transfer, SCNT can be used to generate pluripotent stem cells (ES) from the nucleus of a donor somatic differentiated cell. The nucleus of a mammalian oocyte is replaced by the nucleus of a diploid somatic cell. This gives rise to an embryo that can be cultured in vitro to the blastocyst stage from which ES cell lines can be established. Direct reprogramming of somatic cells by retroviral transduction of four transcription factors (Sox2, Oct3/4, Klf4, and c-Myc) converts somatic differentiated cells to induced pluripotent stem cells (iPS cells), or ES-like cells, without oocyte or embryo destruction. In the initial reprogramming experiments, iPS cells were selected for drug resistance from somatic cells carrying a drug resistance gene transcribed under the control of a promoter selectively active in pluripotent cells such as that of Fbx15, Nanog, or Oct3/4. The selectivity of Nanog or Oct3/4 is higher than that of Fbx5 but all the three approaches have been used. However, selection by drug resistance is not essential and iPS colonies can be isolated using only morphological criteria. Retroviral expression of c-Myc is not essential for reprogramming of fibroblasts, but increases efficiency. The viral LTRs are silenced by long-term culture of pluripotent cells. Reprogramming results in the reactivation of the endogenous expression of Sox2, Oct3/4 and Klf4, which are turned off in somatic cells.

developmental restrictions had to be reversible epigenetic changes rather than irreversible genetic alterations. Moreover, these landmark studies opened a new avenue for the generation of pluripotent cells with the same genetic content (except for mitochondrial DNA) and therefore histocompatible with the adult nuclear cell donor, thus solving the issue of immune rejection in the use of pluripotent cells for therapeutic purposes. However, despite its multiple biotechnological applications, SCNT still has important ethical and legal limitations for clinical application, since it also involves oocyte manipulation and destruction for the isolation of pluripotent cells.

In search of alternative ways to obtain pluripotent cells it was demonstrated that ES cells also have the ability to convert an adult somatic differentiated cell to a pluripotent state by cell fusion [12, 13] (reviewed in [14, 15]). Moreover, further studies demonstrated that factors involved in reprogramming were nuclear factors, suggesting that transcription factors were implicated in the reprogramming process (reviewed in [1]).

20.1.3. Direct Reprogramming: Generation of iPS Cells

These experiments set the basis for the striking breakthrough in stem cell research, published in 2006 by Kazutoshi Takahashi and Shinya Yamanaka. They demonstrated that retrovirus-mediated expression of only four transcription factors in embryonic or adult mouse fibroblasts was sufficient to reprogram these cells to a pluripotent ES-like stage (Fig. 20.1) [16]. In a landmark experiment they tested 24 different genes, understood likely to be involved in the maintenance of pluripotency, for reprogramming capacity. These were either genes specifically expressed in ES cells of unknown function, transcription factors related to pluripotency such as Sox2, Oct3/4, Nanog, etc., and growth/proliferation-related genes potentially involved in ES cell self-renewal such as c-Myc, Stat3, or Klf4. They found out that the retroviral expression of a combination of Oct3/4, Sox2, c-Myc, and Klf-4 was sufficient to directly reprogram mouse fibroblasts of embryonic or adult origin to an ES-like cell stage. These reprogrammed cells were called iPS cells, for *i*nduced *p*luripotent *s*tem cells. The expression of these four factors alone was sufficient for reversal of the epigenetic modifications accumulated in the chromatin of an adult differentiated cell during development and return it to a chromatin configuration resembling that of a noncommitted, pluripotent cell.

In this initial experiment, reprogrammed pluripotent cells were obtained from fibroblasts carrying a neomycin phosphotransferase (G418 resistance) gene knocked into the Fbx15 locus ($Fbx15^{\beta geo/\beta geo}$) such that G418 resistance was expressed under the transcriptional control of the Fbx15 gene [17]. Since Fbx15 is not expressed in fibroblasts, but is expressed in ES cells, G418 selection was used in combination with viral transfection of fibroblasts to identify drug-resistant cells with ES cell characteristics

(Fig. 20.1). However, in this experiment, reversion to a fully pluripotent state was not complete. The iPS cells obtained were similar but not identical to mouse ES cells according to transcriptional and epigenetic criteria. Moreover, these cells fulfilled some of the criteria of pluripotency but not all. They were able to form embryoid bodies in tissue culture and teratomas in nude mice in which cells from the three embryonic cell layers could be found. However, they were not capable of contributing to viable chimeras when injected into mouse blastocysts. Therefore, these results indicated that probably Fbx15 was not the best marker for selection since the endogenous gene could be sufficiently reactivated in partially reprogrammed iPS cells.

20.1.4. Improved iPS Cells from Mouse and Human

One year later three groups showed that selection of reprogrammed cells by drug resistance driven by the promoters of the pluripotency essential genes Nanog or Oct3/4 resulted in selection of iPS cells of better quality than those selected by Fbx15-driven drug resistance (Fig. 20.1) [18–20]. In fact when these reprogrammed cells were carefully analyzed it was shown that they were more similar to ES cells than Fbx15-selected iPS cells according to several criteria. First, retroviral expression of the four factors was silenced in these iPS cells, while they were not in the Fbx15-selected iPS. Retroviral LTR silencing is a characteristic of pluripotent cells, probably due to the activity of Dnmt3a2 methyltransferase [18, 21]. Also endogenous pluripotent cell markers Nanog, Oct3/4, and Sox2 were expressed at similar levels to those of ES cells and the promoters of these genes were fully demethylated, in contrast to that found in the Fbx15-selected iPS cells. In general, the global pattern of ES cell gene expression was practically the same as that found in iPS cells. Finally, the silenced X-chromosome from female somatic cells was reactivated in these iPS cells in contrast to the Fbx15-iPS. But the most rigorous criteria of pluripotency was that these reprogrammed cells were able to contribute to the germ line of iPS cell-mouse chimeras. Moreover, Wernig et al. [20] generated late gestation mouse embryos completely derived from Nanog-selected iPS cells by tetraploid embryo complementation where the tetraploid host embryo only contributes to extraembryonic tissues and the entire embryo is derived from the introduced iPS cells [20]. These experiments also showed that delaying drug selection yielded better-quality iPS cells. In fact, soon after it was shown that wild type fibroblasts could be reprogrammed with four factors, without any selection, following only morphological criteria [22, 23].

One of the problems of iPS cells generated by expression of the four factors, Oct3/4, Sox2, Klf4 and c-Myc, is that mice generated with these iPS cells developed tumors due to the reactivation of retroviral c-Myc expression, thus compromising the clinical application of iPS cells generated by the combination of these four factors. However, soon after it was shown that retroviral c-Myc expression is dispensable for mouse and human fibroblast reprogramming so that pluripotent iPS cells could be generated from fibroblasts by retroviral transduction of only three factors: Oct3/4, Sox2 and Klf4, although at lower efficiency [24, 25]. The elimination of c-Myc from the reprogramming protocol is an advantage in terms of reducing the potential tumorigenicity of the reprogrammed cells if expression of c-Myc is reactivated, and the activation of some relevant Myc targets such as TERT also takes place when only three factors are used [26].

Reprogrammed human somatic cells or human iPS cells would therefore be the source of patient-specific pluripotent cells and would resolve the issues of immune response and the ethical concerns related to the use of other sources of pluripotent cells. Human iPS cells were first obtained by Yu et al. [27] directly from human fibroblasts in 2007 using the combination of factors OCT3/4, SOX2, NANOG, and LIN28.

Although both mouse and human iPS cells were first generated from fibroblasts, more recently other cell types have also been reprogrammed by a combination of transcription factors. Mouse mature B-lymphocytes, pancreatic β-cells, hepatocytes, gastric epithelial cells, neural progenitors as well as human keratinocytes have been successfully reprogrammed to iPS cells (reviewed in [1]).

Reprogramming of mouse embryonic fibroblasts by retroviral expression of three or four transcription factors is still one of the best experimental settings for genetic characterization of iPS cells and for studying the molecular mechanisms involved in cellular reprogramming and maintenance of pluripotency. Using this approach we and others have recently demonstrated that senescence/immortalization-related tumor suppressor genes, such as p53 and the Ink4/Arf locus, are a barrier for somatic cell reprogramming [28–32]. Also by reprogramming MEFs derived from telomerase-deficient mice, we have shown that cells with short telomeres are less efficiently reprogrammed than cells with normal telomere length and that telomerase is required for telomere elongation during reprogramming [26, 29].

Here, we describe a reliable protocol for selection-independent direct reprogramming of mouse embryonic fibroblasts to iPS cells by retroviral expression of Oct3/4, Sox2, and Klf4.

20.2 Protocol for iPS Cells Generation from MEFS

This protocol is based upon that previously described by Blelloch et al. with some modifications [22]. Similar to the original protocol described by Takahashi and Yamanaka [16], it involves retrovirus-mediated expression of the three reprogramming factors. Once MEFs are reprogrammed, iPS cell culture conditions are very similar to those already established for ES cell culture (*see* Chapter 14). It is likely that handler familiarity with the culture of ES cells will contribute to the success of the procedures described here.

20.2.1. Materials

The materials described below are those specifically used for the reprogramming procedure and are mainly related to tissue culture equipment and reagents.

20.2.1.1. Equipment

The equipment is very similar to that used for routine culture of ES cells. All plastic and glass materials for tissue culture should be disposable.

- Laminar flow cabinet for routine tissue culture.
- Horizontal flow cabinet with stereomicroscope for picking iPS colonies.
- CO_2 tissue culture incubator set at 37°C, 5% CO_2, 85% humidity.
- Phase-contrast inverted microscope with 4×, 10×, 20×, and 40× objectives.
- Table top refrigerated centrifuge.
- Refrigerated microcentrifuge.
- Good-quality tissue culture plastic plates (Falcon or Nunc) ranging from 96-well to 6-well format and 35-, 60-, 100-, and 150-mm plates. V-shaped bottom 96-well plates (Costar) are used for trypsinizing individual colonies.
- Sterile plastic centrifuge tubes: 1-, 2-, 15-, and 50-ml
- Sterile plastic pipettes: 1–25 ml
- Cryotubes (freezing vials) (2 ml)
- Glass Pasteur pipettes
- Cellulose acetate microfilters (0.2 and 0.45 μm)
- Micropipettes: 20 μl, 200 μl, and 1,000 μl.
- Multichannel pipettes: 8 or 12 channels (20–200 μl)

20.2.1.2. Retroviral Plasmids

- pMXs-Klf4 (Addgene 13370)
- pMXs-Sox2 (Addgene 13367)
- pMXs-Oct3/4 (Addgene 13366)
- pMXs-c-Myc (Addgene 13375) (optional)
- pBabe-PURO-EGFP (Addgene 14430) (optional)

These plasmids are high copy number plasmids that contain the defective retroviral backbone pMX [33] or pBabe to express the cDNA of the mouse transcription factors Klf4, Sox2, Oct3/4, and c-Myc or the EGFP protein, respectively, under the transcriptional control of the viral LTR (long terminal repeat). The plasmid DNA, once transfected into the cell line HEK293T, together with the ecotropic packaging plasmid pCL-Eco (described below), is converted to viral RNA by the retroviral reverse transcriptase and packaged into infective (nonreplicative) mouse ecotropic viral particles. The viral LTR is transcriptionally active in MEFs but is silenced in pluripotent iPS and ES cells by de novo methylation, after long-term culture [21].

– pCL-Eco (Addgene 12371) [34].

This plasmid expresses ecotropic Moloney mouse leukemia virus retroviral *gag/pol/env* genes required for viral genome reverse transcription and packaging of retroviral particles. The Moloney mouse leukemia env protein encoded by this ecotropic murine retrovirus packaging plasmid results in efficient transfection of mouse cells.

Plasmids are provided in *E. coli* DH5a and can be maintained in this strain or any recA$^-$ bacterial strain by selection in ampicillin 100 μg/ml.

20.2.1.3. Cell Lines

– Both HEK293T/17 (ATCC #CRL-11268) or HEK293T (ATCC #CRL-11268) cell lines can be used. HEK293T/17 is a highly transfectable subclone of HEK293T obtained at ATCC. This cell line is used to produce retroviral particles upon transfection of the plasmids described above. The viral particles are released into the media in which the cells are cultured, from which they can be harvested.
– MEFs (primary mouse embryonic fibroblasts) are routinely obtained from mouse embryos at day E12.5–E13.5 of development by the method described below.

20.2.1.4. Culture Media

– DMEM: Dulbecco's modified Eagle medium (D-MEM) (1×), liquid (High Glucose) with GlutaMAX™ and Sodium Pyruvate (Invitrogen 31966021).
– KO-DMEM: KnockOut™ D-MEM (Invitrogen 10829018)
– Fetal Calf Serum (FCS) (high quality). FCS does not need to be ES cell tested since it is only used for MEFs and HEK293T cells culture. A routine test for toxicity and growth on MEFs is enough in this case. It must be heat inactivated at 56°C for 30 min.
– KSR (Knockout Serum Replacement) (Invitrogen 10828028). KSR is used for iPS cell selection and sub-culture. It is advisable to test each batch of KSR for its capacity to maintain pluripotency of ES cell lines before using it for iPS

cell culture or reprogramming, since batch-to-batch variations may be found. However, previous testing is not as critical for KSR as it is for FCS when used for ES or iPS culture.

- LIF (ESGRO™, Millipore ESG1107)
- GlutaMAX™-I Supplement (Invitrogen 35050061)
- Penicillin/Streptomycin: (Pen/Strep 100×; Invitrogen 15070063). Penicillin-Streptomycin, liquid, contains 5,000 units of penicillin (base) and 5,000 μg of streptomycin (base)/ml. Final concentration 50 units/50 μg/ml
- 2-Mercaptoethanol 50 mM (1,000×) (Invitrogen 31350)
- MEM Non-Essential Aminoacids Solution 10 mM (100×) (Invitrogen 11140035)
- Trypsin-EDTA solution (0.05% Trypsin-EDTA 1×, GIBCO 25300)
- D-PBS: Dulbecco's Phosphate-Buffered Saline (1×) (Invitrogen 14190144)
- DMSO (Dimethylsulfoxide) (Sigma D2650)

20.2.1.5. Reagents

- Kit for plasmid DNA isolation (High Speed Plasmid Maxi Kit Qiagen 12263)
- FuGENE 6 Transfection Reagent (Roche 11815091001)
- Polybrene (Sigma 107689). Stock solution 8 mg/ml in water, filtered through a 0.2 μm filter. Polybrene (hexadimethrine bromide) is a cationic polymer used to increase the efficiency of retroviral infection of cells in culture. Polybrene acts by neutralizing the charge repulsion between virions and sialic acid on the cell surface.
- Gelatin (Sigma G-1890) 0.1% in water and autoclaved
- Mitomycin-C (Sigma M-0503). Stock solution 10 mg/ml in PBS. Filtered through a 0.2 μm filter and used at a final concentration of 10 μg/ml
- Alkaline Phosphatase Detection Kit. Millipore SCR004
- Luria Broth medium (LB): SIGMA L2542-Ampicillin: Sodium salt, SIGMA A9518

20.2.2. MEF Reprogramming Procedure

20.2.2.1. MEF Isolation and Culture

Any procedure to obtain a good-quality primary culture of MEFs can be used. However since MEF quality is important for the success of the reprogramming protocol, we will describe the procedure that we routinely use to prepare cultures of MEFs for reprogramming. We use the same MEF isolation protocol to prepare feeder cells for iPS cell production and culture.

Mice from any genetic background can be used to establish MEF cultures, although different genetic backgrounds may have different reprogramming efficiency or kinetics. MEFs from C57BL/6 embryos are frequently used since this is a commonly used inbred genetic background. We routinely use C57BL/6J. OlaHsd mice from Harlan in our facility. C57BL/6 mice are pigmented so, once MEFs are reprogrammed, C57BL/6 iPS can be easily tested for chimera contribution by routine aggregation with albino CD-1 morulae (we routinely use Harlan Hsd: ICR (CD-1®) mice) or microinjection into albino B6(Cg)-Tyr^{c-2J}/J mice (Jackson Laboratories, USA). The procedure described here provides more MEFs than those needed for a reprogramming session and can be scaled down by reducing the number of harvested embryos.

1. Embryos are harvested from pregnant females at E12.5–E13.5, by hysterectomy.
2. The uterine horns containing the fetuses are collected in PBS containing Pen/Strep. From this stage all the manipulations are performed in the tissue culture hood under sterile conditions.
3. Embryos are washed in PBS containing Pen/Strep (1×) several times until most blood is removed. Head and internal organs are removed with sharp forceps and carcasses washed extensively in PBS-Pen/Strep. From this point onwards, each embryo is processed individually.
4. Embryos are chopped with fine scissors into very small pieces, each in 0.5–1.0 ml of PBS-Pen/Strep in a 60 mm tissue culture plate. Then, 5 ml of Trypsin-EDTA is added to each plate. Plates are incubated at 37°C in a CO_2 incubator for 5 min. After that, the suspension of trypsin-dissociated cells and tissue pieces is pipetted up and down several times with a 10 ml plastic pipette.
5. Incubation and pipetting is repeated one or two more times until practically no pieces of tissue are left and all the tissue is dissociated into single cells.
6. At this point, the suspension is collected and transferred to a 150 mm tissue culture plate containing 25 ml of DMEM + 10% FCS + Pen/Strep (one embryo per plate).
7. Fibroblasts are cultured for 2–4 days until the cells reach confluence. MEFs from each plate are trypsinized, frozen in freezing medium (25% FCS and 10% DMSO in DMEM), and stored in liquid nitrogen. These fibroblasts are at passage 1 and can be further expanded to be used either for reprogramming or for feeder preparation.

If MEFs are going to be used for reprogramming, usually three vials of approximately 4×10^6 cells at passage 1 are frozen in 1 ml of freezing medium from each confluent plate. Each vial is thawed in 5 × 100 mm dishes from which they can be frozen when they reach confluence (passage 2 MEFs) or expanded further 1–5 to obtain MEFs at passage 3. Cells from each plate are frozen in a single vial that will be thawed in a single 100 mm plate for reprogramming at day-2. For reprogramming, passage 2 or 3 MEFs should be used. MEFs at higher passage numbers reprogram less efficiently.

If MEFs are going to be used as feeder cells for iPS culturing, each confluent plate at passage 1 is frozen in one vial containing 1 ml of freezing medium. Each vial is thawed in 5 × 150 mm dishes (passage 2 MEFs) and expanded 1–5 to obtain MEFs at passage 3. MEFs for use as feeder cells should be expanded up to passage 3. Usually 25 × 150 mm plates of MEFs at passage 3 are obtained from a single embryo, by expanding 1–5 in each passage.

At passage 3, the 25 × 150 mm plates are mitotically inactivated by treatment with Mitomycin-C (10 μg/ml final concentration in 25 ml of culture medium, per plate) for 2–3 h. After that, MEFs are washed two times with PBS, trypsinized, counted and centrifuged at 1,000 rpm (1950 × *g*) for 10 min at 4°C. The cell pellet is resuspended in freezing medium at a density of 7×10^6 cells/ml. Vials containing 1 ml of the cell suspension are frozen and stored in liquid nitrogen. A good MEF preparation from a single embryo will result in approximately 1.5×10^8 feeder cells after expansion and Mitomycin treatment and that is enough cells for approximately 80 × 100 mm feeder plates.

Prepare feeder cells for iPS culture on the day prior to reprogramming, Prepare tissue culture plates by gelatin coating. Use enough gelatin (0.1% in water) to cover the bottom surface of the plate, and aspirate immediately or after a few minutes at room temperature. Mitomycin-C-treated MEFs are thawed and plated out in DMEM + 10% FCS + Pen/Strep. Do not use KSR instead of FCS for plating feeder cells, since MEFs do not attach as well in KSR. From one frozen vial of MEFs (7×10^6 Mitomycin-treated cells), 4 × 100 mm plates of feeder cells can be prepared (2×10^5 cells/cm^2).

20.2.2.2. Reprogramming of MEFs

The protocol for MEF reprogramming starts 4 days before the day in which the reprogramming is initiated (day 0), since retrovirus should be freshly prepared from HEK293T cells for each reprogramming experiment. A scheme of the protocol is described in Fig. 20.2.

Day-4:

HEK293T cells are thawed and plated in HEK293T culture medium (DMEM + 10% FCS + Pen/Strep).

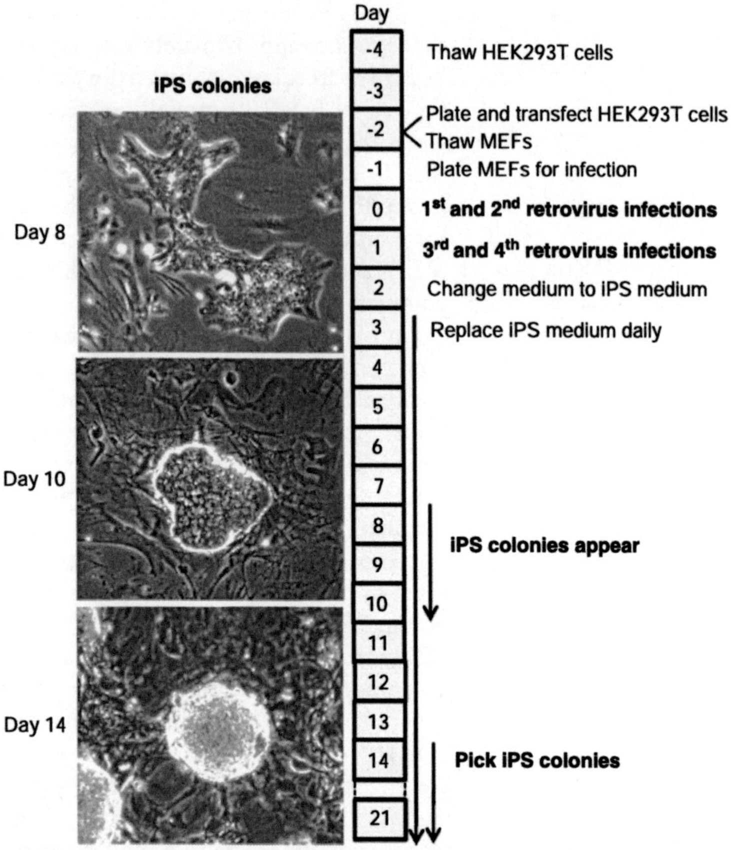

Fig. 20.2 Scheme of MEF reprogramming protocol by retroviral transduction of three transcription factors. The protocol takes 3–4 weeks from retrovirus production to picking of iPS colonies. Day 0 is the day in which reprogramming starts with the two first rounds of retroviral infections. HEK293 cells are used to prepare retrovirus expressing the reprogramming factors. The important landmarks of the reprogramming protocol are represented in *bold*. On the left the morphology of the iPS colonies at different times during the reprogramming process is shown.

Day-3:

Change medium on HEK293T cells.

Day-2:

Transfection of HEK293T cells with retroviral plasmids:

1. HEK293T cells are trypsinized, counted, and plated in 100 mm tissue culture plates (5×10^6 cells/ plate) in 10 ml of HEK293T culture medium. The number of plates to be prepared depends on the number of different retroviruses to be produced. We routinely use only three factors for reprogramming: Sox2, Oct4 and Klf4, omitting c-Myc since its expression is not essential for reprogramming fibroblasts and it may have oncogenic effects if re-activated. Besides,

the expression of c-Myc induces a high rate of apoptosis in the iPS colonies that results in many cells dying and floating in the medium. After plating the cells, the transfection protocol is initiated.

2. Preparation of plasmid cocktails:
   (a) 4 μg of pMX-Sox2 + 4 μg of pCLEco
   (b) 4 μg of pMX-Oct4 + 4 μg of pCLEco
   (c) 4 μg of pMX-Klf4 + 4 μg of pCLEco
   (d) 4 μg of pBabe-PURO-EGFP + 4 μg of pCLEco

   Plasmid DNA is prepared from DH5a bacteria growing in LB medium + ampicillin 100 μg/ml, using the DNA extraction kit described and following the recommendations of the manufacturer. Plasmid DNA is resuspended in TE at a concentration of 1 μg/ml.
3. For each plasmid cocktail, a mix of 24 μl of FuGENE 6 and 576 μl of DMEM (600 μl total volume) is prepared and incubated at room temperature for 5 min.
4. After the incubation each plasmid cocktail (approximately 8 μl) is added to the DMEM + FuGENE 6 mix and incubated 45 min at room temperature. The ratio μg DNA/μl FuGENE 6 should be 1/3 (W/V).
5. The DNA/FuGENE 6 mix is added to the HEK293T cells. Do not remove the medium from the culture plate, simply add the FuGENE 6/DNA mix to the medium dropwise with a 1,000 μl pipettor, distributing it around the plate.

   We routinely do four transfections corresponding to the plasmid cocktails (a), (b), (c), and (d) described above.

   Transfections (1), (2), and (3) will give rise to retroviruses expressing the three reprogramming transcription factors and are essential. Transfection 4 is used as a control of infection efficiency and is optional. If c-Myc is also going to be used for reprogramming, a mix of the plasmids pMX-c-Myc and pCL-Eco should be prepared and treated in the same way as transfections 1–3.
6. On the same day one vial of the primary MEFs (Passage 2–3) is thawed in 2–3 × 100 mm plates.

Day-1:

7. Replace the medium on the transfected HEK293T cells with 10 ml of fresh culture medium.
8. MEFs are plated for reprogramming: 2.5–5 × 10^5 cells in a 100-mm plate in DMEM + 10% FCS + Pen/Strep.
   Four rounds of infection are done, sequentially, during the next 48 h.

Day 0:

First retrovirus infection of MEFs (two rounds)

Morning: first round of infection

9. Conditioned medium from each individual transfection (10 ml from each) is collected. These media contain the retrovirus expressing each of the three reprogramming factors (Klf4, Sox2, and Oct3/4) or EGFP. Replace the medium on the transfected HEK293T with 10 ml of fresh medium.
10. Each conditioned medium is spun separately at 1,000 rpm for 5 min at RT to remove cell debris.
11. Supernatants are individually collected and filtered through 0.45 μm filters.
12. Polybrene 8 μg/ml (final concentration) is added to each supernatant.
13. 1.5 ml of the supernatants (10 ml) containing each of the reprogramming retroviruses (Klf4-, Sox2-, and Oct3/4-) are mixed together (4.5 ml total). For the control plate, 1.5 ml of the supernatants containing Klf4-, Sox2-, Oct3/4-, and EGFP-retroviruses are mixed (6 ml total).
14. The medium of the MEF plates is replaced with the supernatant pools: 4.5 ml for each reprogramming plate and 6 ml for the control plate. As many as 6 × 100-mm plates of MEFs can be infected with the retrovirus supernatant from one plate of HEK293T cells.

    The retroviral-supernatants are prepared fresh for each reprogramming experiment.
15. Afternoon/evening: second round of infection (10–12 h interval)

    The same retrovirus infection procedure performed on the morning of day 0, from HEK293T cells conditioned medium collection to infection of MEFs, is repeated.

Day 1:

16. Morning: third round of infection (repeat the procedure as the day before)
17. Afternoon/evening: fourth round of infection (repeat the procedure as the day before)

Day 2:

18. The medium of the infected MEF plates is replaced with 10 ml of iPS medium (DMEM or KO-DMEM, either media can be used) supplemented with 15% KSR + LIF (1,000 U/ml) + nonessential amino acids (1×) + 2-mercaptoethanol 0.5 mM + Pen/Strep (1×). If KO-DMEM is used it should

be supplemented with Glutamax, since this component is not included by the manufacturer.

Days 3–12:

19. iPS medium should be replaced every day. iPS colonies will start to appear on the monolayer of MEFs around day 8–10. They can be easily recognized by their morphology, similar to that of ES cell colonies (Fig. 20.2).

    Keep changing medium daily.

Days 12–14:

20. Colonies are clearly visible and essentially all have a nice ES cell colony-like morphology: round, shiny, and with well-defined edges (Fig. 20.2).

    Usually individual colonies are picked at day 14 after transfection, under a stereomicroscope but they can be picked between 14 and 21 days according to their size.

20.2.2.3. Picking iPS Colonies

From this step forward, picking and culturing iPS cells is very similar to ES cell culture (*see* Chapter 14).

1. 96-well plates with feeder cells are prepared 1–2 days before picking iPS colonies following the procedure described in **step** 7 of Subheading 20.2.2.1.
2. Medium of reprogrammed plates is replaced 2–3 h before picking the colonies.
3. During picking of iPS colonies, the medium in the plates is replaced with KO-DMEM or DMEM without any additives.
4. Under a stereomicroscope, in a horizontal flow hood, individual colonies are picked using a micropipette (Gilson P20 with yellow tip) set to collect 10 μl. Physically each individual colony is detached from the fibroblast monolayer with the tip of the pipette and aspirated into the tip together with 10 μl of medium. Three-dimensional, sharp-edged round colonies are selected for picking.
5. Colonies are individually transferred to wells of V-shaped 96-well plates containing 30 μl of Trypsin/EDTA.
6. After picking, incubate 96-well plates at 37°C, in a CO_2 incubator, for 5–10 min.
7. With a multichannel pipette, 70 μl of iPS medium is added to each well to neutralize the trypsin. The suspension is pipetted up and down several times (10–20 times) to disaggregate the colonies.
8. Replace the media on the 96-well plates, preplated with feeder cells, in which the picked colonies will be cultured, with 100 μl iPS cell medium.

9. Disaggregated colonies in 100 μl of medium are transferred to the 96-well plates with feeder cells. The total volume in each well will be 200 μl. Incubate overnight in the CO_2 tissue culture incubator.
10. Next day the medium in each well is replaced with 150 μl of iPS medium using the multichannel pipette.
11. Medium is replaced every other day.
12. iPS clones can then be consecutively expanded to 24-well plates, 35 mm plates and 60 mm plates. During this expansion, feeder cells should be used. The same criteria used for ES cell expansion, in terms of cell density and dilution, should be followed.

iPS clones can be frozen from a 60 mm plate. We usually freeze one vial with half of the cells from the 60 mm plate in liquid N_2 and the other half of the plate is expanded to a 100 mm plate and frozen from there.

20.2.2.4. Quantification of Reprogramming Efficiency

The efficiency of reprogramming is dependent, among other factors, upon the efficiency of retroviral infection.

Efficiency of retroviral infection can be calculated from the plate of MEFs infected with the retroviral pool containing 1.5 ml of the Oct4, Sox2, Klf4, and EGFP viruses. At day 3 of the protocol (48 h after the fourth round of viral infection; **step 19**), cells are trypsinized and the proportion of $EGFP^+$ cells is determined by flow cytometry analysis [35]. Typically 40–60% of the total number of cells are positive for EGFP.

To calculate reprogramming efficiency, the control plate is stained using an alkaline phosphatase kit, following the manufacturer's instructions, at day 14 of the protocol, **step 20** of Subheading 20.2.2.2. The total number of iPS colonies stained positive for alkaline phosphatase are counted.

The efficiency is calculated as: [total number of colonies]/[(number of MEFs plated = 2.5×10^5) x (0.4–0.6 = fraction of $EGFP^+$ cells at day 3)].

Usually the efficiency of reprogramming WT MEFs by this method is 0.7–1% of the transfected cells, similar to that previously described [22].

If FACS analysis is not available, another way to obtain an estimate of the reprogramming efficiency is to count iPS colonies per plate and refer to the total number of MEFs plated. In this case, the efficiency of the retrovirus infection is not taken into account and the efficiency should be around 0.3–0.6%.

Practically, 100% of the colonies reprogrammed from wild type MEFs retain ES-like morphology after passaging and can be established as iPS cell lines.

20.2.3. Analysis of iPS Clones. Pluripotency Assays

The main criteria for initial judgment of the pluripotency of isolated iPS clones is, as for ES cell clones, morphological criteria (Fig. 20.3). iPS colonies should look like ES cell colonies, round, compact, shiny and with well-defined edges. This morphology should be retained during iPS cell expansion. However further tests can be used to establish iPS cell pluripotency, both in vitro and in vivo. Here, we describe some of these tests. Karyotype analysis, although not a pluripotency test, is also included since it helps to eliminate those clones containing chromosomal aberrations that may compromise the pluripotency of the iPS clones.

20.2.3.1. Karyotype Analysis

Although not essential it is convenient to karyotype iPS cell lines. Spectral karyotyping (SKY) [36] is one of the best methods for karyotype analysis. However, it is expensive and requires sophisticated equipment and the expertise of a cytogeneticist. Giemsa staining of colcemid-treated cultures and metaphase counting is the method of choice for routine chromosome analysis without karyotyping [37].

In general, if early passage WT MEFS are used the frequency of clones with chromosomal aberrations is low (<10%); however, this frequency may increase for some genetically modified MEFs.

It is important to know the sex of the iPS clones in order to plan chimera production and breeding strategies if germ line transmission is required. If the reprogrammed MEFs derive from a male (XY) embryo, iPS clones obtained from those MEFs will also be XY. Checking for the presence of the Y chromosome by PCR can easily be done, using Y-chromosome-specific primers [38].

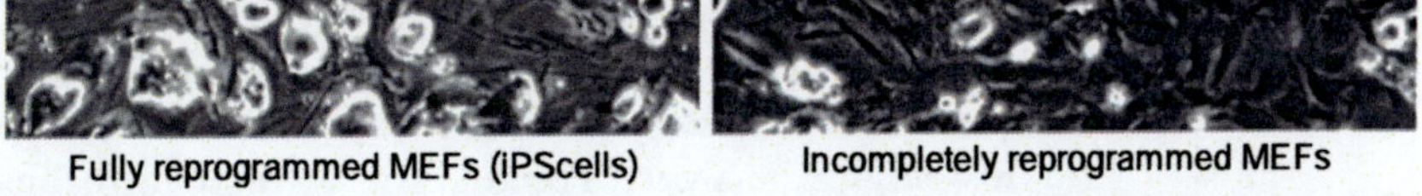

Fig. 20.3 Morphology is a good indication of iPS quality. (a) Culture of good-quality pluripotent iPS cells with the capacity to generate germ line transmitting chimeras. iPS colonies look homogeneous, round, and shiny with morphology similar to that of ES cells. (b) Culture of iPS cells that are not fully reprogrammed. *Arrows* indicate colonies with bad morphology: they appear flat, with no shiny edges and individual cells can be distinguished. These cells do not give rise to chimeras when injected into blastocysts although they may express some pluripotency genes and even give rise to teratomas.

20.2.3.2. Expression of Pluripotency Markers

Different markers can be used to assay iPS pluripotency in vitro. These are genes that are expressed in pluripotent ES cells but not in MEFs. Confirmation of the expression of these genes in iPS colonies is a good measure of pluripotency, especially if pluripotent ES cells are used as a control. However, these criteria are not absolute. Colonies not fully reprogrammed can still be positive for these markers.

Alkaline phosphatase expression can be monitored by cellular staining with available commercial kits (*see* Subheading 20.2.1.5).

SSEA-1 expression can be checked by immunofluorescence using anti-SSEA1 antibody (mouse monoclonal MC-480 antibody; Cell Signaling Technology; Diluted 1:100).

Expression of endogenous Oct3/4, Sox2, and Klf4 can be monitored by Western blot using antibodies anti-Oct3/4 (OCT3/4/H-134; rabbit polyclonal; Santa Cruz Biotechnology sc-9081; diluted 1:250), anti-Sox2 (rabbit polyclonal antibody; Millipore AB5603; diluted 1:250), and anti-Klf4 (rabbit polyclonal antibody GKLF (H-180); Santa Cruz Biotechnology sc20691; diluted 1:250) on RIPA-buffer extracts (see e.g. www.millipore.com for RIPA-buffer recipe and protein extraction protocol) resolved by electrophoresis in 4–12% NuPAGE Novex Bis–Tris precast gels (Invitrogen, Cat. No. NP0329BOX).

Endogenous expression of pluripotency markers such as Nanog and Sox2 can be also checked by reverse transcription and PCR analysis using the primer pairs described in Takahashi and Yamanaka [16].

20.2.3.3. Teratoma Formation

Like ES cells, iPS cells give rise to teratomas when injected into immune deficient mice. The iPS cells obtained from WT MEFs by the protocol described here generate teratomas in which cells derived from the three embryonic lineages, endoderm, mesoderm and ectoderm can be identified (Fig. 20.4).

Basically, once cultures of iPS cells are established, 2×10^6 cells are injected subcutaneously on the dorsal area of nude mice (Crl:Un(Ico)-Foxn1nu from Charles River). Cells growing on feeder cells or on gelatin-coated plates are trypsinized to a single cell suspension and counted. Cells are centrifuged at 1,000 rpm for 10 min at 4°C, and resuspended in PBS at a density of 20×10^6 cells per ml; 100 μl (2×10^6 cells) of the cell suspension are injected subcutaneously in nude mice. When the tumors reach approximately 0.5–1.0 cm of diameter, usually after 14–21 days, they are surgically removed from the skin and processed for histological analysis. Tumors are fixed in formalin at 4°C overnight, embedded in paraffin and sectioned at 3–5 mm. Sections are stained with hematoxylin/eosin or immunostained with antibodies against markers of ES cell differentiation. For basic immunostaining procedures, refer to www.millipore.com under "immunostaining techniques". Anti-neuronal nuclei antibody

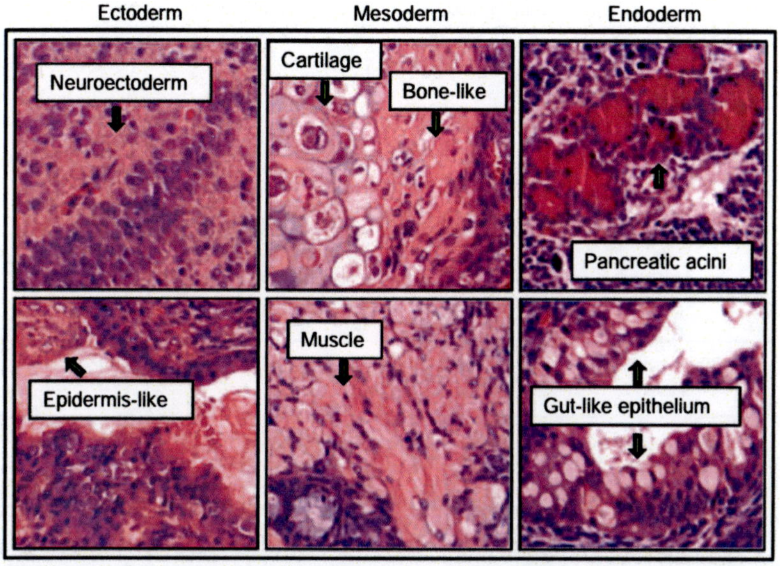

Fig. 20.4 Teratomas generated by MEFs reprogrammed by retroviral expression of Oct3/4, Sox2, and Klf4. iPS cells generated from MEFs reprogrammed by the protocol described here give rise to teratomas when injected subcutaneously in nude mice. The teratomas contain cells of the three different embryonic lineages: ectoderm, mesoderm, and endoderm. *Arrows* indicate areas of the teratomas where mesoderm-derived lineages such as cartilage, bone or muscle, ectoderm-derived lineages like epidermis or neuroectoderm or endoderm-derived pancreatic- or gut-like epithelial cells can be distinguished by hematoxylin/eosin staining of histological sections.

(NeUN; mouse monoclonal; Millipore MAB377; diluted 1:1500) is used for neuroectoderm staining; anti-cytokeratin-19 (CK-19 rat monoclonal antibody; Dev. Stu. Hybridoma Bank DSHB; supernatant, prediluted) is used for ectoderm staining; anti-common-muscle actin (mouse monoclonal HHF-35 antibody; Dako MO635; diluted 1:50) for mesoderm staining and anti-chymotrypsin (mouse monoclonal 4E1 antibody; ABD Serotec 2100-0657; diluted 1:25) for endoderm immunostaining.

20.2.3.4. Chimera Generation and Germ Line Contribution

The most stringent evaluation of iPS pluripotency is, as for ES cells, the testing of their capacity to contribute to the generation of germ line transmitting chimeras. The same techniques and protocols used to generate chimeras from ES cells can be used to generate chimeras from iPS clones. iPS clones reprogrammed from C57BL/6 MEFs can be tested for chimera contribution by aggregation with CD1 morulae or by microinjection into albino B6(Cg)-Tyr$^{c\text{-}2J}$/J blastocysts. In general, better results are obtained by microinjection but chimeras with high contribution of iPS cells can also be obtained by aggregation with outbred embryos. The reprogramming method described here results in the production of germ line competent iPS cells from wild type C57BL/6J MEFs (Fig. 20.5).

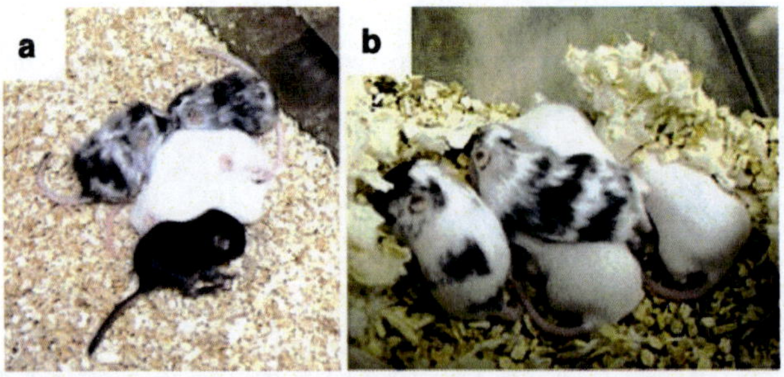

Microinjection in B6(Cg)-*Tyr*$^{c-2J}$/J blastocysts

iPS clone	Blastocysts Injected	Cells injected	Blastocysts transferred	Pups born	Chimeras Sex (% pigmentation)
iPS 3F-7	52	5-8	52	5	**1M(100%)**
iPS 3F-9	43	5-8	43	4	1F(60%)
iPS 4F-4	58	5-6	58	8	**2M (80%)** 1F(100%)

Aggregation with CD-1 morulae

iPS clone	Morulae aggregated	Cells aggregated	Embryos transferred	Pups born	Chimeras Sex (% pigmentation)
iPS 4F-3	171	4-8	171	15	1M(30%) 1M(5%) 1F(50%) 2F(30%)
iPS 718-1.3	71	5-15	66	13	2M (50%) 1M (10%)
iPS 62T.1	153	5-15	153	11	1M(100%) **1M(70%)** 1F(30%)

Fig. 20.5 Chimeras generated by microinjection and aggregation of reprogrammed MEFs. iPS cells generated by this protocol from WT MEFs contribute to chimera generation by microinjection into albino B6(Cg)-Tyrc-2J/J blastocysts (**a**) or by aggregation with CD1 morulae (**b**). In this case, the *black* coat color in chimeric animals indicate the level of iPS contribution. The tables below show representative sessions of iPS microinjection or aggregation with different iPS clones reprogrammed from WT MEFs. *M* male, *F* female. In *bold* are chimeras that transmitted in the germ line.

20.3 Troubleshooting

Problem: The number of iPS colonies after retroviral transfection is zero or very low.

Solution: This indicates a problem with the retroviral infection, expression of any of the transcription factors or both. Check the quality of the retroviral plasmid prep by routine agarose gel electrophoresis since good-quality DNA is essential for efficient transfection. Determine the plasmid DNA concentration by running in parallel a molecular weight marker (MWM) with a known concentration. Run several dilutions of the maxiprep and compare the intensity of the band with the intensity of the MWM after ethidium bromide staining of the gel and UV light exposure.

Check each plasmid pMX-Sox2, pMX-Oct4, or pMX-Klf4 separately for transcription factor expression by transient transfection of HEK293T cells (DNA + FuGENE, as described) with each one of them and checking Sox2, Oct4 and Klf4 expression by RT-PCR or Western blot, 48 h after transfection. If any of them does not express the corresponding factor, use another source of plasmid for that particular one. Check the packaging cell line by co-transfecting the plasmid pBabe-PURO-EGFP together with the plasmid pCLEco as described. Take the supernatant and infect MEFs with it. Count EGFP-positive cells or do FACs analysis. The efficiency of infection should be 70–80%.

Problem: The morphology of iPS colonies is bad.

Solution: Reprogramming is not complete. Check for endogenous transcription factors expression as described in Subheading 20.2.3.2 using as a control another fully reprogrammed iPS cell line or an ES cell line if available.

Check the quality of the MEF culture used for reprogramming. If the cells look senescent (large, flat, fried egg-morphology), or if they grow slowly probably MEFs have been in culture for too long or they went through a large number of cell divisions in culture before reprogramming. Prepare a new batch of MEFs as described and do not expand them beyond passage 2–3.

Problem: iPS cells do not contribute efficiently to chimeras.

Solution: Assuming that there is not a problem with the microinjection procedure itself, check your iPS culture conditions: quality of feeders, LIF and KSR, CO_2% and % humidity of your incubator. Check for potential mycoplasma contamination. Also check endogenous transcription factors expression as described above. Culture the iPS cells for a few more passages in good tissue culture conditions before injecting them.

20.4 Conclusions and Perspective

The generation of pluripotent cells from adult somatic cells by direct reprogramming is one of the most important achievements in stem cell science. Moreover, the methodology for reprogramming cells is relatively simple and independent of embryo or oocyte manipulation which makes it affordable and free of ethical or political restrictions. The protocol we describe here for reprogramming MEFs is still very similar to the one originally described by Kazutoshi Takahashi and Shinya Yamanaka in 2006 and involves retroviral transduction of three essential reprogramming transcription factors Oct3/4, Sox2, and Klf4. Our protocol generates iPS cells from MEFs at a rate close to 1% of the retrovirus transduced cells. These iPS cells fulfill all the criteria of pluripotency, including contribution to germ line chimeras.

However, for clinical use of patient-specific pluripotent stem cells retroviral-independent reprogramming protocols would be desirable, since the integration of retroviral DNA into the genome may result in unwanted genetic alterations. These alternative methods are already being explored, documented by an increasing number of publications. Some of these new approaches involve gene-free reprogramming by the use of small molecules that activate pluripotency genes or the use of genetic systems that do not require DNA integration (reviewed in [1]).

The molecular mechanisms underlying reprogramming by transcription factors remain largely unknown and this is an area for which reprogramming of MEFs will be extremely useful. iPS cells generated from MEFs derived from genetically modified mice will be an excellent tool for the understanding of the genetic basis of reprogramming and the establishment of pluripotency. Genes involved in different aspects of reprogramming can be identified using this approach and this information will help to develop more clinically oriented reprogramming procedures.

Many challenges need to be faced, also, with regard to the clinical applications for iPS cells. These would include finding good protocols for inducing iPS differentiation and delivery, understanding which type of pluripotent or multipotent cells are optimal for transplantation, large-scale production of iPS cells and even gene targeting in reprogrammed cells to rescue genetic defects in the original somatic cells. However, the speed at which this field has progressed in the last 3 years since Takahashi and Yamanaka published their seminal work in 2006 gives us hope that the therapeutic benefits derived from reprogrammed somatic cells are not very far away.

Acknowledgements

The authors thank Carmen Gómez, Marta Riffo, Jaime Muñoz, Maribel Muñoz and Rosa Serrano for their excellent assistance with tissue culture and mouse work, and Marta Cañamero for the histological analysis of teratomas. We also thank the Animal Facility and Comparative Pathology Units at the CNIO for their valuable technical support.

References

1. Hochedlinger K, Plath K (2009) Epigenetic reprogramming and induced pluripotency. Development 136:509–523
2. Evans MJ, Kaufman MH (1981) Establishment in culture of pluripotential cells from mouse embryos. Nature 292:154–156
3. Martin GR (1981) Isolation of a pluripotent cell line from early mouse embryos cultured in medium conditioned by teratocarcinoma stem cells. Proc Natl Acad Sci USA 78:7634–7638
4. Silva J, Smith A (2008) Capturing pluripotency. Cell 132:532–536

5. Ying QL, Wray J, Nichols J, Batlle-Morera L, Doble B, Woodgett J, Cohen P, Smith A (2008) The ground state of embryonic stem cell self-renewal. Nature 453:519–522
6. Jaenisch R, Young R (2008) Stem cells, the molecular circuitry of pluripotency and nuclear reprogramming. Cell 132:567–582
7. Murry CE, Keller G (2008) Differentiation of embryonic stem cells to clinically relevant populations: lessons from embryonic development. Cell 132:661–680
8. Buehr M, Meek M, Blair K, Yang J, Ure J, Silva J, McLay R, Hall J, Ying QL, Smith A (2008) Capture of authentic embryonic stem cells from rat blastocysts. Cell 135:1287–1298
9. Li P, Tong C, Mehrian-Shai R, Jia L, Wu N, Yan Y, Maxson RE, Schulze EN, Song H, Hsieh CL, Pera MF, Ying QL (2008) Germline competent embryonic stem cells derived from rat blastocysts. Cell 135:1299–1310
10. Thomson JA, Itskovitz-Eldor J, Shapiro SS, Waknitz MA, Swiergiel JJ, Marshall VS, Jones JM (1998) Embryonic stem cell lines derived from human blastocysts. Science 282:1145–1147
11. Wilmut I, Schnieke AE, McWhir J, Kind AJ, Campbell KH (1997) Viable offspring derived from fetal and adult mammalian cells. Nature 385:810–813
12. Tada M, Takahama Y, Abe K, Nakatsuji N, Tada T (2001) Nuclear reprogramming of somatic cells by in vitro hybridization with ES cells. Curr Biol 11:1553–1558
13. Cowan CA, Atienza J, Melton DA, Eggan K (2005) Nuclear reprogramming of somatic cells after fusion with human embryonic stem cells. Science 309:1369–1373
14. Hochedlinger K, Jaenisch R (2006) Nuclear reprogramming and pluripotency. Nature 441:1061–1067
15. Geoghegan E, Byrnes L (2008) Mouse induced pluripotent stem cells. Int J Dev Biol 52:1015–1022
16. Takahashi K, Yamanaka S (2006) Induction of pluripotent stem cells from mouse embryonic and adult fibroblast cultures by defined factors. Cell 126:663–676
17. Tokuzawa Y, Kaiho E, Maruyama M, Takahashi K, Mitsui K, Maeda M, Niwa H, Yamanaka S (2003) Fbx15 is a novel target of Oct3/4 but is dispensable for embryonic stem cell self-renewal and mouse development. Mol Cell Biol 23:2699–2708
18. Okita K, Ichisaka T, Yamanaka S (2007) Generation of germline-competent induced pluripotent stem cells. Nature 448:313–317
19. Maherali N, Sridharan R, Xie W, Utikal J, Eminli S, Arnold K, Stadtfeld M, Yachechko R, Tchieu J, Jaenisch R, Plath K, Hochedlinger K (2007) Directly reprogrammed fibroblasts show global epigenetic remodeling and widespread tissue contribution. Cell Stem Cell 1:55–70
20. Wernig M, Meissner A, Foreman R, Brambrink T, Ku M, Hochedlinger K, Bernstein BE, Jaenisch R (2007) In vitro reprogramming of fibroblasts into a pluripotent ES-cell-like state. Nature 448:318–324
21. Cherry SR, Biniszkiewicz D, van Parjis L, Baltimore D, Jaenisch R (2000) Retroviral expression in embryonic stem cells and hematopoietic stem cells. Mol Cell Biol 20:7419–7426
22. Blelloch R, Venere M, Yen J, Ramalho-Santos M (2007) Generation of induced pluripotent stem cells in the absence of drug selection. Cell Stem Cell 1:245–247
23. Meissner A, Wernig M, Jaenisch R (2007) Direct reprogramming of genetically unmodified fibroblasts into pluripotent stem cells. Nat Biotechnol 25:1177–1181
24. Nakagawa M, Koyanagi M, Tanabe K, Takahashi K, Ichisaka T, Aoi T, Okita K, Mochiduki Y, Takizawa N, Yamanaka S (2008) Generation of induced pluripotent stem cells without Myc from mouse and human fibroblasts. Nat Biotechnol 26:101–106
25. Wernig M, Meissner A, Cassady JP, Jaenisch R (2008) c-Myc is dispensable for direct reprogramming of mouse fibroblasts. Cell Stem Cell 2:10–12
26. Marion RM, Strati K, Li H, Tejera A, Schoeftner S, Ortega S, Serrano M, Blasco MA (2009) Telomeres acquire embryonic stem cell characteristics in induced pluripotent stem cells. Cell Stem Cell 4:141–154
27. Yu J, Vodyanik MA, Smuga-Otto K, Antosiewicz-Bourget J, Frane JL, Tian S, Nie J, Jonsdottir GA, Ruotti V, Stewart R, Slukvin II, Thomson JA (2007) Induced pluripotent stem cell lines derived from human somatic cells. Science 318:1917–1920
28. Li H, Collado M, Villasante A, Strati K, Ortega S, Cañamero M, Blasco MA, Serrano M (2009) The Ink4/Arf locus is a barrier for iPS cell reprogramming. Nature 460:1136–1139
29. Marion RM, Strati K, Li H, Murga M, Blanco R, Ortega S, Fernández-Capetillo O, Serrano M, Blasco MA (2009) A p53-mediated DNA damage response limits reprogramming to ensure iPS cell genomic integrity. Nature 460:1149–1153
30. Hong H, Takahashi K, Ichisaka Y, Aoi T, Kanagawa O, Nakagawa M, Okita K, Yamanaka S (2009) Suppression of induced pluripotent stem cell generation by the p53-p21 pathway. Nature 460:1132–1135
31. Kawamura T, Suzuki J, Wang YV, Menéndez S, Morera LB, Raya A, Wahl GM,

Izpisúa-Belmonte JC (2009) Linking the p53 tumour suppressor pathway to somatic cell reprogramming. Nature 460:1140–1144

32. Utikal J, Polo JM, Stadtfeld M, Maherali N, Kulalert W, Walsh RM, Khalil A, Rheinwald JG, Hochedlinger K (2009) Immortalization eliminates a roadblock during cellular reprogramming into iPS cells. Nature 460:1145–1148
33. Kitamura T, Koshino Y, Shibata F, Oki T, Nakajima H, Nosaka T, Kumagai H (2003) Retrovirus-mediated gene transfer and expression cloning: powerful tools in functional genomics. Exp Hematol 31:1007–1014
34. Naviaux RK, Costanzi E, Haas M, Verma IM (1996) The pCL vector system: rapid production of helper-free, high-titer, recombinant retroviruses. J Virol 70:5701–5705
35. Fulwyler MJ (1980) Flow cytometry and cell sorting. Blood 6:173–184
36. Schröck E, Veldman T, Padilla-Nash H, Ning Y, Spurbeck J, Jalal S, Shaffer LG, Papenhausen P, Kozma C, Phelan MC, Kjeldsen E, Schonberg SA, O'Brien P, Biesecker L, du Manoir S, Ried T (1997) Spectral karyotyping refines cytogenetic diagnostics of constitutional chromosomal abnormalities. Hum Genet 101:255–262
37. Nagy A, Gertsenstein M, Vintersten K, Behringer R (2003) Manipulating the mouse embryo, a laboratory manual, 3rd edn. Cold Spring Harbor Laboratory Press, Cold Spring Harbor, NY
38. Koopman P, Gubbay J, Vivian N, Goodfellow P, Lovell-Badge R (1991) Male development of chromosomally female mice transgenic for Sry. Nature 351:117–121

Chapter 21

The Preparation and Analysis of DNA for Use in Transgenic Technology

Anna B. Auerbach, Peter J. Romanienko, and Willie H. Mark

Abstract

The techniques of pronuclear and blastocyst injection are now routine procedures for making transgenic and gene-targeted mice, respectively. For many institutions, Core Facilities are the standard venue for producing such genetically modified animals for use in both basic and biomedical research. While the overall approach to microinjection has changed little over the years, the methodologies for the preparation of DNA for pronuclear injection and electroporation into embryonic stem (ES) cells has seen considerable improvement. Most significantly, procedures for DNA purification have been much simplified. In addition, rapid and cost efficient protocols are now available for genotyping of transgenic, gene-targeted, and mutant mice. In this chapter, we describe protocols that we employ to purify DNA for transgenic and gene-targeting work, as well as procedures for the preparation of DNA for both genotyping mice by PCR and by Southern blot analysis. The major goal behind the revision of these techniques has been to reduce the time and labor needed in order to obtain high quality DNA in sufficient quantities for specific procedures.

Abbreviations

BAC	Bacterial artificial chromosome
BSA	Bovine serum albumin
DNA	Deoxyribonucleic acid
dNTPs	Deoxyribonucleotide triphosphates
EB	Qiagen elution buffer
ES	Embryonic stem
EU	Endotoxin unit
kb	Kilobase
LB	Luria-Bertani
mM	Millimolar
PCR	Polymerase chain reaction
PE	Qiagen wash buffer
pg	Picogram
pl	Picoliter
pmol	Picomole
QG	Qiagen solubilization buffer

S. Pease and T.L. Saunders (eds.), *Advanced Protocols for Animal Transgenesis*, Springer Protocols,
DOI 10.1007/978-3-642-20792-1_21, © Springer-Verlag Berlin Heidelberg 2011

RFLPs	Restriction fragment length polymorphism
RNA	Ribonucleic acid
RNase	Ribonuclease
SDS	Sodium dodecyl sulfate
SSC	Saline sodium citrate
TBE	Tris–borate–EDTA
WT	Wild type

21.1 Introduction

All DNA for introduction into live cells, either zygotes or embryonic stem (ES) cells, should meet certain criteria, by being:

- Free of chemical and particulate contaminants
- Of high physical quality (not degraded)
- At a concentration well tolerated by cells
- In a buffer that will not impede or interfere with ES cell growth and embryonic development

In the sections below, we describe protocols for harvesting and purification of DNA that will result in products that meet the above criteria and be of a quality suitable for pronuclear injection and electroporation into ES cells.

21.1.1. Preparation of DNA for Pronuclear Microinjection and ES Cell Electroporation

The quality of a purified DNA construct and its concentration are critical factors for efficient introduction into mouse zygotes for production of transgenic animals, [1] or ES cell gene targeting [2].

For the cell to survive and divide following introduction of a foreign DNA, the DNA solution must be free of chemical contaminants as well as particulate material often introduced during DNA preparation. (See list of Contaminants and Methods for Their Removal in [3].) Of equal importance is the concentration of DNA used for either pronuclear injection of zygotes or ES cell electroporation. It is critical that the concentration of DNA is accurately measured, particularly if the DNA is to be introduced into mouse zygotes, where the concentration is typically between 2 and 5 ng/μl [1, 4]. Depending on the size of the DNA construct, this corresponds to hundreds of copies of the transgene DNA molecule per pl. The rates at which transgenic mice are produced are lowered when higher concentrations of DNA are used. At 10 ng/μl, numbers of transgenic founders produced will be fewer than at 5 and 7.5 ng/μl, suggesting that too much injected DNA may have a detrimental effect on embryo survival. This may relate to the DNA concentration itself and/or the presence of contaminants in the solution [1, 5, 6]. Van Keuren and colleagues [7] performed a retrospective analysis of the results of microinjection of BAC-based transgenic constructs. A lower and narrower (0.1–0.5 ng/μl) BAC DNA concentration range produced an optimal balance of embryo survival, birth rates and the rate at which transgenic pups were produced. Eggs injected with BAC DNA molecules are even more sensitive to high DNA concentrations than those injected with higher concentrations of small plasmid-based transgenes.[1]

To generate transgenic mice efficiently, the DNA introduced into the zygote/cell must retain its integrity. Often, degradation of purified transgene DNA is caused by improper solution formulation and storage conditions. On an agarose gel, DNA degradation is readily visible as a smear below the band containing the linearized vector. Its presence will impair determination of DNA concentration. Currently, most transgenic facilities integrate strong quality control into their protocols and many of them offer transgene purification as a service.

The quality of the DNA used for electroporation is also important for gene targeting in ES cells. Contaminants might affect survival of electroporated ES cells, as well as lower targeting efficiency. Removal of endotoxins during DNA preparation and the use of phase-lock tubes (separating water from organic phase) during phenol extraction can be helpful. When using PBS-based electroporation buffers, it takes longer to uniformly dissolve DNA than in Tris–EDTA (TE) solution. The amount of time taken would depend on vector size and concentration. It is important to check the concentration, eventually adjust it, and run a gel to verify completion of digestion, as well as the integrity of the sample before electroporation. Again, a significant smear of DNA below the band containing the linearized vector will affect determination of DNA concentration and ultimately may lower targeting efficiency, since the smear (representing smaller DNA molecules) may contain some molecules which have shorter arms of homology and will more likely integrate only by random insertion, thereby increasing the background of non-homologous integration events.

21.1.2. Preparation of DNA for Genotyping

There are many methods for the preparation of genomic DNA from mouse tissues for genotyping. Preparation procedures should be tailored to the genotyping assay performed, as the quality and quantity of the DNA preparation can vary depending upon the procedure used. Also, the amount of time and cost of preparation should be considered. For simple PCR genotyping, small amounts of relatively impure template are adequate, whereas Southern blotting requires intact, clean genomic DNA that can be readily digested by restriction enzymes and is of a quantity suitable for multiple assays and archiving.

[1]One of the reasons might be an overall high osmolarity of the BAC DNA solution containing polyamines. Zygotes start to shrink in solutions with osmolarity greater than 300 mOsm/L [22] and BAC DNA solution might have even higher osmolarities (personal observation).

For PCR genotyping, a simple, cost-effective method of template preparation is the HotSHOT (Hot Sodium Hydroxide and Tris) method [8]. The DNA is less pure and more dilute, but that is countered by using a suitably efficient Taq polymerase and optimized PCR conditions. The method described herein is not suitable for genotyping by Southern blot analysis because the DNA concentration is too low for setting up convenient restriction digestion. Furthermore, the DNA is denatured and more degraded when compared to other methods of preparation. Nonetheless, this type of genomic DNA serves well as a template for PCR genotyping and is very convenient and reliable in our hands for mid-throughput genotyping. For low-to-medium throughput PCR genotyping, it is convenient to consolidate existing allele-specific reactions into a small number of set parameters through optimization, rather than perform many different independent sets of reactions. Of the 300+ assays we use, 90% can be run in either of two cycling parameters, where the difference is simply primer annealing temperature. A compromise annealing temperature is found by running an annealing gradient where template amount, number of cycles and primer concentration remain constant and a range of primer annealing temperatures are tested. This leads to more efficient use of laboratory equipment, personnel, and disposable plastic ware.

For Southern blotting, tail tissue DNA is obtained using a standard method of Proteinase K digestion followed by phenol/chloroform extraction. The resulting DNA is of high quality and quantity, and is suitable for most postpreparation procedures. Again, in Southern blotting or any routine laboratory procedure, we need to find a balance between quality, consistency, and convenience. Although Southern blotting is less commonly used than PCR in routine genotyping, it is still the standard method for identification and analysis of transgenic founder mice, confirming gene targeting in ES cells and further confirmation of targeting in the first generation of offspring from germline chimeras. As it is a lengthy procedure requiring days to obtain results for an average of 30–40 samples per blot, and as there are multiple steps and variables in the entire procedure from quality of genomic DNA, to transfer efficiency and radiolabeling of probes, to name a few, great care should be taken while performing each procedural step.

21.2 Preparation of Construct for Gene Targeting in ES Cells

Cesium chloride gradient centrifugation is the standard approach for the preparation of gene-targeting vector DNA, but it is time consuming and labor intensive. In our Core we use a more efficient method, based on the alkaline lysis approach, to isolate plasmid DNA followed by phenol chloroform extraction. The method provided here is simple and effective, and provides a good yield of clean DNA for electroporation of embryonic stem cells.

Large-scale preparation procedure consists of the following steps:

Culture

Harvest and lysis of the bacterial cells carrying the plasmid of interest

Plasmid purification with commercial kit and linearization

Phenol/chloroform extraction, and finally

Evaluation (quality and quantity) of the purified targeting construct

For DNA plasmid purification we recommend the Qiagen Plasmid Maxi Kit (Cat. No. 12162) and *see* Subheading 21.4.3, (modified from [9]). Alternatively, one may choose to use the Qiagen EndoFree Plasmid Maxi Kit (Cat. No.12362) to obtain an endotoxin-free DNA preparation. These are gravity-flow, anion-exchange-based kits for the purification of plasmid DNA. Bacterial lysate is cleared by centrifugation and applied to the *Qiagen*-column under low-salt conditions, which enables the plasmid DNA to bind to the column matrix. After extensive washing, the bound DNA is eluted with high salt buffer and recovered by isopropanol precipitation and centrifugation. In the case of the EndoFree Maxi Kit, cell lysate is cleared by filtration using QIA-filter Cartridges, endotoxin removal buffer is added and the solution is incubated before applying it to the ion-exchange column. The endotoxin removal step often yields plasmid DNA with less than 0.1 EU/μg DNA. Both Qiagen Plasmid Maxi Kits provide DNA yields from 100 to 500 μg when 100–500 ml of bacterial culture is used.

21.2.1. Materials

First, determine a Screening Strategy by reviewing the targeting construct and the physical map of the genetic locus after the predicted homologous recombination reaction.

21.2.1.1. Physical Maps of the Genomic Locus

A physical map of the wild-type locus to be targeted and a map of the expected targeted allele will be needed. The maps should clearly show the restriction sites and the location of the DNA probe(s) that will be used for screening. In addition, the map should show the exon/intron structure of the targeted gene and the promoter, if appropriate.

21.2.1.2. Physical Map of the Targeting Vector

A physical map of the targeting vector should clearly show both positive and negative selection markers, the flanking genomic

DNA sequences, and any loxP or FRT sites. A scheme of the targeting strategy and the predicted outcome should be determined at the outset of any targeting project. DNA probes intended for use in screening of targeted clones should be tested ahead of time to confirm that the vector is constructed as intended.

21.2.1.3. Targeting Construct

A minimum of 80 μg of linearized DNA at a concentration 0.5–1.0 μg/μl is prepared using a Qiagen Plasmid Maxi Kit (Cat. No. 12162), ethanol precipitated, and left in ethanol until ready to use for electroporation. On the day of electroporation, the DNA is collected by centrifugation and resuspended in sterile PBS without Ca and Mg in a laminar flow hood to maintain sterility of the DNA (see procedure below). Alternatively, purified DNA maybe stored in PBS or TE at 4°C for a couple of weeks before electroporation.

21.2.1.4. Picture of the Targeting Construct Digest

Run a small amount of the linearized DNA on an agarose gel to verify its purity and concentration. The plasmid DNA should also be digested with a few common restriction enzymes such as *Eco*RI, *Bam*HI, *Bgl*II, *Pst*I, or *Xba*I to be sure that linearization is complete and to demonstrate that the targeting vector is constructed correctly as evidenced by the anticipated restriction enzyme fragments.

21.2.1.5. DNA Probe for Screening

Prepare at least 100 ng of 250–300 bp DNA probe(s) to screen the selected clones. Probe(s) should ideally reside in a region *outside* of the homologous arms of the targeting vector (external probe). An autoradiograph of a Southern blot of mouse genomic DNA digested by restriction enzymes hybridized with the DNA probe should also be obtained to verify that the screening strategy might be utilized for identification of targeted ES cell clones.

21.2.1.6. Equipment and Reagents

Incubator shaker.

Sorvall High Speed Centrifuge (or Beckman) with cooling.

Rotor SLA-1500 for 250 ml or SLA-3000 for 500 ml bottles.

Rotor SS-34 for Oak Ridge tubes.

Microcentrifuge.

Spectrophotometer.

Restriction enzymes, if appropriate.

21.2.1.6.1. Solutions and Supplies

Qiagen plasmid DNA purification kit (e.g., Cat. No. 12162).

LB broth EZMix powder (Sigma-Aldrich, Cat. No. L7658) or Terrific Broth (to improve yield in slowly growing plasmid-bearing bacteria) (Sigma-Aldrich, Cat. No. T9179 or T5574) supplemented with an appropriate antibiotic in all steps.

LB Plates with selective antibiotic, 100 mm (Fisher Scientific, Cat. No. 08-757-12), stored at 4°C covered with aluminum foil.

Centrifuge bottles (Thermo Fisher Scientific, Cat. No. 08142–250 ml, or Cat. No. 7149–500 ml).

Selective antibiotics:

Antibiotic stock	Concentration storage	Working concentration (dilution)
Ampicillin (sodium salt)	50 mg/ml in water (500×)	100 μg/ml (2 μl of stock/ml)
Chloramphenicol	34 mg/ml in ethanol (200×)	170 μg/ml (5 μl of stock/ml)
Kanamycin	25 mg/ml in water (500×)	50 μg/ml (2 μl of stock/ml)

Stocks stored at −20°C.

TE buffer, pH 8.0, or 10 mM Tris–Cl, pH 8.5.

Phenol/chloroform/isoamyl alcohol (25:24:1) (Fisher Scientific, BP1752I-400) pH 7.5.

Sodium acetate 3 M, pH 5.2 (Sigma-Aldrich, Cat. No. S7899).

Erlenmeyer flasks.

Oak Ridge tubes, FEP 50 ml with caps (Nalgene, Cat. No. 3114–0050 or PP).

Eppendorf Safe-Lock tubes 1.5–2.0 ml (Eppendorf, Cat. No. 022363204 - 022363352).

Electroporation buffer (Millipore, Cat. No. ES-003-D).

Isopropanol.

Ethanol.

Glycerol, frozen.

Pipette tips.

20 ml flask.

Parafilm.

Notes: Maximum Recommended Culture Volumes for QIAGEN-tip 500:

High-copy plasmids (50–500 copies per bacteria cell) – 100 ml.

Low-copy plasmids (1–10 copies per cell) – 500 ml.

Low-copy plasmids that have been amplified in the presence of chloramphenicol should be treated as high-copy plasmids.

The expected plasmid DNA yields are 300–500 μg for high-copy plasmids and 100–500 μg for low-copy plasmids. If working with low-copy vectors, it may be beneficial to increase the lysis buffer volumes in order to increase the efficiency of alkaline lysis, and thereby the DNA yields. Also, using a rich broth for bacterial growth such as Terrific Broth can increase the density of cells and thus the yield of plasmid.

21.2.2. Procedure

21.2.2.1. Bacterial Culture

- Prepare agar plates using LB or other culture medium and when agar is cool to the touch, add antibiotic.
- Scrape off a small piece of frozen bacterial glycerol stock using a sterile pipette tip (do NOT allow glycerol stock to thaw!) and streak on the agar plate. Incubate overnight at 37°C.
- Select a single colony and transfer, using a pipette tip, into 5 ml of LB or other bacterial culture broth with antibiotic. Use a 20-ml flask. Wrap the plate with parafilm and store in the fridge (4°C) for the short term (days rather than weeks). Culture with shaking for ~8 h (~300 rpm) at 37°C on an incubator shaker.
- Transfer 0.2 ml (100 ml) or 1 ml (500 ml cultures) of your *starter culture* into LB or other culture broth with antibiotic. Inoculate 100 ml (high-copy plasmids) to 500 ml (low-copy plasmids) of medium with bacteria from the starter culture. Use 500 ml or 2,000 ml Erlenmeyer flask for the culture. Grow overnight (12–16 h) at 37°C with vigorous shaking.

21.2.2.2. Harvesting of Bacterial Cells

- *Centrifuge* at 6,000 × *g* (5,000 rpm) for 15 min at 4°C.
- Decant supernatant; turn tube upside down on paper towels to remove as much of the liquid as possible.
- Weigh and record the pellet mass and the culture volume. Store at −20°C if not continuing immediately to the next step. (Collection should be ~3 g/l.)

21.2.2.3. Preparation of Solutions

Centrifuge briefly the RNase A solution (supplied in Qiagen kit in a 10 mg/ml or 100 mg/ml solution) and add to *buffer P1* (resuspension buffer also supplied) before use, to a final concentration of 100 μg/ml. Check *buffer P2* (lysis buffer provided in kit). If Sodium Dodecyl Sulfate (SDS) precipitation is visible, dissolve it by warming at 37°C.

Prechill *buffer P3* (neutralization buffer provided in kit) to 4°C.

Optional: Add the LyseBlue reagent (provided in kit) to buffer P1 and mix before use. Use one vial LyseBlue (centrifuge briefly before use) per bottle of P1 buffer to achieve a 1:1,000 dilution. LyseBlue provides visual identification of optimum buffer mixing thereby preventing the common handling errors that lead to inefficient cell lysis and thus a decrease in yield of plasmid DNA.

21.2.2.4. Resuspension of Bacterial Cells

- Resuspend the bacterial pellet from 21.2.2.2 in 10 ml of *buffer P1* containing RNase A.
- Pipette up and down until suspension is homogenous and no cell clumps are visible. Transfer DNA solution to 50-ml Oak Ridge tubes.

- If LyseBlue reagent is used, vigorously shake the buffer bottle to ensure LyseBlue particles are completely dissolved.

21.2.2.5. Lysis of Bacterial Cells

- Add 10 ml of *buffer P2*, mix thoroughly by gently inverting until the lysate is clear and uniform (Do not vortex!), and incubate at room temperature for 5 min. The lysate should appear viscous when lysis is complete. **Note**: The bottle of buffer P2 should be closed immediately after use to avoid acidification due to exposure to air.
- *If LyseBlue is used, the cell suspension will turn blue upon addition of Buffer P2. Mixing should result in a homogeneously blue-colored suspension. If the uneven colorization is observed or if brownish cell clumps are still visible, continue to mix the solution until a homogeneously colored suspension is achieved.*

21.2.2.6. Neutralization of Bacterial Lysate

- Add 10 ml of chilled *buffer P3*, gently invert tubes four to six times to ensure complete mixing of SDS with cell lysate. Any SDS remaining in solution can inhibit the binding of DNA to the column. Incubate on ice for 20 min.

After addition of buffer P3, a fluffy white precipitate forms and the lysate becomes less viscous. The precipitated material contains bacterial chromosomal DNA, proteins, and other cell debris. The lysate should be mixed thoroughly to ensure complete precipitation of all chromosomal DNA and cellular debris. If the mixture still appears viscous, more mixing is required to completely neutralize the solution.
Note: If LyseBlue reagent is used, the suspension should be mixed until all trace of blue disappears and the suspension is colorless.

21.2.2.7. Clearing of Bacterial Lysate

- *Centrifuge* at ≥20,000 × *g* (15,000 rpm) for 30 min at 4°C.
- Remove the clear supernatant, which contains the plasmid DNA.
- Immediately transfer the supernatant to a clean Oak Ridge tube and centrifuge at ≥20,000 × *g* (15,000 rpm) for 15 min at 4°C to remove any trace amount of contaminating debris (*see* **note** below).
- *Remove supernatant promptly.*

Note: This second centrifugation serves to remove any particular material that may clog the QIAGEN-tip column thereby reducing or even preventing gravity flow.

- *Optional: In order to determine whether growth and lysis conditions were optimal, take a 120 μl aliquot from the cleared lysate supernatant and save for an analysis by gel electrophoresis (sample 1).*

21.2.2.8. Binding of Plasmid DNA to Ion-Exchange Column

- *Equilibrate a QIAGEN-tip by applying 10 ml of buffer QBT* (equilibration buffer, supplied in the kit) and allowing the column to empty by gravity flow. Allow the column to drain completely before applying DNA to the column.
- *Apply the supernatant from Subheading 21.3.2.7 to the column* and allow it to enter the resin by gravity.

The supernatant should be loaded onto the column *promptly.* If it is left too long and becomes cloudy, it must be centrifuged again to remove any precipitate to avoid clogging of the column.

- *Optional: In order to determine the efficiency of DNA binding to the QIAGEN resin, take a 120 μl aliquot from the flow-through and save for analysis by gel electrophoresis (sample 2).*

21.2.2.9. Elution of Plasmid DNA from Column

- After the plasmid DNA solution has passed through the column, *wash the column twice with 30 ml of buffer QC* (wash buffer, supplied), which is the maximum volume the column can hold.
- *Optional: In order to monitor the quality of this washing step, take a 240 μl aliquot from the combined wash fractions and save for an analysis by gel electrophoresis (sample 3).*
- *Add 15 ml buffer QF* (elution buffer, supplied) to the column after the last wash and collect the eluate into a clean 50 ml polypropylene Oak Ridge tube(s).

Note: For constructs larger than 45–50 kb, prewarming the elution buffer to 65°C may help to increase the yield.

- *Optional: In order to check DNA concentration of the eluate, take a 60 μl aliquot of the collected solution and save for an analysis by gel electrophoresis (sample 4).*
- Store the eluate at 4°C or proceed to the next step. *Storage at 4°C longer than overnight is not recommended.*

Note: Re-equilibrated columns can be used more than once for the same plasmid.

21.2.2.10. Precipitation of Eluted DNA

- *Add 10.5 ml of isopropanol* to the eluted DNA at room temperature (to minimize co-precipitation of salts and RNA). Mix well by shaking.
- *Centrifuge* immediately at ≥15,000 × *g* (~9,600 rpm) for 30 min at 4°C.
- *Carefully decant the supernatant.*

Note: The DNA pellets have a smooth, clear appearance and thus may be difficult to see. They are also loosely attached to the bottom of the tube so care must be used in manipulating them. Marking the tube prior to centrifugation may help in finding the pellet.

Resuspend the pellet in 700 μl TE and transfer to an Eppendorf microtube. Precipitate again with an equal volume of isopropanol and 1/10 volume of 3 M sodium acetate.

21.2.2.11. Washing the Plasmid DNA Pellet

- Add 1.0 ml of 70% ethanol at room temperature and centrifuge for 10 min at high speed in a microcentrifuge.
- Carefully decant the supernatant without disturbing the pellet, and carefully remove the remaining ethanol using a pipetman.

The 70% ethanol wash removes the salt introduced in the precipitation of the plasmid DNA and replaces isopropanol with the more volatile ethanol.

- *Repeat the wash.*
- *Air-dry the pellet.* This usually takes 5–10 min at room temperature.

Over-drying the pellet will make the DNA difficult to re-dissolve especially for high molecular weight DNA samples.

- *Re-dissolve the DNA* in a suitable volume (typically 100–200 μl) of *TE buffer* (pH 8.0) at room temperature for 1 h gently mixing occasionally or at 4°C overnight.

Mix well and rinse the walls to ensure complete recovery of the DNA.
Note: Handle DNA gently. Avoid pipetting the DNA up and down, since this may cause shearing. Flicking the tube with fingers works well.

- After resuspension, *spin DNA* in a microfuge for 5 min to remove any particular matter and transfer the DNA to a clean, sterile microfuge tube.

21.2.2.12. Determination of Yield of DNA

- To determine the yield of the recovered DNA, use a *spectrophotometer.* For reliable spectrophotometric readings, the A260 OD values should be between 0.1 and 1.0.

Optional: Removing and saving aliquots during the purification procedure (samples 1–4) is helpful in trouble-shooting. Samples can be analyzed by agarose gel electrophoresis to determine if any unanticipated DNA loss occurred during the purification procedure.

21.2.2.13. Digestion of Plasmid DNA

To linearize the targeting vector in preparation for electroporation, *digest 150 μg of the plasmid DNA* with an appropriate restriction enzyme(s) in three Eppendorf microtubes.

- *Prepare digest mix:*
  - 50 μg DNA
  - 30 μl Buffer appropriate for the enzyme

- 30 μl 10× BSA
- 5 μl Enzyme (20 units/μl)
- Add ddH_2O to a total volume of 300 μl/tube

- *Mix well and incubate* for 4 h to overnight at 37°C.
- Run 2 μl of DNA on an agarose gel to make sure that the *digestion is complete.*

21.2.2.14. Purify the Linearized DNA by Phenol/Chloroform Extraction

After digestion, the DNA must be purified by phenol/chloroform extraction. This step is necessary to remove proteins from nucleic acids. *Wear gloves and work in a chemical fume hood.*

- *Add 300 μl phenol/chloroform/isoamyl alcohol* to each tube and shake to mix well, usually about 5–7 min.
- *Spin tubes* in a microfuge at maximum speed for 5 min to separate the phases.
- *Remove the aqueous phase to a new tube*, being careful not to transfer any of the protein from the interface.
- Repeat the extraction *two more times.*
- *Add an equal volume of chloroform* (no phenol) or chloroform/isoamyl alcohol (24:1) and repeat extraction twice to remove any traces of phenol.
- *Add 1/10th volume of 3 M sodium acetate* (pH 5.2) to each tube. This gives a final concentration of ~0.25 M sodium acetate. Mix well by shaking.
- Add 2 volumes of *ice-cold 100% ethanol.* Mix well by shaking and leave in a −20°C freezer for 1 h or overnight.

21.2.2.15. Evaluation of the Purified Targeting Vector

- It is recommended to use two methods, *spectrophotometry* or fluorometry and *agarose gel electrophoresis,* to determine the yield, purity, and integrity of the vector DNA.
- On the day of electroporation, resuspend the DNA in 1 ml 70% ethanol, to provide 80 μg or more of linearized DNA at a concentration 0.5–1.0 μg/μl

21.2.2.15.1. Just Before Electroporation

- Spin in a microfuge to collect DNA. In a sterile environment, laminar flow hood, open the tube to decant the ethanol, dry in hood, and resuspend in sterile PBS (electroporation buffer). Mix well by shaking and incubate at 50°C to ensure all the DNA is in solution.
- Take a small aliquot to determine quality and quantity of DNA as stated above.

21.3 Protocol for Transgene DNA Preparation for Microinjection of Mouse Zygotes

The quality of the transgene DNA is a critical factor in the efficient production of transgenic mice. Poor quality DNA can influence virtually every step of transgenic mouse production, from lysis of injected zygotes to the impairment of postimplantation development of the manipulated eggs [10]. In addition, particulate contaminants will clog microinjection needles causing difficulties during zygote microinjection. DNA submitted by different laboratories to a Core facility for injection can differ significantly in quality even if the same protocol is used to prepare the transgene DNA. This likely reflects individual differences in the performance of experimental procedure and in the handling of the materials used for DNA purification. Therefore, our Core strongly recommends that investigators use the DNA Purification Service offered by our molecular biology group.

The basic steps for transgene DNA isolation and purification are as follows:

Propagation of bacteria containing the plasmid of interest

Plasmid DNA purification

Restriction digest of plasmid DNA to release the transgene DNA fragment

Separation of transgene DNA from the bacterial vector by Agarose Gel Electrophoresis

Gel extraction and purification of transgene DNA

Precipitate and wash in ethanol, and finally

Analysis and preparation of transgene DNA for microinjection

Our Core uses the QIAquick Gel Extraction Kit to purify the transgene DNA insert after completion of gel electrophoresis. Briefly, the procedure consists of *(a) running of the preparatory agarose gel, (b) excision and solubilization of gel slices containing the transgene DNA, (c) binding of the DNA insert to the column, (d) washing and elution of the DNA from the column, and (e) evaluation of the DNA quality and concentration.* QIAquick Gel Extraction Kit (Qiagen Cat. No. 28704) contains spin columns for silica-membrane-based purification of DNA fragments from agarose, three buffers – one of them containing a chaotropic[2] salt and collection tubes. DNA adsorbs to the silica-membrane under acidic conditions and in the presence of a highly concentrated chaotropic salt [11, 12] such as guanidine thiocyanate (buffer

[2]Chaotropic agents like guanidine and urea are capable of increasing the solubility of some organic molecules in water by weakening hydrophobic and hydrogen bonds within macromolecules such as proteins.

QG – solubilization and binding buffer), which removes water from hydrated molecules in solution [13]. Contaminants like polysaccharides and proteins do not adsorb and pass through the column. Impurities are efficiently washed away (buffer PE diluted with ethanol), and the pure DNA is eluted under slightly basic conditions with low-salt elution buffer (buffer EB – elution buffer, 10 mM Tris–Cl, pH 8.5). The purified DNA can also be eluted in TE (10 mM Tris–Cl, 1 mM EDTA, pH 8.0).

Even though this kit is intended for the purification of DNA ranging in size from 70 bp to 10 kb from agarose gels in TAE (40 mM Tris acetate and 1 mM EDTA, pH 8.3) buffer, in our hands, we have successfully used it to isolate DNA up to 20 kb without significant shearing, if the DNA was handled gently. For gel purification of DNA 40 bp to 50 kb, the QIAEX II Kit (Cat. No. 20021) can be used instead.

21.3.1. Materials

The laboratory requesting microinjection of DNA into mouse zygote provides:

- *Plasmid DNA*

  Minimal amount of 50–100 μg DNA in 50–100 μl is needed. The user is required to provide a physical map of the plasmid DNA with restriction enzyme sites clearly indicated, particularly those used for releasing the transgene from the vector.

- *Picture of the digested DNA*

  The user must provide a picture of the digested plasmid DNA that clearly shows the bands representing the vector backbone and the transgene DNA insert. This information will assist the Transgenic Core staff in knowing what to expect in the bulk digestion of the plasmid DNA.

Note: Users are strongly advised to have a Southern blot analysis protocol to identify transgenic mice. This will avoid delays in genotyping of founder mice once they are generated.

21.3.1.1. Equipment and Supplies

Horizontal Electrophoresis Gel boxes.

Gel-doc XR System (BioRad).

Transilluminator and UV light mask.

Heating block or water bath at 56°C.

Eppendorf microcentrifuge.

Clear 1.7 ml Eppendorf tubes.

1.5 ml Eppendorf tubes.

Scalpel blades.

SeaKem LE or SeaPlaque agarose (Lonza, Cat. No. 50001 or 50101) or

SeaKem GTG agarose (Lonza Cat. No. 50071).

QIAquick Gel Extraction Kit (Qiagen, Cat. No. 28704 or 28706).

21.3.1.2. Solutions

95% Ethanol.

100% Isopropanol.

Sodium acetate 3 M, pH 5.2 (Sigma-Aldrich, Cat. No. S7899).

TE buffer (10 mM Tris–Cl. 1 mM EDTA, pH 8).

EB buffer

TAE.

TBE.

PE buffer.

Control DNA, e.g., lambda *Hin*dlll.

Loading dye.

Embryo tested water (Sigma-Aldrich, Cat. No. W1503).

Microinjection buffer (Millipore, Cat. No. MR-095-10F).

21.3.2. Procedure

21.3.2.1. Grow, Harvest, and Lyse Bacteria

The protocol for bacterial growth and plasmid isolation has been described in Subheading 21.2.2.1.

21.3.2.2. Digest Plasmid with Restriction Enzymes

- Excise most of the prokaryotic vector from the transgene DNA as transgenes are subject to epigenetic regulation when the bacterial plasmid or sequences are present. While transgenic mice have been made without removing plasmid sequences, the presence of prokaryotic sequences may adversely affect expression of some transgenes [14–17]. Further, linearized DNA is more recombinogenic hence increasing the efficiency of integration into the mouse genome.

Note: It is also important *not* to choose a restriction site too close to the promoter of the transgene. The linearized end of the transgene DNA can be digested by nucleases in the zygotic nucleus and if the promoter is impaired, transgene expression will be compromised.

21.3.2.3. Separate Transgene DNA Fragment from Plasmid Vector Sequences

Using the appropriate restriction enzymes, excise the transgene DNA fragment from the plasmid backbone, eliminating as much prokaryotic sequence as possible.

Electrophorese the digested DNA in a 0.5–1% agarose gel, using an appropriate voltage to ensure the restriction fragments are well separated for easy gel purification.

Note: Use a high quality agarose such as SeaKem GTG (Lonza, Cat. No. 50074) or SeaPlaque (Lonza 50101).

21.3.2.4. Solubilize Gel Slice

Using a clean, sharp scalpel and a long-wavelength UV Transilluminator, excise the DNA fragment from the agarose gel.

Minimize the size of the gel slice by trimming excess agarose.

Weigh the gel slice in a clear Eppendorf tube.

- *Add 3 volumes of buffer QG* (solubilization buffer included in kit) to 1 volume of a 1% agarose gel (assume that 100 mg of gel corresponds to 100 μl). For higher concentration agarose gels, increase the volume of buffer QG by one volume of buffer per 1% of agarose.
- *Incubate* the microfuge tube in a 56°C water bath, mixing gently by inverting the tube every 5 min until the agarose has completely dissolved.
- Check the color of the dissolved gel. It should be yellow (similar to buffer QG). If the color of the solubilized gel is even slightly orange or violet, *adjust pH* by adding 5 μl aliquots of 3 M sodium acetate until the color of the mixture returns to yellow, *pH ~7.5*. This is a critical step as adsorption of the DNA to the membrane is pH-dependent.

Note: Buffer QG contains a pH indicator, which is yellow at pH ≤7.5 (orange or violet at higher pH), allowing quick determination of the optimal pH for DNA binding.

- Add 1 gel volume of isopropanol to the sample and mix it by inverting several times.

Note: This step increases the yield of DNA fragments smaller than 0.5 kb and larger than 4 kb.

21.3.2.5. Binding DNA to the Column

- Place a spin column in a 2 ml collection tube (provided in the kit).
- *Load the sample onto the column*, and *centrifuge* for 1 min at 10,000 × *g* (~13,000 rpm).

Note: It is recommended to use no more than 400 mg of melted agarose/column. The maximum volume of the column reservoir is 700 μl. For sample volumes of more than 700 μl, simply load and spin again.

- Discard flow-through.

21.3.2.6. Remove Traces of Agarose

- *Add 0.5 ml of buffer QG to column* and centrifuge for 1 min. in the microfuge. This is an important step for the removal of agarose traces left in the DNA preparation.

21.3.2.7. Wash

- *Wash column* by adding ethanol to buffer PE following the kit's instruction, then add 0.75 ml of buffer PE (binding buffer) to the column and let it stand 2–5 min. before centrifuging. Spin in the microfuge for 1 min.
- *Discard the flow-through*. To remove residual ethanol (introduced with buffer PE), spin the column for an additional 1 min in a new microtube.

21.3.2.8. Elute DNA

- Place column into a clean 1.5 ml Eppendorf tube.
- *Add 50 μl of buffer EB* (elution buffer) to the center of the membrane, centrifuge for 1 min in a microfuge.

Note: DNA yields and concentration depend on the following three factors: the volume of elution buffer, the way in which the buffer is applied to the column, and the incubation time of the buffer on the column.

To increase the transgene DNA concentration, volume of buffer EB may be decreased to 30 μl. Carefully add the elution buffer to the center of the membrane to ensure that the buffer completely covers the membrane. Let stand for 1 min and then centrifuge for 1 min in a microfuge. Expect eluate volumes of about 45 μl and ~28 μl, respectively.

Notes: Elution efficiency is pH-dependent with the maximum efficiency achieved between pH 7.0 and 8.5. The purified DNA can also be eluted in TE (10 mM Tris–HCl, 1 mM EDTA, pH 8.0).

To maximize elution of bound DNA, add EDTA to 1 mM concentration to the buffer EB and add to the center of the membrane.

21.3.2.9. Evaluate DNA Quality and Concentration

- The yield of the DNA should be determined by *spectrophotometry* or fluorometry. For reliable spectrophotometric DNA quantification, the A260 readings should be 0.1 or higher. (See comments regarding absorbance ratios in Fig. 21.1).
- In addition, *agarose electrophoresis* should be used to confirm the yield, purity (single DNA band), and the integrity (lack of a smear beyond the DNA band of interest) of the transgene DNA. Known quantities (100 and 200 ng) of a DNA marker, such as lambda *Hin*dIII, should always be used as a standard for concentration determination.

21.3.2.10. Store DNA

- To protect integrity of the transgene DNA, *store DNA in the −20°C* freezer until ready to inject.

Note: Purification of the transgene DNA fragment may be done using electroelution instead followed by purification with Elutip-D minicolumns (Whatman Cat. No. **10462617**). Our Core finds that the Qiaquick Kit works well for the purification of many different kinds of plasmid DNA.

Transgene DNA should be prepared at a minimum concentration of 30–40 ng/μl. The DNA should be resuspended in either TE or buffer EB with EDTA. A high concentration allows extensive dilution of the transgene DNA as well as any minor contaminants present in the DNA preparation. A greater than tenfold dilution will ensure that low-level chemical contaminants can be sufficiently reduced so that egg lysis and/or developmental blockage will be kept to a minimum. DNA integrity and concentration should be evaluated on an agarose gel, but it is prudent to

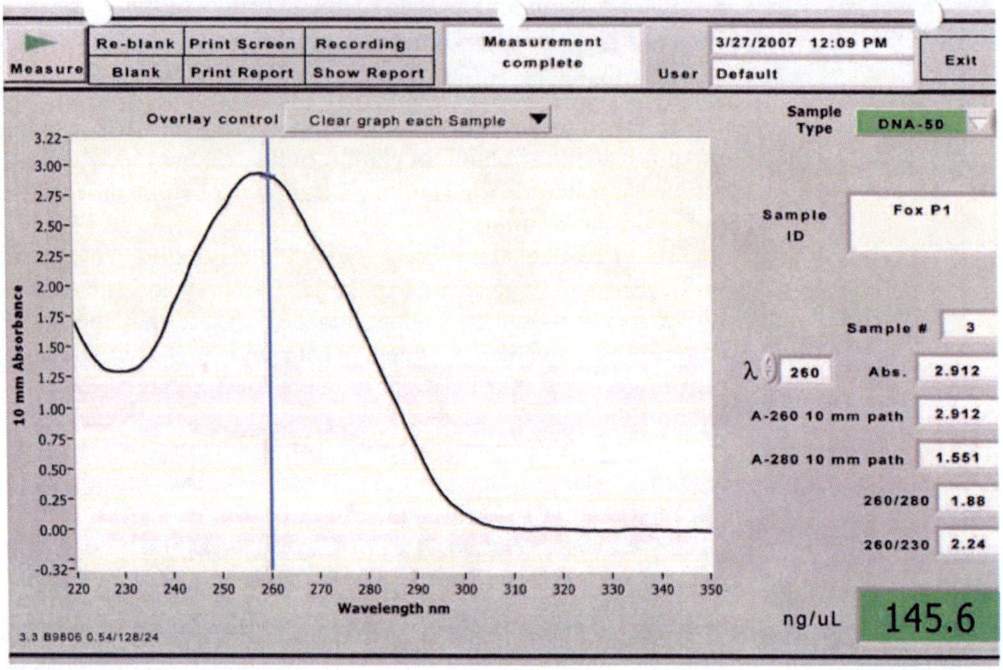

Fig. 21.1 Example of DNA absorbance curve generated by the Nanodrop 2000c spectrophotometer (Thermo Scientific). This snapshot of computerized record allows DNA purity evaluation (typical peak at 260 nm optical density is shown) by reviewing the absorbance ratios at different wavelengths and provides DNA concentration record.

use a second independent method of DNA evaluation like spectrophotometry or fluorometry. Information of DNA purity (Fig. 21.1), integrity (picture of the agarose gel), and concentration should be documented.

Note: A good DNA preparation should have a 260/280 nm absorbance ratio of ≥1.8, an indication of minimal protein contamination. The A260/230 nm absorbance ratio is used as a secondary measure of nucleic acid purity and should be between 2.0 and 2.2. A lower 260/230 nm absorbance ratio would indicate the presence of contaminants, such as salts and/or polysaccharides, which tend to absorb at 230 nm.

21.4 Preparation and Analysis of DNA for PCR Genotyping

The invention of the polymerase chain reaction (PCR) in 1983 is generally credited to Kary Mullis et al. [18]. PCR genotyping is faster and might be better suited than Southern blot analysis for genotyping subsequent generations of newly established permanent transgenic lines. In any case, a transgene detection strategy should be sensitive enough to detect transgene DNA at the single copy level, thus avoiding accidental euthanasia of transgenic animals due to false-positive PCR results.

To extract mouse genomic DNA for PCR genotyping, we use the HotSHOT (*Hot Sodium Hydroxide and Tris*) method [8]. This simple, effective, and inexpensive technique was developed from an alkaline lysis protocol for extraction of genomic DNA from human tissue samples [19]. The Hotshot method is best suited for small tissue samples such as a single ear punch, or 1–2 mm of tail biopsy from pups younger than 21 days of age. For larger tissue samples, higher volume of lysis buffer should be used. DNA obtained with this approach is suitable for use in PCR and is comparable to the traditional proteinase K lysis method [20].

21.4.1. Materials

21.4.1.1. Equipment

Thermal cycler (Eppendorf EPGradient).
Table-top centrifuge.
Gel-doc XR System (BioRad).
Multichannel pipette (optional).
pH paper strips.
0.2 ml PCR tubes.
Barrier pipette tips and pipetter.

21.4.1.2. Solutions and Reagents

Water (Sigma, Cat. No. W4502).

Clontech Advantage 2 Polymerase Mix (Clontech, Cat. No. 639202).

Primers (Operon).

dNTP's (Invitrogen Cat.No.10297-117 or Sigma Cat. No. DNTP100A).

10× Advantage 2 PCR Buffer (Clontech, Cat. No. 639138).

Primers (0.5 μM or 0.5 pmols each).

Alkaline Lysis Solution I: 25 mM NaOH, 0.2 mM EDTA.

Check the pH of Solution I; it should be around 12 without adjusting.

Neutralizing Solution II: 40 mM Tris–HCl (Sigma, Cat. No. T-5941), pH 5.

Dissolve Tris–HCl in water without adjusting pH.

Notes: Mix equal volumes of the Lysis and the Neutralizing solutions and then test the final pH using pH paper. It should be about pH 8.0. To prepare stocks of lysis and neutralizing buffers (Solutions I and II), it is best to use new, unopened bottles of water to minimize contamination by DNA present in laboratory.

PCR Master Mix (in this order):

Water (Sigma-Aldrich, Cat. No. W4502).

10× Advantage 2 PCR Buffer (Clontech, Cat. No. 639138).

0.2 mM, each dNTPs.

Primers (0.5 μM or 0.5 pmols each).

Clontech Advantage 2 Polymerase Mix (Clontech, Cat. No. 639202).

21.4.1.3. Supplies

Individually capped PCR tubes, 0.2 ml (Axygen, Cat. No. PCR-02-C).

96-well nonskirted PCR plates (Fisher Sc., Cat. No. 14230232) or thermal cycler Strip tubes (Axygen, Cat. No. PCR-0208) and strip caps (PCR-02CP).

Barrier tips 20 and 200 μl (Molecular BioProducts, Cat. No. 2749 and 2769).

pH Paper/Sticks.

PCR plate sealing film (Denville Sc., Cat. No. B1212-X).

21.4.2. Procedure

21.4.2.1. Tissue Preparation

- Place a small amount of ***mouse tissue*** (an ear punch, or 1–2 mm of tail biopsy) into a 0.2 ml PCR tube with cap. Using barrier tips, add 75 μl of ***Lysis Solution I*** to each tube containing the tissue sample.
- ***Incubate*** capped tubes at 94°C for 30 min in a thermocycler or in a heating block. For preparation of large quantities of samples, a multichannel pipetter is used for more efficient liquid handling. Alternatively, a computerized liquid handler can also be used.
- After lysis, ***add 75 μl of Neutralizing Solution II*** to each sample. Mix vigorously by shaking the capped tubes and ***centrifuge*** the samples for 2 min at 1,200 rpm (117 × *g*).

Note: Larger pieces of tissue (e.g., tail tissue) do not dissociate completely and small amounts of debris may remain. Under these conditions, remove dissolved tissue solution carefully from the top after centrifugation. While tissue debris does not interfere with the PCR assay, it is advisable to avoid introducing particulate matter into the final PCR reaction tube.

21.4.2.2. PCR Assay

- ***Set up PCR reactions***

In general, 10 μl PCR reactions are set up using 1× (dilute 10× with distilled water) Advantage 2 Polymerase Reaction Buffer, 0.5 μl of template DNA, 0.5 μM of EACH primer, and 0.2 mM of each dNTP.

We use half the manufacturer's recommended amount of Polymerase Mix. We do not use the optional SA (Short Amplicon) buffer provided by Clontech. Water is added to bring the reaction volume to 11 μl before the addition of template DNA. This extra volume of water is to compensate for evaporation during the cycling reaction.

It is recommended that an annealing gradient be performed for every new primer pair used for genotyping. For example, set the lower and upper limits of the annealing temperature to 54°C and 68°C, respectively, and run the PCR for 35 cycles with all other parameters kept constant.

By determining a suitable annealing temperature for each new primer pair, one can consolidate PCR assays of different genetic loci that show similar annealing temperatures so fewer independent reaction conditions need to be used. This promotes efficient usage of thermocyclers in Core facilities where there may be a large workload with limited number of PCR machines for genotyping many different alleles, each using their own parameters.

Notes: Clontech Advantage 2 Polymerase Mix is a mixture of processive and proofreading thermostable polymerases with antibody hot-start capability. We chose the Clontech Advantage 2 Polymerase Mix, because it gave the best PCR products using a given template/primer combination, when compared to eight other commercially available Taq polymerases. Of course, each laboratory should determine what Polymerase best suits their needs as PCR conditions and thermocyclers can have an effect on the outcome of the quality and quantity of PCR product.

- *Set up positive and negative controls*

In genotyping transgenic mice (as opposed to gene-targeted loci or assays involving RFLPs), it is strongly advisable to include an *internal control* to ensure that the DNA sample is of acceptable quality such that amplification would yield a readily detectable PCR product. For this purpose, we use primers for known mouse genes to serve as internal controls. Furthermore, a *positive* control of known transgenic mouse DNA (preferably the same as that to be genotyped) should also be included. This will prevent false-negative results and thus the loss of positive animals. To eliminate false-positive results, we typically perform two types of *negative* controls: a wild-type mouse DNA template and a no template control.

- *Run Analytical Gel*

Following amplification, 1 μl of loading dye is added to each well, and 3–6 μl of the reaction mix are loaded onto a 2% agarose gel. After electrophoresis, gels are stained for about 2 min in TBE containing 10 μg/ml ethidium bromide and immediately imaged under UV light using a BioRad Gel-Doc XR System.

21.4.3. PCR Assays for Common Sequences

21.4.3.1. Neoresistance Gene (Neo Phosphotransferase II)

Neo 5′ primer: GTACTCGGATGGAAGCCGGTCTT
Neo 3′ primer: GCCAAGCTCTTCAGCAATATCACG
Expected product: 280 bp band

Program name: Gen64

Cycling conditions: Initial denaturation is at 94°C for 3 min, followed by denaturation at 94°C for 15 s, annealing at 64°C for 30 s each, and extension at 72°C for 90 s for 35 cycles, then a final extension at 72°C for 7 min.

21.4.3.2. Cre Recombinase

Cre primer 5′: TGATGGACATGTTCAGGGATC

Cre primer 3′: CAGCCACCAGCTTGCATGA

Expected product: 850 bp

Program Name: Gen64

Cycling conditions: same as above for Neo.

21.4.3.3. LacZ (β-galactosidase) Gene

LacZ primer5′: CATCCACGCGCGCGTACATC

LacZ primer 3′: CCGAACCATCCGCTGTGGTAC

Expected product: 360 bp

Program Name: Gen68

Cycling conditions: Initial denaturation is 3 min at 94°C, followed by denaturation at 94°C for 15 s, annealing and extension both at 68°C for 90 s for 35 cycles, then a final extension at 72°C for 7 min.

21.5 Preparation and Analysis of DNA by Southern Blot Genotyping

The Southern blot technique is the standard approach for detecting a specific gene sequence of interest in the mouse genome [21]. Southern blot analysis not only detects the presence of a transgene but can also reveal integrity of the integrated transgene sequences. With appropriate transgene DNA copy number controls, Southern blot analysis can be used to determine copy number per genome of the transgene in the host animal, making it the method of choice for genotyping transgenic founders. At times, Southern blot analysis can reveal the number of integration sites in the founder animal based on the hybridization pattern of the genomic DNA. We also highly recommend that Southern blots be used to analyze the F1 generation for the same reason, as well as to identify segregation of multiple integrants. Independent integration sites may show different transgene expression patterns or levels, and thus the F1 mice should be considered as independent lines. To reduce the number of mice to be analyzed by Southern blotting, we frequently use PCR to reveal the positive animals prior to Southern blot analysis.

Southern blot analysis is also useful in providing a quantitative measurement of reconstitution by ES cells of host tissues in chimeras obtained from blastocyst injection of gene-targeted ES cells. Frequently, the intensity of the hybridizing bands observed in Southern blots corresponds with the extent of chimerism indicated by coat color. Therefore, in rare cases where coat color is not an option for determining chimerism, Southern blot analysis can be used.

21.5.1. Materials

21.5.1.1. Equipment

Incubation oven.

Multichannel pipetter.

Rotating mixer (optional).

Vortex.

SpectraMax Plus[384] Spectrophotometer or similar.

BD-Falcon UV Transparent 96-well plate (Fisher Scientific Cat. No. 08-772-135).

Eppendorf Centrifuge 5415D or similar.

Eppendorf tubes 1.7 ml.

Shaker.

Water bath.

21.5.1.2. Supplies and Solutions

Agarose.

Ethidium bromide.

Milli-Q (Millipore) water.

100 mM Tris, 10 mM EDTA Stock Solution.

10% SDS.

5 M NaCl.

Proteinase K stock 20 mg/ml (Invitrogen, Cat. No. 25530–031).

Phenol/chloroform/isoamyl alcohol (25:24:1), pH 7.5 (Fisher Scientific, BP1752I-400).

Isopropanol.

[α -32P]-dCTP (3000Ci/mmmol specific activity) (Perkin Elmer, Cat. No. BLU013H500UC).

Ethanol.

Lysis Buffer.

50 mM Tris pH 8.0 (Trizma, 1 M stock; Sigma-Aldrich, Cat. No. T2694).

100 mM EDTA pH 8.0.

1% SD100 mM NaCL.

TE buffer (10 mM Tris–HCl pH 8.0, 1 mM EDTA).

21.5.2. Procedure

21.5.2.1. Tissue Lysis

- For each 1 cm *mouse-tail tissue* obtained from a 10–14-day-old animal, use 500 μl of *lysis buffer and Proteinase K* (0.5 mg/ml).
- *Incubate* at 56°C overnight in a 1.7 ml microfuge tube. Occasional shaking of the tube can optimize lysis of the tail tissue. Alternatively, the tube can be rotated continuously during overnight incubation.

21.5.2.2. DNA Purification

- Add 450 μl ***phenol/chloroform/isoamyl alcohol*** to each mouse-tail lysate, and mix by shaking manually or by using a rotating mixer for about 10 min.
- Centrifuge at 12,000 rpm (11,770 × *g*) for 10 min in a microfuge.
- Remove ***the top aqueous phase containing the DNA*** without disturbing the interface (about 400 μl). Place into a new microfuge tube.
- Add 450 μl ***isopropanol*** to the DNA, mix repeatedly by inversion until a DNA precipitate can be seen.
- Centrifuge at 12,000 rpm for 10 min at room temperature to pellet the DNA.
- Remove the isopropanol and add 1 ml 70% ethanol to wash the pellet.
- Centrifuge at 12,000 rpm for 1–2 min to collect the DNA pellet.
- Repeat the ethanol washing to remove any trace amounts of salt in the DNA pellet.
- Spin down the DNA pellet, remove the ethanol, and briefly dry the pellet

  Note: If pellet remains slightly moist, it will dissolve much more readily.
- Add 78 μl *TE*, and incubate the tube at 56°C overnight to dissolve the DNA pellet. Again, occasional shaking of the tube will ensure that high molecular weight genomic DNA is completely dissolved after incubation.

21.5.2.3. Determination of DNA Concentration Using a Plate Reading UV Spectrophotometer

- Prepare a UV Transparent 96-well plate by adding 100 μl water into each well.
- Add 2 μl of dissolved DNA into each well. Use one well for control that contains only 2 μl of TE.
- Use a multichannel pipetter to mix the samples in each well before reading the OD.
- Genomic DNA samples can be used immediately or stored at 4°C.

21.5.3. Southern Blot Genotyping

21.5.3.1. Materials

21.5.3.1.1. Equipment

Power Supply.
Horizontal Electrophoresis gel Apparatus.
Stratagene, Stratalinker UV Crosslinker 2400.
Hot plate.
X-ray developer.
Hybridization Oven (Techne-HB-1D).

Incubator with rotator.
−70°C freezer.
Transfer trays.
Water bath.
Microfuge.

21.5.3.1.2. Solutions and Supplies

Hybond N+ (Nylon) membrane (GE Healthcare, Cat. No. RPN303B).
Whatman 3 mm filter paper.
Rediprime II Random Prime Labeling System kit (GE Healthcare, Cat. No. RPN1633) or BioPrime DNA labeling System (Invitrogen Cat. No. 18094–011) QIAGEN DyeEx^{TM} 2.0 Spin Kit (Qiagene, Cat. No. 63204).
X-ray film and cassette.
Pyrex dish.
Glass plate.
1.5 ml Eppendorf tubes.
TE buffer.
Restriction enzyme and appropriate buffer solution (comes with restriction enzyme).
BSA.
Agarose gel.
Ethidium bromide.
Lambda DNA digested with *Hin*dlll.
*Bam*Hl (or other restriction enzyme).
TBE.
0.2 M EDTA.
Plastic wrap.
Screw cap microfuge tubes.

21.5.3.1.3. Denaturing Solution

1.5 M NaCl.
0.5 M NaOH.

21.5.3.1.4. Neutralization Solution

1.5 M NaCl.
0.5 M Tris–Cl pH 7.5.
Transfer Buffer: 10× SSC; adjust pH to 7.0 using HCl

21.5.3.1.5. Hybridization Buffer

For total ~50 ml:

0.5 M Na_2PO_4 (which gives 1 M of Na^+)	25 ml 1 M Na_2PO_4
1 mM EDTA	100 μl 0.5 M EDTA
1% BSA	0.5 g BSA
7% SDS	17.5 ml 20% SDS
	7.5 ml H_2O

Note: Hybridization buffer should be freshly prepared. Add SDS last because it can precipitate out of solution in high salt.

21.5.3.2. Procedure

Briefly, mouse tissue DNA is digested with restriction enzyme and the digested DNA is separated by gel electrophoresis. The DNA fragments are denatured in situ, neutralized, and transferred onto a nylon membrane by blotting. The denatured DNA fragments are immobilized onto the membrane by UV cross-linking. The membrane is incubated with a gene-specific radiolabeled probe to allow detection of the DNA fragment of interest. After washing, the blot is exposed to X-ray film to reveal the genomic DNA fragment homologous to the probe.

21.5.3.2.1. Enzymatic Digestion

- Based on the OD readings, DNA samples are diluted to a concentration of approximately 1 μg/μl (or 0.5 μg/μl) with TE buffer.
- For each restriction digestion, between 5–8 μg of DNA are used. Add the desired restriction enzyme (1.5–2.5 μl) to 30 μl, which already contains DNA, 1× reaction buffer and BSA, if needed, in a 1.5-ml Eppendorf tube. When using restriction enzymes that exist only at low concentrations, <5 U/μl, the digestion reaction volume should be increased to accommodate the larger volume of enzyme used. Alternatively, longer digestion times can be used.
- *Incubate DNA with restriction enzyme overnight* at 37°C. Occasional agitation of the tube will enhance digestion of the viscous genomic DNA. Alternatively, the tubes can be rotated continuously during overnight digestion. If digestion is not complete, the addition of spermidine to a final concentration of 3 mM in the reaction mix can alleviate this problem (0.1 M spermidine stock stored in −20°C freezer).

21.5.3.2.2. Electrophoresis of Samples

- Prepare a 0.8% *agarose gel* with ethidium bromide to separate the digested DNA samples. To resolve smaller DNA fragments, a higher percentage (1%) gel can be used.
- *Electrophorese* in parallel, a DNA size marker such as lambda DNA digested with *Hin*dIII (commercially available). Also run known quantities of the transgene DNA to serve as copy number controls. A 2.6 pg of a 5 kb transgene DNA equals one haploid copy of the transgene integrated into the genome, when using 8 μg of genomic DNA for a Southern blot. We generally include controls representing 1 copy, 10 copies, and 50 copies per genome with founder tail DNA samples during Southern blot analysis. The linear transgene DNA is mixed with wild-type genomic DNA that is predigested with a restriction enzyme (we use *Bam*HI) to mimic total mouse genomic DNA. We digest large amounts of WT DNA and then phenol:chloroform extract, precipitate and resuspend the digested DNA for this purpose.

1 copy control = WT DNA digested with *Bam*HI + 2.6 pg linear 5 kb DNA

10 copy control = WT DNA digested with *Bam*HI + 26 pg linear 5 kb DNA

50 copy control = WT DNA digested with *Bam*HI + 130 pg linear 5 kb DNA

Run gel overnight at 30 V in 1× TBE. Alternatively, run at 100 V for about 5–6 h.

Note: Mark the gel by cutting one corner of gel after running.

21.5.3.2.3. Transfer DNA Samples to Hybond N+ Membrane

Denaturing Gel

- Soak the gel in Denaturing Solution for 30 min at room temperature on platform rotator.
- Prepare membrane, and wick (during denaturation) for each gel:
  - Cut the Hybond N+ membrane: 13 × 22 cm (or whatever your gel dimensions are)
  - Cut three pieces of Whatman paper: 13 × 22 cm
  - Cut one wick from Whatman paper: 13 × 34 cm

Neutralizing Gel

- Pour off the denaturing solution and add neutralization solution, place on platform rotator at room temperature for 30 min.

Note: The gel may remain in the neutralizing solution until ready to use (but not overnight).

- Label the membrane with pencil and rinse the membrane in water and then equilibrate in 10× SSC Transfer Buffer for 1–2 min.

Assembling Transfer

- Pour 10× SSC Transfer Buffer into transfer tray to a depth of about 2–3 cm.
- Put a support in the center of the tray.
- Put the wick on top of support.
- Put gel on top of wick. Label the membrane on the corner and put the membrane on top of gel.
- Put three layers of Whatman paper on top of membrane.
- Stack paper towel (about 20 cm height) on top of Whatman paper.

- Put a glass plate on top of paper towel and put a 0.5–1 Kg weight on top of glass plate (e.g., 500 ml or 1,000 ml plastic bottle filled with water). If the weight is too much, the gel will crush and capillary transfer will be blocked.
- Seal the four sides of membrane by plastic wrap on top surface of support to prevent shortcut of capillary movement.
- Transfer overnight (about 18 h) or over the weekend.

Notes: Make sure there are no air bubbles between the wick and the gel and between the gel and the membrane.Alkaline transfer cannot be used with nitrocellulose membrane.

- *Immobilizing the DNA on the membrane*

Rinse the membrane in 2× SSC. There are two ways of immobilization:

- UV cross-linking using Stratagene, Stratalinker UV 2400. Touch the membrane on Whatman to get ride of extra liquid and put the membrane on the floor of the UV linker chamber, DNA side facing up. Choose "Auto link." Then store the membrane between two Whatman papers.

  Alternatively,

- Put the membrane between two sheets of Whatman paper and put the sandwich between two glass plates (keep the membrane flat). Bake the membrane at 80°C, for 2 h.

21.5.3.2.4. Label the Probe with ^{32}P Using Rediprime II Random Prime Labeling System Kit Label DNA Probe (Lately We've Had Good Results with the Invitrogen BioPrime DNA Labeling System)

- Heat water bath to about 100°C.
- *Dilute the DNA* to be labeled to a concentration of 25 ng in 45 μl of TE buffer in a microfuge tube with a screw cap.
- *Denature the DNA probe* by heating to 95–100°C for 5 min in a water bath.
- *Snap-cool the DNA probe* by placing on ice for 5 min after denaturation.
- *Centrifuge* briefly to bring the contents to the bottom of the tube.
- *Add the denatured DNA probe to the reaction tube* (provided by the kit), do not mix at this stage.
- *Add 5 μl [α -^{32}P]-dCTP* (3,000 Ci/mmol specific activity) and mix by pipetting up and down and carefully moving the pipette tip around in the solution.
- *Incubate* at 37°C for 10–20 min.
- *Stop the reaction* by adding 5 μl of 0.2 M EDTA.
- *Purify the labeled probe* using the QIAGEN DyeEx™ 2.0 Spin Kit:

- Gently vortex the spin column to resuspend the resin
- Loosen the cap on the column a quarter turn to release any vacuum inside the spin column
- Snap off the bottom closure of the spin column and place the spincolumn in a 2 ml collection tube
- Centrifuge for 3 min at 2,900 rpm (690 × *g*) (in Eppendorf Centrifuge 5415D)
- Carefully transfer the spin column to a clean centrifuge tube and slowly apply the 50 μl labeled DNA probe to the gel bed

  Note: Take care not to put the pipette tip into the gel bed and aim to load the probe into the center of the column.
- Centrifuge for 3 min at 2,900 rpm to separate the DNA probe from the unincorporated 32P-dCTP. Carefully, remove *the eluate* for use in hybridization.

- *Denature the purified DNA* probe by heating to 95–100°C for 5 min, then snap-cool on ice for 5 min.
- *Dilute* this 50 μl labeled probe to 300 ml with TE.

21.5.3.2.5. Hybridization

Prehybridizing the blot with hybridization buffer will reduce nonspecific binding of the probe to the membrane. After hybridization, the blot must be washed thoroughly to remove weakly bound probe in order to reduce background often seen in Southern blotting. Finally, visualize the hybridized DNA fragments by exposure to X-ray film.

- *Prehybridize* the membrane with 10 ml of hybridization buffer for at least 30 min or up to 1 h at 65°C (several hours is alright).
- After this prehybridization step, discard the hybridization buffer.

Add 10 ml of fresh *hybridization* buffer into each cylinder and then add 50 μl of the radiolabeled probe directly into the cylinder and mix well (labeled probe amount may need to be titrated, but if incorporation is 25% or greater using 50–100 μl of the diluted probe is suitable).

- *Hybridize overnight* at 65°C with continuous rotation.
- Next day, discard the hybridization solution in appropriate containers.
- Briefly rinse off residual probe and hybridization buffer with a quick room temp wash using 10–20 ml of low stringency wash, (see below) in the hybridization tube gently rotating the tube by hand.
- *Wash the membrane* with increasing stringency at 65°C:

- First wash – Low stringency: 2× SSC with 0.2% SDS, 20 min,
- Second wash – Medium stringency: 1× SSC with 0.2% SDS, 20 min.
- Last wash – High stringency: 0.2× SSC with 0.2% SDS, 20 min.

- Then ***soak the membrane*** in 2 SSC with 0.2% SDS in a Pyrex dish at room temperature.

21.5.3.2.6. Signal Detection

- Place the membrane on a Whatman paper to remove excess wash buffer.
- Place the membrane in plastic wrap and fold over the edges twice.
- Place the sealed membrane into cassette containing an intensifying screen, put orientation tag at two corners.
- In dark room, assemble film in the cassette.
- *Expose* to film at −70°C overnight.
- *Develop film.*
- ***Analyze*** film to identify tail samples with hybridizing bands and ***determine approximate copy number*** using copy number controls. If band signals are weak, especially if one cannot see the single copy control, expose film for 3 days or longer.

21.6 Isolation of Genomic DNA from ES Cells for Genotyping

The two main objectives of the methodology for isolating DNA from ES cells from 96-well plates are as follows:

1. To harvest sufficient genomic DNA from a single well to perform both a PCR and a Southern blot assay to screen for the targeted allele.
2. To obtain ES cell DNA with a quality that will allow complete digestion of the ES cell DNA by the diagnostic restriction enzymes so genotyping results will be unambiguous.

The method provided below is modified from [2].

21.6.1. Materials

21.6.1.1. Equipment

Rotator – Nutator (optional).

Centrifuge with an adaptor for 96-well plates (Beckman Coulter, Allegra 6R) – *optional.*

37°C water bath.

55°C oven.

21.6.1.2. Supplies

96-well tissue culture plates (Nunc, Cat. No. 165306).

Multichannel pipette.

Tape (UniSeal) for 96-well plates (Whatman, Cat. No. 7704–001 or Nunc, Cat. No. 236366).

Parafilm or Saran wrap.

Agarose.

Restriction enzymes.

Lambda/*Hin*dlll DNA size marker.

21.6.1.3. Solutions

ES Cell Qualified 0.1% Gelatin Solution (Millipore, Cat. No. ES-006-B).

1 M Tris stock (Trizma, 1M stock; Sigma-Aldrich, Cat. No. T2694).

0.5 M EDTA stock.

PBS with Ca/Mg.

5 M NaCl stock.

5% N-Lauroylsarcosine sodium salt stock (Sigma-Aldrich, Cat. No. L9150).

Cell Lysis Buffer: (for two 96-well plates)

10 mM Tris–HCl pH 7.5	0.1 ml 1 M Tris–HCl stock
10 mM EDTA	0.2 ml 0.5 M EDTA stock
10 mM NaCl	20 µl 5 M NaCl stock
0.5% (w/v) sarkosyl 1.0 ml 5% stock of N-lauroylsarcosine sodium salt in ddH2O	
1 mg/ml proteinase K	0.5 ml 20 mg/ml proteinase K stock
Sterile water	To make final volume of 10 ml buffer

Note: Prepare lysis buffer fresh for each experiment.

Proteinase K (stock stored at −80°C) – add fresh each time.

95% Ethanol.

TE (Resuspension buffer) 10 mM Tris–HCl pH 8.0, 1 mM EDTA.

Digest cocktail: restriction enzyme, restriction enzyme buffer, BSA, spermidine (to a final concentration of 1 mM; stored in −80°C freezer).

1% gelatin (sterile) solution.

PBS with Ca/Mg.

21.6.1.4. DNA Precipitation Buffer

10 ml of 95% ethanol with 150 µl of 5 M NaCl (this buffer may appear cloudy).

21.6.2. Procedure

21.6.2.1. Seeding the 96-Well Plates

- Working with a multichannel pipetter, *plate ES cells* into two sets of gelatinized 96-well plates.
- Culture them to high-density and change media daily (*see* Chapter 14).

The ES cells may need between 2–5 days to reach confluency.

One set of plates serves as a back up and should be stored frozen at −80°C (see below). Be sure the plates and their lids are clearly labeled.

Note: The ES cells for analysis may differentiate during culture. This will not affect the genomic DNA under study.

21.6.2.2. Preparing 96-Wells Plates for DNA Isolation

Before harvesting cells for DNA extraction, mark the empty wells or wells with lower cell densities in order to correlate the efficiency of DNA recovery. If necessary, the amount of TE buffer used for resuspending the DNA pellet can be adjusted to give a more consistent DNA concentration.

Nonaseptic technique can be used at this point in processing the cells for DNA extraction and changes of PBS might be discarded directly by inverting the plate.

- *Rinse each well* with 100–200 μl PBS with Ca/Mg.
- Invert the plate and drain the wash onto a stack of paper towels.
- Repeat the wash.
- Seal plate(s) by wrapping them in parafilm and store at −80°C, if not immediately used for DNA extraction. It is advisable to wrap the plate in several layers of "diaper" or absorbent paper to protect them from inadvertent thawing (when the −80 freezer is opened).

21.6.2.3. DNA Extraction

- *Lyse cells* with 50 μl/well of lysis buffer. If the plates are stored at −80°C, remove from the freezer and allow them to reach room temperature before proceeding.
- *Seal the plates* with lids, by wrapping with parafilm, place in a tight container lined with wet paper towels to prevent evaporation, and *incubate* at 55°C overnight.
- Precipitate DNA from the lysed cells by adding 100 μl of Precipitation Buffer, prechilled to −20°C, to each well. Tap the plate to mix well. The use of a rotator to mix buffer and cells can hasten the lysis and precipitation process.
- After ~1 h of incubation at room temperature, a white precipitate of DNA should be evident. Depending on the amount of cells present in the well, the quantity of DNA may be different and thus care should be taken to look for the precipitated DNA, as the white precipitant may be small.

- Gently invert plates onto a stack of paper towels to drain off ethanol solution. The precipitate is attached to the dish.
- *Wash* the wells with cold 70% ethanol, about 150–200 μl.
- *Invert the plate* and gently drain the wash onto paper towels. Alternatively, *spin the plate again as above to make sure DNA is attached.*
- Repeat the ethanol wash two more times.
- After the wash is drained, tilt plates slightly and allow the DNA to *dry* for 10–20 min at room temperature in the cell culture hood.
- *Resuspend DNA* in 30 μl of TE buffer in room temperature.
- Once the DNA is well resuspended, *determine the quantity* of DNA by reading the absorbance.

Note: Over-drying the pellet will make the DNA difficult to re-suspend.

The typical yield of DNA from a confluent well is approximately 10 μg. For Southern blot analysis use the entire preparation of DNA. For PCR use 1–2 μl of the DNA solution.

21.6.2.4. Restriction Enzyme Digestion and Southern Blot Analysis

Note: Check digestion efficiency using described protocol on DNA prepared from parental ES cells before starting your gene-targeting experiment.

- Use DNA plates from 21.6.2.3 to identify the clones that have undergone homologous recombination of the targeting vector. Make up 10 μl per sample of *digest cocktail* containing: restriction enzyme, restriction enzyme buffer, BSA, spermidine (to a final concentration of 3 mM; 0.1 M stock stored in −80°C freezer). Use an excess of restriction enzyme 10 U per 1 μg of DNA except where star activity can result under these conditions.
- Add 10 μl of the enzyme cocktail directly to the 30 μl of ES cell DNA in the 96-well plate.
- *Add the restriction enzyme mixture* for your Southern assay directly to wells, *Mix* well, cover the plate with parafilm to avoid evaporation, and *incubate overnight* at 37°C or the appropriate temperature specified for the enzyme used.
- *Run the digested DNA* on an agarose gel with appropriate size markers (lambda/*Hin*dIII) and wild-type ES cell DNA digested with the same restriction enzyme to serve as a negative control.

Note: Check digestion efficiency as well as the DNA isolation procedure using the described protocol using parental ES cells before starting your gene-targeting experiment.

21.6.2.5. For PCR Screening

- Set up PCR reaction by using 1–2 μl of the isolated ES cell DNA. Conditions of cycling will need to be adjusted depending on the locus to be amplified and the primer pairs used.

References

1. Brinster RL, Chen HY, Trumbauer ME, Yagle MK, Palmiter RD (1985) Factors affecting the efficiency of introducing foreign DNA into mice by microinjecting eggs. Proc Natl Acad Sci USA 82:4438–4442
2. Matise M, Auerbach W, Joyner A (2000) Production of targeted embryonic stem cell clones. In: Joyner AL 2nd (ed) Gene targeting: a practical approach. IRL Press at Oxford University Press, New York, pp 101–132
3. Nagy A, Gertsenstein M, Vintersten K, Behringer R (2003) Production of transgenic mice. In: Manipulating the mouse embryo: a laboratory manual. Cold Spring Harbor, New York, pp 289–358
4. Nottle MB, Haskard KA, Verma PJ, Du ZT, Grupen CG, McIlfatrick SM, Ashman RJ, Harrison SJ, Barlow H, Wigley PL, Lyons IG, Cowan PJ, Crawford RJ, Tolstoshev PL, Pearse MJ, Robins AJ, d'Apice AJ (2001) Effect of DNA concentration on transgenesis rates in mice and pigs. Transgenic Res 10:523–531
5. Page RL, Canseco RS, Russel CG, Johnson JL, Velander WH, Gwazdauskas FC (1995) Transgene detection during early murine embryonic development after pronuclear microinjection. Transgenic Res 4:12–17
6. Martin MJ, Houtz J, Adams C, Thomas D, Freemann B, Keirns J et al (1996) Effect of pronuclear DNA microinjection on the development of porcine ova in utero. Theriogenology 46:695–701
7. Van Keuren ML, Gavrilina GB, Filipiak WE, Zeidler MG, Saunders TL (2009) Generating transgenic mice from bacterial artificial chromosomes: transgenesis efficiency, integration and expression outcomes. Transgenic Res 18:769–785
8. Truett GE, Heeger P, Mynatt RL, Truett-Walker JA, Warman ML (2000) Preparation of PCR-quality mouse genomic DNA with hot sodium hydroxide and tris (HotSHOT). Biotechniques 29:52–54
9. Qiagen Plasmid Purification Handbook, 3rd edn (2005) http://kirschner.med.harvard.edu/files/protocols/QIAGEN_QIAGEN-PlasmidPurification_EN.pdf. Accessed Nov 2005, pp 19–23
10. Auerbach AB, Norinsky R, Ho W, Losos K, Guo Q, Chatterjee S, Joyner AL (2003) Strain-dependent differences in the efficiency of transgenic mouse production. Transgenic Res 12:59–69
11. Vogelstein B, Gillespie D (1979) Preparative and analytical purification of DNA from agarose. Proc Natl Acad Sci USA 76:615–619
12. Boom R, Sol CJ, Salimans MM, Jansen CL, Wertheim-van Dillen PM, van der Noordaa J (1990) Rapid and simple method for purification of nucleic acids. J Clin Microbiol 28:495–503
13. Hatefi Y, Hanstein WG (1969) Solubilization of particulate proteins and non-electrolytes by chaotropic agents. Proc Natl Acad Sci USA 62:1129–1136
14. Chada K, Magram J, Raphael G, Radice E, Lacy E, Costantini F (1985) Specific expression of a foreign beta-globin gene in erythroid cells of transgenic mice. Nature 314:377–380
15. Hammer RE, Brinster RL, Palmiter RD (1985) Use of gene transfer to increase animal growth. Cold Spring Harb Symp Quant Biol 50:379–387
16. Krumlauf R, Hammer RE, Brinster RL, Chapman VM, Tilghman SM (1985) Regulated expression of alpha-fetoprotein genes in transgenic mice. Nucleic Acids Res 14: 9667–9678
17. Townes TM, Lingrel JB, Chen HY, Brinster RL, Palmiter RD (1985) Erythroid-specific expression of human ß-globin genes in transgenic mice. EMBO J 4:1715–1723
18. Mullis K, Faloona F, Scharf SR, Horn G, Erlich H (1986) Specific enzymatic amplification of DNA in vitro: the polymerase chain reaction. Cold Spring Harb Symp Quant Biol 51:263–273
19. Rudbeck L, Dissing J (1998) Rapid, simple alkaline extraction of human genomic DNA from whole blood, buccal epithelial cells, semen and forensic stains for PCR. Biotechniques 25:588–590
20. Blin N, Stafford DM (1976) A general method for isolation of high molecular weight DNA from eukaryotes. Nucleic Acids Res 3:2303–2308
21. Southern EM (1975) Detection of specific sequences among DNA fragments separated by gel electrophoresis. J Mol Biol 98: 503–517
22. Collins JL, Baltz JM (1999) Estimates of mouse oviductal fluid tonicity based on osmotic responses of embryos. Biol Reprod 60:1188–1193
23. QIAquick Spin Handbook (2008) www1.qiagen.com/HB/QIAquickGelExtractionKit_EN. Accessed Mar 2008, pp 25–26

Chapter 22

Colony Management

Karen Brennan

Abstract

Technical advances in the generation of genetically modified (GM) mice and the efforts of large-scale consortia have provided a wealth of resources to the biomedical research community. It has never been easier to obtain a specifically modified allele. However, the establishment and production of GM mouse models using basic breeding methods can still present a unique set of challenges. This chapter aims to equip those new to mouse colony management with some valuable tools. Topics covered include basic reproduction and inheritance, welfare assessment, nomenclature, breeding schemes, and calculators for the production of desired genotypes. Breeding of transgenic founders or chimeras through the first generations is discussed as well as the influence of strain background on this process. Animal records management systems also play an important role in effective colony management.

Abbreviations

BLASTN	Basic local alignment search tool for nucleotide sequence comparison
cM	Centimorgans
CRL	Charles River Laboratory
ES cell	Embryonic stem cell
FISH	Fluorescent in situ hybridization
GM	Genetically modified
GPL	General public license
het	Heterozygous
hom	Homozygous
ILAR	Institute for Laboratory Animal Research
JCMS	The Jackson Laboratory's Colony Management System
MACS	Marker assisted congenic screening
MEF	Mouse embryonic fibroblast
MGD	Mouse Genome Database
MGI	Mouse Genome Informatics
PCR	Polymerase chain reaction
RFID	Radio frequency identification
TJL	The Jackson Laboratory
wt	Wild type

S. Pease and T.L. Saunders (eds.), *Advanced Protocols for Animal Transgenesis*, Springer Protocols,
DOI 10.1007/978-3-642-20792-1_22, © Springer-Verlag Berlin Heidelberg 2011

22.1 Introduction

The deliberate breeding of animals to select for desirable traits is an ancient process. The emergence of new technologies to create specific germline modifications and to examine their effects at a molecular level has made it possible to refine and accelerate these processes. The mouse has provided an ideal model in which to further understanding of gene function in mammals, having a short generation time, relatively simple housing needs, and embryos that could be manipulated in vitro with ease. The technologies described earlier in this manual will lead you through the design and development of novel genetically modified (GM) mouse lines. Having reached this point, you will have successfully generated your founder animals and are now beginning an exciting stage of your study. Presumably you have carefully considered the strategy for your genetic modifications, giving thought to the best strain background to give you a successful outcome, and have a sound protocol for identification of founder animals. In establishing your colony and breeding animals for experimental studies there will also be decisions to be made about strategy. As always, time is critical and you will want to be as productive as possible within the limitations of your resources and of reproductive biology.

At this point it would be wise to discuss your future animal needs for experimental plans with your facility management to devise a breeding plan that takes into account your space and time limitations. Make a commitment to communicate regularly with the animal care staff who have daily contact with your animals, as they can provide valuable insight into the impact of your alterations on animal behavior. Follow closely the progress of breeding and, if you are required to do so, diligently process genotype or phenotype data required for selection of breeding animals. With this approach you will be well on the way to generating sound experimental data. In this chapter, some of the fundamentals of establishing production colonies from GM mouse lines are discussed, with reference to several excellent texts where further information on each topic can be sought.

22.2 Establishing Breeding of Novel GM Lines

22.2.1. Assessment of Welfare in the First Generations

Broadly speaking, there are three purposes for which GM lines are now produced. Firstly, there are novel modifications produced for the study of a particular gene, biological process, or disease, either by gene trapping, targeting approaches, or by insertional transgenesis. Secondly, there are lines produced to be used as tools, in which the effect of the alteration on the biology of the mouse is not of interest in itself, except as part of a strategy to develop a model. The many recombinase-expressing lines fall into this group, as do lines that carry fluorescent tags or other reporter constructs. Thirdly, there are conditional or inducible mutants that carry the potential for generation of phenotypes of interest, but must first be bred with other strains or undergo treatment to complete the genetic alteration.

For all types of mouse models an assessment of the impact of the alterations on general well-being and reproductive performance must be made in the first generations of animals produced. It is clear that if a line is to be established and be useful, the mice must be able to survive to sexual maturity and reproduce. It is important to obtain a complete picture of the characteristics of the line if phenotypic assessments are not to suffer from interference from underlying health problems. Detrimental phenotypes should be identified promptly so that specialized support can be given to the animals in order to prolong survival and to alleviate distress [1].

In the first generations bred, complete records should be generated on breeding performance, growth weights, morphological and behavioral characteristics, as well as how these characteristics relate to genotype. Assessment should begin at the neonatal period and continue up to the most advanced age to which mice will be retained. Comparison should not only be made between the GM and wild-type individuals within a line, but ideally also with mice of the parental strain, or as closely matched as possible, which have been held under the same housing and environmental conditions. It is very often the case that a novel mouse line displays no observable differences to its parental strain. It has been estimated, however, that the welfare of up to 30% of lines is in some way affected by the alterations [2, 3]. Once the unique characteristics of a line have been established and noted in the first generations, then these observations can form the basis of a breeding and husbandry plan. This plan needs to be reviewed at intervals and when significant changes occur as breeding progresses (i.e., segregation of alleles, change of genetic background due to backcrossing or inbreeding). It is also important that such relevant information about strains should be transferred to receiving institutions when the mice are distributed.

Table 22.1 lists qualities that should be monitored systematically in a new mouse model. More details on monitoring for welfare can be found in the Full report of the Mouse Welfare Assessment working group: Assessing the welfare of genetically altered mice [1], with suggested monitoring templates and a template mouse "passport." Details can be found on the www.nc3rs.org.uk site.

Table 22.1
Qualities for initial welfare assessment

Age range	Observable qualities
Neonatal period to weaning	Suckling ability (presence of milk spot)
	Skin color
	Relative sizes within litter
	Growth rates, compared to parental strains
	Skeletal or morphological abnormalities in tooth, eye, ear, or digit development
	Coat appearance and timing of fur growth
	Activity and responsiveness to handling
Weaning to mature adults	Growth rates and adult body weights
	Reproductive performance; fertility, fecundity, and mothering abilities
	Appearance; coat (piloerection, hair loss), eyes (sunken, pale), and skin color
	Behavior and activity; alertness, aggression, and presence of stereotypies
	Posture and gait
	Clinical signs of poor health such as discharge, seizures, weight loss, and lesions

It should be remembered that detrimental effects on the animal's well-being may arise as a desired and predicted outcome of the genetic modification or may be an indirect consequence and unrelated to the expected phenotype. Random insertions of transgenes into the genome can disrupt endogenous gene sequences at the site of insertion [4]. When bred to homozygosity, the effects of the disruption of the endogenous gene may become apparent. If the gene is subject to haploinsufficiency effects, i.e., if two copies of the wild type gene are needed to maintain normal expression, then disruption of the endogenous gene will have consequences for hemizygous mice as well.

Although targeted modifications should overcome the problems associated with random insertions, ES cell lines are unstable in continuous culture [5–7] and can acquire deletions and rearrangements unrelated to the targeted change. An example of this is related by Kumar and colleagues [8], where in the production of a gene-targeted *Zfa* strain, a deletion and transposition event led to the knock-out of the *Nr2e1*gene. The genes are linked as they are located on the same chromosome, although distant. An aggressive phenotype ("fierce" mice) was initially observed as correlating with genotyping for the targeted *Zfa* allele, but was eventually attributed to the $Nr2e1^{frc}$ allele. In due course, by backcrossing to a wild-type strain the two alleles could be segregated.

22.2.2. Nomenclature

With the proliferation of mouse lines now made widely available, a system for naming lines becomes essential for accuracy in communication about mouse models. Rules and guidelines for nomenclature are set by the International Committee for Standardized Genetic Nomenclature for Mice, and are well explained under the Mouse Genome Informatics (MGI) nomenclature homepage (1) and in a tutorial to be found at The Jackson Laboratory (TJL) website (2). The elements of a formal name are depicted in Fig. 22.1.

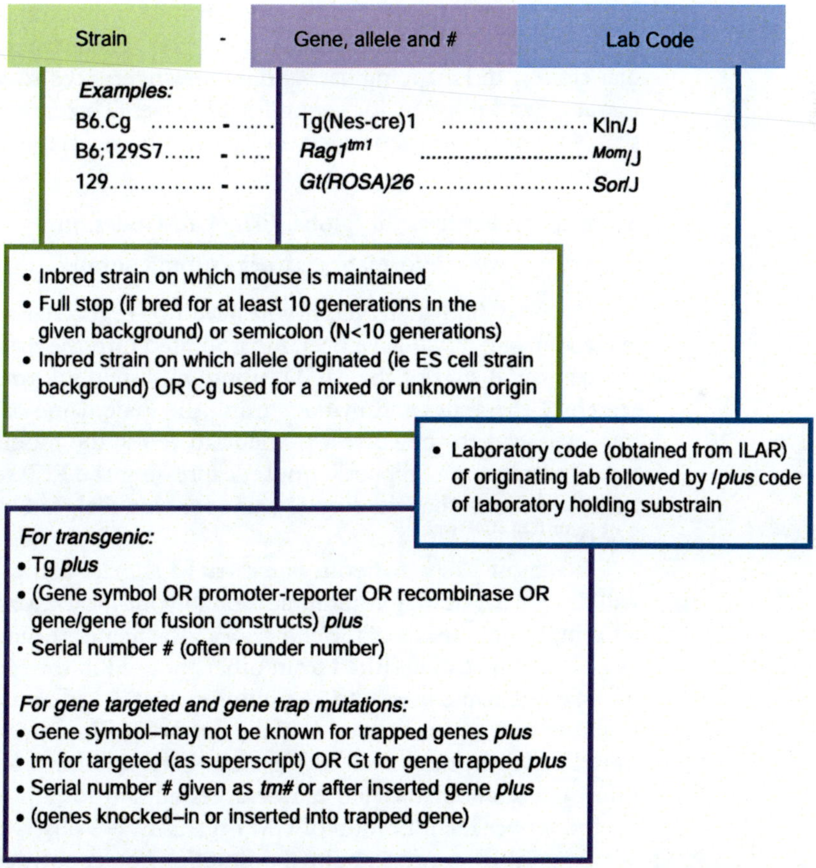

Fig. 22.1 This is a simplified depiction of strain names that holds true for many congenic strains. As many other combinations exist allowing for the description of many types of genetic alterations that now are in existence, refer to the nomenclature guidelines for further information (http://www.informatics.jax.org/mgihome/nomen/index.shtml).

1. http://www.informatics.jax.org/mgihome/nomen/index.shtml
2. http://jaxmice.jax.org/support/nomenclature/tutorial.html

Names in these formats convey detailed information about the mouse line, including the background strain, the gene symbol, the symbol for the modified allele or transgene, whether random insertion, gene targeting, or gene trapping was used and the laboratory of origin. Applying the formal name to your novel mouse line is advantageous before first publication. This can be effected as soon as you have established stable germline transmission of the mutation.

You must use, or first apply for, a Laboratory Registration Code unique to your lab or institution, which is issued by the Institute of Laboratory Animal Research (ILAR) (3). You can then submit a genetically engineered allele to Mouse Genome Database (MGD) (4) and the modified allele will be given an MGI accession number. This database provides a unique comprehensive collection of mouse genetic markers, phenotypes, molecular clones, and mapping data and is cross-referenced to scientific literature and other major genomics databases. Through the same site (4), your novel mouse strain can also be registered under a formal strain name.

3. http://dels.nas.edu/global/ilar/Lab-Codes.html
4. http://www.informatics.jax.org/submit.shtml

You may find that even if you have not registered a mutant allele yourself, it will have been incorporated into the database by the curatorial staff of the MGD, using published information. A search of the genes and markers using the gene name or a synonym will lead to the correct nomenclature for the modification, e.g., a search for a "flipper" mouse, expressing the FLP recombinase, will reveal that the formal name for this allele is Gt(ROSA)26Sor$^{tm1(FLP1)Dym}$.

It is most likely that your line carries a shorter name that you will use for day-to-day communication in your local environment. The important aspects of the choice of a local name are that the line can be readily distinguished from others housed in the same facility, that the name is easy to communicate with, and that it works together with any electronic database employed. For some facilities "Fred" may work as well as Tg(Actb-LacZ) as long as further information about the line can be accessed and local regulations on the proper identification of GM organisms are observed.

A word of caution about the notation for genotypes. While classical geneticists will use a "+" indication for the wild-type gene, it has also been used inappropriately to indicate the presence of a transgene or alteration, which can result in confusion of genotypes and the potential for loss of lines. The use of the allele name to indicate genotype should always be used in preference. For instance, a homozygous mouse carrying transgene nicknamed “hpt” should be indicated by “hpt/hpt,” rather than “+/+,” which indicates a wild-type genotype in standard nomenclature. Even if a system has developed for use locally that is entirely consistent and well understood, be sure to always use standard notations when sharing mice with external parties.

22.2.3. Understanding Mendelian Inheritance

The key to planning effective breeding strategies is an understanding of Mendelian inheritance of genes. Each parent carries two alleles for all genes located on autosomal chromosome pairs. In the production of haploid gametes, the alleles segregate, with gametes having an equal chance of carrying either allele. Offspring will result which carry two alleles, each contributed by either parent, and the combinations which result are governed by the laws of probability, as long as there are no other selection factors involved. Figure 22.2 shows the outcome of the mating of two parents, each carrying one wild-type allele and one altered allele, or in other terms, heterozygous for the mutant allele. The

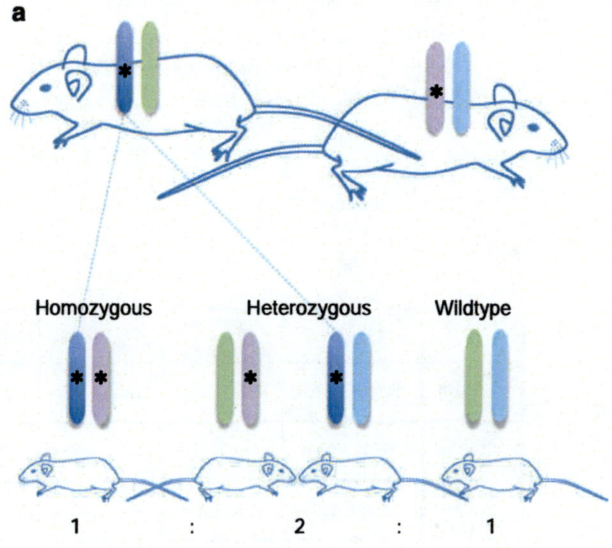

b

	*	+
*	**	*+
+	+*	++

Fig. 22.2 (**a**) Matings of mice each heterozygous for a particular allele, indicated by *, will result in offspring that are homozygous, heterozygous, or wild type with respect to the allele in the ratio of 1:2:1. This can also be depicted using a Punnett square (**b**), where * indicates the transgene and + the wild-type allele.

probable outcome for this combination of alleles results in offspring that are homozygous, heterozygous, and wild type for the genetic alteration at the ratios of 1:2:1. Likewise, breeding of a heterozygous to a wild-type mouse will yield heterozygous and wild-type offspring at a ratio of 1:1, and breeding of a heterozygous to a homozygous mouse will yield heterozygous and homozygous mice at a ratio of 1:1.

When there are two independent mutations to be considered, as long as they are not physically linked by location on the same chromosome, the probabilities of each outcome occurring separately can be multiplied together to determine the expected frequency of a combined outcome. Thus, the mating of two mice heterozygous with respect to two GM alleles, could be expected to produce double homozygous offspring at the rate of 1/16, and offspring that are heterozygous for each allele at 1/8. A Punnett square is used in Fig. 22.3 to predict the genotypes of the offspring of such a mating. This approach can be a useful tool for understanding other complex matings.

Inserted transgenes, strictly speaking, have no allelic pairing, and so carriers are referred to as hemizygous rather than heterozygous. The same Mendelian principles apply, however, when breeding these animals, the absence of the transgene also being a heritable quality. By employing these predictions of breeding outcome to the simplest and the most complex of breeding strategies, breeding plans can be designed to produce the required genotypes of mice with the greatest efficiency.

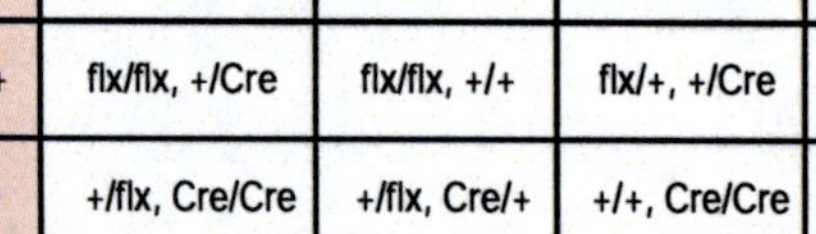

♀ \ ♂	flx Cre	flx +	+ Cre	+ +
flx Cre	flx/flx, Cre/Cre	flx/flx, Cre/+	flx/+, Cre/Cre	flx/+, Cre/+
flx +	flx/flx, +/Cre	flx/flx, +/+	flx/+, +/Cre	flx/+, +/+
+ Cre	+/flx, Cre/Cre	+/flx, Cre/+	+/+, Cre/Cre	+/+, Cre/+
+ +	+/flx, +/Cre	+/flx, +/+	+/+, +/Cre	+/+, +/+

Fig. 22.3 A Punnett square illustrates here how the alleles segregate in the offspring of a mating of two mice, each heterozygous for two mutant alleles, flx and Cre. Possible allelic combinations in gametes are indicated from male and female partners. If this would represent the crossing of a floxed allele with a Cre recombinase-expressing mouse to create a conditional null mutant, then the mice with a flx/flx genotype and either Cre/+ or Cre/Cre could be expected to have both copies of the floxed gene excised (knocked-out). This combination represents one quarter (4/16) of the offspring produced. If there are problems with Cre toxicity in the Cre/Cre mice or recombinase insufficiency this would affect half of these animals.

22.2.3.1. Sex-Linked Alleles

If your transgene or gene trap vector has inserted into a sex chromosome this will become apparent as you follow the transmission of the transgene from founder to offspring. A transgene located on the Y chromosome will not be transmitted by a male breeder to any female offspring but will be inherited by all male progeny (Fig. 22.4a). On the other hand, a transgene on the X chromosome will follow an inheritance pattern identical to somatic chromosomes when transmitted through the female germline (Fig. 22.4b). When hemizygous males with an X-linked transgene are mated, however, all female progeny will inherit the transgene but none of the males. Only females of an X-linked line can be bred to homozygosity for the transgene (Fig. 22.4c), but for functional analysis, the effects of X-inactivation on phenotypic expression need to be taken into account. Early in embryonic development in females one of the X chromosomes is inactivated at random. Adult females will have a mixture of cells that have either an active maternally or paternally derived X chromosome. Therefore, in terms of gene expression, there will only be one functional allele. For further reading on the implications of X-chromosome inactivation in your model, *see* Papaioannou and Behringer [9] and Migeon [10].

An example of an X-linked GM mouse line is the widely used "deleter" mouse ([11], MGI:2176179) used for generating a ubiquitous germline knock-out using the Cre-*loxP* recombinase system. In this strain, the Cre transgene is located on the X chromosome, and is expressed in early embryogenesis, prior to implantation. As Cre expression precedes X-inactivation in this case, the Cre-mediated deletion is effected in every cell, and not subject to X-inactivation mosaicism. One limitation to the use of

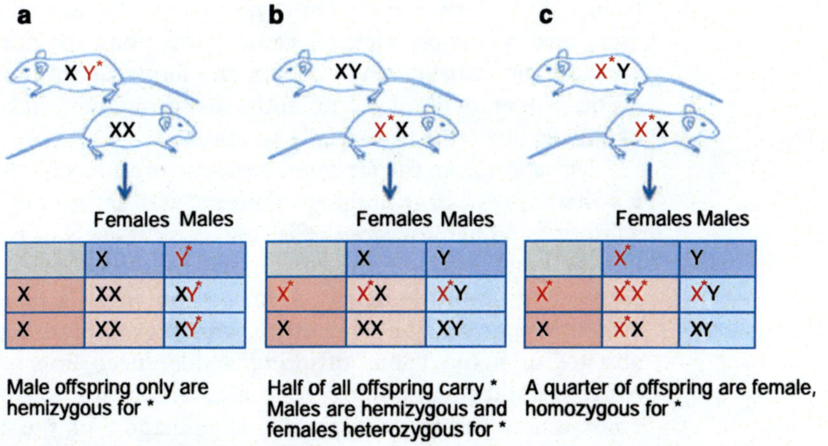

Fig. 22.4 An illustration of the inheritance of alleles located on sex chromosomes. Chromosomes carrying modified alleles are indicated with an asterisk. Combinations shown are: heterozygous male with wild-type female (**a**), wild-type male with heterozygous female (**b**) and both male and female heterozygous (**c**). Punnett squares show the occurrence of transmission of the allele to offspring. Males are indicated by *blue* shading, females by *pink* shading.

this line is that a deletion in male offspring can only be effected if a female deleter is mated to a floxed male (floxed = flanked by *loxP*). If your transgenic line is by chance X or Y linked you will need to take this into account in your breeding strategies and you may encounter some limitations if and when you wish to cross with other lines.

22.2.3.2. Linkage of Two or More Genetic Modifications

When designing breeding strategies with more than one site of modification it is important to verify what is known about the chromosomal location of each. Difficulties will be encountered in producing animals carrying double mutations if the loci occur on the same chromosome. Due to random recombination events, it is not impossible to obtain these animals, but you must wait for a relatively rare event to occur. If the distance between your loci is known, then an estimate of the frequency of a crossover occurring can be made. Two megabases of DNA approximately equates to 1 cM, which is defined as a 1% chance of recombination between two markers at two loci in a single generation ([12], available online at http://www.informatics.jax.org/silverbook). Thus to produce one mouse in which two markers, 5 megabases apart, have segregated due to random recombination, about 250 mice would have to be bred.

22.2.3.3. Embryonic Lethality

Failure of homozygous mice to survive may be an anticipated outcome of your genetic alteration, due to disruption of essential functions by gene knock-out or mutation. A lethal phenotype can also occur in heterozygous mice, for instance due to haploinsufficiency (for null alleles) or production of a hypomorphic protein (from a transgene), with reduced levels of activity. Mouse strains with X and Y-linked mutations are particularly susceptible to the effects of phenotype in the hemizygous state due to the lack of a functional wild-type allele. Because predictions are not always accurate and lethality may occur in any model as an unexpected event, genotypes of mice born in the first generations of breeding of a novel line must be carefully scrutinized.

Deviation from the expected outcome of a breeding plan that is sustained over large numbers of litters, with skewing in favor of wild-type and heterozygous offspring, is most likely to be due to loss of homozygous mutants during embryonic development. Depending upon the degree of expression across a population, or penetrance, of the phenotype, this may lead to complete absence of homozygous offspring and reduced litter sizes, or reduced numbers of homozygous mice surviving to birth. When embryonic lethality is suspected, examination of the uteri of mothers at different stages during pregnancy will help to establish at what stage the homozygous offspring are lost. Papaioannau and Behringer [9], in “Mouse Phenotypes: a Handbook of Mutational Analysis,” have comprehensively detailed the steps to take in such an investigation in Chapter 5: "Prenatal Lethality." If homozygous embryos are not found, even amongst preimplantation embryos, it may be an indication that the mutant allele has affected spermatogenesis or sperm performance, leading to a bias at fertilization in favor of heterozygous or wild-type genotypes [13, 14].

22.2.4. Breeding Schemes for Generating Required Genotypes

In the first breeding and phenotyping of your novel mouse strain you will want to establish whether your modified allele is recessive or dominant, and whether it is lethal in either case. A dominant mutation is one in which a phenotype occurs to the same degree in both heterozygous and homozygous animals. If a phenotype is observed in heterozygous but is more severe in homozygous mice, then it is referred to as a semidominant mutation. If the phenotype is present only in homozygous mice, then the mutation is recessive. Once this is known, you will be able to determine the best breeding strategy for maintenance of your line and the generation of study mice.

Another consideration for the choice of breeding scheme is the requirement for control animals for your study. Depending on the sensitivity of the phenotype to genetic and environmental factors, littermate controls may be preferred to mice of the same inbred background sourced from other colonies. If the line is to be crossed to mice of a different inbred background, then this will also influence the breeding strategy (as discussed in Subheading 22.5).

The following breeding schemes are options for the generation of various genotypes:

Heterozygous mutant × wild type: *Expected ratio of offspring: 1 het: 1 wt*

This will be your best option for continuous breeding if your phenotype is recessive lethal or the production of homozygous pups is deleterious to breeding performance, e.g., if late embryonic stage deaths cause difficulty in parturition. If your modification is dominant and mice heterozygous for the allele are predominantly required, then this approach has the advantage of producing study mice as well as wild-type littermate control animals, in equal numbers. If you are outcrossing or backcrossing to alter the strain background of your line, this type of cross will be necessary. In this case, the generation of equal numbers of wild-type mice, which will not be useful for further breeding, is unavoidable. For this reason, backcross colonies usually consist of one or two breeding pairs only.

Heterozygous mutant × heterozygous mutant: *Expected ratio of offspring: 1 hom: 2 het: 1 wt*

A popular breeding approach as homozygous, heterozygous (or hemizygous), and wild-type offspring are produced within the same litter and are therefore as similar as possible to each other with respect to environmental influences and maternal nutrition. Wild-type offspring will be available for use as littermate controls. Heterozygotes produced in excess of study needs can be used for further breeding as long as the line is maintained on a pure inbred background. This approach may still result in an oversupply of one of the genotypes.

Heterozygous mutant × homozygous mutant: *Expected ratio of offspring: 1 hom: 1 het*

May be desirable in cases where larger numbers of homozygous mice are required, and where wild-type littermates are not needed. For strains that are maintained on a pure inbred background, i.e., isogenic, or congenic strains, (*see* Table 22.2), the need to produce control mice of the same mixed genetic background is not a major concern. If homozygosity for the mutation has a detrimental effect on the survival or fertility of one sex and not the other, this is the best approach. Where both heterozygous and homozygous mice are required to study a dose effect of the mutation then these matings will produce equal numbers of each.

Homozygous mutant × homozygous mutant: *All offspring are homozygous*

This is an ideal way to maintain your strain if you have a nonlethal mutation that does not negatively influence breeding performance. This is the most efficient way to maintain "tool" lines such as recombinase-expressing lines, as the need for molecular screening by PCR can be very much reduced, saving cost and time. It is advisable to screen at regular intervals, however, to confirm genotype as mistakes in breeding may occur. As all mice can

Table 22.2
Terms in use

Term	Meaning
Isogenic	Strain background is genetically identical between modified sequences and the rest of the genome
Co-isogenic	Genetically identical strain background throughout except for the GM allele
Congenic	Genetically identical strain background except for the GM allele and the chromosomal region immediately neighboring it
Outcross	A mating of mice from different strain backgrounds
Backcross	A mating of one mouse to another that is of a strain background identical to one of its parents
Inbred	A strain which has undergone sibling matings for at least 20 generations and is homozygous at more than 99.99% of all loci (except for Y chromosome)
Outbred	A strain in which individuals are genetically diverse and heterozygous at all loci

be used as study animals or for continued breeding, wastage is at a minimum. If the strain is not on a completely inbred background then there is a chance of in-breeding depression occurring [12], as alleles that contribute poor breeding characteristics become homozygous (fixed). The effects of this can be quite sudden and marked, but can be reversed if homozygous animals can be bred to wild-type animals of the reference genetic strain. After one generation, in which heterozygous mice would be produced, heterozygotes can be mated to breed the strain again to homozygosity. It is a good preventative measure against genetic drift in these colonies to regularly mate for one generation to the reference genetic strain; every 5 generations is recommended by TJL.

22.2.5. Production Methods

22.2.5.1. Two-Tiered Colonies: Stem and Expansion

Unless your needs for animal numbers are very small, you will probably benefit from having a stem (or nucleus) colony and an expansion breeding colony, as depicted in Fig. 22.5. This allows you to address the different requirements of maintaining a stable strain and generating animals for experimental use. The priority of your stem colony will be to establish your strain on an ideal

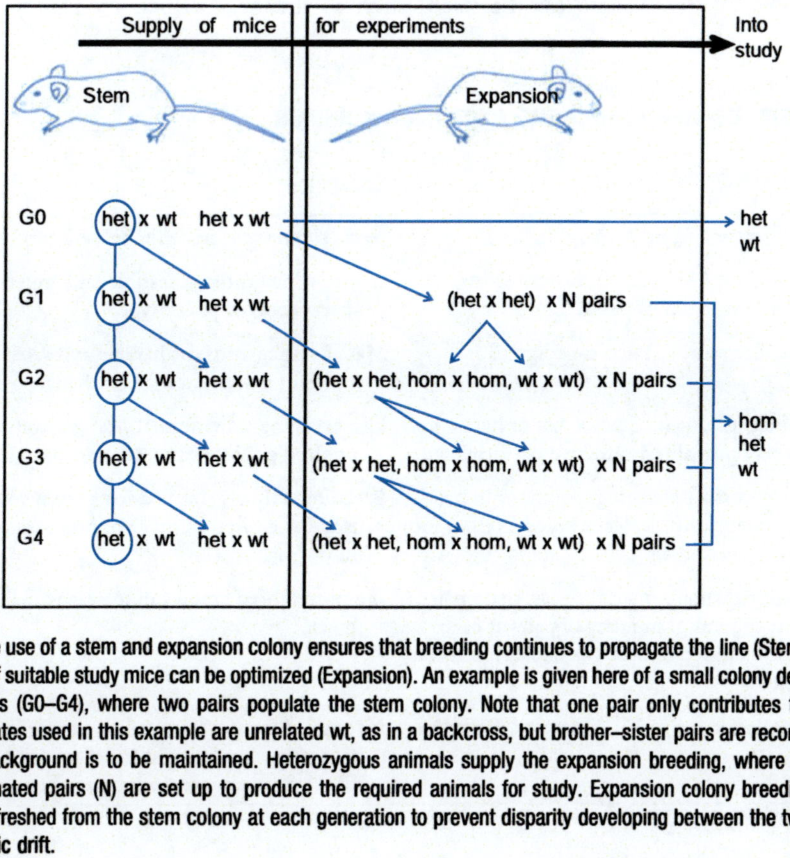

Fig. 22.5 The use of a stem and expansion colony ensures that breeding continues to propagate the line (Stem) while the production of suitable study mice can be optimized (Expansion). An example is given here of a small colony depicted over 4 generations (G0–G4), where two pairs populate the stem colony. Note that one pair only contributes future stem breeders. Mates used in this example are unrelated wt, as in a backcross, but brother–sister pairs are recommended if an inbred background is to be maintained. Heterozygous animals supply the expansion breeding, where a sufficient number of mated pairs (N) are set up to produce the required animals for study. Expansion colony breeding animals should be refreshed from the stem colony at each generation to prevent disparity developing between the two colonies due to genetic drift.

genetic background and then stably maintain it. Protection from genetic drift, inbreeding, genetic contamination, and disease are crucial. The expansion colony can be tailored to meet the needs of ongoing experiments, and a more relaxed approach can be taken to breeding and accessibility if the integrity and continuity of the line is assured in the stem colony. Expansion colonies can make use of trio and harem breedings. Timed matings can be set up specifically to generate numbers for a single study, or stable pairs can be maintained for a constant output. To avoid overproduction of animals, supply should be matched to demand. In order to prevent genetic divergence of the stem and expansion colonies, the number of generations that are supplied by the expansion colony for expansion colony breeding pairs should be limited to three generations. At that point, it is essential to take future breeding stock for the expansion colony from the stem colony. Some have found color systems to be helpful in keeping track of generations. An example of this is the "traffic light" system, in which the first mice supplied from the stem colony are labeled green, their progeny orange, and their progeny in turn red, in order to limit expansion breeding to three generations. Table 22.3 highlights the main differences between stem and expansion breeding. For larger scale production of a strain, a multitier structure can be used, such as those described by Hardy [15] and

Table 22.3
Comparison between stem and expansion colonies

Stem	Expansion
Goal: to preserve and perpetuate the line	Goal: to produce mice for analysis of phenotype
House in highest level barrier for protection against introduction of pathogens	House where most accessible and convenient for observation of phenotype
Maintain stable pairs of known pedigrees	Mate as pairs, trios, or harems and separate when sufficient litter numbers reached
Use breeding scheme (i.e., wt × het or het × het) that favors optimal reproductive performance	Use breeding scheme that best generates mice of required genotypes as test and control animals
Lengthen intervals between generations in order to slow genetic drift. Replace breeders from fifth litters wherever possible	Breeding animals sourced from stem colony and used for limited number of generations (2–3) only
If a pure inbred strain, mate brother–sister pairs. Mate the minimum number of pairs at each generation needed to keep the expansion colony supplied	If a pure inbred strain can mate any male and female
If backcrossing, mate to N10 against reference strain	If supplied from a stem colony that is being backcrossed, take note of N# of any breeding mice used to generate experimental animals

Festing [16]; however, the needs of most GM models can be well served by a two-tier colony.

22.2.5.2. Calculating Number of Breeding Females Required

Making a calculation of the number of mating pairs required for generating experimental animals can sometimes be challenging for those new to managing production colonies. Use of an algorithm like that given below (derived from TJL Breeding Resource Manual) can make this easier. Firstly, you will need to be aware of the following:

The desired number, sex, age, and genotypes that you require

The average output of the breeding females of your line

The usual time interval between litters

Any skewing in the ratios of genotypes produced

Any tendency to preweaning mortality

Examples of the range of these figures are given in Table 22.4

Table 22.4
Reproductive details of mice

Feature	Expected ranges
Age of sexual maturity (males)	6–7 weeks
Age of sexual maturity (females)	7–8 weeks
Estrus cycle frequency	4–5 days
Best age to start mating	7–8 weeks
Gestation period	18–21 days
Generation time	12–14 weeks In general: 3 weeks gestation, 3 weeks suckling, 3–5 weeks until sexual maturity
Litter interval	3–5 weeks
Litter size	3–20
Weaning age	19–28 days Pups need to be weaned before the next litter is born to avoid trampling of the newborns. For some strains, this may be earlier than optimal weaning age
Female fecundity	5–8 litters
Reproductive life span females	2–8 months
Reproductive life span males	2–12 months

Example 1: You require each month, 8 wild-type female and 8 female homozygous for your modification at within two weeks of age. You can set up breeding pairs with heterozygous mice sourced from your stem colony. The following is known about your mouse strain:

Average litter size is 6

Average interval between litters is 3 weeks

Mendelian ratios are normal

Preweaning mortality is at 10% across all genotypes

For calculation purposes, use the following:

Frequency: 1 = weekly, 2 = bi-weekly, 4 = monthly	4
Range of ages in weeks: 4 = 4 weeks, 2 = 2 weeks, 1 = 1 week	2

Calculate as follows:

Divide frequency by age range	2
Multiply by number of usable mice required monthly	8
Multiply by 1 if both sexes required, 2 if only one sex	2
Multiply by 1 if all progeny are desired genotype, 2 if half, 4 if a quarter	4
Multiply by 1 if Mendelian ratios are normal	1

Note: This calculation was based on the requirement to produce 8 mice only. This is because if 18 breeders are required to produce 8 homozygous females monthly, then an equal number of wild-type females will result.

(A) Number of mice required to be produced in total monthly	128
Average number of pups born/litter	6
× Preweaning mortality factor	0.9
÷ Average interval between litters in months	0.75
(B) Productivity per female/month	7.2
Number of female breeders needed to be maintained = A ÷ B	
For this example: 128 ÷ 7.2 = 18	

Additional factors can be added into the calculation, for instance a percentage overproduction to guarantee that sufficient numbers are supplied [e.g., multiply number of mice (A) by 1.1 for a 10% excess]. Note that in addition, 64 males and 32 heterozygous female mice will be produced in excess to the experimental needs. A further 16 homozygote and wild-type mice produced will fall outside the 2-week age range.

An alterative to the above scheme is to use homozygote and wild-type mice sourced from the stem colony (if maintained as het × het breeding pairs), and use wild-type females to generate wild-type progeny, and homozygous females mated with homozygous males to create homozygous progeny. This assumes that the reproductive abilities of each genotype are equivalent and that littermate controls are not required. In this case, there is an advantage to be gained by being able to deliver the study mice with half the number of breeding females, and without the production of excess heterozygote mice.

Example 2: You require each month, 5 homozygous males to be available within two weeks of age. You can set up breeding pairs with heterozygous mice sourced from your stem colony. The following is known about your mouse strain:

Average breeding output is 6 per month

Average interval between litters is 3 weeks

Mendelian ratios are skewed: homozygous are only found at a frequency of 12.5% in progeny of matings of heterozygous mice

Preweaning mortality is at 25%

Calculation is as follows:

Frequency: 1 = weekly, 2 = bi-weekly, 4 = monthly	4
Range of ages in weeks: 4 = 4 weeks, 2 = 2 weeks, 1 = 1 week	2
Divide frequency by age range	2
Multiply by number of usable mice required monthly	5
Multiply by 1 if both sexes required, 2 if only one sex	2
Multiply by 1 if all are desired genotype, 2 if half, 4 if a quarter	4
Multiply by 1 if Mendelian ratios are normal, >1 if a genotype occurs at a reduced frequency	2
(A) Number of mice required to be produced in total monthly	160
Average number of pups/litter	6
× Preweaning mortality factor	0.75
÷ Average interval between litters in months	0.75
(B) Productivity per female/month	6
Number of female breeders needed to be maintained = A ÷ B	
For this example: 160 ÷ 6 = 27	

To set up a constant supply can be very economical if you require only small numbers of mice at any time. It is useful also if you are conducting preliminary studies before developing larger scale experiments and if you are able to work with mice that fall within a range of ages. To produce a larger number of mice less frequently, to use one sex only, or to narrow the usable age range will all result in an increase in the number of breeding pairs required and associated costs. As an example: a weekly supply of 20 mice at 10 weeks of age requires that 16 mating pairs are set up, while a supply of 40 mice every two weeks at the same age requires 32 mating pairs, and generates an equal number of mice which are not used.

22.2.5.3. Timed Matings

Timed matings may be needed when it is important that the exact date of copulation is known. This is the case when fetuses are required for developmental studies and exact Theiler stages must be known as precisely as possible. Additionally, it is useful to be able to schedule parturition dates when perinatal intervention is required. It may also be necessary at times to generate adult mice with a narrow age window (less than 1 week). To produce timed-mated pregnant females, the mated females must be checked daily from the date of pairing with the male for the presence of a copulation plug. Plug checking should occur soon after the onset of the daylight period. The estrus cycle is 4–5 days in female mice, so most females will mate within five days of pairing. Alternatively, if a number of timed mated females are required for experimental use on 1 day, then five times the number of plugged females required should be mated on one night, since 20% of females only are likely to be in estrus. In nature, only 80% of matings result in a pregnancy. This detail should also be taken into account when setting up timed matings, as well as any strain characteristics, such as reduced fertility in males, (C57BL/6 for example). The proportion of females ready for mating can be increased by the use of male pheromones (Whitten effect) or by the selection of females in estrus or diestrus (these techniques are described in [17]). Once a plug is observed, the female is separated from the male and weighed. A plug will show that mating has taken place, but will not guarantee a pregnancy. Significant weight gain in the following 14 days will confirm pregnancy. If timed-mated females are often required, then a group of single-housed stud males can be kept for this purpose. Recording copulation results and particularly the ensuing pregnancies for each stud male will help in identifying and removing ineffectual males and improving the performance of the stud pool as a whole.

22.2.6. The Importance of Robust Genotyping Protocols

As your colony expands, timely and accurate genotyping results become critical. The consequences of poorly identified genotypes in a breeding colony can lead to delays in research progress, unnecessary per diem costs, inconsistent data, and at worst, the unintentional loss of mouse lines. It is wise to consider as early as possible whether your genotyping methods are robust, reliable, and economical enough for your needs.

It is recommended, if using PCR-based screening methods, to have a unique set of primers for each mutant allele. The use of antibiotic resistance genes, recombinases, or other sequences common to many lines will not help to distinguish between mice if accidental mixing occurs in the facility. In the case of gene-targeted modifications, the flanking sequences are known and it is an easy matter to design primers that will identify a wild-type or mutant allele. For a transgene, internal primers can often be used, as the combination of promoter and coding sequences will not usually have an endogenous equivalent. A quantitative PCR method may need to be established to determine heterozygous from homozygous transgenic lines. Distinguishing between point mutations or subtle sequence variations will always be difficult. If strains with similar characteristics and names are unable to be distinguished by routine screening, then think about ways to prevent any confusion of strains, by renaming, using colored cage cards or housing in separate rooms.

22.3 How to Breed Transgenic Founders

Transgenic founders are created by the random insertion of foreign DNA into the genome, whether by pronuclear microinjection or viral transduction. Your founder animals will have been identified by the presence of transgenic sequences in genomic DNA isolated from tissue biopsies. It is likely that you will have produced multiple founders from the injection sessions (typically from 2 to 10 founders), and will need to select those from which you will establish stable colonies. If generated by microinjection of DNA, including use of BAC or YAC constructs [46], there is likely to be only one site of insertion in each founder. Multiple sites of insertion, however, are still a possibility that needs to be ruled out for each line. Multiple copies of the transgene may be inserted at the same locus in tandem arrays, but not all will be transcriptionally active [18]. When lentiviral vectors are used, insertion is very likely to occur at multiple sites in the genome of a founder, with the result that transgenes will then segregate on further breeding [19]. A single copy of the transgene construct, however, will have integrated at each site.

Southern blots with carefully selected restriction enzymes and probe sequences can give you the best information about number of integration sites when characterizing your transgenic founder animals and can also be used to check for any gross rearrangements of your construct. Details on how to establish copy number by

Southern or dot blot can be found in Nagy et al. [17] and Lois [20]. Southern blots should be used to examine both the founder animals and their progeny, as an unrecognized segregation of transgenes at different integration sites (with potentially different expression profiles) can confound interpretation of transgene expression data.

All founder animals are considered unique at F0 and are to be kept and characterized as separate lines, as each has arisen from an embryo that has undergone a unique integration event. Although the transgene is identical in each case, the consequences of the site of integration on expression of the transgene and other neighboring genes may be very different. Integration in areas of varying relative transcriptional activities and the influence of flanking sequences ("position effects") can result in a wide range of expression levels and tissue profiles between lines derived from different founders, all carrying the same transgene [21]. This diversity can sometimes be an advantage in the analysis of phenotype, but more frequently, the number of lines for analysis is narrowed down within the first generations to those that show an adequate level of expression in the primary tissues of interest.

While it is not essential in the characterization of transgene function to determine the site of integration, there are a number of desirable benefits to this. Knowledge of the surrounding sequences allows for the determination of heterozygous and homozygous individuals by PCR by the use of primers to distinguish disrupted from undisrupted alleles. Information about chromosome location can aid in planning breeding of multiple genetic modifications. Finally, awareness of any disruption of endogenous genes can be useful in assessing unexpected phenotypes. By sequencing fragments synthesized from primers internal to the transgene, but extending into the surrounding sequence, a BLASTN search will identify the insertion site [22]. FISH analysis can be used to confirm the predicted chromosomal location [23].

If the transgenic founder is male, then it can be mated to a number of females to generate a large number of F1 offspring for breeding as well as subsequent analysis. If female, production will be slower at one litter at a time and securing breeding animals will have to take precedence initially over analysis of phenotype.

22.3.1. Choice of Strain for Mating Founder Mice

If embryos of an inbred strain have been used for injection (e.g. C57BL/6, FVB/N), then breeding to the same strain of mice will maintain a pure strain background. If desired, another strain can be used concurrently to start backcrossing into an alternative strain background. Quite commonly, the embryos used for injection are the F2 offspring of the mating of two hybrid animals (e.g., B6SJL F1 or B6D2 F1), chosen for their robust embryo yields in response to superovulation, and for ease of visualizing the pronuclei. In this case, the option exists to maintain a hybrid strain by continuing to breed with F1 hybrids. This can have advantages in terms of good fertility and fecundity, which are features of hybrid vigor. In addition, genetic variation between individuals will be stably maintained between generations. Alternatively, you may choose one of the parental strains and begin backcrossing with this strain. Due to the prevalence of C57BL/6 in mouse models, this strain will be the most likely candidate.

22.3.2. Generation of F1 Mice

Establishing germline transmission from transgenic founder animals is usually straightforward as long as the modification itself does not impair reproductive function. Certainly transgenic offspring should be identified and known to be fertile before the founder animals are used for any purpose other than breeding. The mating of founder F0 mice to wild-type mates should produce offspring hemizygous for the transgene at a rate of 50%. Founders that do not produce any transgenic offspring can be eliminated. Transmission of the transgene at a frequency of less than 50% indicates mosaicism in the founder. Integration of the transgene can occur before the first division of the zygote, in which case the transgene will be found in all cells. Frequently, however, it is found that stable integration of the transgene is delayed until after one or more rounds of DNA replication, giving rise to a mosaic founder in which only a proportion of cells, and gametes, carry the transgene [24, 25]. A mosaic founder should not be used for analysis. The F1 progeny of the founder will not be affected by any mosaicism and so are more suitable for use. F1 mice can be used for initial analysis of transgene expression in the hemizygous state, and lines selected for further study by expression level or distribution.

If a high frequency of transgene transmission is seen in the F1 generation, then it is likely that multiple sites of integration are segregating. This could be confirmed by Southern analysis. In this case, further breeding of F1 mice, and possibly subsequent generations, to wild-type mates will be required. It will not be possible to determine which lines are preferred in terms of expression until stable inheritance is achieved. Unless your initial number of founders was small (less than 3), it will usually be necessary to make a selection from all your founders for the lines that have desirable expression levels and tissue distribution, with the minimum copy and insertion numbers. Thus, it is advisable to breed from all of your founders until you have completed this initial analysis and identified the most useful lines. If this analysis extends over a long time period, it can be beneficial to cryopreserve some lines in case they are needed for further investigation.

22.3.3. Generation of F2 Mice

Hemizygous F1 mice will transmit the transgene at normal Mendelian ratios. These mice can be intercrossed to generate F2 mice for study of the effects of a homozygous allele, but care should be taken to watch for recessive lethality effects, especially in the embryonic stages. As discussed in Subheading 22.2.1, effects of insertional mutagenesis on an endogenous gene may become apparent. It is highly desirable to maintain your line as homozygous hereafter, as this reduces the need to genotype and reduces animal numbers. This can only be achieved, however, if the line is on a pure inbred background and has no associated lethality or reproductive impairment. If the line is not on a pure background, then breeding should proceed as hemizygous mating to wild-type mice (if the wild-type strain is one of the parent strains then this is termed backcrossing) in order to avoid fertility problems due to inbreeding depression.

22.4 How to Breed Targeted Mutations

The use of homologous recombination in mouse embryonic stem cells to create site-specific deletions, insertions, and alterations is referred to as gene targeting. In a related approach, gene trapping, recombination takes place selectively in regions of the genome of high transcriptional activity. The founder animals resulting from either gene targeting or trapping are chimeric, consisting of a mixture of cells derived from a host embryo and the ES cell clone carrying the genetic alteration. Chimeras are produced in a number of ways (detailed in Chapters 16 and 17 in this manual), the most common being the injection of ES cells into the blastocoel cavity of a host embryo. Other methods include the injection of ES cells into eight cell embryos, the aggregation of ES cells with zona-denuded diploid embryos, or the incorporation of ES cells into tetraploid embryos by either aggregation or injection. The method used will have some implications for the choice of strains used and subsequent breeding.

Mouse ES cells that can contribute to the germline were initially established using 129 substrains, although there are now ES cell lines available that are derived from a variety of strains, 129 and C57BL/6 strains are still the most widely used. There are two criteria in the selection of host embryo strain: the first being that the combination of ES cell strain and host embryo strain is one that favors ES cell involvement in gamete formation. The second is that coat color can be used to distinguish host embryo from ES cell-derived tissues and identify chimeric mice. Percentage chimerism can be evaluated by relative contribution of the different coat colors and germline transmission can be confirmed when the coat color of the offspring indicates they are of ES cell origin. Table 22.5 lists the coat colors of some common ES cell strains as well as the color combinations that result when chimeras are produced with host embryos of C57BL/6, "albino" C57BL/6 and BALB/c strains. Predictions for coat color in offspring, when test breeding for germline transmission, are also given, although in some cases coat color cannot be used as the primary means of identifying ES cell-derived offspring. Chimeras of ES cells with outbred strains are also included in Table 22.5; however, these are almost exclusively produced by aggregation or by injection of ES cells into the perivitelline space of an eight-cell embryo. Microinjection of outbred blastocysts with ES cells is inefficient for generation of chimeras capable of germline transmission [26], although the combination of outbred eight-cell embryos and ES cells can produce germline chimeras [27, 28].

In the generation of your targeted ES cells you will have selected for those clones that incorporate the genetic modification as designed. You can expect that all chimeras produced with the same ES cell clone should give rise to mouse lines that are essentially identical. You should be cautious, however, to treat mouse lines produced from different clones as distinct, as clonal expansion and treatment in vitro may have resulted in unique genetic changes. Chromosomal abnormalities are common in ES cells, particularly trisomies that in some cases can confer a growth advantage to the clone [6, 7]. Screening for euploidy of candidate clones before producing chimeras (*see* Chapter 17 in this manual) will result in a better outcome overall, but will not be a guarantee that the clone retains full pluripotency. In practice, the ability to "go germline" can differ widely between clones. Mouse lines generated from two or more clones carrying the same genetic alteration can be used to provide confirmation of a phenotype. Ultimately one line will be preferred and retained for further study as there is little value in maintaining separate but essentially identical GM mouse lines.

In order to establish a line, a chimera must have sufficient contribution from ES cells to the gonads that ES cell-derived gametes can be produced. Typically, a single ES cell clone will have been used to generate a number of chimeric mice. The male chimeras with the highest proportion of ES cell coat color are selected for breeding as they have the best chance of transmitting through the germline. Most mouse ES cells commonly used are male, and will therefore interfere with the reproductive development of females if present in the ovaries at a high percentage. You should see a distortion of the sex ratio towards male mice amongst the chimeras, as a high number of male cells incorporated into a female host embryo will result in a mouse that is functionally male. Female chimeras are often infertile, although an ES cell clone that has lost the Y chromosome and thus has become XO can result in the generation of fertile female chimeras that will transmit the mutant allele. This is reported to occur with enough frequency that it is worthwhile retaining chimeric females for breeding if

Table 22.5
Coat colors observed in the generation of mouse chimeras from commonly used ES cell strains

Representative ES cell line(s)	Mouse strain of ES cells [genotype, color gene markers] (phenotype)	Preferred host strain for generation of chimeras [genotype, color gene markers] (phenotype)	Resulting coat colors in chimeras	Test breeding for assessing germline transmission (GLT)
129 ES cell strains:				
R1	(129X1/SvJ × 129S1/SvImJ)F1 [A^w/A^w; $Oca2^p/Oca2^+$; Tyr^{c-ch}/Tyr^+] (white/light-bellied agouti)	C57BL/6J [*a*/*a*] (black, nonagouti)	Agouti patches over black background	C57BL/6J. Any agouti pup demonstrates GLT
		"Albino B6" B6-(Cg)-Tyr^{c-2j}/J [*a*/*a*; Tyr^{c-2j}/Tyr^{c-2j}] (albino) *Or* C57BL/6-Tyr^{c-Brd} [*a*/*a*; Tyr^{c-Brd}/Tyr^{c-Brd}] (albino)	Agouti pigmented patches over albino background	"Albino B6" B6-(Cg)-Tyr^{c-2j}/J (or any albino outbred or inbred strain). Any pigmented pup (various coat colors, from agouti to pink-eyed/chinchilla) demonstrates GLT
		Outbred CD1 (or ICR)[a] [Tyr^c/Tyr^c] (albino)	Pigmented patches (agouti and black) over albino background	Outbred CD1 (or ICR). Any pigmented pup (various coat colors, from agouti to pink-eyed/chinchilla) demonstrates GLT
E14, E14.1, E14TG2a	129P2/OlaHsd [A^w/A^w; $Oca2^p/Oca2^p$; Tyr^{c-ch}/Tyr^{c-ch}] (white/light-bellied, pink-eyed, light chinchilla, light tan)	C57BL/6J [*a*/*a*] (black, nonagouti)	Pale agouti patches over black background	C57BL/6J. Any agouti pup demonstrates GLT
		"Albino B6" B6-(Cg)-Tyr^{c-2j}/J [*a*/*a*; Tyr^{c-2j}/Tyr^{c-2j}] (albino) *Or*	Pale agouti patches over albino background (very light pigmented beige/creamy patches over albino coat color)	"Albino B6" B6-(Cg)-Tyr^{c-2j}/J (or any albino outbred or inbred strain). Any pigmented pup (white/light-bellied agouti, chinchilla) demonstrates GLT

		C57BL/6-$Tyr^{c\text{-}Brd}$ [a/a; $Tyr^{c\text{-}Brd}/Tyr^{c\text{-}Brd}$] (albino)		
W9.5, CJ7… D3, CK35… W4, TC-1… AB1… J1…	129S1/SvImJ 129S2/SvPas 129S6/SvEvTac 129S7/SvEvBrd 129S4/SvJae [A^w/A^w] (white/light-bellied agouti)	C57BL/6J [a/a] (black, nonagouti)	Agouti patches over black background	C57BL/6J. Any agouti pup demonstrates GLT
		"Albino B6" B6-(Cg)-$Tyr^{c\text{-}2j}/J$ [a/a; $Tyr^{c\text{-}2j}/Tyr^{c\text{-}2j}$] (albino) *Or* C57BL/6-$Tyr^{c\text{-}Brd}$ [a/a; $Tyr^{c\text{-}Brd}/Tyr^{c\text{-}Brd}$] (albino)	Agouti patches over albino background	"Albino B6" B6-(Cg)-$Tyr^{c\text{-}2j}/J$ (or any albino outbred or inbred strain). Any pigmented pup (white/light-bellied agouti) demonstrates GLT
		Outbred CD1 (or ICR)[a] [Tyr^c/Tyr^c] (albino)	Pigmented patches (agouti and black) over albino background	Outbred CD1 (or ICR). Any pigmented pup (various coat colors, from agouti to black) demonstrates GLT
RW-4	129X1/SvJ (The "chinchilla" and "albino" alleles of the tyrosinase gene segregate in this strain) [A^w/A^w; $Oca2^p/Oca2^p$; $Tyr^{c\text{-}ch}/Tyr^{c\text{-}ch}$] (white-bellied, pink-eyed, light chinchilla, light tan) *Or* [A^w/A^w; $Oca2^p/Oca2^p$; $Tyr^{c\text{-}ch}/Tyr^c$] (white-bellied, pink-eyed, light chinchilla, off-white) *Or* [A^w/A^w; $Oca2^p/Oca2^p$; Tyr^c/Tyr^c] (albino)	C57BL/6J [a/a] (black, nonagouti)	Pale agouti to albino patches over black background	C57BL/6J. Any agouti pup demonstrates GLT

(continued)

Table 22.5 (continued)

Representative ES cell line(s)	Mouse strain of ES cells [genotype, color gene markers] (phenotype)	Preferred host strain for generation of chimeras [genotype, color gene markers] (phenotype)	Resulting coat colors in chimeras	Test breeding for assessing germline transmission (GLT)
C57BL/6 ES cell strains:				
JM8, JM8.F6, JM8.N4, C2, VGB6	C57BL/6N [*a/a*] (black, nonagouti)	"Albino B6" B6-(Cg)-*Tyr*$^{c\text{-}2j}$/*J* [*a/a*; *Tyr*$^{c\text{-}2j}$/*Tyr*$^{c\text{-}2j}$] (albino) *Or* C57BL/6-*Tyr*$^{c\text{-}Brd}$ [*a/a*; *Tyr*$^{c\text{-}Brd}$/*Tyr*$^{c\text{-}Brd}$] (albino)	Pigmented patches (black) over albino background	"Albino B6" B6-(Cg)-*Tyr*$^{c\text{-}2j}$/*J* (or any albino outbred or inbred strain). Any pigmented pup (various coat colors, from black to agouti) demonstrates GLT
		BALB/c [*A/A*; *Tyrp1*^{b}/*Tyrp1*^{b}; *Tyr*^{c}/*Tyr*^{c}] (albino)	Pigmented patches (agouti and black) over albino background	BALB/c (or any albino outbred or inbred strain). Any pigmented pup (various coat colors, from black to agouti) demonstrates GLT
		Outbred CD1 (or ICR)[a] [*Tyr*^{c}/*Tyr*^{c}] (albino)	Pigmented patches (agouti and black) over albino background	Outbred CD1 (or ICR). Any pigmented pup (various coat colors, from agouti to black) demonstrates GLT
JM8A3, JM8A3.N1, JM8A1.N3	C57BL/6N ("agouti") [*A*^{tm1Brd}/*a*] (Agouti)	C57BL/6J [*a/a*] (black, nonagouti)	Agouti patches over black background	C57BL/6N. Both agouti and black pups demonstrate GLT. Must be distinguished from host-derived black pups by genotype screening
Bruce 4	C57BL/6J – N3 NZB backcross [*a/a*] (black, nonagouti)	"Albino B6" B6-(Cg)-*Tyr*$^{c\text{-}2j}$/*J* [*a/a*; *Tyr*$^{c\text{-}2j}$/*Tyr*$^{c\text{-}2j}$] (albino) *Or*	Pigmented patches (black) over albino background	"Albino B6" B6-(Cg)-*Tyr*$^{c\text{-}2j}$/*J* (or any albino outbred or inbred strain). Any pigmented pup (various coat colors, from black to agouti) demonstrates GLT

		C57BL/6-*Tyr*$^{c\text{-}Brd}$ [*a*/*a*; *Tyr*$^{c\text{-}Brd}$/*Tyr*$^{c\text{-}Brd}$] (albino)		
		BALB/c [*A*/*A*; *Tyrp1*^{b}/*Tyrp1*^{b}; *Tyr*^{c}/*Tyr*^{c}] (albino)	Pigmented patches (agouti and black) over albino background	BALB/c (or any albino outbred or inbred strain). Any pigmented pup (various coat colors, from black to agouti) demonstrates GLT
"Albino B6"	"Albino B6" B6-(Cg)-*Tyr*$^{c\text{-}2j}$/*J* [*a*/*a*; *Tyr*$^{c\text{-}2j}$/*Tyr*$^{c\text{-}2j}$] (albino)	C57BL/6J [*a*/*a*] (black, nonagouti)	Albino patches over black background	"Albino B6" B6-(Cg)-*Tyr*$^{c\text{-}2j}$/*J* (or any albino outbred or inbred strain). Any albino pup demonstrates GLT
F1 and other strains:				
G4... VGF1... V6.5... V6.4...	(129S6/SvEvTac × C57BL/6NCr)F1 (C57BL/6J x129S6/SvEvTac) F1 (129S4/SvJae x C57BL/6J)F1 (C57BL/6J x 129S4/SvJae)F1 [*A*^{w}/*a*] (white/light-bellied agouti)	C57BL/6J [*a*/*a*] (black, nonagouti)	Agouti patches over black background	C57BL/6J. Any agouti pup demonstrates GLT
		"Albino B6" B6-(Cg)-*Tyr*$^{c\text{-}2j}$/*J* [*a*/*a*; *Tyr*$^{c\text{-}2j}$/*Tyr*$^{c\text{-}2j}$] (albino) *Or* C57BL/6-*Tyr*$^{c\text{-}Brd}$ [*a*/*a*; *Tyr*$^{c\text{-}Brd}$/*Tyr*$^{c\text{-}Brd}$] (albino)	Agouti patches over albino background	"Albino B6" B6-(Cg)-*Tyr*$^{c\text{-}2j}$/*J* (or any albino outbred or inbred strain). Any pigmented pup (white/light-bellied agouti) demonstrates GLT
		Outbred CD1 (or ICR)[a] [*Tyr*^{c}/*Tyr*^{c}] (albino)	Pigmented patches (agouti and black) over albino background	Outbred CD1 (or ICR). Any pigmented pup (various coat colors, from agouti to black) demonstrates GLT
TT2				

(continued)

Table 22.5 (continued)

Representative ES cell line(s)	Mouse strain of ES cells [genotype, color gene markers] (phenotype)	Preferred host strain for generation of chimeras [genotype, color gene markers] (phenotype)	Resulting coat colors in chimeras	Test breeding for assessing germline transmission (GLT)
	(C57BL/6 × CBA)F1 [*A/a*] (agouti)	C57BL/6J [*a/a*] (black, nonagouti)	Agouti patches over black background	C57BL/6J. Any agouti pup demonstrates GLT
NOD	NOD/ShiLtJ [*A/A*; Tyr^c/Tyr^c] (albino)	C57BL/6J [*a/a*] (black, nonagouti)	Albino and agouti patches over black background	NOD/ShiLtJ (or any albino outbred or inbred strain). Any albino pup demonstrates GLT

Notes: Prepared by Lluis Montoliu (CNB-CSIC, Madrid, Spain) with modifications by Karen Brennan, July 2010
[a]Outbred host embryos are used primarily for chimera generation as 8-cell embryos in aggregations with ES cells or with injection of ES cells into the perivitelline space

unsuccessful in establishing germline transmission with male chimeras. Your first step however, once chimeric mice have reached breeding age, should be to breed a selection of the highest percentage male chimeras to multiple females in a short space of time to test for germline transmission. Males exhibiting chimerism above 50% are preferable for breeding, but this may vary according to the ES cell line used.

Typically, when breeding chimeras produced by blastocyst injection of ES cells, mice of both ES cell and host embryo origin would be found amongst the offspring, at a ratio that reflects the proportion of chimerism in the gonads. Chimeras produced from aggregation of ES cells with eight-cell to morula stage embryos will have a different pattern of transmission. On the whole, these chimeras will tend to transit to all offspring or to none. This is presumably the result of the higher relative contribution by the ES cells when incorporated into an earlier stage embryo. It is advisable, particularly in the case of chimeras derived from blastocyst hosts, to produce several litters from each chimera as it is not uncommon to find that chimeras that do not transmit in their first litter may quite well do so in the second or even third litters produced.

22.4.1. Choice of Strain for Mating Chimeras

In order to identify the pups that are ES cell-derived, coat color can again be used as an indicator of origin. The coat colors of the offspring will be a result of the combination of coat color genotypes of either the host embryo or the ES cell strain and the strain of mate chosen. Coat color genetics involves genes that encode for pigment produced by both melanocytes and the hair follicle cells, and is a complex phenotype. It is well worth researching beforehand the possible coat color outcomes of your matings so that you are prepared by day 10, when fur appears, to evaluate your litter, particularly if a new ES cell or host embryo strain is to be used. More information can be found in "The coat colors of mice" by W. Silvers [29], which is available online at http://www.informatics.jax.org/wksilvers. Note that nomenclature for coat color alleles has changed, and a key to old and new symbols is also found at this site. A good discussion of this topic is also found in Pease [30].

The choice of mate for breeding the chimera is important as it dictates your expectations for offspring coat color as well as determining the strain background on which your first phenotypic analysis will occur. If you have used 129 strain ES cells and C57BL/6 host embryos to create a chimera, mating this to a C57BL/6 strain mouse will produce your mutant mice on a (C57BL/6 × 129) F1 background. Your ES cell-derived pups will have a black agouti appearance (black hair with yellow banding, as opposed to brown agouti: brown hair with yellow banding) and will be easily distinguished from those that are host embryo derived, which will be C57BL/6 and black. The mutant mice will not be on a pure inbred background for analysis, however, and you may require backcrossing for up ten generations to achieve this. Alternatively, if your selection of mate for the same chimera is a 129 strain mouse (the substrain used should be the same as that of the ES cells) then you produce your mutant mice on a co-isogenic (*see* Table 22.2) background immediately. The coat color of the offspring will depend on the coat color genes of the particular 129 substrain used. Since the host embryo derived pups will have the same color as the ES cell-derived ones, you will need to use molecular screening to distinguish your mutants.

When using an ES cell clone of a C57BL/6 background it is ideal to use an albino host embryo, for instance, BALB/c or an albino C57BL/6 strain such as C57BL/6-$Tyr^{c\text{-}2J}$ or C57BL/6-$A^{c\text{-}Brd}$. If the resulting chimera is mated with an albino mate, ES cell-derived pups are indicated by any coat color other than albino. Specifically, if you mate chimeras generated with C57BL/6 ES cells to albino C57BL/6 females, germline transmission will be indicated by the presence of black pups and if mated to BALB/c females, then the production of black agouti pups will indicate germline transmission (Table 22.5). An advantage to be gained by breeding an albino C57BL/6 mouse with the chimera would be that the F1 generation is already co-isogenic in a C57BL/6 background, with the exception of the allele conferring albino coat color. A similar refinement is possible using an agouti C57BL/6 ES cell strain JM8A3 [31] to create chimeras with a C57BL/6 host embryo. These chimeras display agouti patches on a black background. The mating of such chimeras to C57BL/6 mice will result in both agouti and black pups of ES cell origin as the restored agouti locus exists as a dominant heterozygous in the JM8A3 ES cells. In this case, you would need to screen the pups for the mutant genotype to distinguish those that are black and ES cell origin from those that are black and host embryo derived.

If you are not able to track germline transmission by use of a dominant coat color allele and wish to retain the genetic integrity of your mouse mutation, you can choose to first test-mate your chimeras with a strain that will give a good indication of germline transmission by color, and then follow up with matings of a different strain that will generate offspring of a desirable genetic background. It is not unusual to use molecular screening methods alone to distinguish mutant pups in a litter, by PCR analysis of tail, ear, or blood samples. This is, however, a more labor-intensive approach to screen all offspring from chimeras for germline transmission than using coat color. Once a chimera is known to transmit, it is not essential in subsequent litters to be able to distinguish the ES cell-derived pups by coat color if other means can be used.

If a backcrossing strategy is to be used to transfer the genetic modification onto another strain (*see* Subheading 22.5.2), then the mating of the chimera to a strain unrelated to the ES cell strain will be the first outcross of the process. If you are using a conditional strategy and wish at this point to introduce a recombinase mediated modification to your mutant sequences, you may use the recombinase-producing mouse for mating with chimeras, rather than wait until F1 animals are available for breeding. This method can be used to remove a selection cassette, for instance, that may interfere with regulation or expression of your model. Keep in mind, though, that as this step is usually is irreversible you may benefit from retaining your unmodified line as a separate breeding colony, for mating to other recombinase-producing mouse lines in the future.

22.4.2. Generation of F1 Mice

Recombination by use of gene targeting or trapping usually occurs only on one allele; therefore the ES cell clone used would be heterozygous for the mutant allele. The F1 progeny of the chimera that are ES cell derived, would therefore be either wild type or heterozygous for the mutant allele, at a ratio of 1:1. Screening by Southern, PCR, or phenotypic methods is required to distinguish these genotypes. If a dominant effect is expected from the genetic alteration, then it may become evident in this generation. Once inheritance of the mutant allele has been established, you should set aside heterozygous mice derived from one or more chimeras as foundation breeding animals. If several clones have been used to produce chimeras, then offspring should be maintained as separate colonies, distinguished by clone number.

If the situation arises that ES cell-derived F1 progeny are produced, as detected by coat color, but that no individuals heterozygous for the mutant allele are detected, then there are two possibilities. It is possible that your ES cell clone consisted of a mixed population of cells, and as a result, wild-type ES cells, or a subpopulation that are a result of nonhomologous recombination, have populated your chimeras. In this case, you will either have to subclone to produce a pure population of cells with the correct homologous recombination, or select another positive clone from the same electroporation for injection, or start again with electroporation and selection. Alternatively, it may be that your mutant allele has a dominant lethal effect and you may have to consider a different vector design strategy using conditional alleles.

22.4.3. Generation of F2 Mice

In the F2 generation, the first homozygous mutant mice can be generated for analysis from matings of heterozygous F1 mice and should be closely examined for the influence of phenotype that may affect their development and well-being. Except in the case of a deletion of a gene where the function or significance is unknown, there will be some expectation about the impact of the genetic alteration, and this will be a guide to the expected severity of the outcome. These mice are likely to be examined on a mixed genetic background and this should be taken into account when reporting findings. Phenotypes should be subsequently confirmed on one or more defined inbred backgrounds.

To propagate the mouse line, heterozygous F1 animals produced with a mixed strain background should be mated to wild-type mice of an appropriate strain (*see* comments Subheading 22.5.1). If backcrossing to create a congenic line had been initiated with the chimera mating, crossing individuals of the F1 generation to a wild-type mouse of the desired strain would result in N2 offspring. As the chimera was most likely male, it would be recommended to cross a female F1 to a wild-type male at this point to ensure that the Y chromosome from the desired strain is introduced. Subsequent matings should be of male GM mice to female wild type (*see* Subheading 22.5.2). Once an inbred strain background has been established, and the genetic modification has no implications for the reproductive ability of homozygous mice, then the line can be propagated thereafter in a homozygous state, by brother × sister matings.

22.5 How to Maintain Strain Background, Backcross to Create Congenic Strains or Keep an Outbred Line

The generation of germline genetic alterations has historically made use of particular mouse strains that have facilitated the development of these technologies. Transgenic mice created by pronuclear injection have frequently been made with F2 embryos generated from F1 parents, themselves crosses of C57BL/6 with CBA, DBA/2, or SJL strains. Inbred strains such as FBV/N and the more challenging C57BL/6 have also been used [32]. These have produced large numbers of embryos with easily injectable pronuclei and good survival rates. Targeted transgenesis first employed ES cells derived from 129 substrains, as this strain more readily resulted in germline transmission after manipulation [33]. Other inbred strains have been successfully used, including C57BL/6 and BALB/c [34, 35]. The advent of global projects for mouse knock-outs (6) have recently seen a major shift towards use of the C57BL/6N substrain for gene targeting and gene trapping strategies [31]. (6) http://www.knockoutmouse.org/

The strain in which your founder animals are produced may not be the ideal strain for conducting phenotypic studies. It is frequently advantageous to cross your mutant allele into another genetic background and possibly even multiple backgrounds. The influence of strain background on phenotype is of such significance that confirming a phenotype in two unrelated strains is highly desirable. It is also desirable to achieve a uniform genetic background for the line as quickly as possible, the goal being to maintain the GM allele against a defined inbred background. As making changes to strain backgrounds takes time, it would be good practice to anticipate your future needs by breeding founder animals to (1) the inbred strain to which they are most closely related (e.g., the comparable 129 strain for ES cell-derived mice, and the strain of injected zygotes for transgenic mice), and concurrently to (2) the inbred strain that could be expected to be best for analysis of phenotype. Bear in mind that unless your eventual background is isogenic (*see* Table 22.2) with your genetic alteration, that is, of an identical strain background, there will be present in your mouse model residual genomic sequences from the strain used for production. Current improvements in mouse genetic resources should lead increasingly to new models being generated in an isogenic fashion, from construct design through to establishment of GM colonies.

22.5.1. Working with a Mixed Strain Background

Frequently, the situation arises that the founder mice, or their F1 progeny, are a mixture of two or more strain backgrounds. As the line is expanded, and homozygous mice produced for analysis, it is a temptation to use heterozygous mice for further breeding. Whilst this might be needed for initial analysis of mutations in a homozygous state, for long-term colony maintenance, this should be strongly discouraged as a poor breeding practice [36]. While some hybrid vigor can be helpful in first generations, propagating through breeding related mice in effect creates new inbreeding. Starting from a uniform mixture of alleles at F1, there will be gradual loss of heterozygosity at each generation, which will occur more rapidly the fewer mating pairs are used. Genetic variability between individual mice will increase and inbreeding depression may occur due to fixing of alleles that are detrimental to reproductive performance. If your initial GM mice are of a mixed strain background, it is preferable to maintain the colony by breeding to wild-type mice of a defined inbred background for at least 10 generations. After this the strain will be considered congenic (*see* Table 22.2), i.e., with a genetically identical background to other mice of that inbred strain, with the exception of the sequences immediately surrounding the selected marker (your GM allele). If a hybrid background is most desirable for the line, breeding pairs should be set up with F1 hybrids at each generation, which will not only prevent inbreeding, but will preserve the degree of genetic variability between individuals.

22.5.2. Backcrossing to Create a Congenic Line

In order to backcross to create a congenic line, you will first need to cross your founder, or starting animals to a wild-type mouse from the strain you are backcrossing into, i.e., "the target strain." If this strain is not already contributing to your starting mouse, this is called an outcross (offspring are termed F1). If it is one of the parental strains of the starting mouse, then this is your first backcross (offspring are termed N1). The offspring will contain 50% of the genetic background of the starting mouse and 50% of the wild-type mate. Mice that are heterozygous for the mutant allele are selected as the next generation breeding mice and mated again to wild type. Note that their offspring are termed N2 whether the parents were Fl or N1. Continuing to mate in this way, there will be a reduction of 50% at each generation of the genetic background of the starting mouse, until at N10 there is 99.9% genetic contribution of the desired strain background. This is the point at which further breeding brings insignificant increase of the strain contribution (a chromosome fragment of ~20 cM will persist around the allele you selected for) and so the backcross is considered complete at N10. Note that at N4 there is already more than 90% of the genome of the desired strain so it is commonplace that analysis begins sooner than N10, and the generation used declared on publication. Strains that are in the process of being backcrossed but are not yet at N10 are termed "incipient congenics."

It is important that at least one cross is the mating of a female carrying your mutant allele to a wild-type male. It is usual to begin with this step as the outcross. In the following generations, males should be used for breeding in order to retain the Y chromosome of the target strain. The advantage of using males as well is that they tend to be capable of mating about a week sooner than females and this reduces the overall time for completing the backcross. Depending on the age of sexual maturity of the target strain, it can take from 2 to 2.5 years to complete a backcross, which is a significant investment in time and resources.

22.5.3. Speed and High Speed Backcrossing

There has been much interest in reducing the number of generations required to reach >99% of the target strain, and consequently producing congenic mice sooner for study. A process of "speed congenics" or marker assisted congenic screening (MACS) has been devised to hasten the backcrossing process [37–39]. Polymorphic DNA markers are used to identify the most appropriate breeding animal with the least amount of contaminating donor genome at each generation [40]. By this means it can be possible to reach an equivalence of ten standard backcrossing generations (less than 0.1% contribution from the originating strain) by the fifth backcross generation, "N6" [41, 42]. The efficacy of this technique is related to the number of animals available for screening at each generation and the density of the screen used, [40, 41]. If taking this approach, you should aim to have from 10 to 15 heterozygous male mice at each generation to screen as potential breeders and use a marker panel for which the average interval between marker is about 10 cM. As only 25% of your offspring will be heterozygous males you will actually need to breed 4–5 times this number. There will be residual donor strain sequence around the engineered allele as well as gaps of donor strain persisting in the genome, particularly if a low density (>10 cM intervals) screen is used. This can be overcome by tailoring the marker panel more precisely as breeding progresses, so that the region around the engineered gene is examined at closer intervals, and that markers are not used on regions that have already attained homozygosity. In short, it seems that marker-assisted congenics may not represent simply an acceleration of the backcrossing process, but may result in a congenic strain in which regions of donor strain may persist undetected. It is advisable therefore to continue backcrossing to N10 even after the marker panel indicates that homozygosity with the target strain has been reached.

A further development to speed congenics has been recently proposed under the name "high-speed congenics" [43]. Having used markers to reduce the number of generations required to reach a 99.9% pure strain background, now attention has turned to the time taken to produce a generation by natural breeding. By harvest of round spermatids and intracytoplasmic injection into oocytes, fertilization can occur between an immature male and a fertile female. By collection of round spermatids from males at 22 days of age, approximately 3 weeks per generation can be saved. This reduces the time required for backcross to N10 to about 6 months when combined with marker-assisted selection. This is of course a highly technical and expensive approach, but in terms of research time may well in some cases be considered cost effective.

22.5.4. Maintaining a Line on an Outbred Background

Although the majority of mouse models are studied on an inbred background, because of the obvious advantages of using a defined genetic background, there are some cases where outbred backgrounds are preferred. An outbred strain will be heterozygous at each allele and carry a number of allelic variants. Individuals, therefore, will be genetically nonidentical. Commonly used outbred strains include CD-1, ICR, Swiss Webster, NIH Swiss, and NIMR. The chief desirable properties of these strains are very good fertility, fecundity, and mothering abilities. Studies that use large numbers of animals and particularly require littermates as controls may select an outbred strain for this reason. Cost can also be a factor as these mice are far less expensive to produce than inbred strains. Generally, however, it is disadvantageous to conduct genetic studies of GM mouse models on an outbred background as comparisons are not readily made with phenotypic data from published studies using inbred strains and subtle phenotypes may also go undetected against a highly variable phenotypic background.

When maintaining an outbred strain, care must be taken to preserve the heterogeneity and genetic variability of the population. Creating breeding bottlenecks by using too few pairs at each generation will lead to a rapid loss of alleles and inbreeding. Closed colonies held at separate sites over time will also diverge genetically very quickly from each other. This will also occur in closed inbred colonies, but over a much greater period of time. In an outbred colony, if mating pairs in a closed colony are selected at random, then at least 25 pairs must be maintained at each generation in order to incur less than a 1% inbreeding coefficient [16]. A computer program can be used to generate random matings, e.g., RandoMate [44]. If mating pairs are selected using a system for maximal avoidance of inbreeding (*see* Table 22.6), then a smaller number of pairs can be used, but inbreeding will inevitably occur, albeit at a slower rate, if the colony remains closed (Table 22.7). Heterogeneity is best maintained in a GM colony on an outbred background by mating heterozygotes to wild-type animals acquired from your reference colony. In this case, the best reference colonies are generally those held by the major commercial suppliers of animals (TJL, Taconic, CRL, Harlan) where large breeding numbers can be maintained. Further reading on breeding strategies for outbred strains can be found in Festing [16] and in Hardy [15].

Table 22.6
Selection of breeding mice for maximal avoidance of inbreeding

New mating pair number	Male from old mating pair	Female from old mating pair
1	1	2
2	3	4
3	5	6
4	7	8
5	9	10
6	11	12
7	2	1
8	4	3
9	6	5
10	8	7
11	10	9
12	12	11

[16]

Table 22.7
Percentage of inbreeding per generation for outbred colonies

Number of breeders		Inbreeding per generation-random selection	Inbreeding per generation-maximal avoidance
Male	Female		
4	4	6.25	3.13
13	13	1.92	0.96
25	25	1.00	0.5
80	80	0.31	0.16

Modified from Festing [16]

22.6 Computer-Assisted Colony Management

Undoubtedly the best tool for managing a rodent colony is an effective computer database with specialized software. It not only promotes good management of animal numbers, but also facilitates data collection on reproductive performance and genealogy. A database that is a good fit for the facility can save time for all parties. It can protect the biosecurity of your animals by preventing multiple trips to the animal facility to check information and can be used to schedule tasks and create workflows between yourself and animal care staff. It can benefit communication between different workgroups. As well as the core functions of managing breeding and tracking of animals, some have been designed to manage the needs of high-throughput phenotyping or ENU mutagenesis projects and others can incorporate compliance management.

Many researchers starting out to manage a colony will begin with a spreadsheet system, typically recording mouse data on Excel or similar. Once animal numbers climb, the benefits of a relational database, however, are clear. While excel files can be duplicated, distributed, and then edited in a nonsynchronized manner, a relational database that is centrally administered will protect the integrity of the data. The resulting data quality is higher, as consistency can be checked or even enforced, and editorial privileges can be controlled. Data contained within linked tables can be queried and sorted with greater sophistication and reports can be defined that can allow you to process large numbers of records with ease. A relational database can also work together with other software tools to generate invoices, statistical analyses, or pedigrees. One example of this is an ancestral tree shown in Fig. 22.6 that is generated by a relational database that makes use of CraneFoot pedigree drawing software [45]. You may be in the position that your animal holding facility is already committed to a database, but in the case that you are involved in establishing a system, here are a few points to consider.

The best approach for all would be to build a customized database that can take into account a facility's particular needs; however, this takes a commitment in time and financial resources that not many institutions can afford. Each facility differs not only in species of animals housed, caging types, and animal identification systems used, but also with respect to local requirements for licensing, animal use reporting, and cost accounting. It is also inevitable that practices in a facility change over time due to legislative revisions, advances in animal care, or simply due to the changing nature of research needs. There are a number of commercial software options that come with varying levels of product support and ability to customize to the client's needs, some of which have benefited from years of development. Alternatively, there are also options made available to the scientific community without charge, such as JCMS from Jackson Labs (JAX license) and MausDB from the German Mouse Clinic (open source software available under GPL). All of these require a level of skill and commitment from the client for the installation and administration of the database. If you are looking for a ready-made solution, be prepared to learn that no database solution will be an exact fit to every facility; expect to find a package that will meet up to 80% of your needs and consider the means by which you will bridge the remaining 20%.

Older databases made use of relational database software such as Filemaker Pro and Access, built on client/server architecture, but browser-based enterprise platform independent systems have become favored over these. Web-based systems based on SQL or Oracle offer better access for multiple users and enable access from remote locations. It is important to consider the type of hardware to be used with any system. You may wish to locate computers within the animal rooms or may prefer to use mobile handheld devices such as tablets. Printers and scanners need to be supplied if making use of a bar coding system for cage cards and if implantable RFID or microchip devices are used to identify animals. If the possibility exists for direct uploading of data from body weight balances, and other devices for physiological measurement (temperature, heart rate) to the database system, then automated data capture can save time and minimize error.

Multiple users have to be considered in the choice of a database. The primary point of data entry will be the animal rooms and so the system has to be fast and easy for animal care staff to use, benefiting their workflows and not duplicating record keeping. The interface may need to be available in more than one language

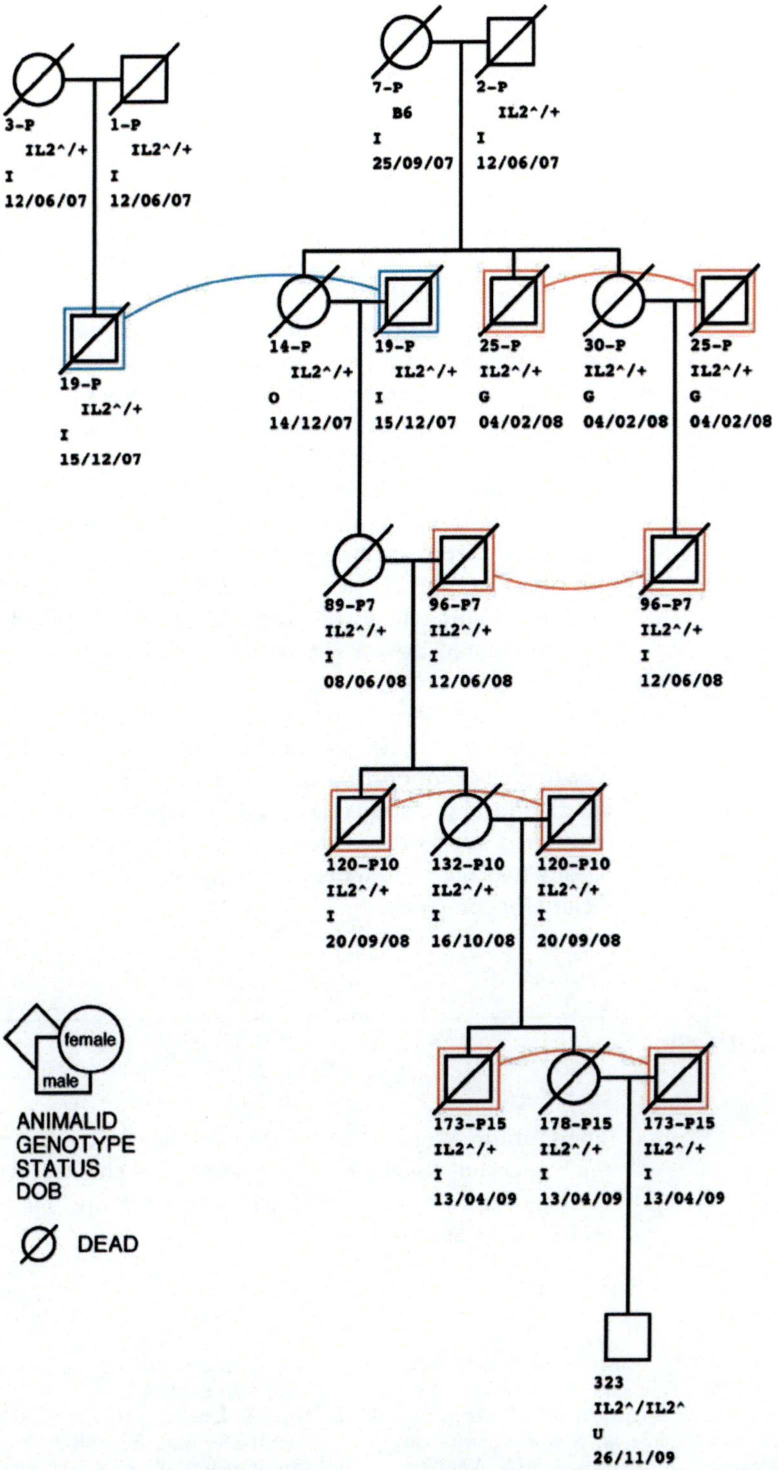

Fig. 22.6 An ancestral tree derived from the animal records management system of the Garvan Medical Research Institute (Stuart) makes use of the CraneFoot open-source pedigree drawing software (http://www.finndiane.fi/software/cranefoot/). Such programs can be used to depict the sometimes complex breeding history of a line and relationships between individual mice.

to cater for a mixed language workforce. For research staff, the displayed information must be relevant and clear to understand without the need to resort to additional documents. The search functions should be flexible enough to deal with many types of queries and actions should be able to be applied in bulk to groups of animals as well as individuals. Data should be easily exported, as Excel files or other compatible files, when it is necessary to link to other applications. For facility management, there are further criteria. A manager will examine whether workflows can be managed efficiently with task setting tools. It will be essential that legislative reporting requirements can be met and that new reports can be designed as requirements change. Costs must be recoverable with ease. The extent to which the database is able to manage compliance and to limit animal numbers and usage to approved levels is also to be considered.

The main factors in the successful adoption of a database would seem to be (1) the ease with which data can be added by the animal care staff or those that have primary responsibility for the record keeping. (2) Having an intuitive interface that requires minimal training is also a factor, as a system perceived as too complex would drive users to other options. (3) The ability of the system to produce custom designed reports for monitoring colony management, breeding statistics, animal tracking, and daily tasks which would be of benefit to both managers and users. (4) The degree of support for users and administrators of the system, whether local or external, will also be important. A strong level of continued support and development will ensure that whatever colony management system is chosen will serve the needs of the facility for some time.

Acknowledgements

The author would like to thank the following people for their generous assistance in preparing this chapter: Ruth Arkell, Michael Dobbie, Holger Maier, Kristina Nagy, Thomas Preiss, and Duncan Sparrow.

References

1. Wells DJ, Playle LC, Enser WEJ, Flecknell PA, Gardiner MA, Holland J, Howard BR, Hubrecht R, Humphreys KR, Jackson IJ, Lane N, Maconochie M, Mason G, Morton DB, Raymond R, Robinson SJA, Watt N (2006) Assessing the welfare of genetically altered mice: working group report. Lab Anim 40:111–114
2. Thon R, Lassen J, Hansen AK, Jegstrup IM, Ritskes-Hoitinga M (2002) Welfare evaluation of genetically modified mice – an inventory of reports to the Danish Animal

Experiments Inspectorate. Scand J Lab Anim Sci 29:45–53
3. Buehr M, Hjorth JP, Hansen AK, Sandøe P (2003) Genetically modified laboratory animals – what welfare problems do they face? J Appl Anim Welf Sci 6:319–338
4. Meisler MH (1992) Insertional mutation of "classical" and novel genes in transgenic mice. Trends Genet 8:341–344
5. Hughes ED, Qu YY, Genik SJ, Lyons RH, Pacheco CD, Lieberman AP, Samuelson LC, Nasonkin IO, Camper SA, Van Keuren ML, Saunders TL (2007) Genetic variation in C57BL/6 ES cell lines and genetic instability in the Bruce4 C57BL/6 ES cell line. Mamm Genome 18:549–558
6. Liu X, Wu H, Loring J, Hormuzdi S, Disteche CM, Bornstein P, Jaenisch R (1997) Trisomy eight in ES cells is a common potential problem in gene targeting and interferes with germ line transmission. Dev Dyn 209: 85–91
7. Sugawara A, Goto K, Sotomaru Y, Sofuni T, Ito T (2006) Current status of chromosomal abnormalities in mouse embryonic stem cell lines used in Japan. Comp Med 56:31–34
8. Kumar RA, Chan KL, Wong AH, Little KQ, Rajcan-Separovic E, Abrahams BS, Simpson EM (2007) Unexpected embryonic stem (ES) cell mutations represent a concern in gene targeting: lessons from "fierce" mice. Genesis 38:51–57
9. Papaioannau VE, Behringer RR (2005) Mouse phenotypes: a handbook of mutational analysis. Cold Spring Harbor Laboratory Press, Cold Spring Harbor, NY
10. Migeon BR (1994) X-chromosome inactivation: molecular mechanisms and genetic consequences. Trends Genet 10:230–235
11. Schwenk F, Baron U, Rajewsky K (1995) A Cre-transgenic mouse strain for the ubiquitous deletion of loxP-flanked gene segments including deletion in germ cells. Nucleic Acids Res 23:5080–5081
12. Silver L (1995) Mouse genetics: concepts and applications. Oxford University Press, Oxford
13. Butler A, Gordon RE, Gatt S, Schuchman EH (2007) Sperm abnormalities in heterozygous acid sphingomyelinase knockout mice reveal a novel approach for the prevention of genetic diseases. Am J Pathol 170:2077–2088
14. Yan W (2009) Male infertility caused by spermiogenic defects: lessons from gene knockouts. Mol Cell Endocrinol 10:24–32
15. Hardy P (2004) Gnotobiology and breeding techniques. In: Hedrich H (ed) The laboratory mouse. Elsevier Academic Press, London, UK
16. Festing M (1999) Introduction to laboratory animal genetics. In: Poole T (ed) The UFAW handbook on the care and management of laboratory animals, 7th edn. Blackwell Science, Oxford, UK, pp 61–93
17. Nagy A, Gertsenstein M, Vintersten K, Behringer R (2003) Manipulating the mouse embryo: a laboratory manual, 3rd edn. Cold Spring Harbor Laboratory Press, Cold Spring Harbor, NY
18. Palmiter RD, Brinster RL (1985) Transgenic mice. Cell 41:343–345
19. Lois C, Hong EJ, Pease S, Brown EJ, Baltimore D (2002) Germline transmission and tissue-specific expression of transgenes delivered by lentiviral vectors. Science 295: 868–872
20. Lois C (2006) Generation of transgenic animals using lentiviral vectors. In: Pease S, Lois C (eds) Mammalian and avian transgenesis-new approaches. Springer, Heidelberg, pp 1–22
21. Clark AJ, Bissinger P, Bullock DW, Damak S, Wallace R, Whitelaw CB, Yull F (1994) Chromosomal position effects and the modulation of transgene expression. Reprod Fertil Dev 6: 589–598
22. Liang Z, Breman AM, Grimes BR, Rosen ED (2008) Identifying and genotyping transgene integration loci. Transgenic Res 17:979–983
23. Matsui S, Sait S, Jones CA, Nowak N, Gross KW (2002) Rapid localization of transgenes in mouse chromosomes with a combined Spectral Karyotyping/FISH technique. Mamm Genome 13:680–685
24. Wilkie TM, Brinster RL, Palmiter RD (1986) Germline and somatic mosaicism in transgenic mice. Dev Biol 118:9–18
25. Whitelaw CB, Springbett AJ, Webster J, Clark J (1993) The majority of G0 transgenic mice are derived from mosaic embryos. Transgenic Res 2:29–32
26. Schwartzberg PL, Goff SP, Robertson EJ (1989) Germ-line transmission of a c-abl mutation produced by targeted gene disruption in ES cells. Science 246:799–803
27. Nagy A, Rossant J, Nagy R, Abramow-Newerly W, Roder JC (1993) Derivation of completely cell culture-derived mice from early-passage embryonic stem cells. Proc Natl Acad Sci USA 90:8424–8428
28. Poueymirou WT, Auerbach W, Frendewey D, Hickey JF, Escaravage JM, Esau L, Doré AT, Stevens S, Adams NC, Dominguez MG, Gale NW, Yancopoulos GD, DeChiara TM, Valenzuela DM (2007) F0 generation mice fully derived from gene-targeted embryonic stem cells allowing immediate phenotypic analyses. Nat Biotechnol 25:91–99

29. Silvers WK (1979) The coat colors of mice: a model for mammalian gene action and interaction. Springer, Heidelberg
30. Pease S (2006) Ancillary techniques. In: Pease S, Lois C (eds) Mammalian and avian transgenesis-new approaches. Springer, Heidelberg
31. Pettitt SJ, Liang Q, Rairdan XY, Moran JL, Prosser HM, Beier DR, Lloyd KC, Bradley A, Skarnes WC (2009) Agouti C57BL/6N embryonic stem cells for mouse genetic resources. Nat Meth 6:493–495
32. Auerbach AB, Norinsky R, Ho W, Losos K, Guo Q, Chatterjee S, Joyner AL (2003) Strain-dependent differences in the efficiency of transgenic mouse production. Transgenic Res 12:59–69
33. Simpson EM, Linder CC, Sargent EE, Davisson MT, Mobraaten LE, Sharp JJ (1997) Genetic variation among 129 substrains and its importance for targeted mutagenesis in mice. Nat Genet 16:19–27
34. Seong ES, Saunders TL, Stewart CL, Burmeister M (2004) To knockout in 129 or in C57BL/6: that is the question. Trends Genet 20:59–62
35. Noben-Trauth N, Köhler G, Bürki K, Ledermann B (1996) Efficient targeting of the IL-4 gene in a BALB/c embryonic stem cell line. Transgenic Res 5:487–491
36. Silva AJ, Simpson EM, Takahashi JS, Lipp H-P, Nakanishi S, Wehner JM, Giese KP, Tully T, Abel T, Chapman PF, Fox K, Grant S, Itohara S, Lathe R, Mayford M, McNamara JO, Morris RJ, Picciotto M, Roder J, Shin H-S, Slesinger PA, Storm DR, Stryker MP, Tonegawa S, Wang Y, Wolfer DP (1997) Mutant mice and neuroscience: recommendations concerning genetic background. Banbury Conference on genetic background in mice. Neuron 19:755–759
37. Lander ES, Schork NJ (1994) Genetic dissection of complex traits. Science 265: 2037–2048
38. Wakeland E, Morel L, Achey K, Yui M, Longmate J (1997) Speed congenics: a classic technique in the fast lane (relatively speaking). Immunol Today 18:472–477
39. Petkov PM, Cassell MA, Sargent EE, Donnelly CJ, Robinson P, Crew V, Asquith S, Haar RV, Wiles MV (2004) Development of a SNP genotyping panel for genetic monitoring of the laboratory mouse. Genomics 83: 902–911
40. Weil MM, Brown BW, Serachitopol DM (1997) Genotype selection to rapidly breed congenic strains. Genetics 146: 1061–1069
41. Armstrong NJ, Brodniki TC, Speed TP (2006) Mind the gap: analysis of marker-assisted breeding strategies for inbred mouse strains. Mamm Genome 17:273–287
42. Markel P, Shu P, Ebeling C, Carlson GA, Nagle DL, Smutko JS, Moore KJ (1997) Theoretical and empirical issues for marker-assisted breeding of congenic mouse strains. Nat Genet 178:280–284
43. Ogonuki N, Inoue K, Hirose M, Miura I, Mochida K, Sato T, Mise N, Mekada K, Yoshiki A, Abe K, Kurihara H, Wakana S, Ogura A (2009) A high-speed congenic strategy using first-wave male germ cells. PLoS ONE 4(3):e4943
44. Schmitt AO, Bortfeldt R, Neuschl C, Brockmann GA (2009) RandoMate: a program for the generation of random mating schemes for small laboratory animals. Mamm Genome 20: 321–325
45. Mäkinen V-P, Parkkonen M, Wessman M, Groop P-H, Kanninen T, Kaski K (2005) High-throughput pedigree drawing. Eur J Hum Genet 13:987–989
46. Giraldo P, Montoliu L (2001) Size matters: use of YACs, BACs and PACs in transgenic animals. Transgenic Res 2:83–103

Chapter 23

Cryopreservation

B. Pintado and J. Hourcade

Abstract

Cryopreservation is the method of choice not only for archiving mouse mutant models at the end of an experiment, but also to create a security back up during their development. Three key factors need to be considered in cryopreservation: why, when, and how. In this chapter, we intend to discuss these factors and to provide an overview of current cryopreservation techniques. We will also refer to some experimental approaches that may become used more widely in the future. We will try to highlight advantages and disadvantages of each method with regard to investment in equipment, skills and technical limitations, and also review factors that compromise the efficiency in each approach. Finally, we will compile specific protocols for freezing mouse spermatozoa and embryos.

23.1 Introduction

In any project designed to generate a genetically modified animal model, archiving has to be included as a relevant aspect of the general picture and cryopreservation is the method of choice. Cryopreservation fulfills several objectives that may arise as an experimental procedure evolves, from the provision of back-up stocks during the development of a mouse model, to a definitive archive at the end of the experiment when no further use of the model is expected in the short term, but where stocks need to remain available for future studies. The strategy in each situation may differ and in order to make the best choice, it is necessary to define the objective and based on that, to select the best approach. In this chapter we intend to provide an overview of current cryopreservation techniques, mentioning some experimental approaches that may become more widely used in the future. We will also try to highlight advantages and disadvantages of each method, consider the necessary investment in equipment, manual skills and technical limitations, and consider also factors that compromise the efficiency of these procedures. Finally, we will

S. Pease and T.L. Saunders (eds.), *Advanced Protocols for Animal Transgenesis*, Springer Protocols,
DOI 10.1007/978-3-642-20792-1_23,

compile specific protocols for freezing mouse spermatozoa and embryos.

23.2 Key Factors in Archiving Mutant Lines

There are three points that need to be addressed in archiving mutant lines: why, when, and how. In order to choose the best strategy, it is important to have a clear picture of the future requirements for the use of a strain, in order to determine how much effort to invest in the procedure. Embryo cryopreservation requires the use of a significant number of donor females. Furthermore, recovery of an archived strain includes embryo transfer and a certain amount of time before one can successfully re-establish breeding stock. However, compared to other archiving strategies such as cryopreservation of gametes, embryo cryopreservation saves time when the line contains more than one genetic modification, or the transgene is maintained in homozygosis.

As we will discuss later, the continued breeding of a transgenic line is not the method of choice for long-term preservation, but it may be the best approach if that strain will need to be expanded within the next 6–12 months. However, be aware that there are several common mistakes that may be made under these circumstances. For instance, in order to reduce the number of animals maintained and therefore reduce associated costs, it might be considered appropriate to maintain the line with a minimum number of animals. This approach poses risks for several reasons. From a practical perspective, the lack of direct interest in the line tends to promote a delay in the renewal of mating pairs. Many transgenic lines have been produced or backcrossed on to inbred backgrounds, which may result in a significant decrease in fertility that will affect males and females perhaps earlier than expected. As a consequence, the line can easily be lost after a period of inattention.

Another important argument for the archiving of transgenic models is the impact of genetic drift. After several generations within a closed colony, it is possible that small genetic changes will accumulate, which may have the potential to modulate the phenotype, effectively modifying it from that exhibited by the original line. This risk increases when the mutant is maintained on a mixed genetic background. In this case, the accumulation of small genetic changes, deleterious to breeding performance, may lead to inbreeding depression, where a decline in reproductive performance becomes very evident after a few generations of intercrossing. Moreover, spontaneous mutations can occur and additionally influence the characteristic of the model. In order to avoid the accumulation of genetic changes in an existing mutant line, it is advisable to backcross the mutation an international to the original parental strain after a certain number of generations. This procedure may be impossible when there is no real way to define the parental wild type background to be used in the procedure. The increasing awareness of this problem has highlighted the need to create congenic lines, where, after a certain number of backcrosses to a pure inbred strain, it is assumed that the congenic line resembles that inbreed strain in all genes except the genetic modification.

Based on all the above points, as a general rule, only active lines that are in constant demand or will be used within the next 12 months should be maintained as a live strain. Under any other circumstances, the best approach is some form of cryopreservation.

23.2.1. Why Should We Preserve a Mutant Strain?

Many research institutions face the question of whether to cryopreserve their mouse models or not. A project finishes and there is a need to preserve a characterized line that eventually may be needed by other research groups. As previously discussed, maintaining the line in a live condition is unwise, except in very specific situations. In addition, many animal facilities are dealing with space issues [1]. The exponential growth of mutant strains of mice obtained either by microinjection or stem cell technology is forcing the maintenance of those transgenic lines upon the laboratory that created them and, as a result, on the animal facility that supports that laboratory. Many institutions involved in work with mutant rodents have to face increasing costs that are difficult to support. Also, a lack of available space within the animal facility has the potential to delay the start of new experimental protocols. Even though public consortiums or private institutions have been created for archiving and maintaining genetically modified animals, a large number of transgenic models that do not represent a genetic mutation of major interest or demand may not be welcomed at a puplic repository. As a rule, such repositories and archives select which strains to archive, based on a scientific review process, in order to preserve those mutants that represent valuable biomedical models, fully characterized genetically and phenotypically, which may be of common use.

The cost to cryopreserve a mouse strain is much lower than the ongoing maintenance of live animals. Nevertheless, the samples require maintenance on a continual basis and at cost. The recovery of live animals made possible by recent technical developments in intracytoplasmic sperm injection (ICSI) of dry spermatozoa (that do not demand any kind of maintenance for storage), and from sperm cells collected from epididymides of dead animals kept at −20°C for several years [2], or by cloning technology in the use of certain tissues from frozen animals [3] are promising technologies for the future, but are not yet available as routine procedures. However, the emergence of these technologies now provides an approach for preserving an important line, with the prospect of future use, even if only the most basic infrastructure is available in the animal facility.

In addition to the practical issues, there is also an ethical issue to consider. The phenotypic nature of a mutation may impair the health of the animals. We should be aware that the impairment is also present even if the animals are not used for experimentation and it is transmitted from one generation to the next by normal breeding as the animal model is maintained. Therefore, for humane reasons, such an animal model would better be stored as cryopreserved embryos or germ cells,

To summarize, given the technical advances, it is possible to plan a selective cryopreservation strategy in which those lines with clear possibilities of being needed in the future should be preserved as embryos, making recovery relatively easy and archiving those other lines with a remote likelihood of being needed in the future, in an economic way. Currently, the latter would be cryopreservation as frozen spermatozoa, or even as dead sperm heads with an almost inexistent cost of maintenance.

23.2.2. When to Preserve a Mutant Strain?

Usually, transgenic strains become cryopreserved as part of a public archive only once they have been determined as important, by having been fully characterized. This typically occurs at the end of an experiment. The mutant lines stored in an international repository represent a small share of the total number of lines generated and characterized. However we should abandon the idea of considering cryopreservation only at the final stages of an experiment, for the purpose of keeping currently unused biomodels available for future demands. Rather, it is very desirable to bank down embryos at an earlier stage, during the development of an experimental animal model, thus quickly creating a back up once transmission and expression have been demonstrated. There are other reasons for the proposing that the cryopreservation of transgenic models should be accomplished as soon as possible. Firstly, for the purpose of avoiding genetic or epigenetic modification of the transgene. In certain conditions, especially with small constructs, where a multicopy integration has taken place, or with those constructs carrying an abundance of prokaryotic DNA, the host genome may recognize the transgene as exogenous and tend to inactivate it [4, 5]. This may occur immediately, or after a number of generations [6]. Secondly, the concurrent use of animals for both experimental purposes and cryopreservation procedures may, in time, reduce the number of animals needed or used in total. Surplus experimental animals may be used for embryo or sperm collection procedures and even animals that have to be maintained throughout the experimental study may be used as stud males in an embryo collection procedure, if this does not interfere with the experimental protocol. Thirdly, the availability of a cryopreserved pool of embryos carrying the mutation of choice would allow for repopulating, in situations that demand the culling of a line due to the unexpected outbreak of an infectious disease, or a natural disaster occur.

23.2.3. How to Preserve a Mutant Strain?

The decision to archive a transgenic model raises questions about what should be cryopreserved and which method is most appropriate. Cryopreservation of mouse embryos [7] has been the standard method for the preservation of a mutant strain for a long time, but it is not the only option. Nowadays it is feasible to preserve gametes, either spermatozoa or oocytes. Spermatozoa preservation remains far from optimal in certain strains but the methodology is achieving a consistent degree of success and when lines with a single genetic modification are to be maintained as heterozygotes or hemizygotes, this approach is an alternative procedure with certain clear advantages in comparison with the traditional approach of embryo freezing [8]. However in those lines where more than one genetic modification is present, embryos are always the strategy of choice, since spermatozoa cryopreservation determines segregation of modifications in the progeny and a consequent need of a number of intercrosses to recover the original genetic profile.

Oocytes have also been successfully preserved but, to date, this approach does not provide any advantage over embryos or spermatozoa, because only matured oocytes have been used. Alternative approaches, such as the preservation of ovarian tissue and nuclear transfer methodology using preserved somatic cells, have proven to be viable options. However, the technical complexity involved in the ultimate recovery of live animals means that these approaches are far from routine procedures and, at the moment, they cannot be considered the strategy of choice.

23.3 Embryo Cryopreservation

Water is the main component of living cells and when it solidifies as a consequence of low temperature, the crystals formed damage cellular structures permanently. Cryopreservation methods seek to minimize or avoid the formation of intracellular ice. Vitrification, one of the three cryopreservation methods that has allowed the recovery of live animals from embryos preserved at subzero temperatures, avoids crystal formation altogether. This method, described in 1985 [9], is based on the solidification of water in a glass-like form. The two remaining cryopreservation methods involve the production of ice crystals. They are the so-called "slow" or "equilibrium" method [7], chronologically the first system described and the "quick freezing" method [10].

23.3.1. Slow-Cooling Method

This method, also known as equilibrium method, seeks to acquire a supercooled state of intracellular water that remains unfrozen in the presence of external ice. Due to the consequent hyperosmotic external environment, the internal water leaves the cell. When the temperature reaches a certain level, intracellular crystallization takes place. However, the crystals formed are of small size, thus minimizing the potential for damage of the cell to occur.

Temperature control is not enough to ensure cell survival and in order to speed up intracellular water reduction, the addition of a cryoprotective agent is necessary. Two different kinds may be used: permeating cryoprotectants such as glycerol, ethyleneglycol or DMSO and nonpermeating cryoprotectants such as sucrose. Permeating cryoprotectants are small molecules able to go through the cell membrane and substitute for intracellular water, thus speeding up the water depletion process. In the slow or "equilibrium" methodology, temperature is decreased slowly from room temperature to −35, −40 or −80°C, depending on the packaging method, before plunging into liquid nitrogen . This freezing technique also requires the presence of permeating cryoprotectants in the freezing medium, at approximately 1.5 M. This procedure requires strict control of temperature decrease, provided by a programmable freezer. This provides its best advantage and also greatest disadvantage. Programmable freezers imply an investment only justified by routine use. On the other hand, because most of the process takes place under controlled conditions, this method ensures highly reproducible results. Another advantage of this method is that different containers can be used, either cryotubes or plastic insemination straws and since cryoprotectants are used at low concentration, minor variations in the exposure time of embryos to freezing solution do not compromise cell viability

As a general profile of the freezing program, samples are equilibrated with cryoprotectant at room temperature. Then they are cooled to −7°C and kept at that temperature for several minutes. Seeding of the sample is induced within this plateau, either manually or automatically, and then temperature starts to diminish gradually at a rate of about 0.3–0.5°C/min until −40°C or −80°C, depending on whether the container used is a straw or a cryotube. Facilities such as the Jackson Laboratory still use the original protocol described by Whittingham et al. [7]. Temperature decreases from −6°C to −80°C at a rate of 0.5°C/min and then samples are plunged into liquid nitrogen. The only change that has been made to this protocol is that cryotubes have substituted the original glass ampoules. This approach implies also a slow and controlled thawing process consisting of the removal of the cryotube from liquid nitrogen and warming of samples at ambient temperature, until thawing of all ice crystals. This process may take 10–15 min.

A second very extended slow freezing protocol shortens freezing time and is based on the use of plastic straws. In this case, temperature reaches −35/−40°C at the same rate of 0.3–0.5°C/min, and then straws are plunged into liquid nitrogen (revised in [11]). This approach does not demand slow thawing. In this case, plastic straws are thawed by allowing the dissipation of liquid nitrogen vapors for 30 s and then by immersion in a water bath at room temperature, till the ice melts.

Regardless of the container employed, cryotube or straw, the slow freezing procedure is time consuming because it takes from 90 to 140 min to completion, depending on the specific freezing curve used. However, it is a very consistently reproducible method and very effective, particularly suitable for people without great expertise in embryo handling. The method can be successfully used with different preimplantation stage embryos, from 2-cell to blastocyst stage.

This system, combined with in vitro fertilization (IVF) can be used to obtain a large number of frozen embryos in a short time. This so-called "speed cryo" depends upon the production of large numbers of oocytes that are fertilized with sperm from males of the mutant line. The resulting 2-cell embryos are frozen following the slow freezing protocol. The advantage of this method is that only three to four mutant males are required in order to obtain a large number of embryos if commercial females are used as embryo donors. If males are homozygous for the modification, all embryos obtained will be heterozygotes, meanwhile if males provided are heterozygotes, then only half of the embryos will carry the genetic modification. For this specific protocol, females are usually obtained from a commercial source. This means that if the mutation is on a mixed background, the embryos obtained by this approach may be considerably different, as compared with the genetics of the founder line. The limitation of the procedure lays in whether the genetic background of the sperm donor lends itself to IVF, in the recovery of sperm samples that are sufficient in number and motility. Strains that have been seen to be very well suited to the procedure are C57Bl/6, FVB/N, and DBA/2.

23.3.2. Quick Freezing Method

The quick freezing method achieves intracellular water depletion by means of a combination of permeating and nonpermeating cryoprotectants. In this procedure, permeating cryoprotectants are used at two to three times the concentration than would be in the equilibrium method. Nonpermeating agents, usually macromolecules like sucrose, are unable to go across the cell membrane. They increase intracellular water efflux by increasing the extracellular osmotic pressure; as a result, blastomeres shrink in an attempt to equilibrate intra- and extracellular osmotic pressure. These combined cryoprotectant concentrations are highly toxic, so the whole procedure has to be performed quickly and in some cases at low temperatures, to minimize cell damage. This allows little margin for variations in exposure times and it complicates handling to some extent. However, the procedure consistently reduces the time required to freeze embryos from hours to minutes, and the use of a programmable freezer is unnecessary. The complete procedure can be performed with simple basic equipment, easily found in any laboratory. Once the handler is familiar with the procedure and has acquired a certain competency, survival rates are similar to that of the classical methodology. The quick freeze method has been used successfully with rats [12] and mice [13]. The only limitation is that not all embryo stages are recoverable from this procedure. The method is best suited to cryopreservation of compacted morulae.

23.3.3. Vitrification

Vitrification implies the use of cryoprotectant in a highly toxic concentration and requires tight control of embryo exposure to such cryoprotectant solutions, with no place for delays. This demands skilled personnel in order to achieve consistently good results. Temperature decrease has to be extremely fast, which implies the use of packaging that allows swift temperature interchange. Improvement of embryo survival has been accomplished by the use of open-pulled straws [14]. The vitrification technique allows the cryopreservation of a wide range of embryonic stages, from oocytes and one cell to expanded blastocyst stage. On the other hand, the technique faces a serious sanitary deficiency. Straws are not sealed, and hence interchange with the LN_2 of the tank is a possibility. Some new improvements of the packaging system have addressed this problem [15], allowing effective sealing of each sample.

Embryo collection should meet certain sanitary requirements specified at the receiving animal facility. Even though embryo rederivation has been used successfully to eradicate infectious diseases [16–18], it should be noted that any cryopreservation process that permits the survival of eukaryotic cells will also permit survival of prokaryotic contaminants like bacteria, Mycoplasma species or viruses. For these reasons, it is important to follow an embryo collection protocol according to the recommendations given by the International Embryo Transfer Society (IETS) and adopted by the International Epizooties Office (IEO) in its International Animal Health Code Appendix 3.3.5. Laboratory Rodent and Rabbit Embryos.

The number of embryos preserved should be as many needed in order to re-establish the strain at least twice. This number will vary, depending on the strain-specific ability to survive cryopreservation and also the genetic status of donors, i.e., homozygous or heterozygous for the mutation. It has to be noted that when donors are heterozygous for the transgene, only half of the embryos obtained will carry the desired genetic information and, for this reason, a greater number of embryos should be preserved. As a general rule, around 400 embryos will guarantee recovery of a mouse strain, but each case is different and should be considered on its own merits.

One important factor to consider in cryopreservation of a given strain is the superovulatory response that will be unique to that strain. There are certain genetic backgrounds with a very low response to superovulatory treatments. This problem is usually associated with inbred strains. Among such strains, 129/J, DBA/2J and FVB/N are considered to be low responders, C57BL/6J females show a higher response, but the fertilization efficiency of males is lower than that of hybrids such as B6CBAF1 or B6D2F1 [19]. The response to superovulation is not the only significant difference among strains; there are also differences in their response to in vitro culture conditions [20, 21] and in the ability of embryos to implant and go to term after embryo transfer [22, 23]. Because all these factors need to be considered, a true measure of banking efficiency is obtained by dividing the number of pups born by the total number of thawed embryos [24]. Unfortunately, this value needs to be established for each mutant strain. Even with mutant mice on a hybrid background, several generations of intercrossing between brothers and sisters in order to achieve homozygosity may also affect fertility. The problem increases when the genetic modification affects reproductive performance.

Most transgenic lines are generated directly in, or backcrossed to, an inbred background and hence a high number of donors need to be superovulated in order to collect enough embryos for manipulation. It is also important to consider the fertility of stud males. Males above 6–9 months of age may have reduced fertility and libido. The availability of homozygous animals for cryopreservation is sometimes restricted or it can only be provided after a breeding period. Cryopreservation can be accomplished in heterozygosity by mating superovulated wild type females purchased from a commercial source and mated with heterozygous transgenic males or by in vitro fertilization, as we mentioned previously in the speed cryo method. These strategies will speed up the cryopreservation process considerably, but if recovery of the strain is required, several weeks will be necessary before homozygous animals are available. Heterozygosity is also the only choice in certain mutations with a homozygous lethal phenotype and in such specific cases other banking approaches, like spermatozoa freezing, should seriously be considered as an alternative.

23.4 Spermatozoa Cryopreservation

Whereas spermatozoa cryopreservation has been a unique tool to promote genetic improvement in livestock, cryopreservation of mouse spermatozoa has represented a technical challenge. The morphology of murine spermatozoa is quite different from the pear-like shape of the sperm of other mammals and this somehow limits membrane elasticity and the ability to freeze sperm in a viable condition. In addition to this, mouse spermatozoa are very sensitive to changes in osmotic conditions [25]. The first successful cryopreservation of murine spermatozoa was achieved in 1990 [26] and since then several authors have reported the low reproducibility of these protocols. The reasons for this lay not only within the freezing protocol itself, but also with the clear influence of the mouse strain used. The strain of origin is a highly influential and sometimes limiting factor in the ability of post-thaw sperm cells to fertilize. Again, inbred strains show a marked disadvantage over outbred and hybrid strains [27]. Amongst the protocols described in the literature, those based on the use of a combination of nonpermeating agents, raffinose and skim milk are becoming more popular, most of them following the protocol described by Nakagata [28]. However there is still room from improvement, new more efficient extenders for sperm dilution have been described recently and also some changes in the standard freezing procedure that enhance results dramatically for certain strains such as C57Bl/6. However, these results still do not reach the levels of post-thaw fertilization that it is possible to accomplish with hybrid strains [29, 30].

When considering sperm cryostorage as an alternative to embryos, it is important to note that spermatozoa cannot be used as a way to eradicate certain diseases. Unlike embryos, which can be washed several times in sterile solutions and even trypsinized to eliminate debris and potential pathogenic agents from the sample, spermatozoa need to be frozen immediately after collection in a diluted form, and it is impossible to get rid of biological residues from the donor male. For this reason, archiving of sperm samples should be used only in the case of known microbiological profiles of the donor colony. We recommend using only donors that are free of all pathogens as specified in the FELASA health monitoring recommendations. However, it is possible, when using for IVF cryopreserved spermatozoa that include mouse pathogens in the media, to produce pathogen-free mice [16]. A recent publication demonstrates that the presence of cumulus cells during in vitro fertilization prevents transmission of Mouse Minute Virus [31]. This result suggests it is possible to recover pathogen-free animals, even if the only source of gametes is infected sperm samples. Since sperm samples from infected animals cannot be washed, there is a risk of inadvertent contamination due to a mistake during sperm handling, at the time of sample collection. Therefore, cryopreservation of sperm from infected males is justifiable only in extraordinary circumstances. In these cases, PCR examination of washing media from the embryos generated by IVF is an additional safety measure that should be adopted [32]. In any case, samples should be clearly marked and stored in a specific tank since it has been shown there is a possibility of cross-contamination between semen samples stored in the same liquid nitrogen tank [33].

The C57BL/6 mouse is the most widely used inbred mouse strain. Besides the efforts directed toward the improvement of the freezing technique, many transgenic lines created or backcrossed to C57BL/6 background have already been cryopreserved by the method described by Nakagata [28]. Significant efforts have been directed toward designing a methodology that may result in improved quality of C57BL/6 sperm samples upon recovery. Apparently, the main problem of C57BL/6 is the inability of its frozen–thawed spermatozoa to capacitate after freezing. Capacitation is a key step during the fertilization process, as a result of which acrosome reaction can be accomplished. In an in vitro situation, this is mediated through the presence of albumin and bicarbonate ions in the in vitro fertilization medium that favors, among other effects, cholesterol efflux, starting a chain of signals that eventually will induce capacitation. Unfortunately, these components fail to trigger capacitation in frozen/thawed C57BL/6 sperm samples. Recently, it has been demonstrated that the inclusion of methyl-beta-cyclodextrin favors cholesterol mobilization from the sperm membrane, increasing substantially the fertilization rate [30]. Monothioglycerol has also shown to improve fertilization rates [29]. Other approaches to improve fertility have been based on the in vitro selection of motile sperm [34] or even in vivo selection through artificial insemination [35]. Whatever the source or previous treatment of sperm, it is essential to have an adequate population of motile spermatozoa after thawing. If this is unattainable, intracytoplasmic sperm injection (ICSI) has proven an efficient route toward fertilization, even if spermatozoa are completely immobile after thawing [36].

From a technical point of view, sperm cryopreservation does not require very expensive equipment. A few males may produce enough samples for fertilization of thousands of oocytes and hence reduced space in the bank is needed, compared to embryo storage. This procedure also reduces significantly the number of animals needed in order to bank down a mutant strain and no hormonal treatment will be required. However, some disadvantages have to be taken into consideration. The first one is the limited ability of frozen/thawed sperm to fertilize oocytes, in very important strains like C57BL/6. Some clear improvements have been made and recently published, as we previously mentioned. But there is still a reduced efficiency compared to other inbred strains or hybrid animals. In addition, the recovery process becomes more complicated than recovery of fertilized embryos, because an in-vitro fertilization procedure will be needed in order to produce the embryos to be transferred into recipient mice, for development to term. A third negative aspect is that with spermatozoa, offspring produced are always heterozygotes, and some months will pass before homozygote offspring are produced. If a mouse model carries more than one transgene, it will take even longer to produce double or triple homozygous mice.

23.5 Oocyte and Ovarian Tissue Cryopreservation

Oocytes are cells that are especially difficult to cryopreserve. The necessity of keeping a functional spindle in the cell allowing activation and meiotic resumption determines that structural damage of the cell should be minimized. In addition to this, oocytes are larger cells than blastomeres and water depletion is more difficult. Only vitrification offers a valid alternative to cryopreservation of oocytes, but the need for a fast decrease in temperature to avoid crystallization implies the use of very specific storage containers. The most effective to date are open-pulled straws. These are essentially the straws used for artificial insemination in livestock, pulled to a smaller diameter, the inner slightly wider than the oocyte. However, preservation of mature oocytes does not provide any true technical advantage since their collection from donors requires exactly the same effort as embryo collection with regard to hormonal treatment and strain-limited superovulatory response.

A true achievement would be the preservation of the thousands of immature oocytes present in a single ovary. The major biological limitation would be achieving maturation of those female gametes. The generation of live offspring from xenografted ovaries in mice [37] shows one possible maturation mechanism. But there are unsolved technical barriers that put this approach far away from becoming a routine procedure for consideration in the near future, as is also true for nuclear transfer from somatic cells [36, 38]

23.6 Genome Resource Banking Management

Appropriate record keeping is essential in any animal facility and this is even more important in a genetic resource bank. It is necessary to keep in mind that the person who freezes the sample may not be the person who thaws it. For this reason, records need to provide extremely detailed and accurate information about the frozen material, including the freezing method, sample location (s), and identification codes. Accurate record keeping and efficient storage methods are key to avoiding unnecessary risks toward sample temperature oscillations, when looking for a specific straw or cryotube, or unnecessary long-term exposure of personnel to cold. Both straws and cryotubes present space to specify this information. Relying on the information written on a plastic goblet or box that contains unmarked samples should be avoided. Even the most diligent of handlers may drop a sample into the storage container, so storing blank straws/tubes in a labeled container presents an unnecessary risk. For this reason the actual container of the biological sample (straw, cryotube, ampoule) should reflect identification of the construct, date and a code which allows access to all crucial information. This means that extensive records should be kept to serve as an inventory, either on written cards or with computer support. The International Embryo Transfer Society provides some extended recommendations in its handbook [39] that should be used as reference. In addition to the written information, it is very helpful to employ a color-coding system for the straws. Some vendors provide colored straws, or different color combinations on the cotton plug at one end of the straw. But this can also be accomplished with colored permanent markers. Liquid nitrogen vapors make it difficult to read the information on the straws, but different color marks are easily recognized.

23.7 Safety Considerations

Finally, there are some safety considerations that should be kept in mind. Throughout the freezing and thawing process there are unique hazards that require specific precautions, such as the use of protective goggles, insulated gloves and coats that protect eyes and skin from exposure and potential cryo-injury. Ampoules or straws may, in certain conditions, explode when retrieved from liquid nitrogen and a face shield should be mandatory. These precautions should be adopted in addition to any other safety measures required by the host Institution.

23.8 Protocols

Three protocols for embryo and sperm freezing, that have worked very reliably in the author's hands, will be described. However, we encourage the reader to check alternatives. Valuable resources for these are the Jackson Laboratory at http://cryo.jax.org/slow.html, the European Mouse Mutant Archive at http://www.emmanet.org/protocols.php, or the Riken Bio-resource Center at http://www.brc.riken.jp/lab/animal/en/protocol.shtml, among others.

23.8.1. Protocol 1: Slow Freezing of 8-Cell Embryos in Plastic Semen Straws

The following protocol was kindly provided by Dr. Jorge Sztein. Even though it was developed for cryopreservation of 8-cell embryos, the method is well suited for use with 2-cell to morula stage embryos.

23.8.1.1. Equipment

This procedure requires a programmable freezer that allows a controlled decrease in temperature to −40°C. It can be based on alcohol (FTS Systems Biocool III) or liquid nitrogen (MTG Medical Technology, Planer Kryo, or Thermo Scientific Cryomed freezers). Embryos are packed in artificial insemination 0.25 cc plastic straws (available from MTG, Irvine Scientific, and IMV Technologies).

23.8.1.2. Materials: Freezing Media

Embryos are frozen in medium M2 (Sigma M-7167) containing 1.5 M propylene glycol (PROH) (Sigma P-1009), prepared as follows:

Pipette 8.8 ml of M2 into a falcon tube and add 1.2 ml of PROH. Filter through a 0.22 μm pore filter to remove contaminants.

23.8.1.3. Diluent Medium

Consists of 1 M sucrose in M2 supplemented with streptomycin sulfate 0.050 g/l (Sigma S9137) and 0.06 g/l penicillin G potassium salt (Sigma P-4687). Weigh 17.1 g of sucrose (Sigma S-7903) and dissolve in 45 ml of M2. Adjust final volume to 50 ml, with M2. Filter sterilize, 0.22 μm pore size (Millex TM GP SLGP033RB, Millipore). The media can be stored for up to 15 days at 4°C, but it is necessary to homogenize it by gently mixing before use.

23.8.1.4. Other Reagents

KSOM+AA (Speciality Media MR 121-D)

23.8.1.5. Freezing Procedure

1. Collect embryos from donors and select those of grade 1 (excellent) or 2 (good) according to morphological appearance. After all have been collected, perform successive washes in at least five 35 mm dishes of M2, changing pipettes between dishes to remove any possible pathogens.
2. Start the programmable freezer and set the hold temperature at −7°C.
3. Take a 133 mm straw and using a metal rod with a stop, push the plug from the end to a position 75 mm from the other end (Fig. 23.1b). Mark the straw with a permanent marker including relevant embryo information and a color code that

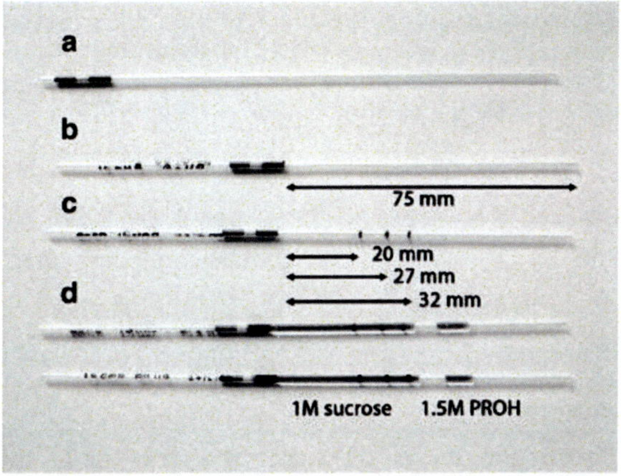

Fig. 23.1 Slow freezing of 8-cell embryos in plastic semen straws. (**a**) Plastic 133 mm semen straw with plug. (**b**) Plastic straw after moving the plug at one end with a metal rod toward the other end, leaving a distance of 75 mm. The space between the original position and the present position of the plug can be used for straw identification. (**c**) Straw with three marks, at 20, 27 and 35 mm of the present position of the plug. (**d**) Loaded straws with 1 M sucrose and the cryoprotectant solution, 1.5 M propylene glycol (PROH) ready to be loaded with equilibrated embryos.

facilitates recognition when the straw is immersed in liquid nitrogen. Alternatively, a wire marker can be used for identification. With the permanent marker, make three marks on each straw (Fig. 23.1c) at 20, 27, and 32 mm of the plug.

4. Load the straws as follows connecting it to an empty 1 ml syringe or a micro-pipetting aid (Brand, Cat No 258 00). Connect the syringe to the straw at the end where embryo information is written.
5. Aspirate diluent medium (1 M sucrose) to mark 3.
6. Aspirate air so that the sucrose meniscus reaches mark 2.
7. Aspirate 1.5 M PROH so that the sucrose meniscus reaches mark 1.
8. Aspirate air until the column of sucrose reaches the plug and seals the straw when it contacts the polyvinyl alcohol in the plug.
9. Pipette embryos collected into a 35 mm dish of 1.5 M PROH. Gently shake them to allow dilution of the washing medium and permeation of the cryoprotectant. Embryos should equilibrate for 15 min at room temperature. Load groups of 10–15 embryos and pipette them into the 1.5 M PROH fraction of each straw. Seal the straw using Crista-seal® (Hawksley cat no 01503), or a heat sealer. There is no fixed number of embryos to be loaded in a straw, we usually group this number to facilitate transfer after thawing and

avoid wasting valuable biological material. Larger numbers could be used if needed.

10. Place the straws inside the freezer and allow 5 min to equilibrate.

    Seed the sucrose fraction by touching near the plug with the tips of metal forceps cooled in liquid nitrogen.

11. Wait 5 min, and then check that the ice has migrated to the embryo fraction.
12. Cool to 35–40°C at a rate of 0.33°C/min.
13. Plunge the straws directly into liquid nitrogen.

23.8.1.6. Thawing Procedure

1. Transfer the straw from the liquid nitrogen container to a smaller liquid nitrogen container, for example a thick wall styrofoam box or Dewar flask of any size that allows the complete immersion of the straw in liquid nitrogen, either vertical or horizontal. The second option is better since it is easier to locate and identify the straw to be thawed if several are carried at the same time.
2. Pick up the selected straw with liquid nitrogen pre-cooled forceps and allow liquid nitrogen vapors to dissipate for 30–40 s, in air.
3. Plunge the straw in a water bath at room temperature for a few seconds till the ice disappears. Wipe the straw.
4. Cut the seal and place the cut end in a 35 mm Petri dish at room temperature. Cut the plug on the other end through the PVA leaving about half the cotton plug in place.
   Use the metal rod with the stop to push the contents of the straw into the 35 mm Petri dish. The sucrose fraction will mix with the embryos in the cryopreservation medium helping to expel the cryoprotectant from inside the cells. As a consequence embryos will shrink noticeably.
5. Wait for 5 min and then transfer the embryos into a dish containing M2 medium at room temperature for another 5 min. They will recover their normal appearance.
6. Transfer to the oviducts of E0.5 pseudopregnant recipients or culture till blastocyst stage in KSOM+AA and transfer into the uterus of E2.5 pseudopregnant recipients.

23.8.2. Protocol 2: Quick Freezing of Embryos

The following procedure has been adapted from the original paper from Abas Mazni et al. [13] and it has been used routinely by the authors [40]. It represents one of the many variants of quick freezing, but this particular procedure can be performed at room temperature and there is some room for a margin of error in embryo exposure time to cryoprotectant. The drop containing the equilibrated embryos is loaded directly avoiding the necessity of placing the embryos into the straw under a stereomicroscope. Among the disadvantages it is necessary to mention that the procedure does not work with all embryo preimplantation stages. Results with 2-cell embryos have proven unsatisfactory and the best results are achieved with compacted morulae collected from the uterotubal junction 72–78 h post hCG. This method requires the use of plastic artificial insemination straws of 0.25 cc, as described in Subheading 23.8.1.

23.8.2.1. Materials

Cryoprotectant Solutions

The original freezing solution consists of a 3 M Ethylene Glycol (Sigma E-9129) in M-2 (Sigma M- 7167) with 2% Fetal Calf Serum and 0.25 M sucrose. To prepare 0.25 M sucrose, add 4.28 g of sucrose (Sigma S-9378, Merck 7651 FW 342.3) to 45 ml of M-2 and stir until completely in solution. Then adjust the volume to 50 ml and filter through 0.22 µm filter (Millex™ GP SLGP033RB, Millipore) to remove contaminants.

To prepare the cryoprotectant solution, in a sterile hood add 8.4 ml of Ethylene Glycol to 39.1 ml of the 0.25 M sucrose in M2 and 2.5 ml of inactivated sterile fetal calf serum (FCS), to obtain a final volume of 50 cc. This freezing solution can be kept at 4°C for 15 days.

Thawing Solution

Consists of 0.5 M sucrose in M2. To prepare it, weigh 8.55 g of sucrose and dissolve in 45 ml of M2. Adjust final volume to 50 ml, with M2. Filter sterilize, 0.22 µm pore size. It can be stored for up to 15 days at 4°C, but it is necessary to homogenize it by gently mixing before use.

Other Reagents

FHM EmbryoMax® FHM HEPES Buffered Medium (1×), liquid, w/Phenol Red from Millipore MR-024-D

Preparation of the Freezing Chamber

This consists of a polystyrene box with a wall of at least 2 cm thickness, with a cover and internal measurements of approximately 15 cm × 25 cm × 15 cm (W×L×D) in which two lines are marked, one at 5 cm from the bottom and another 0.5 cm above the first line. The lower line marks the level of LN2 and the upper one the level where an aluminum plate of 14 cm × 20 cm and 1 mm thick is to be placed. Other metals may be used, but aluminum is preferred since it does not oxidize. (Metal plates can be found in hardware stores or manufactured by a metal working company.) The plate should not cover more than ¾ of the total inner length of the box, because the straws have to be dropped into the liquid nitrogen. To correctly locate the aluminum plate in the box we place it on top of a metal rack smaller than the aluminum plate, but any kind of support that maintains the plate in a horizontal position can be used. The chamber has to be prepared in advance by loading it to the lower mark with LN_2 and then placing the metal plate in position. The box should be closed with the cover for a few minutes, to allow the metal plate to cool and stabilize in temperature. Just before starting to freeze embryos, check the level of LN_2 and if necessary, add more LN_2. In this case, avoid dropping LN_2 on the metal plate, use the free space on one side.

23.8.2.2. Method

All manipulations are performed at room temperature. Ten to 15 embryos are loaded into each straw. A maximum of three straws are processed at a time. One drop of 250–300 μl and three drops of 100–110 μl (*see* Fig. 23.2) of the cryoprotectant solution are placed on the lid of a 6 cm diameter culture dish.

1. After collection from donors 30–45 embryos loaded in the minimum possible volume of the used washing medium (i.e., M-2, FHM) are placed in the big drop of cryoprotectant solution on the dish, a timer is started and the dish is gently shaken mixing the embryos with the cryoprotectant solution.
2. The pipette is washed in cryoprotectant solution to eliminate remaining washing medium and groups of 10–15 embryos are moved into each of the small drops and then loaded into

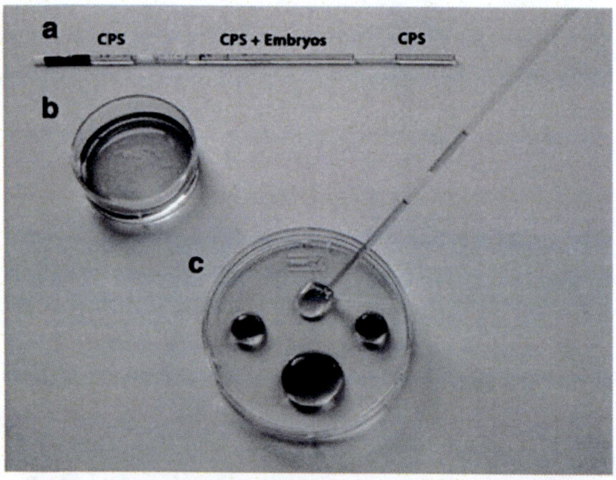

Fig. 23.2 Quick freezing protocol. (**a**) Straw ready for sealing. Embryos are located in the center expanse of cryoprotectant (CPS) solution. At either side of this medium there is an air bubble and then a smaller volume of CPS without embryos. (**b**) Petri dish with cryoprotectant solution used to load the initial and terminal CPS drops. (**c**) A 60 × 15 Petri dish lid is used to help with embryo loading. Three drops of 110 μl plus a large drop of 300–400 μl of CPS are placed in the lid. The total number of embryos that will be frozen at the same time are expelled in the minimum possible volume of medium into the large drop. Then, with a new pipette they are distributed among the three drops. Loading is performed by connection of the straw to an aspiration device. The straw is filled first with a drop from the Petri dish (**b**), then air, then the total volume of one 110 μl drop containing the embryos, then another air bubble, and lastly, more CPS from dish B till the plug seals when it comes in contact with the medium. Three straws can be easily loaded during the 5 min of equilibration time.

plastic straws with a micropippeting aid (Brand cat No 25800) as described below.

3. First, a 1.5 cm of cryoprotectant without embryos, then 10–15 mm of air, next the full drop of 110 μl of cryoprotectant with the embryos, another air space of 10–15 mm and lastly, more cryoprotectant solution till the first drop of cryoprotectant contacts the polyvinyl alcohol in the straw sealing that end.
4. The other end is sealed with heated forceps and straws are left horizontally on the bench till the 5 min equilibriation time is completed. Pick up the straws at the cotton plug end and place them gently on the metal plate inside the freezing chamber. Straws should be separated. The first one is placed closer to the border of the plate.
5. Two minutes later, with the aid of forceps previously immersed in LN_2, the straws are pushed toward the edge of the plate and dropped into LN_2.

23.8.2.3. Thawing

1. Transfer the straw from the container to a smaller container, for example the cover of a Styrofoam container. Pick up the straw with pre-cooled forceps and allow the liquid nitrogen vapors dissipate for 20–30 s.
2. Plunge the straw in a water bath at 37°C for another 20 s till the ice disappears. Take out and dry the straw with tissue paper, cut the seal and place that end inside a 35 mm Petri dish, full of the thawing medium (0.5 M sucrose in M-2) at room temperature.
3. Cut the plug on the other end and with the aid of a syringe or a pipette-aid, expel the entire contents of the straw into the thawing medium. Alternatively, use a metal rod to push the plug down the straw.
4. Gently shake the dish for 5 min, allowing the cryoprotectant to leave the embryos. They will shrink considerably upon exposure to the hyper-osmotic sucrose solution.
5. Place the embryos into M2 media in a dish at room temperature and shake gently for another 5 min. They will recover their normal appearance.
6. Transfer to the oviducts of E0.5 pseudopregnant recipients or culture till the blastocyst stage in KSOM (Speciality Media MR 121-D) and transfer into the uterus of E2.5 pseudopregnant recipients.

23.8.3. Protocol 3: Sperm Cryopreservation

This protocol has been adapted from the procedure described by Refs. [28, 41].

23.8.3.1. Animals

Mature 2- to 6-month-old males housed singly for at least for 1 week prior to sperm recovery.

23.8.3.2. Materials

Cryoprotectant solution: CPA.

18% Raffinose pentahydrate (Sigma R-0250).

3% Dehydrated skim milk (Difco 0032-17-3).

Heat 37 ml of ultrapure (milli-Q) water in a 50 ml conical Falcon (Nunc catalog No 362696) tube in a water bath at 60°C. Add raffinose and vortex till completely in solution. Add skim milk and vortex. Leave till foam disappears and add water to 50 ml.

Centrifuge in 2 ml eppendorf tubes 30 min at 4°C.

Recover supernatant in a Falcon tube and filter through 0.45 μm (Millex HP 0.45 Cat No SLHP033NS).

Check osmolarity. It should be between 460 and 480 mOsm. The supernatant is translucent.

Aliquot in 1.2 ml vols and keep frozen at −80°C. Once thawed, do not re-freeze.

Freezing chamber.

The freezing chamber consists of a styrofoam container with a wall of at least 2 cm thickness and internal measurements of approximately 20 cm × 30 cm × 15 cm (W×L×D) .The container should have a lid and liquid nitrogen, to a depth of 5 cm, in the bottom of the container. Place a styrofoam platform of approximately 18 cm × 20 cm (W×L) that is 2.5 cm thick, floating on LN2, with some toothpicks on it. These will be used to prevent contact between straws. Leave the chamber to cool at least 10 min before use. With this thickness of Styrofoam, a decrease of −40°C/min is achieved.

23.8.3.3. Method: Sperm Collection

1. Collect both epididymi and vas deferens from the donor male and place them in M2 in a Petri dish at room temperature. Carefully remove all the fat and the blood vessels that accompany the vas deferens.
2. Cut the surface of the cauda epididymi and vas deferens and place them in a pre-warmed 35 mm dish containing 1 ml of CPA. Do not leave the CPA solution for too long at 37°C since evaporation will raise osmotic pressure sensitivity and will compromise sperm survival. Express sperm from the vas deferens and cauda epididymi and avoid leaving tissue debris with the sperm.
3. Remove all tissue and leave sperm to dissipate for 10 min at 37°C in a 5% CO_2 humidified incubator.

23.8.3.4. Freezing Procedure

1. Label the freezing straw, at the cotton plug end, with identifying information and connect to a 1 ml syringe or a pipettor.

2. Using the syringe, pull up 100 μl of the fertilization medium (HTF, Cambrex Bio Science BE02-021F) followed by an air bubble of approx. 1 cm and then 100 μl of sperm suspension in CPA into 0.25 ml straws.
3. Close the end with heated forceps or Cristaseal®.
4. Place the straw horizontally on the Styrofoam platform in the pre-cooled freezing chamber for 10 min. Then plunge into liquid nitrogen.

23.8.3.5. Thawing Procedure

1. Take the straw from liquid nitrogen with pre-cooled forceps avoiding touching the sperm. Let vapors dissipate for 15–20 s and then plunge in a water bath at 37°C till all ice crystals disappear (approx. 30 s).
2. Expel the contents into an eppendorf tube and centrifuge at 700 g for 5 min.
3. Remove the supernatant and replace it with in vitro fertilization medium (HTF or similar), a maximum of 80 μl are placed in the tube.
4. Mix gently with a wide bore pipette tip and place in a CO_2 incubator so that the viable sperm can achieve capacitation and swim to the upper part of the medium. This will take around 30 min. Up to 40 μl of the supernatant can be used for fertilization of oocytes.

Note: Based on recent publications, the addition of either 1 mM methyl-beta-cyclodextrin [30] of 477 μM monothioglycerol [29] in the cryoprotectant solution enhances sperm survival in C57BL/6. Also, note that capacitation time in this strain takes longer than others and it is recommended at least 60 min in HTF prior to adding the sperm into the in vitro fertilization dish.

References

1. Knight J, Abbott A (2002) Full house. Nature 417:785–786
2. Ogonuki N, Mochida K, Miki H, Inoue K, Fray M, Iwaki T, Moriwaki K, Obata Y, Morozumi K, Yanagimachi R, Ogura A (2006) Spermatozoa and spermatids retrieved from frozen reproductive organs or frozen whole bodies of male mice can produce normal offspring. Proc Natl Acad Sci USA 103: 13098–13103
3. Wakayama S, Ohta H, Hikichi T, Mizutani E, Iwaki T, Kanagawa O, Wakayama T (2008) Production of healthy cloned mice from bodies frozen at −20 degrees C for 16 years. Proc Natl Acad Sci USA 105: 17318–17322
4. Garrick D, Fiering S, Martin DI, Whitelaw E (1998) Repeat-induced gene silencing in mammals. Nat Genet 18:56–59
5. Houdebine LM (2000) Transgenic animal bioreactors. Transgenic Res 9:305–320
6. Koetsier PA, Mangel L, Schmitz B, Doerfler W (1996) Stability of transgene methylation patterns in mice: position effects, strain specificity and cellular mosaicism. Transgenic Res 5:235–244
7. Whittingham DG, Leibo SP, Mazur P (1972) Survival of mouse embryos frozen to −196 degrees and −269 degrees C. Science 178: 411–414
8. Critser JK, Mobraaten LE (2000) Cryopreservation of murine spermatozoa. ILAR J

41:197–206, National Research Council, Institute of Laboratory Animal Resources
9. Rall WF, Fahy GM (1985) Ice-free cryopreservation of mouse embryos at −196 degrees C by vitrification. Nature 313:573–575
10. Szell A, Shelton JN (1986) Sucrose dilution of glycerol from mouse embryos frozen rapidly in liquid nitrogen vapour. J Reprod Fertil 76:401–408
11. Leibo SP (1989) Equilibrium and nonequilibrium cryopreservation of embryos. Theriogenology 31:85–93
12. Chupin D, De Reviers MM (1986) Quick freezing of rat embryos. Theriogenology 26: 157–167
13. Abas Mazni O, Valdez CA, Takahashi Y, Hishinuma M, Kanagawa H (1990) Quick freezing of mouse embryos using ethylene glycol with lactose or sucrose. Anim Reprod Sci 22:161–169
14. Kong IK, Lee SI, Cho SG, Cho SK, Park CS (2000) Comparison of open pulled straw (OPS) vs glass micropipette (GMP) vitrification in mouse blastocysts. Theriogenology 53: 1817–1826
15. Tsang WH, Chow KL (2009) Mouse embryo cryopreservation utilizing a novel high-capacity vitrification spatula. Biotechniques 46: 550–552
16. Peters DD, Marschall S, Mahabir E, Boersma A, Heinzmann U, Schmidt J, Hrabe de Angelis M (2006) Risk assessment of mouse hepatitis virus infection via in vitro fertilization and embryo transfer by the use of zona-intact and laser-microdissected oocytes. Biol Reprod 74:246–252
17. Schiewe MC, Hollifield VM, Kasbohm LA, Schmidt PM (1995) Embryo importation and cryobanking strategies for laboratory animals and wildlife species. Theriogenology 43:97–104
18. Van Keuren ML, Saunders TL (2004) Rederivation of transgenic and gene-targeted mice by embryo transfer. Transgenic Res 13: 363–371
19. Hogan B, Beddington R, Costantini F, Lacy E (1994) Manipulating the mouse embryo: a laboratory manual. Cold Spring Harbor Laboratory Press, New York
20. Chatot CL, Ziomek CA, Bavister BD, Lewis JL, Torres I (1989) An improved culture medium supports development of random-bred 1-cell mouse embryos in vitro. J Reprod Fertil 86:679–688
21. Suzuki O, Asano T, Yamamoto Y, Takano K, Koura M (1996) Development in vitro of preimplantation embryos from 55 mouse strains. Reprod Fertil Dev 8:975–980
22. Byers SL, Payson SJ, Taft RA (2006) Performance of ten inbred mouse strains following assisted reproductive technologies (ARTs). Theriogenology 65:1716–1726
23. Munoz I, Rodriguez de Sadia C, Gutierrez A, Blanquez MJ, Pintado B (1994) Comparison of superovulatory response of mature outbred mice treated with FSH or PMSG and developmental potential of embryos produced. Theriogenology 41:907–914
24. Rall WF, Schmidt PM, Lin X, Brown SS, Ward AC, Hansen CT (2000) Factors affecting the efficiency of embryo cryopreservation and rederivatoin of rat and mouse models. ILAR J 41:221–227, National Research Council, Institute of Laboratory Animal Resources
25. Willoughby CE, Mazur P, Peter AT, Critser JK (1996) Osmotic tolerance limits and properties of murine spermatozoa. Biol Reprod 55:715–727
26. Tada N, Sato M, Yamanoi J, Mizorogi T, Kasai K, Ogawa S (1990) Cryopreservation of mouse spermatozoa in the presence of raffinose and glycerol. J Reprod Fertil 89: 511–516
27. Sztein JM, Farley JS, Mobraaten LE (2000) In vitro fertilization with cryopreserved inbred mouse sperm. Biol Reprod 63:1774–1780
28. Nakagata N (2000) Cryopreservation of mouse spermatozoa. Mamm Genome 11:572–576
29. Ostermeier GC, Wiles MV, Farley JS, Taft RA (2008) Conserving, distributing and managing genetically modified mouse lines by sperm cryopreservation. PLoS ONE 3:e2792
30. Takeo T, Hoshii T, Kondo Y, Toyodome H, Arima H, Yamamura K, Irie T, Nakagata N (2008) Methyl-beta-cyclodextrin improves fertilizing ability of C57BL/6 mouse sperm after freezing and thawing by facilitating cholesterol efflux from the cells. Biol Reprod 78: 546–551
31. Mahabir E, Bulian D, Needham J, Schmidt J (2009) Lack of transmission of mouse minute virus (MMV) from in vitro-produced embryos to recipients and pups due to the presence of cumulus cells during the in vitro fertilization process. Biol Reprod 81(3): 531–538
32. Janus LM, Smoczek A, Hedrich HJ, Bleich A (2009) Risk assessment of minute virus of mice transmission during rederivation: detection in reproductive organs, gametes, and embryos of mice after in vivo infection. Biol Reprod 81(5):1010–1015
33. Bielanski A, Bergeron H, Lau PC, Devenish J (2003) Microbial contamination of embryos and semen during long term banking in liquid nitrogen. Cryobiology 46:146–152

34. Bath ML (2003) Simple and efficient in vitro fertilization with cryopreserved C57BL/6J mouse sperm. Biol Reprod 68:19–23
35. Pintado B, Hourcade JD, Pérez-Crespo M, Gutiérrez-Adán A (2008) Intraoviductal insemination with frozen C57Bl/6J sperm increases fertility rate compared to standard IVF. Transgenic Res 17:1017
36. Wakayama T, Whittingham DG, Yanagimachi R (1998) Production of normal offspring from mouse oocytes injected with spermatozoa cryopreserved with or without cryoprotection. J Reprod Fertil 112:11–17
37. Snow M, Cox SL, Jenkin G, Trounson A, Shaw J (2002) Generation of live young from xenografted mouse ovaries. Science 297:2227
38. Wakayama S, Kishigami S, Wakayama T (2009) Cloning of ES cells and mice by nuclear transfer. Meth Mol Biol 530:251–265, Clifton, NJ
39. IETS (1998) Manual of the International Embryo Transfer Society: a procedural guide and general information for the use of embryo transfer technology, emphasizing sanitary precautions, 3rd edn. IETS, Champaign, IL
40. Gutierrez A, Garde J, Artiga CG, Munoz I, Pintado B (1993) In vitro survival of murine morulae after quick freezing in the presence of chemically defined macromolecules and different cryoprotectants. Theriogenology 39: 1111–1120
41. Sztein JM, Farley JS, Young AF, Mobraaten LE (1997) Motility of cryopreserved mouse spermatozoa affected by temperature of collection and rate of thawing. Cryobiology 35:46–52

Chapter 24

Shipment of Mice and Embryos

Shirley Pease

Abstract

Following the advent of transgenic technology and the subsequent establishment of international consortia for broad-based genomic analysis of the mouse, the need for sharing of resources in the exchange of mouse models and material has reached an all time high. Live animals, embryos, and gametes are ever more increasingly shipped around the world in an exchange or purchase of research material. In this chapter, we shall look at requirements for shipping of live animals locally, as well as review the current requirements for receipt of live animals by governing authorities in Europe, the UK, the USA, Australia, and China. We shall also review procedures for shipping of cryopreserved embryos and gametes, also embryos at ambient temperatures.

24.1 Introduction

The need to move novel mutant mouse models around the world is increasing, as ever more mutations are generated, either by individual laboratories or more recently by the various international consortia, whose ultimate goal is to generate mutations for every gene in the mouse genome. Today, more than ever, investigators may find their gene of interest already introduced into the mouse or rat genome and all that remains before experiments can start is to import the mutation in some form, to the laboratory. A new mutation may be available either as embryonic stem cells or live animals or cryopreserved embryos. In most cases, we are not free to move animals from A to B at will. There are national importation regulations to abide by and most likely local institutional requirements that need to be met. International regulations for the transportation of animals change frequently. So I shall not try to present those regulations in detail as they stand today, since they likely will be meaningless in a short while. I will, however, try to outline the general considerations that must be given to the shipment of live animals and embryos, with safe arrival at their destination the goal.

S. Pease and T.L. Saunders (eds.), *Advanced Protocols for Animal Transgenesis*, Springer Protocols,
DOI 10.1007/978-3-642-20792-1_24, © Springer-Verlag Berlin Heidelberg 2011

24.2 Shipping of Live Animals

24.2.1. Regulatory Bodies

Transportation is a stressful event for laboratory rodents [1] and it is in everyone's best interests to make a maximum effort to ensure that these valuable animals arrive at their destination in good health. There have been many reviews, assessments and regulations drawn up by various bodies, in order to help the scientific community (and others) meet that goal. The majority of live animals are shipped by air, though some may travel by road. There are international regulations that must be satisfied, as well as local or national requirements, all of which cater to the safety, comfort, and general well-being of the animals. The International Air Travel Association (IATA) has Live Animals Regulations that must be met by its member airlines. Nonmember airlines are not held to these standards, but are subject to local national regulations, such as those developed by the Ministry of Agriculture, Fisheries and Food (UK) and the USDA (USA). The Laboratory Animal Science Association (LASA) convened a transport working group in 2005 to review current regulations relating to animal welfare in transportation [1]. Other international guidelines have been developed by the European Union [2]. All regulations and guidelines are directed toward managing rodent welfare in stipulating details of temperature control, food and water supply, container size, stocking density and ventilation that should/must be provided while the animals are in transit [3–6]

24.2.2. National Shipments

Most shipments, national or international are sent by air, with the assistance of a courier, such as World Courier (http://www.worldcourier.com) that specializes in the transport of live animals. Some carriers are IATA certified, some are not. We recommend you use a carrier that is certified, for the assurance it brings that certain standards for the shipping of live animals will be met. The IATA issues its Live Animals Regulations on an annual basis and live animal shippers who move their animals with an airline that is IATA certified must abide by the most recent regulations as well as any governmental regulations stipulated by the receiving country. Your courier will be able to advise you on airline requirements for the packaging of animals (*see* Subheading 24.2.4), will book the shipment onto prearranged flights, pick up the animals and ensure their safe transport to the airport, plus collect and deliver them to their destination upon arrival.

24.2.3. International Shipments

International shipments are a little more complicated, because in addition to receiving institutional requirements for documentation (*see* Subheading 24.2.4), customs requirements of the receiving country also need to be met. In most cases, a veterinary health certificate stating the origin and health status of the animals, a waybill, and a pro-forma invoice will be required. In addition, veterinary certification may be required from a national authority, for example from the USDA if animals are traveling from the USA, or from the Ministry of Agriculture if traveling from the UK. The national requirements for the receipt of live animals into each country are constantly changing, but your courier will have up-to-date information about the customs requirements for all countries. Some examples are given below. The specified documentation must accompany the shipment and be present and complete when the animals arrive in customs at their destination. Incomplete documentation may result in delay in the delivery of animals or their return to their origin or euthanasia at the port of entry.

International requirements for receipt of animals change quite frequently. So we will give only a general indication, for your information now as to what kind of documentation may be required. This is in addition to the health profile data the receiving institution may have requested, which will have been provided before the animals are shipped (*see* Subheading 24.2.4). Documentation required may take the form of an import permit, a pro-forma invoice, a veterinary health certificate from the institute of origin and a national authority, plus the waybill, as supplied by your courier.

Currently, the most frequent destinations for shipment of laboratory mice are: China, France, Germany, UK, and USA. Of these, China and France require nationally endorsed paperwork (i.e., from USDA in the USA). Germany, the UK, and the USA do not require nationally endorsed paperwork.

For outgoing shipments from the USA, the USDA requires that this endorsement, when required, be prepared in the state of origin, which further complicates timing by not allowing endorsement at the port of departure, such as New York or Los Angeles. Furthermore, in some cases, this endorsement is considered to have only a limited life, requiring that documentation and shipping of animals is acutely coordinated. Current examples of such destinations are:

Netherlands – 10 days

Hong Kong – 10 days

France – 10 days

Czechoslovakia – 10 days

Hungary – 10 days

Singapore – 7 days

Spain – 5 days

Australia – 72 h (prior to flight departure)

Israel – 48 h

Animals shipped to the EU are required to have a UN aligned Certificate of Veterinary Entry Document, http://europa.eu.int/eur-lex/pri/en/oj/dat/2004/l_021/l_02120040128en00110023.pdf.

Some countries, such as Spain, Israel, Sweden, and France have their own health certificates which must be completed and signed by the shipper's institutional veterinarian and/or a USDA official.

The current list of countries that require USDA endorsed paperwork as of December 2010 is:

AKL – New Zealand

AMS – The Netherlands

ATH – Greece

BCN – Spain

BJS – China

BRU – Belgium

BUD – Hungary

BUE – Buenos Aires

CDG – France

DEL – India

DUB – Ireland

FCO – Italy

HKG – Hong Kong

MAD – Spain

MEX – Mexico

MOW – Russia

MXP – Italy

OSL – Norway

SCL – Chile

SHA – China

SEL – Korea

SIN – Singapore

SYD – Australia

TLV – Israel

TPE – Taiwan (China)

TYO – Japan

VIE – Austria

In general, when completing documentation for international shipments, it is important to make sure that *all* the details are correct. The number of animals must be correct, the details of strain name must be used and correct in all cases, changes should not be made to documentation and where possible and appropriate, include "pathogen-free" and "laboratory animals" in the documentation.

24.2.4. Health Status

It is recognized that the microbiological profile of animals used in research is of significant importance, since the presence of pathogens can confound research results. Therefore many institutions have a well-defined list of organisms deemed acceptable as endemic within their rodent colonies, for a variety of reasons (*see* Chapter 25, " Pathogen Free Rederivation of Mouse Strains") and they strive to maintain that profile when considering importation of new mouse strains. Therefore, it is in the best interests of the receiving institution to require documentation from the sending institution, detailing the health profile for the preceding 1 or 2 years of strains to be shipped. They will likely require health profile data in advance, so that they can determine how to handle/house incoming mice upon arrival, without placing resident colonies at risk.

There are many ways in which pathogens may enter a closed mouse colony and incoming animals clearly have the potential to act as vectors. In addition, animals in shipment may come into contact and be contaminated by others of the same species that are not disease-free. In the shipping of animals, laboratory mice from various different institutions, as well as other species, may be held in a common area for a length of time prior to, during, or after shipment. They may be placed adjacent to other animal shipments on loading docks and containers may be left standing in areas that are home to wild rodent colonies. Such possible exposures and resulting transmission of microbes have the potential to change the microbiological status of animals shipped. Disease may be transmitted by contact, aerosols, fomites, urine, and feces. For this reason it is incumbent upon those shipping the animals to be sure not only that they are provided with adequate food, water and warmth if necessary, but also that they are packaged securely in containers that will serve as a barrier between the occupants and the world at large. Upon receipt of animals, it is wise to decontaminate the exterior of shipping containers prior to handling the incoming crates or animals within.

24.2.5. Traveling Conditions and Packaging

IATA regulations stipulate that rodents in transit may be exposed to actual temperatures of no less than 5°C and no more than 35°C. The National Research Council stipulates 4–34°C (39–93°F). Laboratory rodents have a mean body temperature of 33.9°C. In low temperatures, they are able to huddle together in nesting material to conserve heat. But in high temperatures, they are unable to respond effectively. Ambient temperatures of above 37.1°C for more than a short period of time will result in the death of the animal [7]. In a recent survey, rodents en route to other institutions within the USA and on international flights were observed to experience a very wide range in temperature [8]. Almost 50% of shipments were exposed to temperatures above 29°C and nearly 15% of shipments to temperatures below 7.2°C. In addition, 50% of shipments were exposed to temperature ranges of 11°C or more. Most domestic shipments take 24 h, whereas international shipments may take up to 72 h.

24.2.5.1. Containers

Safe containers, designed for the purpose, can be obtained from suppliers such as Simonsen, the Jackson Laboratory or Taconic Farms (Fig. 24.1). Plastic shipping crates rather than cardboard are more suitable for international flights. The containers usually come with filter material placed over all air vents in the container. This serves two purposes. Most importantly it will protect the occupants from air or fomite-born pathogens. Filter material also serves to stabilize and maintain the internal temperature within the container to some degree. As to the number of animals to be packed into one filtered container, the LASA working Group [1] recommends 120 sq cm of floor space be provided per mouse of up to 20 g in body weight (150 sq cm at >20 g) and between 160 and 600 sq cm per rat, from weaning age to 250 g bodyweight.

Rodents in transit will benefit from the addition of plenty of bedding and nesting material, e.g., Nestlets (Fig. 24.2a) which will help them to adjust to the stresses of transportation and maintain body temperature. If cold weather conditions are

Fig. 24.1 Shipping crates, supplied by Simonsen, Taconic, and Jackson Laboratories. Crates are available in different sizes and/or with dividers, for separation of groups of mice. Filter material covers ventilation ports, protected internally by metallic mesh.

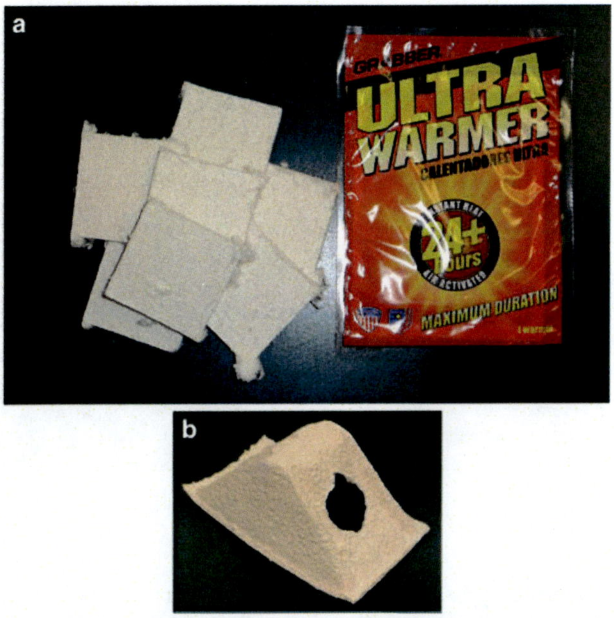

Fig. 24.2 Aids to temperature control during shipping. (a) Nestlets may be shredded by animals for nest-building and conservation of warmth. If extreme low temperatures are anticipated, then the inclusion of activated warm pads, such as "Grabbers," will provide warmth for up to 12 h. (b) Shepherd shacks, for nest-building and warmth conservation.

anticipated, then adding a source of warmth in the form of a hand-warming gel packs, such as "Grabber Hand Warmers" (Fig. 24.2a) is a good idea, even though the effect may not last longer than 12 h. Shepherd shacks (Fig. 24.2b) are also recommended for mice in transit. They will naturally build a nest within this smaller space and the shack will help to conserve warmth.

24.2.5.2. Food and Water Supply

Of course, animals must be provided with enough food and water for the trip. Water may be provided in the form of gelatinized water packs, such as those from Perotech or Shepherd Specialty Papers (Fig. 24.3). Rodents will chew through the outer wrapper to get to the gelatin. Food can be provided as regular laboratory mouse diet pellets on the floor of the traveling container. Young, lactating, or debilitated animals may be provided with a source of diet and water combined in the form of DietGel 76A or DietGel Recovery, manufactured by Hydrogel, distributed by Newco.

Once the animals are placed within, the container needs to be sealed securely with tape, eliminating all sources of air supply other than via the filter material. Use large and clear address labels, including "Live Animals" and "Handle with Care."

Fig. 24.3 Provision of water in the form of a gel pack.

24.2.6. Candidates for Shipping

Although travel is stressful for all animals, adult animals normally will travel well. In general, ship cage mates in groups, but do not mix individuals previously caged separately. Females are much less likely to fight than males, but a new grouping of females will place additional stress on the animals.

Weaning is a stressful time for laboratory rodents. Therefore, it is advisable to ship animals that are at least 4 days post weaning. Alternatively, include the dam in the shipment. Preweaning litters must, of course, be shipped with their mother.

The EU guidelines [2] state that females shall not be shipped during the last 10th of the gestation period or for at least 1 week after parturition. The LASA working group [1] recommends that females not be shipped during the last one-fifth of gestation, that is after E17.5

After transit, it may take between 1 and 7 days for shipped rodents to adapt to their new environment [9]. Circadian rhythms may take up to 2 weeks to return to normal [10]

24.3 Shipment of Cryopreserved Embryos or Sperm

24.3.1. Documentation

Receiving institutions may ask for health data relating to the strain of mouse from which embryos were cryopreserved. However, since embryo washing and implantation is a method used for the pathogen-free rederivation of mouse strains (*see* Chapter 25), this information could be considered less critical to receipt of preimplantation embryos than live animals.

With regard to international shipments, the conditions that apply to the importation of live animals may not alwys apply to preimplantation embryos. Regulations need to be determined on a case by case basis. In Australia for example, imported embryos must be proven free of Haanta virus. New Zealand currently will not accept importation of murine sperm from anywhere other than Australia.

24.3.2. Liquid Nitrogen (LN2) Shipments

Cryopreserved embryos and sperm samples can be shipped safely in a "dry liquid nitrogen shipper," such as the Taylor-Wharton CX100. A dry shipper is a container with a vacuum-sealed jacket, much like a regular LN2 cell-storage tank. Inside the container, there is a central canister (Fig. 24.4a) suitable for the placement of a cane of the type that is used to hold screw-topped LN2 cell-storage vials (Fig. 24.4b). These canes can also be used to hold cryopreservation straws containing either embryos or sperm, held in place by tube bottoms (Fig. 24.4c). Surrounding the central canister is a volume of absorbent material that is designed to

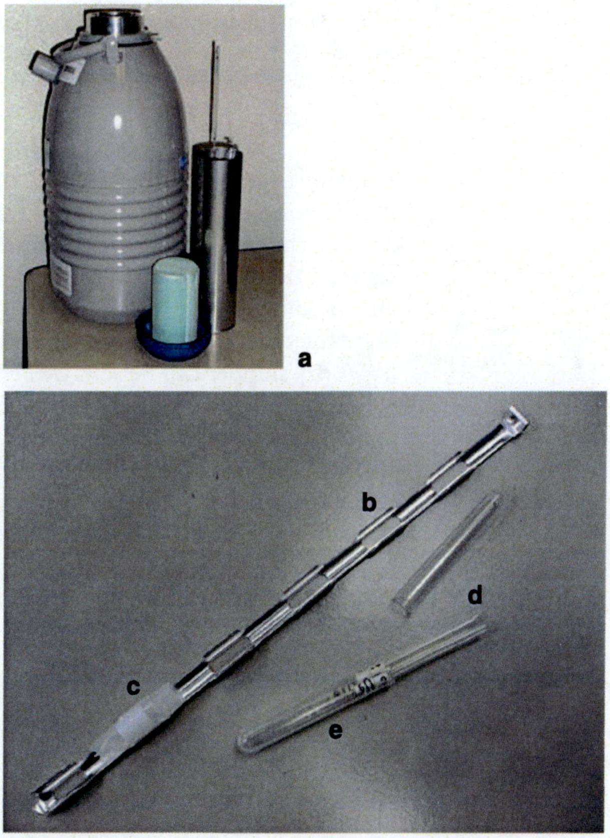

Fig. 24.4 Shipping of frozen embryos. (a) A dry shipper and internal canister. (b) A standard cane for securing of (c), vials, or (d), straws, enclosed in tubes, (e), and placement in the internal canister of a dry shipper.

become saturated with liquid nitrogen. The shipper described absorbs a volume of approximately 4.4 L, which is enough to maintain the interior of properly functioning canister at −180° for 17 days. Thus, the contents are maintained at ultra low temperature and in the event that the container is knocked over, free liquid nitrogen will not be at risk of spilling out of the container.

The shipper must be prepared for use on the day prior to shipping. Instructions come with the canister of course. Briefly, the stopper and central canister should be removed and the shipper weighed. Each shipper, according to size, is able to retain a certain volume of LN2, which will be reflected in the end weight of the container after it has been filled. Reaching this maximum weight gain will ensure that ultra low temperatures are maintained for the maximum amount of time. Fill the container with liquid nitrogen, up to the level of the bottom of the neck. Allow the container to stand for 10 min, to allow the absorbent material to take up the LN2. After about 10 min, top the container up with LN2 again. Repeat this cycle until the LN2 level no longer falls within 10 min. This procedure can take up to 10–15 cycles. It may be convenient to replace the internal canister, canes and stopper, and allow the whole to stand overnight. Prior to loading the shipper with material to be shipped, the free liquid nitrogen in the container must be poured out and the shipper rendered "dry." At this point, the weight of the shipper should have increased by about 8 lbs, for a CX100. Return the internal canister and load the canes with vials or straws to be shipped. Make sure that all shipper components, canes, etc., have been precooled before loading with vials or straws. Replace the stopper, close the lid, and secure with a cable tie (Ty-Rap from Thomas Betts). Place the container inside its external cover.

A dry shipper can be sent by Fedex as nonhazardous material. A label should be placed on the outside of the external cover stating "Dry shipper, not restricted." Better service may be obtained from a courier service such as World Courier. In the event of a delay, a courier service ensure adequate LN2 levels are maintained, whereas non-courier servicing agents will not. The contents of the smallest shipper, i.e., volume of 1.5 L (MVE model SC 2/1V) should remain at ultra low temperature for a period of 8 days.

24.3.3. Shipment of Embryos at Ambient Temperatures

Embryos between two-cell and blastocyst stage can remain at ambient temperatures, in the right media conditions, for periods of up to 48 h and still continue normal development, once returned to the incubator or implanted into recipient females. Embryos at one-cell stage are more sensitive to suboptimal temperatures, so shipping at ambient temperatures is not recommended for this stages of embryo development. Essentially, embryos need to be placed into tubes (such as Nunc cryo tubes catalog 375353 or 1.2 mL Corning cryo vials, catalog #430487) that are filled to the top with M2 media, and sealed [11]. Alternatively, use CO_2 equilibriated ES cell media, with LIF, for the transport of blastocysts. Equilibrated M16 or KSOM can also serve as transport media and may be the best choice for embryos at two-cell stage for prevention of two-cell block. In every case, nonhepes buffered media should be preequilibrated, the transportation tube filled to the top of the vial with media and tightly sealed. The presence of air in the tube may hasten the drift of correct pH in the media, so try to minimize this. The tube should be wrapped in tissue paper and placed inside a larger tube and then in a padded envelope or a box insulated with Styrofoam "peanuts" and shipped overnight. The safest approach is to arrange for overnight delivery. But in the event of a delay, blastocysts should arrive in un-hatched condition after up to 48 h. If low temperatures may be an issue, then place some warm packs in an insulated box, surrounding the tube containing the embryos. This will keep temperatures above freezing point within the box.

If you are routinely shipping embryos, you may wish to consider the purchase of a portable incubator, such as the BioTherm INC-12V portable incubator from Cryologic (http://www.cryologic.com). This incubator keeps a steady temperature of 32.5°C to 40°C for a period of 30 h, but does not provide CO_2 injection. Straws or vials can be accommodated within.

As to the receipt of embryos, those at the four-cell stage offer the opportunity to assess viability by in vitro culture for a period of time before implantation. The disadvantage to receipt of blastocyst stage embryos is that they, optimally, need to be implanted soon after receipt, which may create logistical problems as well as preclude any opportunity to assess the viability of the embryos.

24.4 Suppliers

Hydrogel
117 Preble Street,
Portland,
Maine, 04101

The Jackson Laboratory
600 Main Street
Bar Harbor, Maine 04609
http://www.jax.org

Grabber Inc.
5760 N HawkEye Ct SW
Grand Rapids MI 49509

http://www.warmers.com

MVE Bio-Medical Division
Chart Industries, Inc.
3505 County Road 42 West
Burnsville, MN 55306-3803, USA
http://www.chartbiomed.com

Newco Distributors Inc.
10700 7th St
Rancho Cucamonga, CA 91730
http://www.newcodistributors.com

Perotech
2 Select Ave, Unit 7
Toronto, ON M1V 5J4, Canada
http://www.perotech.com

Shepherd Specialty Papers
8080 Moorsbridge Road
Kalamazoo, MI 49024
http://www.ssponline.com

Simonsen Laboratories
1180 Day Rd Suite C
Gilroy
California 95020-9348
orders@simlab.com

Taconic Farms
One Hudson City Centre
Hudson, NY 12534
http://www.taconic.com

Taylor-Wharton-Cryogenics
4075 Hamilton Boulevard
Theodore, AL 36582
http://www.taylorwharton.com

Thomas + Betts Corporation
8155 T&B Boulevard
Memphis, TN 38125

Acknowledgements

Grateful thanks are due to *Mike Hackett of World Courier, Los Angeles*, for assistance in assembling details of shipping documentation.

References

1. Swallow J, Anderson D, Buckwell A, Harris T, Hawkins P, Kirkwood J, Lomas M, Meacham S, Peters A, Prescott M, Owen S, Quest R, Sutcliffe R, Thompson K (2005) Guidance on the transport of laboratory animals: report of the Transport Working Group established by the Laboratory Animal Science Association (LASA). Lab Anim 39:1–39
2. European Convention on the Protection of Animals during International Transport (2003) http://conventions.coe.int/Treaty/en/Treaties/Html/193.htm
3. Hartung J (2006) The new EU animal transport regulation: improved welfare and health or increased administration? Dtsch Tierärztl Wochenschr 113:113–116
4. International Air Transport Association (2007) Live animals regulations. IATA, 800 Place Victoria, P.O. Box 113 Montreal, QC, Canada H4Z 1M1
5. National Research Council (2006) Guidelines for the humane transportation of research animals. The National Academies Press, 500 Fifth St, NW, Washington, DC
6. National Research Council (2010) Guide for the care and use of laboratory animals. The National Academies Press, 500 Fifth St, NW, Washington, DC
7. Kaplan HB, Brewer NR, Blair WH (1983) Physiology. In: Foster HS, Small JD, Fox JG (eds) The mouse in biomedical research. Academic, New York, NY, pp 248–292
8. Syversen E, Pineda FJ, Watson J (2008) Temperature variations recorded during interinstitutional air shipments of laboratory mice. J Am Assoc Lab Anim Sci 47:31–36
9. Obernier JA, Baldwin RL (2006) Establishing an appropriate period of acclimatization following transportation of laboratory animals. ILAR J 47:364–369
10. Van Ruiven F, Meijer GW, Van Zutphen LFM, Ritskes-Hoitinga J (1996) Adaptation period of laboratory animals after transport: a review. Scand J Lab Anim Sci 23:185–190
11. Kelley KA (2010) Transport of mouse lines by shipment of live embryos. Methods Enzymol 476:25–36

Chapter 25

Pathogen-Free Mouse Rederivation by IVF, Natural Mating and Hysterectomy

J.M. Sztein, R.J. Kastenmayer, and K.A. Perdue

Abstract

The increased popularity of genetically modified animals in collaborative studies has stimulated the widespread interchange of mice among institutions with inconsistent health standards. While the presence of certain organisms may be tolerated at one institution, the impact on the studies at another may demand clean animals. The importance of a well-established veterinary health surveillance program and a list of organisms to be excluded from the facility, i.e., a Specific Pathogen Free (SPF) list, become critical for animal facility management and maintenance of institutional standards.

Most institutions allow direct entry of mice into a holding room only when animals come from production areas of recognized vendors with a known high health status maintained for many years. A different plan must be in place when mice are introduced from areas of a lower or unknown health status. Rederivation is the method by which laboratory animals can be "cleaned" or decontaminated of certain pathogens, including transmissible zoonotic diseases, before being introduced into barrier housing facilities.

25.1 Introduction

Rederivation is the method by which laboratory animals can be "cleaned" or decontaminated of certain pathogens before being introduced into barrier housing facilities. Initially, the rederivation procedure was used to improve colony health status by eradicating sero-positive animals. At that time, the "conventional laboratory animal" potentially harbored a broad collection of organisms, which may have included ectoparasites, endoparasites, fungi, protozoa, bacteria, mycoplasma and viruses [1, 2]. Currently, the rederivation technique is still as important as it was in the past [3], but now it is primarily employed to maintain the pathogen-free status that was so difficult to achieve [4–7].

The increased popularity of transgenic animals in collaborative studies stimulated the widespread interchange of mice among institutions with inconsistent health standards [8]. While the presence of certain organisms may be tolerated at one institution,

S. Pease and T.L. Saunders (eds.), *Advanced Protocols for Animal Transgenesis*, Springer Protocols,
DOI 10.1007/978-3-642-20792-1_25, © Springer-Verlag Berlin Heidelberg 2011

the impact on the studies at another may demand clean animals. For this reason, it is important to stipulate health conditions before introducing imported mice directly into animal holding rooms. The importance of a well-established veterinary health surveillance program and a list of organisms to be excluded from the facility, i.e., a Specific Pathogen Free (SPF) list, become critical for animal facility management and maintenance of institutional standards. An institution may have multiple animal facilities; each with a different health status and its own "SPF" level [9]. Thus, the establishment of footpath patterns and training of all staff entering the facility, on the delineated health status levels between facilities or rooms, are crucial to maintaining the integrity of the desired health status.

Routine receipt of animals from vendors versus importation of animals from external institutions require different levels of attention. Most institutions are permissive – allowing direct entry of mice into a holding room – only when animals come from production areas of recognized vendors with a known high health status maintained for many years. A different plan must be in place when mice are introduced from research laboratories. Rederivation of these animals, prior to introducing them into the colony, is the safest procedure to protect, first, the receiving institution's animal colonies from any possible outbreak and, second, researchers from misinterpretation of study results caused by infected animals [6, 10]. Rederivation also protects the staff and researchers from the rare event of receiving animals carrying a transmissible zoonotic disease [11].

25.1.1. Outline of the Procedure

Laboratory animals become infected via two possible modes of contamination: horizontal or vertical transmission. Horizontal transmission occurs with direct or indirect (fomite) contact with infected animals. Most of the horizontally transmitted contaminants such as mites, endo-parasites, bacteria, mycoplasma and many viruses can be excluded by the rederivation procedures. Vertically transmitted viruses are those that infect the gametes or are transferred through the placenta. Some viruses of this group present a human risk; for example, lymphocytic choriomeningitis virus (LCMV) and hantaviruses.

The original concept for rederivation was conceived in the early 1960s and was accomplished by cesarean section (hysterectomy) [1, 12–14]. Although this technique is not used frequently nowadays, it will be described below. Today, the first choice for rederivation is by embryo transfer [7, 15–20]. It is well demonstrated that pre-implantation embryos from one-cell to blastocyst stage have less possibility of being contaminated than are term pups obtained by hysterectomy. The zona pellucida protects the embryos against infection. Pathogens such as mycoplasma and bacteria that may attach to the embryo can be "cleaned" by treating the zona pellucida with hyaluronidase followed by multiple washes in culture media. The preimplantation embryos float in the oviduct lumen without much contact with the mother's uterus until implantation occurs, whereas the newborns obtained by hysterectomy were connected through the placenta and were sharing body fluids with their mother for several weeks. The rodent placenta is hemochorial, and they share blood and cells with each other. Tracer studies have demonstrated that the pool of amniotic fluid is not a static reservoir, but that there is a continuous interchange between the amniotic sac and the maternal and fetal circulation, with the fluid and its chemical constituents in a constant state of flux [21].

For rederivation, embryos can be collected at any of the preimplantation stages, thoroughly washed at least five times and surgically transferred into a "clean" recipient. At weaning, the surrogate mother should be screened for a list of unacceptable pathogens and, if she tests negative, the pups can be considered to be "clean" or free of those pathogens.

25.1.2. Principles and Applications

Embryo transfer has almost completely replaced the application of hysterectomy for rederivation. Perhaps the evolution of transgenic methodology which brought popularity to the embryo transfer technique was responsible for this change. Despite the reason, it is confirmed that rederivation by embryo transfer can eliminate the following pathogens: mouse parvovirus (MPV), mouse hepatitis virus (MHV), minute virus of mice [(MVM), also seen in the literature as mouse minute virus (MMV)], epizootic diarrhea of infant mice (EDIM) due to a rotavirus, mouse encephalomyelitis virus (GD-VII), mouse adenoviruses, *Helicobacter* spp., endoparasites and ectoparasites [6, 7, 22, 23]. It was also demonstrated that embryo transfer eliminates bacteria such as *Pasteurella pneumotropica* that are not cleared by cesarean rederivation [19, 20, 24].

The original rederivation using an embryo transfer protocol collected embryos on a set day after natural mating [19]. Although it is now known that embryos collected at any of the pre-implantation stages can be used, the one, two, or eight-cell embryos are the most popular. Another option is to produce the embryos using in vitro fertilization (IVF) [20]. Our laboratory performs rederivations based on IVF embryo production. This approach tends to reduce the number of animals used and the required amount of housing space. The increased efficiency of this method of rederivation is reflected in a reduction of the total time frame for the completion of the project.

To perform a rederivation, five females and one or two males of the same genotype are needed when homozygosity must be maintained; one or two founder males if rederivation using commercially purchased wild-type females are acceptable. Generally, importation of a new strain implies receiving only one or two pair of breeders. Unfortunately, this is not a sufficient number of animals to do a homozygous rederivation; therefore it becomes necessary to breed and expand the colony and use the weanlings for oocyte collection. Expanding a colony requires a "dirty" area or facility that breeder pairs and the initial offspring can be housed, or an area where a traditional quarantine and testing for pathogens can occur.

The time frame to complete an uncomplicated rederivation varies depending on the situation as follows:

1. *By IVF* – if there are enough animals available for IVF, the time frame can be counted as:

Two days superovulation ⇒IVF ⇒ 1 day till two-cell stage ⇒ embryo transfer ⇒ 21 days gestation ⇒ 21 days till weaning ⇒ screening surrogate mother and pups released: total of 7–8 weeks.

2. *By natural mating:*

2 days superovulation ⇒ plug (the amount of time here depends on the collection day – 1, 2 or 3) ⇒ embryo transfer ⇒ 21 days gestation ⇒ 21 days till weaning ⇒ screening surrogate mother and pups released: total of 7–8 weeks. However, for this option, the time required to expand the colony and produce mature stud males needs to be considered.

3. *From frozen embryos:*

Transfer upon receiving: ⇒ 21 days gestation ⇒ 21 days till weaning ⇒ screening surrogate mother and pups released: total 6–7 weeks. For frozen sperm add 2 more days.

4. *By hysterectomy:*

Mating ⇒ 19–20 days gestation till hysterectomy ⇒ transfer to foster mother ⇒ 21 days till weaning ⇒ send putative mother for screening and pups released: total 7–8 weeks.

Among these options, IVF has two important advantages:

1. Manipulation of the mating time: Some lines are not good breeders and it could be time consuming to obtain enough animals for a rederivation. While this problem is reflected in any option, with IVF, one male potentially gives enough sperm to fertilize hundreds of oocytes. After IVF, the embryos are washed and cultured overnight before scoring them for fertilization at the two-cell stage; fertile embryos are transferred into a pseudopregnant recipient. Any extra embryos produced, and not used for transfer, are cryopreserved as backup, in case the rederivation needs to be repeated.

2. Housing space in an isolation room: Upon receipt of imported animals they may go to an isolation room or to a quarantine room. If the number of animals received is enough for an IVF procedure, the holding time is approximately 1 week, considering a 3-day rest period after shipping and the 2 days needed to perform the superovulation. If there are only limited breeding pairs (one or two), then a minimum of seven more weeks of housing for natural breeding and weaning of pups must be added before starting the superovulation treatment of the oocyte donors. If only male animals were received, commercially purchased wild-type (WT) females are used as oocyte donors.

25.1.3. Superovulation

Superovulation is the term applied to the hormone treatment used to optimize the number of oocytes that a female will ovulate, therefore reducing the number of animals required [25, 26]. The hormone treatment also synchronizes the estrous cycle of the females to be used, facilitating the experimental design.

Depending on the genetics of the background strain, female mice may naturally release from five to ten oocytes during ovulation. Fowler and Edwards in 1957 described the use of gonadotropins in mice to induce superovulation; a treatment that increased two- to fourfold the number of oocytes recovered. Not every mouse strain responds to the hormone treatment with the same intensity; hybrid strains and outbreds, in general, respond better than inbred strains [27, 28]. Among inbred strains there is a classification that divides the strains into two groups: high ovulators such as C57BL/6J, 129/SvJ, CBA; and low ovulators such as A/J, C3H/HeJ, BALB/cJ (for a more detailed table, *see* [29–31]).

The hormones are administrated by intraperitoneal (i.p.) injection. First, a dose ranging from 2.5 to 10 International Units (IU) of pregnant mare serum gonadotropin (PMSG or eCG, equine chorionic gonadotropin) is given. PMSG prepares the oocytes for ovulation via an action similar to follicle stimulating hormone (FSH). Approximately 48 h later, a second i.p. injection of 2.5 to 10 IU of human chorionic gonadotropin (HCG), with an action similar to luteinizing hormone (LH), is given. HCG stimulates the ovary to produce dehiscence or rupture of follicles resulting in ovulation. The dose of both hormones varies from strain to strain and with the age of the donors. We prefer to use 21–30-day-old females regardless of the strain background. In general, these young females respond better to the superovulation treatment than do older females with a well-established estrous cycle. It is important to note that females can be younger and smaller for an IVF experiment than for natural mating where male acceptance plays an important role in the mating success.

25.1.3.1. Superovulation Protocol

Inject five 21–23-day-old females with 0.1 ml of sterile phosphate buffered saline (PBS) or 0.9% NaCl containing 5 IU PMSG i.p. at 3 p.m. Forty-six to 50 hours later, at approximately 4 p.m., inject the females with 0.1 ml of sterile PBS or 0.9% NaCl containing 5 IU of HCG i.p.

Important Tip: For an IVF experiment, the time that this last injection is given dictates the time the oocytes should be collected. Oocyte collection occurs the following morning between 13 and 15 h after the HCG injection; therefore plan the whole procedure accordingly.

Note: For IVF, some groups find that hormone priming at 3–4 p.m. brings optimal results. Others find that up to a 9 p.m. injection time works equally well. This may be affected by age of egg donors and light cycle timing.

25.1.4. Embryo Culture

It is important to stress that the success of rederivation by any method, is related in part, to the culture media used [32]. The embryos produced by IVF are routinely cultured overnight in the same media used for the IVF procedure and viable embryos are selected for transfer only after the first cleavage. However, if it is necessary to culture the embryos for longer periods, we use KSOM AA as a sequential culture media. There are many choices of mouse embryo culture media, including M16, CZB and KSOM AA which are the most popular. For IVF and overnight culture, we use commercially available media from Cook: K-RVFE-50, but HTF or TYH media also work very well.

For embryo handling outside of the incubator – flushing, collection and washing – the optimal media to use would be that which is HEPES buffered, in a formulation such as FMH or M2 media. The HEPES additive maintains pH more effectively than bicarbonate buffer based culture media while on the bench. The volume of the culture drop also plays an important role in embryo development; it was established that a lower drop volume is better [32]. Although the use of 10–50 μl drops has been described, we consistently use 200–250 μl drops with excellent results. All culture media drops should be covered by mineral oil, light paraffin oil or silicon oil. If the oil is not embryo tested by the manufacturer, a trial run using embryos of known outcome should be exposed to the oil to check for toxicity. Additionally, if the oil is not obtained from a reliable source, it should be cleaned by washing it overnight. Washing the oil is achieved by mixing it with same volume of culture media and placing the mix on a magnetic laboratory stirrer. Discard the media and dispense the oil in amber bottles to protect it from light exposure (or cover the bottle with aluminum foil).

25.2 Rederivation by IVF: Protocol for Fresh and Frozen Sperm [28, 33]

25.2.1. Materials

- Five 21- to 23-day-old females per strain for IVF,
- One or two 3- to 6-month-old males housed singly for 3–4 days prior to the IVF,
- PMSG (eCG) hormone: 5 IU per female Sigma/Aldrich, St Louis, MO C# G4877,
- HCG hormone: 5 IU per female Sigma/Aldrich, St Louis, MO C#CG-10,
- Plastic Falcon dishes 1007 (60 × 15 mm) BD- Falcon 351007,
- Plastic Falcon dishes 1008 (35 × 10 mm) BD- Falcon 353001,
- Media:
  - Holding: M2, Specialty media (Millipore) Billerica, MA. C# MR-015-D,
  - IVF: HTF, Specialty Media (Millipore) Billerica, MA. C#MR-070-D,
  - Cook K-RVFE-50, Cook Medical, www.cookmedical.com, Bloomington, IN,
  - Extended Culture: KSOM AA Specialty media (Millipore) Billerica, MA C# MR-106-D.
- KSOM + AA Specialty Media (Millipore) Billerica, MA. C# MR-106-D,
- CZB Specialty Media (Millipore) Billerica, MA. C# MR-019-D,
- FHM with Hyaluronidase: Specialty Media (Millipore) Billerica, MA. C# MR-056-F,
- TYH: modified Tyrode's medium [34],
- This catalog number: MR-004-D is for acidic Tyrode's medium, which has a very low pH and will dissolve the zona pellucida. TYH is not available as a commercially prepared medium and has to be prepared in the lab. The composition of modified Tyrode's medium is described in the paper by Quinn, referenced here,
- Embryo tested culture grade mineral oil Sigma/Aldrich, St Louis, MI. C#8410,
- Hyaluronidase type IV-5, Sigma. C# H3884,
- Ketamine: KETASET®, Fort Dodge Animal Health, Fort Dodge, IA,
- Xylazine: Rompun® Bayer AG, Leverkusen, Germany,

- Epinephrine: Amphastar Pharmaceuticals Company, El Monte, CA,
- Betadine: Purdue Pharma L.P, Stamford, CT,
- Wide pipette tips (Rainin Instruments, Oakland, CA. C# HR-250W),
- Tuberculin syringe (1 ml) with a 30 G needle. Terumo Medical, Somerset, NJ. C# SS-01T,
- Dumont watchmaker forceps. Roboz Surgical, Gaithersburg, MD C#RS-4976,
- Glass pipettes for embryo manipulation (we pull our transfer pipettes using hematocrit capillary glass tubing). VWR Westchester, PA. Micro capillary pipette 100 μl C#72690-004,
- Mouthpiece and rubber tubing. Sigma/Aldrich, St Louis, MO. C# A5177.

25.2.2. Equipment

- 5% CO_2, 37 C, 95% humidity incubator. Heracell 150i. Air Jacketed. Thermo Scientific, Waltham, MA,
- Dissecting microscope: Leica MZ12. Leica Microsystems, Bannockburn, IL,
- Mouse euthanasia lids –Euthanex Corp, Palmer PA. Euthanex lids C# E20028,
- CO_2 tank with an attached flowmeter regulator for euthanasia–flowmeter regulator M1-320-12FM. Western Medica, Westlake, OH. AVMA: www.avma.org/resources/euthanasia.pdf,
- Optional for cryopreservation: controlled rate freezer or materials for the vitrification method (*see* Chapter 23),
- Hemacytometer: Sigma C# Z359629,
- Makler Chamber: MidAtlantic Diagnostics, Inc. Mount Laurel, NJ.

25.2.3. Suppliers

Media: M2, HTF, KSOM + AA from Millipore (formerly Specialty Media),
Mouse K-RVFE-50 from Cook Australia,
Hormones: PMSG, Sigma G4877 (Sigma-Aldrich),
HCG, Sigma CG10 (Sigma-Aldrich).

25.2.4. Method

The day before IVF, prepare the dishes in the following way [28, 33]:

- Fresh sperm dish: one dish (35 × 10 mm) for each strain with 300 μl of IVF medium (K-RVFE-50) covered with mineral oil.
- Fertilization, washing and culture drops: one dish (60 × 15 mm) with five 250 μl drops of K-RVFE-50 medium. Use the center drop for fertilization and the outer drops for washing the eggs three times; the fourth drop is for culture of the zygotes overnight.

- Cover the drops with mineral oil that has been tested and proven safe for embryos. Equilibrate (incubate) the dishes at 37°C under 5% CO_2, 95% humidity in the incubator overnight.

1. The morning of the IVF, 30–45 min before the females reach the 13–15 h post-HCG, euthanize the male (3–6 months old housed separately at least for 3–4 days before IVF). Following euthanasia, dissect out both epididymides along with the vas deferens. Place both tissues in the sperm dish and cut the epididymis and the vas making 5–7 slashes with the edge of a syringe needle. Return the dish into the incubator and let the sperm swim-out of the epididymis for at least 10 min.
2. Under a dissecting microscope at 30×, visually check the sperm concentration and the motility; with the aid of a wide bore pipette tip pick up 10 μl of sperm from the edge of the drop – or more if the sperm concentration is low and transfer this amount to the fertilization drop. Let the sperm incubate for 30–45 min allowing them to capacitate. After incubation, visually check the sperm drop again for motility, quality and concentration.

Note: On average, the sperm concentration count is about 30 million per ml depending on the mouse strain. After a 10 min swim-out incubation, the collection drop appears very dense in population and full of movement when sperm are healthy and optimal. The edge of the sperm drop should be visually examined under a dissecting scope at higher magnification (40×). Motility is considered to be any movement of the sperm and progressive rectilineal or progressive motility is described when the spermatozoon moves in a forward direction. A sperm sample is considered to be excellent if greater than 80% of the sperm in a given field have forward movement. Concentration is considered to be optimal if the drop appears full of spermatozoa and poor if it appears watery with just a few spermatozoa in the field. The sperm drop is also checked for agglutination. A sample with agglutinated head-to-head sperm is considered to be of poor quality. It is common to find all kinds of variations from the optimal point to lowest point of the scale, since it is not rare to find some males that do not have any sperm. Motility also varies from almost 100% motile to zero. A sample with 60–70% motility is common and very good to work with. The visual evaluation for a trained eye is for the everyday work as accurate as a counting chamber (hemocytometer, Makler Chamber). Follow the same indication for frozen-thawed sperm.

3. Euthanize the donor females using carbon dioxide (CO_2) gas. Dissect out the oviducts and place them in a 500 μl drop of M2

media. One by one tear the ampullae to release the oocytes along with the cumulus cells. Work quickly to avoid leaving embryos too long in the dead body. Work with females in small groups of 3–5 at a time for collection of oviducts in M2.

4. Once you collect all oviducts from the group, then open the ampullae and transfer the clutches of oocytes/cumulus cells to a clean dish. Wash the oocytes once in a M2 drop to eliminate any blood or cell debris that could be attached to the cumulus mass.
5. Repeat the procedure with next group of 3–5 mice until all oocytes are collected. All these steps can be done at room temperature using M2 medium. The use of IVF media for collection of oocytes, however, will require the use of CO_2 for maintenance of the correct pH.

Note: Some groups find that embryo collection in buffered media at room temperature works well. Others find that maintaining the ova at 37°C brings improved results. This can be accomplished by keeping the time of embryo collection on the bench to less than 3 min, or by the use of a warm plate for holding collected oocytes.

6. Wash collected cumulus masses in K-RVFE-50 medium prior to transfer into the fertilization drop. Be careful to transfer the oocytes from drop to drop with as little media as possible.
7. Transfer all collected and washed oocytes and cumulus masses to the fertilization drop, using a wide tip pipette. Add 10 μl of sperm suspension to each fertilization drop. After mixing sperm and eggs together, place the dish into the incubator.
8. Four to six hours later, wash the oocytes in K-RVFE-50 drops to remove excess sperm. Transfer the oocytes from the fertilization drop to one of the side drops for washing. Repeat the wash procedure twice using a fresh drop of medium for each wash. Once the embryos have been washed, return the dishes in the incubator and leave them overnight.
9. Next morning count the number of two-cell embryos produced as an indication that fertilization occurred. Wash these embryos again, through at least five, 200 μl drops of M2 medium before surgically implanting them into a pseudopregnant female or cryopreserving them. The fertilization rate varies depending on the strain and the type of in vitro cross performed. With fresh sperm, the fertility rate, as measured by the number of two-cell embryos produced for C57BL mice, averages 80%. This percentage is altered by the male background, the male-to-male natural variations and the man-made genetic modification(s) the line carries. 129 and BALB/c strains give the poorest results, ranging from 0 to

40%. Hybrid animals in general have better fertility than inbred lines; similarly, an intercross between a homozygous inbred male with WT females has a better outcome.

Note: If only females were received for IVF rederivation, it is advisable to treat the oocytes with hyaluronidase prior to exposing them to the sperm. Hyaluronidase eliminates the cumulus cells and cleans the zona pellucida of possible pathogens attached to it. After the enzyme treatment the oocytes must be washed thoroughly in M2 media alone to remove all of the hyaluronidase.

25.2.5. In Vitro Fertilization Using Frozen Sperm

1. Set up the dishes for IVF the previous day (see protocol for IVF), without the sperm collection dish. Thaw the sperm samples in a 37°C water bath for about 30 s and place the thawed sperm directly into the IVF drop – 250 μl of K-RVFE-50 medium overlaid with mineral oil.
2. If sperm have been cryopreserved in straws, after thawing, push the contents of CPA + sperm column (10 μl) into the IVF drop of K-RVFE-50 overlaid with mineral oil and incubate at 37°C for 45 min. If sperm are in cryotubes, then after thawing, pipette with a wide bore tip 10 μl of the sperm sample directly into the IVF drop and follow same indications as for the straw.
3. After harvesting the oocytes, transfer them to the IVF drop with sperm.
4. Proceed further as IVF with fresh sperm.

Factors to be considered:

- If the IVF rederivation is done using sperm from the strain to be rederived and the females are obtained pathogen free from a recognized vendor, there is no need to treat the oocytes with hyaluronidase.
- The zona pellucida will, with time, become impenetrable to sperm. This phenomenon is called zona hardening. The longer the oocytes sit in the carcass without being removed, or stay in the culture medium, the harder the zona will become. Therefore, work fast when dissecting the oviducts and tearing the ampullae to collect the oocytes. Keep in mind that the success rate for the IVF experiment will be highest when performed as close as possible to 13 h after the HCG injection.
- The temperature and pH of the medium is critical. Keep everything at 37°C as much as possible. The pH of the IVF medium will begin to rise as soon as it is exposed to air. Try to reduce the exposure time outside the CO_2 incubator as much as possible. There are media, such as M2, that were specifically designed for work outside of a CO_2 rich environment, called handling media. These media contain HEPES and the pH of these media will not change when exposed to air. Although

HEPES buffered media is indicated for collecting and washing embryos, it is not a good practice to culture embryos in them. Here we described protocols using the media K-RVFE-50 from Cook; however, HTF media work perfectly well for both fresh and frozen IVF.

- Test the quality of the mineral oil before using it by culturing some fertilized B6D2 eggs, or any other strain that produces embryos that culture well in vitro, in small drops covered by the oil in question and evaluate their development. A good practice is to set up a round of IVF, testing the oil in use vs. the new side by side, expecting to see the same result in both. Using hybrid B6D2 embryos, the percentage of developing embryos to two-cell should be close to 90%. Mineral oil is a petroleum derivative and could carry elements toxic to the embryo. Hence, it is advisable to purchase already embryo tested products. Mineral oil is also light sensitive so it is good practice to keep the bottles from being exposed to light.

25.3 Rederivation by Natural Mating

25.3.1. Natural Mating

Natural is the organic version of IVF. It is the option of choice for difficult to breed strains, for strains that will not respond to superovulation or for those that do not fertilize well in vitro. Many transgenic or knockout (KO) models have impaired fertility and sometimes it is better to do the rederivation by natural mating. For a standard natural mating rederivation at least five stud males and five superovulated females are needed. After the last hormone injection (HCG) the female is introduced into the male cage and the following morning copulation plugs are checked.

The options here are: collect the embryos at the one-cell stage, or let the embryos "incubate" in the female and harvest them the following day at the two-cell stage or on the third day at the eight-cell stage. We collect one-cell embryos, treat the zygote cumulus cell complex with hyaluronidase, wash the embryos thoroughly five times and then culture them overnight. The next morning, score all cleaved embryos and transfer what is needed. Collection of the two-cell and eight-cell embryos by flushing the oviduct is rather complicated compared with tearing the ampulla to obtain the one-cell embryo. Nevertheless, a trained technician should not be inconvenienced by any of these variations, but should choose whatever suits his/her abilities better, because the results are similar for all of the options. The rationale for our choice of harvesting at one-cell stage and culturing to two-cell stage for embryo transfer is:

(a) At the two-cell stage there is clear evidence of development when compared to the zygote or one-cell embryo stage; therefore there are fewer possibilities of mistakenly selecting unfertilized oocytes.

(b) Collecting embryos at one-cell stage and then culturing to two-cell stage or beyond is simpler than flushing the oviducts to collect the two-cell or eight-cell stage embryos.

(c) Transferring at the two-cell stage results in a safety cushion compared with eight-cell embryos. If there are no pseudopregnant females for the transfer, the embryos can go back into culture and be used the following day.

25.3.2. Embryo Washing

Washing the embryos several times in large drops of media is one of the principles for "cleaning" them for rederivation by this method. The manual of the International Embryo Transfer Society (IETS) recommends washing the embryos thoroughly at least ten times [35]. This consideration was established to avoid introducing diseases when transferring imported large animal embryos (especially cattle and pigs) coming from around the globe. Theoretically it was established that the efficacy of the wash had a direct relationship to the size of the drop, i.e., dilution volume, and the size of the embryo. Recently it was demonstrated that five washes is effective in eliminating most pathogens attached to the zona pellucida regardless of the drop size [32].

When collecting mouse ova or one-cell embryos for rederivation, it is advisable to perform the first wash in media with hyaluronidase (300 μg/ml) added. Watch the embryos until the cumulus cells are disrupted and the naked zona pellucida has been exposed for some seconds to the enzyme; then follow with the stipulated washes. The enzyme treatment eliminates pathogens that are attached to the zona pellucida. It is important to use sterile disposable materials and a different pipette between washing drops.

25.4 Embryo Transfer to Infundibulum

25.4.1. Materials

Recipient females: 0.5 day plugged (post coitum) pseudopregnant females

Instruments:

- Scissors, C# RS-5882, Roboz Surgical, Gaithersburg, MD,
- Anatomical forceps (straight or curved with serrated tips) C# RS-5210 and 5211, Roboz Surgical, Gaithersburg, MD,
- Tissue forceps with 1 × 2 teeth C# RS-5150, Roboz Surgical, Gaithersburg, MD,

- Dumont (watchmakers) or No. 5 forceps C#RS-4976, Roboz Surgical, Gaithersburg, MD,
- Serrefine (e.g., Fine Scientific Tools: 18050-28 or 8051-28) or Dietrich microbulldog clamp (Roboz Surgical, Gaithersburg, MD C# RS-7438),
- Autoclip applier (Clay Adams B-D 763007),
- Autoclips (Clay Adams B-D 7631),
- Mouthpiece and rubber tubing: Consists of flexible silicone rubber nosepiece, hard plastic mouthpiece, and 15 in. of latex tubing. C# A5177. Sigma/Aldrich, St Louis, MO,
- Transfer pipette: pipettes are pulled by hand from Micro capillary pipette 100 μl, C#72690-004; VWR, West Chester, PA,
- M2 medium Specialty media (Millipore) Billerica, MA. C# MR-015-D.

25.4.2. Pseudopregnant Recipients

The female recipient for an embryo transfer should be hormonally prepared for the gestation of those embryos. Natural mating of mature females with vasectomized males will produce pseudopregnant recipients suitable for the embryo transfer. A post-coitum vaginal plug is used to identify these females. The vaginal plug will indicate 0.5-day post-coitum and is the only external indication available for synchronization of the recipient with the stage of the embryo. The best surrogate mouse mothers are 2–3-month-old female hybrids such as B6D2F1, B6CBAF1 and popular outbred strains such as CD1, CF1 or Swiss Webster.

To produce enough pseudopregnant females for a working day via natural mating, it is important to keep in mind that the mouse estrous cycle is 5 days long; therefore one out of five females will be in estrous on a certain day and will accept the male. However, if the females were caged together for at least a week, their cycles will synchronize with each other (Whitten effect) [36]. Mice will also synchronize their estrous cycles and even terminate a pregnancy when exposed to the pheromones of an unknown male (Bruce effect) [37]. Females can also be selected for mating by observing aspects of the external genitalia; if a mouse is in estrous the vulva will be pinkish and dilated, indicating that she is ready to accept the male. Another option, suitable for large production labs, is to set up matings of two females per male. The number of plugged females will reach the maximum score on the third day.

It is important to rotate the vasectomized males and retire the males that reach age of 8 months. It is also advisable to discard those vasectomized males that do not produce plugs after three consecutive matings within a period of 1 month. Because it has been demonstrated that the length of the pseudopregnancy in mice is 14 days [38, 39], any pseudopregnant female not used on the day of the experiment may be introduced to a stud male again 2 weeks later, when she should start cycling again.

25.4.3. Anesthesia

Although we recognize the controversies surrounding the use of Avertin as an anesthetic, we have never encountered any problem besides the low body temperature side effect. Therefore, we still recommend its use:

Avertin at 2.5%: 2,2,2,-Tribromethanol 2.5 g (Morre-Tec Industries #1693, Sigma-Aldrich C# T48402); tert-amyl alcohol 5.0 ml (Fisher: A-730-1; Sigma-Aldrich: C#240486 (2 methyl-2 butanol).

Dissolve the tribromethanol in tert-amyl alcohol and then add distilled water until the total volume is 200 ml.

Place on a magnetic stirrer until the solution is mixed into one phase.

Store in a brown bottle and keep refrigerated until use. Warm the solution to room temperature before use.

Dose at 0.2 ml/10 g body weight.

An alternative to Avertin for mice is the mix of 2 ml ketamine hydrochloride solution (50 mg/ml) and 0.8 ml xylazine hydrochloride (20 mg/ml) (Rompun) in water to 10 ml. Inject 0.1 ml of the solution per 10 g of body weight (100 mg/kg ketamine, 16 mg/kg Xylazine).

25.4.4. Surgical Embryo Transfer Method

Anesthetize the plugged recipient with an intraperitoneal injection of Avertin or the anesthesia recommended by your institution's veterinarian. Shave the surgical site (dorsal lumbar area) and prepare the area with an alternating surgical scrub of Betadine® antiseptic followed by 70% alcohol (three times). The NIH-OACU (Office of Animal Care and Use) through the Animal Research Advisory Committee (ARAC) has specific guidelines at http://oacu.od.nih.gov/ARAC/index.htm. Place an appropriately sized (5 cm × 5 cm) fenestrated drape over the surgical field. A 1.0–1.5 cm transverse skin incision using scissors is made and the skin reflected to allow visualization of the ovarian fat pad through the abdominal wall. An incision in the muscles over the fat pad of approximately 1.0 cm is made with a second pair of scissors, and the ovarian fat pad is gently pulled through the incision to allow for visualization of the oviduct union with the ovarian tissue. The ovarian fat pad is clamped with a small serrefine to keep it in place during the intervention. A small tear is made through the bursa that covers the ovary and oviduct, and the bursa is reflected away from the ovary to allow access to the infundibulum. A small pipette containing the embryos to be transferred is introduced into the infundibulum, and the embryos are gently released into the oviduct. The ovarian fat pad is then released from the serrefine clamp and allowed to retract back into the body cavity. Close the abdominal muscles using a 3-0 or 4-0 absorbable suture. A few drops of bupivacaine hydrochloride Injection 0.25% (analgesic) are instilled topically at the incision site and the skin is then closed with 1–2 wound clips and the animal is allowed to recover in a warm place. We normally transfer 12–15 embryos into the left oviduct only; however, the transfer can be done into both sides repeating the sequence described; transfer approximately 8–10 embryos into each side.

Tips

- It is a good practice to check the ovaries of the plugged (0.5 day) recipients for "ovulation points" before the embryo transfer. If pearly-whitish points are there, these will become the corpus lutea after the implantation, and are a good indication that the embryos will implant.
- Avoid using fat females; the ideal weight is between 20 and 30 g. The excess fat will modify the effect of the anesthesia, requiring more than the usual dose; moreover, tearing the ovarian bursa from fat females is bloodier than from thinner females.
- Carefully tear the bursa by taking hold of the tissue while avoiding blood vessels. Avoid touching the ovary because it may cause bleeding that could ruin the embryo transfer. A drop of epinephrine at a concentration of 1.2 mg/ml instilled into the site will help stop any bleeding. Try to sponge any blood in the area before transferring the embryos. Be extremely careful not to touch the blood with the end of your transfer pipette as it will clot in the tip and the embryos will be trapped in the pipette.
- Fire polish to blunt the tip of the transfer pipette for a smooth transfer; a rough or sharp pipette tip will make it difficult to introduce the embryos into the infundibulum.
- Do not blow too hard when releasing the embryos into the infundibulum; they can get smashed against the oviduct wall.
- Do not blow too much air into the oviduct; embryos will adhere to air bubbles and can be lost.
- We routinely do embryo transfers on the left side only. Never use more than 15 embryos per unilateral transfer/female, since litters born with more than ten pups are rare. Remember Grandma's saying: "Don't put all your eggs in one basket".
- As a rule of thumb, 40–50% of the embryos transferred will go to term as pups born. However, the rate could be anywhere from 0 to 90%, with lower percentages being a lot more common.

25.4.5. Protocol Variations

Like many other techniques in biology studies, rederivation does not have a golden rule about the stage of the embryo to transfer, the number of embryos transferred or whether the embryos should be produced in vitro or in vivo. In fact, as we mentioned before, many successful rederivations have been done by cesarean delivery of pups at term. There is also a great variety in the types of culture media, the use of which depends on the preference and experience of the technician. However, steps to follow for achieving successful rederivation include: using at least the minimum number of washes necessary for cleaning pathogens off the embryo, using HEPES buffer media for working outside the incubator, and using a bicarbonate buffered media plus amino acids for culture of embryos.

25.5 Rederivation by Cesarian Section

Cesarean rederivation or hysterectomy was commonly used in the early 1960s to produce gnotobiotic (germ-free) animals or to introduce new pups into SPF containment. The rationale for its use was that the sterile environment of the uterus protected the conceptuses from infection. Now that it is known that the uterus is not a sterile environment, hysterectomy is no longer the first choice for rederivation of rodents. However, it is an option when animals do not respond to other techniques or where cesarean rederivation is still the only option, e.g., Guinea pigs [12].

The methodology requires maintenance of a recipient colony in production on the clean side and a donor colony outside of the barrier. The key to success for this technique is to synchronize the donor's cesarean section with a putative foster mother that recently delivered pups so that litters may be interchanged. It is important to use strains with good maternal behavior to act as surrogate mothers since the final results will depend on their accepting and fostering the pups. Strains considered good mothers are CD1, B6D2F1 and BALB/c among others. As in all methods there are small personal touches needed for adaption of the technique chosen, so that it will best suit your needs. A common dilemma is whether it is better to transfer the uterus into the clean area, along with the conceptuses or leave it behind in the dirty area. The original methodology from the 1960s and the 1970s dictated that at term uteri were transferred through an entry/exit port submerged in a disinfectant solution. However, that practice resulted in potentially infected tissue being introduced into the clean environment. This is why some investigators prefer to aseptically collect the pups in a small sterile container and transfer them enclosed into the restricted area.

Before introducing the pups to the foster mother's nest or cage, it is a good practice to impregnate the operator's gloves with the cage's smell by touching the bedding or the nest prior to handling the newborns. If the foster mother's litter has a different coat color it is also a good idea to leave a couple of them in the nest to help with nursing acceptance.

25.5.1. Equipment and Reagents

- Container with disinfectant solution,
- Hair clippers,
- Sterile scissors and smooth and toothed forceps,
- Sterile container such as a large disposable Petri dish where the newborns can be transferred.

25.5.2. Protocol Description

- The pups should be mature to increase the chances of survival. Palpation can help determine that the pups are at term. At term, a fetus will present as an elongated shape instead of feeling like a sphere-shaped mass. If the mating plug was scored, then euthanize the pregnant donor female at 18 days of gestation.
- Be sure that an adequate recipient foster mother is available.
- Euthanize the female donor according to the institution's Animal Care and Use Committee (IACUC) requirement. Consider combining CO_2 asphyxiation with cervical dislocation to avoid compromising the conceptuses with an excess of CO_2.
- Shave the abdominal area.
- Disinfect the outside of the donor by wiping the animal with an appropriate disinfectant such as Betadine.
- With a pair of forceps, clip the skin and the abdominal muscles and open the abdominal cavity.
- Remove the uterus with care not to puncture the uterine walls. Leave behind all unnecessary tissues such as the mesentery, its adjacent arteries and veins, the ovaries and the vagina.
- With the dirty forceps discard the carcass.
- Over a clean area (diapers or absorbent paper) and with clean instruments, open the uterus to release the pups. Remove the placentas and cut each umbilical cord. If the pups are mature, only a small amount of bleeding will occur from the cord. Blot each pup with a tissue paper and stimulate breathing. If the pups are mature they may squeak upon manipulation, which is a good sign.
- Transfer the newborns to the foster mother's nest.
- At weaning, screen the foster mother for unwanted microorganisms to evaluate the success of the rederivation.

25.6 Frozen Embryos, Frozen Sperm

25.6.1. Sharing Resources

Increasingly, the interchange of mice around the world is in the form of frozen embryos or frozen sperm [8, 18, 40, 41]. Today there are many animal model repositories where investigators can obtain frozen embryos or sperm of a specific mouse model. Both of these cryopreserved materials facilitate the rederivation of new strains into animal facilities. Embryos are shipped from international or national repositories in dry liquid nitrogen shippers directly into the laboratory. Therefore basic training in handling frozen embryos is necessary, not only to safeguard the embryos, but also for the safety of the operators dealing with liquid nitrogen. Every institution should send their thawing protocol along with the embryo shipment. One should never assume that embryos were frozen under the same protocols used by the receiving institution. There are numerous freezing protocols using different cryoprotectants, cooling rates or vitrification methods. Strictly follow the thawing protocol attached to the embryo shipment. After thawing, wash the embryos thoroughly through five drops of media as previously described for the standard rederivation.

In the case of receiving frozen sperm, the rederivation should be performed with the same care. Although the donated oocytes are harvested from females obtained from a recognized vendor, the male gamete is harder to clean than an embryo and potentially can carry excluded pathogens. It is recommended that after the embryos have been generated, washed and transferred that the recipient female be quarantined and screened, as for any other rederivation, before the weaned pups are released to routine colony holding.

25.6.2. Embryo Cryopreservation

When embryos for rederivation are produced, there may be some extra embryos remaining after the embryo transfer. Although not the topic of this chapter, cryopreservation is the tool for managing those resources. Extra embryos should not be wasted, as they may potentially be needed for a second procedure if the first transfer does not work. There are many methods ranging from slow freezing to vitrification, with or without specialized equipment, for preserving surplus embryos (*see* Chapter 23). Each laboratory should select the appropriate method according to their needs. A cryopreservation laboratory is the perfect complement to the embryo laboratory.

25.7 Colony Management Strategies

25.7.1. Determining the Health Status of the Rederived Mice

Surrogate mothers and their offspring should be housed apart from the general facility population until their health status is determined. While a well-executed rederivation procedure, performed under ideal circumstances, will remove most unwanted microbes, the possibility of horizontal contamination of the embryos and surrogate mothers by fomites must be considered. Depending on the amount of risk, the rederived animals present and the impact of a positive animal on other animals in the facility, they should be housed in either isolation areas or microisolator cages. The least desirable option is to house rederived animals in microisolator cages in a room shared with facility animals. If this cannot be avoided, then cages of rederived animals should be opened as little as possible and handled after all other cages in the room.

It is important to develop a surveillance strategy to ensure the detection of excluded agents after rederivation (*see* Subheading 25.7.4). The most commonly used surveillance strategy is quarantining and screening of immune competent recipients.

The agents we routinely test for when only embryos enter are: murine norovirus (MNV), MHV, EDIM, MVM), MPV and GDVII by serology. When live mice are brought into the animal facility for rederivation, fomite transmission is a concern and screening includes: MNV, MHV, EDIM, MVM, MPV and GDVII by serology, and endo-parasites.

Tips

- A seropositive recipient may have cleared a low-level infection and she and the pups may be pathogen free. Immune competent pups can be tested serologically; otherwise PCR methods or sentinels (cohort or dirty bedding) must be used.
- The timing of testing may impact clean quarantine housing availability and therefore rotation of available housing must be considered when devising the test schedule.
- A minimum of 6 weeks, from embryo transfer to testing of the recipient, should elapse before performing serology testing. This amount of time is needed for contaminates to proliferate to a level at which the body recognizes the organism's presence and mounts a detectable antibody response.
- When testing for parasites, the prepatent period of the parasite must be considered.

- Methods for assessing an animal's microbiological status continue to evolve [42] and should be considered. Molecular techniques may be used in lieu of, or in conjunction with, serology to determine rederivation success [43, 44].

25.7.2. Facility Health Status: Excluded Agents

The number and type of agents to test for are decided after consideration of infection risk [8, 45] and risk tolerance. Prevalence and the biology of each microbe of concern determine infection risk. Some viruses listed on commercially available screening panels are rarely encountered today [43, 44, 46]. Other agents are prevalent both within and outside of the United States [43, 44, 47–49]. Transmission of agents is affected by their environmental stability, the type of caging used [50, 51] and handling procedures [52]. Contamination of rederived animals with endo- or ectoparasites is unlikely if only embryos are brought into the rederivation laboratory.

The amount of testing performed is often dependent on budgetary constraints. Risk tolerance and budgetary constraints must balance. While testing for all possible infections may ease fears of contamination, the available budget often limits comprehensive testing. A broad knowledge of agent prevalence and biology along with broad quarantine and testing experience will provide a better understanding of the potential risk presented by imported animals.

Designating a colony as SPF requires stating a list of all agents from which the colony is considered to be free. While there are multiple articles discussing the importance of SPF mice, very few of the articles define what pathogens were tested for and excluded [6, 9].

As an alternative to the SPF designation, our institute maintains a list of acceptable microbial agents for different facilities; all other microbes are considered excluded. The list for most of our facilities follows:

Agents identified in the colony that are not on the accepted list require action. Generally the infected animals are removed but in some cases the microbe is added to the acceptable list. Addition of new agents may occur after consideration of: the agent's impact on the research being performed, the methods available to identify infected animals, the agent's prevalence in the colony, methods available for eliminating the agent, time and resources required to remove the agent, budgetary impact, and the potential for the agent to be reintroduced. Both *Helicobacter* spp. and MNV were found in the majority of our facilities [53] as well as in many facilities throughout the United States [45], North America and Europe [43, 44]. As a result, both agents were added to the acceptable list for most of our facilities.

Tips

- While our institution has an umbrella contract that stipulates required standards for vendor supplied animals, each facility independently decides what agents will be excluded. In many cases, vendor animals have a higher health status than that which is found in our facilities.
- The primary source of recipient animals (generally a vendor) will dictate the maximal possible level of exclusion by rederivation. Infectious agents known to be present at the recipient source facility should be considered to be present in the recipient animals.
- We maintain facilities of varying health status and stipulate a footpath pattern that must be followed when entering more than one facility within a 24-h period.
- Standardized health monitoring recommendations are available [54].
- The list of microbes that a facility decides to exclude varies based on a multitude of factors, to include:
  - The need to accept and transfer mice quickly from outside institutions and facilities.
  - The type of research being performed – breeding colonies, long term studies and immunological research – dictates a more comprehensive exclusion list.
  - Biosecurity of existing facilities, i.e., what is the likelihood that excluded agents will be introduced [55, 56]?
  - Are the financial resources available for testing or handling a break in health status?
  - The number of exports to outside collaborators; collaborations involving the exchange of animals may be facilitated by having animals free of all known microbes that affect research, if the receiving facility is unable to accept embryos.

25.7.3. Limitations of Rederivation to Exclude Pathogens

Available methods vary in their effectiveness at clearing mice of undesirable bacteria, viruses, fungi and parasites. Temporary cessation of breeding [57], cesarean rederivation, cross-fostering [4, 14, 58, 59] and germplasm transfer have all been successfully used to rid rodent colonies of research confounding microbes. When selecting a particular method, the biology of the microbes present and the immune competency of the animals involved must be considered to improve the probability that the method chosen will be successful. To date, the most consistently successful method for clearing all types of microbes is embryo transfer. While microbial proteins and nucleic material have been shown to be present in reproductive tissues [60–62], and experimental infections suggest that microbes can be transferred via embryos [23, 63–65], there are no reports of embryo transfer rederivation failure, under natural conditions when stringent rederivation procedures are followed, endogenous retroviruses being the exception.

25.7.4. Management of Microbes Within Institutional Facilities

The list of excluded agents, the surveillance strategy and the steps to be implemented in the event of detection of an excluded pathogen are best incorporated into a series of Standard Operating Procedures (SOPs) prior to housing any animals.

Ideally, a separate facility to hold incoming animals for breeding and embryo collection should be available. The health status of this facility should allow introduction of known infected rodents. Our program has a facility for this purpose that accepts rodents infected with any agent other than hantaviruses, LCMV and mouse poxvirus.

Our program rederives rodents into two separate facilities. One facility allows introduction of animals infected with the agents in Table 25.1. For the second facility, MNV and *Helicobacter* spp. are removed from the acceptable pathogen list due to the type of research being performed. As the majority of our facilities are endemically infected with MNV and *Helicobacter* spp., most rederivations are performed into the facility that allows these agents. Mice rederived into the other facility are either housed in that stricter barrier facility or, occasionally, they are exported to facilities free of MNV and *Helicobacter* spp.

To maintain the targeted health status, a plan for minimizing entry of excluded agents into a facility is needed. This plan should include, at a minimum, the following:

- *Identify and control points of entry for pathogens.*

Rodent pathogens may enter via the following mechanisms:

Animals. Ideally all animals should be obtained from a reliable vendor, bred within the facility or obtained through embryo rederivation. While quarantine and thorough pathogen screening before entry may ensure that animals are negative for excluded agents, this practice carries a higher risk for introducing an unwanted agent. A pest management program should be in place to prevent entry of feral rodents.

Fomites. Microbes may adhere to people, equipment, supplies and animals. Employees should not come into contact with rodents outside of the rederivation area unless they are of the same health status as the rederived animals. Employees are discouraged from keeping rodents in their homes for any reason. If an employee comes into contact with rodents outside of the animal facility they are required to shower daily at work. Showering and donning of clean scrubs must be done outside of the

Table 25.1
Example acceptable pathogen list

Virus	Murine norovirus (MNV)
Bacteria	*Actinobacillus* sp.
	Bordetella bronchiseptica
	Corynebacterium sp. (not *kutscheri*)
	Enterococcus sp.
	Escherichia coli
	Helicobacter spp.
	Klebsiella oxytoca
	Klebsiella pneumoniae
	Moraxella morganii
	Pasturella pneumotropica
	Proteus mirabilis
	Pseudomonas aeruginosa
	Staphylococcus aureus
	Staphylococcus epidermidis
	Staphylococcus xylosus
	Streptococcus sp. alpha and beta (Group B and Group G) hemolytic
Fungi	*Trichosporon beigelii*
	Pneumocystis murina (formerly *P. carinii*)
Parasites	*Chilomastix* sp.
	Entamoeba muris
	Trichomonas sp.
	Tritrichomonas muris

animal facility before entering the facility. To minimize spread via people, the use of lab coats, gloves, shoe covers and other protective clothing should be mandated. Inanimate objects should be disinfected or sterilized, and sharing of equipment between facilities should be minimized. If animals must be moved from facility to facility, they should be transported in intact closed filtered carriers that can be surface disinfected prior to entering the facility. A flow pattern for objects that could act as fomites should be established to allow for movement from cleaner to dirtier area only. Ideally, only oviducts or embryos should move from the dirty area to the rederivation laboratory.

If research involves the injection of biological material derived from rodents, the materials should be PCR tested for contamination prior to use. Water should be treated to remove or destroy excluded microbes. Bedding and feed should be autoclaved or irradiated.

- *Prevent the spread of established pathogens from room to room.*

Spread of pathogens within an animal facility can be traced to shared procedure or equipment rooms that are not adequately disinfected after use, or to fomites such as used cages, shared equipment or personnel trafficking. While fomites can be minimized as described above, the dedication of procedure and equipment rooms for the rederivation effort is best addressed at the facility design level.

- *Prevent the spread in the animal room of established pathogens.*

Spread of microbes within a room can be minimized through the use of a HEPA filtered change station, aseptic cage change technique and the use of individually ventilated or filtered cages. When removing used cages from the room, removing intact cages from the change station rather than stacked component parts minimizes transmission in the room. The disinfection of work areas, equipment and gloved hands between cages will minimize microbe spread between cages.

- *Prevent the establishment of pathogens in animals.*

This can be achieved by obtaining healthy animals and then minimizing exposure of susceptible populations to pathogens. Neonates and recently weaned animals are more susceptible to pathogens, thus limiting access and experimental manipulation of young animals will minimize exposure to fomites and other sources of infection.

Aseptic cage change: (modified from [52])

- Operate the HEPA filtered change station according to the manufacturer's direction.
- Chemically disinfect or UV irradiate the internal surfaces of the change station immediately prior to use.
- Using disinfected gloved hands, assemble the recipient (new) cage, to include food, bedding/nesting material and water within the change station. Inspect the recipient cage to ensure that all filter materials and all solid surfaces are intact. Ensure that the new cage is properly set up to receive animals by opening the cage top and sliding the wire bar lid to the side or remove and place it on the inverted cage top.

- Clean the outer surfaces of the cage to be changed to remove surface debris prior to placing it in the change station.
- Remove the lid of the cage to be changed. Ensure that the internal surface of this lid does not contact the work surface and that the lid is placed away from the recipient cage.
- Disinfect gloved hands prior to accessing the mice contained within the cage to be changed or use disinfected forceps.
- Transfer the mice to the recipient cage, confirming animal sex, number and individual identification as appropriate. Do not transfer used bedding or nest material.
- Replace the wire bar and lid of the recipient cage, now housing the mice.
- Reassemble the used cage and prepare it for removal from the room.

References

1. Pollard M (1967) Applications of germ-free animals to problems in comparative medicine. Adv Vet Sci 11:139–157
2. Baker DG (1998) Natural pathogens of laboratory mice, rats, and rabbits and their effects on research. Clin Microbiol Rev 11(2): 231–266
3. Franklin CL (2006) Microbial considerations in genetically engineered mouse research. ILAR J 47(2):141–155
4. Artwohl JE, Purcell JE, Fortman JD (2008) The use of cross-foster rederivation to eliminate murine norovirus, helicobacter spp., and murine hepatitis virus from a mouse colony. J Am Assoc Lab Anim Sci 47(6): 19–24
5. Baker HJ (1988) Rederivation of inbred strains of mice by means of embryo transfer. Lab Anim Sci 38(6):661–662
6. Fray MD, Pickard AR, Harrison M, Cheeseman MT (2008) Upgrading mouse health and welfare: direct benefits of a large-scale rederivation programme. Lab Anim 42(2):127–139
7. Van Keuren ML, Saunders TL (2004) Rederivation of transgenic and gene-targeted mice by embryo transfer. Transgenic Res 13 (4):363–371
8. Mahabir E, Bauer B, Schmidt J (2008) Rodent and germplasm trafficking: risks of microbial contamination in a high-tech biomedical world. ILAR J 49:347–355
9. Lindsey JR (1998) Pathogen status in the 1990s: abused terminology and compromised principles. Lab Anim Sci 48(6):557–558
10. Nicklas W, Baneux P, Boot R, Decelle T, Deeny AA, Fumanelli M, Illgen-Wilcke B (2002) Recommendations for the health monitoring of rodent and rabbit colonies in breeding and experimental units. Lab Anim 36:20–42
11. Ike F, Bourgade F, Ohsawa K, Sato H, Morikawa S, Saijo M, Kurane I, Takimoto K, Yamada Y, Jaubert J, Berard M, Nakata H, Hiraiwa N, Mekada K, Takakura A, Itoh T, Obata Y, Yoshiki A, Montagutelli X (2007) Lymphocytic choriomeningitis infection undetected by dirty-bedding sentinel monitoring and revealed after embryo transfer of an inbred strain derived from wild mice. Comp Med 57(3):272–281
12. Glage S, Dorsch M, Hedrich HJ, Bleich A (2007) Rederivation of helicobacter hepaticus-infected mongolian gerbils by caesarean section and cross-fostering to rats and mice. Lab Anim 41(1):103–110
13. Randall D, Cross D (1984) Cesarean rederivation of guinea-pigs by the use of flexible film isolators. Lab Anim 18(4):337–337
14. Watson J, Thompson KN, Feldman SH (2005) Successful rederivation of contaminated immunocompetent mice using neonatal transfer with iodine immersion. Comp Med 55(5):465–469
15. Besselsen DG, Romero-Aleshire MJ, Munger SJ, Marcus EM, Henderson KS, Wagner AM (2008) Embryo transfer rederivation of C.B-17/Icr-*Prkdc(scid)* mice experimentally infected with mouse parvovirus 1. Comp Med 58(4): 353–359
16. Kornblatt AN, Adams RL, Morgenstern SE, Herrup K, Diglio TJ, Sunter K (1984) Embryo transplant rederivation of mice. Lab Anim Sci 34(5):518–519

17. Okamoto M, Matsumoto T (1999) Production of germfree mice by embryo transfer. Exp Anim 48(1):59–62
18. Rall WF, Schmidt PM, Lin X, Brown SS, Ward AC, Hansen CT (2000) Factors affecting the efficiency of embryo cryopreservation and rederivation of rat and mouse models. ILAR J 41(4):221–227
19. Reetz IC, Wullenweber-Schmidt M, Kraft V, Hedrich HJ (1988) Rederivation of inbred strains of mice by means of embryo transfer. Lab Anim Sci 38(6):696–701
20. Suzuki H, Yoruzo K, Watanabe T, Nakura M, Adachi J (1996) Rederivation of mice by means of *in vitro* fertilization and embryo transfer. Exp Anim 45(1):33–38
21. Kaufman MH, Bard JBL (1999) The anatomical basis of mouse development. Academic Press, San Diego, CA, p 30
22. Eaglesome MD, Hare WCD, Singh EL (1980) Embryo transfer: a discussion on its potential for infectious disease control based on a review of studies on infection of gametes and early embryos by various agents. Can Vet J 21:106–112
23. Mahabir E, Bulian D, Needhan J, Schmidt J (2009) Lack of transmission of mouse minute virus (MMV) from *in vitro* produced embryos to recipients and pups due to the presence of cumulus cells during the *in vitro* fertilization process. Biol Reprod 81:531–538
24. Rouleau AM, Kovaks PR, Kunz HW, Amstrong DT (1993) Decontamination of rat embryos and transfer to specific pathogen -free recipients for the production of a breeding colony. Lab Anim Sci 43:611–615
25. Brooke DA, Orsi NM, Aincough JFX, Holwell SE, Markham AF, Coletta PL (2007) Human menopausal and pregnant mare serum gonadothrophins in murine superovulation regimens for transgenic applications. Theriogenology 67:1409–1413
26. Fowler RE, Edwards RG (1957) Induction of superovulation and pregnancy in mature mice by gonadotrophins. J Endocrin 15:374–384
27. Byers SL, Payson SJ, Taft RA (2006) Performance of ten inbred mouse strains following assisted reproductive technologies (ARTs). Theriogenology 65(9):1716–1726
28. Sztein JM, Farley JS, Mobraaten LE (2000) *In vitro* fertilization with cryopreserved inbred mouse sperm. Biol Reprod 63:1774–1780
29. Hogan B, Constantini F, Lacy E (1986) Manipulating the mouse embryo, a laboratory manual. Cold Spring Harbor Laboratory Press, New York
30. Nagy A, Gertsenstein M, Vintersten K, Beringher R (2003) Manipulating the mouse embryo. A laboratory manual, 3rd edn. Cold Spring Harbor Laboratory series, New York
31. Silver LM (1995) Mouse genetics. Concepts and applications. Oxford University Press, New York
32. Gardner DK, Lane M, Watson AJ (eds) (2004) Culture of the Mammalian Embryo. In: A laboratory guide to the mammalian embryo. Oxford University Press, NY, pp 41–61
33. Vasudevan K, Raber J, Sztein J (2010) Fertility comparison between wild type and transgenic mice by *in vitro* fertilization. Transgenic Res 19(4):587–594
34. Quinn P, Barros C, Whittingham DG (1982) Preservation of hamster oocytes to assay the fertilizing capacity of human spermatozoa. J Reprod Fertil 66:161–168
35. Stringfellow DA, Seidel SM (eds) (1990) Manual of the International Embryo Transfer Society. A procedural guide and general information for the use of embryo transfer technology, emphasizing sanitary precautions. IETS. IL
36. Whitten WK, Bronson FH, Greenstein JA (1968) Estrus-inducing pheromone of male mice: transport by movement of air. Science 161(841):584–585
37. Bruce HM (1959) An exteroceptive block to pregnancy in the mouse. Nature 184:105
38. Dewar AD (1959) Observations on pseudopregnancy in the mouse. J Endocrinol 18: 186–190
39. Dewar AD (1973) Effects of hysterectomy on corpus luteum activity in the cyclic, pseudopregnant and pregnant mouse. J Reprod Fertil 33:77–89
40. Glenister PH, Rall WF (2000) Cryopreservation and rederivation of embryos and gametes. In: Jackson IJ, Abbott CM (eds) Mouse genetics and transgenics: a practical approach. Oxford University Press, New York, pp 27–59
41. Peters DD, Marschall S, Mahabir E, Boersma A, Heinzmann U, Schmidt J, Hrabe de Angelis M (2006) Risk assessment of mouse hepatitis virus infection via *in vitro* fertilization and embryo transfer by the use of zona intact and laser -microdissected oocytes. Biol Reprod 74:246–252
42. Weisbroth SH, Peters R, Riley L, Shek W (1998) Microbiological assessment of laboratory rats and mice. ILAR J V 39(4)
43. Pritchett-Corning KR, SheK WR, Henderson SK, Clifford CB (eds) (2009) Companion guide to rodent health surveillance for research facilities. Charles River Laboratories, MA
44. Pritchett-Corning KR, Cosentino J, Clifford CB (2009) Contemporary prevalence of

infectious agents in laboratory mice and rats. Lab Anim 43(2):165–173
45. Livingston RS, Riley LK (2003) Diagnostic testing of mouse and rat colonies for infectious agents. Lab Anim 32:44–51
46. National Research Council, Institute of Laboratory Animal Resources, Committee on Infectious Diseases of Mice and Rats (1991) Objectives, terminology, and overview of pathogen status. In: Infectious diseases of mice and rats. National Academy Press, Washington, DC, pp 3–12
47. Carty AJ (2008) Opportunistic infections of mice and rats: Jacoby and Lindsey revisited. ILAR J 49(3):272–276
48. Clifford CB, Watson J (2008) Old enemies, still with us after all these years. ILAR J 49: 291–302
49. Henderson KS (2008) Murine norovirus, a recently discovered and highly prevalent viral agent of mice. Lab Anim 37(7):314–320
50. Lipman NS, Corning BF, Saifuddin M (1993) Evaluation of isolator caging systems for protection of mice against challenge with mouse hepatitis virus. Lab Anim 27 (2):134–140
51. Compton SR, Homberger FR, Paturzo FX, Clark JM (2004) Efficacy of three microbiological monitoring methods in a ventilated cage rack. Comp Med 54:382–392
52. Kraft LM (1958) Observations on the control and natural history of epidemic diarrhea of infant mice (EDIM). Yale J Biol Med 31: 121–137
53. Perdue K, Green KJ, Copeland M, Barron E, Mandel M, Faucette LJ, Williams EM, Sosnovtsev SV, Elkins WR, Ward J (2007) Naturally occurring murine norovirus infection in a large research institution. J Am Assoc Lab Anim Sci 46(4):39–46
54. FELASA Working Group on Health Monitoring of Rodent and Rabbit Colonies: Nicklas (Convenor) W, Baneux P, Boot R, Decelle T, Deeny AA, Fumanelli M, Illgen-Wilcke B (2002) Recommendations for the health monitoring of rodent and rabbit colonies in breeding and experimental units. Lab Anim 36:20–42
55. England JJ (2002) Biosecurity: safeguarding your veterinarian:client:patient relationship. Vet Clin Food Anim 18:373–378
56. Lipman NS, Homberger FR (2003) Rodent quality assurance testing: use of sentinel animal systems. Lab Anim 32(5):36–43
57. Weir EC, Bhatt PN, Barthold SW, Cameron GA, Simack PA (1987) Elimination of mouse hepatitis virus from a breeding colony by temporary cessation of breeding. Lab Anim Sci 37:455–458
58. Lipman NS, Newcomer CE, Fox JG (1987) Rederivation of MHV and MEV antibody positive mice by cross-fostering and use of the microisolator caging system. Lab Anim Sci 37(2):195–199
59. Singletary KB, Kloster CA, Baker DG (2003) Optimal age at fostering for derivation of helicobacter hepaticus-free mice. Comp Med 53 (3):259–264
60. Scavizzi F, Raspa M (2005) *Helicobacter typhlonius* was detected in the sex organs of three mouse strains but did not transmit vertically. Lab Anim 40:70–79
61. Skinner HH, Knight EH (1974) Factors influencing pre-natal infection of mice with lymphocytic choriomeningitis virus. Arch Gesamte Virusforsch 46(1–2):1–10
62. Dejucq N, Bernard J (2001) Viruses in the mammalian male germinal tract and their effects on the reproductive system. Microbiol Mol Biol Rev 65:208–231
63. Mims CA (1966) Immunofluorescence study of carrier state and mechanism of vertical transmission in lymphocytic choriomeningitis virus infection in mice. J Path Bact 91:395–402
64. Janus LM, Smoczek A, Hedrich HJ, Bleich A (2009) Risk assessment of minute virus of mice transmission during rederivation: detection in reproductive organs, gametes, and embryos of mice after *in vivo* infection. Biol Reprod 81(5):1010–1015
65. Carthew P, Wood MJ, Kirby C (1985) Pathogenicithy of mouse hepatitis virus for preimplantation mouse embryos. J Reprod Fert 73:207–213

Chapter 26

Refinement, Reduction, and Replacement

Jan Parker-Thornburg

Abstract

In their 1959 publication *The Principles of Humane Experimental Technique*, Russell and Burch defined three criteria to be used to alleviate the sources and incidences of inhumanity when performing animal experimentation. These include ***reduction, replacement,*** and ***refinement***. To quote these authors, "Replacement means the substitution for conscious living higher animals of insentient material. Reduction means reduction in the numbers of animals used to obtain information of a given amount and precision. Refinement means any decrease in the incidence or severity of inhumane procedures applied to those animals which still have to be used." In this chapter, we will discuss methods that can be applied in transgenic animal facilities to better adhere to the goals of the three R's, as well as include a fourth "R"-ways of re-using animals.

26.1 Introduction

Animals have been used since the time of ancient Greek scientists to help humans understand many aspects of physiology and genetics. From early Egypt, where dissection of both human cadavers and animals led to an understanding of gross anatomy, to the present, where the creation of genetically modified mice can lead to an understanding of minute molecular structures in the cell, animals have been essential for our progress. We have long recognized the role of animals in biomedicine and our ethical duty to treat them with respects. However, it was not until 1959 that a codified method of addressing this duty was established. A method of defining the humanity of animal experimentation came from the United Kingdom with the publication of *The Principles of Humane Experimental Technique* by W.M.S. Russell and R.L. Burch [47]. These authors clearly defined pain and distress as it applies to "lower" animals. They suggested that, because of the higher level of consciousness in humans, it is our ethical duty as researchers to be exquisitely sensitive to pain and

S. Pease and T.L. Saunders (eds.), *Advanced Protocols for Animal Transgenesis*, Springer Protocols,
DOI 10.1007/978-3-642-20792-1_26, © Springer-Verlag Berlin Heidelberg 2011

distress in animals, much as we would with a baby. To fail to do so would be inhumane.

As a direct result of their analysis, Russell and Burch defined three criteria to be used to alleviate the source and incidence of inhumanity when performing animal experimentation. These include ***reduction, replacement,*** and ***refinement***. To quote these authors, "Replacement means the substitution for conscious living higher animals of insentient material. Reduction means reduction in the numbers of animals used to obtain information of a given amount and precision. Refinement means any decrease in the incidence or severity of inhumane procedures applied to those animals which still have to be used." To these, we could add a fourth "R" – ***reuse***. This would include managing animals such that they could be used for more than one study or purpose, within the limits of current regulations.

We will now examine how these concepts can be applied in a facility that "manufactures" genetically engineered animals. One problem with this type of operation is that large numbers of animals are required at the start to produce a small number of founder animals for any project. What are some methods we can use to reduce, replace, refine, or reuse?

26.2 Reduction

The principle of reduction is perhaps one of the easiest methods of adhering to the ethical standards defined by Russell and Burch. This is where we can make the greatest inroads to humane treatment of animals. Some ideas for doing so include the following:

26.2.1. Use Already Available Resources

One should determine whether an animal or cell line to generate that animal is already available or can be recreated by inter-crossing available mouse models. As described in Chapter 2, various entities are now producing embryonic stem cell lines that knock out every gene in the mouse. These can be accessed using the GeneTrap Consortium website: http://www.genetrap.org. As well, due to the rate at which this technology has been applied, many genetically engineered mice have already been generated world wide, some of which may apply to the project in hand. All investigators should make a practice of conducting a literature search to determine whether or not the animal model is already available. This is a requirement for U.S. investigators submitting an animal protocol [http://www.aphis.usda.gov/animal_welfare/downloads/policy/policy12.pdf].

Transgenic facility managers will generally be aware of the availability of a particular animal model at their institutions.

Managers who learn of research plans that will duplicate effort can manage that information to avoid the duplication, all the while maintaining the strictest confidentiality to protect the investigators involved. As an example, if Investigator A wishes to bring in a new animal model that was recently rederived for Investigator B; the manager can send a discreet inquiry to each investigator to ask whether their possession or need for this model is restricted information, or whether they would be willing to contact someone who has/needs the same model. Thus, in a transgenic facility, a manager can serve as a unique conduit for collaboration.

26.2.2. Take Advantage of Efficiency of Scale

While many transgenic core facilities perform related procedures (pronuclear injection, blastocyst injection, cryopreservation, rederivation), some do not. In some cases these procedures are spread among different entities throughout an institution, say a transgenic facility that does injections and a veterinary medicine department that provides cryopreservation and rederivation services. In other instances, various investigators in an institution may have developed these services specifically for use in their own lab. This duplication of effort requires duplication of colonies and results in unnecessary animal use. Investigators within the same institution should be encouraged to consolidate procedures or share animal colonies to reduce animal use. Due to internal procedures related to animal protocol issues, this may be difficult, but it is worth exploring and developing a means to do so. As an example, one institution (Baylor College of Medicine, Houston, TX, USA) established colonies to provide pseudopregnant females for all investigators needing them at the institution. In this case, the animals simply need to be ordered on the day before or day of use. It would be fairly easy to institute similar procedures for donor females, and one could consider whether this could be made into a billable service.

26.2.3. Eliminate Wasteful Protocols

Researchers who conduct animal research need to assess whether protocols can be changed to eliminate unnecessary use of animals. As an example, some investigators have held to the belief that oocytes produced by superovulation are inferior to those produced by natural matings. However, it is difficult to produce the requisite numbers of oocytes for injections using natural matings, thus requiring many more females for donors. After many years of injections to produce both transgenic and gene-targeted mice, it is now clear that the use of oocytes produced by superovulation does *not* compromise the project, and reduces animal use significantly. Thus, failure to use superovulation techniques would violate the "reduce" tenet of the three R's.

It is imperative that all protocols that affect the success of a project be assessed for efficacy. One common procedure in transgenic facilities is to accept a purified DNA fragment from an investigator for injection. Every time this is done, the project is entirely dependent upon the DNA isolation capabilities of the personnel in that investigator's lab. The best way of alleviating this dependency is for the transgenic core lab to isolate all DNA fragments for injection (*see* Chapter 21). This would insure that the project is being done with consistently clean, high-quality DNA. However, if the use of DNA prepared in a contracting lab is necessary (due to the inability of the transgenic core lab to isolate DNA fragments), then each sample should be checked for toxicity prior to setting up a full injection. DNA can be tested by injecting five to ten oocytes and then incubating them overnight to two-cell, and preferably to blastocyst stage. Determination of DNA toxicity prior to initiation of a full injection will easily save 10–12 animal lives, and probably more (considering that the project would be re-done after failure to get live births).

The Core manager should constantly be evaluating the efficacy of standard protocols that are used. In addition, literature searches, web searches, e-mail questions (using the online sites such as the Transgenic List [http://www3.imperial.ac.uk/lifesciences/services/research/transgeniclist]) and face-to-face contact during meetings are all valuable in determining whether a new protocol might prove more effective at reducing the numbers of animals used.

26.2.4. Optimize Standard Protocols

In order to reduce the numbers of animals, it is essential not only that protocols that work are used, but that the BEST protocols available are used.

Make sure your superovulation is super! Perhaps you are now convinced that you should move from natural matings to superovulation to produce oocytes for injection. It is not enough to simply decide to superovulate. Now, you must choose the best protocol for superovulating the females you have. Here, one needs to assess the strain of the animal (as different strains superovulate differently) [2, 8, 57], the age and/or weight of the animal, the source of the chemicals used, the amount of the chemicals used [14], the timing of the injections, and the housing conditions of the animals (specifically, the light cycle). A difference in any of these parameters can change the outcome of the superovulation procedure, so it is essential to determine what works best for your conditions. Table 26.1 describes some superovulation protocols used at the author's facility.

Choice of strains: There are two considerations to make when choosing a strain for transgenic models. Some investigators prefer to use outbred stocks presuming that they mimic the human condition more closely. However, due to the uncontrolled variation of outbred strains, they may actually confound the analysis [16]. Thus, it is preferable to use inbred strains if at all possible. In addition, there are instances where an investigator would like

Table 26.1
Superovulation protocols for four commonly used mouse strains

Strain	Age	Results
C57BL/6	3–3.5 weeks	31 oocytes/female
B6D2F1	6–8 weeks	33 oocytes/female
BALB/c	9–10 weeks	23 oocytes/female
FVB	6–8 weeks	20 oocytes/female

The superovulation protocols noted above are used routinely at the M. D. Anderson Cancer Center's Genetically Engineered Mouse Facility. The light cycle for this facility is a 12/12 cycle (on at 7 a.m., off at 7 p.m.). Female mice are dosed with 5 IU of PMS and 5 IU of HCG (47–49 h after PMS). Hormone administration occurs between 11 a.m. and 1 p.m. for each strain. PMS is obtained from the National Hormone & Peptide Program (NHPP) at UCLA. The powder is diluted in sterile isotonic saline (0.9%) to a final concentration of 50 IU/ml. 1.2 ml aliquots are stored at −80°C and thawed no sooner than 0.5 h prior to use. HCG (10,000 IU) is obtained from Sigma Chemicals. A stock solution (2,000 IU) is made by resuspending the powder in 5 ml of sterile water. 25 μl of the stock solution is placed into a 1.5 ml eppendorf tube and lyophilized to generate working tubes that are stored at −80°C. Immediately prior to injection, 1 ml of isotonic saline (0.9%) is added to each tube for a final concentration of 50 IU/ml

transgenic production done in a specific strain – say C57BL/6. Most injectionists find it difficult to inject into these oocytes, as the pronucleus is smaller and develops later. When one looks at animal lives, it may take double the number of donor females to produce the transgenic founders needed (perhaps using 20 females vs. 10 females). However, considering the numbers of animals it would take to backcross the gene from an easily injected strain into a C57BL/6 background (five mouse generations if doing speed congenics, ten mouse generations by standard crosses), the numbers of animal lives saved by doing the injection directly into C57BL/6 (or any other inbred background) would be significant.

Don't take anything for granted. It is relatively standard to reassess procedures when a project fails. However, it is beneficial to head off problems before they occur. For that reason, it is helpful to test every batch of new media and oil using an oocyte incubation test (set up five to ten fertilized oocytes in a drop of old [previously tested] media and new media, or old media and new oil; incubate to blastocyst stage). These tests can be done on a regular basis – weekly or every other week – or upon receipt of new media and oil. Also, one should test the %CO_2 in your incubator at least monthly using Fyrite analysis.

26.2.5. Eliminate Repeat Procedures

The biggest waste of animals happens when a project fails and is then repeated. This happens with some regularity at all transgenic core facilities. However, it is critical that an assessment is made at the time of failure to determine the cause, and then eliminate that cause during the next attempt. What are some reasons that a project would fail? Numerous possible causes for a project to fail are discussed below. To narrow down the reason, the first question should always be: Is the failure limited to this project, or does it affect all projects in the animal room? If the failure is limited to a single project, then it is much easier to identify the problem, as it is likely to relate to DNA or cells. On the other hand, if projects have been working well, and there is suddenly a failure of numerous projects at the same time, then one would need to examine many other possible causes.

1. Outside of the building/environmental causes for failure: Many observations have been made to indicate that nearby construction projects can interfere with gestation, birth, and nursing behavior in mice [43]. Thus, whether there is nearby construction might be the first question when projects fail. Unfortunately, there is not much that can be done other than to wait out the construction. Rarely, weather events can also affect the building such that the mice are disturbed – for example, we documented increased building vibration during a recent hurricane, and subsequent decreases in number of pups born. Low humidity levels can be caused by atmospheric drying in Northern climates during the cold months or by the warm Santa Ana winds (for those who live on the west coast of the United States). Prolonged exposure to low humidity levels can reduce numbers of pups due to chronic dehydration of the gestating females.
2. Facility causes for failure: Malfunction of the light timers which cause changes in the light cycle will have a negative effect on pup numbers and fertilization rate of donor oocytes. As well, one needs to be aware of possible problems with mechanical equipment. The HVAC venting system can emit high frequency sound that can affect animals, as can cage-change hoods in the room. Laptops in the room, carts that are noisy, and even movement sensors that automatically unlock the doors have been shown to have negative effects on the animals due to the emission of high frequency sound [58]. A simple bat detector maybe useful in determining whether such problems exist in an animal housing facility.
3. Housing causes for failure: In general, it is recommended that transgenic and gene-targeted animals be housed using the most microbiologically safe methods available. This is because the animals will be shipped from the transgenic animal room

to other rooms in the animal facility so that investigators have access to them. Thus, microisolator housing is recommended at the very least. Infections of these sensitive animals can easily result in project failure [29, 30]. In addition, one should check the food available to the animal – breeder diet (we use diet that is ≥20% protein, ≥10% fat) is recommended for gestating females. The added proteins and fats in breeder diet improve the birth and survival rates of mouse pups.

While not a common problem, the bedding and any enrichment devices in the cage can also be suspect when a project fails. In general, bedding and enrichment will not adversely affect immunocompetent mice. However, cotton microfibers from nesting enrichment has been shown to cause irritation and infection in nude mice [6, 46]

When a project fails, very careful consideration should also be given to the integrity of the water supply system. If animals are housed in a unit with automatic water, the failure of a water valve to close properly could cause the adults and pups to drown; as well, leakage from the valve onto the bedding can cause the pups to become hypothermic, at which time, the mother will likely cannibalize the pups. Also, consider whether there is sufficient water getting to the animals. If the watering system is inefficient at delivering water to the animals, or if there is forced air into the cage around the watering valve, there can be dehydration of the animals. This may result in a gradual loss of fertility in the studs and donors, in addition to late-stage pregnancy failure and cannibalism. Table 26.2 demonstrates an instance of where oocyte numbers and birth rates were affected by housing in a specific type of ventilated microisolator (now discontinued).

4. Personnel: The simple act of picking up a mouse has been shown to increase stress hormones in the animal [12, 18]. In our experience, over time mice will acclimatize to handling by one person. We and others have noted that if the familiar animal care staff member leaves for vacation, or is switched out for another person, the animals will often respond by losing pregnancies or cannibalizing pups. Highly trained technicians will practice gentle handling of the animals and will understand the devastating effects that raised voices and loud noises can have in an animal facility. Failure has not been documented due to singing, playing a radio, etc., probably because these sounds are consistent and not sudden or loud. Mice hear in the range of 0.5–120 kHz [61, 64]. Human noises (voices, whistling, singing) generally range from 0.5 to 5 kHz [32, 42, 64]. Thus, noises we make fall in the very lower spectrum of the hearing range of mice.

Table 26.2
Egg production and birth rate can be affected by housing conditions

Egg production rates

Pre/ Post	Date	Avg. fertilized oocytes	Avg. total oocytes	Avg. no. of females	Avg. no of plugged females	Avg. no of oocytes/ female	Avg. no of fertile oocytes/ female
Pre	4/2001–8/2001	146	214	11	10	19.5	14.6
Post	8/2001–6/2002	187	285	12	8	23.8	23.4

Birth rates

Pre/ Post	Date	Avg. no. of fertilized oocytes	Avg. no. of implanted embryos	Avg. no. of pseudopregnant females	Avg. no. of pregnant females	Avg. no. of births	Avg. no. of births/ pregnant female
Pre	4/2001–8/2001	146	107	3.2	1.2	3.9	3.3
Post	8/2001–6/2002	187	128	4.3	3.2	18.4	5.7

Unpublished data collected at The Ohio State University Transgenic Core Facility between April 2001 and June 2002. "Pre" refers to the time period during which all animals were housed in ventilated microisolators with forced air being supplied around the water source. Due to low numbers of pregnancies and births, all animals were moved to static microisolators in August of 2001. "Post" refers to the time period during which animals were housed in static microisolators with water supplied from bottles. As shown above, we noted a 60% increase in the average number of fertilized oocytes per female and a 73% increase in birth rate after changing to static microisolators. Later tests of ventilated microisolation caging from other sources suggest that this effect was most likely due to the placement of the air supply in relation to the watering system

26.2.6. If Feasible, Substitute Alternate Methods That Will Produce the Necessary Results While Using Fewer Animals

Not all investigators need an established genetically propagated line to perform their studies. Protocols have been published where a transgenic construct is directly injected into the tissue of interest – skeletal muscle for example [23, 51]. The gene is expressed for a period of time, but will not be passed on to subsequent generations, obviating the need for the many animals associated with transgenic mating schemes.

Recently, a protocol has been developed where a transgene-containing vector can be injected through the tail vein. The use of hydrodynamic injection (a very large and fast bolus of saline containing the transgenic vector into the tail vein) can result in uptake of the transgene in some cell types [26, 63, 65]. The transgene is not transmitted to subsequent generations, but sufficient expression may be seen in the injected animal to obtain publishable results.

If expression is expected in lung epithelium, it is possible to deliver a transgenic DNA through inhalation. This procedure, however, generally requires the use of a viral entry system (such as adenovirus), making it a BSL-2 procedure.

As described in Chapter 10 on Lentiviral Transgenesis, it is possible to produce many transgenic founders from one initial founder using lentiviral infection of the oocyte. This infection can result in insertion of several transgenes in the same oocyte, which can then be separated through standard mating procedures [28]. Use of this technique could result in the use of fewer female donors and requires BSL-2 handling. However, one also needs to account for extra animals required for outcrossing to separate the insertions, which may nullify the savings in animal usage.

In our facility, we have identified two areas that we deemed wasteful of mice. The first was in the generation of blastocysts for ES cell injection. Typically, we would superovulate, but there were occasions when the superovulation was not good, resulting in few blastocysts for injection. Other times, we would not use our stud males for several weeks, and their response to the donor females was poor, resulting in few blastocysts for injection. We solved this problem by performing weekly matings of superovulated females with the males, obtaining day 2.5 morulae and then doing a slow-cool method of embryo cryopreservation. As a result, we now have banks of embryos, ready for injection that simply need to be thawed the day before the injection procedure. This has resulted in our ability to inject immediately upon receiving a request and eliminated days with small numbers of injectable blastocysts [41]. Others have used this same procedure to store pronuclear stage oocytes for pronuclear injections, although better results may be obtained if one uses vitrification in this instance [3, 24]. A great advantage of using cryopreserved embryos for injections is that many embryos can be stored over a very short period of time, reducing the need for maintaining large donor and stud colonies. This would be especially beneficial when rarely used strains are requested.

A second area where we (and others) have been able to reduce animal numbers is for cryopreservation. For years, this technique has generally been limited to the cryopreservation of embryos, due, in large part, to poor sperm survival upon thawing. Highly technical and more expensive-assisted fertilization techniques, such as intracytoplasmic sperm injection [53], would need to be applied if low-quality sperm were to be useful in re-establishing a strain of mouse. More recently, use of the agent monothiolglycerol (MTG) has been shown to increase sperm survival and motility to the point where sperm freezing is a viable method of preserving a mouse line, irrespective of background strain [39]. However, only-freezing sperm results in reconstitution of a mouse line that is, by genetic definition, maximally heterozygous for the desired gene. The problem arises when an investigator needs to cryopreserve a line in a homozygous state, or where multiple transgenes need to be maintained. This problem can be solved by performing an in vitro fertilization followed by cryopreservation. In this case, multiple female embryo donors will be used, but IVF using sperm from one or two donor males eliminates the need for multiple stud males for mating. Two-cell embryos that result are vitrified and archived. This method can also be used to generate many embryos from only one or two males and 10–15 wild-type females to freeze/vitrify heterozygous embryos. Typically, 200–300 embryos can be archived in one session using sperm from one male. Use of an IVF-cryopreservation procedure can thus save time, animal lives, shelf space, and per diem costs to the investigator. Cryopreservation protocols are described in detail in Chapter 23.

26.2.7. Verify the Integrity of Your ES Cells

Poor-quality embryonic stem (ES) cells will not make germline-transmitting mice. Thus, it is critical to know how good your ES cells are prior to targeting and prior to injection. This can be done by sending a sample out for karyotyping, or by making chromosome spreads for counting. Cells that are trisomic for chromosome 8 and, less commonly, chromosome 11, will grow robustly, but will not make mice that transmit through the germline [27, 52]. We, and others, have found that an ES cell line can deteriorate over time [15, 34]. This can only be detected by assessing the ability of the cells' contribution to chimeras, and then by determining how many high-percentage chimeras are sterile, and how many will go germline. In our case, only 30% of high-percentage chimeras from one ES cell line would go germline, and this was independent of the lab that made the initial clone (clones from the same ES cell line generated by three different labs). We subsequently rejected this cell line for use, replacing it with a tested line that went germline easily (70% of high-percentage chimeras to germline). By switching to a more competent cell line, we reduced the number of animals used for testing for germline transmission, as well as avoided repeat injections with alternate clones. Also, be aware of the number of cells you are injecting into a blastocyst. We have found that with an excellent hybrid cell line we need to inject very few cells (five to eight) for optimal survival of the embryo, whereas other cell lines may require the injection of greater numbers of cells.

26.3 Replacement

As the procedures of generating mice are the premise of what we do, there are few areas where replacement is possible. However, we should always be aware that we can ask investigators whether they have considered the use of other methods that do not involve the use of animals. Three methods immediately come to mind.

First, an age-old method of gene analysis is to examine gene expression and its subsequent effects in vitro using tissue culture cells. Cells are available from a wide variety of tissues, and for both "normal" (as far as immortalized cells can be considered normal) and pathogenic states (such as tumor cell lines). Often, preliminary studies are performed using tissue culture cells. For example, expression of transgenes can be verified in cell culture prior to pronuclear injection. The caveats of cell culture are that one type of cell cannot mimic the physiology of an entire animal, and that analysis done in tissue culture is not always indicative of what would occur in vivo [37, 45]. However, in an effort to replace animals, we should be willing to address the possibility of performing an in vitro analysis for projects where this could result in publishable data.

A second possibility is that certain developing tissues such as cardiac tissue or neurons can be obtained by making use of embryoid bodies (EBs) – ES cells that have differentiated in culture using techniques that optimize for cell-type-specific differentiation [25, 31, 38, 48]. While not replacing studies that require analysis of many physiological systems (which would require an intact animal) EBs and lineage-defined differentiated cells can provide a limited analysis of several systems.

Recently, induced pluripotent stem cells were obtained from differentiated somatic cells using several embryonic induction genes (including Oct 3/4, Sox 2, c-Myc, and KLF4) [54]. The use of somatic cells to generate induced pluripotent stem cells (iPSCs) obviates the need to euthanize mice for that purpose. In addition, iPSCs have been shown capable of differentiation into neural cells, macrophages, cardiac cells, embryoid bodies, and teratomas [49, 60]. Thus, processes in these systems could be studied without using mice. Unfortunately, this system is limited to use of these cell types only. The study of interactions among cell types, or analyses of processes such as metastasis require an intact animal system.

26.4 Refinement

Those of us "raised" as scientists will see the word "refinement" and automatically presume that this refers to streamlining our experiments to make them more efficient for obtaining repeatable and reliable results. However, in the animal care field, refinement refers to a process that will reduce stress and trauma in the animals we work with. As such, methods of refinement will generally involve improvements in surgical procedures and euthanasia.

Many of us were trained in surgical procedures before mice and rats were added to the list of regulated species. Thus, our standard surgical procedures may be near and dear to our hearts, but not necessarily procedures that are optimal for the animal or recommended by oversight bodies [13]. Prior to dismissing "new" techniques summarily, however, it may be wise to do the experiment. If the new technique is instituted and well executed, is the animal better off? Are more pups born? Below are listed the major areas where transgenic specialists can refine techniques.

The Guide for the Care and Use of Laboratory Animals [36] lists the following criteria as important for performing surgery in animals: "presurgical planning, personnel training, aseptic and surgical technique, animal well-being, and animal physiologic status." Each of these areas should be assessed to determine if refinement techniques can be applied.

Presurgical planning for mouse surgeries should first include prepping the physical areas for anesthesia and surgery in a manner deemed adequate by veterinary standards. Set up will include laying out surgical instruments (that have been properly sterilized) and any additional items used during surgery (sponges, surgical spears, etc.). Sterilization of surgical instruments is easily done using a glass-bead sterilizer located in the surgical area. Figure 26.1 displays a properly organized surgical area. Organization prior to a surgery can prevent mistakes that could result in a failed pregnancy or infection of the animal.

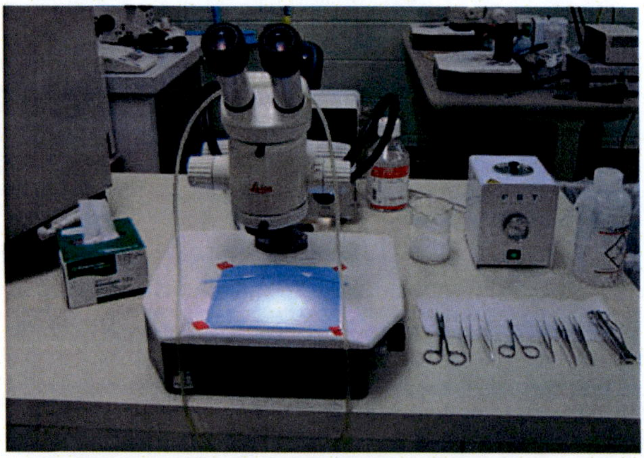

Fig. 26.1 Example of a prepared surgical table for implantations: the surgical area shown was prepared by first wiping all surfaces with 70% ethanol. A square of presterilized SpaceDrape® (Locus Technology Inc.) was placed on the surgical microscope to assist with keeping the animal warm. Sterile surgical spears (Medtronic-Solan) used for tissue manipulation are placed onto the drape. Surgical instruments were sterilized by placing them into the glass-bead sterilizer for 10–30 s. After sterilization, they are placed onto a sterile gauze pad located on the work surface in the order that they will be used.

A second area that is critical for surgical success is in the area of training. Someone who is new to embryo transfer surgery should be trained on mice that are not part of a project. A new trainee should start with observation of a trained surgeon using a double-headed microscope. Video observation of implant techniques generally fail to show the detail required for isolation of the infundibulum. The second stage of training should be hands-on, done on one or two mice that are euthanized prior to waking up from the anesthesia and using colored beads (Sephadex™ beads) instead of embryos. Once the trainer is satisfied that the trainee can isolate the oviduct and implant the embryos, then females implanted with embryos can be taken to term to determine whether a pregnancy results. These steps can be gone through rapidly for a very quick and skilled learner. However, even after months of training, there are trainees who never develop to the point where they can be trusted to implant an animal. These technicians should be given other duties in the lab, rather than subjecting the animals to additional distress. Once competent, a very valuable method of intensive surgical training is to have the new surgeon responsible for rederivations for the facility. Successful training can be monitored by numbers of pups born as well as by assessing the surgical records. Surgical records that include type, method of administration and amount of anesthesia and analgesia and documentation of postsurgical monitoring of the surgical site (often for up to 7 days) are now being required by many Animal Care and Use Committees. Finally, according to the Guide, "good surgical technique includes asepsis, gentle tissue handling, minimal dissection of tissue, appropriate use of instruments, effective hemostasis, and correct use of suture materials and patterns." Both new and "old" surgeons should exhibit these good surgical techniques.

Another area that should be tested for refinement is that of anesthesia. The anesthesia used for implant procedures may differ from institution to institution. Avertin has been a standard injectable anesthesia for implant surgery. It is still used at many institutions, but is often recommended against due to the observations of abdominal adhesions with repeat administration [62], although this may be due to the particular method of anesthetic preparation used by the authors [59]. As well, avertin causes a drastic reduction in heart rate in the animal, which can lead to death by hypothermia during surgery. However, since avertin provides 7–15 min of anesthesia at a surgical plane, it is still widely used by transgenic technicians (with guidelines for preparation and storage determined often by the institution's Animal Care and Use Committee) [40].

Many institutions are now requiring their mouse surgeons to use a ketamine mix or an inhalable anesthetic such as isoflurane. Each of these also has pluses and minuses. On a switch from avertin to ketamine/xylazine, this author found that more pups were born after the surgery. In addition, the animals would become mobile more quickly, thus reducing the possibility of hypothermia. However, a major drawback with ketamine is that a DEA license is required to obtain the drug, and careful records of use and disposal must be maintained for regulatory requirements. Isoflurane is an inhalable anesthetic that allows the animal to reach a surgical plane of anesthesia very rapidly – within 30 s. Isoflurane requires a machine with a controlled-rate vaporizer for proper administration; and, to avoid danger to the surgeon, an effective scavenging system. This may necessitate many more tubes going into and out of the surgical area. Isoflurane is also quite expensive. But, if the proper equipment is available and the expense can be absorbed, use of this anesthetic is an excellent refinement for mouse surgery, as the effects of isoflurane on heart rate in the mouse are minimal, especially when compared to those seen due to ketamine and avertin (Fig. 26.2).

Another surgical refinement is to treat the mouse with an analgesic such as buprenorphine at the time of surgery. Treatment with analgesics is an important refinement even when mice do not show the signs of pain [1]. This is a minor addition to surgical procedure that provides pain relief to the mouse after the recovery, and should be a part of all survival surgeries.

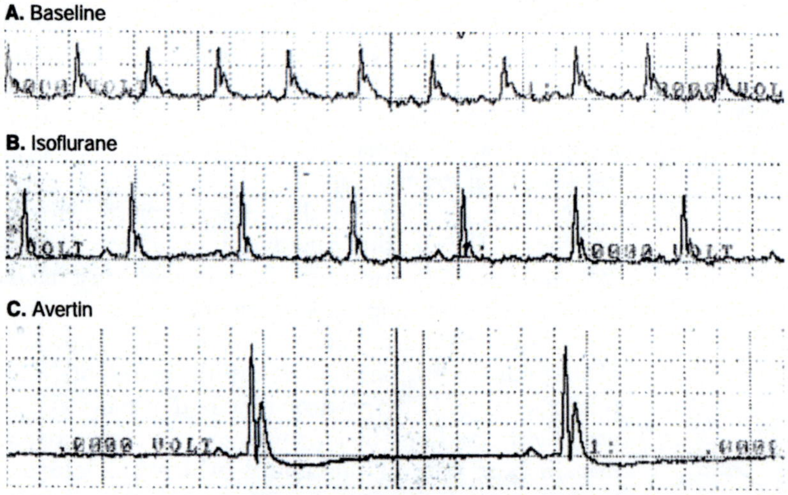

Fig. 26.2 Electrocardiographic analysis of a mouse treated with two different anesthetics: (**a**) a single mouse wild-type mouse was used at the end of an experiment to examine how different anesthetics affect heart rate. For the previous experiment, the mouse had been implanted with a telemetric electrocardiographic transmitter (Data Sciences International) for remote monitoring of heart rate. For this analysis, a baseline was established in the mouse prior to administering anesthetic. The mouse was then treated with isoflurane (2.5% in O_2) to reach a surgical plane of anesthesia (as determined by loss of pedal reflex). After recording the ECG, the mouse was allowed to recover, at which time it re-established a normal baseline ECG. The mouse was then treated with Avertin (240 mg/kg) and the ECG was recorded while at a surgical plane of anesthesia. The mouse was humanely euthanized prior to recovery.

A very easy refinement to make during and after mouse surgeries is to provide warmth to the animal, typically by administration of warmed fluids and by placing the anesthetized animal on a warm plate for surgery and recovery. Not only does post-surgical warmth reduce the possibility of hypothermia, it also results in increased implantation rates [4]. It is helpful to the animal to provide a subcutaneous injection of approximately 500 μl of pre-warmed isotonic saline solution prior to or just after the surgery. This injection can also help dilute out the effects of injected anesthesia, allowing the animal to recover more quickly.

Removal of hair at the site of surgery... whether or not to do so is a question that is often raised with animal oversight committees. The rodent surgical procedures for aseptic surgery published by the Animal Welfare Information Center of the USDA in 1993 include the use of clipping the surgical site, or removing hair by plucking, followed by careful disinfection [13]. However, several studies of human brain surgery have shown that carefully parting the hair over the surgical site causes less trauma and infection [7, 22, 33]. Typically, mice are clipped for surgery and the skin disinfected with 70% ethanol and wiped with a sterile gauze pad. Alternatively, betadyne or chlorhexidine preparations may be used for sanitization of skin, prior to surgery. However, many surgeons do not clip and also achieve high pup birth rates (which is the gold standard for aseptic surgery). Care must be taken, though, to avoid getting the fur into the surgical site upon closure [11].

The last refinement we will address is that of euthanasia. Unfortunately, due to the nature of our work, many animals are used and eventually euthanized. The best manner of performing euthanasia is presently a matter of discussion among animal welfare groups throughout the United States and Europe [10, 19]. For many years, the typical recommendation for euthanasia from the veterinary staff has been to asphyxiate the animals using CO_2 [http://avma.org/resources/euthanasia.pdf]. However, many (most) transgenic technicians have been well-trained in the use of cervical dislocation, and prefer to use that technique due to observations that it provides a less-painful death when performed correctly. Studies have been done regarding various methods of euthanasia in mice with conflicting conclusions, and, in fact, support can be found for both positions [9, 20, 50]. Many previous studies relied on small numbers of animals and slight differences in euthanasia protocols. Determination of the best method of euthanasia will take additional research using carefully controlled studies that test statistically significant numbers of animals. For this to occur, it is imperative to have effective communication between veterinarians and research personnel.

26.5 Reuse

We would all like to reduce the large numbers of animals used in transgenic and targeting procedures. However, after instituting some of the ideas listed previously, we are still using many animals. The question then becomes, are there ways to generate additional use of the animals? How can we recycle? Some methods are described below. However, prior to instituting these, it is essential that you have permission from your institutional Animal Care and Use Committee.

Discussion with various transgenic facilities shows that the methods of increasing use of the animals can vary. Some ideas include (1) offering wild-type animal tissues to investigators (free) that can be collected from the females euthanized as oocyte donors or old stud males and unused recipient females; (2) offering older, wild-type breeders to investigators; (3) where choices of strain permit (i.e., C57BL/6 × C57BL/6 but not C57BL/6 × B6D2F1) one can use wild-type offspring from blastocyst injections for breeding or as subsequent embryo donors; (4) one can donate frozen carcasses of wild-type mice to people with snakes, to be used as a food source; (5) one can donate frozen carcasses to institutions that house raptors to be used as a food source. (6) On the rare occasions where a project cannot be performed on schedule and superovulated and mated females are available, wild-type embryos can be isolated as morulae and frozen for later blastocyst injection, or, the females can be allowed to go to term and the progeny used for subsequent projects. The concept of wise use of animals dictates that we not allow an animal to go to waste; an animal should rarely, if ever, fail to have its life have meaning.

26.6 Conclusions

In summary, I have presented the concepts of humane and responsible use of animals as described by Russell and Burch. It is essential that users of large numbers of laboratory animals adhere to the concepts of ***reduction, replacement,*** and ***refinement***, and that, if possible, we also include the possibility of ***reuse***. As stewards of the animals that provide such a valuable service, it is our responsibility to manage our use of them with a thoughtful and informed approach. Most of us are very proud of the work we do to create animal models that mimic human and animal diseases in a genetically malleable system. As a result of our work, the animals produced can be used to decipher how disease begins, progresses, and hopefully, can be interrupted and cured.

References

1. Arras M, Rettich A, Cinelli P, Kasermann HP, Burki K (2007) Assessment of post-laparotomy pain in laboratory mice by telemetric recording of heart rate and heart rate variability. BMC Vet Res 3:16
2. Auerbach AB, Norinsky R, How W, Losos K, Guo Q, Chatterjee S, Joyner AL (2003) Strain-dependent differences in the efficiency of transgenic mouse production. Transgenic Res 12:59–69
3. Bagis H, Odaman H, Sagirkaya H, Dinnyés A (2002) Production of transgenic mice from vitrified pronuclear-stage embryos. Mol Reprod Dev 61(2):173–179
4. Bagis H, Odaman Mercan H, Dinnyes A (2003) Exposure to warmer postoperative temperatures reduces hypothermia caused by anesthesia and significantly increases the implantation rate of transferred embryos in the mouse. Lab Anim 38:50–54
5. Baumans V (2005) Science-based assessment of animal welfare: laboratory animals. Rev Sci Tech 24(2):503–513
6. Bazille PG, Walden SD, Koniar BL, Gunther R (2001) Commercial cotton nesting material as a predisposing factor for conjunctivitis in athymic nude mice. Lab Anim 30(5):40–42
7. Bekar A, Korfali E, Dogan S, Yilmazlar S, Baskan Z, Aksoy K (2001) The effect of hair on infection after cranial surgery. Acta Neurochir (Wien) 143(6):533–536
8. Byers SL, Payson SJ, Taft RA (2006) Performance of ten inbred mouse strains following assisted reproductive technologies (ARTS). Theriogenology 65:1716–1726
9. Cartner SC, Barlow SC, Ness TJ (2007) Loss of cortical function in mice after decapitation, cervical dislocation, potassium chloride injection, and CO2 inhalation. Comp Med 57 (6):570–573
10. Conlee KM, Stephens ML, Rowan AN, King LA (2004) Carbon dioxide for euthanasia: concerns regarding pain and distress, with special reference to mice and rats. Lab Anim 39:137–161
11. Cooper DM, McIver R, Bianco R (2000) The thin blue line: a review and discussion of aseptic technique and postprocedural infections in rodents. Contemp Top Lab Anim Sci 39 (6):27–32
12. Croft AP, O'Callaghan MJ, Shaw SG, Connolly G, Jacquot C, Little HJ (2008) Effects of minor laboratory procedures, adrenalectomy, social defeat or acute alcohol on regional brain concentrations of corticosterone. Brain Res 1238:12–22
13. Cunliffe-Beamer TL (1993) Applying principles of aseptic surgery to rodents. AWIC Newsletter 4(2):3–6
14. Edgar DH, Whalley KM, Mills JA (1987) Effects of high-dose and multiple-dose gonadotropin stimulation on mouse oocyte quality as assessed by preimplantation development following in vitro fertilization. J In Vitro Fert Embryo Transf 4:273–276
15. Federov LM, Haegel-Kronenberger H, Hirchenhain J (1997) A comparison of the germline potential of differently aged ES cell lines and their transfected descendants. Transgenic Res 6:223–231
16. Festing MF (2004) The choice of animal model and reduction. Altern Lab Anim 32 (Suppl 2):59–64
17. Flecknell P (2002) Replacement, reduction and refinement. ALTEX 19(2):73–78
18. Hale KD, Weigent DA, Gauthier DK, Hiramoto RN, Ghanta VK (2003) Cytokine and hormone profiles in mice subjected to handling combined with rectal temperature measurement stress and handling only stress. Life Sci 72(13):1495–1508
19. Hawkins P, Playle L, Golledge H, Leach M, Banzett R, Coenen A, Cooper J, Danneman P, Flecknell P, Kirkden R, Niel L, Raj M (2006) Report from the Newcastle consensus meeting on carbon dioxide euthanasia of laboratory animals. University of Newcastle upon Tyne, UK, pp 1–17, 27–28 Feb 2006
20. Howell RL, Donegan CL, Pinkert CA (2003) Mouse embryo yield and viability after euthanasia by CO_2 inhalation or cervical dislocation. Comp Med 53(5):510–513
21. International Gene Trap Consortion (2009) http://www.genetrap.org. Cited 19 June 2009
22. Iwami K, Takagi T, Arima T, Takayasu M (2006) [Cranial surgery without shaving: practice and results in our hospital]. No Shinkei Geka 34(9):901–905
23. Kawano R, Ishizaki M, Maeda Y, Uchida Y, Kimura E, Uchino M (2008) Transduction of full-length dystrophin to multiple skeletal muscles improves motor performance and life span in utrophin/dystrophin double knockout mice. Mol Ther 16(5):825–831
24. Keskintepe L, Agca Y, Pacholoczyk GA, Machnicka A, Critser JK (2001) Use of cryopreserved pronuclear embryos for the production of transgenic mice. Biol Reprod 65 (2):407–411
25. Kurosawa H (2007) Methods for inducing embryoid body formation: in vitro

differentiation system of embryonic stem cells. J Biosci Bioeng 103(5):389–398
26. Liu F, Song Y, Liu D (1999) Hydrodynamics-based transfection in animals by systemic administration of plasmid DNA. Gene Ther 6(7):1258–1266
27. Liu X, Wu H, Loring J, Hormuszi S, Disteche CM, Bornstein P, Jaenisch R (1997) Trisomy eight in ES cells is a common potential problem in gene targeting and interferes with germ line transmission. Dev Dyn 209 (1):85–91
28. Lois C (2005) Generation of transgenic animals with lentiviral vectors. In: Pease S, Lois C (eds) Mammalian and avian transgenesis: new approaches (principles and practice). Springer, Berlin
29. Mahabir E, Reindl K, Mysliwietz J, Needham J, Bulian D, Markoullis K, Schmidt J (2008) Absence of germline transmission after blastocyst injection with murine embryonic stem cells cultured with mouse hepatitis virus and mouse minute virus. Platform presentation at the 2008 AALAS National Meeting, Indianapolis, IN
30. Markoullis K, Bulian D, Holzlwimmer G, Quintanilla-Fend L, Zitzelberger H, Heiliger K, Mysliwietz J, Uphoff CC, Drexler HG, Adler T, Busch DH, Schmidt J, Mahabir E (2008) Mycoplasmas affect cell parameters of murine embryonic stem cells, germline transmission and the resulting chimeric progeny. Platform presentation at the 2008 AALAS National Meeting, Indianapolis, IN
31. McKiernan E, O'Driscoll L, Kasper M, Barron N, O'Sullivan F, Clynes M (2007) Directed differentiation of mouse embryonic stem cells into pancreatic-like or neuronal- and glial-like phenotypes. Tissue Eng 13 (10):2419–2430
32. Meyer J (2004) Bioacoustics of human whistled languages: an alternative approach to the cognitive processes of language. An Acad Bras Cienc 76(2):405–412
33. Miller JJ, Weber PC, Pael S, Ramey J (2001) Intracranial surgery: to shave or not to shave? Otol Neurotol 22(6):908–911
34. Nagy A, Rossant J, Nagy R, Abramow-Newerly W, Roder JC (1993) Derivation of completely cell-culture derived mice from early-passage embryonic stem cells. Proc Natl Acad Sci USA 90:8424–8428
35. Nagy A, Gertsenstein M, Vintersten K, Behringer R (2003) Manipulating the mouse embryo: a laboratory manual. Cold Spring Harbor Laboratory Press, Cold Spring Harbor, NY
36. National Research Council (1996) Guide for the care and use of laboratory animals. National Academy Press, Washington, DC
37. Nguyen QG, Buskin JN, Himeda CL, Shield MA, Hauschka SD (2003) Differences in the function of three conserved E-boxes of the muscle creatine kinase gene in cultured myocytes and in transgenic mouse skeletal and cardiac muscle. J Biol Chem 278 (47):46494–46505
38. Osakada F, Ikeda H, Sasai Y, Takahashi M (2009) Stepwise differentiation of pluripotent stem cells into retinal cells. Nat Protoc 4 (6):811–824
39. Ostermeier GC, Wiles MV, Farley JS, Taft RA (2008) Conserving, distributing and managing genetically modified mouse lines by sperm cryopreservation. PLoS ONE 3(7):e2792
40. Papaioannou VE, Fox JG (1993) Efficacy of tribromoethanol anesthesia in mice. Lab Anim Sci 43(2):189–192
41. Parker-Thornburg JV, Alana JL, Smith CN, Detry M, Rojas R, Baskin KK (2005) Cryopreserved morulae can be used to efficiently generate germline-transmitting chimeras by blastocyst injection. Transgenic Res 14:685–690
42. Raphael LJ, Borden GJ, Harris KS (2007) Speech science primer: physiology, acoustics, and perception of speech. Lippincott Williams & Wilkins, Philadelphia, PA
43. Rasmussen S, Glickman GM, Norinsky R, Quimby FW, Tolwani R (2009) Construction noise decreases reproductive efficiency in mice. J Am Assoc Lab Anim Sci 48 (4):363–370
44. Report of the American Veterinary Medical Medical Association Panel on Euthanasia (2000) http://avma.org/resources/euthanasia.pdf. Cited 27 Aug 2009
45. Rindt H, Gulick J, Knotts S, Neumann J, Robbins J (1993) In vivo analysis of the murine beta-myosin heavy chain gene promoter. J Biol Chem 268(7):5332–5338
46. Rowson KEK, Michaels L (1980) Injury to young mice caused by cottonwool used as nesting material. Lab Anim 14:187
47. Russell WMS, Burch RL (1959) The principles of humane experimental technique. Charles C. Thomas, Springfield, IL
48. Schroeder IS, Wiese C, Truong TT, Rolletschek A, Wobus AM (2009) Differentiation analysis of pluripotent mouse embryonic stem (ES) cells in vitro. Methods Mol Biol 530:219–250
49. Senju S, Haruta M, Matsunaga Y, Fukushiima S, Ikeda T, Takahashi K, Okita K, Yamanaka S, Nishimura Y (2009) Characterization of dendritic cells and macrophages generated by directed differentiation from mouse induced pluripotent stem cells. Stem Cells 27 (5):1021–1031

50. Springell DA, Costin NS, Pilowsky PM, Goodchild AK (2005) Hypotension and short-term anaesthesia induce ERK1/2 phosphorylation in autonomic nuclei of the brainstem. Eur J Neurosci 22:2257–2270
51. Springer ML, Rando TA, Blau HM (2002) Gene delivery to muscle. Curr Protoc Hum Genet Chapter 13:Unit 13.4
52. Sugawara A, Goto K, Sotomaru Y, Sofuni T, Ito T (2006) Current status of chromosomal abnormalities in mouse embryonic stem cell lines used in Japan. Comp Med 56(1):31–34
53. Szczygiel MA, Kusakabe H, Hanigimachi R, Whittingham DG (2002) Intracytoplasmic sperm injection is more efficient than in vitro fertilization for generating mouse embryos from cryopreserved spermatozoa. Biol Reprod 67(4):1278–1284
54. Takahashi K, Yamanaka S (2006) Induction of pluripotent stem cells from mouse embryonic and adult fibroblast cultures by defined factors. Cell 126(4):663–676
55. Transgenic-list (2009) http://www3.imperial.ac.uk/lifesciences/services/research/transgeniclist. Cited 18 June 2009
56. United States Department of Agriculture (2000) APHIS Animal Care Policy Manual. www.aphis.usda.gov/animal_welfare/policy.shtm. Cited 18 June 2009
57. Vergara GJ, Irwin MH, Moffatt RJ, Pinkert CA (1997) In vitro fertilization in mice: strain differences in response to superovulation protocols and effect of cumulus cell removal. Theriogenology 47:1245–1252
58. Weigler BJ, Carte AD, Martineau JM, Noson D, Soriano P (2007) Motion detectors as the cause of poor reproductive performance in mice. Poster presentation at the 2007 AALAS National Meeting, Charlotte, NC
59. Weiss J, Zimmermann F (1999) Letters to the Editor: Tribromoethanol (Avertin) as an anaesthetic in mice. Lab Anim 33:192–193
60. Wernig M, Lengner CG, Hanna J, Lodato MA, Steine E, Foreman R, Staerk J, Markoulaki S, Jaenisch R (2008) A drug-inducible transgenic system for direct reprogramming of multiple somatic cell types. Nat Biotechnol 26(8):916–924
61. Willott JF, Demuth RM, Lu SM, Van Bergem P (1982) Abnormal tonotopic organization in the ventral cochlear nucleus of the hearing-impaired DBA/2 mouse. Neurosci Lett 34 (1):13–17
62. Zeller W, Meier G, Bürki K, Panoussis B (1997) Adverse effects of tribromoethanol as used in the production of transgenic mice. Lab Anim 32:407–413
63. Zhang G, Budker V, Wolff JA (1999) High levels of foreign gene expression in hepatocytes after tail vein injections of naked plasmid DNA. Hum Gene Ther 10 (10):1735–1737
64. Zheng QY, Johnson KR, Erway LC (1999) Assessment of hearing in 80 inbred strains of mice by ABR threshold analyses. Hear Res 130(1–2):94–107
65. Zhu HZ, Wang W, Feng DM, Sai Y, Xue JL (2006) Conditional gene modification in mouse liver using hydrodynamic delivery of plasmid DNA encoding Cre recombinase. FEBS Lett 580(18):4346–4352

Index

S. Pease and T.L. Saunders (eds.), *Advanced Protocols for Animal Transgenesis*, Springer Protocols,
DOI 10.1007/978-3-642-20792-1, © Springer-Verlag Berlin Heidelberg 2011

Z